METHODS IN ENZYMOLOGY

EDITORS-IN-CHIEF

John N. Abelson   Melvin I. Simon

DIVISION OF BIOLOGY
CALIFORNIA INSTITUTE OF TECHNOLOGY
PASADENA, CALIFORNIA

FOUNDING EDITORS

Sidney P. Colowick and Nathan O. Kaplan

Methods in Enzymology

Volume 184

Avidin–Biotin Technology

EDITED BY

Meir Wilchek

DEPARTMENT OF BIOPHYSICS
THE WEIZMANN INSTITUTE OF SCIENCE
REHOVOT, ISRAEL

Edward A. Bayer

DEPARTMENT OF BIOPHYSICS
THE WEIZMANN INSTITUTE OF SCIENCE
REHOVOT, ISRAEL

ACADEMIC PRESS, INC.
Harcourt Brace Jovanovich, Publishers
San Diego New York Boston
London Sydney Tokyo Toronto

This book is printed on acid-free paper. ♾

ACADEMIC PRESS, INC.
San Diego, California 92101

United Kingdom Edition published by
ACADEMIC PRESS LIMITED
24-28 Oval Road, London NW1 7DX

LIBRARY OF CONGRESS CATALOG CARD NUMBER: 54-9110

ISBN 0-12-182085-8 (alk. paper)

PRINTED IN THE UNITED STATES OF AMERICA
90 91 92 93 9 8 7 6 5 4 3 2 1

Table of Contents

Section I. Introduction

Section II. Biotin–Binding Proteins

Section III. General Methodology

A. Preparation of Biotin Derivatives

B. Preparation of Avidin and Streptavidin Derivatives

C. Assays for Avidin and Biotin

Section IV. Applications

A. Isolation and Purification

B. Localization

C. Protein Blotting

D. Immunoassays

E. Gene Probes

F. Composite and Special Applications

Contributors to Volume 184

Article numbers are in parentheses following the names of contributors.
Affiliations listed are current.

MARGIE ADAMS (37), *Department of Biology, University of California, San Diego, La Jolla, California 92093*

PAUL ANZIANO (35), *Department of Biochemistry, University of Texas Health Sciences Center, Dallas, Texas 75235*

HIROAKI ASOU (45), *Department of Physiology, School of Medicine, Keio University, Tokyo 160, Japan*

STRATIS AVRAMEAS (55), *Immunocytochimie, Institut Pasteur, 75724 Paris, Cedex 15, France*

ETIENNE-EMILE BAULIEU (30), *Faculté de Médecine, Université Paris XI, and INSERM U 33 Communications Hormonales, Hôpital de Bicêtre, 94275 Bicêtre Cedex, France*

EDWARD A. BAYER (2, 3, 4, 6, 8, 13, 14, 15, 18, 22, 23, 26, 31, 36, 47, 48, 54, 65, 71, 73), *Department of Biophysics, The Weizmann Institute of Science, Rehovot 76100, Israel*

L. DAWSON BEALL (28), *Department of Immunology, Otsuka America Pharmaceutical Co., Rockville, Maryland 20850*

STEPHAN BECK (72), *MilliGen/Biosearch, Burlington, Massachusetts 01803*

HAYA BEN-HUR (8, 23, 47, 48), *Department of Biophysics, The Weizmann Institute of Science, Rehovot 76100, Israel*

MELVIN BERGER (74), *Department of Pediatrics, Case Western Reserve University School of Medicine, Cleveland, Ohio 44106*

MELVIN L. BILLINGSLEY (53), *Department of Pharmacology, Pennsylvania State University College of Medicine, Hershey, Pennsylvania 17033*

CLAUDE BONNARD (75), *NESTEC SA, Centre de Recherche Nestlé, CH-1800 Vevey, Switzerland*

CHRISTINE L. BRAKEL (51), *ENZO Biochem, Inc., New York, New York 10013*

MARK S. BROWER (51), *Division of Hematology/Oncology, Department of Medicine and the Specialized Center for Research in Thrombosis, The New York Hospital-Cornell Medical Center, New York, New York 10021*

J. WILLIAM BUCKIE (32), *Department of Anatomy, University of Cambridge, Cambridge CB2 3DY, England*

GYAN CHANDRA (7), *Department of Biochemistry, Glaxo Research Laboratories, Research Triangle Park, North Carolina 27709*

HSHI-CHI CHANG (57), *Department of Veterinary Public Health, Faculty of Veterinary Medicine, Hokkaido University, Sapporo 060, Japan*

JASBIR CHAUHAN (10), *Department of Biochemistry and Molecular Biology, Faculty of Medicine, University of Manitoba, Winnipeg, Manitoba R3E 0W3, Canada*

LING-MEI CHEN (77), *Department of Forensic Medicine, Shanghai Medical University, Shanghai 200032, China*

GWEN V. CHILDS (44), *Department of Anatomy and Neurosciences, The University of Texas Medical Branch, Galveston, Texas 77550*

MARY K. CONRAD (76), *National Institute on Drug Abuse, Baltimore, Maryland 21224*

EVERLY CONWAY DE MACARIO (58), *Wadsworth Center for Laboratories and Research, New York State Department of Health and School of Public Health, State University of New York, Albany, New York 12201*

GEOFFREY M. W. COOK (32), *Department of*

Anatomy, University of Cambridge, Cambridge CB2 3DY, England

KRISHNAMURTI DAKSHINAMURTI (10, 12), *Department of Biochemistry and Molecular Biology, Faculty of Medicine, University of Manitoba, Winnipeg, Manitoba R3E 0W3, Canada*

CORNELIS DE GROOT (33), *Laboratory of Cell Biology and Histology, Cellular Immunology Group, University of Amsterdam, Academic Medical Center, 1105 AZ Amsterdam, The Netherlands*

RAYMOND C. DUHAMEL (21), *Bard Cardiosurgery Division, C.R. Bard Incorporated, Billerica, Massachusetts 01821*

ANNE H. DUTTON (37), *Department of Biology, University of California, San Diego, La Jolla, California 92093*

BARBARA J. FENN (68), *Department of Biochemistry, Medical College of Wisconsin, Milwaukee, Wisconsin 53226*

EVE FINKELSTEIN (49), *Syva Co., Palo Alto, California 94303*

FRANCES M. FINN (27), *Protein Research Laboratories and Department of Medicine, University of Pittsburgh School of Medicine, Pittsburgh, Pennsylvania 15261*

RICHARD FISHEL (35), *Laboratory of Chromosome Biology, BRI-Basic Research Program, National Cancer Institute, Frederick Cancer Research Facility, Frederick, Maryland 21701*

ANTHONY C. FORSTER (69), *Department of Biology, Yale University, New Haven, Connecticut 06520*

GLYN R. FOX (52), *Department of Biology, University of Victoria, Victoria, British Columbia V8W 2Y2, Canada*

STANLEY C. FROEHNER (50), *Department of Biochemistry, Dartmouth Medical School, Hanover, New Hampshire 03756*

BARBARA FUDEM-GOLDIN (17), *Department of Molecular Pharmacology, Albert Einstein College of Medicine, Bronx, New York 10461*

KIMBERLY GARRY (51), *Division of Hematology/Oncology, Department of Medicine and the Specialized Center for Research in Thrombosis, The New York Hospital-Cornell Medical Center, New York, New York 10021*

GULILAT GEBEYEHU (66), *Life Technologies Incorporated, Gaithersburg, Maryland 20877*

ALEXANDER N. GLAZER (19), *Division of Biochemistry and Molecular Biology, Department of Molecular and Cell Biology, University of California, Berkeley, California 94720*

PAULA J. GRABOWSKI (34), *Section of Biochemistry, Brown University, Providence, Rhode Island 02912*

JOHN G. GRAY (7), *Department of Biochemistry, Glaxo Research Laboratories, Research Triangle Park, North Carolina 27709*

N. MICHAEL GREEN (5), *Laboratory of Protein Structure, National Institute for Medical Research, London NW7 1AA, England*

ERNEST V. GROMAN (22), *Advanced Magnetics Incorporated, Cambridge, Massachusetts 02138*

NOBUO HASHIMOTO (57), *Department of Veterinary Public Health, Faculty of Veterinary Medicine, Hokkaido University, Sapporo 060, Japan*

ELI HAZUM (29), *Department of Endocrinology, Glaxo Research Laboratories, Research Triangle Park, North Carolina 27709*

GREGORY S. HEARD (11), *Department of Neurology, Medical College of Virginia, Virginia Commonwealth University, Richmond, Virginia 23298*

TIMOTHY M. HERMAN (68), *Department of Biochemistry, Medical College of Wisconsin, Milwaukee, Wisconsin 53226*

YAFFA HILLER (6), *Department of Biophysics, The Weizmann Institute of Science, Rehovot 76100, Israel*

KLAUS HOFMANN (27), *Protein Research Laboratories, University of Pittsburgh School of Medicine, Pittsburgh, Pennsylvania 15261*

PAUL HOROWITZ (25), *Department of Biochemistry, University of Texas Health Science Center at San Antonio, San Antonio, Texas 78284*

SU-MING HSU (39), *Department of Pathology and Laboratory Medicine, University of Texas Health Science Center at Houston, Houston, Texas 77225*

WALTER L. HURLEY (49), *Department of Animal Sciences, University of Illinois, Urbana, Illinois 61801*

KENNETH R. HUSKINS (64), *E. I. Du Pont De Nemours & Company, Inc., Medical Products Department, Diagnostics Systems Research and Development Division, Newark, Delaware 19714*

JEANNE HYMES (11), *Department of Human Genetics, Medical College of Virginia, Virginia Commonwealth University, Richmond, Virginia 23298*

KENNETH A. JACOBSON (79), *Laboratory of Chemistry, National Institute of Diabetes, Digestive, and Kidney Diseases, National Institutes of Health, Bethesda, Maryland 20892*

ROBERT J. JOVELL (58), *Wadsworth Center for Laboratories and Research, New York State Department of Health and School of Public Health, State University of New York, Albany, New York 12201*

RANDALL L. KINCAID (53), *Section on Immunology, National Institute of Alcohol Abuse and Alcoholism, Rockville, Maryland 20852*

LEONARD KLEVAN (66), *Life Technologies Incorporated, Gaithersburg, Maryland 20877*

RONALD A. KOHANSKI (20), *Department of Biochemistry, Mt. Sinai School of Medicine, New York, New York 10029*

FORTUNE KOHEN (56), *Department of Hormone Research, The Weizmann Institute of Science, Rehovot 76100, Israel*

JEAN-PIERRE KRAEHENBUHL (75), *Swiss Institute for Experimental Cancer Research and Institute of Biochemistry, University of Lausanne, CH-1066 Epalinges, Switzerland*

ULRICH KRÜGER (60), *Department of Clinical Chemistry and Pathobiochemistry, Medical Faculty, University of Technology, D-5100 Aachen, Federal Republic of Germany*

WILLIAM J. LAROCHELLE (50), *Laboratory of Cellular and Molecular Biology, National Cancer Institute, National Institutes of Health, Bethesda, Maryland 20892*

M. DANIEL LANE (20), *Department of Biological Chemistry, The Johns Hopkins University, School of Medicine, Baltimore, Maryland 21205*

MERLIN M. L. LEONG (52), *Department of Biology, University of Victoria, Victoria, British Columbia, V8W 2Y2 Canada*

COREY LEVENSON (67), *Department of Chemistry, Cetus Corporation, Emeryville, California 94608*

ODED LIVNAH (9), *Department of Structural Chemistry, The Weizmann Institute of Science, Rehovot 76100, Israel*

MATHEW M. S. LO (76), *ICI Americas, Inc., Wilmington, Delaware 19897*

ALBERTO J. L. MACARIO (58), *Wadsworth Center for Laboratories and Research, New York State Department of Health and School of Public Health, State University of New York, Albany, New York 12201*

JAMES L. MCINNES (69), *Department of Biochemistry, University of Adelaide, Adelaide, South Australia 5000, Australia*

KATHRYN E. MEIER (78), *Department of Pharmacology, Howard Hughes Medical Institute, University of Washington, Seattle, Washington 98118*

DONALD M. MOCK (24, 25), *Department of Pediatrics, University of Iowa Hospitals and Clinics, Iowa City, Iowa 52242*

RANDAL E. MORRIS (42), *Department of Anatomy and Cell Biology, University of Cincinnati, College of Medicine, Cincinnati, Ohio 45267*

LATA S. NERURKAR (63), *Laboratory of Tumor Cell Biology, National Cancer Institute, National Institutes of Health, Bethesda, Maryland 20892*

JEANNE E. NEUMANN (64), *E. I. Du Pont De Nemours & Company, Inc., Medical Products Department, Diagnostics Systems Research and Development Division, Newark, Delaware 19714*

WALTER NEWMAN (28), *Department of Immunology, Otsuka America Pharmaceutical Co., Rockville, Maryland 20850*

IZHAK NIR (41), *Department of Pathology, University of Texas Health Science Center at San Antonio, San Antonio, Texas 78284*

DANIEL J. O'SHANNESSY (16), *SmithKline Beecham Pharmaceutical, Research and Development, King of Prussia, Pennsylvania 19406*

GEORGE A. ORR (17), *Department of Molecular Pharmacology, Albert Einstein College of Medicine, Bronx, New York 10461*

DAVID S. PAPERMASTER (41), *Department of Pathology, University of Texas Health Science Center at San Antonio, San Antonio, Texas 78284*

KEITH R. PENNYPACKER (53), *Department of Anatomy, Medical College of Pennsylvania, Philadelphia, Pennsylvania 19129*

RICHARD M. PINO (43), *Department of Anatomy, Louisiana State University Medical Center, New Orleans, Louisiana 70112*

JOSEPH W. POLLI (53), *Department of Pharmacology, Pennsylvania State University College of Medicine, Hershey, Pennsylvania 17033*

ZAFAR I. RANDHAWA (28), *Protein Chemistry Section, Otsuka America Pharmaceutical Co., Rockville, Maryland 20850*

KEVIN J. REAGAN (64), *E. I. Du Pont De Nemours & Company, Inc., Medical Products Department, Diagnostics Systems Research and Development Division, Newark, Delaware 19714*

EDWARD S. RECTOR (12), *Department of Immunology, Faculty of Medicine, University of Manitoba, Winnipeg, Manitoba R3E 0W3, Canada*

GÉRARD REDEUILH (30), *Faculté de Médecine, Université Paris XI, and INSERM U 33 Communications Hormonales, Hôpital de Bicêtre, 94275 Bicêtre Cedex, France*

AVI REISFELD (71), *Department of Biophysics, The Weizmann Institute of Science, Rehovot 76100, Israel*

ALEXANDER RICH (35), *Department of Biology, Massachusetts Institute of Technology, Cambridge, Massachusetts 02138*

FREDERIC M. RICHARDS (1), *Department of Molecular Biophysics and Biochemistry, Yale University, New Haven, Connecticut 06511*

JEFFREY M. ROTHENBERG (22, 71), *Department of Biophysics, The Weizmann Institute of Science, Rehovot 76100, Israel*

ARNOLD E. RUOHO (78), *Department of Pharmacology, University of Wisconsin, School of Medicine, Madison, Wisconsin 53706*

CATHARINE B. SAELINGER (42), *Department of Microbiology and Molecular Genetics, University of Cincinnati College of Medicine, Cincinnati, Ohio 45267*

RACHEL SCHNEERSON (62), *Laboratory of Developmental and Molecular Immunity, National Institute of Child Health and Human Development, National Institutes of Health, Bethesda, Maryland 20892*

BARBARA G. SCHNEIDER (41), *Department of Pathology, University of Texas Health Science Center at San Antonio, San Antonio, Texas 78284*

ALEXANDER SCHWARZ (15), *Institute of Biotechnology 2, Nuclear Research Center, D-5170 Julich, Federal Republic of Germany*

CLAUDE SECCO (30), *Faculté de Médecine, Université Paris XI, and INSERUM U 33 Communications Hormonales, Hôpital de Bicêtre, 94275 Bicêtre Cedex, France*

MING-CHUAN SHAO (77), *Department of Biochemistry, Shanghai Medical University, Shanghai 200032, China*

EDWARD L. SHELDON (67), *Molecular Devices Corporation, Menlo Park, California 94025*

JOHN E. SHIVELY (59, 60), *Division of Immunology, Beckman Research Institute of the City of Hope, Duarte, California 91010*

S. J. SINGER (37), *Department of Biology, University of California, San Diego, La Jolla, California 92093*

DEREK C. SKINGLE (69), *Department of Biochemistry, University of Adelaide, Adelaide, South Australia 5000, Australia*

EHUD SKUTELSKY (36), *Department of Pathology, Sackler School of Medicine, Tel Aviv University, Ramat Aviv, Israel*

FRANK STIEKEMA (33), *Laboratory of Cell Biology and Histology, Cellular Immunology Group, University of Amsterdam, Academic Medical Center, 1105 AZ Amsterdam, The Netherlands*

CHRISTIAN J. STRASBURGER (56), *Department of Internal Medicine, University of Lübeck, Lübeck D-2400, Federal Republic of Germany*

LUBERT STRYER (19), *Department of Cell Biology, Stanford University School of Medicine, Stanford, California 94305*

JOEL L. SUSSMAN (9), *Department of Structural Chemistry, The Weizmann Institute of Science, Rehovot 76100, Israel*

KAZUO SUTOH (46), *Department of Pure and Applied Sciences, College of Arts and Sciences, University of Tokyo, Komaba, Tokyo 153, Japan*

ANN SUTTON (62), *Center for Biologics Evaluation and Research, Food and Drug Administration, Bethesda, Maryland 20892*

ROBERT H. SYMONS (69), *Department of Biochemistry, University of Adelaide, Adelaide, South Australia 5000, Australia*

ARTHUR J. SYTKOWSKI (38), *Laboratory for Cell and Molecular Biology, Division of Hematology and Oncology, New England Deaconess Hospital, Department of Medicine, Harvard Medical School, Boston, Massachusetts 02115*

IKUO TAKASHIMA (57), *Department of Veterinary Public Health, Faculty of Veterinary Medicine, Hokkaido University, Sapporo 060, Japan*

THÉRÈSE TERNYNCK (55), *Immunocytochimie, Institut Pasteur, 75724 Paris, Cedex 15, France*

WILLIE F. VANN (62), *Laboratory of Bacterial Polysaccharides, Center for Biologics Evaluation and Research, Food and Drug Administration, Bethesda, Maryland 20892*

RAPHAEL P. VISCIDI (70), *Eudowood Division of Infectious Diseases, Department of Pediatrics, The Johns Hopkins Hospital, Baltimore, Maryland 21205*

CHRISTOPH WAGENER (59, 60), *Department of Clinical Chemistry and Pathobiochemistry, Medical Faculty, University of Technology, D-5100 Aachen, Federal Republic of Germany*

CHRISTIAN WANDREY (15), *Institute of Biotechnology 2, Nuclear Research Center, D-5170 Julich, Federal Republic of Germany*

ROGER A. WARNKE (40), *Department of Pathology, Stanford University Medical Center, Stanford, California 94305*

ROBERT WATSON (67), *PCR Division, Cetus Corporation, Emeryville, California 94608*

JOHN A. WEHRLY (64), *E. I. Du Pont De Nemours & Company, Inc., Imaging Systems Department, Research and Development Division, Brevard, North Carolina 28712*

LAWRENCE M. WEISS (40), *Division of Pathology, City of Hope National Medical Center, Duarte, California 94305*

JAMES S. WHITEHEAD (21), *Vector Laboratories, Incorporated, Burlingame, California 94010*

LUCIA WICKERT (59), *Department of Clinical Chemistry and Pathobiochemistry, Medical Faculty, University of Technology, D-5100 Aachen, Federal Republic of Germany*

MEIR WILCHEK (2, 3, 4, 6, 8, 13, 14, 15, 18, 22, 23, 26, 31, 36, 47, 48, 54, 65, 71, 73), *Department of Biophysics, The Weizmann Institute of Science, Rehovot 76100, Israel*

FINN WOLD (77), *Department of Biochemistry and Molecular Biology, The University of Texas Medical School, Houston, Texas 77225*

BARRY WOLF (11), *Departments of Human Genetics and Pediatrics, Medical College of Virginia, Virginia Commonwealth University, Richmond, Virginia 23298*

JAN WORMMEESTER (33), *Laboratory of Cell Biology and Histology, Cellular Immunology Group, University of Amsterdam, Academic Medical Center, 1105 AZ Amsterdam, The Netherlands*

ROBERT H. YOLKEN (61), *The Johns Hopkins University, School of Medicine, Baltimore, Maryland 21205*

Preface

The development of avidin–biotin technology has unified and simplified a variety of technologies in which antibodies, lectins, hormones, and other binders have been used to localize, isolate, and identify the corresponding antigen, sugar moiety, receptor, etc. In all of these studies, in order to detect the latter, the binder counterpart had to be modified in some manner with a specific marker. In many cases, the preparation of protein–protein conjugates were required. The use of such conjugates resulted in a number of problems, often due to the modified physical and chemical state of the binder, which could lead to an alteration in the specificity and activity of the resultant conjugate.

The contribution of the avidin–biotin system is manifold. Foremost, of course, is the remarkable affinity between the two biomolecules. An added dimension of the system is the presence on egg white avidin (or its bacterial relative, streptavidin) of four biotin-binding sites which often generate an amplified signal. Moreover, avidin and streptavidin are exceptionally stable proteins, amenable to a wide variety of modifications and conjugations which only negligibly disturb the activity or specificity of their biotin-binding properties. Thus, protein–protein conjugates between the binder and marker are unnecessary, and only the biotinylation of the binder is required. Subsequent interactions are accomplished using the appropriate avidin- or streptavidin-associated probe.

In organizing this volume, we would have been delighted to have been able to present definitive studies which could be used as a basis for the design of new systems and applications. Unfortunately, this was not possible. In this explosive phase of the development of avidin–biotin technology for different applications, we chose to present the currently established uses from which (combined perhaps with a touch of imagination) the experienced investigator should be able to extrapolate information for the design of new and exciting uses for this complex.

There is much repetition, particularly with respect to the biotinylation procedures and the preparation of avidin-associated probes. Since the former is critical to most applications, we could not check the validity of each procedure, but have provided a general detailed chapter describing protocols and guidelines for the biotinylation of proteins based on our own research. A definitive opinion on whether a spacer group is required between the biotin moiety and the binder or probe is not given. In some cases the spacer group is definitely required and an enhanced or stabilized signal is received, but in others there appears to be no advantage to its use. Again, the properties of each system must be determined empirically. A clear stand has not been taken on whether an avidin-conjugated probe should be used or whether a two-step approach (using either sequential treatment or preformed complexes of native avidin and biotinylated probe) should be employed. Again, each researcher must determine experimentally which approach is most pertinent for the project involved.

We would like to acknowledge Mrs. Dvorah Ochert for her indispensable assistance with the manuscripts, the correspondence, the editing, typing, and retyping, all this in addition to her heavy departmental responsibilities. We also would like to thank our wives and families for their patience and understanding during the periods of heightened activity. And, of course, we are grateful to all of the contributors for their cooperation in preparing their manuscripts for this volume.

MEIR WILCHEK
EDWARD A. BAYER

METHODS IN ENZYMOLOGY

VOLUME I. Preparation and Assay of Enzymes
Edited by SIDNEY P. COLOWICK AND NATHAN O. KAPLAN

VOLUME II. Preparation and Assay of Enzymes
Edited by SIDNEY P. COLOWICK AND NATHAN O. KAPLAN

VOLUME III. Preparation and Assay of Substrates
Edited by SIDNEY P. COLOWICK AND NATHAN O. KAPLAN

VOLUME IV. Special Techniques for the Enzymologist
Edited by SIDNEY P. COLOWICK AND NATHAN O. KAPLAN

VOLUME V. Preparation and Assay of Enzymes
Edited by SIDNEY P. COLOWICK AND NATHAN O. KAPLAN

VOLUME VI. Preparation and Assay of Enzymes (*Continued*)
Preparation and Assay of Substrates
Special Techniques
Edited by SIDNEY P. COLOWICK AND NATHAN O. KAPLAN

VOLUME VII. Cumulative Subject Index
Edited by SIDNEY P. COLOWICK AND NATHAN O. KAPLAN

VOLUME VIII. Complex Carbohydrates
Edited by ELIZABETH F. NEUFELD AND VICTOR GINSBURG

VOLUME IX. Carbohydrate Metabolism
Edited by WILLIS A. WOOD

VOLUME X. Oxidation and Phosphorylation
Edited by RONALD W. ESTABROOK AND MAYNARD E. PULLMAN

VOLUME XI. Enzyme Structure
Edited by C. H. W. HIRS

VOLUME XII. Nucleic Acids (Parts A and B)
Edited by LAWRENCE GROSSMAN AND KIVIE MOLDAVE

VOLUME XXVII. Enzyme Structure (Part D)
Edited by C. H. W. HIRS AND SERGE N. TIMASHEFF

VOLUME XXVIII. Complex Carbohydrates (Part B)
Edited by VICTOR GINSBURG

VOLUME XXIX. Nucleic Acids and Protein Synthesis (Part E)
Edited by LAWRENCE GROSSMAN AND KIVIE MOLDAVE

VOLUME XXX. Nucleic Acids and Protein Synthesis (Part F)
Edited by KIVIE MOLDAVE AND LAWRENCE GROSSMAN

VOLUME XXXI. Biomembranes (Part A)
Edited by SIDNEY FLEISCHER AND LESTER PACKER

VOLUME XXXII. Biomembranes (Part B)
Edited by SIDNEY FLEISCHER AND LESTER PACKER

VOLUME XXXIII. Cumulative Subject Index Volumes I–XXX
Edited by MARTHA G. DENNIS AND EDWARD A. DENNIS

VOLUME XXXIV. Affinity Techniques (Enzyme Purification: Part B)
Edited by WILLIAM B. JAKOBY AND MEIR WILCHEK

VOLUME XXXV. Lipids (Part B)
Edited by JOHN M. LOWENSTEIN

VOLUME XXXVI. Hormone Action (Part A: Steroid Hormones)
Edited by BERT W. O'MALLEY AND JOEL G. HARDMAN

VOLUME XXXVII. Hormone Action (Part B: Peptide Hormones)
Edited by BERT W. O'MALLEY AND JOEL G. HARDMAN

VOLUME XXXVIII. Hormone Action (Part C: Cyclic Nucleotides)
Edited by JOEL G. HARDMAN AND BERT W. O'MALLEY

VOLUME XXXIX. Hormone Action (Part D: Isolated Cells, Tissues, and Organ Systems)
Edited by JOEL G. HARDMAN AND BERT W. O'MALLEY

VOLUME XL. Hormone Action (Part E: Nuclear Structure and Function)
Edited by BERT W. O'MALLEY AND JOEL G. HARDMAN

Volume XLI. Carbohydrate Metabolism (Part B)
Edited by W. A. Wood

Volume XLII. Carbohydrate Metabolism (Part C)
Edited by W. A. Wood

Volume XLIII. Antibiotics
Edited by John H. Hash

Volume XLIV. Immobilized Enzymes
Edited by Klaus Mosbach

Volume XLV. Proteolytic Enzymes (Part B)
Edited by Laszlo Lorand

Volume XLVI. Affinity Labeling
Edited by William B. Jakoby and Meir Wilchek

Volume XLVII. Enzyme Structure (Part E)
Edited by C. H. W. Hirs and Serge N. Timasheff

Volume XLVIII. Enzyme Structure (Part F)
Edited by C. H. W. Hirs and Serge N. Timasheff

Volume XLIX. Enzyme Structure (Part G)
Edited by C. H. W. Hirs and Serge N. Timasheff

Volume L. Complex Carbohydrates (Part C)
Edited by Victor Ginsburg

Volume LI. Purine and Pyrimidine Nucleotide Metabolism
Edited by Patricia A. Hoffee and Mary Ellen Jones

Volume LII. Biomembranes (Part C: Biological Oxidations)
Edited by Sidney Fleischer and Lester Packer

Volume LIII. Biomembranes (Part D: Biological Oxidations)
Edited by Sidney Fleischer and Lester Packer

Volume LIV. Biomembranes (Part E: Biological Oxidations)
Edited by Sidney Fleischer and Lester Packer

VOLUME LV. Biomembranes (Part F: Bioenergetics)
Edited by SIDNEY FLEISCHER AND LESTER PACKER

VOLUME LVI. Biomembranes (Part G: Bioenergetics)
Edited by SIDNEY FLEISCHER AND LESTER PACKER

VOLUME LVII. Bioluminescence and Chemiluminescence
Edited by MARLENE A. DELUCA

VOLUME LVIII. Cell Culture
Edited by WILLIAM B. JAKOBY AND IRA PASTAN

VOLUME LIX. Nucleic Acids and Protein Synthesis (Part G)
Edited by KIVIE MOLDAVE AND LAWRENCE GROSSMAN

VOLUME LX. Nucleic Acids and Protein Synthesis (Part H)
Edited by KIVIE MOLDAVE AND LAWRENCE GROSSMAN

VOLUME 61. Enzyme Structure (Part H)
Edited by C. H. W. HIRS AND SERGE N. TIMASHEFF

VOLUME 62. Vitamins and Coenzymes (Part D)
Edited by DONALD B. MCCORMICK AND LEMUEL D. WRIGHT

VOLUME 63. Enzyme Kinetics and Mechanism (Part A: Initial Rate and Inhibitor Methods)
Edited by DANIEL L. PURICH

VOLUME 64. Enzyme Kinetics and Mechanism (Part B: Isotopic Probes and Complex Enzyme Systems)
Edited by DANIEL L. PURICH

VOLUME 65. Nucleic Acids (Part I)
Edited by LAWRENCE GROSSMAN AND KIVIE MOLDAVE

VOLUME 66. Vitamins and Coenzymes (Part E)
Edited by DONALD B. MCCORMICK AND LEMUEL D. WRIGHT

VOLUME 67. Vitamins and Coenzymes (Part F)
Edited by DONALD B. MCCORMICK AND LEMUEL D. WRIGHT

VOLUME 81. Biomembranes (Part H: Visual Pigments and Purple Membranes, I)
Edited by LESTER PACKER

VOLUME 82. Structural and Contractile Proteins (Part A: Extracellular Matrix)
Edited by LEON W. CUNNINGHAM AND DIXIE W. FREDERIKSEN

VOLUME 83. Complex Carbohydrates (Part D)
Edited by VICTOR GINSBURG

VOLUME 84. Immunochemical Techniques (Part D: Selected Immunoassays)
Edited by JOHN J. LANGONE AND HELEN VAN VUNAKIS

VOLUME 85. Structural and Contractile Proteins (Part B: The Contractile Apparatus and the Cytoskeleton)
Edited by DIXIE W. FREDERIKSEN AND LEON W. CUNNINGHAM

VOLUME 86. Prostaglandins and Arachidonate Metabolites
Edited by WILLIAM E. M. LANDS AND WILLIAM L. SMITH

VOLUME 87. Enzyme Kinetics and Mechanism (Part C: Intermediates, Stereochemistry, and Rate Studies)
Edited by DANIEL L. PURICH

VOLUME 88. Biomembranes (Part I: Visual Pigments and Purple Membranes, II)
Edited by LESTER PACKER

VOLUME 89. Carbohydrate Metabolism (Part D)
Edited by WILLIS A. WOOD

VOLUME 90. Carbohydrate Metabolism (Part E)
Edited by WILLIS A. WOOD

VOLUME 91. Enzyme Structure (Part I)
Edited by C. H. W. HIRS AND SERGE N. TIMASHEFF

VOLUME 92. Immunochemical Techniques (Part E: Monoclonal Antibodies and General Immunoassay Methods)
Edited by JOHN J. LANGONE AND HELEN VAN VUNAKIS

VOLUME 106. Posttranslational Modifications (Part A)
Edited by FINN WOLD AND KIVIE MOLDAVE

VOLUME 107. Posttranslational Modifications (Part B)
Edited by FINN WOLD AND KIVIE MOLDAVE

VOLUME 108. Immunochemical Techniques (Part G: Separation and Characterization of Lymphoid Cells)
Edited by GIOVANNI DI SABATO, JOHN J. LANGONE, AND HELEN VAN VUNAKIS

VOLUME 109. Hormone Action (Part I: Peptide Hormones)
Edited by LUTZ BIRNBAUMER AND BERT W. O'MALLEY

VOLUME 110. Steroids and Isoprenoids (Part A)
Edited by JOHN H. LAW AND HANS C. RILLING

VOLUME 111. Steroids and Isoprenoids (Part B)
Edited by JOHN H. LAW AND HANS C. RILLING

VOLUME 112. Drug and Enzyme Targeting (Part A)
Edited by KENNETH J. WIDDER AND RALPH GREEN

VOLUME 113. Glutamate, Glutamine, Glutathione, and Related Compounds
Edited by ALTON MEISTER

VOLUME 114. Diffraction Methods for Biological Macromolecules (Part A)
Edited by HAROLD W. WYCKOFF, C. H. W. HIRS, AND SERGE N. TIMASHEFF

VOLUME 115. Diffraction Methods for Biological Macromolecules (Part B)
Edited by HAROLD W. WYCKOFF, C. H. W. HIRS, AND SERGE N. TIMASHEFF

VOLUME 116. Immunochemical Techniques (Part H: Effectors and Mediators of Lymphoid Cell Functions)
Edited by GIOVANNI DI SABATO, JOHN J. LANGONE, AND HELEN VAN VUNAKIS

VOLUME 117. Enzyme Structure (Part J)
Edited by C. H. W. HIRS AND SERGE N. TIMASHEFF

VOLUME 118. Plant Molecular Biology
Edited by ARTHUR WEISSBACH AND HERBERT WEISSBACH

VOLUME 119. Interferons (Part C)
Edited by SIDNEY PESTKA

VOLUME 120. Cumulative Subject Index Volumes 81–94, 96–101

VOLUME 121. Immunochemical Techniques (Part I: Hybridoma Technology and Monoclonal Antibodies)
Edited by JOHN J. LANGONE AND HELEN VAN VUNAKIS

VOLUME 122. Vitamins and Coenzymes (Part G)
Edited by FRANK CHYTIL AND DONALD B. MCCORMICK

VOLUME 123. Vitamins and Coenzymes (Part H)
Edited by FRANK CHYTIL AND DONALD B. MCCORMICK

VOLUME 124. Hormone Action (Part J: Neuroendocrine Peptides)
Edited by P. MICHAEL CONN

VOLUME 125. Biomembranes (Part M: Transport in Bacteria, Mitochondria, and Chloroplasts: General Approaches and Transport Systems)
Edited by SIDNEY FLEISCHER AND BECCA FLEISCHER

VOLUME 126. Biomembranes (Part N: Transport in Bacteria, Mitochondria, and Chloroplasts: Protonmotive Force)
Edited by SIDNEY FLEISCHER AND BECCA FLEISCHER

VOLUME 127. Biomembranes (Part O: Protons and Water: Structure and Translocation)
Edited by LESTER PACKER

VOLUME 128. Plasma Lipoproteins (Part A: Preparation, Structure, and Molecular Biology)
Edited by JERE P. SEGREST AND JOHN J. ALBERS

VOLUME 129. Plasma Lipoproteins (Part B: Characterization, Cell Biology, and Metabolism)
Edited by JOHN J. ALBERS AND JERE P. SEGREST

VOLUME 130. Enzyme Structure (Part K)
Edited by C. H. W. HIRS AND SERGE N. TIMASHEFF

VOLUME 131. Enzyme Structure (Part L)
Edited by C. H. W. HIRS AND SERGE N. TIMASHEFF

VOLUME 132. Immunochemical Techniques (Part J: Phagocytosis and Cell-Mediated Cytotoxicity)
Edited by GIOVANNI DI SABATO AND JOHANNES EVERSE

VOLUME 133. Bioluminescence and Chemiluminescence (Part B)
Edited by MARLENE DELUCA AND WILLIAM D. MCELROY

VOLUME 134. Structural and Contractile Proteins (Part C: The Contractile Apparatus and the Cytoskeleton)
Edited by RICHARD B. VALLEE

VOLUME 135. Immobilized Enzymes and Cells (Part B)
Edited by KLAUS MOSBACH

VOLUME 136. Immobilized Enzymes and Cells (Part C)
Edited by KLAUS MOSBACH

VOLUME 137. Immobilized Enzymes and Cells (Part D)
Edited by KLAUS MOSBACH

VOLUME 138. Complex Carbohydrates (Part E)
Edited by VICTOR GINSBURG

VOLUME 139. Cellular Regulators (Part A: Calcium- and Calmodulin-Binding Proteins)
Edited by ANTHONY R. MEANS AND P. MICHAEL CONN

VOLUME 140. Cumulative Subject Index Volumes 102–119, 121–134

VOLUME 141. Cellular Regulators (Part B: Calcium and Lipids)
Edited by P. MICHAEL CONN AND ANTHONY R. MEANS

VOLUME 142. Metabolism of Aromatic Amino Acids and Amines
Edited by SEYMOUR KAUFMAN

VOLUME 143. Sulfur and Sulfur Amino Acids
Edited by WILLIAM B. JAKOBY AND OWEN GRIFFITH

VOLUME 144. Structural and Contractile Proteins (Part D: Extracellular Matrix)
Edited by LEON W. CUNNINGHAM

VOLUME 145. Structural and Contractile Proteins (Part E: Extracellular Matrix)
Edited by LEON W. CUNNINGHAM

VOLUME 146. Peptide Growth Factors (Part A)
Edited by DAVID BARNES AND DAVID A. SIRBASKU

VOLUME 147. Peptide Growth Factors (Part B)
Edited by DAVID BARNES AND DAVID A. SIRBASKU

VOLUME 148. Plant Cell Membranes
Edited by LESTER PACKER AND ROLAND DOUCE

VOLUME 149. Drug and Enzyme Targeting (Part B)
Edited by RALPH GREEN AND KENNETH J. WIDDER

VOLUME 150. Immunochemical Techniques (Part K: *In Vitro* Models of B and T Cell Functions and Lymphoid Cell Receptors)
Edited by GIOVANNI DI SABATO

VOLUME 151. Molecular Genetics of Mammalian Cells
Edited by MICHAEL M. GOTTESMAN

VOLUME 152. Guide to Molecular Cloning Techniques
Edited by SHELBY L. BERGER AND ALAN R. KIMMEL

VOLUME 153. Recombinant DNA (Part D)
Edited by RAY WU AND LAWRENCE GROSSMAN

VOLUME 154. Recombinant DNA (Part E)
Edited by RAY WU AND LAWRENCE GROSSMAN

VOLUME 155. Recombinant DNA (Part F)
Edited by RAY WU

VOLUME 156. Biomembranes (Part P: ATP-Driven Pumps and Related Transport: The Na,K-Pump)
Edited by SIDNEY FLEISCHER AND BECCA FLEISCHER

VOLUME 157. Biomembranes (Part Q: ATP-Driven Pumps and Related Transport: Calcium, Proton, and Potassium Pumps)
Edited by SIDNEY FLEISCHER AND BECCA FLEISCHER

Volume 171. Biomembranes (Part R: Transport Theory: Cells and Model Membranes)
Edited by Sidney Fleischer and Becca Fleischer

Volume 172. Biomembranes (Part S: Transport: Membrane Isolation and Characterization)
Edited by Sidney Fleischer and Becca Fleischer

Volume 173. Biomembranes [Part T: Cellular and Subcellular Transport: Eukaryotic (Nonepithelial) Cells]
Edited by Sidney Fleischer and Becca Fleischer

Volume 174. Biomembranes [Part U: Cellular and Subcellular Transport: Eukaryotic (Nonepithelial) Cells]
Edited by Sidney Fleischer and Becca Fleischer

Volume 175. Cumulative Subject Index Volumes 135–139, 141–167 (in preparation)

Volume 176. Nuclear Magnetic Resonance (Part A: Spectral Techniques and Dynamics)
Edited by Norman J. Oppenheimer and Thomas L. James

Volume 177. Nuclear Magnetic Resonance (Part B: Structure and Mechanism)
Edited by Norman J. Oppenheimer and Thomas L. James

Volume 178. Antibodies, Antigens, and Molecular Mimicry
Edited by John J. Langone

Volume 179. Complex Carbohydrates (Part F)
Edited by Victor Ginsburg

Volume 180. RNA Processing (Part A: General Methods)
Edited by James E. Dahlberg and John N. Abelson

Volume 181. RNA Processing (Part B: Specific Methods) (in preparation)
Edited by James E. Dahlberg and John N. Abelson

Volume 182. Guide to Protein Purification
Edited by Murray P. Deutscher

VOLUME 183. Molecular Evolution: Computer Analysis of Protein and Nucleic Acid Sequences
Edited by RUSSELL F. DOOLITTLE

VOLUME 184. Avidin–Biotin Technology
Edited by MEIR WILCHEK AND EDWARD A. BAYER

VOLUME 185. Gene Expression Technology (in preparation)
Edited by DAVID V. GOEDDEL

VOLUME 186. Oxygen Radicals in Biological Systems (Part B: Oxygen Radicals and Antioxidants) (in preparation)
Edited by LESTER PACKER AND ALEXANDER N. GLAZER

VOLUME 187. Arachidonate Related Lipid Mediators (in preparation)
Edited by ROBERT C. MURPHY AND FRANK A. FITZPATRICK

VOLUME 188. Hydrocarbons and Methylotrophy (in preparation)
Edited by MARY E. LIDSTROM

Section I

Introduction

[1] Reflections

By FREDERIC M. RICHARDS

Avidin was recognized as a biological factor in egg white in the late 1920s during the discovery and isolation of the vitamin biotin.[1] The latter compound joined the other known water-soluble vitamins, which received interest both because of the related deficiency diseases and because of the emerging identification of vitamins as cofactors in various metabolic processes. Avidin activity was found in various avian eggs and egg jelly of invertebrates but could not be found in many other tissues or organisms. Its function in the egg was presumed to be that of an antibiotic along with several other proteins of similar function but very different mechanisms, lysozyme, for example. With no general biological function, and no apparent clinical use as an antibiotic, avidin was essentially ignored for about two decades.

In the late 1950s the laboratories of Wakil and Lynen both reported the discovery of covalently bound biotin with a coenzyme function.[2] Avidin at once took on the role of a tool in the investigation of this new class of enzymes. The use of avidin in the study of the biotin enzymes appeared suitable for that purpose but of limited general application. Shortly afterward Green began studies of the detailed protein chemistry of avidin, focusing on its general structure and activity rather than its use as a reagent.[3]

Of particular interest since the earliest days has been the remarkable strength of the interaction between avidin and biotin. The binding is characterized by a dissociation constant of the order of 10^{-15} *M*. This value corresponds to a free energy of association of about 21 kcal/mol, a staggeringly large value for the noncovalent interaction of a protein with a molecule as small as biotin. In general such binding is found only in systems involving liganded metal ions either as partial covalent bonds or chelates. No metal ions are involved in the avidin–biotin binding. Of special importance is the very slow off-rate that accompanies such a tight association.

[1] F. Kögl and B. Tönnis, *Z. Physiol. Chem.* **242,** 43 (1936).
[2] F. Lynen, J. Knappe, E. Lorch, G. Jutting, and E. Ringelmann, *Angew. Chem.* **71,** 481 (1959); S. J. Wakil and D. M. Gibson, *Biochim. Biophys. Acta* **41,** 122 (1960).
[3] N. M. Green, *Adv. Protein Chem.* **29,** 85 (1975).

While these chemical studies were proceeding, Chaiet and colleagues[4] reported avidinlike activity from various species of *Streptomyces*, one more, apparently, in the series of antibiotic materials produced by these bacteria. While similar to egg avidin in many ways, streptavidin has a much more acidic isoelectric point. Again, with no clinical application this substance appeared to be a curiosity, and interest was slow in developing. The future importance of this form of avidin was certainly not clear at the beginning.

In the early to mid-1970s a series of events occurred which appear to have changed the perception of the avidin–biotin system. John Edsall, Chris Anfinsen, and I, as editors of *Advances in Protein Chemistry*, asked Michael Green to write a review article on avidin. The article contained all of the detailed protein chemical information then available on avidin. The evidence for a tetramer with the four biotin-binding sites arranged in two clusters was presented. The clever cross-linking experiments with bisbiotin samples of varying chain length defined the deep binding pocket and recessed surface area in which it lies. The binding constants for a large number of biotin derivatives were collected. The review[3] was published in 1975 and served as a background for subsequent applications.

The major published studies of Green on the properties of the avidin–biotin system suggested to Heitzmann and myself the possibility of developing a general tool for labeling membranes,[5] while others were applying the system for various other purposes.[6] The convenient single carboxyl group on biotin and the relatively low reactivity of the fused ring system made the preparation of reagents for the chemical modification of proteins straightforward. Because of the enormous binding constant we thought of the whole system as sort of a chemist's version of an antigen–antibody labeling procedure, to which, in fact, it is a complement. We did note that the very slow off-rate would allow sequential labeling of different reactive groups using different heavy metal tags, but we did not envision the use of this system for separations. Because of the close association of our laboratories, Wallace and Engelman used this labeling procedure, shortly after its development, to study general protein distribution and redistribution during the phase transition in *Acholeplasma* membranes.[7]

[4] L. Chaiet, T. W. Miller, F. Tausig, and F. J. Wolf, *Antimicrob. Agents Chemother.* **3,** 28 (1963); L. Chaiet and F. J. Wolf, *Arch. Biochem. Biophys.* **106,** 1 (1964).

[5] H. Heitzmann and F. M. Richards, *Proc. Natl. Acad. Sci. U.S.A.* **71,** 3537 (1974).

[6] E. A. Bayer, M. Wilchek, and E. Skutelsky, *FEBS Lett.* **68,** 240 (1976); E. A. Bayer and M. Wilchek, *Trends Biochem. Sci.* **3,** N257 (1978); K. Hofmann, F. M. Finn, and Y. Kiso, *J. Am. Chem. Soc.* **100,** 3585 (1978).

[7] B. A. Wallace, F. M. Richards, and D. M. Engelman, *J. Mol. Biol.* **107,** 255 (1976).

Completely independently, Tom Broker and Norman Davidson at the California Institute of Technology had the same general ideas for using the avidin–biotin system, and they developed procedures for gene mapping by labeling RNA molecules, especially tRNA, and then using them as hybridization probes to locate genes in double-stranded DNA.[8] After hearing about this work at meetings, we did have some casual discussions with David Ward in Human Genetics at Yale, University.[9] It was clear that the biotin-avidin system as a general cytochemical labeling procedure had arrived.[10]

I doubt that any of us at the time imagined the explosion in applications that would occur over the 1980s[11] and that is documented in this volume. With the basic underpinnings for the technology now in hand, as well as cloned genes for the avidins[12] and crystal structures of the proteins,[13] the future is bright indeed.

[8] J. E. Manning, N. D. Hershey, T. R. Broker, M. Pellegrini, H. K. Mitchell, and N. Davidson, *Chromosoma* **53,** 107 (1975); T. R. Broker, L. M. Angerer, P. Yen, N. D. Hershey, and N. Davidson, *Nucleic Acids Res.* **5,** 363 (1978).

[9] P. R. Langer, A. A. Waldrop, and D. C. Ward, *Proc. Natl. Acad. Sci. U.S.A.* **78,** 6633 (1981).

[10] E. A. Bayer and M. Wilchek, *Methods Biochem. Anal.* **26,** 1 (1980); M. Wilchek and E. A. Bayer, *Immunol. Today* **5,** 39 (1984).

[11] M. Wilchek and E. A. Bayer, *Anal. Biochem.* **171,** 1 (1988).

[12] M. L. Gope, R. A. Keinänen, P. A. Kristo, O. M. Conneely, W. G. Beattie, T. Zarucki-Schulz, B. M. O'Malley, and M. S. Kulomaa, *Nucleic Acids. Res.* **15,** 3595 (1987); G. Chandra and J. G. Gray, this volume [7]; C. E. Argarana, I. D. Kuntz, S. Birken, R. Axel, and C. R. Cantor, *Nucleic Acids Res.* **14,** 1871 (1986).

[13] E. Pinn, A. Pähler, W. Saenger, G. Petsko, and N. M. Green, *Eur. J Biochem.* **123,** 545 (1982); A. Pähler, W. A. Hendrickson, M. A. G. Kolks, C. E. Argarana, and C. R. Cantor, *J. Biol. Chem.* **262,** 13933 (1987).

[2] Introduction to Avidin–Biotin Technology

By MEIR WILCHEK and EDWARD A. BAYER

The avidin–biotin system has many uses in both research and technology. This general introductory chapter presents an overview of the principles and advantages of this system, and illustrates the numerous applications of avidin–biotin technology. It also explains in general and explicit terms the basic idea behind avidin–biotin technology.

Principle

The major distinguishing feature of the avidin–biotin system is the extraordinary affinity ($K_a = 10^{15}\ M^{-1}$) that characterizes the complex formed between the vitamin biotin and the egg-white protein avidin (or streptavidin, its bacterial relative from *Streptomyces avidinii*).[1] The interaction is so strong that even biotin coupled to proteins (such as native biotin-requiring carboxylases) is available for binding by avidin.

The rationale in using the avidin–biotin system is based on the premise that if one chemically modifies any biologically active compound with biotin through its valeric acid side chain, the biological and physicochemical properties of the biotin-modified molecule will not be changed significantly. Likewise, the biotin moiety of the derivatized molecule would still be available for interaction with avidin. Another distinctive feature of this system is the multiple (four) binding sites of avidin for biotin. This provides us with the possibility of cross-linking between different biotin-containing molecules and adds another dimension to the use of this multifaceted system.

If a reporter group of some sort is attached to the avidin molecule, the conjugate can be used for many different purposes (Fig. 1). The avidin-conjugated probe can be added as a single chemically conjugated entity, or avidin (by virtue of its four biotin-binding sites) can be applied in native form together with a biotinylated probe. The latter alternative can be instituted either sequentially in a stepwise fashion or in a single step as preformed complexes. Figure 2 (which was first published in an earlier volume in this series[2]) presents these two alternatives for labeling.

Figure 1 also provides a list of some of the types of binders that can be used to label a given target site, as well as a list of probes that can be either conjugated with avidin (for direct interaction with the biotinylated binder) or derivatized with biotin (for complex formation with underivatized avidin). The various applications of avidin–biotin technology reported to date are also presented in Fig. 1.

Advantages

Figure 1 demonstrates why the avidin–biotin system has become so popular: a simplified and unified approach can be taken. Some of the major advantages in using avidin–biotin technology can be summarized as follows:

1. The exceptionally high affinity and stability of the avidin–biotin complex ensures the desired conjunction of binder and probe.

[1] N. M. Green, *Adv. Protein Chem.* **29,** 85 (1975).

[2] E. A. Bayer, E. Skutelsky, and M. Wilchek, this series, Vol. 62, p. 308.

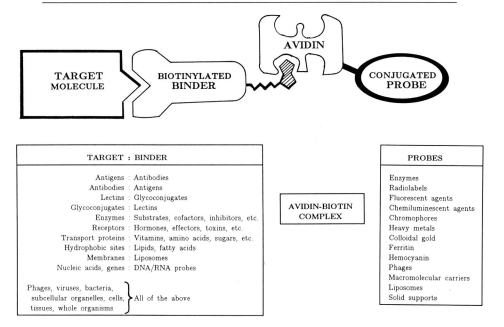

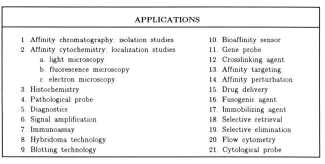

TARGET	BINDER
Antigens	Antibodies
Antibodies	Antigens
Lectins	Glycoconjugates
Glycoconjugates	Lectins
Enzymes	Substrates, cofactors, inhibitors, etc.
Receptors	Hormones, effectors, toxins, etc.
Transport proteins	Vitamins, amino acids, sugars, etc.
Hydrophobic sites	Lipids, fatty acids
Membranes	Liposomes
Nucleic acids, genes	DNA/RNA probes
Phages, viruses, bacteria, subcellular organelles, cells, tissues, whole organisms	All of the above

AVIDIN-BIOTIN COMPLEX

PROBES
Enzymes
Radiolabels
Fluorescent agents
Chemiluminescent agents
Chromophores
Heavy metals
Colloidal gold
Ferritin
Hemocyanin
Phages
Macromolecular carriers
Liposomes
Solid supports

APPLICATIONS	
1. Affinity chromatography: isolation studies	10. Bioaffinity sensor
2. Affinity cytochemistry: localization studies	11. Gene probe
a. light microscopy	12. Crosslinking agent
b. fluorescence microscopy	13. Affinity targeting
c. electron microscopy	14. Affinity perturbation
3. Histochemistry	15. Drug delivery
4. Pathological probe	16. Fusogenic agent
5. Diagnostics	17. Immobilizing agent
6. Signal amplification	18. Selective retrieval
7. Immunoassay	19. Selective elimination
8. Hybridoma technology	20. Flow cytometry
9. Blotting technology	21. Cytological probe

FIG. 1. Generalized scheme illustrating the essentials of avidin–biotin technology. A biologically active target molecule is recognized by a biotinylated binder that is subsequently recognized by avidin conjugated to an appropriate probe. Lists of target–binder pairs, the various probes that have been used, and the broad spectrum of applications of avidin–biotin technology are also presented.

2. Biotin can readily be attached to most binders and probes, and, following biotinylation, the biological activity and physical characteristics are commonly retained. In fact, in many cases, a binder can be biotinylated without extensive purification, since its interaction with the target molecule serves to remove "irrelevant" biotinylated substances.

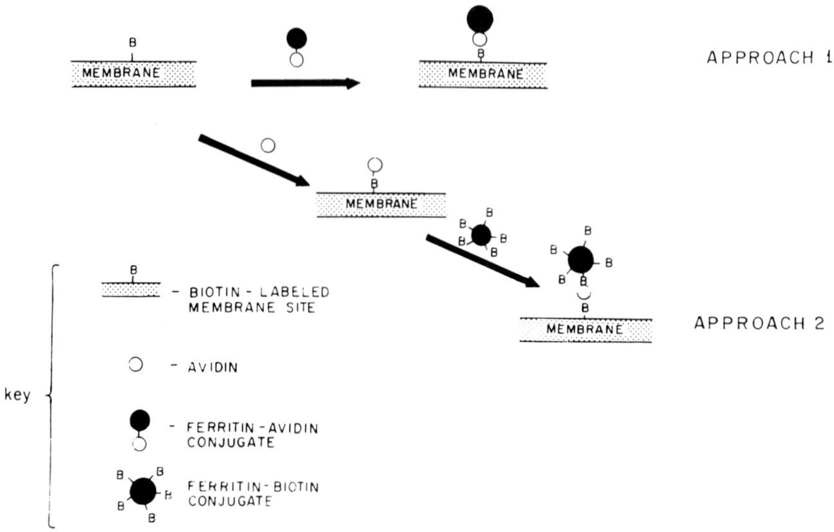

FIG. 2. Schematic of two alternative approaches for labeling biotin-modified cell membrane sites: (1) one-step method, using ferritin-conjugated avidin, and (2) two-step method, employing sequential treatment with avidin and biotin-labeled ferritin.

3. The multiplicity of biotin groups per binder combined with the tetrameric structure of avidin leads to amplification of the desired signal.
4. The system is amenable to double-labeling and kinetics studies.
5. The system is extremely versatile. A given molecular target can interact with a single type of biotinylated binder that can then be analyzed by various means using different avidin-conjugated probes. Conversely, the status of different target molecules in a given system can be compared using a variety of biotinylated binders and a single avidin-conjugated probe. Of course, the versatility is further extended through the combined use of different biotinylated binders and avidin-associated probes.
6. A wide spectrum of different biotinylating reagents, biotinylated binders, and both biotinylated and avidin-containing probes is available from a variety of commercial sources.

Application

The development of avidin–biotin technology was not the outcome of a single event, nor was it the brainchild of a single individual or research group. Rather, it developed through an evolutionary process (the collec-

TABLE I
MILESTONES IN DEVELOPMENT OF AVIDIN–BIOTIN TECHNOLOGY

Year	Historical highlights	Ref.
1916	"Toxic substances" in egg white	3
1936	Biotin isolated	4
1942	Structure of biotin determined	5,6
	Synthesis of biotin hydrazide	7
1943	Biotin synthesized	8–10
1952	Avidin purified from egg white	11,12
1958	Coenzyme function of biotin described	13,14
1963	Extensive protein chemical studies on avidin	15–18
	Streptavidin and stravidin isolated	19
1965	Retardation of avidin on biotinyl cellulose	20
1968	Affinity chromatography of avidin on biocytin–Sepharose	21
1969	Structure of stravidin determined	22
1970	Avidin crystallized	23
	Biotin-containing peptides isolated on avidin column	24
1971	Avidin sequenced	25
	N-Hydroxysuccinimide and *p*-nitrophenyl esters of biotin synthesized	26
1972	Biotinylated phages used for assay of avidin and biotin	27
1973	Avidin monomer column	28
1974	Localization of biotinylated sites on purified membranes; ferritin–avidin conjugates	29
1975	Biotinylated RNA	30
	Spacer-containing biotinylating reagents	31
	Comprehensive review on avidin	1
1976	Biotinylated antibodies and lectins (affinity cytochemistry)	32
1977	Biotinylated hormones	33–35
1978	*TIBS* review: "Emerging Techniques"	36
1979	ABC approach for enzyme immunoassay	37
	Biotinylated lipids	38
1980	First comprehensive review of applications	39
1981	Biotinylated DNA probes	40
	Premade complexes	41
	Iminobiotin columns and reagents	42,43
1982	Insulin receptor isolated using biotinylated insulin	44–46
1983	Protein and DNA blotting using biotinylated binders	47–49
1986	Cloning of streptavidin gene	50
	Nonglycosylated avidin	51
1988	Crystallographic structure of streptavidin	52,53
1990	*Methods in Enzymology*, "Avidin–Biotin Technology"	This volume

tive product of curiosity, necessity, observation, imagination, and creative thought) which began with the early discovery of toxic substances in egg white and culminated with the commercial availability of literally hundreds of "avidin–biotin" products from dozens of companies. Some of the historical milestones in the development of avidin–biotin technology[3–53] are presented in Table I.

[3] W. G. Bateman, *J. Biol. Chem.* **26,** 263 (1916).
[4] F. Kögl and B. Tönnis, *Z. Physiol. Chem.* **242,** 43 (1936).
[5] V. du Vigneaud, K. Hofmann, and D. B. Melville, *J. Am. Chem. Soc.* **64,** 188 (1942).
[6] V. du Vigneaud, D. B. Melville, K. Folkers, D. E. Wolf, R. Mozingo, J. C. Kereszteolf, and S. A. Harris, *J. Biol. Chem.* **146,** 475 (1942).
[7] K. Hofmann, D. B. Melville, and V. du Vigneaud, *J. Biol. Chem.* **144,** 513 (1942).
[8] S. A. Harris, D. E. Wolf, R. Mozingo, and K. Folkers, *Science* **97,** 447 (1943).
[9] S. A. Harris, D. E. Wolf, R. Mozingo, R. C. Anderson, G. E. Arth, N. R. Easton, D. Heyl, A. N. Wilson, and K. Folkers, *J. Am. Chem. Soc.* **66,** 1756 (1944).
[10] S. A. Harris, N. R. Easton, D. Heyl, A. N. Wilson, and K. Folkers, *J. Am. Chem. Soc.* **66,** 1757 (1944).
[11] H. Fraenkel-Conrat, N. S. Snell, and E. D. Ducay, *Arch. Biochem. Biophys.* **92,** 80 (1952).
[12] H. Fraenkel-Conrat, N. S. Snell, and E. D. Ducay, *Arch. Biochem. Biophys.* **92,** 97 (1952).
[13] S. J. Wakil, E. B. Titchener, and D. M. Gibson, *Biochim. Biophys. Acta* **29,** 225 (1958).
[14] F. Lynen, J. Knappe, E. Lorch, G. Jutting, and E. Ringelmann, *Angew. Chem.* **71,** 481 (1959).
[15] N. M. Green, *Biochem. J.* **89,** 585 (1963).
[16] N. M. Green, *Biochem. J.* **89,** 599 (1963).
[17] N. M. Green, *Biochem. J.* **89,** 609 (1963).
[18] M. D. Melamed and N. M. Green, *Biochem. J.* **89,** 591 (1963).
[19] L. Chaiet, T. W. Miller, F. Tausig, and F. J. Wolf, *Antimicrob. Agents Chemother.* **3,** 28 (1963).
[20] D. M. McCormick, *Anal. Biochem.* **13,** 194 (1965).
[21] P. Cuatrecasas and M. Wilchek, *Biochem. Biophys. Res. Commun.* **33,** 235 (1968).
[22] K. H. Baggaley, B. Blessington, C. P. Falshaw, and W. D. Ollis, *Chem. Commun.,* 101 (1969).
[23] N. M. Green and E. J. Toms, *Biochem. J.* **118,** 67 (1970).
[24] A. Bodanszky and M. Bodanszky, *Experientia* **26,** 327 (1970).
[25] R. J. DeLange and T.-S. Huang, *J. Biol. Chem.* **246,** 698 (1971).
[26] J. M. Becker, M. Wilchek, and E. Katchalski, *Proc. Natl. Acad. Sci. U.S.A.* **68,** 2604 (1971).
[27] J. M. Becker and M. Wilchek, *Biochim. Biophys. Acta* **264,** 165 (1972).
[28] N. M. Green and E. J. Toms, *Biochem. J.* **133,** 687 (1973).
[29] H. Heitzmann and F. M. Richards, *Proc. Natl. Acad. Sci. U.S.A.* **71,** 3537 (1974).
[30] J. E. Manning, N. D. Hershey, T. R. Broker, M. Pellegrini, H. K. Mitchell, and N. Davidson, *Chromosoma* **53,** 107 (1975).
[31] E. A. Bayer, T. Viswanatha, and M. Wilchek, *FEBS Lett.* **60,** 309 (1975).
[32] E. A. Bayer, M. Wilchek, and E. Skutelsky, *FEBS Lett.* **68,** 240 (1976).
[33] K. Hofmann and Y. Kiso, *Proc. Natl. Acad. Sci. U.S.A.* **73,** 3516 (1976).
[34] K. Hofmann, F. M. Finn, H.-J. Friesen, C. Diaconescu, and H. Zahn, *Proc. Natl. Acad. Sci. U.S.A.* **74,** 2697 (1977).
[35] D. Atlas, D. Yaffe, and E. Skutelsky, *FEBS Lett.* **95,** 173 (1978).
[36] E. A. Bayer and M. Wilchek, *Trends Biochem. Sci.* **3,** N257 (1978).
[37] J.-L. Guesdon, T. Ternynck, and S. Avrameas, *J. Histochem. Cytochem.* **27,** 1131 (1979).
[38] E. A. Bayer, B. Rivnay, and E. Skutelsky, *Biochem. Biophys. Acta* **550,** 464 (1979).

We initially intended to include in this volume all the available applications which have thus far appeared in the primary literature, since each contribution has its own special message to convey. We soon discovered that this would be impossible, since the literature is simply too vast. Indeed, many of the methodologies involving avidin and/or biotin have already found their way into this series (Table II), although their contribution to the emerging technology has not generally been emphasized. A more extensive view of the widespread usage of avidin–biotin technology is presented elsewhere in this volume.[54] In assembling this volume, we therefore attempted to include only papers that either set a historical precedence or present a special method or message.

Judging from the articles and tables[54] that appear in this volume, the avidin–biotin system is extremely versatile, the applications are numerous, and the scope of its future use is virtually limitless. We cite the following from our first major review on this subject[39]:

> Naturally, we will be unable to cover all possible applications; it seems that the potential of the avidin–biotin complex as a tool in molecular biology is unlimited, and that its successful implementation is directly dependent on the needs and imagination of the user. . . . We cannot, of course, foresee all its possible uses, given our own limited imagination and/or restricted knowledge. We do hope, however, that this review will serve to stimulate the imagination of the readers and convert some of them to users of the method.

[39] E. A. Bayer and M. Wilchek, *Methods Biochem. Anal.* **26,** 1 (1980).

[40] P. R. Langer, A. A. Waldrop, and D. C. Ward, *Proc. Natl. Acad. Sci. U.S.A.* **78,** 6633 (1981).

[41] S.-M. Hsu, L. Raine, and H. Fanger, *J. Histochem. Cytochem.* **29,** 577 (1981).

[42] K. Hofmann, S. W. Wood, C. C. Brinton, J. A. Montibeller, and F. M. Finn, *Proc. Natl. Acad. Sci. U.S.A.* **77,** 4666 (1980).

[43] G. A. Orr, *J. Biol. Chem.* **256,** 761 (1981).

[44] F. M. Finn, G. Titus, H. Nemoto, T. Noji, and K. Hofmann, *Metabolism* **31,** 691 (1982).

[45] F. M. Finn, G. Titus, D. Horstman, and K. Hofmann, *Proc. Natl. Acad. Sci. U.S.A.* **81,** 7328 (1984).

[46] R. A. Kohanski and M. D. Lane, *J. Biol. Chem.* **260,** 5014 (1985).

[47] B. B. Gordon and S. D. J. Pena, *Biochem. J.* **208,** 351 (1982).

[48] K. Ogata, M. Arakawa, T. Kasahara, K. Shioiri-Nakano, and K. Hiraoka, *J. Immunol. Methods* **65,** 75 (1983).

[49] J. J. Leary, D. J. Brigati, and D. C. Ward, *Proc. Natl. Acad. Sci. U.S.A.* **80,** 4045 (1983).

[50] C. E. Argarana, I. D. Kuntz, S. Birken, R. Axel, and C. R. Cantor, *Nucleic Acids Res.* **14,** 1871 (1986).

[51] Y. Hiller, J. M. Gershoni, E. A. Bayer, and M. Wilchek, *Biochem. J.* **248,** 167 (1987).

[52] A. Pähler, W. A. Hendrickson, M. A. Gawinowicz Kolks, C. E. Argarana, and C. R. Cantor, *J. Biol. Chem.* **262,** 13933 (1987).

[53] P. C. Weber, D. H. Ohlendorf, J. J. Wendoloski, and F. R. Salemme, *Science* **243,** 85 (1989).

[54] See M. Wilchek and E. A. Bayer, this volume [3].

TABLE II
EARLIER COVERAGE OF AVIDIN–BIOTIN TECHNIQUES IN THIS SERIES

Subject	Volume [Chapter][a]
Biotin	
Assays for biotin	18A[62], 18A[63], 18A[74], 57[37], 62[49], 62[50], 62[51], 122[12], 122[13]
Biotin analogs	18A[66], 18A[67], 18A[68], 18A[70], 18A[71], 109[37],
Biotin derivatives	18A[64], 18A[65], 34[20], 46[72], 57[37], 62[54], 62[55], 62[63], 70[13], 83[12], 92[37], 109[37], 138[35], 151[39]
Biotinylated proteins	62[55], 62[57], 83[12], 92[20], 92[37], 121[43], 121[68], 133[25], 149[11], 150[38], 159[56]
Biotinylated hormones	98[22], 109[37], 124[5]
Biotinylated glycoconjugates	62[55], 83[12], 138[35]
Biotinyl lipids	138[35], 149[11]
Biotinylated nucleic acids	164[25], 168[54], 170[4]
Immobilized biotin	34[20], 71[8], 122[13], 170[4]
Iminobiotin and derivatives	109[37], 122[14], 122[15], 135[7]
Avidin	
Assays for avidin	18A[74], 18A[75], 18A[76], 57[37], 62[51], 62[52], 62[53], 122[12], 122[13]
Modified avidins	92[20], 109[37], 133[25]
Fluorescent avidin	85[48], 92[37], 122[12], 150[38]
Radioactive avidin	62[53], 70[13]
Avidin conjugates	62[55], 83[12], 92[37], 93[20], 149[11]
Immobilized avidin	112[6], 121[68], 122[13], 124[5], 125[40]
Antiavidin antibodies	18A[76], 122[15]
Preparation of biotin-binding proteins	
Avidin	18A[73], 34[20], 122[14]
Streptavidin	109[37]
Egg-yolk protein	62[56]
Antibiotin antibodies	62[57]
Holocarboxylase synthetase	107[15]
Applications	
Affinity cytochemistry	62[55], 83[12], 85[48], 103[11], 129[19], 151[39], 168[54]
Affinity labeling	62[63], 164[25]
Immunoassay	73[33], 92[20], 103[27], 121[43], 133[25]
Flow cytometry	103[13], 150[38]
Affinity chromatography	34[20], 109[37], 122[14], 122[15], 124[5], 125[40], 135[7], 170[4]
Blotting	138[35], 159[56]

[a] *Methods in Enzymology,* Vol. 18A (1970) [62] A. F. Carlucci; [63] D. B. McCormick and J. A. Roth; [64] H. Ruis, D. P. McCormick, and L. D. Wright; [65] J. E. Christner and M. J. Coon; [66] K. Ogata; [67] J. P. Tepper, H.-C. Li, D. B. McCormick, and L. D. Wright; [68] K. Ogata; [70] S. Iwahara, D. B. McCormick, and L. D. Wright; [71] H. Ruis, R. N. Brady, D. B. McCormick, and L. D. Wright; [73] N. M. Green; [74] N. M. Green; [75] R.-D. Wei; [76] S. G. Korenman and B. W. O'Malley
Vol. 34 (1974) [20] E. A. Bayer and M. Wilchek

References to TABLE II (*continued*)

Vol. 46 (1977) [72] E. A. Bayer and M. Wilchek
Vol. 57 (1978) [37] H. R. Schroeder, R. C. Boguslaski, R. J. Carrico, and R. T. Buckler
Vol. 62 (1979) [49] R. L. Hood; [50] K. Dakshinamurti and L. Allan; [51] H. J. Lin and J. F. Kirsch; [52] H. A. Elo and P. J. Tuohimaa; [53] M. S. Kulomaa, H. A. Elo, and P. J. Tuohimaa; [54] B. K. Sinha and C. F. Chignell; [55] E. A. Bayer, E. Skutelsky, and M. Wilchek; [56] H. W. Meslar and H. B. White III; [57] M. Berger; [63] E. A. Bayer and M. Wilchek
Vol. 70 (1980) [13] J. J. Langone
Vol. 71 (1981) [8] M. L. Ernst-Fonberg and J. S. Wolpert
Vol. 73 (1981) [33] J.-L. Guesdon and S. Avrameas
Vol. 83 (1982) [12] E. A. Bayer, E. Skutelsky, and M. Wilchek
Vol. 85 (1982) [48] K. Wang, J. R. Feramisco, and J. F. Ash
Vol. 92 (1983) [20] C. Stähli, V. Miggiano, J. Stocker, Th. Staehelin, P. Häring, and B. Takacs; [37] L. Wofsy
Vol. 93 (1983) [20] T. I. Ghose, A. H. Blair, and P. N. Kulkarni
Vol. 98 (1983) [22] H. T. Haigler
Vol. 103 (1983) [11] R. D. Broadwell and M. W. Brightman; [13] J. C. Cambier and J. G. Monroe; [27] G. E. Trivers, C. C. Harris, C. Rougeot, and F. Dray
Vol. 107 (1984) [15] N. H. Goss and H. G. Wood
Vol. 109 (1985) [37] F. M. Finn and K. H. Hofmann
Vol. 112 (1985) [6] L. Illum and P. D. E. Jones
Vol. 121 (1986) [43] J. E. Shively, C. Wagener, and B. R. Clark; [68] T. V. Updyke and G. L. Nicolson
Vol. 122 (1986) [12] R. D. Nargessi and D. S. Smith; [13] L. Goldstein, S. A. Yankofsky, and G. Cohen; [14] G. A. Orr, G. C. Heney, and R. Zeheb; [15] R. Zeheb and G. A. Orr
Vol. 124 (1986) [5] E. Hazum
Vol. 125 (1986) [40] P. Dimroth
Vol. 129 (1986) [19] C.-T. Lin and L. Chan
Vol. 133 (1986) [25] G. Barnard, E. A. Bayer, M. Wilchek, Y. Amir-Zaltsman, and F. Kohen
Vol. 135 (1987) [7] M. T. W. Hearn
Vol. 138 (1987) [35] M. Wilchek and E. A. Bayer
Vol. 149 (1987) [11] B. Rivnay, E. A. Bayer, and M. Wilchek
Vol. 150 (1987) [38] D. M. Segal, J. A. Titus, and D. A. Stephany
Vol. 151 (1987) [39] G. H. Smith
Vol. 159 (1988) [56] R. L. Kincaid, M. L. Billingsley, and M. Vaughan
Vol. 164 (1988) [25] J. Ofengand, R. Denman, K. Nurse, A. Liebman, D. Malarek, A. Focella, and G. Zenchoff
Vol. 168 (1989) [54] P. C. Emson, H. Arai, S. Agrawal, C. Christodoulou, and M. J. Gait
Vol. 170 (1989) [4] T. M. Herman

Indeed, the prediction of future breakthroughs in the use of avidin–biotin technology is beyond our imaginative capacity, and we leave development of new directions to the readers of this volume and future users of this universally applicable technology.

[3] Applications of Avidin–Biotin Technology: Literature Survey

By MEIR WILCHEK and EDWARD A. BAYER

Because of the extensive literature on avidin–biotin technology, we are presenting a survey of reference material to date in tabular form (see Ref. 1 in list at the end of this article). The tables are divided more or less according to the sections in this volume. In order to assist readers in understanding the details therein, a list of pertinent abbreviations is given in Table I.

Table II presents the experimental details from studies which have used avidin columns for the isolation of target material. Native biotin-containing systems (e.g., biotin-requiring enzymes) could be isolated directly on such columns. The isolation of other materials, such as membrane proteins and glycoconjugates, is dependent on the mediation of a biotinylated binder, e.g., antibody or lectin. One of the problems in the use of avidin–biotin technology for isolation purposes is the difficulty encountered in eluting the bound material from the column. Therefore, elution conditions used in a given study are also listed in Table II. In some cases, avidin columns have been used for the selective elimination of a given protein or cell type, such that elution from the column was not required.

Tables III, IV, and V provide information relating to cytochemical localization studies. These types of studies usually involve the microscopic visualization of membrane-based sites mediated through the avidin–biotin interaction. In some cases, target sites have been directly biotinylated using group-specific biotinylating reagents. In others, biotinylated lectins, hormones, antibodies, and other binders were employed to mediate between the target molecule and the avidin–probe conjugate or complex. In order to give readers an overall view of the various approaches used by different groups, we summarize in Tables III–V details of the biotinylated binder as well as the avidin–probe system.

The use of protein blotting for identification and analysis of target molecules in complex mixtures of biological material centers on two basic approaches: (1) direct labeling of blotted material using biotinylating reagents and (2) indirect labeling using biotinylated binders. The staining methodologies developed have been used for blot transfers of material separated by SDS–PAGE or isoelectric focusing, as well as for direct detection of target material on dot blots. The biotinylation step may be carried out either in solution (before transfer to nitrocellulose) or directly on the blotted material. Details regarding the reagents, binders, and probes that have been used are listed in Tables VI and VII.

One of the most prevalent uses of avidin–biotin technology in recent years has been for immunoassays. In many cases, the signal is enhanced 10-fold or more, owing to the four biotin-binding sites of avidin and the multiple biotinyl groups on the derivatized antibody. The approach is suitable for both monoclonal and polyclonal antibody preparations; any immunoglobulin class may be employed; the biotinylation step can be carried out either on the primary antibody or on a biotinylated secondary antibody system; both avidin–probe conjugates and complexes have been used; and sequential as well as homogeneous assays have been reported. Details of such immunoassay systems are given in Table VIII.

Finally, one of the major reasons that avidin–biotin technology has gained prominence in the 1980s is its application to gene probing. Using a suitable probe, an individual nucleic acid segment can be detected from a large heterogeneous population. Today, most of the work in gene probe development deals with probe design and labeling procedures, and biotin can be introduced into the growing DNA chain through the use of a biotinylated nucleotide triphosphate in conjunction with nick-translation techniques. Alternatively, some of the reactive biotin-containing derivatives can be used for direct introduction of the biotin moiety into DNA. Synthetic oligonucleotides (20 to 30 bases in length), to which biotin residues have been introduced either on the 3′ or 5′ terminus, have also been successfully applied. Details of the various approaches for biotinylation of DNA and their application for hybridization are summarized in Table IX.

It is clear that we could not include all of the "avidin–biotin" literature in Tables II–IX (even with the availability of computerized literature surveys, etc.), owing to (1) the many contributions in which the terms avidin or biotin are not included in the title or key words and (2) the unavailability of certain journals and books in the libraries at our disposal. Even if we could have compiled all of the pertinent literature, one would probably question the worth in the end, since the important aspects or relationships among the various entries would have been masked as a consequence of the volume of material presented. These tables thus serve as a starting point for the interested researcher.

TABLE I
ABBREVIATIONS USED IN TABLES II–IX

Abbreviation	Definition
Ab	Antibody
MAb	Monoclonal antibody
Ab_1, B–Ab_2	Sequential use of polyclonal primary Ab and biotinylated secondary Ab
MAb_1, B–Ab_2	Sequential use of monoclonal primary Ab and biotinylated secondary Ab
ABC	Complexes formed by mixing native avidin with biotinylated probe (identified parenthetically)
ABC (B–AP)	Complexes containing native avidin and biotinylated alkaline phosphatase
ABC (B–HRP)	Complexes containing native avidin and biotinylated peroxidase
acidP	Acid phosphatase
AP	Alkaline phosphatase
Av	Avidin
Av–AP	Avidin-conjugated alkaline phsophatase
Av–Fer	Avidin-conjugated ferritin
Av–gold	Avidin-conjugated colloidal gold
Av–HRP	Avidin-conjugated peroxidase
B–	Covalently coupled biotin
BcapHZ	*N*-Biotinyl-6-aminocaproylhydrazide
BcapNHS	*N*-Biotinyl-6-aminocaproyl-*N*-hydroxysuccinimide ester
BCHZ	Biocytin hydrazide (N^6-biotinyl-L-lysine hydrazide)
BHZ	Biotin hydrazide
BNHS	Biotin *N*-hydroxysuccinimide ester
BNHS (SO_3)	Biotin *N*-hydroxy(sulfo)succinimide ester
B–dUTP	5-(*N*-Biotinyl-3-aminoallyl)deoxyuridine 5′-triphosphate
B–11-dUTP	5′[*N*-(3-Aminoallyl)-*N*′-biotinyl-6-aminocaproyl]deoxyuridine 5′-triphosphate
B–16-dUTP	5′[*N*-(3-Aminoallyl)-*N*-(*N*-biotinyl 6-aminocaproyl)-4-aminobutyryl]deoxyuridine 5′-triphosphate
CHAPS	3′[(3-Chloramidopropyl)dimethylammonio]-1-propane
Con A	Concanavalin A
DBB	*p*-Diazobenzoylbiocytin
ELISA	Enzyme-linked immunosorbent assay
EM	Electron microscopy
FACS	Fluorescence-activated cell sorting (cell cytometry)
Fer	Ferritin
FITC-Av	Fluorescein-derivatized avidin
FM	Fluorescence microscopy
βG	β-Galactosidase
Gal ox.	Galactose oxidase
Glc ox.	Glucose oxidase
Gold	Colloidal gold
GSI	Lectin from *Griffonia simplicifolia*

TABLE I (*continued*)

Abbreviation	Definition
HRP	Peroxidase from horseradish
IL-2	Interleukin 2
LM	Light microscopy
MPB	3-(*N*-Maleimidopropionyl)biocytin
PC	Penicillinase
PE	Phycoerythrin
Photobiotin	*N*-(4-Azido-2-nitrophenyl)-*N*′-(*N*-biotinyl 3-aminopropyl)-*N*′-methyl-1,3-propanediamine
RIA	Radioimmunoassay
RITC-Av	Rhodamine-derivatized avidin
SBA	Soybean agglutinin
SDS	Sodium dodecyl sulfate
StABC	Complexes formed by mixing native avidin with biotinylated probe (identified parenthetically)
StABC (B–AP)	Complexes containing native streptavidin and biotinylated alkaline phosphatase
StABC (B–HRP)	Complexes containing native streptavidin and biotinylated peroxidase
StAv	Streptavidin
StAv–AP	Streptavidin-conjugated alkaline phosphatase
StAv–Fer	Streptavidin-conjugated ferritin
StAv–gold	Streptavidin-conjugated colloidal gold
StAv–HRP	Streptavidin-conjugated peroxidase
WGA	Wheat germ agglutinin

TABLE II
ISOLATION OF BIOLOGICALLY ACTIVE MATERIALS USING AVIDIN COLUMNS

Material purified	Binder	Elution conditions	Ref.[a]
Native biotin-containing systems			
B–transcarboxylase peptides	None	6 *M* Guanidine-HCl, pH 1.5	2
Acetyl-CoA carboxylase aposubunits	None	6 *M* Guanidine-HCl, pH2	3,4
Apo(acetyl-CoA carboxylase)	None	Effluent fractions pooled	5,6
Transcarboxylase subunits	None	Differential pH	7
B–polypeptides	None	SDS–urea	8
B–enzymes	None	Biotin	9,10
Propionyl-CoA carboxylase	None	Biotin gradient	11
Acetyl-CoA carboxylase	None	Biotin	12
Mitochondrial carboxylases	None	Biotin gradient	13
Sodium transport enzyme	None	Biotin	14

(*continued*)

TABLE II (*continued*)

Material purified	Binder	Elution conditions	Ref.[a]
Membrane components and receptors			
Iminobiotin-labeled membrane proteins and glycoproteins	None	Acetate, pH 4	15,16
Surface glycoproteins	B–Con A	2% SDS	17
Membrane antigens	B–MAb	SDS (boiling)	18
Terminal deoxynucleotidytransferase	B–MAb	2 *M* $MgCl_2$	19
IgE receptor	B–IgE	None	20
	IgE, B–MAb	SDS (boiling)	21
Estrogen receptor	B–estradiol	Estradiol	22
Glucocorticoid receptor	B–dexamethasone	Dexamethasone	23
Gonadotropin-releasing hormone receptor	B–GnRH	CHAPS, Tris/glycerol, pH 5.5	24
Insulin receptor	B–insulin	Acetate, pH 5.0	25
	B–insulin	1 *M* NaCl, biotin	26
Lactogenic and somatogenic receptors	B–growth hormone	5.0 *M* $MgCl_2$ or SDS	27
Opioid receptor	B–enkephalin	Enkephalin	28
Cytomegalovirus hydrophobic proteins or glycoproteins	(1) B–MAb (2) MAb_1, B–Ab_2	Glycine-HCl, pH 2.2	29
Selective retrieval or elimination of target material			
B–lectin or antibody	None	None	30
B–thymocytes	None	None	31
B–subunit of transcarboxylase	None	None	32
B–mitogen or inducer	None	None	33
B–enzymes	None	None	34
Daudi lymphoblasts	B–MAb	None	35
Lymphocyte subpopulations	B–Ab	(1) Rosette formation, erythrocyte lysis	36
	B–MAb	(2) None (rosette formation)	37–39
	B–MAb	(3) Mechanical agitation	40
Human B lymphocytes (MAb-producing cells)	B–antigen	Selection by FACS	41
Nucleic acids and associated complexes			
Drosophila DNA, sea urchin DNA, *Leishmania* DNA	B–RNA	None	42
Drosophila DNA	B–RNA	1 *M* NaOH	43,44
B–tRNA	None	6 *M* Guanidine-HCl, pH 2.5	45
B–rRNA fragment (affinity-labeled)	None	70% formic acid	46
R loop structures	B–tRNA	99% formamide	47
B–DNA	None	None	48
B–RNA	None	6 *M* Guanidine-HCl, pH 2.5	49
B–nucleosomes	None	SS-reduction of cleavable biotin analog	50,51
Plasmid DNA	B–DNA	0.1 *M* NaOH	52
Spliceosomes	B–RNA	90° in SDS	53

TABLE II (*continued*)

Material purified	Binder	Elution conditions	Ref.
Transcriptionally active B–DNA	None	SS reduction of cleavable biotin analog	54
Transcription factors	B–DNA	Differential salt concentration	55
Sequence-specific DNA-binding factor	B–DNA	Differential salt concentration	56
Recombinant plasmids	B–DNA	(1) SS reduction of cleavable biotin analog (2) Heat, low ionic strength, phenol extraction	57
Neisseria DNA	B–DNA, B–RNA	EDTA, 1 *M* NaCl, 0.1% SDS	58

[a] See reference list at end of article.

TABLE III
DIRECT LABELING OF BIOLOGICALLY ACTIVE MATERIALS USING AVIDIN–BIOTIN TECHNOLOGY

Target material	Biotin reagent	Probe	Comments	Ref.[a]
Biotin-tagged sites				
Acholeplasma laidlawii membranes	(1) BNHS (2) Periodate/BHZ (3) Gal ox./BHZ	Av–Fer	EM	59
Erythrocyte ghosts				
Membrane sialic acids				
Intact erythrocytes and lymphocytes	Periodate/BHZ	Av–Fer	EM	60–63
Thalassemic erythrocytes				64
Membrane proteins				
Acholeplasma laidlawii membranes	BNHS	Av–Fer	EM	65
Halobacterium halobium membranes				66
Surface sites				
Intact yeast cells	Periodate/BHZ	Av–Fer	EM	67
Membrane galactose residues				
Intact erythrocytes and lymphocytes	Gal ox./BHZ	Av–Fer	EM	68

[a] See reference list at end of article.

TABLE IV
IMMUNOCYTOCHEMICAL LOCALIZATION OF BIOLOGICALLY ACTIVE MATERIAL

Target material	Binder system	Probe	Comments	Ref.[a]
Erythrocyte surface antigens	B–Ab	Av–Fer	EM	30
Human cell surface antigens	Ab_1, B–Ab_2	ABC (B–HRP)	LM	69
Blood group antigens	Ab_1, B–Ab_2	ABC (B–HRP)	LM	70
Urothelial blood group isoantigens	B–Ab	ABC (B–HRP)	LM, biotinylation of affinity-purified Ab directly on column	71
Urothelial antigens	Ab_1, B–Ab_2	ABC (B–HRP)	LM	72
Various antigens (hormones, effectors, and enzymes)	Ab_1, B–Ab_2	Av, [^{3}H]biotin	LM, EM	73
Ia antigen	MAb_1, B–Ab_2	ABC (B–HRP)	EM	74
Various antigens	(1) MAb_1, B–Ab_2 (2) Ab_1, B–Ab_2	ABC (B–Glc ox.)	LM, double labeling	75
Tac antigen	MAb_1, B–Ab_2	Av–Texas red	FM, double labeling	76
Opsin	(1) B–Ab (2) Ab_1, B–Ab_2 (3) MAb_1, B–Ab_2	(1) Av–Fer (2) StAv–gold (3) StAv–Texas red	EM, FM	77–79
S-100 protein	Ab_1, B–Ab_2	ABC (B–HRP)	LM	80
Lactoferrin	Ab_1, B–Ab_2	ABC	LM	81
ras gene product	MAb_1, B–Ab_2	ABC (B–HRP)	LM	82
Laminin	MAb_1, B–Ab_2	ABC (B–HRP)	LM, EM	83
Keratin	Ab_1, B–Ab_2	ABC (B–HRP)	LM	84
Odorant-binding protein	Ab_1, B–Ab_2	ABC (B–HRP)	LM	85
Bone phosphoprotein	Ab_1, B–Ab_2	(1) ABC (B–HRP) (2) Av–gold	LM	86
Albumin-conjugated erythrocytes	B–Ab	StABC (B–fluorescent microspheres)	FACS	87
Intracellular myosin	B–Ab	FITC-Av	FM	88
Intermediate filament proteins (kidney tumors)	MAb_1, B–Ab_2	ABC (B–HRP)	LM	89

Golgi proteins	B–MAb	Av–HRP	EM	90
Fibronectin	Ab_1, B–Ab_2	ABC (B–HRP)	LM	91
Neurofilament protein and galactocerebroside	B–Ab	FITC-Av, RITC-Av	FM	92
Platelet factor 4	B–Ab	(1) FITC-Av (2) Av–Fer	FM, EM	93
Cathepsin B	Ab_1, B–Ab_2	ABC (B–HRP)	LM	94
Terminal deoxynucleotidyl-transferase	B–Ab	(1) FITC-Av (2) Av–HRP	FM, LM	95 96
Enolase	Ab_1, B–Ab_2	Av–HRP	EM	97
γ-Glutamyltransferase	Ab_1, B–Ab_2	ABC (B–HRP)	LM	98
Relaxin	Ab_1, B–Ab_2	ABC (B–HRP)	LM	99,100
Kallikrein-like enzymes	Ab_1, B–Ab_2	ABC (B–HRP)	LM	101
Thioesterase	Ab_1, B–Ab_2	ABC (B–HRP)	LM	102
Chymotrypsinogen	B–Ab	Av–HRP	LM	103
Calmodulin and phospholipase A_2	Ab_1, B–Ab_2	Av–Fer	EM	104
Protein kinase	Ab_1, B–Ab_2	StAv–gold	EM	105
Choline acetyltransferase	Ab_1, B–Ab_2	ABC (B–HRP)	LM	106
Myoglobulin and creatine kinase	Ab_1, B–Ab_2	ABC (B–HRP)	LM	107
Tyrosine monooxygenase	Ab_1, B–Ab_2	ABC (B–HRP)	LM, EM, double labeling	108
Epidermal growth factor	Ab_1, B–Ab_2	ABC (B–HRP)	LM	109
Insulin, glucagon-like peptides	Ab_1, B–Ab_2	ABC (B–HRP)	LM	110
Gonadotropin-releasing hormone	(1) Ab_1, B–Ab_2 (2) B–Ab	(1) ABC (B–HRP) (2) FITC-Av	LM, FM, double labeling amplified signal	111
Human chorionic gonadotropin	Ab_1, B–Ab_2	ABC (B–HRP)	EM	112
ACTH	Ab_1, B–Ab_2	ABC (B–Glc ox.)	LM, double labeling	113
Substance P	Ab_1, B–Ab_2	ABC (B–HRP)	EM	114
LH and FSH, ACTH	Ab_1, B–Ab_2	ABC (B–HRP)	LM, EM	115
Serotonin	Ab_1, B–Ab_2	ABC (B–HRP)	LM	116
CRF	MAb_1, B–Ab_2	ABC (B–HRP)	LM	117

(continued)

TABLE IV (*continued*)

Target material	Binder system	Probe	Comments	Ref.[a]
Pituitary hormones	Ab_1, B–Ab_2	ABC (B–HRP)	LM, EM	118,119
Glucagon-like peptide	Ab_1, B–Ab_2	ABC (B–HRP)	LM	120
Anti-Müllerian hormone	Ab_1, B–Ab_2	StABC (B–HRP)	EM	121
B–IgG	None	(1) ABC (B–HRP) (2) ABC (B–Fer)	LM, EM	122
Intracellular immunoglobulins	Ab_1, B–Ab_2	ABC (B–HRP)	EM	123
Intracellular IgG	B–Ab	ABC (B–HRP)	LM, EM	124
IgG_3 (K isotope)	B–MAb	ABC (B–HRP)	LM	125
IgG	(1) Ab_1, B–Ab_2 (2) Ab, B–protein A	ABC (B–HRP)	LM	126
Membrane-bound MAb	B–Ab	ABC (B–HRP)	EM	127
Immune complexes	B–IgG	Av–HRP	LM	128
Fc receptors	Ab_1, B–Ab_2	ABC (B–HRP)	LM	129
CR1 complement receptors	B–MAb	StAv–PE, B–Ab, StAv–PE	FM, FACS, enhanced signal	130
Platelet glycoprotein IIb–IIIa complex	B–MAb	(1) FITC-Av (2) Av–PE	FM	131
Polymeric immunoglobulin receptors	B–IgA dimer	Av–gold		132
Estrogen receptors	MAb_1, B–Ab_2	ABC (B–HRP)	LM	133
Progesterone receptor	Ab_1, B–Ab_2	ABC (B–HRP)	LM	134
Synaptic associations	Ab_1, B–Ab_2	ABC (B–HRP)	EM, double labeling	135
HIV antigen	Ab_1, B–Ab_2	(1) ABC (B–HRP) (2) Av–HRP	LM, double labeling	136
AIDS antigen in macaques	(1) MAb_1, B–Ab_2 (2) Ab_1, B–Ab_2	ABC (B–HRP)	LM	137
Rift Valley fever virus	B–Ab	ABC (B–HRP)	LM	138
Papillomavirus	Ab_1, B–Ab_2	ABC (B–HRP)	LM	139
Herpes simplex virus	B–Ab	FITC-Av	FM	140
Flavivirus	B–MAb, B–Ab	FITC-StAv	FM	141
Bacterial surface agglutinogen	B–Ab	Av–Fer	EM	142
Cellulosome	B–Ab	Av–Fer	EM	143
Flagellar antigens in *Euglena*	Ab_1, B–Ab_2	(1) FITC-Av (2) Av–Fer	FM, EM	144
Trophoblasts	MAb_1, B–Ab_2	ABC (B–HRP)	LM, double labeling	145

		(2) Av–erythrocytes		
T and B cell subsets	(1) B–MAb (2) MAb_1, B–Ab_2	(1) Av–HRP (2) ABC (B–HRP)	LM, double labeling	149–152
NK cells	MAb_1, B–Ab_2	ABC (B–HRP)	LM, EM, double labeling	153
Suppressor/cytotoxic lymphocytes	B–Ab	ABC (B–HRP)	EM	154
Gastrin-producing cells	Ab_1, B–Ab_2	ABC (B–HRP)	LM	155
IL-2 receptor-bearing cells	MAb_1, B–Ab_2	ABC (B–HRP)	LM	156
Liver-specific protein	Ab_1, B–Ab_2	ABC (B–HRP)	LM	157
Individual cells	MAb_1, B–Ab_2	ABC (B–HRP)	LM	158
Lymphocyte infiltration	MAb_1, B–Ab_2	ABC (B–HRP)	LM	159
γ-Aminobutyric acid axons	Ab_1, B–Ab_2	ABC (B–HRP)	LM, EM	160
Human platelets	MAb_1, B–Ab_2	ABC (B–HRP)	LM	161
Macrophage antigen	MAb_1, B–Ab_2	ABC (B–HRP)	LM	162
Lymphoid cell subpopulations	MAb_1, B–Ab_2	ABC (B–HRP)	LM	163
Granulocyte surface antigens	Ab_1, B–Ab_2	ABC (B–HRP)	LM	164
Epidermal dendritic cells	B–MAb	Av–PE	FACS	165
Quail cells in embryonic chimeras	Ab_1, B–Ab_2	Av–HRP	LM	166
Thy-1.2 antigen on thymocytes	B–Ab	Av–(fluorescent microspheres)	FACS	167
Lymphocyte Leu antigens	B–Ab	ABC (B–PE)	FACS	168
Human B lymphocytes	B–antigen	FITC-Av	FACS	41
Gastrointestinal carcinoma-associated antigen	MAb_1, B–Ab_2	ABC (B–HRP)	LM	169
Melanoma-associated antigens, HLA antigens, Ia antigens	MAb_1, B–Ab_2	(1) ABC (B–HRP) (2) FITC-Av	LM, FM	170
Carcinoembryonic antigen	B–MAb	ABC (B–HRP)	LM	171
Tumor cells	B–MAb	(1) FITC-Av (2) Av, FITC-Ab (3) FITC-Av, FITC-Ab	FACS, enhanced signal	172
Mammary carcinoma cells	MAb_1, B–Ab_2	ABC (B–HRP)	LM	173,174
Estrogen-induced protein (normal and malignant cells and tissues)	MAb_1, B–Ab_2	ABC (B–HRP)	LM	175
Melanoma cells	MAb_1, B–Ab_2	ABC (B–HRP)	LM	176

[a] See reference list at end of article.

TABLE V
NONIMMUNOCYTOCHEMICAL LOCALIZATION OF BIOLOGICALLY ACTIVE MATERIAL

Target material	Binder	Probe	Comments	Ref.[a]
Lectin-induced labeling				
Erythrocytes	Various B–lectins	Av–Fer	EM	30
Microsomal glycoproteins	B–Con A	Av–Fer	EM	177
Macrophages	B–Con A	Av–HRP	EM	178
Skeletal muscle	Various B–lectins	Av–HRP	LM	179
Normal and malignant tissue	B–peanut agglutinin	ABC (B–HRP)	LM	180
Skin	Various B–lectins	ABC (B–HRP)	LM	181
Lymphoma cells	Various B–lectins	ABC (B–HRP)	LM	182
Retinal pigment epithelium	Various B–lectins	(1) ABC (B–HRP) (2) Av–Fer (3) Av B–ferritin	EM	183
Germ cell tumors	Various B–lectins	ABC (B–HRP)	LM	184
Hepatocytes	Various B–lectins	ABC (B–HRP)	LM, EM	185
Prostatic epithelial cells	Various B–lectins	ABC (B–HRP)	LM	186
Glycoprotein storage diseases	Various B–lectins	ABC (B–HRP)	LM	187
Vascular endothelia in brain tumors	B–agglutinin I from *Ulex europaeus*	ABC (B–HRP)	LM	188
Capillary endothelia of cerebral cortex	Various B–lectins	ABC (B–HRP)	EM	189
Parietal cells of gastric mucosa	Various B–lectins	ABC (B–HRP)	LM, EM	190
Retinal rod outer segments	B–Con A	StAv–Texas red	FM	77
Urinary bladder epithelium	Various B–lectins	ABC (B–HRP)	EM	191
Cat duodenum	Various B–lectins	Av–gold	LM	192
Pancreatic carcinomas	Various B–lectins	ABC (B–HRP)	LM	193
Normal vascular epithelium	Various B–lectins	ABC (B–HRP)	LM	194
Effector–receptor interactions				
β-Adrenergic receptor	B–propranolol	Av–Fer	EM	195
Acetylcholine receptors	B–α-bungarotoxin	RITC-Av	FM	196
Acetylcholine receptor	B–toxin	Native avidin	EM	197
Epidermal growth factor receptor	B–EGF	Av–gold	EM	198
Tetanus toxoid receptors (on B cells)	B–tetanus toxin	FITC-Av	FM, FACS	199

TABLE V (*continued*)

Target material	Binder	Probe	Comments	Ref.[a]
Sodium channel	ABC (B–scorpion toxin)	B–HRP	LM, EM	200,201
Gonadotropin-releasing hormone receptor	B–GnRH	ABC (B–HRP)	LM	202
CRF receptor	B–CRF	ABC (B–HRP)	LM	203
Receptors for *Pseudomonas* endotoxin A (internalized by fibroblasts)	B–toxin	Av–gold	EM	204
Parathyroid hormone receptor	B–PTH	ABC (B–HRP)	LM	205
Erythropoietin receptors on myeloma cells	[B–peptide:StABC:B–(fluorescent microspheres)]		FACS	87
Miscellaneous binder systems				
B–subunit of transcarboxylase	None	(1) Native avidin (2) Antibiotin Ab	EM EM	206 32
Nonmuscle contractile proteins	B–heavy meromyosin	FITC-Av	FM	88,93, 207,208
Biotin transport components	None	B–Fer	EM	209
Hydrophobic sites (liposome fusion)	B–gangliosides	Av–Fer	EM	63
	B–lipids (liposomes)	(1) Av–Fer (2) ABC B–(fluorescent liposomes)	EM FM	210 211
G_{M1} gangliosides	B–choleragen	FITC-Av or RITC-Av	FM	212,213
Ornithine decarboxylase	B–α-difluoromethyl-Orn	Av–HRP	LM	214
C3 receptors	B–C3	(1) FITC-Av (2) Av–Fer	(1) FACS (2) EM	215
Avidin	None	B–undecagold	EM	216
Hyaluronic acid	B–link protein	Av–HRP	LM, EM	217
Heavy meromyosin	B–ADP	Avidin	EM	218
B–IgG	None	ABC (B–HRP)	LM, EM	219

[a] See reference list at end of article.

TABLE VI
DIRECT LABELING OF PROTEIN BLOTS

Target material	Biotin reagent	Probe	Comments	Ref.[a]
Various protein standards	BNHS	Av–HRP	Prelabeled protein solutions	220,221
	BNHS	[^{125}I]StAv	Prelabeling of cells	222
	BNHS	StAv, B–HRP	Prelabeled protein solutions	223
	BNHS	Bispecific MAb, HRP	Luminescent detection	224
	BNHS(SO_3)	(1) Av–HRP (2) StAv, anti-StAv, Ab_2–HRP	BNHS staining of blots, amplified signal	225
	Photobiotin	(1) Av–AP (2) StAv–HRP	Prelabeled protein solutions	226
	MPB	ABC (B–HRP)	Staining of blots	227,228
	DBB	ABC (B–AP)	Staining of blots	229
	Periodate/BHZ	Av–HRP	Prelabeled glycoproteins	230,231
Biotin-containing polypeptides	None	FITC-Av	Direct labeling of gels	8
Respiratory syncytial viral proteins	BNHS	Av–HRP	Selective immunoprecipitation of B–proteins using MAb	220
Erythrocyte membrane proteins				
SH, S—S	MPB	ABC (B–HRP)	(1) MPB staining in solution (2) MPB staining of blots (3) MPB staining of membranes	228
Tyr, His	DBB	ABC (B–AP)	DBB staining of blots	229
Erythrocyte membrane glycoproteins				
Sialic acids	Periodate	ABC-HZ (AP)	Staining of blots	232
	Periodate/BcapHZ	(1) ABC (B–HRP) (2) Av–HRP or StAv–HRP (2) Av–HRP or StAv–HRP	Staining of blots	233
Sialyl and/or Gal moieties	Periodate/BCHZ or Gal ox./BCHZ	(1) StABC (B–AP) (2) [^{125}I]StAv	(1) Staining of blots (2) Staining in solution (3) Staining of membranes	222,234
Leukocyte membrane proteins	BNHS(SO_3)	Av–HRP	Prelabeling of cells	235

[a] See reference list at end of article.

TABLE VII
INDIRECT LABELING OF PROTEIN BLOTS

Target molecule	Binder	Probe	Ref.[a]
Neoglycoproteins	B–SBA	Av–HRP	236
Glycosphingolipids	B–GSI	ABC (B–HRP)	237
Glycoproteins of parotid saliva	(1) WGA (2) Various B–lectins	Av–HRP	236
Salivary glycoproteins	B–Con A	ABC (B–HRP)	238
Glycoconjugates in human milk fat globule	Various B–lectins	Av–HRP	239
Fibroblast surface glycoproteins	Various B–lectins	ABC (B–HRP)	240
Myoblast glycoproteins	(1) Various B–lectins (2) Ab_1, B–Ab_2	ABC (B–HRP)	241
Calmodulin-binding proteins	B–calmodulin	ABC (B–enzymes)	242
Complement proteins (crossed immunoelectrophoresis)	MAb_1, B–Ab_2	StABC (B–HRP)	243
Fibronectin	Ab_1, B–Ab_2	ABC (B–HRP)	244
Hapten-labeled proteins	Ab_1, B–Ab_2	Av–HRP	245
Lectins	B–neoglycoproteins	Av–HRP	236
β_2-Microglobulin	Ab_1, B–Ab_2	Av–HRP	246
Myosin	B–ADP	Av–HRP	218
Native B–proteins in plants	None	[^{125}I]StAv	247
Neurofilament polypeptides	Ab_1, B–Ab_2	ABC (B–HRP)	248
Odorant-binding proteins	Ab_1, B–Ab_2	ABC (B–HRP)	85
Postsynaptic muscle cell protein	B–Ab	StAv, anti-StAv, Ab_2–AP	249
Proteins (trypsin)	B–Ab	StAv–acidP	250
Transferrin/Tau proteins in cerebrospinal fluid	Ab_1, B–Ab_2	ABC (B–HRP)	251
Uromodulin	Ab_1, B–Ab_2	ABC (B–HRP)	252
Vimentin	Ab_1, B–Ab_2	ABC (B–HRP)	253
Nerve growth factor	Ab_1, B–Ab_2	StAv–βG	254
Interferon	B–Ab	Av–AP	255
Allotype antibodies (isoelectric focusing)	B–Ab	Av–HRP	256
Human IgG (isoelectric focusing)	Ab_1, B–Ab_2	ABC (B–HRP)	257
Immunoglobulins (isoelectric focusing)	Ab_1, B–Ab_2	ABC (B–HRP)	258
DNA-binding proteins	B–DNA	StAv–HRP	259
5-Methylcytosine in DNA	Ab_1, B–Ab_2	ABC (B–HRP)	260
Low-density lipoprotein (LDL) receptor	B–LDL	StABC (B–HRP)	261
Poliovirus type 1 capsid protein	Ab_1, B–Ab_2	ABC (B–AP)	262
Lysed bacterial colonies	Ab_1, B–Ab_2	ABC (B–HRP)	263
Toxoplasma membrane antigens	B–MAb	Av–HRP	264
Carcinoembryonic antigen (isoelectric focusing)	Ab_1, B–Ab_2	Av, B–HRP	265

[a] See reference list at end of article.

TABLE VIII
IMMUNOASSAYS MEDIATED BY AVIDIN–BIOTIN TECHNOLOGY

Target molecule	Binder	Probe	Method, comments	Ref.[a]
Various proteins	B–Ab	(1) Av–enzymes (2) ABC–enzymes	ELISA	266
Apolipoprotein B	B–MAb	Av–HRP	Competitive ELISA	267
Complement proteins	MAb_1, B–Ab_2	StABC (B–HRP)	Quantitative crossed immunoelectrophoresis	243
Factor VIII	B–MAb	ABC (B–HRP)	ELISA (dot immunoblotting)	268
Ferritin	B–Ab	Av–HRP	ELISA (polystyrene beads)	269
Keratin	Ab_1, B–Ab_2	ABC (B–HRP)	ELISA	270
Lactoferrin	(1) B–antigen (2) B–Ab	(1) Av–HRP (2) ABC (B–HRP)	(1) Competitive ELISA (2) Noncompetitive ELISA	271
Lysozyme	(1) B–Ab (2) B–antigen	Av–HRP	(1) ELISA (2) Competitive ELISA	272
Myelin basic protein	B–Ab	Av–HRP	ELISA	273
Thyrotropin	B–MAb	[^{125}I]MAb	Av-coated beads for adsorption, double-MAb sandwich-type assay, RIA	274
Trout protein	Ab_1, B–Ab_2	StAv, B–[^{32}P]plasmid	RIA	275
Glycosphingolipids	(1) B–lectins (2) MAb_1, B–Ab_2	Av–HRP ABC (B–AP)	"ELISA" ELISA	276 277
Fluorescein	B–MAb	Antigen–HRP	StAv-coated B–plates for adsorption, competitive ELISA	278
5,5-Diphenylhydantoin	Ab, DPH–Av	Pyruvate carboxylase	Competitive assay, homogeneous assay	279
Soluble peptidoglycan	Ab_1, B–Ab_2	ABC (B–HRP)	"Mixed-sandwich ELISA"	280
Angiotensin-1 converting enzyme	(1) Ab_1, B–Ab_2 (2) Ab, B–protein A	ABC (B–HRP)	Antigen-bound plates, competitive ELISA	281
Staphylococcal enterotoxins	B–Ab	Av–AP	ELISA	282
Heat-stable enterotoxin from *E. coli*	Ab_1, B–Ab_2	Av–HRP	Competitive ELISA	283
Nerve growth factor	Ab_1, B–Ab_2	StAv–βG	ELISA (fluorometry)	254

Insulin	Ab_1, B–Ab_2	Ab–HRP	Av-coated beads for adsorption, ELISA	284
Human chorionic somatmammotropin	B–hCS	Av–AP	Competitive ELISA	285
Human chorionic gonadotropin	B–MAb	Av–HRP	ELISA	286
Abscisic acid	Ab_1, B–Ab_2	Av–AP	Competitive ELISA	287
Phytohormones	Ab_1, B–Ab_2	Av–AP	Competitive ELISA	288
Interferon	B–Ab	Av–AP	ELISA	255
Soluble immune response suppressor	Ab_1, B–Ab_2	ABC (B–HRP)	ELISA	289
Interleukin 2	B–MAb	Av–AP	ELISA	290
IgA	B–Ab	Av–HRP	ELISA	291
IgD	B–Ab	Av–HRP	ELISA	292
IgE	(1) B–Ab	Av–HRP	(1) ELISA	293–295
	(2) B–antigen		(2) Competitive ELISA	296
	(3) B–Ab	Ab–HRP	(3) Av-coated slide for adsorption	297
IgE, IgG	Av–Ab	B–erythrocytes	Agglutination	298
IgG allotypes	B–Ab	Av–HRP	ELISA	299
IgG secretion in humans	Ab_1, B–Ab_2	ABC (B–HRP)	ELISA	300
Mouse IgG	B–Ab	(1) StABC (B–HRP)	(1) Hybridoma screening	301
		(2) StAv–acridinium ester	(2) Chemiluminescence immunoassay	302
Mouse MAb	B–Ab	Av–HRP	ELISA, hybridoma screening	303
Ab production in cells	B–Ab	ABC (B–HRP)	ELISA	304
Antimicrosomal Ab production	Ab_1, B–Ab_2	StAv–HRP	ELISA	305
Ab to hepatitis B surface antigen	B–antigen	(1) Av–HRP (2) Anti-biotin Ab–HRP	ELISA	306,307
Encephalitis Ab	(1) B–antigen	Av–HRP	ELISA	308
	(2) B–protein A	Av–HRP		309
Antitetanus toxoid Ab	Ab_1, B–Ab_2	Av, B–AP	ELISA	199
UV lesions in DNA	Ab_1, B–Ab_2	ABC (B–HRP)	ELISA	310
Methylated DNA	Ab_1, B–Ab_2	Av–HRP	ELISA	311
Adenovirus antigen	B–Ab	StABC (B–HRP)	ELISA	312
Bovine leukemia virus	Ab_1, B–Ab_2	ABC (B–HRP)	LM	313
Flavivirus	B–Ab	StAv–HRP	ELISA	141

(continued)

TABLE VIII (*continued*)

Target molecule	Binder	Probe	Method, comments	Ref.[a]
Hantavirus serotypes	B–Ab	ABC (B–HRP)	ELISA	314
Hepatitis B surface antigen	Ab_1, B–Ab_2	(1) Av–AP (2) ABC (B–AP)	ELISA	315
Herpes simplex virus antigen	B–MAb	Av–HRP	ELISA	316
	B–Ab	StAv–AP	ELISA	317
Respiratory syncytial virus	B–Ab	Av–HRP	ELISA	318
Rotavirus	B–Ab	Av, [^{3}H]biotin	RIA	319
Simian viruses	Ab, B–protein A	Av–AP	ELISA	320
Soybean mosaic virus	B–Ab	Av–AP	ELISA	321
Virus antigens in fowl	B–Ab	StABC (B–AP	ELISA, Marek's disease	322
Bacterial antigens	B–Ab	(1) ABC (B–HRP) (2) Av–AP	ELISA	323
Microbial antigens	B–Ab	(1) ABC (B–PC) (2) ABC (B–HRP)	ELISA	324
Bacterial B–polysaccharides	B–antigen	Ab–AP	Av-coated beads for adsorption, ELISA	325
Parasite antigen (*Trypanosoma cruzi*)	Ab_1, B–Ab_2	[^{3}H]Avidin	RIA, Chagas' disease	326,327
Cell surface antigens	B–MAb	Av–AP	ELISA, cell-coated plates	328
Lymphocyte subpopulations	MAb_1, B–Ab_2	ABC (B–HRP)	Cell enumeration	329
Stem-cell antigen	B–Ab	(1) FITC-Av (2) Av, FITC-Ab (3) FITC-Av, FITC-Ab	Fluorescence intensity, enhanced signal	172
Rheumatoid factors	B–IgG (human)	Av–AP	ELISA	330,331
Myeloma-associated antigens (human)	(1) Ab_1, B–Ab_2 (2) Av–Glc ox.	Av–HRP	ELISA	332
Carcinoembryonic antigen	B–Ab	(1) Av, B–serum, [^{131}I]antigen (2) Av–HRP	(1) Av-induced precipitation, homogeneous competitive RIA (2) Competitive ELISA	171, 333–335

[a] See reference list at end of article.

TABLE IX

USE OF BIOTINYLATED GENE PROBES

Type or source of DNA or RNA	Reagent	Mode of incorporation	Mode of detection	Comments	Ref.[a]
Polytene chromosomes (*Drosophila*)	BNHS	Cytochrome *c* bridge, formaldehyde coupling to RNA	(1) ABC (B–polymer spheres) (2) Av–Fer	EM, gene mapping	43,336
	B–dUTP	Nick translation	(1) FITC-antibiotin Ab (2) Antibiotin Ab–HRP	FM, EM	337
	B–11-dUTP	Nick translation	Antibiotin Ab–gold		338
Satellite DNA (mouse)	B–dUTP	Nick translation	(1) FITC-antibiotin Ab (2) Antibiotin Ab–enzyme (3) Av–Fer	FM, LM, EM	339,340
Y chromosome (human)	B–dUTP, B–dCTP	Nick translation	(1) StABC (B–HRP) (2) Antibiotin Ab, FITC-Ab_2	LM, FM, double labeling	341–343
Human and mouse chromosomes	B–11-dUTP	Nick translation	(1) Antibiotin Ab, B–Ab_2, ABC (B–HRP) (2) StABC (B–HRP)	LM	344
Phage DNA	BNHS	Periodate/diamine labeling of RNA	Av–Fer	Gene mapping	45
λ phage DNA	B–dUTP	Nick translation	(1) StABC (B–AP) (2) Bispecific MAb (anti-biotin, anti-HRP)	DNA blotting	345 224
	BcapNHS	Transamination of cytosines	StABC (B–AP)	Enzyme assay	346
Plasmid DNA	B–dUTP	Enzymatic extension at 3′ terminus, oligo-nucleotide probes	ABC (B–HRP)	DNA blotting	347

(*continued*)

TABLE IX (*continued*)

Type or source of DNA or RNA	Reagent	Mode of incorporation	Mode of detection	Comments	Ref.[a]
	B–11-dUTP	Nick translation	(1) StABC (B–HRP) (luminol substrate)	(1) Chemiluminescence on dot blots	348
			(2) StAv–Fer	(2) EM, isolation	52
	5′-B–oligonucleotide	Solid-phase synthesis	StAv–HRP	Blotting	349
	BHZ	Transamination of cytosines	StAv–AP	Blotting	350
Escherichia coli DNA	BNHS	Periodate/cytochrome *c* bridge, labeling of RNA	Av–spheres	EM	44
Cauliflower mosaic virus DNA	B–dUTP	Nick translation	Av–Fer	EM	351
Viral DNA	B–11-dUTP	Nick translation	Av–AP	Enzyme assay	352
Cytomegalovirus DNA	B–dUTP	Nick translation	(1) StAv–AP (2) FITC-Av	(1) Enzyme assay (2) Fluorescence assay	353
			(3) StABC (B–AP)	(3) DNA blotting	354
	B–11-dUTP	Nick translation	(1) Antibiotin Ab, B–Ab_2, ABC (B–HRP)	LM	355
			(2) Av, B–Ab_2, ABC (B–HRP)		356
Viral genomes (infected cell cultures)	(1) B–4-dUTP (2) B–11-dUTP (3) B–16-dUTP	Nick translation	(1) Antibiotin Ab–HRP (2) ABC (B–HRP) (3) FITC-antibiotin Ab	FM, EM	357
Orientation of viral DNA	B–dUTP	DNA polymerase (end labeling)	Av–Fer	EM	358
Flavivirus RNA	Photobiotin	Photoactivation	Av–AP	Spot hybridization	359

Adenovirus	B–dUTP	Nick translation	StABC (B–HRP)	LM	360
	BcapNHS	Transamination of cytosines	StABC (B–AP)	Enzyme assay	346
Papovavirus DNA	B–dUTP	Nick translation	StABC (B–HRP)	Enzyme assay	361
Viral DNA and RNA in tumor cells	B–dUTP	Nick translation	ABC (B–HRP)	Enzyme assay	362
Papillomavirus	B–11-dUTP B–11-dCTP	Nick translation	(1) StABC (B–HRP) (2) ABC (B–HRP)	Colony blotting	363
Hepatitis B virus	B–dUTP	Nick translation	(1) ABC (B–βG) (2) StAv–AP	DNA blotting	364 365
Cowpea mosaic virus	BNHS	Homobifunctional cross-linking reagents, cytochrome *c*, histone, or polyethylamine bridge, RNA probe	(1) Av–HRP (2) Av–AP	Blotting	366
Avocado viroid	Photobiotin	Photoactivation	(1) Av–AP (2)StABC (B–AP)	Blotting	367
Wheat chromosomes	B–dUTP	Nick translation	ABC (B–HRP)	LM	368
Mitochondrial DNA (HeLa cells)	BNHS	Periodate/diamine labeling of RNA	Av–Fer	EM	369
Muscle protein genes	B–dUTP	Nick translation	Antibiotin Ab, RITC-Ab_2	FM, double labeling	370
Ribosomal genes	B–dUTP	Nick translation	Antibiotin Ab, RITC-Ab_2	FM	371
Neurofilament-specific mRNA	B–dUTP	Nick translation	StAv–gold	LM	372
Adenosine deaminase (DNA and mRNA)	B–11-dUTP	Enzymatic extension at 3′ terminus, oligo-nucleotide probes	Av, B–HRP	Blotting	373
Metallothionein gene (human)	Photobiotin	Photoactivation	Av–AP	Blotting	374

(continued)

TABLE IX (*continued*)

Type or source of DNA or RNA	Reagent	Mode of incorporation	Mode of detection	Comments	Ref.[a]
Placental DNA (human)	B–4-dUTP B–11-dUTP B–16-dUTP	Nick translation	ABC (B–AP)	Blotting	375
β-Globin G (human)	B–dATP and B–dCTP analog	Nick translation	StAv–AP	Blotting	376
Restriction fragment (human)	B–11-dUTP	Nick translation	StABC (B–AP)		377
Synthetic oligonucleotides	BNHS	DNA modified at 5′ terminus	(1) Av–AP (2) ABC (B–HRP)		378–381
	BHZ	Periodate	(1) Antibiotin Ab, Ab–HRP (2) StAv–AP	Blotting	381
DNA sequencing bands	BNHS	Oligonucleotides modified at 5′ terminus	StAv, B–AP	Blotting	382
Midivariant mutant RNA	BNHS	DNA modified at 5′ terminus	Avidin	Cleavable spacer	383
DNA repair sites (human diploid fibroblasts	B–dUTP	Nick translation	Av–Fer	EM	384
HLA DNA	B–psoralen	Photoactivation	StAv–HRP	Enzyme assay	385
Colon tumor DNA (mouse)	B–dUTP	Nick translation	(1) Antibiotin Ab, FITC-Ab_2 (2) Av–HRP	FM, EM	386
Sickle cell anemia	B–11-dUTP	Nick translation	StABC (B–AP)	Blotting	387

[a] See reference list at end of article.

References

[1] The tables presented in this chapter are modified from those published in our recent review: M. Wilchek and E. A. Bayer, *Anal. Biochem.* **171**, 1 (1988).
[2] A. Bodanszky and M. Bodanszky, *Experientia* **26**, 327 (1970).
[3] M. D. Lane, J. Edwards, E. Stoll, and J. Moss, *Vitam. Horm.* (*N.Y.*) **28,** 345 (1970).
[4] R. B. Guchhait, S. E. Polakis, P. Dimroth, E. Stoll, J. Moss, and M. D. Lane, *J. Biol. Chem.* **249,** 6633 (1974).
[5] A. D. Landman and K. Dakshinamurti, *Anal. Biochem.* **56,** 191 (1973).
[6] A. D. Landman and K. Dakshinamurti, *Biochem. J.* **145,** 545 (1975).
[7] M. Berger and H. G. Wood, *J. Biol. Chem.* **250,** 927 (1975).
[8] J. A. Swack, G. L. Zander, and M. F. Utter, *Anal. Biochem.* **87,** 114 (1978).
[9] K. P. Henrikson, S. H. G. Allen, and W. L. Maloy, *Anal. Biochem.* **94,** 366 (1979).
[10] K. P. Henrikson and S. H. G. Allen, *J. Biol. Chem.* **254,** 5888 (1979).
[11] R. A. Gravel, K. F. Lam, D. Mahuran, and A. Kronis, *Arch. Biochem. Biophys.* **201,** 669 (1980).
[12] N. B. Beaty and M. D. Lane, *J. Biol. Chem.* **257,** 924 (1982).
[13] J. Oei and B. H. Robinson, *Biochim. Biophys. Acta* **840,** 1 (1985).
[14] P. Dimroth, *FEBS Lett.* **141,** 59 (1982).
[15] G. A. Orr, *J. Biol. Chem.* **256,** 761 (1981).
[16] C. A. Beard, R. A. Wrightsman, and J. E. Manning, *Mol. Biochem. Parasitol.* **16,** 199 (1985).
[17] J. W. Buckie and G. M. W. Cook, *Anal. Biochem.* **156,** 463 (1986).
[18] T. V. Updyke and G. L. Nicolson, *J. Immunol. Methods* **73,** 83 (1984).
[19] S. A. Fuller, A. Philips, and M. S. Coleman, *Biochem. J.* **231,** 105 (1985).
[20] W. T. Lee and D. H. Conrad, *J. Exp. Med.* **159,** 1790 (1984).
[21] T. Nakajima and G. Delespesse, *Eur. J. Immunol.* **16,** 809 (1986).
[22] G. Redeuilh, C. Secco, and E.-E. Baulieu, *J. Biol. Chem.* **260,** 3996 (1985).
[23] B. Manz, A. Heubner, I. Köhler, H.-J. Grill, and K. Pollow, *Eur. J. Biochem.* **131,** 333 (1983).
[24] E. Hazum, I. Schvartz, Y. Waksman, and D. Keinan, *J. Biol. Chem.* **261,** 13043 (1986).
[25] F. M. Finn, G. Titus, D. Horstman, and K. Hofmann, *Proc. Natl. Acad. Sci. U.S.A.* **81,** 7328 (1984).
[26] R. A. Kohanski and M. D. Lane, *J. Biol. Chem.* **260,** 5014 (1985).
[27] M.-T. Haeuptle, M. L. Aubert, J. Djiane, and J.-P. Kraehenbuhl, *J. Biol. Chem.* **258,** 305 (1983).
[28] H. Nakayama, H. Shikano, T. Aoyama, T. Amano, and Y. Kanaoka, *FEBS Lett.* **208,** 278 (1986).
[29] D. R. Gretch, M. Suter, and M. F. Stinski, *Anal. Biochem.* **163,** 270 (1987).
[30] E. A. Bayer, M. Wilchek, and E. Skutelsky, *FEBS Lett.* **68,** 240 (1976).
[31] M. L. Jasiewicz, D. R. Schoenberg, and G. C. Mueller, *Exp. Cell Res.* **100,** 213 (1976).
[32] F. R. Harmon, M. Berger, H. Beegen, H. G. Wood, and N. G. Wrigley, *J. Biol. Chem.* **255,** 9458 (1980).
[33] B. R. Clark, B. J. Mills, K. Horikoshi, J. E. Shively, and C. W. Todd, *J. Immunol. Methods* **67,** 371 (1984).
[34] C. S. Chandler and F. J. Ballard, *Biochem. J.* **237,** 123 (1986).
[35] R. J. Berenson, W. I. Bensinger, D. Kalamasz, and P. Martin, *Blood* **67,** 509 (1986).
[36] J. Wormmeester, F. Stiekema, and K. De Groot, *J. Immunol. Methods* **67,** 389 (1984).
[37] D. Levitt and R. Danen, *J. Immunol. Methods* **89,** 207 (1986).
[38] J. Fürfang and S. Thierfelder, *J. Immunol. Methods* **91,** 123 (1986).

[39] P. J. Lucas and R. E. Gress, *J. Immunol. Methods* **99,** 185 (1987).
[40] R. J. Berenson, W. I. Bensinger, and D. Kalamasz, *J. Immunol. Methods* **91,** 11 (1986).
[41] P. Casali, G. Inghirami, M. Nakamura, T. F. Davies, and A. L. Notkins, *Science* **234,** 476 (1986).
[42] M. Pelligrini, D. S. Holmes, and J. Manning, *Nucleic Acids Res.* **4,** 2961 (1977).
[43] A. Sodja and N. Davidson, *Nucleic Acids Res.* **5,** 385 (1978).
[44] J. Manning, M. Pellegrini, and N. Davidson, *Biochemistry* **16,** 1364 (1977).
[45] T. R. Broker, L. M. Angerer, P. Yen, N. D. Hershey, and N. Davidson, *Nucleic Acids Res.* **5,** 363 (1978).
[46] D. J. Eckermann and R. H. Symons, *Eur. J. Biochem.* **82,** 225 (1978).
[47] B. Wittig and S. Wittig, *Biochem. Biophys. Res. Commun.* **91,** 554 (1979).
[48] P. R. Langer, A. A. Waldrop, and D. C. Ward, *Proc. Natl. Acad. Sci. U.S.A.* **78,** 6633 (1981).
[49] R. W. Richardson and R. I. Gumport, *Nucleic Acids Res.* **11,** 6167 (1983).
[50] M. Shimkus, J. Levy, and T. Herman, *Proc. Natl. Acad. Sci. U.S.A.* **82,** 2593 (1985).
[51] M. L. Shimkus, P. Guaglianone, and T. M. Herman, *DNA* **5,** 247 (1986).
[52] H. Delius, H. van Heerikhuizen, J. Clarke, and B. Koller, *Nucleic Acids Res.* **13,** 5457 (1985).
[53] P. J. Grabowski and P. A. Sharp, *Science* **233,** 1294 (1986).
[54] B. Roseman, J. Lough, E. Houkom, and T. Herman, *Biochem. Biophys. Res. Commun.* **137,** 474 (1986).
[55] M. S. Kasher, D. Pintel, and D. C. Ward, *Mol. Cell. Biol.* **6,** 3117 (1986).
[56] R. P. Singh and V. Natarajan, *Biochem. Biophys. Res. Commun.* **147,** 65 (1987).
[57] B. Rigas, A. A. Welcher, D. C. Ward, and S. M. Weissman, *Proc. Natl. Acad. Sci. U.S.A.* **83,** 9591 (1986).
[58] A. A. Welcher, A. R. Torres, and D. C. Ward, *Nucleic Acids Res.* **14,** 10027 (1986).
[59] H. Heitzmann and F. M. Richards, *Proc. Natl. Acad. Sci. U.S.A.* **71,** 3537 (1974).
[60] E. A. Bayer, E. Skutelsky, D. Wynne, and M. Wilchek, *J. Histochem. Cytochem.* **24,** 933 (1976).
[61] E. Skutelsky, D. Danon, M. Wilchek, and E. A. Bayer, *J. Utrastruct. Res.* **61,** 325 (1977).
[62] T. Suganuma, T. Ohta, S. Tsuyama, T. Kamada, S. Otsuji, and F. Murata, *Virchows Arch. B.* **49,** 1 (1985).
[63] S. Spiegel, E. Skutelsky, E. A. Bayer, and M. Wilchek, *Biochim. Biophys. Acta* **687,** 27 (1982).
[64] I. Kahane, A. Polliack, E. A. Rachmilewitz, E. A. Bayer, and E. Skutelsky, *Nature (London)* **271,** 674 (1978).
[65] B. A. Wallace, F. M. Richards, and D. M. Engelman, *J. Mol. Biol.* **107,** 255 (1976).
[66] R. Henderson, J. S. Jubb, and S. Whytock, *J. Mol. Biol.* **123,** 259 (1978).
[67] E. Skutelsky and E. A. Bayer, *Exp. Cell Res.* **121,** 331 (1979).
[68] E. Skutelsky and E. A. Bayer, *J. Cell Biol.* **96,** 184 (1983).
[69] M. J. Mattes, J. G. Cairncross, L. J. Old, and K. O. Lloyd, *Hybridoma* **2,** 253 (1983).
[70] S. Hirohashi, Y. Ino, T. Kodama, and Y. Shimosat, *JNCI, J. Natl. Cancer Inst.* **72,** 1299 (1984).
[71] C.-W. Lin, M. Fujime, S. D. Kirley, and G. R. Prout Jr., *J. Histochem. Cytochem.* **32,** 1339 (1984).
[72] G. M. Seal, R. G. Rowland, J. V. Thomalla, R. A. Rudolph, D. S. Pfaff, M. Kamer, and J. N. Eble, *J. Urol.* **133,** 513 (1985).
[73] S. P. Hunt and P. W. Mantyh, *Brain Res.* **291,** 203 (1984).

[74] K.-C. Feng-Chen, B.-F. Chen, Z. Liu, and A.-K. Ng, *J. Histochem. Cytochem.* **34,** 1495 (1986).
[75] A. M. Gown, R. Garcia, M. Ferguson, E. Yamanaka, and D. Tippens, *J. Histochem. Cytochem.* **34,** 403 (1986).
[76] S. J. Korsmeyer, W. C. Greene, J. Cossman, S.-M. Hsu, J. P. Jensen, L. M. Neckers, S. L. Marshall, A. Bakhshi, J. M. Depper, W. J. Leonard, E. S. Jaffe, and T. A. Waldmann, *Proc Natl. Acad. Sci. U.S.A.* **80,** 4522 (1983).
[77] A. S. Polans, L. G. Altman, and D. S. Papermaster, *J. Histochem. Cytochem.* **34,** 659 (1986).
[78] D. S. Papermaster, B. G. Schneider, M. A. Zorn, and J. P. Kraehenbuhl, *J. Cell Biol.* **77,** 196 (1978).
[79] I. Nir, B. G. Schneider, and D. S. Papermaster, *J. Histochem. Cytochem.* **32,** 643 (1984).
[80] A. Carbone, A. Poletti, R. Manconi, R. Volpe, and L. Santi, *Hum. Pathol.* **16,** 1157 (1985).
[81] C. Charpin, A. Lachard, N. Pourreau-Schneider, I. Jacquemier, M. N. Lavaut, C. Andonian, P. M. Martin, and M. Toga, *Cancer* **55,** 2612 (1985).
[82] M. Noguchi, S. Hirohashi, Y. Shimosato, A. Thor, J. Schlom, Y. Tsunokawa, M. Terada, and T. Sugimura, *JNCI, J. Natl. Cancer Inst.* **77,** 379 (1986).
[83] R. Jaffe, B. Bender, M. Santamaria, and A. E. Chung, *Lab. Invest.* **51,** 88 (1984).
[84] H. Battifora, K. Sheibani, R. R. Tubbs, M. I. Kopinski, and T.-T. Sun, *Cancer* **54,** 843 (1984).
[85] J. Pevsner, P. B. Sklar, and S. H. Snyder, *Proc. Natl. Acad. Sci. U.S.A.* **83,** 4942 (1986).
[86] M. P. Mark, C. W. Prince, T. Oosawa, S. Gay, A. L. J. J. Bronckers, and W. T. Butler, *J. Histochem. Cytochem.* **35,** 707 (1987).
[87] D. M. Wojchowski and A. J. Sytkowski, *Biochim. Biophys. Acta* **857,** 61 (1986).
[88] M. H. Heggeness and J. F. Ash, *J. Cell Biol.* **73,** 783 (1977).
[89] A. M. Vogel, A. M. Gown, J. Caughlan, J. E. Haas, and J. B. Beckwith, *Lab. Invest.* **50,** 232 (1984).
[90] Y. Chicheportiche, P. Vassalli, and A. M. Tartakoff, *J. Cell Biol.* **99,** 2200 (1984).
[91] H. Sariola, P. Kuusela, and P. Ekblom, *J. Cell Biol.* **99,** 2099 (1984).
[92] H. Asou, N. Iwasaki-Mutou, S. Hirano, T. Turumizu, and Y. Horibe, *Acta Histochem. Cytochem.* **18,** 383 (1985).
[93] M. H. Ginsberg, L. Taylor, and R. G. Painter, *Blood* **55,** 661 (1980).
[94] F. M. Graf and P. Sträuli, *J. Histochem. Cytochem.* **31,** 803 (1983).
[95] T. J. Fetterhoff and R. C. McCarthy, *Am. J. Clin. Pathol.* **83,** 565 (1985).
[96] B. Racklin, R. Bearman, K. Sheibani, C. Winberg, and H. Rappaport, *Leukemia Res.* **7,** 431 (1983).
[97] O. K. Langley and M. S. Ghandour, *Histochem. J.* **13,** 137 (1981).
[98] Y. Suzuki, H. Ishizuka, H. Kaneda, and N. Taniguchi, *J. Histochem. Cytochem.* **35,** 3 (1987).
[99] H. Yki-Järvinen and T. Wahlström, *Acta Endocrinol.* (*Copenhagen*) **106,** 544 (1984).
[100] M. B. Anderson, M. Collado-Torres, and M. R. Vaupel, *J. Histochem. Cytochem.* **34,** 945 (1986).
[101] M. Schachter, D. J. Longridge, G. D. Wheeler, J. G. Mehta, and Y. Uchida, *J, Histochem. Cytochem.* **34,** 927 (1986).
[102] S. Smith, D. Pasco, and S. Nandi, *Biochem. J.* **212,** 155 (1983).
[103] K. Yasuda, N. Yamamoto, and S. Yamashita, *Experientia* **37,** 306 (1981).
[104] S. Weinman, C. Ores-Carton, D. Rainteau, and S. Puszkin, *J. Histochem. Cytochem.* **34,** 1171 (1986).

[105] J. G. Linner, S. A. Livesey, D. S. Harrison, and A. L. Steiner, *J. Histochem. Cytochem.* **34,** 1123 (1986).
[106] F. Eckenstein and H. Thoenen, *EMBO J.* **1,** 363 (1982).
[107] M. M. Moran, R. J. Siegel, J. W. Said, and M. C. Fishbein, *J. Histochem. Cytochem.* **33,** 1110 (1985).
[108] A. N. Van den Pol, *Science* **228,** 332 (1985).
[109] E. C. Salido, L. Barajas, J. Lechago, N. P. Laborde, and D. A. Fisher, *J. Histochem. Cytochem.* **34,** 1155 (1986).
[110] P. H. Smith and B. B. Toms, *J. Histochem. Cytochem.* **34,** 627 (1986).
[111] H. S. Phillips, K. Nikolics, D. Branton, and P. H. Seeburg, *Nature (London)* **316,** 542 (1985).
[112] D. W. Morrish, H. Marusyk, and O. Siy, *J. Histochem. Cytochem.* **35,** 93 (1987).
[113] D. T. Piekut, *J. Histochem. Cytochem.* **35,** 261 (1987).
[114] S. R. Vincent, H. Kimura, and E. G. McGeer, *J. Comp. Neurol.* **199,** 113 (1981).
[115] G. V. Childs (Moriarty), D. G. Ellison, and J. A. Ramaley, *Endocrinology* **110,** 1676 (1982).
[116] K. N. Westlund and G. V. Childs, *Endocrinology* **111,** 1761 (1982).
[117] E. Mezey, J. Z. Kiss, L. R. Skirboll, M. Goldstein, and J. Axelrod, *Nature (London)* **310,** 140 (1984).
[118] G. V. Childs (Moriarty) and G. Unabia, *J. Histochem. Cytochem.* **30,** 713 (1982).
[119] G. V. Childs (Moriarty) and G. Unabia, *J. Histochem. Cytochem.* **30,** 1320 (1982).
[120] C. R. Vaillant and P. K. Lund, *J. Histochem. Cytochem.* **34,** 1117 (1986).
[121] D. Tran, J. Y. Picard, J. Campargue, and N. Josso, *J. Histochem. Cytochem.* **35,** 733 (1987).
[122] R. J. Seitz, K. Heininger, G. Schwendemann, K. V. Toyka, and W. Wechsler, *Acta Neuropathol.* **68,** 15 (1985).
[123] G. Viale, P. Dell'Orto, P. Braidotti, and G. Coggi, *J. Histochem. Cytochem.* **33,** 400 (1985).
[124] N. K. Gonatas, J. O. Gonatas, A. Stieber, T. Ternynck, and S. Avrameas, *J. Histochem. Cytochem.* **35,** 189 (1987).
[125] J. U. Alles and K. Bosslet, *J. Histochem. Cytochem.* **34,** 209 (1986).
[126] S.-M. Hsu and L. Raine, *J. Histochem. Cytochem.* **29,** 1349 (1981).
[127] S. G. Volsen, *J. Immunol. Methods* **72,** 119 (1984).
[128] D. M. Boorsma, J. Van Bommel, and J. V. Heuvel, *Histochemistry* **84,** 333 (1986).
[129] V. M. Elner, A. J. Hass, H. R. Davis, and S. Glagov, *J. Histochem. Cytochem.* **31,** 1139 (1983).
[130] J. H. M. Cohen, J. P. Aubry, M. H. Jouvin, J. Wijdenes, J. Bancherau, M. Kazatchkine, and J. P. Revillard, *J. Immunol. Methods* **99,** 53 (1987).
[131] V. L. Woods, Jr., L. E. Wolff, and D. M. Keller, *J. Biol. Chem.* **261,** 15242 (1986).
[132] L. C. Kühn and J.-P. Kraehenbuhl, *J. Biol. Chem.* **256,** 12490 (1981).
[133] M. C. McClellan, N. B. West, D. E. Tacha, G. L. Greene, and R. M. Brenner, *Endocrinology* **114,** 2002 (1984).
[134] J.-M. Gasc, J.-M. Renoir, C. Radanyi, I. Joab, P. Tuohimaa, and E.-E. Baulieu, *J. Cell Biol.* **99,** 1193 (1984).
[135] M. Maegawa, S. Hisano, Y. Tsuruo, S. Katoh, J. Nakanishi, M. Chikamori-Aoyama, and S. Daikoku, *J. Histochem. Cytochem.* **35,** 251 (1987).
[136] C. A. Wiley, R. D. Schrier, J. A. Nelson, P. W. Lampert, and M. B. A. Oldstone, *Proc. Natl. Acad. Sci. U.S.A.* **83,** 7089 (1986).
[137] L. V. Chalifoux, N. W. King, and N. L. Letvin, *Lab. Invest.* **51,** 22 (1984).

[138] M. E. Faran, W. S. Romoser, R. G. Routier, and C. L. Bailey, *Am. J. Trop. Med. Hyg.* **35,** 1061 (1986).
[139] L. M. Alessandri, G. F. Sterrett, E. C. Pixley, and J. K. Kulski, *Pathology* **18,** 382 (1986).
[140] L. S. Nerurkar, A. J. Jacob, D. L. Madden, and J. L. Sever, *J. Clin. Microbiol.* **17,** 149 (1983).
[141] E. A. Gould, A. Buckley, and N. Cammack, *J. Virol. Methods* **11,** 41 (1985).
[142] E. A. Bayer, E. Skutelsky, S. Goldman, E. Rosenberg, and D. L. Gutnick, *J. Gen. Microbiol.* **129,** 1109 (1983).
[143] E. A. Bayer and R. Lamed, *J. Bacteriol.* **167,** 828 (1986).
[144] A. A. Rogalski and G. B. Bouck, *J. Cell Biol.* **93,** 758 (1982).
[145] B. H. Butterworth, T. Y. Khong, Y. W. Loke, and W. B. Robertson, *J. Histochem. Cytochem.* **33,** 997 (1985).
[146] A. K. Bhan, M. C. Mihm, Jr., and H. F. Dvorak, *J. Immunol.* **129,** 1578 (1982).
[147] A. Karlsson-Parra, U. Forsum, L. Klareskog, and O. Sjöberg, *J. Immunol. Methods* **64,** 85 (1983).
[148] R. J. Mason K. N. Steward, D. A. Power, A. M. Macleod, and G. R. D. Catto, *Immunol. Lett.* **7,** 157 (1983).
[149] R. Warnke and R. Levy, *J. Histochem. Cytochem.* **28,** 771 (1980).
[150] L. Si, T. L. Whiteside, D. H. Van Thiel, and B. S. Rabin, *Lab. Invest.* **50,** 341 (1984).
[151] S. S. Tabibzadeh and M. A. Gerber, *J. Immunol. Methods* **91,** 169 (1986).
[152] H. Sako, Y. Nakane, K. Okino, K. Nishihara, M. Kodama, M. Kawata, and H. Yamada, *Histochemistry* **86,** 1 (1986).
[153] Y.-H. Kang, M. Carl, L. P. Watson, and L. Yaffe, *J. Immunol. Methods* **84,** 177 (1985).
[154] T. Löning, D. Schmitt, W. M. Becker, J. Weiss, and M. Jänner, *Arch. Dermatol. Res.* **272,** 177 (1982).
[155] W. R. Gower, Jr., E. M. McSweeney, T. E. Dyben, and P. J. Fabri, *J. Histochem. Cytochem.* **33,** 1087 (1985).
[156] L. Takacs, H. Osawa, I. Törö, and T. Diamantstein, *Clin. Exp. Immunol.* **59,** 37 (1985).
[157] H. Helin, R. Uibo, I. Paronen, and K. Krohn, *Clin. Exp. Immunol.* **59,** 371 (1985).
[158] P. H. Nibbering, P. C. J. Leijh, and R. Van Furth, *J. Histochem. Cytochem.* **33,** 453 (1985).
[159] Y. T. Konttinen, V. Bergroth, K. Visa-Tolvanen, S. Reitamo, and L. Förström, *Clin. Immunol. Immunopathol.* **28,** 441 (1983).
[160] A. N. Van den Pol, A. D. Smith, and J. F. Powell, *Brain Res.* **348,** 146 (1985).
[161] C.-T. Deng, P. I. Terasaki, Y. Iwaki, F. M. Hofman, P. Koeffler, L. Cahan, N. El Awar, and R. Billing, *Blood* **61,** 759 (1983).
[162] D. A. Hume, D. Halpin, H. Charlton, and S. Gordon, *Proc. Natl. Acad. Sci. U.S.A.* **81,** 4174 (1984).
[163] E. Vasak, K. Atkinson, V. Munro, and J. Biggs, *Transplantation* **16,** 1027 (1984).
[164] M. Henke, L. M. Yonemoto, G. S. Lazar, L. Gaidulis, T. Hecht, S. Santos, and K. G. Blume, *J. Histochem. Cytochem.* **32,** 712 (1984).
[165] C. A. Picut, C. S. Lee, E. P. Dougherty, K. L. Andersen, and R. M. Lewis, *J. Histochem. Cytochem.* **35,** 745 (1987).
[166] C. C. Lance-Jones and C. F. Lagenaur, *J. Histochem. Cytochem.* **35,** 771 (1987).
[167] M. R. Kaplan, E. Calef, T. Bercovici, and C. Gitler, *Biochim. Biophys. Acta* **728,** 112 (1983).
[168] V. T. Oi, A. N. Glazer, and L. Stryer, *J. Cell Biol.* **93,** 981 (1982).
[169] L. B. Olding, J. Thurin, C. Svalander, and H. Koprowski, *Int. J. Cancer* **34,** 187 (1984).

[170] P. G. Natali, A. Bigotti, R. Cavaliere, M. R. Nicotra, and S. Ferrone, *JNCI, J. Natl. Canc. Inst.* **73,** 13 (1984).
[171] C. Wagener, Y. H. J. Yang, F. G. Crawford, and J. E. Shively, *J. Immunol.* **130,** 2308 (1983).
[172] J. W. Berman and R. S. Basch, *J. Immunol. Methods* **36,** 335 (1980).
[173] M. Nuti, Y. A. Teramoto, R. Mariani-Costantini, P. H. Hand, D. Colcher, and J. Schlom, *Int. J. Cancer* **29,** 539 (1982).
[174] P. H. Hand, M. Nuti, D. Colcher, and J. Schlom, *Cancer Res.* **43,** 728 (1983).
[175] D. R. Ciocca, D. J. Adams, D. P. Edwards, R. J. Bjercke, and W. L. McGuire, *Cancer Res.* **43,** 1204 (1983).
[176] D. O. Chee, R. H. Yonemoto, S. P. L. Leong, G. F. Richards, V. R. Smith, J. L. Klotz, R. M. Goto, R. L. Gascon, and M. M. Drushella, *Cancer Res.* **42,** 3142 (1982).
[177] E. R. Boulan, G. Kreibich, and D. D. Sabatini, *J. Cell Biol.* **78,** 874 (1978).
[178] R. J. Walter, R. D. Berlin, J. R. Pfeiffer, and J. M. Oliver, *J. Cell Biol.* **86,** 199 (1980).
[179] S. D. J. Pena, B. B. Gordon, G. Karpati, and S. Carpenter, *J. Histochem. Cytochem.* **29,** 542 (1981).
[180] S.-M. Hsu and L. Raine, *J. Histochem. Cytochem.* **30,** 157 (1982).
[181] Y. Dokusa, K. Takata, M. Nagashima, and H. Hirano, *Histochemistry* **79,** 1 (1983).
[182] H. J. Ree, L. Raine, and J. P. Crowley, *Cancer* **52,** 2089 (1983).
[183] R. M. Pino, *J. Histochem. Cytochem.* **32,** 862 (1984).
[184] S. Teshima, S. Hirohashi, Y. Shimosato, K. Kishi, Y. Ino, K. Matsumoto, and T. Yamada, *Lab. Invest.* **50,** 271 (1984).
[185] P. N. McMillan, L. S. Ferayorni, C. O. Gerhardt, and H. O. Jauregui, *Lab. Invest.* **50,** 408 (1984).
[186] U. Orgad, J. Alroy, A. Ucci, and F. B. Merk, *Lab. Invest.* **50,** 294 (1984).
[187] J. Alroy, U. Orgad, A. A. Ucci, and M. E. A. Pereira, *J. Histochem. Cytochem.* **32,** 1280 (1984).
[188] T. Weber, R. J. Seitz, U. G. Liebert, E. Gallasch, and W. Wechsler, *Acta Neuropathol.* **67,** 128 (1985).
[189] D. Z. Gerhart, M. S. Zlonis, and L. R. Drewes, *J. Histochem. Cytochem.* **34,** 641 (1986).
[190] N. Kessimian, B. J. Langner, P. N. McMillan, and H. O. Jauregui, *J. Histochem. Cytochem.* **34,** 237 (1986).
[191] K. Takai, T. Kakizoe, T. Sekine, S. Sato, and T. Niijima, *J. Urol.* **137,** 136 (1987).
[192] E. Skutelsky, V. Goyal, and J. Alroy, *Histochemistry* **86,** 291 (1987).
[193] E. Skutelsky, J. Alroy, A. C. Ucci, J. L. Carpenter, and F. M. Moore, *Am. J. Pathol.* **126,** 25 (1987).
[194] J. Alroy, V. Goyal, and E. Skutelsky, *Histochemistry* **86,** 603 (1987).
[195] D. Atlas, D. Yaffe, and E. Skutelsky, *FEBS Lett.* **95,** 173 (1978).
[196] D. Axelrod, *Proc. Natl. Acad. Sci. U.S.A.* **77,** 4823 (1980).
[197] E. Holtzman, D. Wise, J. Wall, and A. Karlin, *Proc. Natl. Acad. Sci. U.S.A.* **79,** 310 (1982).
[198] C. R. Hopkins, B. Boothroyd, and H. Gregory, *Eur. J. Cell Biol.* **24,** 259 (1981).
[199] D. Kozbor and J. C. Roder, *J. Immunol.* **127,** 1275 (1981).
[200] H. Darbon, E. Jover, F. Couraud, and H. Rochat, *Int. J. Pept. Protein Res.* **22,** 179 (1983).
[201] T. Nakayama and S. Furuya, *Biomed. Res.* **6,** 175 (1985).
[202] G. V. Childs, Z. Naor, E. Hazum, R. Tibolt, K. N. Westlund, and M. B. Hancock, *J. Histochem. Cytochem.* **31,** 1422 (1983).
[203] K. N. Westlund, P. C. Wynn, S. Chmielowiec, T. J. Collins, and G. V. Childs, *Peptides* **5,** 627 (1984).

[204] R. E. Morris and C. B. Saelinger, *J. Histochem. Cytochem.* **32,** 124 (1984).
[205] A. Niendorf, M. Dietel, H. Arps, J. Lloyd, and G. V. Childs, *J. Histochem.Cytochem.* **34,** 357 (1986).
[206] N. M. Green, R. C. Valentine, N. G. Wrigley, F. Ahmad, B. E. Jacobson, and H. G. Wood, *J. Biol. Chem.* **247,** 6284 (1972).
[207] M. H. Heggeness, K. Wang, and S. J. Singer, *Proc. Natl. Acad. Sci. U.S.A.* **74,** 3883 (1977).
[208] J. B. Geiduschek and S. J. Singer, *Cell* **16,** 149 (1979).
[209] E. A. Bayer, E. Skutelsky, T. Viswanatha, and M. Wilchek, *Mol. Cell. Biochem.* **19,** 23 (1978).
[210] E. A. Bayer, B. Rivnay, and E. Skutelsky, *Biochim. Biophys. Acta* **550,** 464 (1979).
[211] D. L. Urdal and S. Hakomori, *J. Biol. Chem.* **255,** 10509 (1980).
[212] H. Asou, E. G. Brunngraber, and I. Jeng, *J. Histochem. Cytochem.* **31,** 1375 (1983).
[213] H. Asou, N. Iwasaki-Mutou, and S. Hirano, *Neurosci. Res.* **2,** 399 (1985).
[214] G. M. Gilad and V. H. Gilad, *J. Histochem. Cytochem.* **29,** 687 (1981).
[215] M. Berger, T. A. Gaither, R. M. Cole, T. M. Chused, C. H. Hammer, and M. M. Frank, *Mol. Immunol.* **19,** 857 (1982).
[216] D. Safer, J. Hainfeld, J. S. Wall, and J. E. Reardon, *Science* **218,** 290 (1982).
[217] J. A. Ripellino, M. M. Klinger, R. U. Margolis, and R. K. Margolis, *J. Histochem. Cytochem.* **33,** 1060 (1985).
[218] K. Sutoh, K. Yamamoto, and T. Wakabayashi, *Proc. Natl. Acad. Sci. U.S.A.* **83,** 212 (1986).
[219] R. J. Seitz, K. Heininger, G. Schwendemann, and K. V. Toyka, *J. Histochem. Cytochem.* **34,** 547 (1986).
[220] L. J. Anderson, R. A. Coombs, C. Tsou, and J. C. Hierholzer, *J. Clin. Microbiol.* **19,** 934 (1984).
[221] M. Neumaier, U. Fenger, and C. Wagener, *Anal. Biochem.* **156,** 76 (1986).
[222] E. Roffman, L. Meromsky, H. Ben-Hur, E. A. Bayer, and M. Wilchek, *Biochem. Biophys. Res. Commun.* **136,** 80 (1986).
[223] D. Della-Penna, R. E. Christoffersen, and A. B. Bennett, *Anal. Biochem.* **152,** 329 (1986).
[224] M. M. L. Leong, C. Milstein, and R. Pannell, *J. Histochem. Cytochem.* **34,** 1645 (1986).
[225] W. J. LaRochelle and S. C. Froehner, *J. Immunol. Methods* **92,** 65 (1986).
[226] E. Lacey and W. N. Grant, *Anal. Biochem.* **163,** 151 (1987).
[227] E. A. Bayer, M. G. Zalis, and M. Wilchek, *Anal. Biochem.* **149,** 529 (1985).
[228] E. A. Bayer, M. Safars, and M. Wilchek, *Anal. Biochem.* **161,** 262 (1987).
[229] M. Wilchek, H. Ben-Hur, and E. A. Bayer, *Biochem. Biophys. Res. Commun.* **138,** 872 (1986).
[230] D. J. O'Shannessy, M. J. Dobersen, and R. H. Quarles, *Immunol. Lett.* **8,** 273 (1984).
[231] D. J. O'Shannessy, and R. H. Quarles, *J. Appl. Biochem.* **7,** 347 (1985).
[232] E. A. Bayer, H. Ben-Hur, and M. Wilchek, *Anal. Biochem.* **161,** 123 (1987).
[233] D. J. O'Shannessy, P. J. Voorstad, and R. H. Quarles, *Anal. Biochem.* **163,** 204 (1987).
[234] E. A. Bayer, H. Ben-Hur, and M. Wilchek, *Anal. Biochem.* **170,** 271 (1988).
[235] W. L. Hurley, E. Finkelstein, and B. D. Holst, *J. Immunol. Methods* **85,** 195 (1985).
[236] R. Rohringer and D. W. Holden, *Anal. Biochem.* **144,** 118 (1985).
[237] J. Buehler and B. A. Macher, *Anal. Biochem.* **158,** 283 (1986).
[238] E. A. Azen and P.-L. Yu, *Biochem. Genet.* **22,** 1 (1984).
[239] P. Ashorn, P. Vilja, R. Ashorn, and K. Krohn, *Mol. Immunol.* **23,** 221 (1986).
[240] B. B. Gordon and S. D. J. Pena, *Biochem. J.* **208,** 351 (1982).
[241] P. C. Holland, S. D. J. Pena, and C. W. Guerin, *Biochem. J.* **218,** 465 (1984).

[242] M. L. Billingsley, K. R. Pennypacker, C. G. Hoover, D. J. Brigati, and R. L. Kincaid, *Proc. Natl. Acad. Sci. U.S.A.* **82,** 7585 (1985).
[243] S. Bhakdi, D. Jenne, and F. Hugo, *J. Immunol. Methods* **80,** 25 (1985).
[244] M. W. Devouge, B. B. Mukherjee, and S. D. J. Pena, *Virology* **121,** 327 (1982).
[245] J. M. Kittler, N. T. Meisler, D. Viceps-Madore, J. A. Cidlowski, and J. W. Thanassi, *Anal. Biochem.* **137,** 210 (1984).
[246] H. Vissing and O. D. Madsen, *Electrophoresis* **5,** 313 (1984).
[247] B. J. Nikolau, E. S. Wurtele, and P. K. Stumpf, *Anal. Biochem.* **149,** 448 (1985).
[248] J. S. Pachter and R. K. H. Liem, *Dev. Biol.* **103,** 200 (1984).
[249] W. J. LaRochelle and S. C. Froehner, *J. Biol. Chem.* **262,** 8190 (1987).
[250] M. S. Brower, C. L. Brakel, and K. Garry, *Anal. Biochem.* **147,** 382 (1985).
[251] P. Gallo, F. Bracco, S. Morara, L. Battistin, and B. Tavolato, *J. Neurol. Sci.* **70,** 81 (1985).
[252] C. Hession, J. M. Decker, A. P. Sherblom, S. Kumar, C. C. Yue, R. J. Mattaliano, R. Tizard, E. Kawashima, U. Schmeissner, S. Heletky, E. P. Chow, C. A. Burne, A. Shaw, and A. V. Muchmore, *Science* **237,** 1479 (1987).
[253] S. D. J. Pena, M. Opas, K. Turksen, V. I. Kalnins, and S. Carpenter, *Eur. J. Cell Biol.* **31,** 227 (1983).
[254] L. Lärkfors and T. Ebendal, *J. Immunol. Methods* **97,** 41 (1987).
[255] P. H. Van der Meide, M. Dubbeld, and H. Schellekens, *J. Immunol. Methods* **79,** 293 (1985).
[256] K. A. Knisley and L. S. Rodkey, *J. Immunol. Methods* **95,** 79 (1986).
[257] T. Olsson, V. Kostulas, and H. Link, *Clin. Chem.* (*Winston-Salem, N.C.*) **30,** 1246 (1984).
[258] F. Chiodi, A. Siden, and E. Ösby, *Electrophoresis* **6,** 124 (1985).
[259] G. Gabor and R. M. Bennett, *Biochem. Biophys. Res. Commun.* **12,** 1034 (1984).
[260] C. W. Achwal and H. S. Chandra, *FEBS Lett.* **150,** 469 (1982).
[261] D. P. Wade, B. L. Knight, and A. K. Soutar, *Biochem. J.* **229,** 785 (1985).
[262] M. Chow, R. Yabrov, J. Bittle, J. Hogle, and D. Baltimore, *Proc. Natl. Acad. Sci. U.S.A.* **82,** 910 (1985).
[263] P.-L. Hsu, S.-M. Hsu, and E. Appella, *Gene Anal. Technol.* **2,** 30 (1985).
[264] K. Ogata, M. Arakawa, T. Kasahara, K. Shioiri-Nakano, and K. Hiraoka, *J. Immunol. Methods* **65,** 75 (1983).
[265] C. A. Saravis, W. Cantarow, P. V. Marasco, B. Burke, and N. Zamcheck, *Electrophoresis* **1,** 191 (1980).
[266] J.-L. Guesdon, T. Ternynck, and S. Avrameas, *J. Histochem. Cytochem.* **27,** 1131 (1979).
[267] T. S. Watt and R. M. Watt, *Proc. Natl. Acad. Sci. U.S.A.* **80,** 124 (1983).
[268] E. S. Tackaberry, P. R. Ganz, and G. Rock, *J. Immunol. Methods* **99,** 59 (1987).
[269] M. Lee, J. Change, D. Carlson, and M. Burgett, *Clin. Chim. Acta* **147,** 109 (1985).
[270] J. A. Madri and K. W. Barwick, *Lab. Invest.* **48,** 98 (1983).
[271] P. Vilja, K. Krohn, and P. Tuohimaa, *J. Immunol. Methods* **76,** 73 (1985).
[272] A. Francina, H. Cloppet, R. Guinet, M. Rossi, D. Guyotat, O. Gentilhomme, and M. Richard, *J. Immunol. Methods* **87,** 267 (1986).
[273] L. Spatz, L. Whitman, M. J. Messito, G. Nilaver, S. Ginsberg, and N. Latov, *Immunol. Commun.* **12,** 31 (1983).
[274] W. D. Odell, J. Griffin, and R. Zahradnik, *Clin. Chem.* **32,** 1873 (1986).
[275] J. Blanco and G. H. Dixon, *J. Immunol. Methods* **89,** 265 (1986).
[276] K. Molin, P. Fredman, and L. Svennerholm, *FEBS Lett.* **205,** 51 (1986).
[277] J. Buehler, U. Galili, and B. A. Macher, *Anal. Biochem.* **164,** 521 (1987).

[278] M. Suter and J. E. Butler, *Immunol. Lett.* **13,** 313 (1986).
[279] C. Bacquet and D. Y. Twumasi, *Anal. Biochem.* **136,** 487 (1984).
[280] H. Park, A. R. Zeiger, and H. R. Schumacher, *Infect. Immun.* **43,** 139 (1984).
[281] J. J. Lanzillo and B. L. Fanburg, *Anal. Biochem.* **126,** 156 (1982).
[282] I. F. Hahn, P. Pickenhahn, W. Lenz, and H. Brandis, *J. Immunol. Methods* **92,** 25 (1986).
[283] D. E. Lockwood and D. C. Robertson, *J. Immunol. Methods* **75,** 295 (1984).
[284] T. Kohno and I. Ishikawa, *Biochem. Biophys. Res. Commun.* **147,** 644 (1987).
[285] R. Rappuoli, P. Leoncini, P. Tarli, and P. Neri, *Anal. Biochem.* **118,** 168 (1981).
[286] S. K. Gupta, J. L. Guesdon, S. Avrameas, and G. P. Talwar, *J. Immunol. Methods* **83,** 159 (1985).
[287] B. Leroux, R. Maldiney, E. Miginiac, L. Sossountzov, and B. Sotta, *Planta* **166,** 524 (1985).
[288] R. Maldiney, B. Leroux, I. Sabbagh, B. Sotta, L. Sossountzov, and E. Miginiac, *J. Immunol. Methods* **90,** 151 (1986).
[289] T. M. Aune, *J. Immunol. Methods* **84,** 33 (1985).
[290] R. C. Budd and K. A. Smith, *Bio/Technology* **4,** 983 (1986).
[291] K. Adler-Storthz, G. R. Dreesman, D. Y. Graham, and D. G. Evans, *J. Immunoassay* **6,** 67 (1985).
[292] A. Plebani, L. D. Notarangelo, V. Monafo, L. Nespoli, and A. G. Ugazio, *Clin. Allergy* **14,** 373 (1984).
[293] P. V. S. Rao, N. L. McCartney-Francis, and D. D. Metcalfe, *J. Immunol. Methods* **57,** 71 (1983).
[294] A. Plebani, M. A. Avanzini, M. Massa, and A. G. Ugazio, *J. Immunol. Methods* **71,** 133 (1984).
[295] Y. G. Alevy and C. M. Blynn, *J. Immunol. Methods* **87,** 273 (1986).
[296] A. Plebani, A. G. Ugazio, A. M. Avanzini, V. Monafo, and G. R. Burgio, *J. Immunol. Methods* **90,** 241 (1986).
[297] E. C. De Macario, A. J. L. Macario, and R. J. Jovell, *J. Immunol. Methods* **90,** 137 (1986).
[298] J.-L. Guesdon and S. Avrameas, *Ann. Immunol. (Paris)* **131C,** 389 (1980).
[299] S. Jackson, J. A. Sogn, and T. J. Kindt, *J. Immunol. Methods* **48,** 299 (1982).
[300] M. Rosenkoetter, A. T. Reder, J. J.-F. Oger, and J. P. Antel, *J. Immunol.* **132,** 1779 (1984).
[301] R. J. Versteegen and C. Clark, *in* "Monoclonal Antibodies and Functional Cell Lines: Progress and Applications" (R. H. Kennett, K. B. Bechtol, and T. J. McKearn, eds.), p. 393. Plenum, New York, 1984.
[302] R. C. Hart and L. R. Taaffe, *J. Immunol. Methods* **101,** 91 (1987).
[303] M. R. Ziai, L. Imberti, and S. Ferrone, *J. Immunol. Methods* **82,** 233 (1985).
[304] P. Atkinson, B. Bennett, and R. L. Hunter, *J. Immunol. Methods* **76,** 365 (1985).
[305] M. Harigai, M. Kawagoe, W. Hirose, M. Hara, A. Kitani, T. Hirose, K. Norioka, K. Suzuki, and H. Nakamura, *J. Immunol. Methods* **91,** 129 (1986).
[306] I. K. Mushahwar and K. S. Spiezia, *J. Virol. Methods* **16,** 45 (1987).
[307] I. K. Mushahwar, *J. Virol. Methods* **16,** 1 (1987).
[308] H.-C. Chang, I. Takashima, J. Arikawa, and N. Hashimoto, *J. Immunol. Methods* **72,** 401 (1984).
[309] H.-C. Chang, I. Takashima, J. Arikawa, and N. Hashimoto, *J. Virol. Methods* **9,** 143 (1984).
[310] B. Leipold and W. Remy, *J. Immunol. Methods* **66,** 227 (1984).
[311] P. G. Foiles, N. Trushin, and A. Castonguay, *Carcinogenesis* **6,** 989 (1985).
[312] K. Mortensson-Egnund and E. Kjeldsberg, *J. Virol. Methods* **14,** 57 (1986).

[313] E. N. Esteban, R. M. Thorn, and J. F. Ferrer, *Cancer Res.* **45,** 3231 (1985).
[314] D. Goldgaber, C. J. Gibbs, Jr., D. C. Gajdusek, and A. Svedmyr, *J. Gen. Virol.* **66,** 1733 (1985).
[315] C. Kendall, I. Ionescu-Matiu, and G. R. Dreesman, *J. Immunol. Methods* **56,** 329 (1983).
[316] K. Adler-Storthz, C. Kendall, R. C. Kennedy, R. D. Henkel, and G. R. Dreesman, *J. Clin. Microbiol.* **18,** 1329 (1983).
[317] L. S. Nerurkar, M. Namba, G. Brashears, A. J. Jacob, Y. J. Lee, and J. L. Sever, *J. Clin. Microbiol.* **20,** 109 (1984).
[318] A. Hornsleth, B. Friis, and P. A. Krasilnikof, *J. Med. Virol.* **18,** 113 (1986).
[319] R. H. Yolken, *J. Infect. Dis.* **148,** 942 (1983).
[320] D. Katz, J. K. Hilliard, R. Eberle, and S. L. Lipper, *J. Immunol. Methods* **14,** 99 (1986).
[321] R. Diaco, J. H. Hill, E. K. Hill, H. Tachibana, and D. P. Durand, *J. Gen. Virol.* **66,** 2089 (1985).
[322] I. Davidson, M. Malkinson, C. Strenger, and Y. Becker, *J. Virol. Methods* **14,** 237 (1986).
[323] R. H. Yolken, F. J. Leister, L. S. Whitcomb, and M. Santosham, *J. Immunol. Methods* **56,** 319 (1983).
[324] R. H. Yolken and S.-B. Weem, *J. Clin. Microbiol.* **19,** 356 (1984).
[325] A. Sutton, W. F. Vann, A. B. Karpas, K. E. Stein, and R. Schneerson, *J. Immunol. Methods* **82,** 215 (1985).
[326] R. L. Tarleton and R. E. Kuhn, *J. Immunol. Methods* **60,** 213 (1983).
[327] R. L. Tarleton, C. L. Schulz, M. Grögl, and R. E. Kuhn, *Am. J. Trop. Med. Hyg.* **33,** 34 (1984).
[328] S. L. Epstein and J. K. Lunney, *J. Immunol. Methods* **76,** 63 (1985).
[329] I. L. Paradis, E. J. Merrall, J. M. Krell, J. H. Dauber, R. M. Rogers, and B. S. Rabin, *J. Histochem. Cytochem.* **32,** 358 (1984).
[330] M. Kemp, S. Husby, J. C. Jensenius, G. G. Rasmussen, and S.-E. Svehag, *Acta Pathol. Microbiol. Immunol. Scand. Sect. C* **93,** 217 (1985).
[331] M. Zrein, G. De Marcillac, and M. H. V. Van Regenmortel, *J. Immunol. Methods* **87,** 229 (1986).
[332] A. C. Morgan, Jr., and R. F. McIntyre, *Cancer Res.* **43,** 3155 (1983).
[333] B. R. Clark and C. W. Todd, *Anal. Biochem.* **121,** 257 (1982).
[334] C. Wagener, B. R. Clark, K. J. Rickard, and J. E. Shively, *J. Immunol.* **130,** 2302 (1983).
[335] C. Wagener, U. Fenger, B. R. Clark, and J. E. Shively, *J. Immunol. Methods* **68,** 269 (1984).
[336] J. E. Manning, N. D. Hershey, T. R. Broker, M. Pellegrini, H. K. Mitchell, and N. Davidson, *Chromosoma* **53,** 107 (1975).
[337] P. R. Langer-Safer, M. Levine, and D. C. Ward, *Proc. Natl. Acad. Sci. U.S.A.* **79,** 4381 (1982).
[338] H. Kress, E. M. Meyerowitz, and N. Davidson, *Chromosoma* **93,** 113 (1985).
[339] L. Manuelidis, P. R. Langer-Safer, and D. C. Ward, *J. Cell Biol.* **95,** 619 (1982).
[340] N. J. Hutchison, P. R. Langer-Safer, D. C. Ward, and B. A. Hamkalo, *J. Cell Biol.* **95,** 609 (1982).
[341] S. Adolph and H. Hameister, *Hum. Genet.* **69,** 117 (1985).
[342] T. L. J. Boehm and D. Drahovsky, *Clin. Genet.* **30,** 509 (1986).
[343] Y.-F. Lau, *Cytogenet. Cell Genet.* **39,** 184 (1985).
[344] M. L. Murer-Orlando and A. C. Peterson, *Exp. Cell Res.* **157,** 322 (1985).
[345] V. T.-W. Chan, K. A. Fleming, and J. O'D. McGee, *Nucleic Acids Res.* **13,** 8083 (1985).
[346] R. P. Viscidi, C. J. Connelly, and R. H. Yolken, *J. Clin. Microbiol.* **23,** 311 (1986).
[347] A. Murasugi and R. B. Wallace, *DNA* **3,** 269 (1984).
[348] J. A. Matthews, A. Batki, C. Hynds, and L. J. Kricka, *Anal. Biochem.* **151,** 205 (1985).
[349] T. Kempe, W. I. Sundquist, F. Chow, and S.-L. Hu, *Nucleic Acids Res.* **13,** 45 (1985).

[350] A. Reisfeld, J. M. Rothenberg, E. A. Bayer, and M. Wilchek, *Biochem. Biophys. Res. Commun.* **142,** 519 (1987).
[351] J. Menissier, D. J. Hunting, and G. De Murcia, *Anal. Biochem.* **148,** 339 (1985).
[352] E. R. Unger, L. R. Budgeon, D. Myerson, and D. J. Brigati, *Am. J. Surg. Pathol.* **10,** 1 (1986).
[353] J. A. McKeating, W. Al-Nakib, P. J. Greenaway, and P. D. Griffiths, *J. Virol. Methods* **11,** 207 (1985).
[354] N. S. Lurain, K. D. Thompson, and S. K. Farrand, *J. Clin. Microbiol.* **24,** 724 (1986).
[355] D. Myerson, R. C. Hackman, and J. D. Meyers, *J. Infect. Dis.* **150,** 272 (1984).
[356] D. Przepiorka and D. Myerson, *J. Histochem. Cytochem.* **34,** 1731 (1986).
[357] D. J. Brigati, D. Myerson, J. J. Leary, B. Spalholz, S. Z. Travis, C. K. Y. Fong, G. D. Hsiung, and D. C. Ward, *Virology* **126,** 32 (1983).
[358] B. Theveny and B. Revet, *Nucleic Acids Res.* **15,** 947 (1987).
[359] A. M. Khan and P. J. Wright, *J. Virol. Methods* **15,** 121 (1987).
[360] S. A. Gomes, J. P. Nascimento, M. M. Siqueira, M. M. Krawczuk, H. G. Pereira, and W. C. Russell, *J. Virol. Methods* **12,** 105 (1985).
[361] A. J. Aksamit, P. Mourrain, J. L. Sever, and E. O. Major, *Ann. Neurol.* **18,** 490 (1985).
[362] J. K. McDougall, D. Myerson, and A. M. Beckmann, *J. Histochem. Cytochem.* **34,** 33 (1986).
[363] R. Neumann, B. Heiles, C. Zippel, H. J. Eggers, H. H. Zippel, L. Holzmann, and K. D. Schulz, *Acta Cytol.* **30,** 603 (1986).
[364] H. Yokota, K. Yokoo, and Y. Nagata, *Biochim. Biophys. Acta* **868,** 4533 (1986).
[365] J. A. Saldanha, P. Karayiannis, H. C. Thomas, and J. P. Monjardino, *J. Virol. Methods* **16,** 339 (1987).
[366] A. H. Alhakim and R. Hull, *Nucleic Acids Res.* **14,** 9965 (1986).
[367] A. C. Forster, J. L. McInnes, D. C. Skingle, and R. H. Symons, *Nucleic Acids Res.* **13,** 745 (1985).
[368] A. L. Rayburn and B. S. Gill, *J. Hered.* **76,** 78 (1985).
[369] L. Angerer, N. Davidson, W. Murphy, D. Lynch, and G. Attardi, *Cell* **9,** 81 (1976).
[370] D. G. Albertson, *EMBO J.* **4,** 2493 (1985).
[371] D. G. Albertson, *EMBO J.* **3,** 1227 (1984).
[372] P. Liesi, J.-P. Julien, P. Vilja, F. Grosveld, and L. Rechardt, *J. Histochem. Cytochem.* **34,** 923 (1986).
[373] L. K. Riley, M. E. Marshall, and M. S. Coleman, *DNA* **5,** 333 (1986).
[374] J. L. McInnes, S. Dalton, P. D. Vize, and A. J. Robins, *Bio/Technology* **5,** 269 (1987).
[375] J. J. Leary, D. J. Brigati, and D. C. Ward, *Proc. Natl. Acad. Sci. U.S.A.* **80,** 4045 (1983).
[376] G. Gebeyehu, P. Y. Rao, P. SooChan, D. A. Simms, and L. Klevan, *Nucleic Acids Res.* **15,** 4513 (1987).
[377] D. Dykes, J. Fondell, P. Watkins, and H. Polesky, *Electrophoresis* **7,** 278 (1986).
[378] B. C. F. Chu and L. E. Orgel, *DNA* **4,** 327 (1985).
[379] A. Chollet and E. H. Kawashima, *Nucleic Acids Res.* **13,** 1529 (1985).
[380] L. Wachter, J.-A. Jablonski, and K. L. Ramachandran, *Nucleic Acids Res.* **14,** 7985 (1986).
[381] S. Agrawal, C. Christodoulou, and M. J. Gait, *Nucleic Acids Res.* **14,** 6227 (1986).
[382] S. Beck, *Anal. Biochem.* **164,** 514 (1987).
[383] B. C. F. Chu, F. R. Kramer, and L. E. Orgel, *Nucleic Acids Res.* **14,** 5591 (1986).
[384] D. J. Hunting, S. L. Dresler, and G. de Murcia, *Biochemistry,* **24,** 5729 (1985).
[385] E. L. Sheldon, D. E. Kellogg, R. Watson, C. H. Levenson, and H. A. Erlich, *Proc. Natl. Acad. Sci. U.S.A.* **83,** 9085 (1986).
[386] M. E. Royston and L. H. Augenlicht, *Science* **222,** 1339 (1983).
[387] G. J. Garbutt, J. T. Wilson, G. S. Schuster, J. J. Leary, and D. C. Ward, *Clin. Chem. (N.Y.)* **31,** 1203 (1985).

Section II

Biotin-Binding Proteins

[4] Biotin-Binding Proteins: Overview and Prospects

By EDWARD A. BAYER and MEIR WILCHEK

In this section we include chapters that deal with avidin[1] from egg white and streptavidin[2] from *Streptomyces avidinii,* as well as information concerning a variety of other biotin-binding proteins which may eventually become very useful for avidin–biotin technology. The studies that deal with the properties of the very strong biotin-binding proteins (i.e., avidin and streptavidin, both of which boast affinity constants of about 10^{15} M^{-1}) are all relevant to the application-oriented chapters that appear later in this volume. The nuances in both the physicochemical and biochemical characteristics of avidin and streptavidin are important for their appropriate applications. In addition, biotin-binding polyclonal antibodies[3] have already proved useful for many studies,[4] and the availability of biotin-binding monoclonals[5] (especially those which exhibit affinity constants greater than 10^9 M^{-1}) should broaden the use of the system. The biotin-binding proteins in egg yolk[6,7] will also prove to be a useful addition to this family of proteins.

Serious disadvantages in the use of biotin-binding antibodies are their high molecular weight (about twice that of avidin and the other biotin-binding proteins) and the fact that they possess only two binding sites per molecule (versus four for avidin, streptavidin, and the egg-yolk biotin-binding proteins). This means that the efficiency of binding (ligand/protein ratio) by the antibodies is much lower than the avidinlike proteins (over and above the several orders of magnitude differences in affinity constants). Another problem of antibodies is the presence of the Fc portion which may cause extraneous binding with certain cell types (e.g., macrophages and lymphocytes) that have receptors for this part of the antibody molecule. The oligosaccharide residues on the Fc portion may also be a source of nonspecific binding that would interfere with the use of antibodies in avidin–biotin technology. It should also be kept in mind that the affinity of an antibody for a biotinylated protein may be reduced

[1] N. M. Green, this volume [5].
[2] E. A. Bayer, H. Ben-Hur, and M. Wilchek, this volume [8].
[3] M. Berger, this series, Vol. 62 [57].
[4] P. R. Langer, A. A. Waldrop, and D. C. Ward, *Proc. Natl. Acad. Sci. U.S.A.* **78,** 6633 (1981).
[5] K. Dakshinamurti and E. S. Rector, this volume [12].
[6] K. Dakshinamurti and J. Chauhan, this volume [10].
[7] H. W. Meslar and H. B. White, this series, Vol. 62 [56].

compared to that for the free ligand; in many cases the antibody–biotin interaction may therefore be included with the low-affinity biotin-binding proteins. Despite these potential problems, the use of antibodies in avidin–biotin technology has advantages that can be enjoyed by most research groups.

The applicability of the egg-yolk biotin-binding protein, which exhibits a slightly lower affinity for biotin ($\sim 10^{12}$ M^{-1}) than that of avidin or streptavidin, is essentially unknown. Its size (~73 kDa) is similar to that of avidin and streptavidin, and, like the latter, it is a tetramer composed of identical subunits, each of which bears a single biotin-binding site. Unlike the egg-white and bacterial proteins, however, the egg-yolk relative is saturated with free biotin in the native state, a fact that suggests its possible function as a storage protein for biotin. Nonetheless, compared to either avidin or streptavidin, biotin, bound to the active site of the egg-yolk protein, is relatively easy to remove. It is anticipated that the applicability of this protein as a probe in avidin–biotin technology will be extensively explored in the near future.

Another untapped source for use as a replacement probe for avidin is the variety of biotin receptors in nature. To date, none has been isolated in sufficient quantity for application in avidin–biotin technology. However, receptor affinity chromatography has been described as an efficient procedure for the purification of certain ligands,[8] and this approach may eventually be useful for immobilization of biotinylated proteins. Many receptors are monomeric, low molecular weight proteins, and these properties may provide a desirable option in certain applicative systems.

An interesting prospect is the potential use of the enzyme biotin holocarboxylase synthetase[9] in avidin–biotin technology. This enzyme apparently exists as a monomer in the native state. Its size is about half that of avidin, and the affinity for biotin is about 10^7 M^{-1}. In addition to the use of biotin holocarboxylase synthetase as a substitute for avidin, the enzyme may prove an exciting alternative for the biotinylation of other proteins (in addition to carboxylases). In this regard, it is known that the enzyme is specific for a conserved sequence in all biotin-requiring apoenzymes, in which the lysine residue of Ala-Met-Lys-Met is biotinylated. This tetrapeptide can then be incorporated into any protein (e.g., at the C terminus) and the synthetase can be used to incorporate biotin at this specific site. This approach may find use in systems where biotinylation is particularly detrimental to biological activity. Finally, the enzyme biotinidase[6,10] (EC

[8] P. Bailon, D. V. Weber, R. F. Keeney, J. E. Fredricks, C. Smith, P. C. Familletti, and J. E. Smart, *Bio/Technology* **5,** 1195 (1987).

[9] N. H. Goss and H. G. Wood, this series, Vol. 107 [15].

[10] B. Wolf, J. Hymes, and G. S. Heard, this volume [11].

3.5.1.12) may eventually be extensively exploited in avidin–biotin technology, as a replacement for avidin, for biotinylation of binders, or as an analytical tool for releasing the biotin moiety from biotinylated material.

The fact that biotin is a vitamin required in very small quantities for a variety of cellular processes reflects the very high affinity constants exhibited by the broad number of naturally occurring biotin-binding proteins which have been described. As time goes on, it would not be surprising if many more biotin-binding proteins will be described, many of which may prove suitable for application in avidin–biotin technology. In certain instances, some may even supercede the use of avidin, owing to improved molecular characteristics. The use of avidin–biotin technology would undoubtedly be increased enormously if other non-biotin-binding proteins which exhibit very high affinities for unrelated ligands would be described for complementary applications.

[5] Avidin and Streptavidin

By N. MICHAEL GREEN

Introduction

The discovery of the protein avidin resulted from intensive nutritional investigations into the vitamin B complex.[1] Avidin proved to be a minor constituent of egg white that could induce a nutritional deficiency in rats by forming a very stable noncovalent complex with what was subsequently proved to be the B vitamin biotin. The biological role of avidin in egg white appeared to be that of a scavenger, inhibiting bacterial growth. A variant of lower affinity for biotin has been found in egg yolk, which may be important in regulating the supply of biotin during development.[2] Avidin is a highly specialized protein that is only rarely expressed. One might have expected that it was a vertebrate protein of fairly recent lineage were it not for the fact that a close relative, streptavidin, is expressed in a species of *Streptomyces*.[3] However, like proteins have not been found elsewhere in microorganisms, and the possibility remains that its occurrence in *Streptomyces avidinii,* a strain of *Streptomyces lavendulae,* reflects a chance transfection rather than an ancient lineage. The relation of the avidins to a recently discovered family of binding proteins is considered below.

[1] P. Gyorgy, *in* "The Vitamins" (W. H. Sebrell, Jr., and R. S. Harris, eds.), Vol. 1, p. 527. Academic Press, New York, 1954.
[2] H. W. Meslar, S. A. Camper, and H. B. White, *J. Biol. Chem.* **253,** 6979 (1978).
[3] L. Chaiet and F. J. Wolf, *Arch. Biochem. Biophys.* **106,** 1 (1964).

Current interest in avidin derives mainly from its applications in biotechnology based on its rapid and almost irreversible binding of any molecule to which biotin can be linked. This has led to a proliferation of labeling and affinity purification methods, which are the subject of this volume.

The biochemical basis for the very high affinity was first investigated by chemical modification.[4] Interest was further stimulated by the discovery of the coenzyme function of biotin in CO_2 transfer[5] and by the subsequent use of avidin to identify biotinyl enzymes. More extensive characterization of the molecule and its binding sites followed, mainly from the author's laboratory, and the results of this were reviewed in 1975.[6] At that time there was only one report of the use of avidin as a cytochemical label,[7] although from its general properties and its use as a high-resolution electron microscopic label for biotinyl enzymes,[8] it was clear that it had considerable potential in this field. Most work on avidin since the mid-1970s has been on the development of such applications based on avidin labeled with gold, peroxidase, alkaline phosphatase, ferritin, or radioactive or fluorescent labels, using a variety of synthetic bridging agents to provide appropriate biotinyl sites for its attachment.[9,10]

During this period there has been relatively little new biochemical work on the structure and binding properties of avidin or streptavidin. Considerable efforts have been devoted to an X-ray crystallographic approach to the structure, but it proved very difficult to grow acceptable crystals of sufficient size,[11] possibly because avidin is a glycoprotein and its carbohydrate is characteristically heterogeneous.[12] In this respect, streptavidin, which is carbohydrate free, has proved advantageous. Its structure is now determined, but the results were not available for the preparation of this chapter.[13,14] Another main advance has been the cloning and sequencing of the streptavidin gene.[15]

[4] H. Fraenkel-Conrat, N. S. Snell, and E. D. Ducay, *Arch. Biochem. Biophys.* **39,** 97 (1952).

[5] S. J. Wakil, E. B. Titchener, and D. M. Gibson, *Biochem. Biophys. Acta* **29,** 225 (1958).

[6] N. M. Green, *Adv. Protein Chem.* **29,** 85 (1975).

[7] H. Heitzmann and F. M. Richards, *Proc. Natl. Acad. Sci. U.S.A.* **71,** 3537 (1974).

[8] N. M. Green, R. C. Valentine, N. G. Wrigley, F. Ahmad, B. E. Jacobson, and H. G. Wood, *J. Biol. Chem.* **247,** 6284 (1972).

[9] E. A. Bayer and M. Wilchek, *Methods Biochem. Anal.* **26,** 1 (1980).

[10] M. Wilchek and E. A. Bayer, *Anal. Biochem.* **171,** 1 (1988).

[11] E. Pinn, A. Pahler, W. Saenger, G. A. Petsko, and N. M. Green, *Eur. J. Biochem.* **123,** 545 (1982).

[12] R. C. Bruch and H. B. White, *Biochemistry* **21,** 5334 (1982).

[13] P. C. Weber, D. H. Ohlendorf, J. J. Wendoloski, and F. R. Salemme, *Science* **243,** 85 (1989).

[14] W. A. Hendrickson, A. Pähler, J. L. Smith, Y. Satow, E. A. Merritt, and R. P. Phizackerley, *Proc. Natl. Acad. Sci. U.S.A.* **86,** 2190 (1989).

In this chapter, I consider only fundamental molecular properties and those which are important for applications in biotechnology. Particular emphasis is placed on differences between avidin and streptavidin, since although streptavidin has become widely used as a label because of low nonspecific binding, its binding characteristics have not been studied quantitatively. The alternative use of succinylavidin[16] to diminish nonspecific avidin binding is less effective, probably because the carbohydrate of avidin also contributes to the nonspecific binding.[17] Carbohydrate-free avidin can be obtained by use of deglycosylating enzymes, and it has been shown to be present in significant amounts in some commercial preparations from which it may be separated by use of lectin columns.[18] Its biotin-binding properties were unimpaired.

Affinity Methods for Purification of Avidins and Biotinyl Proteins

Improvement on the original use of biotinyl cellulose[19] came with the introduction of iminobiotinyl derivatives of Sepharose[20–23] which utilized the pH dependence of the binding[24] to achieve efficient elution. In iminobiotin, the ureido group becomes a guanidinium group, and only the form in which this is uncharged is strongly bound. The apparent affinity, therefore, should fall by a factor of 10 per pH unit provided that the protonated form does not bind [$K_{app} = K_B K/(H^+ + K)$, where K_B is the dissociation constant for uncharged iminobiotin (10^{-13} M) and K is the acid dissociation constant of the guanidinium group (pK = 11.9)]. The dissociation constant of the avidin–iminobiotin complex increases from 0.03 μM at pH 9 to 13 μM at pH 6, by a factor of 430, rather than the theoretical factor of 10,000, the discrepancy being least above pH 8. It may be that the dissociation constant changes less at lower pH because of a contribution from weak binding of the protonated form of iminobiotin.

---

[15] C. E. Argarana, I. D. Kuntz, S. Birken, R. Axel, and C. R. Cantor, *Nucleic Acids Res.* **14,** 1871 (1986).

[16] F. M. Finn, G. Titus, J. A. Montibeller, and K. Hofmann, *J. Biol. Chem.* **255,** 5742 (1980).

[17] T. V. Updyke and G. L. Nicolson, *J. Immunol. Methods* **73,** 83 (1984).

[18] Y. Hiller, J. M. Gershoni, E. A. Bayer, and M. Wilchek, *Biochem. J.* **248,** 167 (1987).

[19] D. B. McCormick, *Anal. Biochem.* **13,** 194 (1965).

[20] K. Hofmann, S. W. Wood, C. C. Brinton, and F. M. Finn, *Proc Natl. Acad. Sci. U.S.A.* **77,** 4666 (1980).

[21] G. A. Orr, G. C. Heney, and R. Zeheb, this series, Vol. 122, p. 83.

[22] G. A. Zeheb and G. A. Orr, this series, Vol. 122, p. 87.

[23] E. A. Bayer, H. Ben Hur, G. Gitlin, and M. Wilchek, *J.Biochem. Biophys. Methods* **13,** 103 (1986).

[24] N. M. Green, *Biochem. J.* **101,** 774 (1966).

The principle was extended by Hofmann's group, who made iminobiotinyl derivatives of hormones to allow efficient elution of hormone receptors,[25,26] but unfortunately the affinities proved slightly too low to be useful. The dissociation constants for the aminohexanoate and lysine derivatives of iminobiotin at pH 6.8 ($K_d = 10^{-5}\ M$)[25] were 5- to 10-fold greater than for the parent iminobiotin. The effects of using dethiobiotin and of incorporating spacer arms were also studied. A useful summary of the synthesis and properties of these derivatives is available.[27]

The very high affinity of avidin for biotin, although ideal for specific labeling, is a disadvantage for affinity purifications of biotinyl enzymes, where it is not possible to substitute an iminobiotinyl residue. It is possible to weaken the binding by selective oxidation of avidin tryptophans with periodate ($K_d = 10^{-9}\ M$),[6] but the affinity is still too high. A more successful approach has been the use of monomeric avidin–Sepharose, prepared by stripping off noncovalently bound subunits with guanidinium chloride.[6,28–30] The dissociation constant of the biotinyl complex of the monomer is about $10^{-7}\ M$; however, other species with lower dissociation constants are present in these preparations, and these should be blocked with biotin before using the column.

Lipoic acid has some stereochemical similarity to biotin,[6] and lipoyl Sepharose provides another matrix of moderate affinity which has been used for the purification of antibiotin antibodies.[31] It has also been used to purify complexes of myosin S_1, with avidin, using the spare binding sites of an avidin molecule bound to biotinyl S_1.[32] Conversely, avidin-Sepharose columns can, in principle, be used to purify lipoyl peptides.

Streptavidin

The gene for streptavidin has been cloned and sequenced with the ultimate objective of using it in general expression systems for detecting and isolating fusion proteins.[15] It coded for a sequence of 159 amino acids, some 30 residues longer than avidin and longer than expected from molec-

[25] F. M. Finn and K. Hofmann, *Proc. 7th Am. Peptide Symp.* (Pierce Chemical Co., Rockford, IL), 1 (1981).

[26] K. Hofmann, G. Titus, J. A. Montibeller, and F. M. Finn, *Biochemistry* **21,** 978 (1982).

[27] F. M. Finn and K. Hofmann, this series, Vol. 109, p. 418.

[28] K. P. Henrikson, S. H. Allen, and W. L. Maloy, *Anal. Biochem.* **94,** 366 (1979).

[29] R. A. Kohanski and M. D. Lane, *J. Biol. Chem.* **260,** 5014 (1985).

[30] P. Dimroth, *FEBS Lett.* **141,** 59 (1982).

[31] F. Harmon, *Anal. Biochem.* **103,** 58 (1980).

[32] K. Yamamoto and T. Sekine, *J. Biochem.* **101,** 519 (1987).

TABLE I
PROPERTIES OF AVIDINS

Property	Avidin egg white[a]	Streptavidin: Unprocessed[b]	Streptavidin: Processed[c]	Avidin, egg yolk[d]
Amino acid residues	128	159	125–127	—
Subunit size				
From sequence	15,600	16,473	13,400	—
SDS gels	16,400	19,000	14,500	19,000
Subunits	4	4	4	4
Isoelectric pH	10	—	5–6	4.6
ε_{280}	24,000	—	34,000	—
$\Delta\varepsilon_{233}$ (+ biotin)	24,000	—	8,000	7,000
Fluorescence				
λ_{max} (nm)	338 (328)[e]	—	—	—
τ (nsec)	1.8 (0.8)[e]	—	—	—
Binding of HABA				
K_d (μM)	6	—	100	—
ε_{500}	35,000	—	35,000	7,000
K_d biotin (M) (pH 7, 25°)	0.6×10^{-15}	—	4×10^{-14}	1.7×10^{-12}
$t_{1/2}$ (days)	200	—	2.9	0.07

[a] From Ref. 6.
[b] From Refs. 13 and 23.
[c] From Refs. 3 and 6.
[d] From Refs. 2 and 33.
[e] Fluorescence lifetimes were taken from Ref. 34. Numbers in parentheses refer to the avidin–biotin complex.

ular weight measurements. It was found that subunits of both low and high molecular weight were present and that the smaller one, the main constituent of most commercial preparations, was the result of processing at both the N and C termini to give "core" streptavidin of 125–127 residues.[13,23] It had a much higher solubility in water than the unprocessed precursor. This core is identical to avidin at 33% of its residues, including the four tryptophan residues involved in the biotin-binding site. It also resembles avidin in its predicted secondary structure, predominantly β strands and bends,[15] and in other features (Table I).[33,34] Some commercial preparations contain unprocessed streptavidin,[13] and a recently published purification based on iminobiotin columns yields this as the main product.[23]

[33] C. V. R. Murthy and P. R. Aliga, *Biochem. Biophys. Acta* **786,** 222 (1984).
[34] B. P. Maliwal and J. Lakowicz, *Biophys. Chem.* **19,** 337 (1985).

Two differences from avidin are of some importance. Streptavidin contains no carbohydrate, the heterogeneity of which is the probable cause of the poor quality of avidin crystals,[11] and it has a slightly acid isoelectric point[3] (pH 5–6), which minimizes nonspecific adsorption to nucleic acids and negatively charged cell membranes. Quantitative data on its affinity for biotin and its analogs are sparse (see Table II below).

General Properties of Avidins

The main common features of biotin binding by both avidin and streptavidin (Table I) can be summarized briefly. The avidins are stable tetramers with 2-fold symmetry, the binding sites being arranged in two pairs on opposed faces of the molecule (see below). The stability is greatly enhanced by biotin binding, since the total free energy of binding is about 330 kJ/mol of tetramer. The dissociation constant for biotin is so low that it can be estimated only from the ratio of the rate constants for binding and exchange. The binding is accompanied by a red shift of the tryptophan spectrum and by a decrease in fluorescence, either of which can be used as the basis for quantitative assays.[35,36] The spectral changes in the tryptophan residues are accompanied by a marked reduction in their accessibility to reagents such as a *N*-bromosuccinimide. In avidin, the four tryptophans of each subunit are protected when biotin is bound. In contrast, fluorescence quenching by oxygen is not diminished in the avidin–biotin complex; if anything, the rate constant for quenching is increased.[34]

The binding of biotin can be blocked by oxidation of any of several tryptophan residues[6] or, in the case of avidin, by the dinitrophenylation of what appears to be a single lysine residue.[6] Isolation and sequencing of dinitrophenyl (DNP) peptides from avidin in which a single lysine is blocked showed that reaction at any of three lysines inactivates the avidin.[37] Similarly, two essential tryptophans were identified by isolation of peptides after labeling with hydroxynitrobenzyl bromide.[38] The location of these residues is shown in Fig. 1. These results show that inactivation by incorporation of 1 mol of reagent does not necessarily imply that a single specific group is involved. Reaction of any one of two or three different lysines or tryptophans led to inactivation and blocked further reaction. Furthermore, dinitrophenylation of the lysines blocked reaction of the tryptophans with the Koshland reagent, just as it had previously

[35] N. M. Green, this series, Vol. 18A, p. 418.

[36] H. J. Lin and J. F. Kirsch, this series, Vol. 62, p. 287.

[37] G. Gitlin, E. A. Bayer, and M. Wilchek, *Biochem. J.* **242,** 923 (1987).

[38] G. Gitlin, E. A. Bayer, and M. Wilchek, *Biochem. J.* **250,** 291 (1988).

```
                        10        20        30        40        50
Streptav.   12  AAEAGITGTWYNQLGST        FIVTAGAD  GALTGTYESAVGNAES
Avidin       1  ARKCSLTGKWTNDLGSN        MTIGAVNSR GEFTGTYITAVT ATS
Beta pred.          -----                -------      --------   -
Conserved         - - +.+  =   =*        = = -=   . =       =-- .
Fatty acid   1     MAFDGTWKVDRNENYEKFMEKMGINVVKRKLGAHDNLKLTITQEGNK
Beta strands           ----A-----======= ========  ----B-----  --
Ret.cell.    1    PVDFNGYWKMLSNENFEEYLRALDVNVALRKIANLLKPDKEIVQDGDH


                        60        70        80        90       100
Streptav.   53  RYVLTG RYDSAPATDGS  GTALGWTVAWKNNYRNAHSATTWS    QY
Avidin      42  NEIKES PLHGTQNTINKRTQPTFGFTVNWKF    SESTTVFT    GQC
Beta pred.      ---*-               ------  *        ------      -
Conserved         *===  *      .     .  * *.-      .= =..-=      -
Fatty acid 48   FTVKESSNFRNIDVVFEL  GVDFAYSLAD     GTELTGTWTMEGNKL
Beta strands    -C----  ----D-----   ---E---       ----F----   ---
Ret.cell.   49  MIIRTLSTFRNYIMDFQV  GKEFEETGID     DRKCMTTVSWDGDKL


                       110       120       130       140       150
Streptav.   97  VG      GAEARINTQWLL TSGTTEANAWKS TLVGHDTFTKVKPSAA
Avidin      84  FIDR    NGKEVLKTMWLLRSSVNDIGDDWKA TRVGINIFTRLRTQKE
Beta pred.      ---        --*--*--          *   -----------
Conserved       .-      ====.- * .-     -=.-    * *  . - *=+ ..
Fatty acid 91   VGKFKRVDNGKELIAVREIS    GNELIQTYTYEGVEAKRIFKKE
Beta strands    G---   ------H------     ----I----  ----J---
Ret.cell.   94  QCVQK  GEKEGRGWTQWIE    GDELHLEMRAEGVTCKQVFKKVH
```

FIG. 1. Alignment of avidin and streptavidin with fatty acid- and cellular retinol-binding proteins. The avidins[15] are aligned with the other two binding proteins,[39] giving weight to the matching of predicted bends and strands (---) with the actual bends and strands (--*X*--) of the fatty acid-binding protein.[40] === indicates α helix. The serum retinol-binding protein and other members of the β-lactoglobulin subclass are not shown because the percent identity is very low (Fig. 2) except for the N-terminal motif. Conservation is shown as follows: +, complete identity; =, identity of avidin and fatty acid-binding protein; −, conserved hydrophobic site; ., other limited conservation. An asterisk indicates either a contact residue (bound fatty acid) or a site at which modification blocks biotin binding.[37–40]

been shown to protect them from periodate oxidation,[6] emphasizing the complex interactions which can occur in a restricted site.

A gene for avidin has been cloned from chick oviduct.[41] It corresponds to the variant with isoleucine at position 34. Its expression is induced in

[39] J. Sundelin, S. R. Das, U. Eriksson, L. Rask, and P. A. Peterson, *J. Biol. Chem.* **260,** 6494 (1985).

[40] J. C. Sacchetini, J. I. Gordon, and L. J. Banaszak, *J. Biol. Chem* **263,** 5815 (1988).

[41] M. L. Gope, R. A. Keinanen, P. A. Kristo, O. M. Conneely, W. G. Beattie, T. Zarucki-Schulz, B. W. O'Malley, and M. S. Kulomaa, *Nucleic Acids Res.* **15,** 3595 (1987).

other tissues under inflammatory conditions, probably caused by a rise in progesterone levels.[42]

Relation of Avidins to Proteins of Known Structure

A search of the Protein Information Resource (PIR) database revealed no protein with significant similarity to avidin. A more detailed examination of potentially similar β-structured proteins proved more informative. It is clear that avidin is composed almost entirely of β strands and bends, from analysis of its Raman spectrum,[43] from circular dichroism (CD) measurements,[6,44] and from the secondary structure predicted from the sequences.[15,43] Proteins with known antiparallel β structure fall into two main classes: (1) β sandwiches of 6–10 strands such as transthyretin (prealbumin) and immunoglobulins[45] and (2) orthogonal β barrels such as serum retinol-binding protein[46] and β-lactoglobulin.[47] Comparison with the sequences of avidins showed no similarity to any of the first class but did reveal a short N-terminal motif (Fig. 1) common to all members of the second class.[39,48]

The class of β-barrel proteins has been extended by the discovery of several homologous proteins (~40% identity) which bind fatty acids, retinol, or steroids and share both the N-terminal motif and a β-barrel structure[40] with the lactoglobulin subclass, although overall the sequences of the two subclasses show less than 20% identity. The alignments (Fig. 1) and the derived matrix (Fig. 2)[49] show that avidin and streptavidin have greater sequence identity with this subclass of orthogonal β-barrels than they have with the lactoglobulin subclass. The similarity is also greater than that between the two subclasses, although less than that between members of the same subclass. Taken in conjunction with the predicted secondary structure and the physical evidence for β strands, the evidence for a common structure is good. Note that when the percent identity is as low as it is here (<20%), there are many possible alignments which give almost equivalent scores and that the highest score does not necessarily correspond to the structurally most significant alignment. The alignments

[42] H. A. Elo, M. S. Kulomaa, and A. O. Niemela, *Comp. Biochem. Physiol.* **68A,** 323 (1981).

[43] R. B. Honzatko and R. W. Williams, *Biochemistry* **21,** 6201 (1982).

[44] N. M. Green and M. D. Melamed, *Biochem. J.* **100,** 614 (1966).

[45] J. S. Richardson, this series, Vol. 115, p. 341.

[46] M. E. Newcomer, T. A. Jones, J. Aquist, J. Sundelin, U. Eriksson, L. Rask, and P. A. Peterson, *EMBO J.* **3,** 1451 (1984).

[47] M. Z. Papiz, L. Sawyer, E. Eliopoulos, A. C. T. North, J. B. C. Findlay, R. Sivraprasadarao, T. A. Jones, M. E. Newcomer, and P. J. Kraulis, *Nature (London)* **324,** 383 (1986).

[48] J. Godovac-Zimmerman, *Trends Biochem. Sci.* **13,** 64 (1988).

[49] R. Huber, M. Schneider, O. Epp, I. Mayr, A. Messerschmidt, J. Pflugrath, and H. Kayser, *J. Mol. Biol.* **195,** 423 (1987).

Streptavidin	30	12	12	9	8
Avidin		18	9	8	5
Fatty acid (rat)			28	7	9
Retinol (cellular)				8	4
Retinol (serum)					21
					Lactoglobulin

FIG. 2. Identity matrix between orthogonal β-barrel proteins arranged in three subclasses. The percent identities were calculated for the alignments shown in Fig. 1. The percent identity for the lactoglobulin subclass was taken from a structurally based alignment,[49] which was then matched with Fig. 1. The greatest resemblances outside the subclasses are between avidin and the fatty acid-binding protein. Structures of the underlined proteins have been determined.[40,46,47]

in Fig. 1 were obtained by matching the N-terminal motifs and as far as possible matching predicted strands and bends with corresponding features of the known structure. Several lysine and tryptophan residues marked with an asterisk in Fig. 1 contribute to the biotin-binding site[37,38] and are mostly located in regions homologous to those in the fatty acid-binding protein which make contact with ligand.

One feature of the structure of the fatty acid-binding protein which is not present in the predicted secondary structures of the avidins is the pair of short α helices between the A and B strands. These are also absent from the lactoglobulin subclass, where they are replaced by a short β hairpin which extends from the A and B strands.

Although uncertainties of both the alignment and the secondary structure prediction prevent detailed structural conclusions, it can be said that avidins are probably orthogonal β barrels with binding sites enclosed between β sheets. Such a deep binding site is consistent both with the very high affinity and with the deep burial of the binding sites, implied by the results with bifunctional ligands.[50] However, deep burial of a site does not necessarily correlate with a very high affinity. The dissociation constants for fatty acids and retinol, for example, are between 10^{-6} and 10^{-8} *M*.

[50] N. M. Green, L. Konieczny, E. J. Toms, and R. C. Valentine, *Biochem. J.* **125,** 781 (1971).

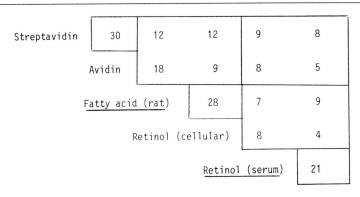

Dissociation Constants for Biotin

The rate constant for biotin binding has been measured only once ($7 \times 10^7 M^{-1} sec^{-1}$ for avidin and biotin at pH 5, 25°), so that values quoted for dissociation constants of other ligands or other conditions, which are based on this value, are only approximate. Direct measurements of the much higher dissociation constants of the D and L isomers of hexylimidazolidone have been compared with those calculated from the rate constants[6] and were found to be higher by a factor of about 4. This could result partly from the rate of binding for the uncharged ligand being lower than that for biotin, since avidin and biotin carry opposite charges at pH 5. Other factors are also involved so that absolute values of most binding constants estimated from rate constants could be in error by an order of magnitude.

Since no measurements of rate constants for either binding or dissociation for the widely used biotin–streptavidin system have been published, some results of recent measurements of the exchange rate are given here (Table II).[51,52] They are compared with previous results for avidin[52] and show a considerably faster release of biotin, in agreement with an earlier comment referring to unpublished work.[53] The rate constants increased as the pH rose, opposite to the behavior of avidin. This could be a consequence of the different isoelectric points. Streptavidin would be expected to show maximum stability near pH 5 and to release its ligand more readily at extremes of pH, where the net charge on the protein increases. In contrast, the net charge on avidin would fall with rising pH up to pH 10.5. The results in Table II are consistent with this hypothesis.

Avidin Polymers

Current knowledge of the spatial relations between the four binding sites of avidin comes from an electron microscopic study of polymers made with bifunctional biotinyldiamines.[50] Now that the structure of streptavidin is known at high resolution it is worth comparing its behavior with these reagents, since this could reveal differences in subunit relationships.

In the reaction with avidin, at least 9 methylene groups between biotinamides are required to produce polymers, and 12 are required to make them sufficiently stable to resist depolymerization by biotin, implying binding sites of considerable depth (15 Å). A further increase in length

[51] R. K. Garlick and R. W. Giese, *J. Biol. Chem.* **263,** 210 (1988).

[52] N. M. Green and E. J. Toms, *Biochem. J.* **133,** 687 (1973).

[53] F. M. Finn, G. Titus, and K. Hofmann, *Biochemistry* **23,** 2554 (1984).

TABLE II
DISSOCIATION RATES OF AVIDIN–BIOTIN COMPLEXES[a]

Rate and half-life	pH 1.7	2.0	3.0	5.0	7.0	9.2	10.5
Avidin							
k ($\text{sec}^{-1} \times 10^7$)	—	200	9	0.9	0.4	—	—
$t_{1/2}$ (days)	—	0.4	9	90	200	—	—
Streptavidin							
k ($\text{sec}^{-1} \times 10^7$)	35	—	19	8.7	28	64	100
$t_{1/2}$ (days)	2.3	—	4.2	9.2	2.9	1.25	0.8

[a] Exchange rates for streptavidin (Celltech, Slough, U.K.) were measured at 25° using [^{14}C]biotin that had been purified by HPLC.[51] The exchange followed a simple first-order course apart from the first 10% which, as with avidin, exchanged more rapidly. The methods and the results for avidin are taken from Ref. 52.

of linking chain (19–22 atoms) greatly *decreases* the length of the polymers, showing that intramolecular bridging occurs and that the shortest path between biotin carboxyls (not necessarily linear) is about 25 Å (Fig. 3 and Ref. 50). Pairs of sites are positioned at opposite ends of the short (40 Å) axis of the avidin molecule, each pair being located in a depression in the protein surface. Similarly situated sites have recently been observed in a tetrameric biliverdin-binding protein,[49] a member of the lactoglobulin family.

The presence of a depression in the protein surface could account for the observation that the length of linking chain required to produce an effective heterobifunctional reagent, such as those introduced by Hof-

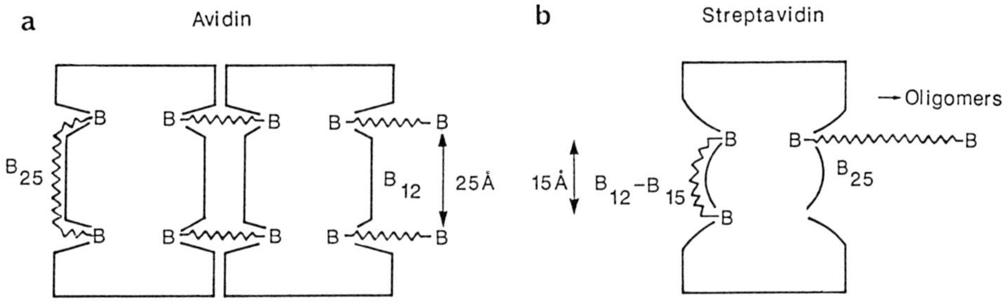

FIG. 3. Spatial relations between neighboring binding sites in avidin and streptavidin. Avidin polymers form when there are more than 8 methylene groups between biotins. Intramolecular bifunctional binding of bisbiotinyl compounds requires a linking chain of at least 19 atoms for avidin (a), whereas a chain of 12 is sufficient for streptavidin (b).

mann for isolating hormone receptors,[27] can be shorter than that required to give avidin homopolymers. It suggests that a small or flexible second protein can approach more closely to the biotin site than can a second avidin molecule. A further example of this is provided by the DNP group of DNP biotin hydrazide, which when bound to avidin is accessible to

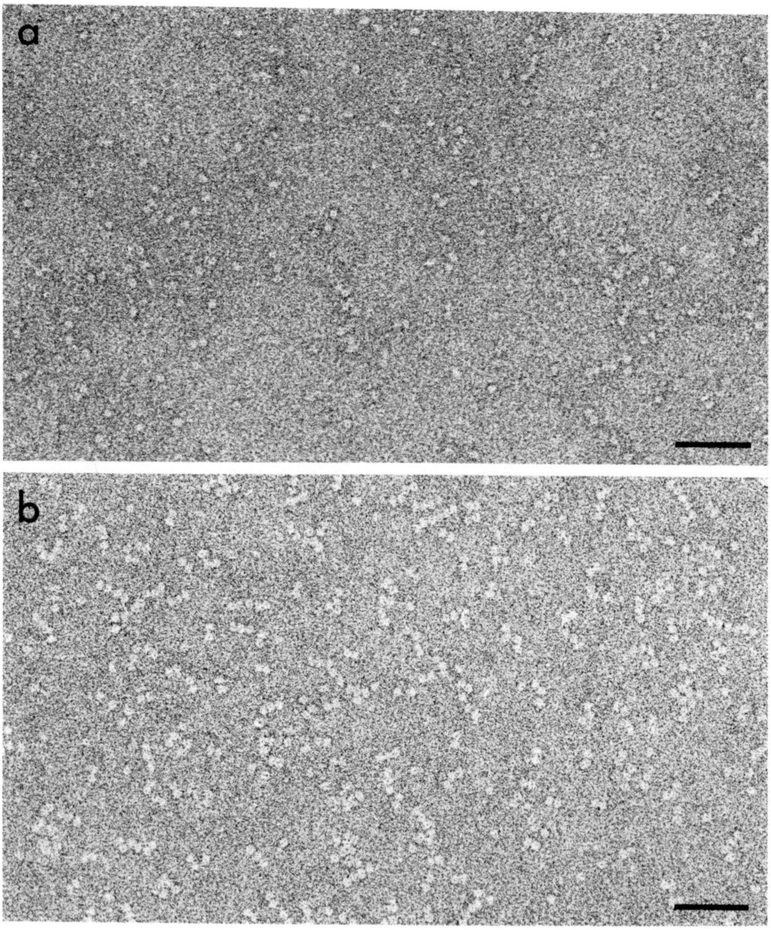

FIG. 4. Avidin polymers. Electron micrographs of streptavidin combined with (a) bis-(biotinamido)dodecane (B_{15}) and (b) bis-[biotinamidobutyramido]decane (B_{25}). In both cases 2 mol of the reagent is bound to each tetramer. The sparsity of polymeric species in (a) implies that most of the bifunctional ligand molecules are bridging two sites within a tetramer (Fig. 3). Only the long-chain reagent gave significant amounts of polymer (b). The repeat distance for this polymer was 46 Å, identical to that observed with avidin and the same B_{25} reagent.[50] The results shown were obtained with streptavidin from Celltech. A second set of experiments with unprocessed streptavidin gave similar results. Bar, 4 nm.

anti-DNP antibody, as measured by fluorescence quenching, although there are only two nitrogen atoms between the DNP group and the carboxyl group of biotin (N. M. Green, unpublished).

Experiments on the reactions between streptavidin and the same series of bisbiotinyl compounds showed interesting differences. Titrations in the presence of the dye 2-(4′-hydroxyazobenzene)benzoic acid (HABA) showed some bifunctional behavior when there were 8 methylene groups on the linking chain (compound B_{11}), and this increased gradually as the chain length was increased to 12 methylene groups (B_{15}). This is similar to previous observations with avidin, except that the change from monofunctional to bifunctional behavior is not so sharply defined. Electron microscopy of the products from B_{12} to B_{15} showed them to be 90% monomeric (Fig. 4a) with a few dimers and trimers. This differs markedly from avidin, which gave long polymers[6,44] (n = 10–20). The streptavidin products resembled those given by avidin with the long-chain B_{25} reagent, suggesting that binding sites of streptavidin are much closer together and that in this case intramolecular bifunctional binding is strongly preferred over polymer formation, even at short chain lengths. Again, in contrast to avidin, the long-chain B_{25} reagent gave the longest polymers (n = 7, Fig. 4b), probably reflecting differences in local topography of the sites, which made it difficult for biotin at the end of a long chain to find its way back to the neighboring site (Fig. 3b).

We conclude that while the individual subunits and binding sites of avidin and streptavidin have similar structures, based on sequence similarity and conservation of tryptophan involvement, the angular relations between subunits in the tetramer is such as to bring the openings into the binding clefts of streptavidin closer together by about 10 Å.

The very high affinity and the 2-fold symmetry of the tetramer form the basis of one of the most useful attributes of the avidin–biotin system. The initial binding is almost irreversible, provided that the biotinyl residue is accessible, and the 2-fold symmetry ensures that if the avidin is binding to a surface molecule many outwardly directed biotin sites will remain vacant and can be saturated with a second biotinyl ligand without displacement of the avidin from the labeled site. This provides the basis for the variety of sandwich and amplification techniques.

The polymers themselves can be prepared with their terminal binding sites vacant, and these have been used as markers for both nucleotide sites and reactive thiol groups on myosin.[54] The polymers can be extensively labeled with concanavalin A (Con A) (Fig. 5), showing that the carbohydrate is not located on the surfaces which carry the binding sites,

[54] K. Sutoh, K. Yamamoto, and T. Wakabayashi, *Proc. Nat. Acad. Sci. U.S.A.* **83,** 212 (1986).

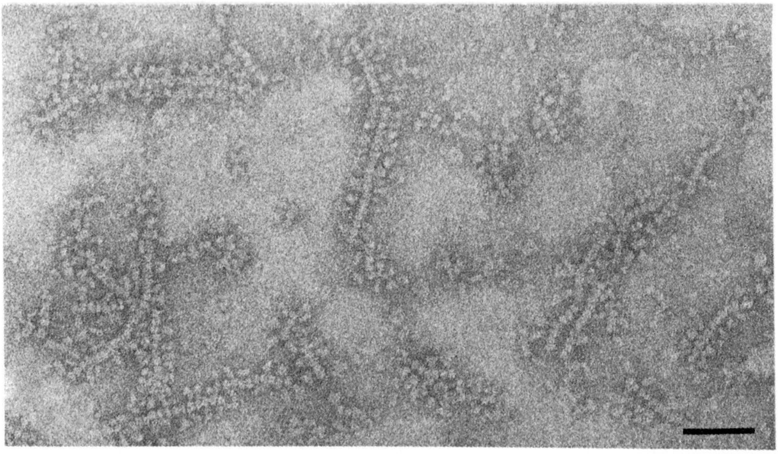

FIG. 5. Avidin polymers (B_{15}), labeled with concanavalin A. Avidin polymers (15 μg/ml) were adsorbed on carbon film. The film was transferred to a solution of Con A (20 μg/ml, 0.15 *M* NaCl, pH 7.5) and left for 0.5–5 min to adsorb. The field shown was labeled for 2 min. Con A tetramers (MW 100,000) are larger than avidin, and their profiles are sometimes triangular. They can be seen in rows, sometimes on both sides of an avidin polymer. Bar, 4 nm.

since these would be inaccessible. Linear polymers have also been observed when avidin binds to oligomeric biotinyl enzymes, such as pyruvate carboxylase,[55,56] from which conclusions could be drawn about the locations of the catalytic sites.

Kinetics of Displacement of Biotinyl Ligands

There have been a number of studies of the effects of a variety of substituents at the nitrogen of biotin amide on the rate of displacement by biotin under different conditions (Table III). The rates increase rapidly with rise in temperature (×2 for 5°), which is consistent with the high ΔH for biotin binding.[6] They also increase when a bulky substituent is present, and this effect can be eliminated by incorporating a six-carbon spacer arm.[26,27] Small hydrophobic groups such as iodophenol can decrease the dissociation rate. Effects of pH are variable (see also Table II) and are especially marked when charged groups are introduced into biotin itself.[6] Some of these effects are discussed in detail in an earlier report on

[55] W. Johannssen, P. V. Attwood, J. C. Wallace, and D. B. Keech, *Eur. J. Biochem.* **133,** 201 (1983).

[56] M. Rohde, F. Lim, and J. C. Wallace, *Eur. J. Biochem.* **156,** 15 (1986).

TABLE III

HALF-TIMES FOR EXCHANGE OF AVIDIN COMPLEXES[a]

Complex	Avidin complexes: AB (slow), 4°	AB (slow), 25°	AB$_4$ (fast), 4°	AB$_4$ (fast), 25°	Succinylavidin Complexes: AB (slow), 25°	AB$_4$ (fast), 25°	Streptavidin, AB, 25°	Refs.
Biotin				*	127		2.9[d]	6,52
Dethiobiotin (DTB)		0.20		*	0.58			6,27,52
DTB–OMe		0.028		*				6
Iminobiotin (free-base form)		0.007[b]		*				6
Corticotropin–biocytin					20	+		57
Corticotropin–dethiobiocytin					0.9	+		57
B–insulin					0.11	+	11	27,53
B–amidohexanoylinsulin					74	+	90	27,53
DTB–insulin					No complex			27,53
DTB–amidohexanoylinsulin					0.25	+		27,53
B–substance P	0.58		0.10					58
B-NH–(*N*-oxo-tetramethyl-piperidin-4-yl)		0.67		0.032				59
B-NH–(*N*-oxo-tetramethyl-piperidin-4-amidoacetyl)		0.21		0.21				59
B-NH–(acetamidoethyl)	5.0	2.4[c]		1.8[c]				51
B-NH–(3-iodo-4-hydroxy-phenylaminoethyl)	380	27[c]		2.7[c]				51
B-NH–(3,5-diiodo-4-hy-droxyphenylaminoethyl)	2000	80[c]		13[c]				51

[a] In all experiments 5–10% of the ligand exchanged more rapidly, an effect which is ignored in this table. The remainder exchanged with a single first-order constant when the ligand had no bulky substituent. When bulky substituents were present and all sites were occupied (AB$_4$), one-half of the ligand exchanged rapidly (+, ~50% of rapid exchange with undetermined rate constant; *, absence of this exchange). The subsequent slow exchange proceeded at the same rate as the exchange of ligand from AB complexes (see text). Data are given in days; the measurements were all made between pH 7.0 and 7.6.

[b] Exchange rate estimated from K_d.[6]

[c] Measurements made at 20°.

[d] Taken from Table II.

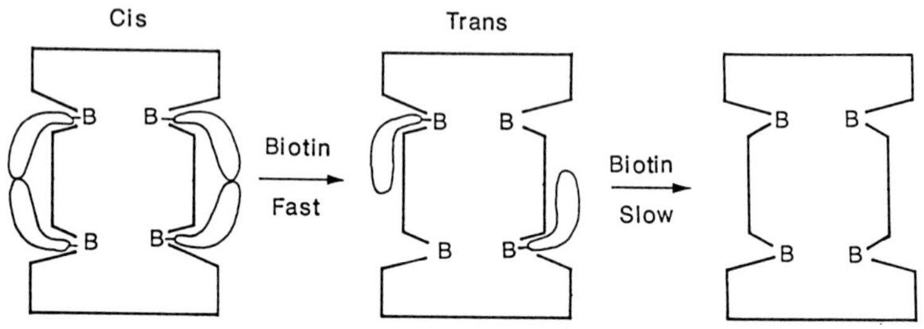

FIG. 6. Interpretation of slow and rapid stages of the exchange process for bulky biotinyl ligands. When bulky ligands occupy neighboring sites, one is displaced relatively rapidly by biotin, leaving firmly bound ligands in the trans position that do not interfere and exchange more slowly.

the use of iodinated biotin derivatives.[52] Succinylation of the avidin has relatively little effect on the exchange rate.[51]

A surprising observation of slow- and fast-exchanging ligands has been made in several sets of measurements,[52,56–59] in apparent conflict with earlier conclusions (from titrations of weakly binding analogs) that the four biotin-binding sites are equivalent and noninteracting.[6] The effect is apparent only in complexes with bulky substituents on the biotin carboxyl and is most marked with fully saturated AB_4 complexes. AB complexes show only a single, slow exchange rate.

The effect can be interpreted in terms of the 2-fold symmetry of the subunit arrangements. The binding sites are organized in pairs on opposite faces of the avidin tetramer, with the biotin carboxyls separated by 15–25 Å (Fig. 6). After saturation of the first site, the second mole of ligand can enter either a cis or a trans position relative to an occupied site. If the biotin carries a bulky substituent, then the cis arrangement will be less stable, and in a fully saturated avidin (AB_4), one-half of the ligands will exchange relatively rapidly.

These conclusions were originally drawn by Sinha and Chignell[59] from a study of the binding of the spin-labeled derivatives such as biotin amido tetramethylpiperidine *N*-oxide. The ESR spectrum of the complex with two sites occupied changed slowly from one which included a major component characteristic of interacting nitroxides (implying a separation of <15 Å) to one in which there was no detectable interaction. The rate constant was the same as the rate of displacement of the first two biotinyl

[57] H. Romovacek, F. M. Finn, and K. Hofmann, *Biochemistry* **22,** 904 (1983).

[58] S. Lavielle, G. Chassaing, and A. Marquet, *Biochem. Biophys. Acta* **759,** 270 (1983).

[59] B. K. Sinha and C. F. Chignell, this series, Vol. 62, p. 295.

nitroxides from a 4 : 1 complex (Table III). This interesting observation is consistent with random initial binding followed by dissociation of nitroxide from the less stable cis sites and recombination at trans sites (Fig. 6).

Conclusions

This chapter provides a summary of recent advances in understanding the interactions between avidin and biotin derivatives, some of which may prove useful for the development of new techniques. Undoubtedly many novel approaches will emerge, stimulated by knowledge of the detailed structure of streptavidin and by the cloning of the avidin genes. Combined with new methods for making biotinyl nucleotides,[60] a range of applications in molecular genetics can be foreseen.

The system could also be expanded by employing a hapten such as DNP, which induces antibodies of very high affinity, to provide a second type of noncovalent linker. This would allow controlled coupling of different macromolecules and could serve to simulate mechanisms or test hypotheses of transmembrane signaling. Another area yet to be developed is the use of avidin monolayers as a basis for biosensor technology. An important new biological role for avidin appears likely to emerge from a recent observation of an avidinlike domain (30% identity) in a gene expressed in the developing ectoderm of the sea urchin embryos.[61,62] The gene product also includes multiple repeats of an epidermal growth factor-like sequence, and it seems likely that it has some biotin-dependent control function.

Acknowledgments

I thank Dr. N. G. Wrigley for electron micrographs, Dr. Meir Wilchek and Dr. R. M. Buckland (Celltech) for samples of streptavidin, and Mr. B. J. Trinnaman for skilled assistance.

[60] J. M. Rothenberg and M. Wilchek, *Nucleic Acids Res.* **16,** 7197 (1988).
[61] D. A. Hursh, M. E. Andrews, and R. A. Raff, *Science* **237,** 1487 (1987).
[62] W. C. Barker, L. T. Hunt, and D. G. George, *Protein Sequence Data Anal.* **1,** 363 (1988).

[6] Nonglycosylated Avidin

By YAFFA HILLER, EDWARD A. BAYER, and MEIR WILCHEK

We have recently found[1] that a commercial preparation of egg-white avidin[2] contains about 30% of a nonglycosylated form of the tetrameric protein. During the course of studies on this protein, we developed a method for the isolation of the nonglycosylated tetramer based on the removal of glycosylated forms[3] using appropriate lectin columns. Since avidin is known to contain both mannose and *N*-acetylglucosamine residues,[4] affinity columns containing concanavalin A (Con A) or wheat germ agglutinin can be used for this purpose. Con A is the lectin of choice because it is relatively inexpensive and because immobilized forms are obtainable from a variety of commercial sources.

Even though no method is currently available for the experimental deglycosylation of avidin,[5] we describe here a method for the separation and isolation of the fully nonglycosylated tetramer from mixtures of glycosylated and partially nonglycosylated forms of the protein. This method will undoubtedly be useful when deglycosylation procedures are eventually described.[6]

[1] Y. Hiller, J. M. Gershoni, E. A. Bayer, and M. Wilchek, *Biochem. J.* **248,** 167 (1987).

[2] Batch 1283, Belovo Soc. Coop. (Bastogne, Belgium).

[3] Glycosylated forms of avidin refer to tetramers containing either fully glycosylated monomers, monomers bearing truncated oligosaccharides, or mixtures of glycosylated and nonglycosylated monomers.

[4] R. J. DeLange and T.-S. Huang, *J. Biol. Chem.* **246,** 698 (1971).

[5] In this chapter, a distinction is made between "nonglycosylation" and "deglycosylation"; in the former case the mechanism by which egg-white avidin has "lost" its oligosaccharide chains is unknown, wheras in the latter case the oligosaccharide residues of glycosylated forms of avidin are removed experimentally. Preliminary results using endoglycosidase F gave promising results in the deglycosylation of certain commercial preparations of avidin (e.g., Belovo) but not with others (e.g., avidin^{EX}, STC Laboratories Inc., Winnipeg, MB), suggesting qualitative differences in the oligosaccharide content of the two products (T. Viswanatha, personal communication).

[6] In our laboratory, we have isolated from nature bacterial strains that are capable of producing deglycosylated avidin. We are currently in the process of isolating the enzyme(s) responsible for this activity.

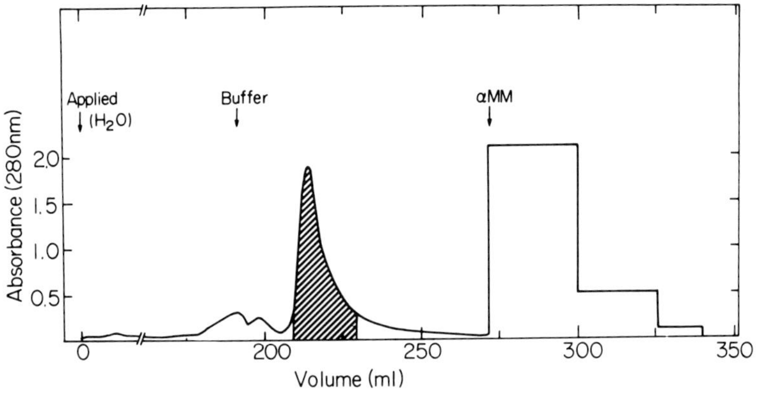

FIG. 1. Preparative purification of nonglycosylated avidin by chromatography on Con A–agarose. A solution of Belovo avidin (~0.5 mg/ml in distilled water) was applied to a 25-ml Con A–agarose column until protein appeared in the effluent (thus indicating saturation of the column). The nonglycosylated avidin (shaded peak) was eluted by washing the column with Tris buffer. The glycosylated forms of avidin were eluted with three sequential volumes (~30 ml each) of the same buffer containing α-methyl-D-mannoside (αMM).

Affinity Chromatography

Reagents

Avidin[2]

Con A–agarose (7–10 mg/ml resin, Sigma Chemical Co., St. Louis, MO)

Tris buffer (25 m*M* Tris-HCl, pH 7.4, containing 150 m*M* NaCl, 1 m*M* $CaCl_2$, and 1 m*M* $MgCl_2$)

α-Methyl-D-mannopyranoside (5% aqueous solution)

Procedure

Belovo avidin (87 mg) is dissolved in 192 ml distilled water, and the solution is centrifuged (10,000 *g* for 15 min). The clarified solution is applied to a Con A column (25 ml). Tris buffer is used to wash the column, after which the nonglycosylated avidin peak appears. Elution of the adsorbed glycosylated forms is accomplished with 5% α-methyl-D-mannopyranoside in the same buffer.

The affinity chromatographic pattern of a representative preparation of nonglycosylated avidin is shown in Fig. 1. The fractions that contain nonglycosylated avidin (shaded area, Fig. 1) and those that are eluted with the competing sugar are pooled separately.[7] The two preparations are dialyzed exhaustively and lyophilized. Yield of nonglycosylated avidin, 24.7 mg (30%). Near-quantitative elution (54 mg, 65%) of the glycosylated forms of avidin can also be achieved.

Properties of Nonglycosylated Avidin

The purified nonglycosylated avidin preparation is characterized on SDS–PAGE by only one band which migrates at a position coincident with that of the low molecular weight band of Belovo avidin (M_r 15,500). The isolated protein is essentially devoid of sugar, as indicated by the lack of interaction of blotted samples with Con A or wheat germ agglutinin. The biotin-binding characteristics are virtually unchanged.

Comments

The nonspecific adsorption of egg-white avidin is usually attributed to two inherent characteristics of the molecule, namely, the high p*I* and the presence of sugar residues. For this reason, many laboratories and commercial enterprises have employed the neutral, nonglycosylated bacterial counterpart streptavidin as a substitute for the egg-white protein, despite the high cost of the bacterial protein. The availability of a nonglycosylated preparation of egg-white avidin would eliminate at least one of the abovementioned disadvantages. This is particularly important when applying avidin–biotin technology to cells that may express lectins on their surfaces. The nonglycosylated avidin has the same biotin-binding properties as the fully glycosylated form, which indicates that the sugars are not important for the biological activity. Thus, nonglycosylated avidin may eventually prove to be the preferred substitute for native avidin.

[7] Effluent fractions are first examined spectrophotometrically and then by SDS–PAGE to ensure that only those containing the nonglycosylated forms of the protein are pooled.

[7] Cloning and Expression of Avidin in *Escherichia coli*

By GYAN CHANDRA and JOHN G. GRAY

The gene coding for the biotin-binding protein avidin was isolated from a chicken oviduct cDNA, λgt11 phage library using an oligonucleotide probe derived from the published avidin protein sequence.[1] In

[1] R. J. DeLange and T. S. Huang, *J. Biol. Chem.* **246,** 698 (1971).

METHODS IN ENZYMOLOGY, VOL. 184

this chapter we describe the screening of the λ phage library and the isolation of the gene coding for avidin protein. The successful expression of a functional form of this protein in *Escherichia coli* is also described.

Molecular Cloning of Avidin

Bacterial Strains, Phage, and Plasmids

A λgt11 chicken oviduct cDNA library was obtained from B. W. O'Malley. Bacterial strain *E. coli* Y1088[2] is used as a host to propagate the λ phage. *Escherichia coli* strain JM103[3] is used for propagation of the M13 mp19 sequencing bacteriophage and for expression of the protein with plasmid pkk233-2.[4]

Deoxyoligonucleotide Probe

The deoxyoligonucleotide probe is synthesized on a DNA synthesizer using phosphoramidite chemistry.[5] The probe is selected from the least degenerate codon region in the protein. The probe sequence used is determined from a tract of 8 amino acid (425–447 bp of the gene sequence) as indicated in Fig. 1. The probe is a 23-mer (23 bases) mixed-base probe and has the following sequence:

5′ TTCCA(A,G)TC(A,G)TC(G,T)CC(A,G)AT(A,G)TCGTT 3′

Solutions and Media

20× SSC: 3 *M* NaCl, 0.3 *M* sodium citrate
Denaturation: 0.5 *N* NaOH, 1.5 *M* NaCl
Neutralization: 0.5 *M* Tris-HCl (pH 7.4), 1.5 *M* NaCl
5× SSCPE: 2.4 *M* NaCl, 0.3 *M* sodium citrate, 130 m*M* KH_2PO_4, 20 m*M* EDTA
100× Denhardt's: 2% (w/v) bovine serum albumin, 2% (w/v) polyvinylpyrrolidone, 2% (w/v) Ficoll (400,000 MW)
Hybridization: 25% (v/v) formamide, 5× SSC, 2× Denhardt's, 150 μg/ml sonicated salmon sperm DNA, 5 m*M* EDTA, 0.1% (w/v) sodium dodecyl sulfate (SDS)
Wash solution: 2× SSC, 0.1% (w/v) SDS
LB broth (per liter): 10 g Bacto-tryptone, 5 g yeast extract, 5 g NaCl
SM buffer: 50 m*M* Tris-HCl (pH 7.5), 100 m*M* NaCl, 10 m*M* $MgSO_4$, 0.01% (w/v) gelatin

[2] R. A. Young and R. W. Davis, *Science* **222,** 778 (1983).
[3] J. Messing, Recomb. *DNA Tech. Bull.* (NIH Publ. No. 79-99,2) **2,** 43 (1979).
[4] E. Amann and J. Brosius, *Gene* **40,** 183 (1985).
[5] S. L. Beaucage and M. H. Caruthers, *Tetrahedron Lett.* **22,** 1859 (1981).

```
          10          20          30          40          50
           *           *           *           *           *
G AAT TCC GCA AGG AGC ACA CCC GGC TGT CCA CCT GCT GCA GAG ATG GTG CAC GCA ACC
                                                          Met Val His Ala Thr

  60          70          80          90          100         110
   *           *           *           *           *           *
  TCC CCG CTG CTG CTG CTG CTG CTG CTC AGC CTG GCT CTG GTG GCT CCC GGC CTC TCT
  Ser Pro Leu Leu Leu Leu Leu Leu Leu Ser Leu Ala Leu Val Ala Pro Gly Leu Ser

      120         130         140         150         160         170
        *           *           *           *           *           *
  GCC AGA AAG TGC TCG CTG ACT GGG AAA TGG ACC AAC GAT CTG GGC TCC AAC ATG ACC
  Ala Arg Lys Cys Ser Leu Thr Gly Lys Trp Thr Asn Asp Leu Gly Ser Asn Met Thr

          180         190         200         210         220
            *           *           *           *           *
  ATC GGG GCT GTG AAC AGC AGA GGT GAA TTC ACA GGC ACC TAC ATC ACA GCC GTA ACA
  Ile Gly Ala Val Asn Ser Arg Gly Glu Phe Thr Gly Thr Tyr Ile Thr Ala Val Thr

230         240         250         260         270         280
  *           *           *           *           *           *
  GCC ACA TCA AAT GAG ATC AAA GAG TCA CCA CTG CAT GGG ACA GAA AAC ACC ATC AAC
  Ala Thr Ser Asn Glu Ile Lys Glu Ser Pro Leu His Gly Thr Glu Asn Thr Ile Asn

    290         300         310         320         330         340
      *           *           *           *           *           *
  AAG AGG ACC CAG CCC ACC TTT GGC TTC ACC GTC AAT TGG AAG TTT TCA GAG TCC ACC
  Lys Arg Thr Gln Pro Thr Phe Gly Phe Thr Val Asn Trp Lys Phe Ser Glu Ser Thr

        350         360         370         380         390         400
          *           *           *           *           *           *
  ACT GTC TTC ACG GGC CAG TGC TTC ATA GAC AGG AAT GGG AAG GAG GTC CTG AAG ACC
  Thr Val Phe Thr Gly Gln Cys Phe Ile Asp Arg Asn Gly Lys Glu Val Leu Lys Thr

            410         420         430         440         450
              *           *           *           *           *
  ATG TGG CTG CTG CGG TCA AGT GTT AAT GAC ATT GGT GAT GAC TGG AAA GCT ACC AGG
  Met Trp Leu Leu Arg Ser Ser Val Asn Asp Ile Gly Asp Asp Trp Lys Ala Thr Arg

  460         470         480         490         500         510
    *           *           *           *           *           *
  GTC GGC ATC AAC ATC TTC ACT CGC CTG CGC ACA CAG AAG GAG TGA GGA TGG CCC CGC
  Val Gly Ile Asn Ile Phe Thr Arg Leu Arg Thr Gln Lys Glu ---

      520         530         540         550         560         570
        *           *           *           *           *           *
  AAA GCC AGC AAC AAT GCC GGA GTG CTG ACA CTG CTT GTG ATA TTC CTC CCA ATA AAG

          580         590
            *           *
  CTT TGC CTC AGA CAA AAA AAA AAA A
```

FIG. 1. Complete cDNA sequence of the avidin gene. The underlined sequence indicates the signal peptide of the protein. The 8 amino acids within the box were used to select the deoxyoligonucleotide probe involved in screening the cDNA library.

Breaking buffer: 200 mM Tris-HCl (pH 7.6), 250 mM NaCl, 10 mM $Mg(CH_3COO)_2$, 5% (v/v) glycerol

Nitrocellulose filters: BA/85, 0.45 μm (Schleicher & Schuell, Keene, NH)

Hybridization unit: OmniBlot processing system [American Bionetics (Fisher Scientific Dist., Pittsburg, PA)]

Transfection

A single colony of *E. coli* Y1088 cells is inoculated into 40 ml of LB broth containing 50 μg/ml ampicillin and 0.2% (w/v) maltose and incubated at 37° with vigorous shaking overnight. The culture is centrifuged at 5000 rpm for 5 min, and the bacterial pellet is resuspended in one-half the original volume (20 ml) of SM buffer.

The λ phage cDNA library is diluted such that 10,000 plaque-forming units (pfu)/100 mm petri plate are obtained.[2] The *E. coli* Y1088 cells (prepared as described above) and properly diluted phage are incubated at room temperature for 15 min to allow for phage adsorption. After adsorption, 3 ml of top agar (LB media containing 0.6% w/v agarose), prewarmed to 45°, is added to each phage dilution and poured onto prewarmed (37°) LB plates containing 50 μg/ml ampicillin. Plates are allowed to solidify at room temperature for 15 min prior to incubation overnight at 37°.

Plaque Lift (Transfer of λ Phage DNA onto Nitrocellulose Filters)

LB plates containing phage plaques are transferred to 4° for 2 hr before proceeding further. Chilling the plates to 4° helps prevent tearing of the top agar overlay when blotting. Nitrocellulose filters (82 mm in diameter) are gently placed over the soft top agar containing the phage plaques. The position of the filter is marked using a 20-gauge syringe needle dipped in India ink. A unique asymmetric pattern is used on each filter to ensure that orientation and individual filters can be identified. Each plate containing phage is blotted in duplicate to insure against false-positive clones.

The first nitrocellulose filter is left in contact with the top agar for approximately 1 min. A second or duplicate filter is allowed to remain in contact for approximately 2 min, before gently removing and placing the DNA side up on a tray containing Whatman 3MM filter paper saturated with a denaturing solution. Filters are transferred successively to neutralizing and to 2 × SSCPE solutions, keeping the filters approximately 2 min in each of these solutions. As the filters are transferred between solutions, excess liquid is removed from the underside of each filter with a paper towel. The filters are transferred to Whatman 3MM filter paper and allowed to air dry before being vacuum dried at 80° for 2 hr.

Hybridization of ^{32}P-Labeled Oligonucleotide Probe to Nitrocellulose Filters

Prehybridization, hybridization, and washing of filters are done in the same Omniblot hybridization bag. Nitrocellulose filters (maximum of 20/bag) are placed back to back in a hybridization bag, for 10 filters. For 20 filters, a sheet of mesh is placed in between as described by the manufacturer. The bag containing the filters is sealed, taking care that no mesh fibers extend into the sealing crease. A vacuum is created inside, using a 30- or 50-ml syringe, to check for leaks. After determining that no leaks are present, 60–80 ml of hybridization buffer is added to the bag containing 20 filters and incubated at 37° for a minimum of 2 hr. Before adding the probe, excess hybridization buffer is removed, leaving approximately two-thirds of the original volume. Approximately 1×10^6 cpm/filter of the end-labeled oligonucleotide probe[6] is diluted with hybridization buffer and added to the bag. Hybridization is accomplished by incubation at 37° for at least 12 hr.

Washing of Filters

Washing of filters is accomplished initially using wash solution at room temperature. For 20 filters a wash volume of 1 liter is sufficient. The first 500-ml wash is performed at room temperature and is accomplished by passing the wash solution in a port on one side of the hybridization bag and collecting the wash in a flask as it exits the port on the opposite side. The whole process is vacuum driven. The second wash is more critical and is accomplished by washing as described above but with wash buffer prewarmed to 42°. The entire washing procedure is monitored with a Geiger counter and is stopped when the majority of counts is effectively removed. The final critical wash may also be done in a 1-liter beaker containing 500 ml of the wash buffer prewarmed to the desired temperature. Less than 5 min of exposure at this temperature is required. Filters are air dried on Whatman 3MM paper, arranged with duplicates side by side, covered with Saran wrap, and autoradiographed at −70° with Kodak X-OMAT XAR film using an intensifying screen (Du Pont, Cronex Lightning Plus) for 24–48 hr.

Plaque Purification

Signals are considered positive when signals on duplicate filters are observed to be in the same position relative to the orientation markings.

[6] A. M. Maxam and W. Gilbert, this series, Vol. 65, p. 499.

Regions approximately 1.5 mm in diameter and thought to contain positive signals are picked using a Pasteur pipet and transferred to 500 μl of SM buffer in a 1.5-ml microcentrifuge tube. After 20 min the phage suspension is used for replating each clone at a lower density of 1000 pfu/petri plate. The replating and rescreening is repeated, thus enhancing the ratio of positive to negative clones, until 100% of the replated plaques give a positive signal.

Remarks

After screening of 100,000 pfu, approximately 50 positive signals were found. The areas corresponding to 27 positive signals were picked and processed for rescreening to obtain individual pure plaques. λ phage DNA was extracted[7] and digested with *Eco*RI restriction enzyme to excise the insert from each clone. The *Eco*RI-digested DNA was electrophoresed on a 1.5% (w/v) agarose gel containing 0.7 μg/ml ethidium bromide in order to determine the size of each insert. Of the 27 clones processed, only 1 contained the entire coding sequence. The largest clone resulted in two bands of approximately 0.4 and 0.2 kbp. The two *Eco*RI fragments were isolated, subcloned into the M13 mp19 phage sequencing vector, and sequenced by the Sanger dideoxy method.[8] The nucleotide sequence as well as the corresponding amino acid sequence are shown in Fig. 1.

Expression of Avidin in *Escherichia coli*

Construction of Expression Plasmid

Utilizing the restriction enzyme information obtained from the DNA sequence analysis, it was determined that the simplest method of achieving avidin expression is to use the expression vector pkk233-2. The pkk233-2 vector contains the *trc* promoter (a hybrid of the *trp* and *lac* promoters with the consensus 17-bp spacing between the *trp* −35 region and *lac UV5* −10 region) and the LacZ ribosome-binding site followed by an ATG initiation codon which is within a unique *Nco*I restriction site. The cloning cassette of pkk233-2 contains three unique restriction sites, *Nco*I, *Pst*I, and *Hin*dIII (Fig. 2). The avidin cDNA gene has two unique restriction sites, *Pst*I and *Hin*dIII. The *Pst*I site of avidin is 2 bases before its ATG initiation codon and is in the same reading frame as the *Pst*I site in pkk233-2. The *Hin*dIII site is located in the 3′-noncoding region of the cDNA gene (Fig. 3).

[7] R. M. Lawn, E. F. Fritsch, R. C. Parker, G. Blake, and T. Maniatis, *Cell* **15,** 1157 (1978).
[8] F. Sanger, S. Nicklen, and A. R. Coulson, *Proc. Natl. Acad. Sci. U.S.A.* **74,** 5463 (1977).

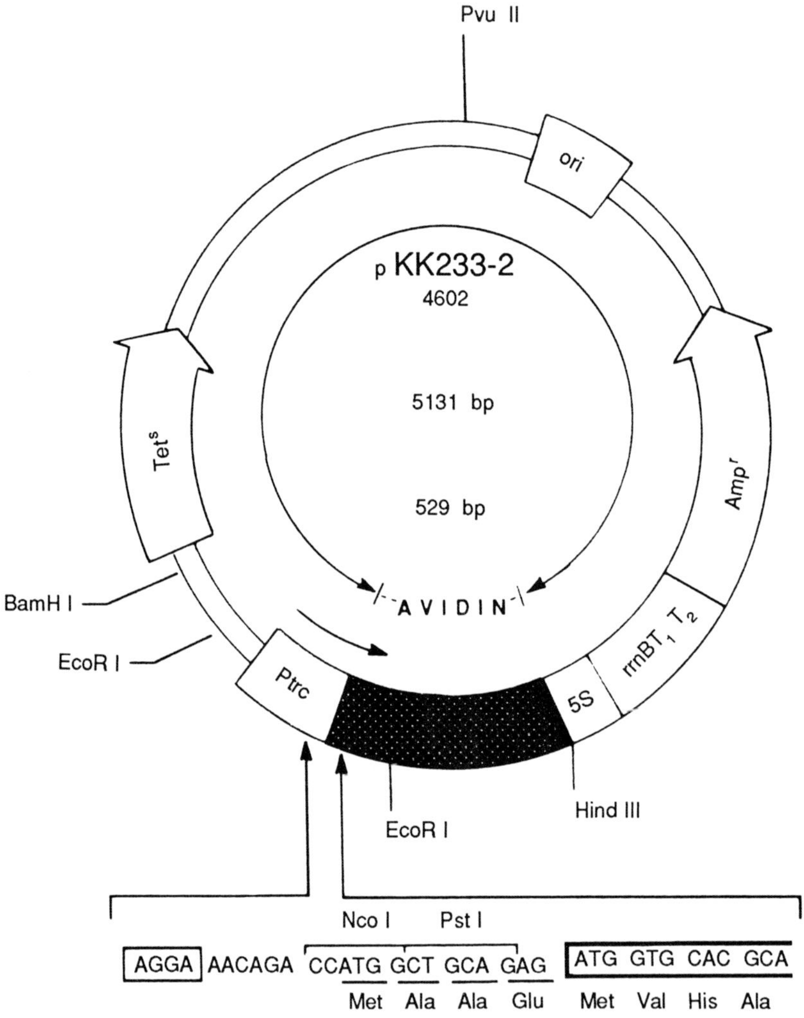

FIG. 2. Construction of the recombinant plasmid expressing the avidin protein under a *trc* promoter (see text). The dark area of the vector represents 529 bp of avidin in the reading frame of the promoter. Amp and Tet stand for ampicillin- and tetracycline-resistance genes, respectively. Met, Ala, Ala, and Glu, before the box, are the four amino acids added to the N terminus of the protein, whereas Met inside the box represents the first methionine (start codon) of the cDNA gene. Details of plasmid construction are described in the text.

Plasmid pkk233-2 is prepared by digesting with restriction enzymes *Hin*dIII and *Pst*I. Avidin cDNA, which had been isolated from λgt11 by *Eco*RI restriction enzyme digestion as two fragments (200- and 400-bp *Eco*RI fragments), is individually digested with *Pst*I (200-bp fragment) or

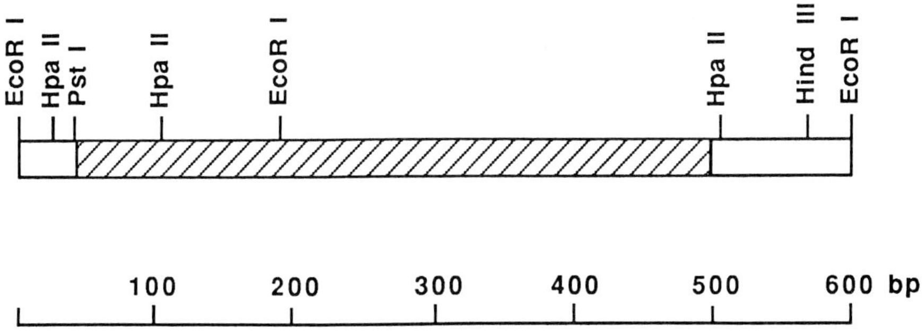

FIG. 3. Restriction map of avidin cDNA. The hatched box represents the coding region of the cDNA, and the open box on either side represents noncoding regions of the avidin cDNA.

*Hin*dIII (400-bp fragment). The subsequent *Pst*I–*Eco*RI and *Eco*RI–*Hin*dIII fragments are gel purified and ligated together to generate a 529-bp *Pst*I–*Hin*dIII fragment. The *Pst*I–*Hin*dIII fragment is ligated into pkk233-2 that had been digested with *Pst*I and *Hin*dIII as described above. *Escherichia coli* JM103 cells transformed[9] with the recombinant plasmid produce individual colonies when plated on LB plates containing 50 μg/ml ampicillin. The plasmid DNA from six colonies was isolated and subjected to restriction enzyme analysis. All colonies tested contained the 529 bp *Pst*I–*Hin*dIII avidin insert. This construction has 4 extra amino acids added to the N-terminal end of the avidin protein (Fig. 2).

Procedure for Expression

Escherichia coli JM103, containing the expression plasmid pkk233-2 plus the avidin gene, is inoculated into 20 ml of LB broth containing 25 μg/ml of ampicillin and grown overnight with shaking at 37°. One-half milliliter of the culture is inoculated into 100 ml of LB broth containing 25 μg/ml ampicillin and incubated with vigorous shaking until a Klett reading of 80 (Klett–Summerson photoelectric colorimeter) is reached. At this stage, the culture is induced by the addition of β-isopropyl-D-thiogalactopyranoside (IPTG) to a final concentration of 1 m*M*.

After a 2-hr induction at 37°, the bacteria are pelleted by centrifuging at 5000 rpm for 5 min. The cell pellet is resuspended in 1/25 of the original volume with cold (4°) osmotic shock buffer [10 m*M* Tris-HCl (pH 7.6), 1 m*M* EDTA, and 20% w/v sucrose], and the suspension is incubated for 30 min on ice. The cell suspension is again centrifuged at 5000 rpm for 5

[9] J. Messing, this series, Vol. 101, p. 20.

min. The cell pellet is resuspended in cold water (4°) at 1/25 the original volume and incubated on ice for 30 min with occasional stirring.

The cell suspension is centrifuged at 8000 rpm for 10 min, and the supernatant is saved for further analysis. The cell pellet is resuspended in breaking buffer and treated with lysozyme (0.2 mg/ml) on ice for 30 min. The lysozyme-treated cells are frozen and thawed 2 times, being careful to keep the temperature below 4°. The cell preparation is sonicated (Artek Sonic Dismembrator, Model 300) 3 times for 30 sec at 1-min intervals using a small probe at maximum setting. The cell lysate is centrifuged at 20,000 *g* for 30 min to clarify. DNA is removed by adding 5 mg/ml of streptomycin sulfate with stirring and subsequent incubation at 4° for 4–6 hr, followed by centrifugation at 20,000 rpm for 15 min. The osmotic shock and cell lysate supernatants are precipitated by adding an equal volume of 80% saturated ammonium sulfate and incubating overnight at 4°. The protein pellets obtained after centrifugation (8000 rpm for 15 min) are resuspended in 1/100 the original volume of resuspension buffer [50 m*M* Tris-HCl (pH 7.4), 1 m*M* EDTA, and 1 m*M* phenylmethylsulfonyl fluoride (PMSF)]. Samples of each preparation (osmotic shock and cell lysate) are boiled for 3 min in Laemmli's denaturing buffer, electrophoresed on a 12.5% Laemmli gel,[10] transferred to nitrocellulose filter paper (Western blot), and probed using rabbit antiavidin antibody.[11]

The Western blot can also be probed with biotinylated alkaline phosphatase[12] in order to determine whether the recombinant avidin is capable of binding biotin. To test for biotin binding by the recombinant avidin, the Western blot containing the avidin protein is first blocked by soaking in a solution of TBS [10 m*M* Tris-HCl (pH 7.4) and 150 m*M* NaCl] containing 0.2% (w/v) Tween 20 for 2 hr. Biotinylated alkaline phosphatase from Sigma (0.5 mg protein containing 5.2 mol of *d*-biotin per mole of alkaline phosphatase) is diluted 1 : 10,000 in TBST (TBS containing 0.05% v/v Tween 20) and added to the filter (about 0.5 ml of solution per cm^2 filter) and incubated for 1 hr at room temperature. The blot is washed 3 times for 5 min in TBST buffer. The indicator is prepared by mixing 66 μl of nitroblue tetrazolium (NBT) substrate with 10 ml of alkaline phosphatase buffer [100 m*M* Tris-HCl (pH 9.5), 100 m*M* NaCl, 5 m*M* $MgCl_2$] and 33 μl of 5-bromo-4-chloro-3-indolyl phosphate (BCIP) substrate as described in the manufacturer's instructions (Promega Biotech, Madison, WI). The indicator is added and the color allowed to develop for 10 min before

[10] U. K. Laemmli, *Nature* (*London*) **227,** 680 (1970).

[11] Z. Wojtkowiak, R. C. Briggs, and L. S. Hnilica, *Anal. Biochem.* **129,** 486 (1983).

[12] Y. Hiller, J. M. Gershoni, E. A. Bayer, and M. Wilchek, *Biochem. J.* **248,** 167 (1987).

being stopped by washing with several changes of water at 5-min intervals. The results of the Western blots (antibody probe and biotin probe) demonstrate that avidin is expressed in *E. coli,* is secreted into the periplasmic space, and is still capable of binding biotin.

Concluding Comments

Using an oligonucleotide probe to the published protein sequence of avidin, we were able to obtain a cDNA clone from a chicken oviduct cDNA library encoding for the entire protein. Gope *et al.*[13] have also reported the molecular cloning of chicken avidin cDNA, and their published sequence is virtually the same as that reported here for the 596-bp clone. The clone of 596 bp contained the published protein sequence of amino acids plus an additional 24 amino acids on the N-terminal end thought to be a signal peptide, which is probably responsible for the avidin protein being found in the periplasmic fraction (osmotic shock) of *E. coli.*

The avidin–biotin studies, performed for the expressed protein secreted in the periplasm of *E. coli,* revealed that recombinant avidin is still capable of biotin binding. Therefore, it will now be possible to study the mechanism responsible for its high-affinity binding to biotin. In addition, cloning of the gene gives the opportunity (1) to tag genes of interest with avidin for study *in vivo* and (2) to make hybrid proteins through genetic engineering that can be easily purified using the avidin–biotin interaction.

Acknowledgments

We are thankful to Ms. Jennifer Lorenz and Mr. Stephen Haneline for their valuable assistance on this project.

[13] M. L. Gope, R. A. Keinänen, P. A. Kristo, O. M. Conneely, W. G. Beattie, T. Zarucki-Schulz, B. W. O'Malley, and M. S. Kulomaa, *Nucleic Acids Res.* **15,** 3595 (1987).

[8] Isolation and Properties of Streptavidin

By EDWARD A. BAYER, HAYA BEN-HUR, and MEIR WILCHEK

In recent years we have advocated the use of avidin–biotin technology, which has since become a universal probe for general use in the biological sciences.[1-3] Use of this system, however, is sometimes restricted due to the high basicity (p*I* ~10.5) and presence of sugar moieties on the avidin molecule, which may lead to nonspecific or otherwise undesired interactions.[1] In this chapter, we describe the isolation and characteristics of an alternative biotin-binding protein produced by the bacterium *Streptomyces avidinii*.[4,5]

The bacterial protein displays remarkable similarity to egg-white avidin in its biotin-binding properties but is not a glycoprotein and is not basic.[6] Owing to its advantageous physical properties, streptavidin has been used either as a replacement for or to complement egg-white avidin in avidin–biotin technology.

Isolation of Streptavidin

The classic method for the isolation of streptavidin from culture broth of *S. avidinii* involves ammonium sulfate precipitation, ion-exchange chromatography, and crystallization.[6] An affinity-based procedure using a CNBr-activated iminobiotin column has also been described.[7] However, only low amounts of streptavidin could be obtained in the latter case, and such columns could be used only a limited number of times because of the unstable nature of the isourea bond under the alkaline conditions required for iminobiotin binding.[8,9]

[1] E. A. Bayer and M. Wilchek, *Methods Biochem. Anal.* **26,** 1 (1980).
[2] E. A. Bayer and M. Wilchek, *Trends Biochem. Sci.* **3,** N237 (1978).
[3] M. Wilchek and E. A. Bayer, *Anal. Biochem.* **171,** 1 (1988).
[4] E. O. Stapley, J. M. Mata, I. M. Miller, T. C. Demny, and H. B. Woodruff, *Antimicrob. Agents Chemother.* **3,** 20 (1963).
[5] L. Chaiet, T. W. Miller, F. Tausig, and F. J. Wolf, *Antimicrob. Agents Chemother.* **3,** 28 (1963).
[6] L. Chaiet and F. J. Wolf, *Arch. Biochem. Biophys.* **106,** 1 (1964).
[7] K. Hofmann, S. W. Wood, C. C. Brinton, Jr., J. A. Montibeller, and F. M. Finn, *Proc. Natl. Acad. Sci. U.S.A.* **77,** 4668 (1980).
[8] M. Wilchek, T. Oka, and Y. J. Topper, *Proc. Natl. Acad. Sci. U.S.A.* **72,** 1055 (1975).
[9] J. Kohn and M. Wilchek, *Enzyme Microb. Technol.* **4,** 161 (1982).

Here, we describe the purification of relatively large quantities of streptavidin using an improved iminobiotin column.[10] The resin is prepared by the chloroformate activation method, which gives stable carbamates.[11,12]

Preparation of Affinity Column

4-Nitrophenyl chloroformate Activation of Sepharose CL-4B

Reagents

Sepharose CL-4B (Pharmacia, Piscataway, NJ), 200 ml
4-Nitrophenyl chloroformate (Aldrich, Milwaukee, WI), 12 g
4-Dimethylaminopyridine (Sigma, St. Louis, MO), 8.5 g
Acetone (absolute)
2-Propanol
Distilled water

Procedure. All steps in the activation procedure are performed at 4° or in an ice bath. Sepharose CL-4B is washed batchwise through a sintered glass funnel with increasing concentrations of acetone in water.[13] To the washed gel, an acetone solution (200 ml) containing 4-nitrophenyl chloroformate (0.3 mmol/ml) is added. To this suspension, 200 ml of an acetone solution containing 4-dimethylaminopyridine (0.35 mol/ml) is added dropwise with stirring, and the resin is stirred for 1 hr.[14] The activated Sepharose is then washed exhaustively with generous quantities of the following solutions in the given order, acetone, acetone–2-propanol (1 : 1), 2-propanol, 2-propanol–water (1 : 1), and distilled water, until all traces of reagent disappear.[15]

Determination of Extent of Reaction. A sample of the activated resin (~200 mg) is weighed, and 5 ml of 0.2 *N* NaOH is added. The resultant solution is diluted 10-fold, and the extent of activation (micromoles 4-nitrophenol per gram of resin) is calculated from the following equation:

$$R = 2.9A_{400}/\text{weight of resin (g)}$$

[10] E. A. Bayer, H. Ben-Hur, G. Gitlin, and M. Wilchek, *J. Biochem. Biophys. Methods* **13,** 103 (1986).

[11] M. Wilchek and T. Miron, *Biochem. Int.* **4,** 629 (1982).

[12] M. Wilchek, T. Miron, and J. Kohn, this series, Vol. 104 [1].

[13] Successive washes are carried out using 0, 25, 50, 75, and 100% (absolute) acetone solutions. The final wash is performed using generous quantities of absolute acetone.

[14] During this time, the solution turns yellow.

[15] The presence of reagent is checked by adding several drops of 0.2 *N* NaOH to the wash solution. The appearance of a yellow color indicates residual amounts of the reagent.

For example, if the absorbance (400 nm) of a sample (220 mg) treated with NaOH in the above-described manner is 0.62, then there would be about 8.2 μmol reactive 4-nitrophenyl groups per gram of resin.

Preparation of Aminohexyl-Sepharose CL-4B

Reagents

4-Nitrophenyl chloroformate-activated Sepharose CL-4B, 200 ml
Diaminohexane (Sigma), 20 g
Hydrochloric acid (concentrated)
0.2 *M* acetic acid
0.05 *N* NaOH
Distilled water

Procedure. Dissolve the diaminohexane in 200 ml of water, and bring the pH of the solution to 10 with HCl. Add the activated resin and stir overnight at 4°.[14] Wash successively with water, acetic acid, water, and NaOH until the yellow color disappears.[16] Then wash the resin with distilled water until the pH of the solution is neutral.

2-Iminobiotin-6-aminohexyl-Sepharose CL-4B

Reagents

Aminohexyl-Sepharose CL-4B, 100 ml
Iminobiotin *N*-hydroxysuccinimide bromide (Sigma),[17] 375 mg (1.1 mmol)
Dimethylformamide
Dioxane
0.1 *N* NaOII
Distilled water

Procedure. The resin is washed successively with NaOH, water, dioxane : water (1 : 3), dioxane : water (3 : 1), dioxane, absolute dioxane, dimethylformamide : dioxane (1 : 3), dimethylformamide : dioxane (3 : 1), and dimethylformamide (dried). To the resultant suspension, a dimethylformamide solution (150 ml) containing iminobiotin *N*-hydroxysuccinimide bromide is added, and the resin is stirred gently overnight at room temperature. The iminobiotin-containing affinity matrix is washed successively with dimethylformamide, dimethylformamide : water (3 : 1),

[16] Washing with alkaline solution results in the release of large amounts of unreacted *p*-nitrophenol.

[17] B. F. Goldin and G. A. Orr, this volume [17].

dimethylformamide : water (1 : 3), and water. The product is resuspended in 100 ml of water containing 0.1% (w/v) sodium azide and stored at 4°.

Determination of Capacity of Iminobiotin Affinity Resin for Avidin. Egg-white avidin[18] (1 mg/ml in 50 m*M* sodium carbonate, pH 11, containing 0.5 *M* NaCl) is added to a small affinity column containing a measured sample (0.5 ml) of 2-iminobiotin-6-aminohexyl-Sepharose CL-4B. The column is washed with the same buffer, and the absorbance (280 nm) of the effluent fractions is measured. The above procedure is repeated until the total absorbance in the effluent fractions is equal to that applied to the column. The column is washed with the same buffer, and the avidin bound to the column is eluted with 50 m*M* sodium acetate (pH 4) containing 2 *M* NaCl. Additional amounts of bound avidin may be eluted with distilled water, brought to pH 4 with HCl. The amount of avidin released from the column is calculated from the published extinction coefficient ($E_{282}^{1\%}$ 15.4).[19]

Preparation of Native Streptavidin

Organism and Growth Conditions

Streptomyces avidinii (ATCC 27419) is grown in liquid culture (1-liter batches) at 30° in malt medium, medium A, or synthetic medium as described previously.[4,20–22] The amount of streptavidin produced in the cul-

[18] For financial reasons, commercially obtained egg-white avidin is used instead of streptavidin for determining the capacity of iminobiotin columns.

[19] N. M. Green, *Adv. Protein Chem.* **29,** 85 (1975).

[20] "Malt medium" consists of 3 g malt extract, 3 g yeast extract, 5 g tryptone, and 10 g dextrose in 1 liter of distilled water. The pH of the solution is brought to 6.4 before sterilization.

[21] Medium A contains 10 g dextrose, 1 g DL-asparagine, 0.1 g K_2HPO_4, 0.5 g $MgSO_4 \cdot 7H_2O$, 10 mg $FeSO_4 \cdot 7H_2O$, and 0.5 g yeast extract dissolved in 1 liter of distilled water. The solution is brought to pH 7.2 before sterilization.

[22] Two types of synthetic media have recently been described [J. Cazin, Jr., M. Suter, and J. E. Butler, *J. Immunol. Methods* **113,** 75 (1988)] which yield relatively large quantities (100–120 mg/liter) of streptavidin. Synthetic liquid medium A consists of 7.0 g L-asparagine, 10 g dextrose, 1 g K_2HPO_4, and 1.7 g "trace element and basic salt mixture" (1.0 g KH_2PO_4, 0.5 g $MgSO_4 \cdot 7H_2O$, 0.1 g NaCl, 0.1 g $CaCl_2 \cdot 2H_2O$, 0.5 mg H_3BO_3, 0.4 mg $MnSO_4 \cdot H_2O$, 0.4 mg $ZnSO_4 \cdot 7H_2O$, 0.2 mg $NaMoO_4 \cdot 2H_2O$, 0.2 mg $FeCl_3 \cdot 6H_2O$, 0.1 mg KI, and 40 μg $CuSO_4 \cdot H_2O$). The mixture is brought to 1 liter with distilled water, the solution is sterilized by autoclaving at 15 psi for 15 min, and the pH of the medium is brought to 6.5. Synthetic liquid medium B is the same as the above, except that the concentrations of KH_2PO_4, $MgSO_4 \cdot 7H_2O$, NaCl, and $CaCl_2 \cdot 2H_2O$ are doubled. The pH is 6.1 after autoclaving.

ture fluids is determined daily by enzyme assay.[23] When this amount reaches a plateau (usually between 5 and 7 days), the culture is harvested.

Affinity Chromatography

Reagents

2-Iminobiotin-6-aminohexyl-Sepharose CL-4B, 30 ml
pH 4 buffer [50 m*M* sodium acetate (pH 4) containing 2 *M* NaCl]
pH 7 buffer [0.1 *M* sodium phosphate (pH 7) containing 0.15 *M* NaCl]
pH 11 buffer [50 m*M* sodium carbonate (pH 11) containing 0.5 *M* NaCl]
6 *N* NaOH
Distilled water

Procedure. The culture broth (1 liter) is passed through gauze, and the cells are removed by centrifugation (7000 *g*). The filtrate is brought to pH 11 with NaOH, and the affinity resin is added directly to the clear solution. The suspension is stirred for 1 hr at 25° and then passed through a sintered glass filter. The resin is washed with 100-ml aliquots of pH 11 buffer until the absorbance (280 nm) of the effluent fractions is minimal. The washed iminobiotin-containing resin is collected and poured into a suitable column (2.5 × 80 cm). The adsorbed streptavidin is eluted using pH 4 buffer.[24] The column is washed with pH 11 buffer, the resin is transferred back to the cultural broth, and the affinity chromatographic process is repeated until the streptavidin content of the culture filtrate is depleted.

Throughout the elution procedure, the absorbance (280 nm) of the eluted fractions is monitored continually. Protein-containing fractions are collected, dialyzed at 4° first against pH 7 buffer and then against triple-distilled water until the conductivity of the solution reaches a nominal value. The contents of the dialysis bag are lyophilized, and the resultant protein is stored at 4°. The affinity column is washed with pH 7 buffer and stored in the same buffer with the addition of 0.1% (w/v) sodium azide.

Comments. Growth of *S. avidinii* on malt medium[4,20] usually yields maximum levels of extracellular streptavidin after 5–7 days. Streptavidin production usually ranges from 40 to 60 mg/liter (up to 20 times that reported by Hofmann *et al.*).[7] Accordingly, we can purify up to 40 mg streptavidin per liter of culture, using the affinity resin and the procedure described here. We have experienced difficulties in using asparagine-

[23] E. A. Bayer, H. Ben-Hur, and M. Wilchek, this volume [23].

[24] In some cases, better elution can be achieved using distilled water brought to pH 4 with HCl, and a second elution step is usually performed using this low ionic strength solution.

based synthetic or semisynthetic media[21,22] with iminobiotin columns, and most of our streptavidin preparations have thus been produced from cultures grown on malt medium.

Biotin-Binding Activity on Blots

The activity of the streptavidin subunit and tetramer can be detected by enzyme assay on nitrocellulose transfers using biotinylated alkaline phosphatase (B–AP).[25]

Reagents

Streptavidin-containing blots (either dot blots or following electrophoretic separation)
B–AP,[26] 3 units/ml in PBS
Quenching solution: 2% (w/v) bovine serum albumin
Phosphate-buffered saline, pH 7.4 (PBS)
Substrate solution: 10 mg naphthol AS-MX phosphate (Sigma, free acid dissolved in 200 μl dimethylformamide) is mixed with a solution containing 30 mg Fast Red TR salt (Sigma) dissolved in 100 ml 0.1 *M* Tris-HCl (pH 8.4)

Procedure. The blot is treated with quenching solution[27] for 1 hr at room temperature and rinsed with PBS. The B–AP solution is applied to the blot, which is then incubated for 30 min at room temperature. The blot is rinsed 3 times with PBS, substrate solution is added, and the reaction is allowed to proceed until the desired bands are visible.[28]

Properties of Streptavidin

In contrast to earlier data which suggested a molecular weight of 60,000 for the tetramer,[4,7] the relative mobility on SDS–PAGE of the native protein purified by our affinity-based procedure appears to consist of a major band exhibiting an M_r of about 75,000. Under extreme denaturing conditions, a major subunit band with a relative molecular mass of about 18,000 is observed. Recently, the streptavidin gene has been

[25] Y. Hiller, J. M. Gershoni, E. A. Bayer, and M. Wilchek, *Biochem. J.* **248,** 167 (1987).

[26] E. A. Bayer and M. Wilchek, this volume [14].

[27] Blots are treated with the required solution at 0.5–1 ml/cm^2 in suitable-sized polystyrene compartmented boxes (Althor Products, Wilton, CT).

[28] Color development usually takes about 15–30 min for visualizing biotin-binding activity in the tetramer. For the activity of the monomer in denaturing SDS–polyacrylamide gels, color development usually requires hours or overnight incubation.

cloned[29]; the results indicate a molecular weight of about 66,000 for the tetramer, thus confirming our contention that the previously reported molecular mass of the tetramer is significantly lower than that of the native protein.

Streptavidin, isolated by our procedure, is capable of binding 4 mol of biotin per mole of subunit. Chemical modification studies indicate that tryptophan residues are involved in biotin binding,[30] similar to those which comprise the binding site of egg-white avidin.[31] There are some indications that tyrosines are also important for binding in both proteins.[32] Unlike lysines of the egg-white protein,[33] the role of lysine in streptavidin is not completely clear.[31]

In addition to the major M_r 75,000 band, streptavidin tends to produce different bands corresponding to two types of postsecretory modifications: proteolytic digestion of the intact M_r 18,000 subunit to a minimal molecular size (~14,000) and aggregation of the native tetramer into higher-order (di-, tri-, etc.) oligomeric forms.[34] All of the observed protein bands (detectable by Coomassie blue staining) coincide with the corresponding biotin-binding bands as determined by blotting experiments,[10,25] indicating that all are derived from streptavidin and that none are contaminating proteins.

The thermal stability of streptavidin is remarkable (Fig. 1). In the presence of SDS, the octamers and tetramers begin to break up into monomers and dimers only at temperatures above 60°. As shown in Fig. 1, streptavidin monomers, formed in this manner, are still capable of binding biotin.

Preparation of Truncated Streptavidin

We have recently shown[34] that the truncated form of streptavidin is produced by the action of low quantities of endogenous proteases which clip the N- and C-terminal appendages of the native protein. Earlier procedures for the isolation of streptavidin from the growth medium of *S. avidinii* have included an initial step wherein the culture fluids are either concentrated or precipitated with ammonium sulfate[4,7]; in either case, both the streptavidin and the proteases are coconcentrated, thus resulting in a more rapid degradation of the molecule. The protein is degraded to a

[29] C. E. Argarana, I. D. Kuntz, S. Birken, R. Axel, and C. R. Cantor, *Nucleic Acids Res.* **14,** 1871 (1986).

[30] G. Gitlin, E. A. Bayer, and M. Wilchek, *Biochem. J.* **256,** 279 (1988).

[31] G. Gitlin, E. A. Bayer, and M. Wilchek, *Biochem. J.* **250,** 291 (1988).

[32] G. Gitlin, unpublished results.

[33] G. Gitlin, E. A. Bayer, and M. Wilchek, *Biochem. J.* **242,** 923 (1987).

[34] E. A. Bayer, H. Ben-Hur, Y. Hiller, and M. Wilchek, *Biochem. J.* **259,** 369 (1989).

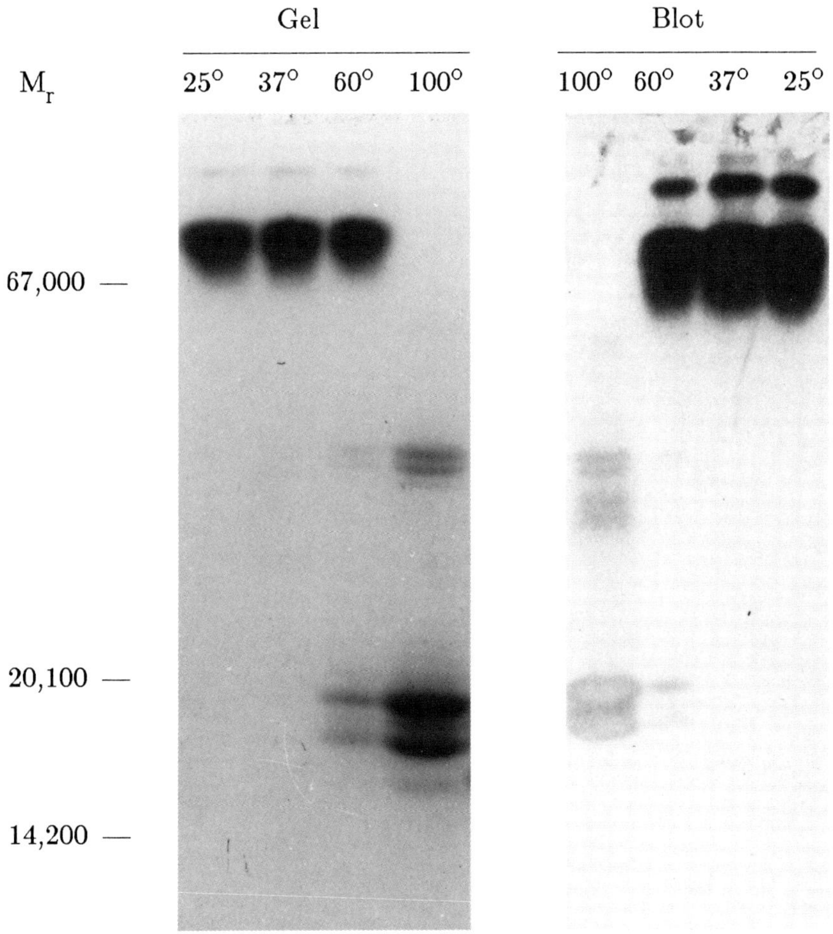

FIG. 1. Temperature-dependent dissociation of native streptavidin into subunits. Samples were incubated at the indicated temperature for 20 min or boiled (100°) for 5 min prior to SDS–PAGE. The gel was stained with Coomassie Brilliant Blue; the nitrocellulose blot transfer was treated with biotinyl alkaline phosphatase and assayed for enzymatic activity using naphthol AS-MX phosphate as substrate. Note the relatively low biotin-binding activity (for the biotinylated enzyme) of the monomers and dimers compared to that of the intact tetramer. Note also the presence of higher-order oligotetramers in the streptavidin preparation.

minimal molecular size (subunit M_r ~13,200), which has been termed "core" streptavidin.[35] This form of streptavidin is the commercially avail-

[35] A. Pähler, W. A. Hendrickson, M. A. Gawinowicz Kolks, C. E. Argarana, and C. R. Cantor, *J. Biol. Chem.* **262,** 13933 (1987).

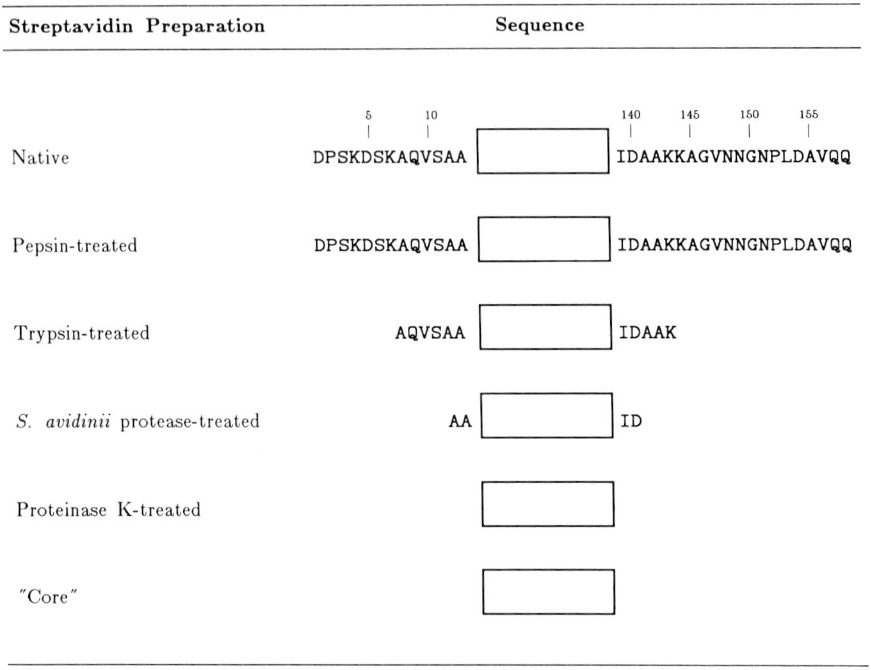

FIG. 2. Comparison of the terminal sequences of various protease-treated streptavidin preparations. The sequence for native streptavidin is that given by Argarana *et al.*[29] For the sake of clarity, the sequence of "core" streptavidin (i.e., the residues between Glu-14 and Ala-138) has been omitted and is designated by boxes.

able protein, and it is remarkably resistant against further degradation. The past difficulties in preparing the intact form of the protein (i.e., inhibiting proteolysis of the N and C termini during purification) have favored the preparation of core streptavidin (total hydrolysis of the termini) in order to produce an electrophoretically uniform product. Using our methodology (described above),[10] the affinity resin is applied directly to the growth medium, thereby impeding endogenous proteolytic action on the terminal appendages.

Many commercial proteases are also capable of converting the native molecule to core streptavidin. The specificity of the protease determines the status of the final product (Fig. 2). To prepare the truncated form, which is similar to the commercially available protein, the best enzymes to use are proteinase K and subtilisin. Other proteases (including papain, thermolysin, and elastase) are also capable of successfully converting streptavidin to a truncated form similar to core streptavidin. However, pepsin and chymotrypsin fail to hydrolyze native streptavidin, owing to the absence of aromatic amino acids in the protease-sensitive terminal stretches.

Reagents

Native streptavidin, 10 mg
Proteinase K (Sigma), 100 μg/ml
Tris-HCl buffer, pH 7.5, containing 5 m*M* EDTA

Procedure. Dissolve the streptavidin in 4.5 ml of Tris-HCl buffer. Add the proteinase K solution (0.5 ml) and incubate for 1 hr at 37°. Bring the solution to pH 11 and purify the product on a 3-ml iminobiotin affinity column as described above for the native form.

Comments. In addition to the truncated monomer producing a single band in denaturing SDS–polyacrylamide gels, the *in vitro* interaction of truncated streptavidin with biotinylated enzymes is superior to that of the native protein.[23] It is thus postulated that the terminal appendages of the intact molecule may interfere sterically with the binding.

Preparation of Streptavidin Oligotetramers

In some preparations of the native protein, enriched amounts of higher-order aggregates (oligotetramers) have been observed. It is not yet certain whether covalent bonding accounts for the aggregation.[36] Nevertheless, the oligotetrameric forms are stable in solution and can be purified by gel filtration.

Reagents

Native streptavidin, 2 mg
Sephacryl S-300 column (15 × 760 mm)
50 m*M* Tris–HCl buffer, pH 7.5

Procedure. Streptavidin (2 mg/ml) is applied to the Sephacryl S-300 column, equilibrated, and eluted with Tris buffer at a flow rate of 10 ml/hr. Fractions of 1 ml are collected, and the absorbance (280 nm) is determined. The small initial peak contains oligotetramer-enriched fractions. These fractions are pooled, concentrated, and rechromatographed under the same conditions. The fractions are examined by SDS–PAGE (10% gels) under nondenaturing conditions (samples are not boiled). The relevant fractions are then pooled, dialyzed against distilled water, and lyophilized.

[36] The appearance of elevated amounts of subunit dimers in oligotetramer-enriched samples analyzed under denaturing conditions (boiling of samples) suggests that the tetramer aggregates are indeed interlinked by covalent bonds.

[9] Crystal Forms of Avidin

By ODED LIVNAH and JOEL L. SUSSMAN

Although many varied and sophisticated applications have been developed[1] for the tight binding of biotin to avidin, the nature and mechanism of the interaction between them are not yet fully understood. High-resolution X-ray crystallographic studies could give us a direct answer to the nature of this strong interaction. It may also be possible to establish a correlation between the structure of bound biotin analogs and their differential affinities toward avidin.

Crystal Forms of Egg-White Avidin

Type I

The first single crystals suitable for X-ray diffraction studies were reported by Green and Joyson in 1970.[2] These crystals were obtained using the dialysis method[3] at room temperature by equilibrating a solution containing 10–20 mg/ml of protein versus a sodium phosphate buffer at pH 5.2. Crystals appeared when the buffer concentration was increased to about 3.0 *M*. Two crystal forms were obtained: prisms with a maximum dimension of 0.2 mm and needles $0.5 \times 0.05 \times 0.05\ mm^3$ in size. The prism-shaped crystals were not studied crystallographically.

The space group and cell dimensions of the needle-shaped crystals were determined, via precession photographs, to be *C*222 with cell dimensions $a = 62$ Å, $b = 107$ Å, $c = 43$ Å. The crystals diffracted up to 2.5 Å and did not appear to be particularly sensitive to prolonged radiation. On soaking biotin into the crystals, there appeared to be no morphological changes in the crystals; however, no crystallographic studies were reported for the complex.

Assuming the molecular mass of the avidin subunit as 15,600 daltons, if there were one subunit per asymmetric unit, then $V_M = 2.28$ Å^3/dalton,[4] whereas if there were two subunits per asymmetric unit, $V_M = 1.14$ Å^3/

[1] M. Wilchek and E. A. Bayer, *Anal. Biochem.* **171,** 1 (1988).

[2] N. M. Green and M. A. Joyson, *Biochem. J.* **118,** 71 (1970).

[3] A. McPherson, "Preparation and Analysis of Protein Crystals," Chap. 4. Wiley, New York, 1982.

[4] V_M is defined as the volume of the asymmetric unit/molecular weight of protein per asymmetric unit.

dalton. Matthews[5] has shown that V_M normally lies within the range 1.68–3.53 for proteins up to 70,000 daltons, so that the asymmetric subunit most probably consists of a single subunit.

Another crystal form was obtained[2] via the dialysis method at room temperature using 3 *M* ammonium sulfate (pH 5.1) as the precipitating reagent. Thin square plates up to 0.4 mm were obtained but not examined crystallographically.

Type II

A different crystal form was obtained[6] from polyethylene glycol (PEG) 6000. These crystals were grown using the batch method[3] at room temperature. The protein solution contained 20–45 mg/ml protein, 10 m*M* sodium phosphate (pH 6.0), and 150 m*M* NaCl. It was reported[6] that to 100 μl of the protein solution PEG 6000 was added until a faint precipitate appeared. The precipitate was then redissolved in 5–15 μl of the buffer. Oil drops developed in the solution and crystals grew out of them after 16–17 days, reaching a final size of 0.8 × 0.2 × 0.2 mm^3 within 1 week. These crystals diffracted to 1.6 Å on still photographs and were stable to irradiation for more than 150 hr at room temperature. Although the crystals diffracted well, the structure determination of avidin could not be advanced owing to a failure to reproduce this crystal form.

The space group was determined by precession photographs, and the cell constants were determined by diffractometer measurements to be $P2_12_12$ with a = 71.6 Å, b = 80.0 Å, c = 43.1 Å. On soaking biotin into the crystals, the a axis undergoes a significant alteration[7,8] to a = 74.7 Å, whereas the b and c axes are essentially unchanged. Again, taking 15,600 as the molecular mass of the avidin monomer, and assuming two avidin molecules in the unit cell, one obtains a value of V_M = 2.06 Å^3/dalton, implying that the crystals contain two subunits in the asymmetric unit.

Type III

A tetragonal crystal form was obtained from ammonium sulfate.[9] The crystals were grown using microdialysis buttons[3] at room temperature.

[5] B. W. Matthews, *J. Mol. Biol.* **33,** 491 (1968).

[6] E. Pinn, A. Pahler, W. Saenger, G. A. Petsko, and N. M. Green *Eur. J. Biochem.* **123,** 545 (1982).

[7] Note that Ref. 6 incorrectly attributes the published value for the *a* axis to native avidin instead of the avidin–biotin complex (see Ref. 8).

[8] A. Pahler, W. A. Hendrickson, M. A. Gawinowicz-Kolks, C. E. Argarana, and C. R. Cantor, *J. Biol. Chem.* **262,** 13933 (1987).

[9] G. Gatti, M. Bolognesi, A. Coda, F. Chiolerio, E. Filippini, and M. Malcovati, *J. Mol. Biol.* **178,** 787 (1984).

The protein solution contained 20 mg/ml protein and 50 m*M* phosphate buffer (pH 5.2). The protein solution was equilibrated versus a solution of 2.7 *M* ammonium sulfate in 50 m*M* phosphate buffer (pH 5.7). Under these conditions crystals grew to the size of $0.4 \times 0.15 \times 0.15$ mm^3 within 2–4 weeks.

It was established via X-ray analysis precession photographs that this crystal form of avidin belongs to tetragonal space group $P4_22_12$ with cell constants $a = b = 79.6$ Å, $c = 84.3$ Å. Reflections were observed to a resolution of 2 Å on still photographs. Considering the molecular mass of the avidin monomer and the V_M values of the other crystal forms, the tetragonal crystal form should contain two avidin subunits per asymmetric unit, giving rise to a V_M value of 2.14 Å/dalton.

Crystal Form of Nonglycosylated Avidin

The carbohydrate moiety of the avidin is not essential for its biotin-binding activity.[10] The biotin-binding properties of the nonglycosylated avidin are equivalent to those obtained for the native (glycosylated) avidin molecule. Owing to the unsuccessful attempts at reproducing some avidin crystal forms, we felt that the avidin with no carbohydrate moiety (thus deleting a flexible part from protein surface) might crystallize more readily than the glycosylated avidin, and well-diffracting crystals would be obtained.

Crystals are grown using the hanging drop method[3] at a constant temperature of 19°. The protein is dissolved at 4 mg/ml[11] into a solution containing 10 m*M* phosphate buffer (pH 6.0) and 18 m*M* sodium azide. Crystals are obtained using PEG 6000 as the precipitating agent. The 8-μl drops contain 2 mg/ml protein, 5.6% PEG 6000, 55 m*M* phosphate buffer (pH 5.4), and 5 m*M* sodium azide. The 1-ml reservoir contains 20% PEG 6000. Crystals grew after 4–5 days and reached the final size of $0.18 \times 0.14 \times 0.04$ mm^3 within 8 days. To obtain larger crystals we applied macroseeding techniques.[12] By reseeding the crystals at least 5 times, they reached a size of $0.4 \times 0.3 \times 0.18$ mm^3.

Our first attempt to determine the unit cell of the nonglycosylated avidin crystals via X-ray crystallography failed within a few hours as the resolution decreased drastically after 6 hr on X-ray irradiation. Using

[10] Y. Hiller, J. M. Gershoni, E. A. Bayer, and M. Wilchek, *Biochem. J.* **248,** 167 (1987); Y. Hiller, E. A. Bayer, and M. Wilchek, this volume [6].

[11] It should be noted that the solubility of nonglycosylated avidin is decreased by factor of 5 compared to avidin, apparently owing to the loss of the carbohydrate chain.

[12] C. Thaller, G. Eichele, L. H. Weaver, E. Wilson, R. Karlson, and J. N. Jansonius, this series, **114,** p. 132.

techniques of Hope[13] for extreme low-temperature X-ray data collection, we were able to preserve the lifetime of the crystal in the X-ray beam practically indefinitely. The method consists of first coating the crystal with a viscous oil in the crystallization droplet and removing all traces of mother liquor. The crystal is then picked up with a thin glass spatula, putting it directly under a stream of boiled liquid N_2 at a temperature of approximately 90 K.

The cell constants and space group were determined by measurements on a Rigaku AFC5-R rotating anode diffractometer operated at 10 kW. The crystal was orthorhombic, space group $P2_12_12$, with cell constants $a = 71.24$ Å, $b = 79.21$ Å, $c = 43.12$ Å, and contained two subunits in the asymmetric unit. On soaking biotin into the crystals the cell constants changed to $a = 73.62$ Å, $b = 79.84$ Å, $c = 43.27$ Å.

When comparing these results with those obtained from the glycosylated avidin orthorhombic crystal,[6] one can see that the crystals appear to be isomorphous, i.e., with a unit cell volume change of about 1%. It may be concluded that the absence of the carbohydrate chain appears not to affect the molecular packing in the crystal.

[13] H. Hope, *Acta Crystalogr.* **B44,** 22 (1988).

[10] Nonavidin Biotin-Binding Proteins

By KRISHNAMURTI DAKSHINAMURTI and JASBIR CHAUHAN

Introduction

The role of biotin as the prosthetic group of the carboxylases and transcarboxylases in various organisms is well established.[1] Apart from these proteins, in which biotin is attached covalently, there are a group of proteins that bind to biotin noncovalently. Avidin, the biotin-binding protein of raw egg white of birds, and streptavidin, the bacterial analog, have an exceedingly high affinity for biotin, with a K_d of about 10^{-15} M, the strongest noncovalent binding known between a protein and a low molecular weight ligand. In addition to these two examples of exceptionally strong affinity, there are other biotin-binding proteins [such as the egg-yolk biotin-binding proteins, biotinidase (EC 3.5.1.12), biotin holocar-

[1] H. G. Wood and R. E. Barden, *Annu. Rev. Biochem.* **46,** 385 (1977).

boxylase synthetase, and nuclear biotin-binding protein] with progressively lower affinities for biotin.

Biotin-binding proteins vary considerably along the evolutionary scale. The biotin biosynthetic pathway of *Escherichia coli* has been well defined. Biotin holocarboxylase synthetase, a biotin-binding protein, has been shown to regulate the synthesis of these enzymes.[2–4] The higher organisms have an absolute requirement for biotin.[5] In the laying hen there are three biotin-binding proteins: avidin (a bacteriostatic protein) present in albumen of egg[6] and two egg-yolk biotin-binding proteins which are thought to supply biotin to the developing embryo.[7–9] Other vitamins (such as riboflavin) have only one species of binding protein[8] which is present in both the egg albumen and the yolk. The occurrence of three distinct biotin-binding proteins suggests a specialized role for biotin in the regulation of embryonic development.[8,9] In mammals, only two proteins have so far been shown to bind biotin: biotin holocarboxylase synthetase and biotinidase.

Recently, attention has been focused on the clinical significance of biotin holocarboxylase synthetase and biotinidase in relation to multiple carboxylase deficiency. Two distinct types of multiple carboxylase deficiencies have been recognized based on the age of onset as well as the nature of the clinical presentation. This disease manifests itself in either a neonatal or infantile form (deficiency of holocarboxylase synthetase) and a late-onset or juvenile form (deficiency of biotinidase).

Biotin Holocarboxylase Synthetase

Biotin carboxylases are synthesized in the form of apoproteins which undergo posttranslational modification by the addition of biotin, the prosthetic group, to the ε-amino group of a lysine residue in the apoproteins. This covalent attachment of biotin to a specific lysine in the apoenzyme is catalyzed by biotin holocarboxylase synthetase in a two-step reaction [Eqs. (1) and (2)].

$$\text{ATP} + \text{biotin (B)} + \text{holoenzyme synthetase (HS)} \rightarrow \text{B–AMP–HS} + \text{PP}_i \quad (1)$$

$$\text{B–AMP–HS} + \text{apoenzyme} \rightarrow \text{holoenzyme} + \text{AMP} + \text{HS} \quad (2)$$

$$\text{B–AMP–HS} + \text{operator} \rightarrow \text{B–AMP–HS–operator} \quad (3)$$

[2] M. A. Eisenberg, *Ann. N.Y. Acad. Sci.* **447,** 335 (1985).
[3] M. A. Eisenberg, O. Prakash, and S. C. Hsiung, *J. Biol. Chem.* **257,** 15167 (1982).
[4] M. A. Eisenberg, *Adv. Enzymol. Relat. Areas Mol. Biol.* **38,** 317 (1973).
[5] K. Dakshinamurti, L. Chalifour, and R. P. Bhullar, *Ann. N.Y. Acad. Sci.* **447,** 38 (1985).
[6] N. M. Green, *Adv. Protein Chem.* **29,** 85 (1975).
[7] R. D. Mandella, H. W. Meslar, and H. B. White III, *Biochem. J.* **175,** 629 (1978).
[8] H. B. White III, *Ann. N.Y. Acad. Sci.* **447,** 202 (1985).
[9] H. B. White III and C. C. Whitehead, *Biochem. J.* **241,** 677 (1987).

Eisenberg *et al.*[3] have purified the enzyme from *E. coli* to homogeneity and have shown that, apart from biotinylating the apocarboxylases, it regulates the gene(s) for biotin biosynthesis. The reaction sequence for the biosynthesis of biotin by *E. coli* and the genes involved in this synthesis have been reported by Eisenberg.[2] The biotin locus in *E. coli* is located in the same region as *gal* and *λatt*, and the order was determined to be *gal, λatt, bio*. The biotin locus is composed of five consecutive genes in the order A, B, F, C, and D (representing the genes coding for various enzymes of biotin biosynthesis) and three unlinked genes *bioH, bioR* (a regulatory gene coding for biotin holocarboxylase synthetase), and *bioP* (a permeability gene).

Similarities between the reactions involved in aminoacyl-tRNA synthesis and biotin holocarboxylase synthetase were demonstrated, in light of the regulatory role recognized for the aminoacyl-tRNA.[2] It was suggested that the holoenzyme synthetase with the bound biotinyl-5′-AMP functions as the corepressor for the biotin operon.[3] In the *trp* operon, the most extensively studied system,[10] tryptophanyl-tRNA acts as a negative modulator at the attenuator site, terminating transcription of the *trp* operon under conditions of tryptophan excess, or as a substrate for protein synthesis when the levels are low. In the biotin operon, the biotinyl-5′-AMP complex either inhibits initiation of transcription by binding to the operator site or utilizes the active form of the biotin complex which is common to both reactions for holocarboxylase synthetase [Eqs. (2) and (3)].

The biotin operon from *E. coli* has been extensively studied and has been sequenced.[2,3,11,12] Eisenberg *et al.*[3] have shown that *E. coli* holocarboxylase synthetase is a single polypeptide of M_r 34,000 which binds biotin with a K_d value of 1.3×10^{-7} *M*. The activated form of biotin, biotinyl 5′-adenylate, binds much more tightly to the holocarboxylase synthetase (K_d 1.1×10^{-9} *M*). Purification, assays, and properties of the bacterial biotin holocarboxylase synthetase are described in detail by Goss and Wood.[13]

Biotinidase

Biotinidase releases covalently attached biotin (possibly from biocytin) during the proteolytic degradation of carboxylases. In the gut this enables the absorption of dietary protein-bound biotin, and in cells the

[10] C. Yanofsky, *Nature (London)* **289,** 751 (1981).

[11] A. Otsuka and J. Abelson, *Nature (London)* **276,** 689 (1978).

[12] P. K. Howard, J. Shaw, and A. J. Otsuka, *Gene* **35,** 321 (1985).

[13] N. H. Goss and H. G. Wood, this series, Vol. 107, p. 261.

recycling of the enzyme-bound biotin is facilitated. Biotinidase has been detected in most mammalian tissues, with high activities being present in liver, kidney, serum, intestine, and adrenal glands.[14] It has been localized in the microsomal fraction of hepatocytes.[14,15] Biotinidase has recently been purified to homogeneity from human serum.[16,17] The p*I* of this enzyme is 4.6. We have shown that [^{3}H]biotin is not released by the action of biotinidase on ^{3}H-biotinylated ribonuclease and polylysines (M_r 3,000–100,000).[16] Craft *et al.*[17] studied the action of biotinidase on polypeptides of various lengths (decreasing in number of residues from 123 to 5). The activity of biotinidase increased as the peptide chain length decreased but was significantly lower than that observed for biocytin. These results suggest that biotin-containing proteins have to be digested by proteases and peptidases to biocytin before the biotin can be released by the action of biotinidase.

Purified human serum biotinidase gives a single band of about M_r 80,000 on the SDS–PAGE system. Because biotinidase is a glycoprotein[16] and glycoproteins frequently behave anomalously on SDS–PAGE,[18] the migration of biotinidase on SDS–PAGE as a function of the pore size of the gel (percent acrylamide) was compared to that of a group of standard proteins. The corrected M_r for biotinidase was calculated by plotting the square root of the slopes for individual proteins (K_r, retardation coefficient) against the cube root of their known M_r. A similar M_r for biotinidase was obtained using sedimentation analysis of the enzyme in a 5–20% (w/v) sucrose density gradient.[16] From these results it was concluded that biotinidase is a single polypeptide of M_r 68,000.

Various synthetic substrates can be used for the assay of biotinidase. An assay based on the colorimetric determination of 4-aminobenzoate released from the synthetic substrate *N*-(*d*-biotinyl) 4-aminobenzoate by the action of biotinidase has been widely used.[19] In an attempt to make the assay more sensitive for clinical application, Wolf and McVoy[20] developed a radioassay in which the release of [^{14}C]carboxyl-4-aminobenzoate from *N*-(*d*-biotinyl) [^{14}C]carboxyl-4-aminobenzoate is measured after the removal of both free biotin and excess unreacted substrate using avidin

[14] J. Pispa, *Ann. Med. Exp. Biol. Fenn Suppl.* **43**(5), 5 (1965).
[15] G. S. Heard, R. E. Grier, D. L. Weiner, J. R. Secor McVoy, and B. Wolf, *Ann. N.Y. Acad. Sci.* **447,** 400 (1985).
[16] J. Chauhan and K. Dakshinamurti, *J. Biol. Chem.* **261,** 4268 (1986).
[17] D. V. Craft, N. H. Goss, N. Chandramouli, and H. G. Wood, *Biochemistry* **24,** 2471 (1985).
[18] S. P. Grefrath and J. A. Reynolds, *Proc. Natl. Acad. Sci. U.S.A.* **71,** 3913 (1974).
[19] J. Knappe, W. Brummer, and K. Biederbick, *Biochem. Z.* **338,** 599 (1963).
[20] B. Wolf and J. R. Secor McVoy, *Clin. Chim. Acta* **135,** 275 (1983).

followed by bentonite precipitation. A fluorometric assay measuring the release of 6-aminoquinoline from *N*-(*d*-biotinyl)-6-aminoquinoline has also been described.[21] More recently, a fluorometric assay measuring release of lysine from the natural substrate (biocytin) was reported.[22]

Biotinidase Assay. The assay based on the hydrolysis of *N*-(*d*-biotinyl) 4-aminobenzoate and the colorimetric determination of the liberated 4-aminobenzoate is the one most widely used. The assay is initiated by the addition of 0.1 ml of enzyme solution to 1.9 ml of a mixture containing 0.1 *M* potassium phosphate buffer (pH 7.0), 10 m*M* EDTA, 0.5 mg bovine serum albumin, and 0.15 m*M* *N*-(*d*-biotinyl) 4-aminobenzoate. The mixture is incubated for 30 min at 37°, and the reaction is terminated by the addition of 0.2 ml of 30% (w/v) trichloroacetic acid. The supernatant, 1.5 ml, is added to 0.5 ml of water. The following reagents are then added successively at room temperature at 3-min intervals: 0.2 ml of sodium nitrite (0.1% w/v), 0.2 ml of ammonium sulfamate (0.5% w/v), and 0.2 ml of *N*-(1-naphthyl)ethylenediamine dihydrochloride (0.1% w/v). After incubation for 10 min, absorbance is measured at 546 nm. An assay mixture without enzyme is used as a blank. A standard curve relating absorbance units to nanomoles of 4-aminobenzoate is established and is linear to an optical density of 0.8.

Biotin-Binding Assay. The biotin-binding assay relies on the strong noncovalent binding of [^{3}H]biotin to biotin-binding proteins. The assay (in a final voume of 0.3 ml) contains the following components: 100 m*M* potassium phosphate buffer (pH 6.0), 150 m*M* NaCl, 1 m*M* EDTA, 1 m*M* 2-mercaptoethanol, 0.1 μCi [^{3}H]biotin, and the appropriate amount of biotin-binding protein. When more highly purified biotinidase preparations are used, 1.0 mg of γ-globulin is also included in the assay to aid in subsequent precipitation of proteins. In the control, unlabeled biotin at a final concentration of 1 m*M* is included to correct for nonspecific binding. The reaction mixture is incubated at room temperature for 5 min, and the protein is then precipitated by the addition of 0.9 ml of cold saturated ammonium sulfate in phosphate buffer (pH 6.0). After standing for 10 min on ice, the precipitated protein is collected by centrifugation and washed once with 1.0 ml of cold 70% saturated ammonium sulfate in phosphate buffer. The precipitate is again collected by centrifugation, then dissolved in 0.1 ml 0.01 *N* NaOH, and the radioactivity is determined with 1 ml of Scintiverse II cocktail.

Biotin-Binding Activity of Human Serum Biotinidase. We have previously shown that biotin inhibits biotinidase activity in the millimolar

[21] H. Wastell, G. Dale, and K. Bartlett, *Anal. Biochem.* **140,** 69 (1984).

[22] H. Ebrahim and K. Dakshinamurti, *Anal. Biochem.* **154,** 282 (1986).

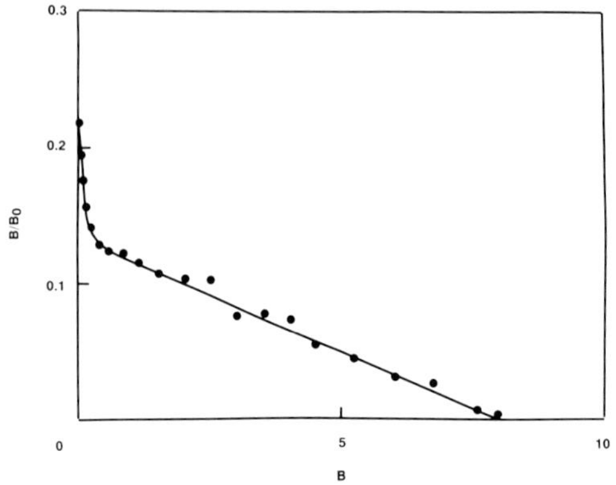

FIG. 1. Concentration-dependent binding of d-[^{3}H]biotin to purified human serum biotinidase. Samples of purified biotinidase were incubated with increasing concentrations of d-[^{3}H]biotin (0.1 nM–10 μM) as described in the text. A parallel set of samples contained the same amount of d-[^{3}H]biotin plus a 1000-fold excess of cold d-biotin for the determination of nonspecific binding. Each point is the mean of duplicate determinations. Linear regression of the points gave K_d values of 3 and 59 nM and B_{max} values of 0.065 and 0.79 mol/mol of biotinidase.

range.[16] However, the physiological significance of this inhibition appears to be very little, as biotin is present only in the nanomolar range in human serum.[23] In this context, it has been shown that the K_m of biotinidase for biocytin is 10 μM. We have determined the binding of free biotin to pure human serum biotinidase. Figure 1 shows the concentration-dependent binding of d-[^{3}H]biotin to biotinidase and the Scatchard transformation of the data.[24] The Scatchard plot shows two linear parts, indicating that two classes of sites with high and low affinities for biotin are present (3 and 59 nM, respectively). Biotin binding has also been investigated using displacement curve analysis. The displacement curve also exhibits both low- and high-affinity binding sites.

The high-capacity low-affinity binding corresponds to that of other vitamin-binding proteins. For example, the thiamine-binding protein binds thiamin with a K_d for thiamin of 5.5×10^{-7} M.[25] Dethiobiotin and iminobiotin displace the bound biotin in the micromolar range. The biotin-binding activity of biotinidase can be completely inhibited with N-bromosuccinimide and p-chloromercuribenzoate, suggesting that tryptophan and cysteine residues are required for biotin binding. It is likely that the

[23] H. Baker, *Ann. N.Y. Acad. Sci.* **447,** 129 (1985).
[24] G. Scatchard, *Ann. N.Y. Acad. Sci.* **51,** 660 (1949).
[25] H. Nishimura, K. Yoshioka, and A. Iwashima, *Anal. Biochem.* **139,** 373 (1984).

cysteine residues in biotinidase are required for the formation of an enzyme–acyl complex for the high-affinity low-capacity binding. The formation of an enzyme–acyl complex has been shown for biotinidase.[14]

Fractionation of Human Serum on Sephadex G-150 Column. Human serum can be fractionated on Sephadex G-150 columns and analyzed for biotinidase and biotin-binding activity. Both activities coincide, suggesting that biotinidase is the only protein which binds biotin in human serum. These findings have been corroborated by binding studies performed on fractionated human serum analyzed on DEAE-Sephacel, hydroxylapatite, and octyl-Sepharose columns. On all column separations, biotin-binding activity coincided with biotinidase activity. Biotin-binding studies with human serum support the notion that biotinidase might be a biotin carrier protein in human serum.

Egg-Yolk Biotin-Binding Protein

It has been demonstrated that biotin in egg yolk is bound to a nondialyzable, heat-denatured component.[26] White and co-workers have isolated this heat-denaturable protein (BBP1) and have shown it to be distinct from avidin.[27] The purified protein has an M_r of 74,300 and is a tetramer of identical subunits. It is a glycoprotein, with a p*I* of 4.6. BBP1 binds biotin with a K_d of 1×10^{-12} *M*. This high affinity for biotin could be used in instances where avidin– or streptavidin–biotin interactions result in nonspecific binding.

The biotin content of eggs has been shown to be directly related to the amount of biotin in the diet.[28] After 2 weeks on a biotin-deficient diet, a hen lays eggs whose hatchability is reduced to zero and whose biotin content is 10% that of a normal hen's egg.[29] A hen's egg contains about 10 μg of biotin. Of this about 1 μg is in the albumen and bound to avidin, where it occupies 15% or less of the binding sites. The rest of the biotin is in egg yolk, bound to the two biotin-binding proteins BBP1 and BBP2. The plasma of the laying hen also contains biotin-binding proteins that are identical to egg-yolk biotin-binding proteins.[9] Both BBP1 and BBP2 are transported by the bloodstream, where these proteins bind biotin, to the ovary and finally are deposited in the developing oocyte. Since most of the biotin in egg yolk is bound to biotin-binding proteins, the physiological

[26] P. Gyorgy and C. S. Rose, *Proc. Soc. Exp. Biol. Med.* **49,** 294 (1949).
[27] H. W. Meslar, S. A. Camper, and H. B. White III, *J. Biol. Chem.* **253,** 6979 (1978).
[28] L. E. Brewer and H. M. Edwards, Jr., *Poultry Sci.* **51,** 619 (1972).
[29] J. R. Couch, W. W. Craven, C. A. Elvehjem, and J. G. Halpin, *Arch. Biochem. Biophys.* **21,** 77 (1949).

function of these biotin-binding proteins is undoubtedly to transport biotin to the egg for future use by the developing embryo.[9] It was further demonstrated that the concentrations of BBP1 and BBP2 in egg yolk are directly related to dietary biotin content. At low concentrations of biotin BBP1 is the major transporter, while at higher concentrations of biotin BBP2 predominates. It has been shown that both avidin[30] and BBP1[31] can be induced by various sex hormones. The induction of two egg-yolk biotin-binding proteins by biotin[9] may suggest a role for biotin in regulating these genes.

BBP1 can be purified by two procedures. Murthy and Adiga[31] have used a combination of the following: DEAE-cellulose chromatography, adsorption of avidin on an antiavidin–Sepharose column, and release of bound biotin by guanidine-HCl followed by chromatography on a biotin–Sepharose column. The protein purified by this method was essentially free of bound biotin. The second procedure developed by Meslar *et al.*[27] can be used if apoprotein is not required. This procedure involves delipidification of egg yolk by butanol extraction, chromatography on a phosphocellulose column, followed by affinity chromatography on a biotin–Sepharose column and finally gel filtration on an Ultrogel AcA 44 column. The first procedure gives a yield of 6%, while the second procedure gives a yield of 1.5%. There is no reported procedure for the purification of BBP2.

Assay for Egg-Yolk Biotin-Binding Protein. The method of White and Whitehead[9] has been used to determine the biotin-binding activity of two egg-yolk biotin-binding proteins. The assay is initiated by the addition of *d*-[^{3}H]biotin (0.1 μCi, specific activity 10 Ci/mmol) and 100 μl of biotin-binding protein in a total volume of 1.0 ml of phosphate buffer (pH 7.0). The mixture is incubated for 30 min at 65° for BBP1 and 45° for BBP2. The cooled incubation mixtures are quantitatively transferred to a small phosphocellulose column (0.25-ml bed volume) and washed with 2 ml of phosphate buffer (0.1 *M*, pH 7.0). Biotin-binding proteins are eluted with 2 *M* NaCl (4 ml). A portion of the eluate is counted with 10 ml Scintiverse II counting fluid.

Nuclear Biotin-Binding Protein

Apart from the role of biotin in carboxylases, biotin has been implicated in other areas of metabolism where its role cannot be explained on the basis of its function as the prosthetic group of known biotin-containing

[30] J. Korpela, *Med. Biol.* **62,** 5 (1984).
[31] C. V. R. Murthy and P. R. Adiga, *Mol. Cell. Endocrinol.* **40,** 79 (1985).

enzymes. As mentioned earlier, biotin plays an important role in *E. coli* by participating in the biotin operon. In avian species, biotin induces egg-yolk biotin-binding proteins. Furthermore, we have presented evidence that biotin enhances levels of cGMP, guanylate cyclase,[32] and glucokinase[33,34] in mammalian tissues, and this has been substantiated recently.[35,36] Various enzymes and proteins can be induced by biotin, but the mechanism by which biotin regulates these enzymes is not yet known. It is likely that transcriptional control, as for hormones, may be involved in regulation. Biotin is taken up by the nuclei *in vivo* even though nuclei do not possess any CO_2-fixing capability, suggesting that the biotin in nuclei does not function as the prosthetic group of biotin carboxylases.[37–40] During biotin deficiency, nuclear biotin seems to be conserved whereas it is preferentially lost from other cellular organelles.[39,40] Biotin in nuclei has been shown to be protein bound. The nuclear protein binds biotin *in vitro* in a reversible manner with a maximal binding of 3.54 pmol/μg protein with a dissociation constant for biotin of 2.2×10^{-7} *M*. Polyacrylamide gel electrophoresis in the presence of sodium dodecyl sulfate indicates an apparent subunit molecular weight of 60,000 for this protein.

Homology around Tryptophan Residues in Biotin-Binding Proteins

Biotin is covalently linked to the ε-aminolysine group in all carboxylases. A conserved sequence, Ala-Met-Bct-Met (Bct is the abbreviation for biocytin), occurs in all carboxylases isolated from various sources, *E. coli* to humans.[1,41,42] It has been suggested that this conservation provides evidence that these biotin-dependent enzymes may have evolved from a common ancestor.[42]

Human serum biotinidase and egg-yolk biotin-binding protein require tryptophan residues in the active center for biotin-binding activity. Green[6] has reviewed the avidin–biotin interaction. Importance has been attached both to tryptophanyl residues (4 per subunit) and hydrogen bonding to the

[32] I. Singh and K. Dakshinamurti, *Mol. Cell. Biochem.* **79,** 47 (1988).

[33] K. Dakshinamurti and C. Cheah-Tan, *Arch. Biochem. Biophys.* **127,** 17 (1968).

[34] K. Dakshinamurti and C. Cheah-Tan, *Can. J. Biochem.* **46,** 75 (1968).

[35] D. L. Vesely, *Science* **216,** 1329 (1982).

[36] J. T. Spence and A. P. Koudelka, *J. Biol. Chem.* **259,** 6393 (1984).

[37] K. Dakshinamurti and S. P. Mistry, *J. Biol. Chem.* **238,** 294 (1963).

[38] K. Dakshinamurti and S. P. Mistry, *J. Biol. Chem.* **238,** 297 (1963).

[39] R. L. O. Boeckx and K. Dakshinamurti, *Biochem. J.* **140,** 549 (1974).

[40] R. L. O. Boeckx and K. Dakshinamurti, *Biochim. Biophys. Acta* **383,** 282 (1975).

[41] H. G. Wood and G. K. Kumar, *Ann. N.Y. Acad. Sci.* **447,** 1 (1985).

[42] A. Lamhonawah, T. J. Barankiewicz, H. F. Willard, D. J. Mahuran, F. Quan, and R. A. Gravel, *Proc. Natl. Acad. Sci. U.S.A.* **83,** 4864 (1986).

ureido ring in the binding process. Biotin binding to streptavidin results in changes in the absorption spectrum, and the tryptophan reactivity of streptavidin is similar to that of avidin.[43] It is possible that for biotinidase and egg-yolk biotin-binding proteins, tryptophan residues in the active center may be required for hydrophobic binding of biotin to these enzymes.

The sequence for streptavidin has recently been published and shows considerable similarity to avidin around the tryptophan residues.[44] Although the sequence for *E. coli* biotin holocarboxylase synthetase has been determined, it is not known whether this protein requires the tryptophan residues for biotin binding. However, the other biotin-binding proteins (biotinidase, avidin, streptavidin, and two egg-yolk biotin-binding proteins) require tryptophan residues for biotin binding. It is therefore possible that there may be homology in the amino acid sequence around the tryptophan residues in all the known biotin-binding proteins.

Conclusions

The best understood role of biotin in various organisms is its function as the prosthetic group of a small number of enzymes, the biotin-dependent carboxylases. Biotin serves as a covalently bound "CO_2 carrier" in these carboxylases. Apart from these enzymes, biotin holocarboxylase synthetase and biotinidase are enzymes of biotin metabolism. Both enzymes are biotin-binding proteins. Biotin holocarboxylase synthetase has been shown to act at the biotin operon in *E. coli* and to regulate the synthesis of various enzymes required for biotin synthesis. Biotinidase has been indicated to act as a biotin carrier protein in human serum. Furthermore, considerable progress has been made in understanding the role of egg-yolk biotin-binding proteins in depositing biotin in eggs. Two proteins, BBP1 and BBP2, have been identified in egg yolk and can be induced by biotin.

Most of the biotin-binding proteins appear to require tryptophan residues for biotin binding, and there is considerable homology of the amino acid sequence around certain tryptophans in both avidin and streptavidin. However, the sequences for the other biotin-binding proteins are not known at present. Furthermore, biotin can regulate various enzymes, and the existence of biotin and a biotin-binding protein in nuclei suggests that there might be a transcriptional control of various genes by biotin similar to that known for hormones.

[43] L. Chaiet and F. Wolf, *Arch. Biochem. Biophys.* **106,** 1 (1964).

[44] C. E. Argarana, I. D. Kuntz, S. Birken, R. Axel, and C. R. Cantor, *Nucleic Acids Res.* **14,** 1871 (1986).

[11] Biotinidase

By BARRY WOLF, JEANNE HYMES, and GREGORY S. HEARD

Biotinidase (biotin-amide amidohydrolase, EC 3.5.1.12) hydrolyzes biocytin (ε-*N*-biotinyllysine) to the vitamin biotin and lysine. Biotinidase does not release biotin that is covalently bound to intact holocarboxylases.[1] Carboxylases must be proteolytically degraded to biocytin or small biotinyl peptides before hydrolysis can occur. Biotinidase is necessary for the endogenous recycling of the vitamin and probably for the liberation of the vitamin from dietary protein-bound sources.[2] Biotinidase activity is deficient in serum and other tissues of most children with the inherited disorder late-onset multiple carboxylase deficiency.[3-5]

In 1954, enzymes in hog liver[6] and in human plasma[7] were described that released biotin from the products of proteolytic degradation of hog liver and from biocytin, respectively. The enzymes appeared to be identical and were called biotinidase or biocytinase. Biotinidase has since been studied in various prokaryotes and in the tissues of many eukaryotes. Some microorganisms can grow only if biotin is included in the medium,[6,8,9] whereas others grow in medium containing either free biotin or biocytin; the latter have biotinidase activity. In animals, the highest specific activity of biotinidase is found in serum.[10,11] Activities are high in liver, kidney, and adrenal glands and low in brain.[11,12] Biotinidase is also

[1] D. V. Craft, N. H. Goss, N. Chandramouli, and H. G. Wood, *Biochemistry* **24,** 2471 (1985).
[2] B. Wolf, G. S. Heard, J. R. Secor McVoy, and H. M. Raetz, *J. Inherited Metab. Dis.* **7** (Suppl. 2), 121 (1984).
[3] B. Wolf, R. E. Grier, W. D. Parker, S. I. Goodman, and R. J. Allen, *N. Engl. J. Med.* **308,** 161 (1983).
[4] B. Wolf, R. E. Grier, R. J. Allen, S. I. Goodman, and C. L. Kien, *Clin. Chim. Acta* **131,** 272 (1983).
[5] B. Wolf, R. E. Grier, R. J. Allen, S. I. Goodman, C. L. Kien, W. D. Parker, D. M. Howell, and D. L. Hurst, *J. Pediatr. (St. Louis)* **103,** 233 (1983).
[6] R. W. Thoma and W. H. Peterson, *J. Biol. Chem.* **210,** 569 (1954).
[7] L. D. Wright, C. A. Driscoll, and W. P. Boger, *Proc. Soc. Exp. Biol. Med.* **86,** 335 (1954).
[8] L. D. Wright, H. R. Skeggs, and E. L. Cresson, *Am. J. Chem.* **73,** 4144 (1951).
[9] M. Koivusalo, C. Elorriaga, Y. Kaziro, and S. Ochoa, *J. Biol. Chem.* **238,** 1038 (1963).
[10] M. Koivusalo and J. Pispa, *Acta Physiol. Scand.* **58,** 13 (1963).
[11] J. Pispa, *Ann. Med. Exp. Biol. Fenn.* **43** (Suppl. 5), 1 (1965).
[12] S. F. Suchy, J. R. Secor McVoy, and B. Wolf, *Neurology* **35,** 1510 (1985).

present in secretory cells, including fibroblasts and leukocytes, and in pancreatic juice and zymogen granules.[2,13] In general, carnivores and omnivores have higher specific activities of biotinidase in serum than do herbivores; the sheep is an exception.[11]

Biotinidase in blood probably originates from the liver.[14] Biotinidase in serum is sialylated, but in rat liver from which blood is removed by perfusion with physiological saline, biotinidase is asialylated.[13] Results of subcellular fractionation of rat hepatocytes suggest that the enzyme is enriched in the microsomal fraction and, to a lesser extent, in the lysosomal fraction.[11,13] Microsomal biotinidase is localized in the rough endoplasmic reticulum and Golgi apparatus.

Biotinidase has been partially purified from *Streptococcus faecalis*,[9] *Lactobacillus casei*,[15] and from hog kidney,[15] liver,[11] and serum.[11] The enzyme from human plasma[1,16] and serum[17] has been purified to homogeneity.

Assay Methods

Both natural and artificial substrates have been used for determining biotinidase activity in various assays using secondary enzymatic,[18] radiometric,[19,20] colorimetric,[4,15] and fluorometric[21–23] measurement. Some methods require derivatization[21] or chromatographic separation of the reaction products.[19,22]

The simplest and most commonly used method for measuring biotinidase activity uses *N*-(*d*-biotinyl) 4-aminobenzoate (BPABA) as substrate. Hydrolysis of BPABA results in the liberation of biotin and 4-aminobenzoate (PABA). We describe two methods based on the use of BPABA. The first method is a modification of the method of Knappe *et al.*,[15] and is used clinically for the diagnosis of biotinidase deficiency.[4] A procedure,

[13] G. S. Heard, R. E. Grier, D. L. Weiner, J. R. Secor McVoy, and B. Wolf, *Ann. N.Y. Acad. Sci.* **447,** 400 (1985).

[14] D. L. Weiner, R. E. Grier, P. Watkins, G. S. Heard, and B. Wolf, *Am. J. Hum. Genet.* **34,** 56A (1983).

[15] J. Knappe, W. Brummer, and K. Biederbick, *Biochem. Z.* **338,** 599 (1963).

[16] B. Wolf, J. B. Miller, J. Hymes, J. R. Secor McVoy, Y. Ishikawa, and E. Shapira, *Clin. Chim. Acta* **164,** 27 (1987).

[17] J. Chauhan and K. Dakshinamurti, *J. Biol. Chem.* **261,** 4268 (1986).

[18] D. L. Weiner, R. E. Grier, and B. Wolf, *J. Inherited Metab. Dis.* **8,** 101 (1985).

[19] L. P. Thuy, B. Zielinska, L. Sweetman, and W. L. Nyhan, *Ann. N.Y. Acad. Sci.* **447,** 434 (1985).

[20] B. Wolf and J. R. Secor McVoy, *Clin. Chim. Acta* **135,** 275 (1983).

[21] H. Ebrahim and K. Dakshinamurti, *Anal. Biochem.* **154,** 282 (1986).

[22] K. Hayakawa and J. Oizumi, *J. Chromatogr.* **383,** 148 (1986).

[23] H. Wastell, G. Dale, and K. Bartlett, *Anal. Biochem.* **140,** 69 (1984).

based on this method, for semiquantitatively measuring biotinidase activity in blood-soaked filter paper disks has been developed.[24] It is used to screen newborn infants for biotinidase deficiency.[25,26] The second method, that of Wolf and Secor McVoy,[20] is specific and more sensitive. Therefore, it is applicable to leukocytes, fibroblasts, and amniotic cells, which have low activities, and to tissues with high activities, but for which the availability is limited.

Modified Method[15] *of Knappe et al.*[4]

Principle. Biotinidase is assayed by measuring the hydrolysis of BPABA. The PABA released is diazotized, coupled to a naphthol reagent, and determined by its absorbance at 546 nm.

Because of its limited sensitivity, this method is applicable only for specimens with high activity, including serum, plasma, and homogenates of liver, kidney, and adrenal glands. The presence of compounds other than PABA that contain a free primary aromatic amino group, such as sulfonamides, can result in the nonspecific formation of colored compounds. Therefore, the appropriate controls must be used.

Reagents

Potassium phosphate buffer, 50 m*M* (pH 6.0)
Trichloroacetic acid solution (TCA), 30% (w/v)
Sodium nitrite solution, 0.1% (w/v), prepared on each day of assay
Ammonium sulfamate solution, 0.5% (w/v)
N-1-Naphthylethylenediamine dihydrochloride solution, 0.1% (w/v); this solution is light sensitive and should be stored in a dark bottle
Buffer A, prepared by dissolving potassium ethylenediaminetetraacetic acid (EDTA) in 50 m*M* potassium phosphate buffer to a final concentration of 50 m*M* and adjusting the pH to 6.0
Buffer B, prepared by dissolving BPABA in 0.1–0.2 ml sodium bicarbonate, 1.0 *M*, and adding to Buffer A to achieve a final concentration of 0.15 m*M* BPABA
Standard solution, prepared by dissolving 27.42 mg of analytically pure PABA in 100 ml of Buffer A; the solution is refrigerated, and for use it is diluted 10-fold, so that 0.1 ml corresponds to 20 nmol of PABA

[24] G. S. Heard, J. R. Secor McVoy, and B. Wolf, *Clin. Chem.* (*N.Y.*) **30,** 125 (1984).

[25] B. Wolf, G. S. Heard, L. G. Jefferson, V. K. Proud, W. E. Nance, and K. A. Weissbecker, *N. Engl. J. Med.* **313,** 16 (1985).

[26] G. S. Heard, B. Wolf, L. G. Jefferson, K. A. Weissbecker, W. E. Nance, J. R. Secor McVoy, A. Napolitano, P. L. Mitchell, F. W. Lambert, and A. S. Linyear, *J. Pediatr.* (*St Louis*) **108,** 40 (1986).

Procedure. One hundred microliters of serum, plasma, or tissue extract is added to 1.9 ml of Buffer B in 12 × 75 mm glass tubes at 37°; once the enzyme source has been added, the tube is vortexed, returned to the water bath, and incubated for 30 min. The following control tubes are also carried through the assay in a final volume of 2.0 ml: (1) buffer blank containing 0.1 ml of buffer A and 1.9 ml of buffer B, (2) PABA standard containing 0.1 ml of standard solution (20 nmol PABA) in 1.9 ml of Buffer B, and (3) sample blank containing 0.1 ml of sample and 1.9 ml of buffer A.

The enzymatic reaction is stopped by adding 0.2 ml of TCA at 4°. The acidified samples are centrifuged at 2000 *g* for 10 min at 25°. A 1.5-ml aliquot is removed from each sample and control, added to a 12 × 75 mm glass test tube, and diluted with 0.5 ml of water. The PABA in each tube is determined by the following procedure. To each diluted aliquot, 0.2 ml of sodium nitrite solution is added, and the solution is mixed. After 3 min, 0.2 ml of ammonium sulfamate solution is added, and the solutions are mixed. After 3 min, 0.2 ml of *N*-1-naphthylethylenediamine dihydrochloride solution is added. The solution is mixed, and after 10 min (during which time a purple color develops in all tubes that contain PABA) the optical density (OD) of the solution is read at 546 nm against distilled water. Optical density increases linearly up to 30 nmol of PABA.

Biotinidase activity is calculated according to the following formula:

$$\text{Activity} = \frac{(\text{OD}_{\text{sample}} - \text{OD}_3) \times 20 \times 10}{(\text{OD}_2 - \text{OD}_1) \times 30}$$

in which the subscripts sample, 1, 2, and 3 signify the optical densities of the sample, buffer blank, standard, and sample blank, respectively; 20 is the number of nanomoles of PABA per tube; 10 is the reciprocal of the sample volume in milliliters; and 30 is the incubation time in minutes. Activity is expressed as PABA released, in nanomoles per minute per milliliter of serum. The biotinidase activity in normal human serum ranges from 4.4 to 10.0 nmol/min/ml serum.[4]

Method of Wolf and Secor McVoy[20]

Principle. Biotinidase is assayed by measuring the hydrolysis of *N*-(*d*-biotinyl) [*carboxyl*-^{14}C]PABA. The [*carboxyl*-^{14}C]PABA released is determined by liquid scintillation counting after separation from unhydrolyzed radioactive substrate.

Reagents

N-(*d*-Biotinyl)-4-[*carboxyl*-^{14}C]aminobenzoic acid solution, 2.3 mCi/mmol, 0.6 m*M*[27]

Potassium phosphate buffer, 0.1 m*M* (pH 6.0)

p-Hydroxymercuribenzoate solution, 2 m*M*

Avidin solution, 1% (w/v)

Bentonite suspension, 4.2% (w/v), in potassium phosphate buffer, 0.1 m*M* (pH 6.0)

Liquid scintillation counting fluid suitable for aqueous solutes

Procedure. Serum is diluted to 1% (v/v) with phosphate buffer. Extracts of leukocytes,[28] fibroblasts, and other tissues are prepared by homogenizing in phosphate buffer in the presence of Triton X-100.[20] Extracts should contain 2–3 mg protein/ml. One hundred microliters of each diluted serum (1%, v/v), leukocyte extract, or fibroblast extract is added to 18 μl of potassium phosphate buffer in a 1-ml polystyrene centrifuge tube. After preincubation for 5 min at 37° in a water bath, the reaction is started by the addition of 15 μl of *N*-(*d*-biotinyl) [*carboxyl*-^{14}C]PABA to each tube. The solution is mixed, and the tube is returned to the water bath. A sample blank containing 100 μl of enzyme solution, 18 μl of *p*-hydroxymercuribenzoate solution, and 15 μl of substrate solution is also carried through the assay.

After incubation for 90–240 min,[29] the reactions are stopped by the addition of 18 μl of *p*-hydroxymercuribenzoate (18 μl of phosphate buffer is added to blanks). Twenty-five microliters of avidin solution is added, and the solution is mixed and then incubated at 25° (ambient) for 30 min. To each tube is added 75 μl of bentonite suspension. The suspension is mixed, incubated for 10 min, centrifuged at 2000 *g* for 2 min (Fisher Microfuge), and then 125 μl of the supernatant is transferred to a scintillation vial. Eight milliliters of liquid scintillation fluid is added, and radioactivity is determined in a liquid scintillation spectrometer. The activity in

[27] Radioactive substrate with a specific activity of 2.3 mCi/mmol may be prepared, as described in Ref. 4, from biotin *N*-hydroxysuccinimide ester (17 mg) and a mixture of [*carboxyl*-^{14}C]PABA (0.2 mg, 58.5 mCi/mmol; Research Products International Corp., Mount Prospect, IL) and unlabeled PABA (7.1 mg).

[28] B. Wolf and L. E. Rosenberg, *J. Clin. Invest.* **62,** 931 (1978).

[29] The duration of the incubation interval over which the release of [*carboxyl*-^{14}C]PABA increases linearly depends on the source of enzyme. One hundred microliters of diluted serum, leukocyte extract, or fibroblast extract can be incubated satisfactorily for 180, 120, and 300 min, respectively. We have found incubation intervals of 120, 90, and 240 min to be practical for serum, leukocytes, and fibroblasts, respectively.

each sample is calculated according to the following formula:

$$\text{Activity} = \frac{(\text{cpm}_{\text{sample}} - \text{cpm}_{\text{blank}}) \times V_2}{\text{cpm}_{\text{substrate}} \times V_1 \times t \times SA}$$

in which $\text{cpm}_{\text{sample}}$ and $\text{cpm}_{\text{blank}}$ are the counts per minute associated with the sample and blank tubes; $\text{cpm}_{\text{substrate}}$ is counts per minute per 15 μl of substrate (typically ~3000 cpm); V_1 and V_2 are volumes (milliliters) of the counted aliquot and the total stopped reaction solution, respectively; t is incubation time (minutes); and SA is the specific activity of the radioactive substrate (counts per minute per nanomole) added to each tube.

Physical Properties

Human serum biotinidase migrates as an α_1-protein on serum electrophoresis.[10,13,25] The p*I* of the enzyme is between 4.0 and 4.2, which reflects the fact that the enzyme is sialylated in serum.[13] After neuraminidase treatment the enzyme migrates to the β region on electrophoresis.[13] The asialylated enzyme has a p*I* of 5.1–5.2 on isoelectric focusing.[30] The enzyme is asialylated in homogenates of human liver and rat liver and kidney.[30]

Estimates of the molecular weight of biotinidase determined by gel filtration chromatography in different laboratories vary greatly. The molecular weight of the human plasma enzyme is about 76,000 on Sephacryl S-200 chromatography.[1] We obtained similar results for the human plasma enzyme.[16] The molecular weights of the hog liver and serum enzymes are 115,000 on Sephadex G-100 and Sephadex G-200.[11] The human serum enzyme is about 110,000 on both Sephadex G-100 and Sephacryl S-200.[17]

There is better agreement for the molecular weight of biotinidase determined by sodium dodecyl sulfate–polyacrylamide gel electrophoresis and sedimentation analysis. The enzyme appears to be a single polypeptide, with a molecular weight of about 66,000–76,000.[1,16,17]

Catalytic Properties

The acyl–enzyme appears to be an intermediate in the biotinidase reaction based on the formation of biotinyl hydroxamate when the enzyme is incubated in the presence of high concentrations of hydroxylamine with either BPABA or biocytin as the substrate.[11,15]

Human serum biotinidase has a K_m for BPABA of 10–34 μM[4,17] and a K_m for biocytin of 5–7.8 μM.[10,17] Human plasma biotinidase has a K_m for

[30] B. Wolf, unpublished data (1985).

BPABA of 10 μM and a K_m for biocytin of 6.2 μM.[1] Hog serum enzyme has a K_m for BPABA of 55 μM and a K_m for biocytin of 8 μM.[11] *Lactobacillus casei* and *S. faecalis* enzymes have similar K_m values.[9,15] The K_m values do not vary significantly over the range pH 5.0–7.0. The V_{max} of the human serum enzyme does change with pH.[17] The V_{max} for biocytin at acid pH is higher for biocytin than for BPABA, whereas the V_{max} for BPABA is higher at neutral pH. The V_{max} for each substrate is similar at pH 7.

Specificity

Biotinidase specifically cleaves *d*-biotinylamides and esters.[11,15] The carboxyl group of biotin is linked through an amide bond with the ε-amino group of a lysine in a highly conserved region of the carboxylases (-ala-met-biocytin-met-).[31] Biotinidase does not cleave biotin from intact holocarboxylases.[1,17] The rate of hydrolysis of biotin from natural and synthetic biotinylated peptides decreases with increasing size of the peptide, suggesting that biocytin or small biotinyl peptides are the primary substrates of biotinidase *in vivo*.[1]

N-d-BPABA is hydrolyzed by hog liver and serum biotinidase at pH 6.5 and by hog kidney biotinidase at pH 6.0 at the same rate as is biocytin.[11,15] *N-l*-BPABA is cleaved at 4–15% of the rate of the *d* isomer, and its K_m is 100-fold higher.[11,15] *d*-Biotin methyl ester is cleaved by the hog liver and serum enzymes at about half the rate of biocytin, whereas hog kidney biotinidase cleaves the methyl ester slightly more rapidly than biocytin.[11,15]

The specificity of hydrolysis of the biotinyl substrates by biotinidase lies in the ureido portion of the biotin molecule. Removal of the sulfur (to yield dethiobiotin) barely alters the rate of hydrolysis.[11,17] However, oxidation of the sulfur in sulfo-BPABA or shortening of the aliphatic side chain (e.g., in nor-BPABA), decreases the activity to 10 or 2% of that of BPABA, respectively.[11] The amide portion is also important. *N*-Biotinyl 3-aminobenzoate is cleaved at 10% of the rate of the *para* compound, whereas the *ortho* derivative is not cleaved.[17] Human serum biotinidase is specific for the lysine portion of the substrate and does not cleave compounds such as biotinylphenylalanine, biotinyl-ε-aminocaproic acid or biotinyl-γ-amino-*n*-butyric acid.[17] Biotinidase cannot cleave biocytin that is bound to avidin.[11]

Partially purified biotinidases from *S. faecalis* and *L. casei* have different specificities from those described above.[9,15] Both biotinamide and

[31] D. B. Rylatt, D. B. Keech, and J. C. Wallace, *Arch. Biochem. Biophys.* **183,** 113 (1977).

BPABA are cleaved at much greater rates than biocytin. Biotinidase from *S. faecalis* can cleave biotinyl-β-alanine and biocytin at approximately equal rates.

Inhibition

Biotin is a competitive inhibitor of biotinidase at acidic pH when either biocytin or BPABA is the substrate. Biotin is not inhibitory with BPABA as the substrate at pH 7.4,[11] and the K_i for biotin is increased markedly with biocytin as substrate at alkaline pH.[17] Biotinidase activity is not competitively inhibited by *dl*-dethiobiotin, *l*-biotin, *l*-lysine, ε-aminocaproic acid, fatty acids, *d*-norbiotin, *d*-biotin sulfone, or PABA.[11,17]

Sulfhydryl inhibitors inactivate biotinidase. Hog liver and serum biotinidases are completely inactivated by 0.1 m*M* *p*-chloromercuribenzoate,[11] and the enzymes in human plasma[1] and hog kidney[15] are totally inactivated by 10 μ*M* *p*-chloromercuribenzoate.[11] Monoiodoacetate also inhibits enzyme activity, whereas *N*-ethylmaleimide or arsenite does not inhibit.[11] Partially purified biotinidases from *L. casei* and *S. faecalis* are not inhibited by 0.1 m*M* *p*-chloromercuribenzoate.[9,15]

Evidence for the presence of a serine hydroxyl group in or near the active site of biotinidase is equivocal. Hog kidney biotinidase has only 15% of normal activity when treated with 1 μ*M* diisopropyl fluorophosphate,[15] and the activity of the human plasma enzyme is inhibited 99% by 0.1 μ*M* phenylmethylsulfonyl fluoride.[1] Phenylmethylsulfonyl fluoride does not inhibit human serum biotinidase at 1 m*M*.[17] Diisopropyl fluorophosphate at 1 m*M* does not completely inactivate the hog serum enzyme and inhibits hog liver enzyme only about 30%.[11] Partially purified biotinidase from *L. casei* is inhibited 50% by 0.4 m*M* diisopropyl fluorophosphate.[15]

The K_i for biotin using BPABA as substrate is 100 μ*M* at pH 5.0 and 1.3 m*M* at pH 6.5 for hog serum biotinidase.[11] The K_i for the hog liver enzyme is 40 μ*M* at pH 5.0 and 0.9 m*M* at pH 6.5.[11] The K_i for the human serum enzyme is 87 μ*M* at pH 5.5 and 1.3 m*M* at pH 7.[17] The K_i for biotin for human serum enzyme when biocytin is the substrate is 20 μ*M* at pH 5.5 and 0.5 m*M* at pH 7.[17] The lack of inhibition by biotin at pH 7.4 is thought to be due to slower dissociation of biotin from the enzyme. The slow turnover of product results in less available, inhibitable enzyme.[11,17]

The effect of pH on the inhibition of biotinidase by biotin appears not to be due to changes in the affinity of the enzyme for the substrate. The K_m values for BPABA and biocytin are virtually unchanged between pH 5.5 and 7.0.[11,17]

Stability

Biotinidase loses activity during freezing and thawing but is stable when stored at −70°.[1,17] After ammonium sulfate precipitation the human plasma biotinidase loses 67% of its initial activity in 9 months.[1] Dialyzing hog serum biotinidase at between pH 5.0 and 6.0 stabilizes the enzyme.[11] Buffers with a salt concentration of at least 0.1 *M* prevent loss of activity.[17] Human serum biotinidase can be stored at 4° in 0.1 *M* phosphate buffer (pH 6.0) containing 1 m*M* 2-mercaptoethanol and 1 m*M* EDTA for 1 month without loss of activity.[17]

The addition of mercaptoethanol (0.1–0.2 m*M*) or dithiothreitol (1–2 m*M*) appears to protect or stabilize the enzyme.[1,11,17,32] Heavy metals, such as copper and zinc, inhibit biotinidase.[11] Therefore, EDTA is included in storage buffers at 1–10 m*M* to prevent this inhibition.

Biotinidase from hog serum and from hog liver is stable both during the enzyme assay and during preincubation for up to 30 min at 37°.[11] Denaturation of the enzyme occurs after 30 min at 60°.[11] The human serum enzyme is denatured 60% after 15 min at 60° and 100% after 15 min at 70°.[17]

[32] K. Hayakawa and J. Oizumi, *Clin. Chim. Acta* **168,** 109 (1987).

[12] Monoclonal Antibody to Biotin

By KRISHNAMURTI DAKSHINAMURTI and EDWARD S. RECTOR

Introduction

The high affinity between the glycoprotein avidin and the vitamin biotin has been exploited in various areas of biological research, particularly in affinity cytochemistry.[1,2] The restrictions of this method have also been recognized.[1] Various cell lines, such as HeLa, baby hamster kidney, and human skin fibroblasts, have been shown to bind and internalize avidin.[3,4] In view of this we have explored the possibility of producing monoclonal antibodies to biotin for use in studies on the cellular functions

[1] E. A. Bayer and M. Wilchek, *Methods Biochem. Anal.* **26,** 1 (1980).
[2] J. Korpela, *J. Med. Biol.* **62,** 5 (1984).
[3] K. Dakshinamurti and L. E. Chalifour, *J. Cell. Physiol.* **107,** 427 (1981).
[4] L. E. Chalifour and K. Dakshinamurti, *Biochem. Biophys. Res. Commun.* **104,** 1047 (1982).

of biotin. Antibodies produced by traditional methods are heterogeneous and hence are not true analytical reagents. We report here the production and characterization of monoclonal antibodies to biotin.

Methods

Materials. Eagle's minimal essential medium (EMEM), fetal bovine serum, penicillin G, streptomycin, fungizone, hypoxanthin, and Freund's complete adjuvant were purchased from Grand Island Biological Company (St. Louis, MO). L-[4,5-^{3}H]Leucine (62 Ci/mmol) was purchased from Amersham Corp. (Oakville, ON). Keyhole limpet hemocyanin (KLH), biotin, carbodiimide, acetyl-CoA, *N*-hydroxysuccinimidobiotin, ATP, bovine serum albumin (BSA), and propionyl-CoA were obtained from Sigma Chemical Co. (St. Louis, MO). [^{3}H]Biotin (10 Ci/mmol) was a gift from Hoffman-LaRoche Inc. (Nutley, NJ). Affi-Gel protein A MAPS kit and reagents for polyacrylamide gel electrophoresis were obtained from Bio-Rad Laboratories (Mississauga, ON). $NaH^{14}CO_3$ (40–60 mCi/mmol) and Aquasol-2 scintillant were purchased from New England Nuclear Corp. (Montreal, QE). Horseradish peroxidase-conjugated affinity-purified goat antimouse IgG + IgM (H + L) was obtained from Kirkeguard & Perry Labs (Gaithersburg, MD). Pristane (1,6,10,14-tetramethylpentadecane) was purchased from Aldrich Chemical Co. (Milwaukee, WI). All other chemicals used were of reagent grade.

Antigen Preparation. Biotin is covalently attached to KLH. Activated biotin (*N*-hydroxysuccinimidobiotin, 10 mg) is added to 62 mg of KLH dissolved in 5.0 ml of 0.2 *M* borate buffer (pH 9.4). This is stirred for 6 hr at room temperature and dialyzed for 24 hr against four changes of 4 liters each of 0.9% (w/v) NaCl. The biotin–KLH is used as the antigen.

Biotin is covalently attached to BSA according to the procedure of Berger.[5] Biotin (500 mg) and 5 μCi of [^{3}H]biotin are suspended in 7.5 ml of 50% aqueous pyridine with constant stirring at room temperature. Carbodiimide (2.5 g) is dissolved in 12.5 ml of 50% aqueous pyridine, and this mixture is added dropwise to the biotin suspension. Stirring is continued for an additional 30 min at room temperature during which time a clear solution is formed. Albumin (250 mg), dissolved in 6.25 ml of water, is added dropwise to the biotin–carbodiimide reaction mixture and stirring is continued at room temperature for 4.5 hr. The mixture is dialyzed for 24 hr against four changes of 4 liters each of 0.9% NaCl. Determination of the radioactivity before and after dialysis indicated approximately 8 mol of biotin bound per mole of BSA.

[5] M. Berger, this series, Vol. 62, p. 319.

Immunization. Female BALB/c mice are immunized by two subcutaneous injections of 5 μg of biotin–KLH emulsified in Freund's complete adjuvant (0.1 ml of emulsion per injection per mouse) at a 1-month interval. At 3 weeks after the second injection and 3 days before fusion, the mice are injected intraperitoneally with 5 μg of biotin–KLH in 0.5 ml of phosphate-buffered saline (PBS). At the time of killing, serum is collected to provide a positive control in subsequent immunoassays.

Cell Fusion and Cloning Procedures. The cell fusion procedure using NS-I myeloma cells and polyethylene glycol as the fusing agent has been described in detail elsewhere.[6] Six individual fusions were performed, generating a total of 576 primary cultures. Supernatants are tested for the mouse antibiotin antibodies 11 days following fusion. Selected cultures are subsequently expanded and retested for antibody activity. Of 150 cultures we produced, 4 were selected for the derivation of monoclonal cell lines. Using the limiting dilution procedure, a total of 188 clones were obtained subsequently, of which 20 were selected for expansion and storage in liquid N_2. Four of these cell lines, each derived originally from a different fusion, were chosen for the production of the monoclonal antibodies reported in this chapter.

Enzyme-Linked Immunosorbent Assay (ELISA). The initial screening and selection of culture supernatants containing antibodies to biotin are accomplished by ELISA. Biotin–BSA (0.1 ml), at a concentration of 10 μg/ml in Na_2CO_3 buffer (pH 9.0), is used to coat the wells of 96-well flat-bottomed polystyrene microtiter plates (Costar, Cambridge, MA). After incubation at 4° overnight, the wells are washed once and incubated for 5 min with a solution of 1% (w/v) BSA in phosphate-buffered saline (PBS/BSA). The plates are then inverted and shaken; and immediately thereafter culture supernatants or control serum samples are placed into the wells (0.1 ml/well), and the plates are incubated at room temperature for 2 hr. Following removal of the samples and one wash with 20 mM imidazole-buffered saline (IBS) containing 0.02% (v/v) Tween 20 (IBS/Tween), peroxidase-conjugated affinity-purified goat antimouse IgG + IgM (H + L) diluted 1 : 100 in PBS/BSA is placed into the wells (0.1 ml/well). The plates are then incubated at room temperature for 2 hr with gentle shaking. The wells are washed 5 times with IBS/Tween, and substrate solution containing 2,2′-azinodi(3-ethylbenzthiazoline sulfonate) and H_2O_2 in cacodylate buffer is added (0.1 ml/well). The apparent absorbance at 405 nm generated in each well is determined after a final incubation for 10 min at room temperature. Antiserum obtained from

[6] E. Rector, T. Nakajima, C. Rocha, D. Duncan, D. Lestourgeon, R. S. Mitchell, J. Fischer, A. H. Sehon, and G. Delespesse, *Immunology* **55,** 481 (1985).

immunized mice provides a positive control for all assays. Normal mouse serum and tissue culture medium are used as negative controls.

Saturation Analysis by ELISA. Monoclonal antibody-containing supernatants from hybridoma cultures are mixed with 100 μl of varying concentrations of biotin–BSA (0–1000 μg/ml), biotin–KLH (0–1000 μg/ml), biotin (10^{-2}–10^{-10} *M*), and biocytin (10^{-2}–10^{-10} *M*). After incubation at 4° for 24 hr, the antibody–antigen mixtures are transferred into 96-well microtiter plates that had been previously coated with biotin–BSA. After incubation for 24 hr at 4°, unbound proteins are removed by three 2-min washes with IBS/Tween. The second antibody is added and the ELISA completed as described above.

Production and Purification of Monoclonal Antibodies. Female BALB/c mice are primed by intraperitoneal injection (0.5 ml/mouse) of pristane. After 2 weeks, monoclonal hybridoma cells (5×10^6) are injected intraperitoneally in 0.5 ml of culture medium. Ascites fluids are collected after approximately 10 days, placed in a tube containing heparin, and clarified by centrifugation.

Immunoglobulins from the ascites fluid are purified using the Affi-Gel protein A MAPS system according to the manufacturer's instructions. The Affi-Gel protein A column (0.5 × 10 cm) is equilibrated with 5 bed volumes of binding buffer. Ascites fluid (2 ml) is diluted with an equal volume of binding buffer and applied to the column. The column is washed with 10 volumes of binding buffer. The IgG is eluted by washing the column with 15 volumes of elution buffer. Protein-containing fractions are pooled and dialyzed against 10 m*M* Tris-HCl buffer (pH 7.5). Purity of the antibody is estimated by electrophoresis on 10% (w/v) sodium dodecyl sulfate (SDS)–polyacrylamide gels according to the method of Laemmli.[7]

Inhibition of Propionyl-CoA Carboxylase by Monoclonal Antibodies to Biotin. The ability of monoclonal antibodies raised against biotin to inhibit the activity of the biotin-containing enzyme propionyl-CoA carboxylase (EC 6.4.1.3) is investigated by using a 30% $(NH_4)_2SO_4$ fraction of rat liver cytosol.[8] A sample of rat liver cytosol is incubated for 15 min at 37° with ascites fluid containing monoclonal antibody to biotin. Propionyl-CoA carboxylase activity is then measured according to procedure of Lane and Halenz.[9]

One unit of enzyme activity is defined as that amount which catalyzes

[7] U. K. Laemmli, *Nature (London)* **227,** 680 (1970).

[8] P. M. Gillevet and K. Dakshinamurti, *Biosci. Rep.* **2,** 841 (1982).

[9] M. D. Lane and D. R. Halenz, this series, Vol. 5, p. 576.

the carboxylation of 1 μmol of substrate per minute at 37° into acid-stable product.

Protein Measurement. Protein content is determined according to the procedure of Lowry *et al.*[10] Bovine serum albumin is used to construct the standard curve.

Development of Monoclonal Antibodies

Fusion of spleen cells obtained from mice immunized with biotin–KLH with the murine myeloma cell line NS-1 have yielded a highly successful fusion. The supernatants from initial fusion cultures were all strongly positive for antibiotin antibodies, and of the crude cultures chosen for secondary screening all retained the ability to produce antibiotin antibodies. Limiting dilutions of crude cultures yielded approximately 50% of the possible number of single clones (188/384) and of these 75% were positive for antibiotin antibodies. Monoclonal antibody was obtained by ascites production in mice using four different clones (Nos. 28, 33, 38, and 142). One clone, No. 33, produces antibody of high affinity that binds both free and haptenic biotin antigens (Fig. 1A–C) as well as biocytin (Fig. 1D). Figure 1 shows that both biotin–BSA and biotin–KLH inhibit antibody binding to the solid phase at concentrations several orders of magnitude lower than that of free biotin.

The profiles of the binding of biotin–BSA and biotin–KLH are similar, and on that basis they should have an equal number of binding sites. It is not possible to calculate the molar ratios of binding between free biotin and biotin–KLH as can be done for free biotin and biotin–BSA, in view of the wide range (3×10^5–9×10^6) given for the molecular mass of hemocyanin. However, the fact that protein-bound biotin inhibits antibody binding to the solid phase more effectively, on a molar basis, than does free biotin should facilitate biological use of this antibody, since cellular and circulating biotin is mostly protein bound. Antibody produced by clone No. 33 is also highly effective at inhibiting the activity of the biotin-containing enzyme propionyl-CoA carboxylase (Table I). The other clones produce antibody of much lower affinity than that derived from clone 33 and are 1000-fold less effective at inhibiting propionyl-CoA carboxylase activity. The activity of acetyl-CoA carboxylase is inhibited in a fashion similar to that for propionyl-CoA carboxylase [4.25×10^3 milliunits (mU) of acetyl-CoA carboxylase is inhibited by 1 ml of ascites fluid

[10] O. H. Lowry, N. J. Rosebrough, A. L. Farr, and R. J. Randall, *J. Biol. Chem.* **193,** 265 (1951).

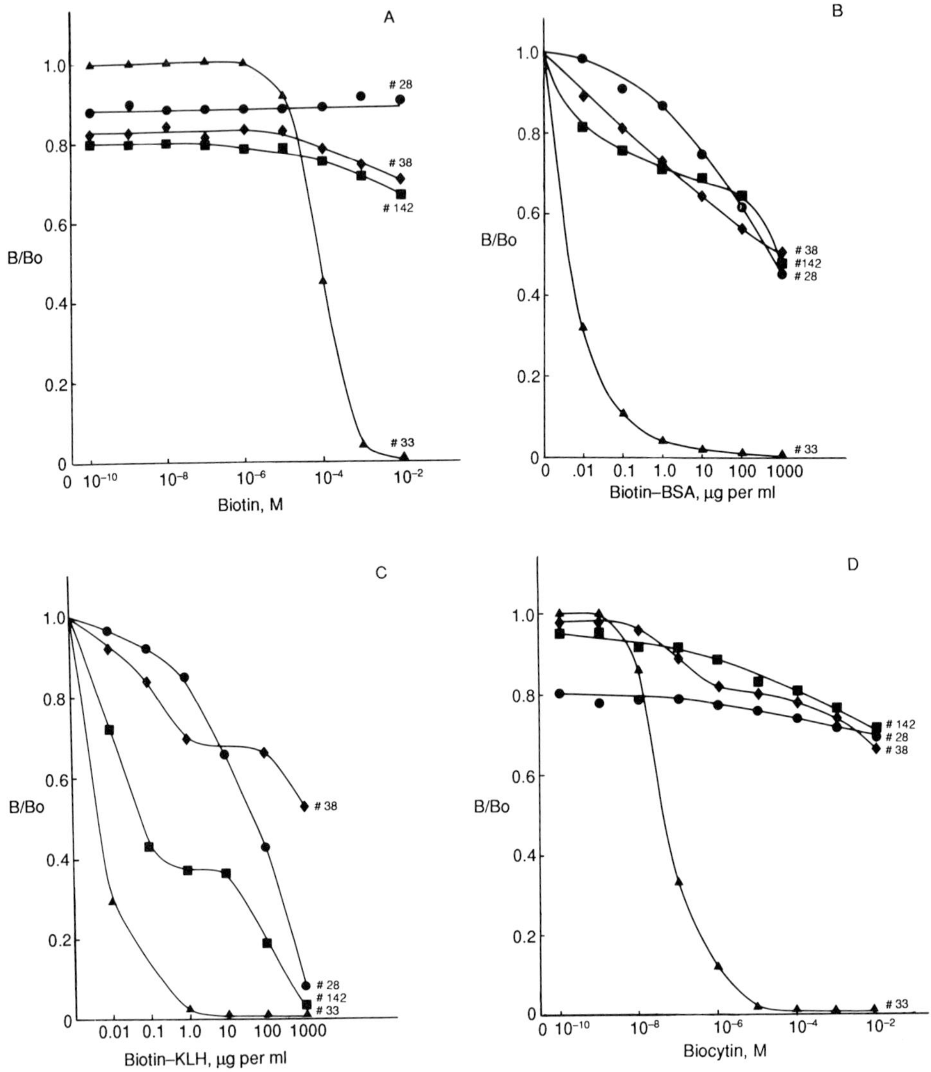

FIG. 1. Relative affinities of monoclonal antibiotin antibodies for biotin (A), biotin–BSA (B), biotin–KLH (C), and biocytin (D). The relative affinity of monoclonal antibodies (clone numbers indicated) for different biotin-containing antigens was determined by competition saturation analyses. The monoclonal antibodies were incubated overnight at 4° with varying concentrations of antigen prior to performing an ELISA with biotin–BSA as a plate-coating antigen. B/B_o represents the amount of monoclonal antibody bound to the plates in the presence of competing antigen (B) relative to the amount bound in the absence of competing antigen (B_o). [From K. Dakshinamurti, R. P. Bhullar, A. Scoot, E. S. Rector, G. Delespesse, and A. Sehon, *Biochem. J.* **237,** 477 (1986), with permission of the publishers.]

TABLE I
INHIBITION OF PROPIONYL-CoA CARBOXYLASE BY MONOCLONAL ANTIBODIES DIRECTED TOWARD BIOTIN[a]

Clone no.	Propionyl-CoA carboxylase (mU) inhibited by 1 ml of ascites fluid
28	14.6
33	15.2×10^3
38	10.4
142	14.9

[a] Rat liver propionyl-CoA carboxylase was incubated for 15 min at 37° with ascites fluid containing monoclonal antibody to biotin. At the end of the incubation, enzyme activity was measured as described in the text. From K. Dakshinamurti, R. P. Bhullar, A. Scott, E. S. Rector, G. Delespesse, and A. Sehon, *Biochem. J.* **237,** 477 (1986), with permission of the publishers.

from clone No. 33]. There is no inhibition of either carboxylase when they were incubated with ascites fluid which did not contain antibiotin antibody.

Purification of Antibody from Ascites Fluid

Monoclonal antibody is purified from ascites fluid produced from clone 33 using the Affi-Gel protein A MAPS system. This system results in a one-step purification of antibody from the ascites fluid. SDS–polyacrylamide gel electrophoretic analysis showed only two major protein bands, corresponding to the immunoglobulin heavy and light chains, thus confirming the high purity of the final antibody preparation. The purified antibody was effective at inhibiting both acetyl-CoA carboxylase and propionyl-CoA carboxylase activities [2.59 mU of acetyl-CoA carboxylase and 2.33 mU of propionyl-CoA carboxylase, respectively, inhibited per microliter of antibody (0.13 μg protein)].

Comments

Immune serum containing antibodies to the haptenic group biotin has previously been prepared by immunization of rabbits with biotin–BSA.[11]

[11] F. Ahmad, P. M. Ahmad, R. Dickstein, and E. Greenfield, *Biochem. J.* **197,** 95 (1981).

However, the multivalent specificity of these antibodies makes them a less attractive reagent in assessing the role of biotin in metabolism. This led us to develop a monoclonal antibody to biotin.

The monoclonal antibody can be used in assessing the role of the biotinyl group in biotin-containing enzymes. Acetyl-CoA carboxylase of mammalian origin exists in the cell as an inactive protomer, which is converted to the enzymically active form in the presence of citrate.[12] As the monoclonal antibody to biotin can be taken up by HeLa cells in culture, the antibody can be used to investigate the cellular protomer-to-polymer ratio of acetyl-CoA carboxylase under different nutritional and hormonal conditions. Assay *in vitro* for acetyl-CoA carboxylase has shown that the polymerized enzyme is not inactivated by the monoclonal antibody whereas it is inactivated when treated with antibody prior to polymerization.

The high affinity between biotin and avidin has been utilized in various areas of biological research. Of much significance is the use of the avidin–biotin complex for the localization and evaluation of cell surface receptors. In these studies, biotin is attached to cell surface functional groups through an appropriate reactive derivative. It has been assumed that coupling a small molecule such as biotin to the receptor would only slightly affect binding characteristics. In some instances this has been verified prior to the use of this technique. However, this is not always so. Thus, the addition of avidin to the medium of biotinylated lymphocytes has been used to mimic lectin-induced stimulation. Whereas the lectin-induced stimulation of lymphocytes is reversible, the stimulation observed with avidin was irreversible.[1]

There are other possible limitations in using this system. Before using the versatile avidin–biotin complex method, it must be determined whether the interaction in the experimental system under study comprises a biotin-containing, biotin-recognizing, or biotin-free system. Moreover, avidin is a basic glycoprotein[13]; the oligosaccharide moiety or its ion-exchange properties might thus contribute to side reactions. In this regard, it has been shown that avidin, perhaps mimicking a natural ligand, binds to isolated rat liver plasma membranes[14] and to a variety of cells studied.[3,4] This emphasizes the need for proper controls for applying the avidin–biotin complex technique to affinity cytochemical studies. The use of monoclonal antibody in these studies would obviate such problems.

[12] M. D. Lane, J. Moss, and S. E. Polakis, *Curr. Top. Cell. Regul.* **8,** 139 (1974).

[13] N. M. Green, *Adv. Protein Chem.* **29,** 85 (1975).

[14] L. E. Chalifour and K. Dakshinamurti, *Biochem. J.* **210,** 121 (1983).

Biotin transport as well as the nonprosthetic group function of biotin could be studied by using the specific antibody. Antibiotin antibody could also be used in isolation and purification procedures where the affinity between biotin and avidin proves to be too strong to be useful.

Acknowledgment

This work was supported by grants from the Medical Research Council of Canada.

Section III

General Methodology

A. Preparation of Biotin Derivatives
Articles 13 through 17

B. Preparation of Avidin and Streptavidin Derivatives
Articles 18 through 21

C. Assays for Avidin and Biotin
Articles 22 through 25

[13] Biotin-Containing Reagents

By MEIR WILCHEK and EDWARD A. BAYER

In order to increase the versatility and utility of the avidin–biotin system, a variety of reagents that can label different functional groups on proteins, saccharides, nucleic acids, and other biologically active compounds are required. In addition to promoting the covalent attachment of the biotin moiety to the desired molecule, such reagents can be used for double labeling via biotinylation either of different sites (target, binder, and/or probe) or of different functional groups on the same site. To date, reagents that can label amino groups, thiols, imidazoles and phenols, carboxyls, and aldehydes (carbohydrates), general electrophilic reagents, and photoreactive reagents for indiscriminant labeling have been described in the literature.

Figure 1 provides some representative examples of the many reagents currently in use. Although the vast majority of past applications used the *N*-hydroxysuccinimide ester of biotin (BNHS) for routine biotinylation of proteins, the other reagents are slowly entering the literature for special, interesting, or general applications. In this context, we originally suggested in an earlier review[1] the use of disulfide- and tartrate-containing derivatives of biotin as cleavable reagents. Indeed, such biotinylating reagents have since been described in the literature, and some are commercially available today. In addition, a new class of reagents, namely, homo- and heterobifunctional, cleavable and noncleavable, biotin-containing cross-linking reagents, is currently being considered for the next generation of biotinylating reagents. Examples of such reagents include maleimido derivatives of either biocytin hydrazide or biocytin-*N*-hydroxysuccinimide ester for the vectorial cross-linking of thiol groups of one macromolecule with carbohydrate or amino groups of another (Fig. 2). Another such reagent, biotinylglutamic acid dihydrazide, can be used to link two oxidized glycoproteins. These reagents should eventually prove useful for the identification and retrieval of cross-linked proteins using avidin and its derivatives.

In this chapter, we describe the synthesis of many of the above-mentioned biotin-containing reagents, the original description and/or syntheses of which originated in our laboratory. Although it would have been appropriate to describe the synthesis of a given reagent in the chapter

[1] E. A. Bayer and M. Wilchek, *Methods Biochem. Anal.* **26,** 1 (1980).

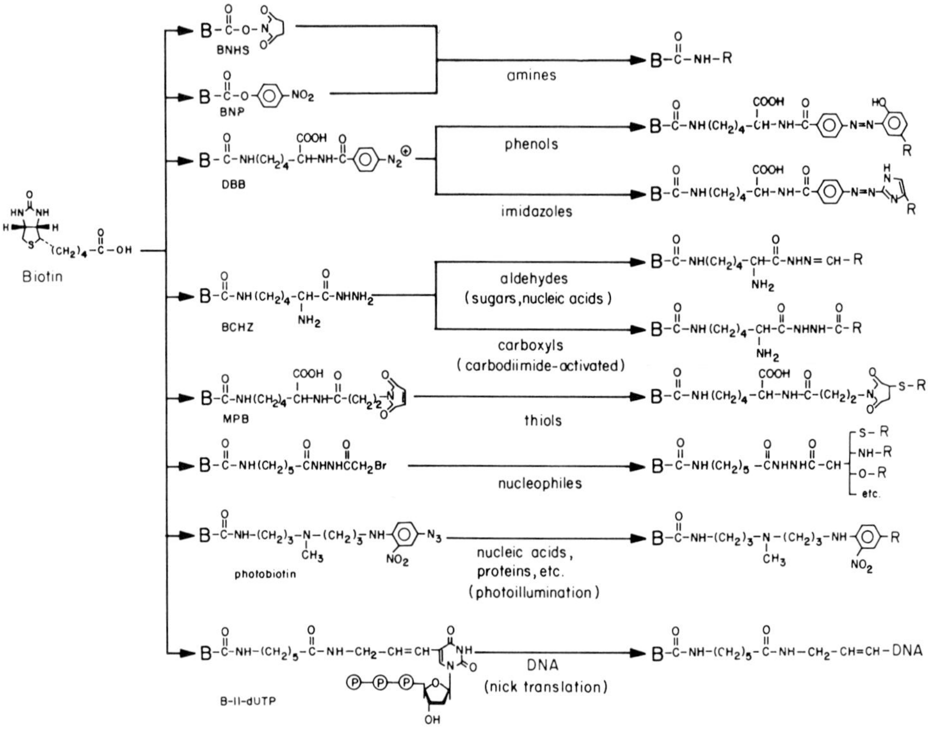

FIG. 1. Biotin and examples of some reactive derivatives. **B** represents the biotin moiety (without the carboxyl group); **R** represents a biologically active compound that possesses the designated functional group.

describing its application, we decided to present the syntheses of all the reagents together in order to emphasize the fact that they comprise one of the two major "partners" of the system.[2]

Short-Chain Biotinylating Reagents

It is not surprising that the first biotinylating reagents were simple derivatives of the biotin molecule. It may, however, be surprising to many researchers to learn that the original synthesis of reactive biotin derivatives was not intended for the currently accepted usage, i.e., a general means for the biotinylation of macromolecules. The synthesis of biotin

[2] The detailed use of these reagents for the biotinylation of proteins is described elsewhere in this volume (E. A. Bayer and M. Wilchek [14]).

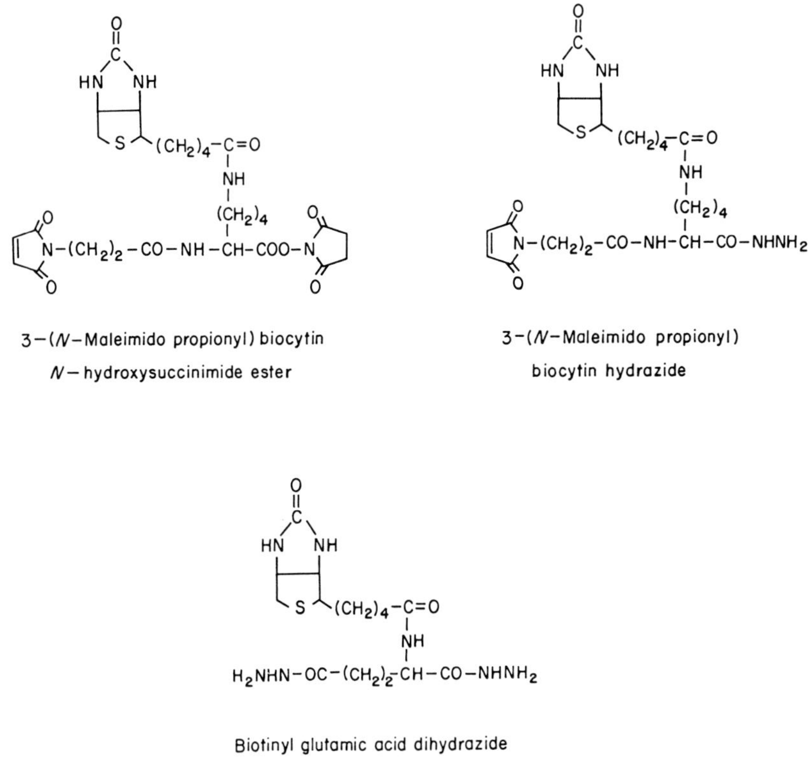

FIG. 2. Structures of some biotin-containing, bifunctional reagents referred to in the text.

hydrazide was reported as early as 1942[3] in one of the series of articles that led to the original determination of the structure of biotin. The *p*-nitrophenyl and *N*-hydroxysuccinimide esters of biotin (BNP and BNHS, respectively) were also synthesized several years before the era of avidin–biotin technology.[4] These reagents were originally designed as potential affinity labels for the site-directed inhibition of the biotin transport system in yeast; BNP successfully inhibited the transport of biotin in these cells, but BNHS did not. It is thus intriguing to note that BNHS, the basic biotinylating reagent for avidin–biotin technology, was originally introduced into the literature as a "negative control" for an entirely different purpose.

[3] K. Hofmann, D. B. Melville, and V. du Vigneaud, *J. Biol. Chem.* **144,** 513 (1942).

[4] J. M. Becker, M. Wilchek, and E. Katchalski, *Proc. Natl. Acad. Sci. U.S.A.* **68,** 2604 (1971).

Preparation of Biotinyl-N-hydroxysuccinimide Ester (BNHS)[5]

Reagents

Biotin (Sigma Chemical Co., St. Louis, MO), 10 g
N-Hydroxysuccinimide[6] (Sigma), 6 g
Dicyclohexylcarbodiimide (Fluka AG, Buchs, Switzerland), 8 g
Dimethylformamide
Distilled water
Ether
2-Propanol

Procedure. Dissolve the biotin in 100 ml of hot dimethylformamide, allow to cool to room temperature, and add the *N*-hydroxysuccinimide with stirring. In a separate flask, dissolve the dicyclohexylcarbodiimide in 20 ml of dimethylformamide and add to the other reactants. The suspension is stirred overnight at room temperature, and the dicyclohexylurea is filtered. The filtrate is evaporated *in vacuo* to a minimum volume, and the residue is precipitated with ether, filtered, and washed well with the same solvent. The product is washed further with 2-propanol. Yield, about 10 g (70%). TLC of the product on silica plates shows a single spot, R_f 0.7 [chloroform : methanol (1 : 2), using dimethylaminocinnamaldehyde spray],[7] compared with R_f 0.3 for biotin using the same system.

Preparation of Biotinyl-p-nitrophenyl Ester (BNP)

Reagents

Biotin (Sigma), 10 g
p-Nitrophenol (Sigma), 8 g
Dicyclohexylcarbodiimide (Fluka), 8.8 g
Dichloromethane
Ether
2-Propanol

Procedure. The biotin is suspended in 120 ml of dichloromethane at room temperature. To the suspension, the *p*-nitrophenol and dicyclohexylcarbodiimide are added successively. The flask is covered, and the suspension is stirred overnight at room temperature. The precipitate is

[5] In addition to being prepared by the procedure presented in this chapter, BNHS is now available from dozens of companies. Several companies also offer long-chain homologs of BNHS.

[6] The *N*-hydroxysulfosuccinimide ester and other active esters (e.g., *N*-hydroxyphthalimido and *p*-nitrophenyl esters) of the above compound may be prepared in a similar manner (with equimolar substitution of the required reactant).

[7] D. B. McCormick and J. A. Roth, this series, Vol. 18 [63].

filtered, and the filtrate is dried under reduced pressure. The yellow gummy residue is washed well with ether, and the product is recrystallized with 2-propanol. The crystals are washed with ether and dried over sulfuric acid in a desiccator. Yield, 8.8 g (60%); mp 156–158°.

Hydrolysis of the product in 0.2 *N* NaOH yields an equivalent of *p*-nitrophenol (determined by A_{420}). TLC on silica plates shows a single spot, R_f 0.45, chloroform : methanol (37 : 3), using dimethylaminocinnamaldehyde[7] or NaOH spray.

Preparation of Biotin Hydrazide (BHZ)

Reagents

Biotin (Sigma), 10 g
Hydrazine hydrate
Thionyl chloride (Fluka)
Methanol
Dimethylformamide

Procedure. To a chilled solution of methanol (100 ml), thionyl chloride (10 ml) is added. The biotin is then added, and the reaction is allowed to continue overnight at room temperature. The solvent is evaporated to dryness; methanol (100 ml) is again added and reevaporated. The residue is dissolved in 50 ml of methanol, and hydrazine hydrate (12 ml) is added. The reaction is allowed to proceed overnight at room temperature, and the product is filtered and washed with ether. The samples are recrystallized from dimethylformamide. A second crop can be obtained from the concentrated filtrate and recrystallized from dimethylformamide. Total yield, 8.2 g (78%).

Long-Chain Biotin-Containing Intermediates

In recent years, it has become apparent that, for steric reasons, the interaction between avidin and a biotinylated macromolecule can be dramatically improved by extending the spacer arm which links the biotin moiety to the surface of the biomolecule. Two derivatives of biotin have found widespread use as extended intermediates for further derivatization into reactive group-specific biotinylating reagents. These are biocytin and biotinyl-*N*-ε-aminocaproic acid. Biocytin (*N*-ε-biotinyl-L-lysine) is a naturally occurring breakdown product of biotin-requiring enzymes. Its synthesis was first described in this connection in 1952.[8] The synthesis of the

[8] D. E. Wolf, J. Valiant, R. L. Peck, and K. Folkers, *J. Am. Chem. Soc.* **74,** 2002 (1952).

aminocaproic acid derivative of biotin was also described before the advent of avidin–biotin technology.[9]

Preparation of Biocytin[10] *Hydrochloride*

Reagents

BNHS, 6.5 g
L-Lysine hydrochloride, 3.5 g
Cupric carbonate, 12 g
Dimethylformamide
Sodium bicarbonate, 400 mg
Distilled water
Ethanol
Ether
0.1 *N* HCl, 10 ml
Hydrogen sulfide

Procedure. L-Lysine hydrochloride is dissolved in 100 ml of distilled water, and cupric carbonate is added. The solution is boiled in a beaker for 5 min, and the precipitate is filtered. The blue filtrate is cooled in an ice bath, and BNHS (dissolved in 50 ml dimethylformamide) is added. Sodium bicarbonate crystals are introduced to the solution, and the reaction is allowed to continue for 4 hr over ice. The blue precipitate is centrifuged, and the supernatant fluids are decanted. The product is washed successively with water and ethanol and dried with ether. Yield, 7 g (90%).

The blue powder is dissolved in 0.1 *N* HCl, and H_2S is bubbled into the solution until no further precipitation occurs. The mixture is heated briefly in a hood until the CuS solid "coagulates." The precipitate is filtered, the residual H_2S is evaporated, and the resultant clear solution is lyophilized. Yield, 6 g (75%).

Amino acid analysis of the hydrolyzed product yields an equivalent of lysine. TLC on silica gel gives a single spot, R_f 0.2, butanol : acetic acid : water (4 : 1 : 1), with no visible contaminants using ninhydrin or dimethylaminocinnamaldehyde spray.[7]

Preparation of Biotinyl-N-ε-aminocaproic Acid

Reagents

BNHS, 10 g
ε-Aminocaproic acid (Merck, Darmstadt, FRG), 5 g in 60 ml of 0.1 *M* sodium bicarbonate

[9] E. A. Bayer, T. Viswanatha, and M. Wilchek, *FEBS Lett.* **60,** 309 (1975).
[10] Biocytin is now also available commercially from Sigma and Calbiochem.

Dimethylformamide, 60 ml
HCl, 1 and 0.1 *N* solutions
Ether

Procedure. Dissolve BNHS in hot dimethylformamide and cool to room temperature. Add dropwise to a bicarbonate solution of ε-aminocaproic acid with vigorous magnetic stirring, and continue stirring for 4 hr at room temperature. The solution is acidified with 1 *N* HCl; the precipitate is washed with cold 0.1 *N* HCl and dried with ether. The crystals are dried *in vacuo* over NaOH. Yield, 7.8 g (74%).

TLC on silica gel gives a single spot, R_f 0.5, chloroform : methanol : acetic acid (9 : 1 : 1), which is easily distinguished from biotin and the reactants (BNHS R_f 0.8, biotin R_f 0.7, ε-aminocaproic acid R_f 0.15). There are no visible contaminants using ninhydrin and biotin-specific[7] sprays.

Long-Chain Biotinylating Reagents

Owing to the improved interaction between avidin and biotinylated macromolecules which bear extended spacer arms, there has been a recent trend to use long-chain derivatives of biotin as reagents for incorporating the biotin moiety into the binder of choice. Of the two intermediates used for synthesis of the long-chain reagents, biocytin is preferred where applicable[11] since the product is generally more soluble in aqueous solutions because of the free amino or carboxyl group that remains after derivatization.

Preparation of Biotinyl-N-ε-aminocaproyl-N-hydroxysuccinimide Ester (BcapNHS)

Reagents

Biotinyl-*N*-ε-aminocaproic acid, 3.6 g
N-Hydroxysuccinimide[6] (Sigma), 1.4 g
Dicyclohexylcarbodiimide (Fluka), 2.3 g
N-Methylpyrrolidone
Ether

Procedure. Dissolve biotinyl-*N*-ε-aminocaproic acid in 70 ml of *N*-methylpyrrolidone.[12] Add the *N*-hydroxysuccinimide followed by the

[11] One should always keep in mind that the remaining group of the biocytin derivative is not necessarily "innocent." Thus, *N*-hydroxysuccinimide derivatives of biocytin cannot be synthesized, since the free amino group would react with the active ester. In another example, the p*K* of biocytin hydrazide is substantially higher than that of biocytin; this may, in some cases, affect its performance as a biotinylating reagent.

[12] *N*-Methylpyrrolidone is more suitable than dimethylformamide as a solvent in this reaction. In order to facilitate the solubilization, warm the solvent to 80°, then cool immediately to room temperature.

dicyclohexylcarbodiimide, and stir gently overnight at room temperature. After filtering the dicyclohexylurea which forms, the product is precipitated by ether (~500 ml) and collected on a sintered glass funnel. Rinse thoroughly with ether and dry in a vacuum desiccator over phosphorus pentoxide. Yield, 3.2 g (70%). TLC on silica gel gives a single spot, R_f 0.3, chloroform : methanol : acetic acid (17 : 2 : 1).

Preparation of Biotinyl-N-ε-aminocaproyl-p-nitrophenyl Ester (BcapNP)

Reagents

Biotinyl-*N*-ε-aminocaproic acid, 1 g
p-Nitrophenol (Fluka), 0.6 g
Dicyclohexylcarbodiimide (Fluka), 0.62 g
N-Methylpyrrolidone
Ether

Procedure. Dissolve biotinyl-*N*-ε-aminocaproic acid in 20 ml *N*-methylpyrrolidone.[12] Add the *p*-nitrophenol followed by the dicyclohexylcarbodiimide and stir gently overnight at room temperature. Filter off the dicyclohexylurea which forms. The product is precipitated by ether and collected on a sintered glass funnel. Rinse thoroughly with ether and dry in a vacuum desiccator.

Preparation of Biocytin Hydrazide (BCHZ)

Reagents

Biocytin, 1 g
Hydrazine hydrate (Merck)
Thionyl chloride (Fluka)
Methanol
Ether
Dimethylformamide
Distilled water
Porapak Type Q column (Waters Associates Inc., Milford, MA), 10 × 1 cm

Procedure. Thionyl chloride (2.5 ml) is added to a solution of methanol (25 ml) cooled in an ice water–acetone bath. Biocytin (1 g) is then added, and the mixture is allowed to react overnight with constant stirring. The solvent is reduced to a minimal volume, and ether is added. The precipitate is filtered, washed with ether, and dried over NaOH under reduced pressure. TLC of the product (biocytin methyl ester) on silica gel gives one spot [R_f 0.48, butanol : acetic acid : water (4 : 1 : 1)] with no visible contaminants using dimethylaminocinnamaldehyde[7] and ninhydrin sprays.

The methyl ester of biocytin (1 g) is dissolved in 20 ml of methanol, and hydrazine hydrate (1 ml) is added. After 48 hr at room temperature, the solvent is concentrated to dryness; drying is continued further in a desiccator under reduced pressure in the presence of H_2SO_4 until the odor of hydrazine can no longer be detected.

The product is dissolved in water (10 ml) and passed through a column containing Porapak Type Q. The column is eluted with water, and 10-ml fractions are collected.[13] Fractions containing biocytin hydrazide are detected on TLC (R_f 0.11, using methanol as the solvent system) by dimethylaminocinnamaldehyde[7] and ninhydrin sprays.[14]

The fractions containing the product are pooled, concentrated to dryness, dissolved in a minimal amount of methanol, and precipitated with ether. Biocytin hydrazide can be crystallized from hot ethanol. On reaction with acetone, three reaction products are obtained (TLC, using methanol as solvent) with R_f values of 0.24, 0.55, and 0.65, representing the anticipated reaction products.

We have also prepared the same compound from *t*-Boc biocytin[15] and *tert*-butyl carbazate in the presence of dicyclohexylcarbodiimide in a dimethylformamide solution. The protecting groups are removed with 4 *N* HCl in dioxane.

Preparation of 3-(N-Maleimidopropionyl)biocytin (MPB)[16]

Reagents

- Biocytin hydrochloride, 1 g in 10 ml of 0.5 *M* sodium bicarbonate
- 3-(*N*-Maleimidopropionyl)-*N*-hydroxysuccinimide ester (Calbiochem, San Diego, CA), 0.7 g
- Dimethylformamide
- 1 *N* HCl
- Ether

Procedure. 3-(*N*-Maleimidopropionyl)-*N*-hydroxysuccinimide ester is dissolved in dimethylformamide (10 ml) and added dropwise to the biocy-

[13] The first several fractions usually contain residual hydrazine hydrochloride. It is essential to remove all of the latter since hydrazine reacts well with aldehydes and, as an impurity, would reduce the final efficiency of the hydrazide reagent.

[14] To hasten elution of the hydrazide, the column is eluted with 80% aqueous methanol following the hydrazine peak. The fractions are then eluted as a sharp peak.

[15] E. Bayer and M. Wilchek, this series, Vol. 34 [20].

[16] MPB and similar maleimido analogs of biocytin are now available from Sigma and Calbiochem.

tin solution. The reaction is carried out for 2 hr, after which the solution is acidified with 1 *N* HCl. The product is allowed to crystallize at 4°. The crystals are washed with cold 0.1 *N* HCl and then dried with ether. Yield, 0.8 g (60%).

MPB appears on TLC as a single spot, R_f 0.3, chloroform : methanol (1 : 1). The product can be detected by dimethylaminocinnamaldehyde,[7] iodine vapors, or by burning the TLC plate. In order to test the reactivity of the spot to sulfhydryls, a 5-fold molar excess of dithiothreitol is interacted with MPB. The free sulfhydryl-containing derivative, MPB-SH, is formed. Unreacted dithiothreitol (R_f 0.9) is separated from the sulfhydryl adduct (R_f 0.35) using the above TLC system. Both dithiothreitol and MPB-SH are detected using 5,5′-dithiobis(2-nitrobenzoic acid) spray (0.1% in ethanol : 0.1% $NaHCO_3$, 1 : 1, v/v). Only the biotin-containing derivative is visible using dimethylaminocinnamaldehyde spray.[7]

Preparation of Other Maleimidobiocytin Derivatives

Other maleimido derivatives of biocytin can be prepared similarly, using commercially available maleimido aliphatic or aromatic *N*-hydroxysuccinimide esters. The aromatic derivatives are less soluble and hence less effective. 4-(*N*-Maleimidobutyryl)-*N*-hydroxysuccinimide ester and 6-(*N*-maleimidocaproyl)-*N*-hydroxysuccinimide ester are commercially available (Calbiochem, Sigma). The biocytin derivatives prepared from these compounds behave similarly to MPB.

Preparation of p-Aminobenzoylbiocytin

N-t-Boc-p-aminobenzoic Acid

Reagents

p-Aminobenzoic acid (Fluka), 13.7 g
BOC-ON (Fluka), 20 g
1 *N* NaOH
Dioxane
Petroleum ether
Distilled water
10% Citric acid

Procedure. *p*-Aminobenzoic acid is dissolved in 1 *N* NaOH (120 ml), and BOC-ON (dissolved in 50 ml dioxane) is added. The reaction mixture is stirred vigorously overnight. Water is added, and the solution is extracted with petroleum ether to remove the dioxane. The aqueous layer is acidified with citric acid, and the precipitate is collected, washed with water, and air dried. Yield, 15 g (65%); mp 185°. TLC on silica plates, ether : petroleum ether (17 : 3), yields a single spot (R_f 0.44) which could be distinguished from the starting material (R_f 0.40) by its insensitivity to ninhydrin staining.

t-Boc-p-aminobenzoyl-N-hydroxysuccinimide Ester

Reagents

N-*t*-Boc-*p*-aminobenzoic acid (6 g)
N-Hydroxysuccinimide (Sigma), 3 g
Dicyclohexylcarbodiimide (Fluka), 5.5 g
Dioxane
Ether

Procedure. To the *N*-hydroxysuccinimide in dioxane (100 ml), solid *N*-*t*-Boc-*p*-aminobenzoic acid is added. The dicyclohexylcarbodiimide (dissolved in 20 ml dioxane) is added, and the solution is stirred for 1 hr. The filtrate is concentrated to minimal volume, and the product is crystallized from ether. Yield, 6.5 g (78%); mp 165°. TLC, ether : petroleum ether (19 : 1), gives a single spot with R_f 0.47 compared to 0.57 for the starting material.

t-Boc-p-aminobenzoylbiocytin

Reagents

Biocytin hydrochloride, 1 g, dissolved in 11 ml of 0.5 *N* NaOH
t-Boc-*p*-aminobenzoyl-*N*-hydroxysuccinimide ester, 0.85 g
Dimethylformamide
Ether
Petroleum ether
Distilled water
10% Citric acid

Procedure. *t*-Boc-*p*-aminobenzoyl-*N*-hydroxysuccinimide ester is dissolved in 10 ml of dimethylformamide, and biocytin is added. Additional dimethylformamide is added if a precipitate is formed, and the reaction is carried out overnight at room temperature. Water is then added, and the reaction mixture is extracted successively with ether and petroleum ether in order to remove the dimethylformamide. The solution is acidified with citric acid, and the precipitate is collected by filtration. The precipitate is washed with ether to remove traces of the reactants. Yield, 1.2 g (85%); mp 171°. TLC, butanol : acetic acid : water (4 : 1 : 1), gives a single spot with an R_f of 0.6 compared to 0.14 for biocytin.

p-Aminobenzoylbiocytin (TFA salt)

Reagents

t-Boc-*p*-aminobenzoylbiocytin, 1.2 g
Trifluoroacetic acid (TFA)
Dichloromethane
Ether

Procedure. *t*-Boc-*p*-Aminobenzoylbiocytin is dissolved in 20 ml of a 1 : 1 mixture of TFA and dichloromethane and left at room temperature for 30 min. Ether is added, and the precipitate is filtered and washed with ether until dry. Yield, 1.05 g (88%). TLC, butanol : acetic acid : water (4 : 1 : 1), gives a single spot (R_f 0.4). On reaction with $NaNO_2$, the product reacts with tyrosine.

p-Diazobenzoylbiocytin (DBB)[17]

Reagents

p-Aminobenzoylbiocytin, 2 mg
2 *N* HCl
$NaNO_2$, 7.7 mg/ml (ice-cold aqueous solution)
1 *N* NaOH
0.1 *M* Borate buffer, pH 8.4

Procedure. *p*-Aminobenzoylbiocytin is dissolved in ice-cold 2 *N* HCl (40.7 μl). An equivalent volume of $NaNO_2$ is added, and the reaction is allowed to continue for 5 min at 4°. The reaction is terminated on addition of 1 *N* NaOH (35 μl). The DBB solution is brought to the desired concentration by dilution with borate buffer, and the solution is used immediately.

Biotin-Containing Cross-Linking Reagents

A new class of probes includes a series of trifunctional reagents in the general form

```
   B
   |
X—C—Y
```

where B includes the bicyclic ring system of biotin, C is a bifurcated spacer arm, and X and Y are reactive functional groups. The latter group-specific functions can either be homofunctional (e.g., two *N*-hydroxysuccinimide groups) or heterofunctional (e.g., a maleimide and hydrazide

[17] Diazobenzoyl derivatives are chemically unstable and cannot be stored for long periods of time. The preparation of DBB from the aminobenzoyl derivative is therefore performed immediately prior to use. The precursor is now available commercially from Calbiochem.

group). The spacer may be extended in any direction by addition of ω-amino acids, in order to facilitate vectorial interaction with target molecules or with the avidin-containing probe. The spacer either can be uncleavable or can theoretically be cleavable in one, two, or three directions; the number and site(s) of cleavage on a biotinylated cross-linking reagent would be determined by the ultimate goal of the researcher, i.e., whether cleavage of the biotin moiety is required to facilitate isolation (release from an avidin column) or whether cleavage of cross-linked macromolecules is desired for identification purposes (e.g., on two-dimensional SDS–PAGE). Suitable spacers may include mono-, di-, oligo-, and polysaccharides and their substituted derivatives. Alternatively, certain synthetic and naturally occurring amino acids (e.g., lysine, ornithine, aspartic and glutamic acids), dipeptides, and oligopeptides are also suitable as bifurcated spacers. Tartryl (periodate-cleavable) and disulfide (thiol-cleavable) derivatives can be used as cleavable extender arms. A wide variety of such reagents can be designed, some of which are presented in Fig. 3.

The bifurcated biotin-containing reagents can be used for preparing conjugates of two different protein species, such as a monoclonal antibody and an enzyme. Alternatively, these reagents can be used for combined cross-linkage and biotinylation of molecular components of a complex biological system, such as the cell membrane. The presence of the biotin moiety allows subsequent isolation, identification, or targeting of the cross-linked molecules of interest. Some examples in the use of these reagents are presented later in this volume.[2]

Preparation of 3-(N-Maleimidopropionyl)biocytin N-Hydroxysuccinimide Ester

Reagents

- MPB, 523 mg
- *N*-Hydroxysuccinimide,[6] 125 mg
- Dicyclohexylcarbodiimide, 225 mg
- Dimethylformamide
- Ether

Procedure. 3-(*N*-Maleimidopropionyl)biocytin is dissolved in 5 ml of dimethylformamide, and the *N*-hydroxysuccinimide is added. The solution is cooled to 4°, and dicyclohexylcarbodiimide is added. The reaction mixture is maintained for 24 hr at 4°. The dicyclohexylurea precipitate is filtered and discarded, and the product is precipitated directly from the filtrate with ether. The precipitate is collected, washed with ether, and the amount of *N*-hydroxysuccinimide ester is determined spectrophotometrically (ε_{261} 10,000 M^{-1} cm^{-1}) after hydrolysis with 0.5 M NH_4OH.

Functional groups for X and y	Spacers
$-O-N$ (N-hydroxysuccinimide)	**Noncleavable**
	$HOOC-(CH_2)_n-COOH$
	$H_2N-(CH_2)_n-NH_2$
$-O-\bigcirc-NO_2$	$H_2N-(CH_2)_n-COOH$
	$H_2N-(CH_2)_4-\overset{NH_2}{\underset{\vert}{CH}}-COOH$
$-\overset{O}{\overset{\Vert}{C}}-(CH_2)_n-N$ (maleimide)	**Cleavable**
	$HOOC-(CH_2)_2-S-S-(CH_2)_2-COOH$
$-NHNH_2$	$H_2N-(CH_2)_2-S-S-(CH_2)_2-NH_2$
$-\overset{O}{\overset{\Vert}{C}}-\bigcirc-N_2^+$ or $-NH-\bigcirc-N_2^+$	$HOOC-\overset{NH_2}{\underset{\vert}{CH}}-CH_2-S-S-CH_2-\overset{NH_2}{\underset{\vert}{CH}}-COOH$
	$HOOC-\overset{HO}{\underset{\vert}{CH}}-\overset{OH}{\underset{\vert}{CH}}-COOH$
$-\overset{O}{\overset{\Vert}{C}}-\bigcirc-N_3^+$ or $-NH-\bigcirc-N_3^+$	$HOOC-(CH_2)_n-HNOC-\overset{OH}{\underset{\vert}{CH}}-\overset{OH}{\underset{\vert}{CH}}-CONH-(CH_2)_n-COOH$

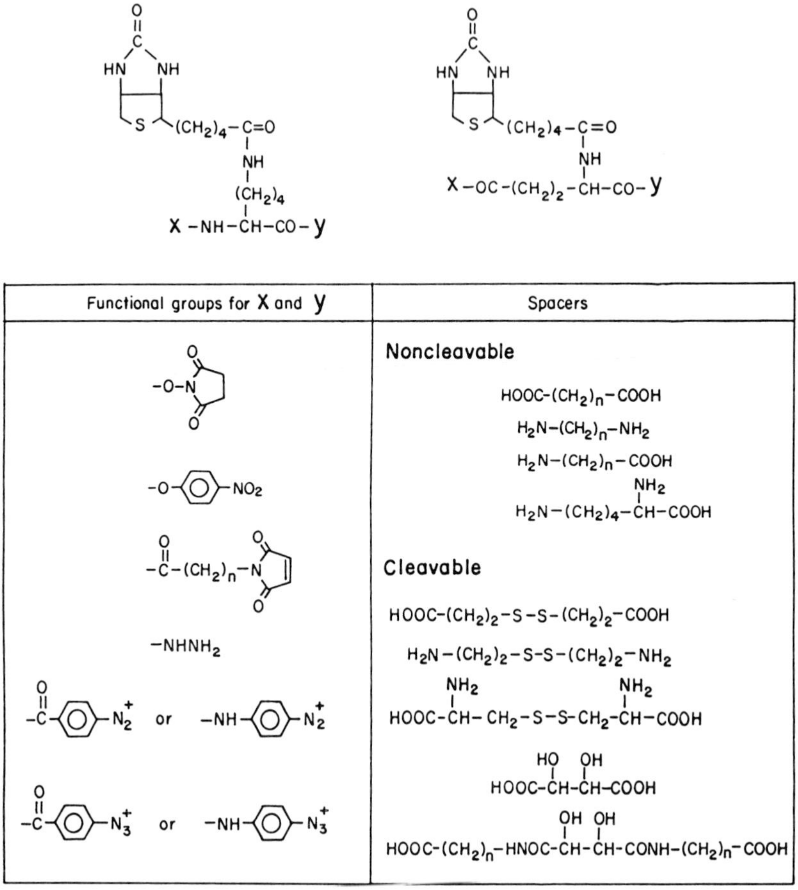

FIG. 3. Examples of biotin-containing cross-linking reagents. The examples shown are based either on biocytin, which provides both an amine and a carboxyl group for further derivatization with reactive groups X and Y, or on biotinyl glutamate, which provides two free carboxyl groups for the same purpose. The reactive functional groups X and Y can either be identical (in the case of homobifunctional reagents) or different (heterobifunctional reagents). In some cases, a given functional group is incompatible with another (e.g., a hydrazide derivative on one side and a p-nitrophenyl or N-hydroxysuccinimide group on the other), since the reagent would polymerize. The reactive groups and/or the biotin moiety can be extended further from the point of bifurcation by the incorporation of a suitable spacer group. If the latter is a cystinyl or tartryl derivative, the resultant reagent would be cleavable at the corresponding position.

Preparation of 3-(N-Maleimidopropionyl)biocytin Hydrazide

Reagents

3-(*N*-Maleimidopropionyl)biocytin *N*-hydroxysuccinimide ester, 125 mg in 1 ml of dimethylformamide
Hydrazine hydrate, 10 μl in 1 ml of 0.5 *M* sodium bicarbonate
Methanol
Ether

Procedure. The solution of 3-(*N*-maleimidopropionyl)biocytin *N*-hydroxysuccinimide is added dropwise to the hydrazine hydrate solution. The reaction is carried out for 3 hr at room temperature. The solution is concentrated to dryness, dissolved in a minimal amount of methanol, and the product is precipitated with ether. Purity is determined by TLC, and the product is identified by its reaction with acetone (and consequent change in R_f value).

Other hydrazide derivatives bearing long spacer arms are prepared by substituting 3 to 5 molar equivalents of the respective dihydrazide (e.g., 100–200 mg adipic acid dihydrazide). These derivatives can also be prepared directly from 3-(*N*-maleimidopropionyl)biocytin using either hydrazine or a desired dihydrazide derivative in the presence of dicyclohexylcarbodiimide.

Biotinyl-L-glutamic Acid

Reagents

BNHS, 3.4 g in 30 ml of dimethylformamide
L-Glutamic acid, 1.5 g in 30 ml of 0.5 *M* sodium bicarbonate
1 *N* HCl
Distilled water

Procedure. The BNHS and glutamic acid solutions are mixed, and the reaction is carried out for 2 hr at room temperature. The solution is acidified with HCl; the resultant precipitate is washed with cold water and dried *in vacuo*. Other aminodicarboxylic acids, peptides, etc., are prepared by substituting equimolar amounts for glutamic acid.

Biotinyl-L-glutamic Acid Dihydrazide

Reagents

Biotinyl-L-glutamic acid, 1 g in 10 ml of methanol
Thionyl chloride, 1 ml
Methanol
Sodium hydroxide

Hydrazine hydrate, 1 ml
2-Propanol

Procedure. The thionyl chloride and the methanolic solution of biotinyl-L-glutamic acid are cooled separately in an acetone–dry ice bath before mixing. The reaction is left at room temperature for 24 hr. The methanol is removed under vacuum; an equivalent volume of methanol is added and removed again by evaporation, and this step is repeated twice again. The residue is dried in a desiccator over sodium hydroxide (to remove traces of HCl). The residue is again dissolved in methanol, hydrazine hydrate is added, and the reaction is left at room temperature for 24 hr. A portion of the biotinylglutamic acid dihydrazide precipitates out and is filtered. The remainder is obtained by evaporation of the methanol and crystallization from 2-propanol.

[14] Protein Biotinylation

By EDWARD A. BAYER and MEIR WILCHEK

An essential prerequisite in the successful application of avidin–biotin technology is the incorporation of the biotin moiety into the experimental system. This may be accomplished by direct biotinylation of the target system using one of the many examples of group-specific biotin-containing reagents described elsewhere in this volume.[1] However, another alternative, one that contributes toward the virtually unlimited potential of the system, is the capacity to biotinylate biologically active molecules which maintain their recognition properties toward a given target molecule.[2-4] In this case, biotin may be introduced chemically via a variety of functional groups on the binder using the same group-specific reagents.

In this chapter, we present some of the procedures by which we have incorporated biotin moieties into proteins. Although in earlier works[4,5] we have professed that biotinylation of a macromolecule is relatively harmless to its activity and only minimally affects its physical and chemical properties, this is not necessarily true, and a given biotinylated binder should always be thoroughly characterized prior to use. We therefore

[1] M. Wilchek and E. A. Bayer, this volume [13].
[2] E. A. Bayer, M. Wilchek, and E. Skutelsky, *FEBS Lett.* **68,** 240 (1976).
[3] E. A. Bayer and M. Wilchek, *Trends Biochem. Sci.* **3,** N257 (1978).
[4] E. A. Bayer and M. Wilchek, *Methods Biochem. Anal.* **26,** 1 (1980).
[5] M. Wilchek and E. A. Bayer, *Immunol. Today* **5,** 39 (1984).

include in this chapter some of the methods that we use in our laboratory for examining the efficiency of the biotinylation reaction, the biological activity of the biotinylated binder, and its propensity to undergo nonspecific interactions. The examples detailed here should serve as guidelines for the biotinylation of other proteins and analysis thereof, especially for cases that have not yet been described in the literature.

Biotinylation of Proteins via Amino Groups

In our initial studies in which the concepts for biotinylating proteins were first instituted, we used biotinyl-*N*-hydroxysuccinimide ester (BNHS) for incorporating the biotin moiety via lysines into several bacteriophages, lectins, and antibodies.[2,6] BNHS is still the most popular biotinylating reagent and has been used almost exclusively for biotinylation of proteins. One of the reasons for its widespread usage is that lysine residues are numerous in most proteins and characteristically occupy an "exposed" position. Moreover, the lysine residues are usually not directly involved in the binding activity, and their modification generally has a nominal effect on the interaction of a protein with its substrate. The utility and efficiency of action of this reagent are underscored by the dozens of commercial enterprises that include BNHS in their catalogs.

Nevertheless, there has been a trend in recent years to replace BNHS with longer-chain homologs, e.g., biotinyl-ε-aminocaproyl-*N*-hydroxysuccinimide ester. Another trend has been to substitute water-soluble analogs (i.e., *N*-hydroxysulfosuccinimide derivatives of biotin) for the conventional biotinylating reagent.[7] In addition to BNHS, other active esters of biotin [e.g., biotinyl-*p*-nitrophenyl ester (BNP) and its homologs][8] are also efficient reagents for incorporating biotin into proteins via amino groups of lysine.

In the following analysis, we present a case study on the biotinylation of a simple binding protein, namely, the galactose-specific lectin peanut agglutinin. Simple procedures are described for determining the average number of biotin groups per protein (on a molar basis), the availability of

[6] J. M. Becker and M. Wilchek, *Biochim. Biophys. Acta* **264,** 165 (1972).

[7] The use of water-soluble derivatives are advantageous when working with proteins that are labile to the organic solvent (e.g., dimethylformamide and dimethyl sulfoxide) used to solubilize BNHS before adding to the protein solution. Such water-soluble reagents may, however, have some disadvantages. For example, if they are predissolved in aqueous solution, they may be subject to substantial levels of autohydrolysis before coming into contact with the protein. In addition, BNHS and its homologs can be stored at −20° for long periods of time in organic solvents.

[8] E. A. Bayer, T. Viswanatha, and M. Wilchek, *FEBS Lett.* **60,** 309 (1975).

biotin groups in a given preparation (the fraction of biotinylated protein that interacts with an avidin column), the relative binding activity of the biotinylated preparation, and the tendency of the preparation to interact nonspecifically with extraneous compounds. Such analyses are highly recommended when commencing work with a new biotinylated binding system. Moreover, such analyses should be performed periodically on stored (aliquoted) material, in order to verify its stability.

Biotinylation of Proteins Using BNHS

Biotinylation of Peanut Agglutinin: A Model Study

Reagents

Peanut agglutinin (PNA), 60 mg
BNHS,[1,9] 20 m*M* (6.8 mg/ml in dimethylformamide)
Sodium bicarbonate, 0.1 *M* containing 0.2 *M* NaCl[10]
Dimethylformamide
0.15 *M* NaCl
Phosphate-buffered saline (pH 7.4) (PBS)

Procedure. PNA is dissolved at a concentration of about 2.5 mg/ml in a 0.1 *M* $NaHCO_3$ solution containing 0.2 *M* NaCl. The solution is centrifuged (1500 *g*, 10 min) to remove solids that fail to dissolve. The supernatant fluids are collected, the optical density (280 nm) is measured, and the solution is brought to 20 μM (2.4 mg/ml, $E_{280}^{1\%} = 7.7$, MW = 120,000).

In order to determine the optimal conditions for incorporating biotin into the protein, preliminary experiments are performed using varying molar ratios of the reagent. Thus, aliquots (0, 8, 20, 40, 80, and 200 μl, representing reagent-to-protein molar ratios of 0, 2, 5, 10, 20, and 50, respectively) of the BNHS solution are added to 4-ml samples of the above-described PNA solution (80 nmol PNA per sample). The final level of dimethylformamide is brought to 5% in each sample.[11]

[9] Other active ester derivatives of biotin, including long-chain, water-soluble, and/or cleavable reagents, may be substituted for BNHS.

[10] Bicarbonate solutions are convenient and appropriate for BNHS-induced biotinylation. Other buffers (e.g., PBS) can also be used provided that extraneous amines are not present. Amine-containing buffers (such as Tris or glycine) are obviously unsuitable.

[11] For proteins that are sensitive to organic solvents, the level of dimethylformamide can be reduced in the sample by using a more concentrated BNHS stock (e.g., 50 mg/ml). At such high concentrations of BNHS, the dimethylformamide solution must be heated to approximately 60° in order to dissolve the reagent. For extreme solvent-sensitive proteins, the use of a water-soluble derivative may be required.

Each reaction mixture is allowed to stand for 2 hr at room temperature. The samples are dialyzed separately, first against 0.15 *M* NaCl and then against PBS. The samples are collected (in each case ~90% of the initial protein can be recovered), brought to 1 mg/ml, and stored at −20° in 1-ml aliquots.

Analysis of Biotinylated Preparation

Average Number of Biotin Groups per Protein Molecule. Several laboratories have reported the use of an avidin-based assay for detecting the extent of biotinylation. However, the biotinylated protein preparation should not be directly assayed for biotin, since the tetrameric nature of the avidin molecule and steric hindrance (the fixed orientation of the protein-bound biotin moieties) would interfere with the precision and accuracy of the assay. A more effective method for detecting the average number of biotin groups per protein molecule would be to first remove biotin from the protein and then to determine the amount of biotin in the sample. Unfortunately, biotinidase (the only known biotin-specific hydrolase)[12] is capable of removing biotin only from biocytin and small biotin-containing peptides; this enzyme is essentially inactive with respect to biotinylated proteins. Thus, until a better enzyme system is described, it is best to hydrolyze the protein extensively either by chemical means or through the action of a commercially available protease. An example of such a procedure is provided as follows.

A measured amount of biotinylated protein (e.g., 10 μg) is added to a solution (0.1 ml) containing 10 m*M* Tris-HCl buffer (pH 7.8), 5 m*M* EDTA, and 0.5% sodium dodecyl sulfate (SDS). The solution is boiled for 3 min, cooled to room temperature, and proteinase K is added to a final ratio of 1 : 50 versus the amount of biotinylated protein in the sample. In this case, a 10-μl aliquot from a stock solution of proteinase K (1 mg/ml in distilled water, stored at −20°) is diluted with 1 ml distilled water, and 10 μl is added to the biotinylated protein sample. The reaction is allowed to proceed at 37° for 3 hr, after which the solution is boiled for 10 min and assayed for biotin content.[13] As a control, a sample of the native (underivatized) protein is similarly treated. The molar ratio of biotin moieties per protein molecule is determined from the following equation:

$$(B/P)_{av} = \frac{0.0041 \text{ [amount of biotin in sample } (\mu g)]}{\text{amount of protein in sample } (\mu g)/\text{MW of protein}}$$

[12] K. Dakshinamurti and J. Chauhan, this volume [10]; B. Wolf, J. Hymes, and G. S. Heard, this volume [11].

[13] Biotin is determined by one of the assays described in this volume ([22]–[25]) and other volumes in this series (for list, see Table II, this volume [2]).

Thus, in our example for biotinylated PNA (MW 120,000), if the amount of biotin detected in a 10-μg protease-degraded sample is determined to be 130 ng, the average number of biotin moieties per protein molecule would be calculated to be 6.4. If the initial BNHS-to-PNA ratio is 10, the calculated value would also indicate the efficiency of the biotinylation reaction.

Another possibility that we and others have tried is to use a radioactive form of the biotinylating reagent for assessing the extent of biotinylation. A tritiated form of BNHS is available commercially from Amersham (Buckinghamshire, England). The radioactive reagent is usually diluted with varying amounts of "cold" BNHS; it should be pointed out that this may be a critical source of error, since the two may differ both in their purity and in their corresponding reactivity with the protein. Consequently, the results obtained using this approach may be inaccurate and misleading.

Availability of Biotin Moiety for Interaction with Avidin. The percentage of biotinylated PNA molecules that are accessible for binding avidin in a given sample is assessed by quantitative affinity chromatography on an avidin column. A measured amount (1 ml, A_{280} ~0.3)[14] of the biotinylated protein is applied to a small (0.5–1 ml) avidin–Sepharose or streptavidin–Sepharose affinity column (1.5–3 mg avidin/ml resin). Fractions of 1 ml are collected, and the column is washed with buffer (e.g., PBS) until the A_{280} reaches a minimal value. The total amount of protein applied to the column and the sum total of protein contained in the effluent fractions are determined spectrophotometrically, and the fraction (%) of biotinylated protein that is available for interaction with avidin is computed from the following equation:

$$\text{B–P/P}_{\text{total}} = \frac{[(A_{280})_{\text{applied}} - (A_{280})_{\text{effluent}}]}{(A_{280})_{\text{applied}}} \times 100$$

The above-described procedure can be modified for reduced format by using a radiolabeled form of the biotinylated protein and lesser amounts of affinity resin. In this case, one can determine the total amount of radioactivity introduced to the affinity resin, the total amount contained in the

[14] The actual amount added to the column should be determined according to the molar ratio of protein added to that of avidin on the column, the latter always being in excess. The four binding sites on the immobilized avidin combined with the multiple biotin moieties on the derivatized protein should be taken into account when calculating the amount of protein to be used. The amount of avidin–Sepharose used should be kept to a minimum, but the amount of protein applied should be high enough to generate a meaningful result. In practical terms, we usually keep the molar ratio of avidin to at least double that of the biotinylated protein.

effluent fractions, as well as the amount associated with the washed affinity resin. In an alternative to chromatography, applied and effluent fractions can be separated by repeated centrifugation using a microcentrifuge.

One of the drawbacks of this procedure for determining the efficiency of biotinylation is that the affinity resin can be used only once and cannot be regenerated. Nevertheless, the availability of the biotin moiety for interaction with avidin is one of the most important characteristics of the biotinylated binding protein; the above-described procedure is simple and effective.

Activity of Biotinylated Protein. The activity of the biotinylated PNA preparation relative to that of the native protein is determined by affinity chromatography on an appropriate column.[15] For this purpose, α-lactose–agarose (Sigma Chemical Co., St. Louis, MO) or an asialofetuin–agarose column can be used.[16] The procedure is similar to that described above for determining the efficiency of biotinylation using an avidin column. In this case, the agarose column can also be regenerated by releasing the bound PNA (biotinylated or native) using the competing sugar (0.2 *M* galactose or lactose). Quantification of the released protein also provides additional data concerning the specificity of the biotinylated molecule for its binding substrate. If the amount of biotinylated PNA released from the column is less than that of the native lectin, other nonaffinity interactions (hydrophobic, electrostatic, etc.) may also contribute to the binding.

Determination of Specificity. An easy and effective way to determine nonspecific interactions of the biotinylated lectin is to use dot blots.[17] A variety of target materials can be employed for this purpose. Since PNA recognizes the exposed penultimate galactose residues in desialylated sialoglycoconjugates, asialofetuin is an excellent positive control for this assay. Other purified proteins (fetuin, albumin, globulin, etc.), which are theoretically unreactive with the biotinylated probe, can be used as negative standards. Other purified macromolecules (polysaccharides, nucleic acids) and complex biological systems (cell extracts, solubilized membranes, etc.) can also serve as target material for determination of nonspecific binding.

Target material, including both negative and positive controls, are applied in serial dilutions to dot blots. The blots are treated with an

[15] The binding activity of other proteins can be determined similarly using appropriate columns (e.g., immobilized antigen for antibody and immobilized inhibitors for enzyme). Alternatively, ELISA or enzyme assays may be used where appropriate.

[16] Commercially obtained asialofetuin (Sigma) can be coupled to activated Sepharose. Alternatively, fetuin–agarose (Sigma) can be treated with neuraminidase in order to remove the terminal sialyl moieties, thereby exposing penultimate galactosyl groups.

[17] E. A. Bayer, H. Ben-Hur, and M. Wilchek, this volume [47].

appropriate quencher,[18] and the biotinylated binder is added.[19] The blots are then treated with an appropriate avidin-containing detection system and stained accordingly.

Negative controls should always include an inactivated or blocked biotinylated binder.[20] As a second type of negative control, one should substitute an unbiotinylated (native) binder in the initial reaction with the target material. Yet another negative control is to pretreat the avidin-containing system with biotin. Each type of negative control provides information concerning potential sources of nonspecific binding in the reaction protocol.[17]

A protocol for testing nonspecific binding of a biotinylated PNA preparation is now given in more detail to exemplify the logic behind this technique. The following purified proteins and membrane extracts are suggested targets that can serve as positive and negative controls for the galactose-binding PNA: asialofetuin (a terminal galactose-bearing glycoprotein, *positive control*), fetuin (galactose in penultimate position, blocked by terminal sialic acid, *negative control*), ovalbumin (high-mannose-containing glycoprotein, *negative control*), bovine serum albumin (BSA) (an example of a nonglycosylated, negatively charged protein, *negative control*), lysozyme (an example of a nonglycosylated, positively charged protein, *negative control*), neuraminidase-treated human erythrocyte ghosts (galactose-exposed complex model target system, *positive control*), and untreated human erythrocyte ghosts (galactose-blocked complex model target system, *negative control*).

The samples are applied in serial dilutions (1000, 300, 100, 30, 10, 3 μg/ml, etc.) to four dot blots. The blots are all treated with 2% BSA for 1 hr at room temperature and rinsed with PBS. To two of the blots, a solution of biotinylated PNA (3 μg/ml in 2% BSA) is added.[19] To one of the blots, the biotinylated PNA solution is added after being preheated to 80° for 10 min. To the fourth, 0.2 *M* lactose is included in the biotinylated PNA solution. The blots are incubated for a further 1 hr, rinsed several

[18] For discussion of appropriate quenching solutions, see R. C. Duhamel and J. S. Whitehead, this volume [21]. For lectins, it is advisable to use an appropriate nonglycosylated quencher (e.g., albumin, gelatin, lysozyme) or deglycosylated quencher (treated with a suitable glycosidase or otherwise modified at the competing sugar component). The system selected should be subjected to rigorous preexamination using appropriate positive and negative controls.

[19] Conditions and amounts of materials are determined in preliminary experiments. Several concentrations of biotinylated PNA can be compared in the final experiment.

[20] For determining the nonspecific binding of lectins, one type of negative control should include the competing sugar (e.g., 0.2 *M* lactose for PNA) in the initial incubation. Heat-inactivated binders (e.g., pretreatment of antibodies at 65°) should also be used as a negative control.

times with PBS, and treated with preformed complexes containing streptavidin and biotinylated alkaline phosphatase (final concentrations, 5 μg/ml and 2.5 units/ml in 2% BSA, respectively).[17] To one of the biotinyl-PNA-treated blots, streptavidin–alkaline phosphatase complexes, which have been pretreated with an excess of free biotin (1 μM final concentration), are added. After a 30-min incubation period at room temperature, the blots are rinsed well with PBS and treated with substrate solution [10 mg naphthol AS-MX phosphate dissolved in 200 μl of dimethylformamide is mixed with a solution containing 30 mg Fast Red dissolved in 100 ml of 0.1 M Tris-HCl (pH 8.4)].

Comments

Although it is certainly of academic interest to determine the average number of biotin residues per protein molecule [$(B/P)_{av}$], it is of greater practical relevance for avidin–biotin technology to determine the effectiveness of interaction of the biotinylated preparation with avidin ($B–P/P_{total}$). For this reason, we have developed the simple, quantitative model system, based on affinity chromatography on an avidin column as described above. Once the $B–P/P_{total}$ value is obtained for a given biotinylated sample, it is important to determine whether the biological activity and specificity of the biotinylated protein have been affected. For biotinylated binders (e.g., lectins and antibodies), a simple binding assay using an appropriate affinity column or enzyme-linked assay can be instituted. For other biologically active probes (e.g., hormones, toxins, and enzymes), activity levels can be determined using the appropriate biological assay. If possible, the activity associated with the biotinylated fraction should also be determined. In most cases, this can best be accomplished by determining the biological activity of the pooled, unbound (effluent) fractions that have passed through the avidin column. This value should then be subtracted from the total activity of the biotinylated preparation. In rare cases (e.g., with certain biotinylated enzymes), the biological activity of the immobilized fraction can be determined *in situ*.

Figure 1 shows the effect of biotinylation on the binding of PNA to avidin- and galactose-containing affinity columns. It can be seen that optimal results are achieved with an initial BNHS-to-PNA ratio of 10. At this ratio, binding to the avidin column reaches a plateau, indicating maximum availability of the biotin groups on the lectin; the biological activity is only minimally affected, as is evident from the near-total binding to the asialofetuin column.

Another essential characteristic of each biotinylated preparation is the ability to determine the degree of nonspecific binding. The dot-blot assay

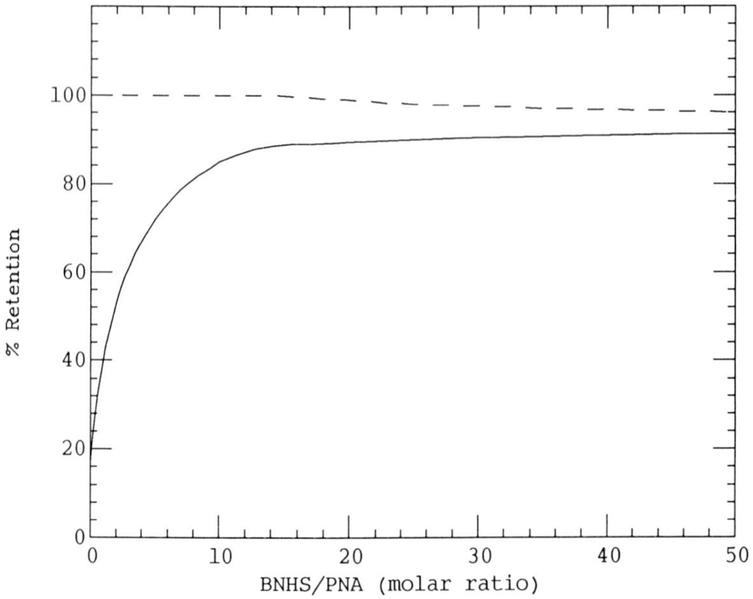

FIG. 1. Reactivity with avidin- and galactose-binding activity of biotinylated PNA samples. Samples, treated with different molar ratios of BNHS, were examined for interaction with avidin- (——) and galactose-containing (– – –) columns.

(using the appropriate battery of positive and negative controls as outlined above) provides a simple semiquantitative method for assessing this property. In the case of the biotinylated PNA preparation, the binding and specificity characteristics are maintained exquisitely at the preferred 10 : 1 BNHS-to-PNA ratio. Thus, in subsequent preparations, we have continued to use this ratio for preparing biotinylated PNA. When biotinylating a protein for the first time, we routinely examine the biotinylated sample for the three critical functions (i.e., interaction with avidin columns, biological activity, and nonspecific binding). If the latter tests suggest that the system is not optimal, then the initial ratio of biotinylating reagent to protein is adjusted accordingly.

Biotinylation of Purified Antibodies

In this section, we present the biotinylation protocol we routinely use for incorporating biotin into purified antibodies using BNHS. Using the following procedure, we have generally obtained highly active, highly selective antibodies that exhibit only low (if any) levels of nonspecific binding.

Reagents

BNHS, 2 mg/ml in dimethylformamide
Antibody, 2 mg protein in 1 ml of 0.1 *M* $NaHCO_3$[10]
0.15 *M* NaCl
Dialysis buffer: PBS or other buffer suitable for antibody

Typical Procedure. Add 25 μl of BNHS solution (2 mg/ml) to antibody solution (2 mg protein in 1 ml of 0.1 *M* $NaHCO_3$).[21] The solution is allowed to stand at room temperature for 1 hr without stirring. The solution is dialyzed overnight against 0.15 *M* NaCl with several changes. The last dialysis is performed against the desired buffer (e.g., PBS). The biotinylated antibody is stable and may be stored in any buffer under appropriate conditions (e.g., −20°).

Comments. The conditions presented in the above-described procedure have been shown to give optimal results for many antibody types and, as presented above, provide about 10 biotin moieties (average) per antibody molecule. Altering the relative concentration or ratio of reagent and/or antibody will alter the extent or efficiency of biotinylation. In general, we try to keep the concentration of antibody between 0.5 and 2.0 mg/ml, although we have biotinylated preparations of antibodies at protein concentrations as high as 50 mg/ml and as low as 50 μg/ml. In every case, we have kept the molar ratio of BNHS to antibody between 10 and 20. Much higher BNHS-to-antibody ratios have been reported in the literature, but we have generally experienced excellent results using the typical procedure as outlined above.

In any case, the three critical functions (i.e., interaction with avidin columns, biological activity, and nonspecific binding) should always be performed with any new biotinylated preparation. The availability of the biotin moiety on biotinylated antibodies can be checked on an avidin column precisely as described above for biotinylated PNA. The biological activity of the biotinylated antibody preparation is best examined in an ELISA system employing the adsorbed antigen in the immobile phase. Nonspecific binding is detected by dot blotting. In this case, it is preferable to use the purified antigen as the positive control, but complex systems (e.g., cell-associated antigen) can also be used. Appropriate negative controls should include purified nonantigenic proteins, and nonreacting related cell systems (in the case of complex positive controls). Other negative controls should include heat-inactivated (65°) biotinylated antibody preparation, substitution of the native (underivatized) antibody for

[21] The concentration and buffer are representative. Using appropriate ratios of BNHS, antibody concentrations of 10 mg/ml and higher can be biotinylated, and dilute solutions containing 10 μg/ml antibody and lower may also be biotinylated.

the biotinylated form in the incubation with the antigen, and use of a biotin-blocked avidin–enzyme conjugate in the second incubation step. The experiment is designed and carried out as described above for determining the nonspecific binding characteristics of PNA, but it is modified as mentioned for the antibody system.

Biotinylation of Whole Antiserum

In some cases, the binding protein does not have to be purified before the biotinylation step. Although all of the proteins in a crude mixture would be expected to undergo biotinylation, the "separation" would occur at a later stage, i.e., when the active biotinylated binding protein recognizes and attaches to the target molecule. The remainder of the biotinylated protein species would be removed during the washes. An effective biotinylation procedure for whole antiserum is presented below.

Reagents

Whole antiserum
BNHS, 34 mg/ml in dimethylformamide
0.1 *M* Sodium bicarbonate
0.15 *M* NaCl
PBS or other buffer suitable for antibody

Procedure. The antiserum (1 ml) is diluted to 20 ml with the bicarbonate solution, and 200 μl of the BNHS solution is added. After 1 hr at room temperature, the solution is dialyzed exhaustively against saline and then once against PBS. The dialyzed biotinylated protein is stored in aliquots at −20°.

Comments. The use of biotinylated antiserum or other crude extracts in avidin–biotin technology should always be subject to strict controls. One of these controls should employ a similarly biotinylated sample of preimmune serum. Another should consist of a biotinylated antiserum with a specificity different from that of the system under study. Other types of controls are similar to those discussed above in the section on purified antibodies.

Biotinylation of Enzymes

Unlike the biotinylation of antibodies, which can more or less be standardized because of the structural similarity of the molecules, the biotinylation of enzymes is dependent on the physical and chemical properties of the given enzyme species. Certain enzymes are exceptionally stable, and the biotinylation can be performed under relatively harsh conditions. Others are so sensitive that an effective biotinylation protocol (one that results in a biologically active biotinylated enzyme) has yet to be described. We present two procedures which result in biotinylated forms of two very useful enzymes for blotting and immunoassay studies, namely, bovine intestine alkaline phosphatase and horseradish peroxidase.

Biotinylation of Alkaline Phosphatase

Reagents

BNHS, 34 mg/ml in dimethylformamide (0.1 *M*)
Alkaline phosphatase [Sigma Type VII-NT, 5000 units (U), 3.66 mg in 0.43 ml][22]
1 *M* Sodium bicarbonate
0.15 *M* NaCl
PBS

Procedure. The enzyme is brought to 3.7 ml with distilled water, and 0.4 ml of the bicarbonate solution is added. An aliquot (9 μl) of the BNHS solution is added, and the reaction is carried out without stirring for 2 hr at room temperature. The solution is dialyzed exhaustively against 0.15 *M* NaCl and then once against PBS. The final concentration of enzyme is brought to 500 U/ml.

Comments. We have repeated this procedure dozens of times with essentially identical results. The biotinylated enzyme can be used to form soluble complexes with either avidin or streptavidin (or their derivatives), which can then be employed for the detection and quantification of a target molecule mediated by a biotinylated binder. The behavior of the biotinylated enzyme in the formation of such complexes reflects both the number and accessibility of biotin moieties as well as its remaining enzymatic activity. These qualities can be identified on dot blots using a suitable biotinylated target protein.[23]

[22] Sodium chloride solutions of alkaline phosphatase are preferred over ammonium sulfate suspensions or lyophilized powders since they can be directly biotinylated with BNHS (or other biotinylating reagents) without preliminary dialysis or solubilization of the enzyme.
[23] E. A. Bayer and M. Wilchek, this volume [18]. See pp. 181–184.

Biotinylation of Horseradish Peroxidase

Reagents

Horseradish peroxidase (RZ = 3.0, from STC Laboratories, Winnipeg, Manitoba, or from Sigma, Type VI), 11 mg
BNHS, 34 mg/ml in dimethylformamide (0.1 *M*)
0.1 *M* Sodium bicarbonate
0.15 *M* NaCl
PBS

Procedure. The enzyme is dissolved in 1.1 ml of sodium bicarbonate solution, and 55 μl of the BNHS solution is added. After 3 hr at room temperature, the biotinylated protein is dialyzed exhaustively against saline and once against PBS. The protein is stored in aliquots at −20°. Alternatively, the biotinylated peroxidase is sterilized by passage through a Millipore filter and stored under sterile conditions at 4°.

Comments. On storage at −20°, the biotinylated peroxidase is less stable than the biotinylated alkaline phosphatase. Following cryostorage for several years, elevated levels of biotinylated peroxidase were required for the formation of effective complexes with avidin. We have not explored other methods for storing the biotinylated preparation of this enzyme.

Alternative Biotinylation Procedures

For most applications, the biotinylation of proteins via the amino groups of lysines is the preferred approach, for reasons which have been outlined above. In recent years, however, we have developed other methods for biotinylating proteins via other functional groups.

New reagents that have been described in the literature have generally been based on the types or classes of biotin-containing reagents which were initially described in our early reviews.[3,4] The need for such reagents may arise in the event that a reactive lysine residue is essential either to the biological activity or to the structural stability of the protein of interest; biotinylation via the lysine residues in such cases would result in the inactivation of the molecule.

In other cases, the biotinylation via lysines may cause an undesirable alteration in the physical characteristics of the molecule. For example, the biotinylation of lysines neutralizes the charge and lowers the p*I* of the protein; this may be unacceptable for certain proteins (e.g., acidic proteins) and may cause autoprecipitation, elevated levels of nonspecific binding, or enhanced susceptibility to proteolytic action in complex cellular systems. Consequently, other group-specific, biotin-containing reagents have been synthesized as potential substitutes for the amine-specific reagents (i.e., BNHS, BNP, and their derivatives).

In the remainder of this chapter, we present a variety of alternative procedures for introducing the biotin moiety into proteins via different functional groups.

Biotinylation of Proteins Using DBB

One such alternative biotinylating reagent, *p*-diazobenzoylbiocytin (DBB),[1] is specific for tyrosyl and histidyl amino acid side chains. In the following analysis, we present a detailed account which demonstrates how a new biotinylation procedure is carried out. In this example, the biotinylation of alkaline phosphatase by DBB is presented.

Reagents

DBB,[24] in 0.1 *M* sodium borate buffer (pH 8.4), at various concentrations (e.g., 20, 50, 100, 200, and 600 μg/ml)
Alkaline phosphatase (Sigma Type VII-NT, 1 mg/ml)[22]
0.15 *M* NaCl or buffer (e.g., PBS)

Procedure. To samples containing 0.5 ml of protein solution is added an equivalent volume of the respective concentration of DBB.[25] The reaction is allowed to take place for 2 hr at room temperature, after which the reaction mixture is dialyzed extensively against 0.15 *M* NaCl or a suitable buffer. The volume of the dialyzed protein solution is adjusted to 2 ml, and the samples are stored at 4° (in the presence of 0.1% sodium azide).

Comments. The relative activities of the DBB-treated enzyme samples are determined by assaying in an appropriate substrate system in microtiter plates. Samples of DBB-labeled alkaline phosphatase are diluted to predetermined levels,[26] and their corresponding hydrolytic action on *p*-nitrophenyl phosphate is determined.[27]

It is interesting that the incorporation of biotin into alkaline phosphatase has a profound effect on its activity (Table I). At an initial ratio of 50

[24] M. Wilchek and E. A. Bayer, this volume [13]. See p. 134.

[25] In this example, the final concentrations of DBB are 10, 25, 50, 100, and 300 μg/ml and the final enzyme concentration is 0.5 mg/ml in all cases. The final molar ratios of DBB per enzyme molecule are approximately 5, 12, 24, 46, and 140. An untreated control (0.5 ml borate buffer is added instead of one of the DBB solutions) is included for comparison.

[26] Dilutions (e.g., 1 : 500, 1 : 1000, 1 : 2000, 1 : 5000, and 1 : 10,000) of the underivatized enzyme are examined for enzymatic activity using *p*-nitrophenyl phosphate. This provides a working range of enzyme concentrations that can then be used for examination of the DBB derivatives.

[27] The substrate solution for alkaline phosphatase consists of *p*-nitrophenyl phosphate (1 mg/ml) in 1 *M* diethanolamine (pH 9.8) containing 0.5 m*M* $MgCl_2$.

TABLE I
RELATIVE ACTIVITY OF DBB-LABELED ALKALINE PHOSPHATASE PREPARATIONS AND THEIR RESPECTIVE INTERACTION WITH AVIDIN–SEPHAROSE

Initial ratio (μg DBB per mg enzyme)	Activity (%)[a]	Bound/applied (%)	Unbound/applied (%)	Total activity (bound + unbound) (%)
0	100	0	26	26
20	153	20	46	66
50	173	49	43	92
100	107	80	44	124
200	72	140	15	155
600	23	152	17	169

[a] Percent activity versus underivatized control sample.

μg DBB per milligram enzyme (equivalent to about 14 mol DBB per mole of enzyme in the initial reaction), the activity of the resultant conjugate is about 60% *greater* than that of the underivatized substrate. As the ratio increases, there is a dramatic reduction of enzymatic activity.

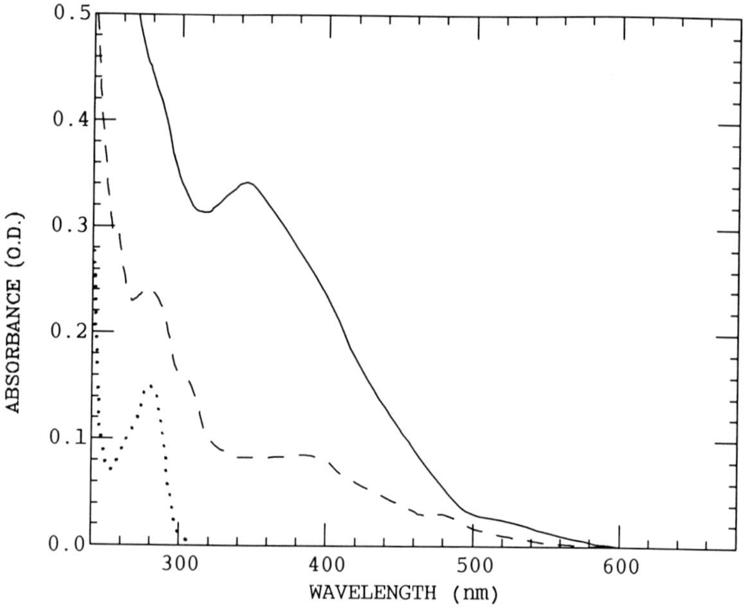

FIG. 2. Absorbance spectra of bovine serum albumin (BSA) and derivatives, showing equimolar concentrations of DBB–BSA (——), dithionite-treated DBB–BSA (– – –), and underivatized BSA (···).

As can be seen in Fig. 2, the average number of biotin moieties per enzyme molecule can be determined spectroscopically (ε_{340} 20,000 M^{-1} cm^{-1}). The DBB-labeled tyrosines are cleavable on reduction with dithionite; DBB-labeled histidines are stable under such conditions. The partial removal of the DBB label from a biotinylated protein is also shown.

In order to analyze further the effect of modification on the biotinylated enzyme, the interaction of the different DBB-modified samples with avidin–Sepharose is compared. Avidin–Sepharose is washed well with PBS, and a sample (50 μg in 100 μl) is added to an Eppendorf centrifuge tube that contains 0.9 ml of DBB-modified alkaline phosphatase (0.5 μg/ml). After intermittent shaking for 1 hr at room temperature, the suspension is centrifuged, and 0.9 ml of the supernatant is removed. The pellet is washed 3 times with 0.9 ml of PBS and resuspended to 1 ml. Aliquots (10 μl) of both the effluent and the resuspended column are mixed separately with 100 μl of substrate solution, and the corresponding activities are examined in triplicate using microtiter plates. The activity of the supernatant (unbound) and avidin–Sepharose-associated (bound) fractions is compared to that of the original (applied) sample.

The results (Table I) show that the underivatized enzyme is inhibited in the presence of avidin–Sepharose. The reason(s) for this inhibition is unknown, but it is clear that there is no enzyme activity associated with the column. Introduction of biotin (via DBB) to the enzyme results in the association of a substantial amount of the enzyme activity with the affinity resin. In fact, at high DBB-to-enzyme ratios, there appears to be an enhancement of enzymatic activity, since the amount of activity bound to the column exceeds that originally applied.

Biotinylation of Thiol-Containing Proteins Using MPB

Another functional group that is appropriate for attaching biotin moieties is the sulfhydryl group. Many proteins are considered SH proteins and in their native state bear free cysteines in the reduced form. These may be directly biotinylated through SH-specific reagents, such as 3-(*N*-maleimidopropionyl)biocytin (MPB) and its analogs. Other proteins contain cystines in the native state, such that two individual cysteines are covalently linked in the oxidized disulfide form. These can be biotinylated only after conversion of the disulfide to sulfhydryls with reducing agents (e.g., mercaptoethanol or borohydride). In cases where a protein lacks cysteines, the protein can be thiolylated by a variety of reagents [including *N*-acetylhomocysteine thiolactone,[28] 2-mercapto(*S*-acetyl)acetic acid

[28] F. H. White, Jr., this series, Vol. 25 [48].

N-hydroxysuccinimide ester,[29] and 2-iminothiolane].[30] The thiolylated protein can then be treated with a maleimido derivative of biotin.

In the following example, we present a simple procedure for the biotinylation of β-galactosidase, an enzyme known to contain high levels of free SH groups, most of which are not critical for biological activity.

Biotinylation of β-Galactosidase Using MPB

Reagents

MPB,[31] 26 μg/ml in PBS

β-Galactosidase (Sigma, Grade VIII), 1 mg dissolved in 0.1 ml 0.15 *M* NaCl

0.15 *M* NaCl or buffer (e.g., PBS)

Procedure. A 0.9-ml aliquot of MPB is added to the β-galactosidase solution, and the reaction is allowed to proceed for 2 hr at room temperature. The biotinylated enzyme is dialyzed exhaustively against PBS, sterilized by passage through a Millipore filter (HA 0.2 μm), and stored at 4°.

Comments. The properties of MPB-labeled β-galactosidase were found to be superior to those of the BNHS-derivatized enzyme.[32] The former retains about 75% of the hydrolytic activity compared to approximately 60% for the latter. Both enzymes are unstable in storage at −20°; however, the MPB-labeled enzyme is only partially inhibited whereas the BNHS-modified enzyme is totally inactivated by freezing. Underivatized β-galactosidase is stable in solution under these conditions. As a consequence, we recommend storage of the MPB-labeled enzyme at 4° under sterile conditions.

MPB-labeled β-galactosidase can be attached to an avidin–Sepharose column, and the immobilized enzyme is completely active (Fig. 3). The biotinylated enzyme is also suitable for use together with avidin or streptavidin as a detection system for either blotting or immunoassay.

Biotinylation of Proteins via Carboxyl Groups

The biotin moiety can also be attached to proteins via carboxyl groups of aspartic and glutamic acids. This can be accomplished using an amino- or hydrazide-containing derivative of biotin. For this purpose, we again recommend using biocytin hydrazide since it bears an extended spacer and is highly soluble in aqueous solutions.

[29] R. J. S. Duncan, P. D. Weston, and R. Wrigglesworth, *Anal. Biochem.* **132,** 68 (1983).

[30] H. J. Schramm and T. Dülffer, *Hoppe-Seyler's Z. Physiol. Chem.* **358,** 137 (1977).

[31] M. Wilchek and E. A. Bayer, this volume [13]. See p. 131.

[32] E. A. Bayer, M. G. Zalis, and M. Wilchek, *Anal. Biochem.* **149,** 529 (1985).

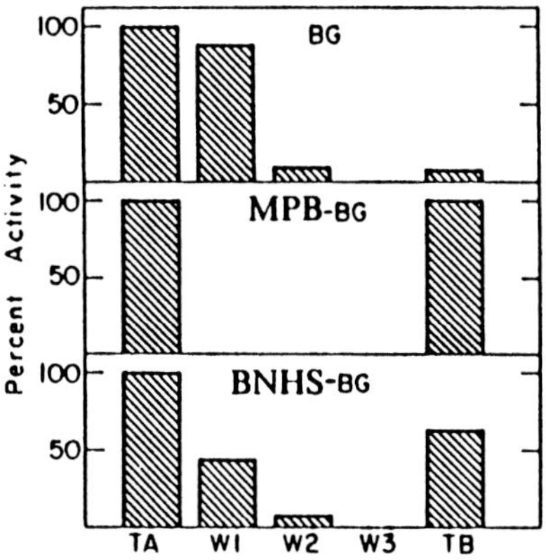

FIG. 3. Immobilization of biotinylated β-galactosidase preparations on avidin–Sepharose. The underivatized enzyme (BG), the MPB-derivatized enzyme (MPB–BG), and the BNHS-derivatized enzyme (BNHS–BG) were incubated in separate experiments with avidin–Sepharose. The total applied enzymatic activity (TA), the activity of the respective washes (W1, W2, and W3), and the total activity bound to the column (TB) were measured. Note the complete adsorption of MPB–BG compared with the partial association of BNHS–BG with the avidin-containing matrix. Only nominal attachment of the underivatized enzyme was observed.

Reagents

Protein sample, between 1 and 10 mg
Biocytin hydrazide,[1] 20 mg/ml in distilled water[33]; the solution is brought to pH 5 with 1 *N* HCl
Water-soluble carbodiimide (WSC)[34]
0.5 *N* HCl
0.15 *M* NaCl
Suitable buffer

Procedure. The protein is dissolved in the biocytin hydrazide solution to a final concentration of 1 mg/ml, and solid WSC is added (2 mg/mg protein). The reaction is carried out at room temperature, and during its course the solution is maintained at pH 5 by the periodic addition (every 30 min or so, when and if necessary) of 0.5 *N* HCl. After 6 hr, the reaction mixture is dialyzed, first against saline and then against an appropriate buffer in the final dialysis step. The concentration of the biotinylated protein is determined; its interaction with avidin, its biological activity,

[33] Solutions of biocytin hydrazide may be heated in a glass test tube over a flame in order to dissolve them completely.

[34] 1-Ethyl-3-(3-dimethylaminopropyl)carbodiimide hydrochloride (Sigma).

and its nonspecific-binding characteristics are assessed. The protein is stored by suitable means.

Comments. It is important to keep the amino derivative of biotin (biocytin hydrazide in this case) in great excess over the protein carboxyl groups. Biocytin, which contains a free carboxyl group, would be unsuitable for labeling acidic residues of proteins since it, too, would react with the WSC. The amount of WSC added to the reaction should be pretested for each protein in order to find the optimal value; too little would result in only partial biotinylation of the protein, and too much may inactivate the protein. It is also important to remember that biotinylation of carboxyl groups of proteins significantly alters their p*I*. In some cases, a highly alkaline protein derivative may result, which would adversely affect many properties, notably nonspecific binding.

Biotinylation of Glycoproteins Using Biocytin Hydrazide

Glycoproteins (and other glycoconjugates) may be biotinylated in several ways. In the following sections, we present both chemical and enzymatic methods that can be used to insert biotin via appropriate sugar residues. In both cases, the sugars are converted to aldehyde derivatives, and the latter are alkylated through an appropriate hydrazide. All procedures result in the alteration of the saccharide moieties of the macromolecule, and this must be considered when employing this mode of biotinylation. On the other hand, in contrast to BNHS derivatizations, the p*I* of the biotinylated protein is essentially unchanged.

Periodate Oxidation

Reagents

Glycoprotein, between 0.5 and 3 mg/ml in PBS
Sodium periodate, 0.1 *M* aqueous solution (freshly made)
Biocytin hydrazide,[1] 20 mg/ml in PBS[33]
Buffer (e.g., PBS)

Procedure. The glycoprotein solution (0.9 ml) is mixed with 0.1 ml of the periodate solution, and the reaction is allowed to proceed for 30 min at room temperature. The oxidized glycoprotein is either dialyzed for 4 hr against 2–4 liters PBS or passed over a Sephadex G-10 column (10 ml). An aliquot (0.1 ml) of biocytin hydrazide is added, and the solution is incubated for 2 hr. The reaction mixture is then dialyzed exhaustively against PBS (or any other suitable buffer) and stored at −20°.

Comments. Periodate oxidation works on vicinal hydroxyl groups of sugar residues. It is therefore assumed that the glycoconjugate of study indeed bears such residues.[35] In our experience, biocytin hydrazide (*N*-ε-biotinyl L-lysine hydrazide) is the aldehyde-specific reagent of choice. The lysine spacer provides improved interaction with avidin (over that of biotin hydrazide), and the extra amino group (compared with biotin ε-aminocaproic acid hydrazide) provides both increased solubility in aqueous solutions and enhanced interaction with aldehydes.

Enzyme-Mediated Oxidation

The following procedure is suitable for the mild introduction of biotin into glycoproteins which contain terminal galactosyl groups or for sialoglycoproteins which have penultimate galactose residues. In the latter case, the procedure is based on the removal of the terminal sialic acid, followed by the conversion of the exposed galactose to the 6-aldehydo derivative by galactose oxidase. The biotin moiety is incorporated by performing the reaction in the presence of biocytin hydrazide.

Reagents

Glycoprotein, between 0.5 and 3 mg/ml in PBS containing 1 m*M* $CaCl_2$ and 1 m*M* $MgCl_2$ (PBS–Ca,Mg)
Galactose oxidase (from *Dactylium dendroides*, Sigma), 100 U/ml in PBS–Ca,Mg, stored in aliquots at −20°
Neuraminidase (from *Vibrio cholerae*, Behringwerke AF, Marburg, FRG), 1 U/ml solution, stored under sterile conditions at 4°
Biocytin hydrazide,[1] 20 mg/ml in PBS[33]
0.15 *M* NaCl, containing 10 m*M* EDTA and 0.1% sodium azide

Procedure. To the glycoprotein solution (0.9 ml) are added 30 μl each of the neuraminidase,[36] galactose oxidase, and biocytin hydrazide stock solutions. The reaction is allowed to proceed at 37° for 2 hr, after which the reaction mixture is dialyzed exhaustively against 0.15 *M* NaCl containing 10 m*M* EDTA and 0.1% sodium azide (the purpose of the additives is to inhibit further enzyme action on the substrate). The biotinylated glycoprotein is stored in the same solution.

[35] Many oligosaccharides are substituted or linked in such a manner that there are no free vicinal hydroxyls available (e.g., chitin). In such cases, the sugars would not be oxidized to the respective aldehyde derivatives and biocytin hydrazide would therefore not be incorporated.

[36] For glycoproteins bearing exposed *terminal* galactosyl residues in the native state, neuraminidase can be omitted from the reaction mixture.

Comments. The above reaction can be performed using immobilized forms of neuraminidase and galactose oxidase.[37] The resin can then be removed from the reaction mixture by centrifugation.

Cross-Linking Experiments

Homo- and heterotri(biotinyl)functional reagents can be used to cross-link two different macromolecular species, and the biotinylated cross-linked macromolecules can be isolated, detected, quantified, etc., by using an appropriate avidin-containing probe. The cross-linking agents are suitable for a wide variety of target macromolecules or cell systems. The cross-linked protein pair can be interacting or neighboring components of a complex biological system (e.g., two macromolecular components of a cell membrane) or can be experimentally designed conjugates (e.g., antibody–antibody, antibody–enzyme, enzyme–enzyme, lectin–lectin, lectin–enzyme, or antibody–toxin). These conjugates would be useful in such processes as immunoassays, enzyme reactions, and site-directed studies, especially where consecutive reactions (i.e., cascades) are applicable.

Cross-Linking of Internal Erythrocyte Plasmalemmal Proteins

Inside-out resealed erythrocyte ghosts[38] (0.1 ml packed volume, washed and resuspended in 1 ml PBS) are treated with a solution (50 μg in 20 μl of dimethylformamide) of 3-(*N*-maleimidopropionyl)biocytin *N*-hydroxysuccinimide. After 1 hr at room temperature, the ghosts are washed with PBS, solubilized in a solution containing 1% sodium dodecyl sulfate (SDS), and analyzed by the desired means (e.g., SDS–PAGE and immunoblotting or isolation on avidin column).

Cross-Linking of Macromolecules on Intact Erthrocyte Membranes

Whole erthrocytes are treated with 1 m*M* sodium periodate for 30 min at 4°. The oxidized cells are then washed with PBS and treated with a 50 μg/ml solution (in PBS) of 3-(*N*-maleimidopropionyl)biocytin hydrazide. The cells are washed, packed, and lysed in 40 volumes of ice-cold lysis buffer [5 m*M* phosphate solution (pH 8.2), containing 2 m*M* phenylmethylsulfonyl fluoride]. The ghosts are washed with the same buffer,

[37] Both enzymes can be attached to appropriate matrices by a variety of means (see K. F. O'Driscoll, this series, Vol. 44 [12]; W. H. Scouten, Vol. 135 [2]; M. Wilchek, T. Miron, and J. Kohn, this series, Vol. 104 [1]).

[38] T. L. Steck and J. A. Kant, this series, Vol. 31 [16].

the membrane proteins are solubilized in 1% SDS, and the biotinylated cross-linked proteins are isolated on an avidin–Sepharose column. The pattern of isolated proteins (removed from the column by boiling the column in sample buffer) is analyzed by SDS–PAGE and compared to that of the original membrane extract.

Vectorial Cross-Linking and Biotinylation of Two Purified Proteins

One of the potentially useful applications of trifunctional reagents is the preparation of protein–protein conjugates with simultaneous inclusion of the biotin moiety into the conjugated species. In the following examples, we describe two different methods to combine an antibody molecule with a second protein. Both approaches are based on the lack of free sulfhydryl groups on the antibody molecule. In one approach, a purified antibody is reacted with a bifurcated, heterofunctional biotin derivative which contains maleimido and *N*-hydroxysuccinimide groups. In this case, the derivative is bound to the antibody molecule via the amino groups of lysine, and the maleimido group remains free for interaction with a thiol-containing protein. In the second approach, the sugar moieties of the antibody are oxidized, and the resultant aldehyde is reacted with a hydrazide-containing maleimidobiotin reagent. Again, the free maleimido group is available for interaction with a thiol-containing protein.

In addition to antibody–enzyme conjugates, other protein pairs (including enzyme–enzyme and antibody–toxin couples) can be prepared in a similar manner. The protocols provided below are, of course, only exemplary, and it is clear that this general approach will spawn a variety of new reagents and procedures for the simultaneous biotinylation and conjugation of any two protein species.

Preparation of Biotinylated Conjugate via Amino and Sulfhydryl Groups. Purified antibody (1.5 mg IgG in 1 ml of PBS) is mixed with 50 μl of a solution of 3-(*N*-maleimidopropionyl)biocytin *N*-hyroxysuccinimide (2 mg/ml in dimethylformamide). The reaction is allowed to proceed at room temperature for 1 hr, after which the labeled antibody is separated from the free reagent by dialysis or by gel filtration on Sephadex G-25. The maleimidobiotinyl IgG is mixed with an equimolar amount (10 nmol; i.e., 1 mg of an M_r 100,000 protein) of a native thiol-containing protein (e.g., β-galactosidase, A or B chain of ricin) or a thiolylated protein (e.g., *N*-acetylhomocysteine thiolactone-modified enzyme). After 1 hr at room temperature, the conjugate is separated from the unconjugated protein species by gel filtration on an appropriate matrix. Alternatively, isoelectric focusing, affinity chromatography, or any other separation technique may be employed in order to remove the unreacted protein species from the conjugated molecules.

Preparation of Biotinylated Conjugate via Carbohydrate and Sulfhydryl Groups. IgG (1 mg in 1 ml of PBS) is brought to 5 m*M* sodium periodate. After 30 min at 4°, the protein is dialyzed (or subjected to gel filtration on Sephadex G-25). 3-(*N*-Maleimidopropionyl)biocytin hydrazide (2 mg in 1 ml of PBS) is added, and the reaction is allowed to proceed for 1 hr at room temperature. The maleimido-biotinylated antibody[39] is dialyzed again in order to remove residual reagent, and an equimolar portion of an appropriate native thiol-containing or artificially thiolylated protein is introduced as described above. After 1 hr at room temperature, the conjugates are separated from the unconjugated protein species by appropriate means.

[39] The antibody molecules are now biotinylated via the oligosaccharide residues and the free maleimido group is now available for secondary interaction with a sulfhydryl protein.

[15] Enzymatic C-Terminal Biotinylation of Proteins

By ALEXANDER SCHWARZ, CHRISTIAN WANDREY, EDWARD A. BAYER, and MEIR WILCHEK

An effective method has yet to be reported for the selective incorporation of a single biotin molecule into proteins at a predetermined site. The biotin-containing labeling reagents described in [13] in this volume are all group specific; such residue-specific biotinylation would therefore be contingent on the presence of a single modifiable amino acid group in the desired target protein. Likewise, the selective modification of C- or N-terminal amino acids is also complicated by the presence of aspartic and glutamic acids and lysines in proteins.

One possible way of modifying a protein selectively would be through enzymatic mediation, either by transpeptidation of the C-terminal amino acid or by coupling a desired amino acid to the C terminus.[1-3] In applying this approach to avidin–biotin technology, biotinylation of the C-terminal amino acid of a target protein would not be expected to change drastically

[1] K. Rose, C. Herrero, A. E. I. Proudfoot, R. E. Offord, and C. J. A. Wallace, *Biochem. J.* **249,** 83 (1988).

[2] K. Morihara, Y. Keno, and Q. Sakima, *Biochem. J.* **240,** 803 (1986).

[3] K. Breddam, F. Widmer, and J. T. Johansen, *Carlsberg Res. Commun.* **46,** 121 (1981).

the biological activity or the three-dimensional structure of the protein, because only one biotin molecule per C terminus is attached. In this chapter, we describe a method for the selective C-terminal biotinylation of proteins using biocytinamide and carboxypeptidase Y (serine carboxypeptidase, EC 3.4.16.1), a nonspecific protease.

Reagents

Carboxypeptidase Y (Carlsberg Biotechnology, Ltd., Copenhagen, Denmark), protein content about 10% by weight
Biocytinamide, synthesized according to a modification of the procedure reported by Hofmann *et al.*[4]
Ethanol
Dimethylformamide
5 *N* NaOH
1 *M* Acetic acid
Distilled water
Sephadex G-50 fine (Pharmacia Fine Chemicals, Uppsala, Sweden), 2 × 20 cm

Procedure. Biocytinamide (100 mg/ml) is dissolved in a mixture of water : dimethylformamide : ethanol (2 : 1 : 1), and the pH is adjusted to 8.5 with NaOH. To this solution, the desired target protein (10 mg/ml) and the carboxypeptidase Y preparation (to a final protein concentration of 0.1 mg/ml) are added, and the reaction is carried out overnight with shaking at 4°. The reaction mixture is applied to a Sephadex G-50 column and eluted with acetic acid. The solution is dialyzed and concentrated by suitable means (lyophilization or ultrafiltration).

Comments. The proteins that we have successfully biotinylated to date include insulin, trypsin, myoglobin, cytochrome *c*, ribonuclease, and lysozyme. The incorporation of biotin into the target protein is demonstrated using dot blots stained with avidin-complexed alkaline phosphatase.[5,6] The extent of biotinylation (i.e., the percentage of target molecules modified with biotin) is determined using an avidin-containing affinity column.[5,7] The biotinylated protein is further characterized on native PAGE by a single band which migrates at the same position as the unmodified protein, indicating that the peptidase has not extensively

[4] K. Hofmann, F. M. Finn, and Y. Kiso, *J. Am. Chem. Soc.* **100,** 3585 (1978). The procedure is modified in the last step. N^{α}-Boc-biocytinamide is dissolved in glacial acetic acid, and 4 *N* HCl in dioxane is added. The solution is kept at room temperature for 30 min, and the product is precipitated with ether, washed with ether, and dried.

[5] E. A. Bayer and M. Wilchek, this volume [14].

[6] E. A. Bayer, H. Ben-Hur, and M. Wilchek, this volume [47].

[7] E. A. Bayer and M. Wilchek, this volume [18].

cleaved the protein. Blot transfer and staining of samples with a suitable avidin-containing probe show that only the modified protein is labeled.

One limitation of the above-described method is that proteins bearing a C-terminal proline are not susceptible to enzymatic modifications using carboxypeptidase Y. For example, in our experiments we were unable to couple biotin to the C terminus (proline) of ovalbumin.

In enzyme-mediated synthesis using carboxypeptidase Y, the use of a good nucleophile results in more efficient coupling to the target protein. Therefore, an amino acid amide is superior to the corresponding free amino acid. An amino acid ester should not be used in combination with carboxypeptidase Y owing to multiple coupling and resultant oligomerization.[8]

[8] K. Breddam, F. Widmer, and J. T. Johansen, *Carlsberg Res. Commun.* **48,** 231 (1983).

[16] Antibodies Biotinylated via Sugar Moieties

By DANIEL J. O'SHANNESSY

Antibodies are a diverse class of glycoproteins that specifically bind with antigen and elicit a number of secondary responses *in vivo*. The structure and composition of antibodies have been well described.[1] However, in the context of this chapter it should be noted that they consist of 82–96% polypeptide and 4–18% oligosaccharide including both N-linked and O-linked oligosaccharide moieties. Most of the applications of antibodies in basic research, diagnostics, and immunotherapy rely on labeling of the antibody with reporter molecules such as biotin. Until recently, most biotinylation procedures involved the covalent modification of the ε-amino groups of lysines with such compounds as biotin-*N*-hydroxysuccinimide ester (BNHS).[2] However, this chemistry is nonspecific in that the biotin label cannot be directed to a particular site on the antibody. In many instances, particularly when labeling monoclonal antibodies, the result of such labeling is low recovery of both antibody protein and activ-

[1] J. R. Clamp and I. Johnson, *in* "Glycoproteins" (A. Gottschalk, ed.), p. 612. Elsevier, Amsterdam, 1972.

[2] E. A. Bayer, E. Skutelsky, and M. Wilchek, this series, Vol. 62, p. 308.

ity.[3] The latter may be due to modification of amino acids in or in close proximity to the antigen combining site.[4]

The fact that antibodies are glycoproteins offers an alternative labeling procedure whereby the biotin can be directed specifically to the oligosaccharide moieties of the antibody. The oligosaccharide moieties of antibodies are not involved in antigen binding and are most often situated at sites far removed from the antigen combining region. Hydrazide derivatives of biotin and its analogs are the reagents of choice for this purpose.[5]

Hydrazide Derivatives of Biotin

A number of hydrazide derivatives of biotin have been described including biotin hydrazide (BHZ),[2] biotin aminocaproylhydrazide (BACH),[6] and biocytin hydrazide.[7] The synthesis of these compounds is not described here. BHZ is available from a number of commercial sources, and BACH is available from Calbiochem (San Diego, CA).

The use of spacer arm-derivatized biotin hydrazides such as BACH and biocytin hydrazide is recommended since the kinetics of labeling antibodies or glycoproteins in general are faster than with BHZ[6] and because the interaction with avidin/streptavidin is more efficient.[6,7] Since BACH is commercially available, the procedure described below uses BACH as the example although the other biotin hydrazides may be substituted if desired. Unlike biotinylation employing BNHS, which needs to be solubilized in organic solvents such as dimethylformamide, BACH has the added advantage of being sufficiently soluble in aqueous solutions so as to avoid the use of organic solvents during the labeling procedure.

Preparation of Antibodies and Labeling with BACH

Hydrazides are very efficient nucleophiles for condensation with aldehydes. Therefore, the reaction scheme involves the generation of aldehydes specifically on the oligosaccharide moieties of antibodies. Although galactose oxidase in conjunction with neuraminidase treatment can be used for this purpose, by far the most versatile and easily performed reaction is oxidation of vicinal diols with sodium metaperiodate.[5]

Procedure. The antibody is prepared in 0.1 *M* sodium acetate, 0.15 *M* sodium chloride (pH 5.5) at up to 50 mg/2.5 ml. Primary amino-containing

[3] T. R. Burkot, R. A. Wirtz, and J. Lyon, *J. Immunol. Methods* **84,** 25 (1985).

[4] C. Schneider, R. A. Newman, D. R. Sutherland, U. Asser, and M. F. Greaves, *J. Biol. Chem.* **257,** 10766 (1982).

[5] D. J. O'Shannessy and R. H. Quarles, *J. Immunol. Methods* **99,** 153 (1987).

[6] D. J. O'Shannessy, P. J. Voorstad, and R. H. Quarles, *Anal. Biochem.* **163,** 204 (1987).

[7] M. Wilchek and E. A. Bayer, this series, Vol. 138, p. 429.

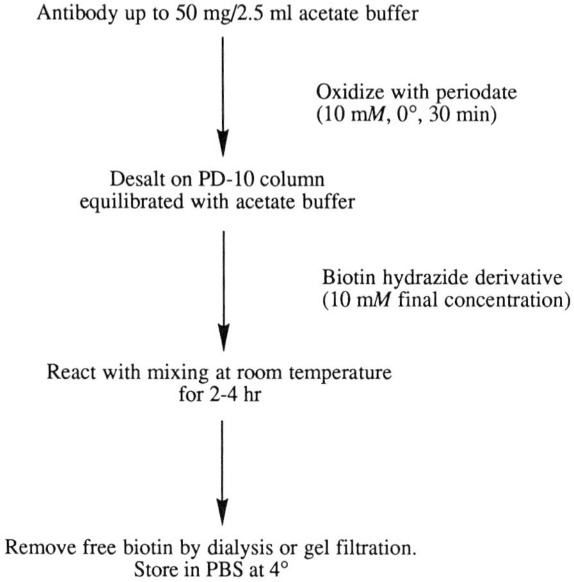

FIG. 1. Specific biotinylation of the sugar moieties of antibodies.

buffers such as Tris should be avoided. The solution is chilled on ice, and 50 μl of a 0.5 *M* aqueous solution of sodium periodate (prepared immediately before use) is added to give a final periodate concentration of 10 m*M*. The solution is quickly mixed, returned to the ice bath, and placed in the dark for 20–30 min.

While the oxidation reaction is proceeding, a Pharmacia PD-10 gel filtration column is equilibrated with the same acetate–sodium chloride buffer used to prepare the antibody. Immediately after the oxidation is complete, the oxidized antibody solution is applied to the PD-10 column and subsequently eluted with 3.5 ml of the same buffer. The sample is collected into a tube containing solid BACH to give a final concentration of 10 m*M* (13 mg). The tube is capped and mixed gently for 2–4 hr. Unreacted BACH is then removed by gel filtration, centrifugal filtration, or dialysis against PBS [phosphate-buffered saline: 0.1 *M* sodium phosphate, 0.15 *M* sodium chloride (pH 7.4)].

Antibodies biotinylated by this procedure (summarized in Fig. 1) have been shown to retain specific avidin-binding characteristics for at least 6 months when stored in PBS at 4°. For small-scale biotinylation reactions, the PD-10 column procedure, which is used primarily to remove unreacted periodate and nonprotein-associated aldehydes generated during the oxidation, can be substituted with a periodate-quenching reagent such as sodium arsenite or glycerol.[6] In this case, once the oxidation reaction is complete, the periodate quenching reagent is added to a final concentration of 40 m*M* and allowed to react for an additional 5 min before addition of BACH.

Notes. The degree of oxidation of the antibody and subsequent labeling with BACH is dependent on a number of factors. Increasing the temperature of oxidation or the concentration of periodate will result in more extensive oxidation of the antibody oligosaccharide. In addition, the reaction between BACH and an aldehyde is a specific acid-catalyzed reaction and is increased as the pH of conjugation is decreased. Therefore, variation of these conditions will affect the degree of labeling obtained and the ultimate rate of biotinylation. However, it should be stressed that antibodies are differentially susceptible to periodate and pH so some compromises must be made. A general and rapid procedure for the optimization of biotinylation of the oligosaccharide moieties of glycoproteins was recently described and is recommended as a prescreen for the biotinylation of antibodies.[6]

Biotinylation has been shown to be dependent on prior oxidation of the antibody.[8] Also, antibodies deglycosylated with trifluoromethanesulfonic acid do not undergo biotinylation, indicating that oxidation with periodate is specific for the oligosaccharide moieties.[8] Further evidence that this procedure results in the biotinylation of sugar moieties is given by the fact that the biotin label can be removed by subsequent treatment of the biotinylated antibody with endoglycosidases.[9]

Applications

To date, few articles have appeared using this methodology. However, the uses of antibodies biotinylated via their sugar moieties should obviously not differ from those presently in use employing antibodies biotinylated via protein moieties. The advantages of the present methodology over those previously described using BNHS are that the biotin label is site directed to the sugar moieties and that the antibodies retain full immunological activity. This is due to (1) the chemistry involved in the labeling procedure and (2) the fact that the sugar moieties of antibodies are most often situated in the Fc portion of the molecule far removed from the antigen-binding Fab region. Biotinylation of the sugar moieties

[8] D. J. O'Shannessy, M. J. Dobersen, and R. H. Quarles, *Immunol. Lett.* **8,** 273 (1984).

[9] D. J. O'Shannessy and R. H. Quarles, *J. Appl. Biochem.* **7,** 347 (1985).

of an antibody is a good alternative procedure when biotinylation with BNHS results in significant or even total loss of immunological activity.[3,8] Also, because of the relative percent oligosaccharide, IgM antibodies are particularly suited to biotinylation of sugar moieties.

The vast array and ready availability of avidin and streptavidin conjugates make biotinylation of antibodies amenable to any detection, quantification, or purification system currently in use employing avidin–biotin technology. Antibodies biotinylated by the procedure presented here have been used in fluorescence-activated cell sorting of peripheral blood lymphocytes,[10] in enzyme-linked immunosorbent assays,[8] immunoblotting techniques,[11,12] and epitope mapping of pollen antigens.[13] In all of these cases, full immunological activity of the antibody preparations was retained after the biotinylation procedure. In a recent report,[14] monoclonal antibodies biotinylated either through the ε-amino groups or through the sugar moieties were compared in a streptavidin–biotin immunoaffinity system. The authors state that biotinylation through the sugar moieties was less effective in this particular system. However, it should be noted that spacer arm-derivatized biotins were used to label the protein but not the sugar moieties, thus making the comparison somewhat invalid. Specific biotinylation of sugar moieties may also be used to assess the purity of an antibody preparation with respect to glycoprotein contamination and to analyze the oligosaccharide distribution of the antibody. Using such a strategy, the existence of monoclonal antibodies with oligosaccharide exclusively on the light chains has been identified.[8,9]

[10] M. J. Dobersen, P. Gascon, S. Trost, J. A. Hammer, S. Goodman, A. B. Noronha, D. J. O'Shannessy, R. O. Brady, and R. H. Quarles, *Proc. Natl. Acad. Sci. U.S.A.* **82,** 552 (1985).

[11] H. J. Willison, D. J. O'Shannessy, and R. H. Quarles, *IRCS Med. Sci.* **14,** 201 (1986).

[12] H. J. Willison, A. I. Ilyas, D. J. O'Shannessy, M. Pulley, B. D. Trapp, and R. H. Quarles, *J. Neurochem.* **49,** 1853 (1987).

[13] A. K. M. Ekramoddoullah, F. T. Kisil, R. T. Cook, and A. H. Sehon, *J. Immunol.* **138,** 1739 (1987).

[14] D. R. Gretch, M. Suter, and M. F. Stinski, *Anal. Biochem.* **163,** 270 (1987).

[17] 2-Iminobiotin-Containing Reagent and Affinity Columns

By BARBARA FUDEM-GOLDIN and GEORGE A. ORR

Introduction

The avidin–biotin complex has an extremely low dissociation constant (K_D) of approximately 10^{-15} *M*.[1] The $t_{1/2}$ for dissociation of this complex is approximately 200 days. The tightness of this noncovalent interaction is the basis for its use as a general detection method in many areas of biochemistry, molecular biology, and immunocytochemistry.[2–4] Streptavidin, a bacterial biotin-binding protein with a similar low dissociation constant, has also been employed in similar studies. In many cases, streptavidin is superior to avidin since nonspecific binding is reduced owing to its less basic p*I* value.[5] An integral component of this technology is the necessity of preparing biologically active avidin/streptavidin probes, i.e., radiolabeled, fluorescent, electron-dense, or enzyme-conjugated probes, uncontaminated by damaged protein or unconjugated reporter molecules.

Cuatrecases and Wilchek developed an affinity isolation procedure for avidin using biocytin (biotinyl-ε-*N*-lysyl)–Sepharose 4B.[6] As a consequence of the tight interaction between avidin and biotin, the conditions required for elution of the specifically bound protein from the affinity matrix were a combination of low pH (1.5) and 6 *M* guanidine-HCl; either alone did not effect elution. Although avidin is not inactivated by these conditions, it is unclear whether certain avidin derivatives, e.g., enzyme conjugates, would retain their desired biological activity. We have overcome this problem by replacing biotin with its cyclic guanidino analog, 2-iminobiotin (Fig. 1).[7–9] In agreement with the observations of Green,[1] we have found that the free base form of 2-iminobiotin forms a stable complex with avidin, but that the salt form interacts poorly.[7] Our studies indicate that the decrease in affinity at neutral and acidic pH

[1] N. M. Green, *Adv. Protein Chem.* **29,** 85 (1975).
[2] E. A. Bayer and M. Wilchek, *Methods Biochem. Anal.* **26,** 1 (1980).
[3] E. A. Bayer and M. Wilchek, this series, Vol. 62, p. 308.
[4] M. Wilchek and E. A. Bayer, *Immunol. Today* **5,** 39 (1984).
[5] L. Chaiet and F. J. Wolf, *Arch. Biochem. Biophys.* **106,** 1 (1964).
[6] P. Cuatrecasas and M. Wilchek, *Biochem. Biophys. Res. Commun.* **33,** 235 (1968).
[7] G. A. Orr, *J. Biol. Chem.* **256,** 761 (1980).
[8] G. Heney and G. A. Orr, *Anal. Biochem.* **114,** 92 (1981).
[9] G. A. Orr, G. C. Heney, and R. Zeheb, this series, Vol. 122, p. 83.

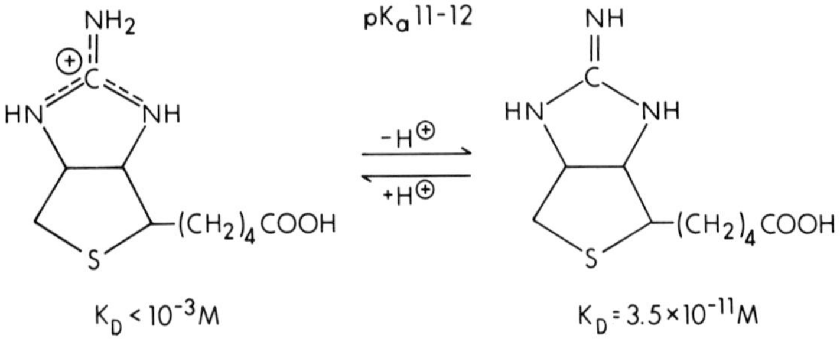

FIG. 1. Structure of 2-iminobiotin showing ionization of the cyclic guanidino group.

values is due to the combined protonation of the cyclic guanidino group of 2-iminobiotin and the ionization of some residue on avidin.[7] We have exploited this pH-dependent alteration in binding to develop a rapid and efficient affinity isolation procedure for avidin and its derivatives. The method is also applicable to streptavidin.[10,11]

Methods

Synthesis of 2-Iminobiotin Hydrobromide

The two-step synthesis involves the alkaline hydrolysis of biotin with barium hydroxide at 140° to yield 5-(3,4-diaminothiophan-2-yl)pentanoic acid.[12] 2-Iminobiotin is subsequently prepared from this diaminocarboxylic acid derivative of biotin by reaction with cyanogen bromide.[13]

5-(3,4-Diaminothiophan-2-yl)pentanoic acid. Biotin (1 g) and barium hydroxide (3 g) are mixed together and placed into a Pyrex hydrolysis tube (15 × 250 mm). Water (7 ml) is added and the tube sealed under mild vacuum (house vacuum line, ~30 mm Hg). After heating at 140° for 21 hr, the tube contents are removed, CO_2 is bubbled into the suspension, and the insoluble precipitate ($BaCO_3$) is removed by filtration. The filtrate is acidified to pH 4 using 2 *N* H_2SO_4, filtered, and concentrated *in vacuo*. Addition of methanol induces crystallization of the diaminocarboxylic acid sulfate, and this process is allowed to continue overnight at 4° (56% yield). Thin-layer chromatography on silica gel ($CHCl_3 : H_2O$: concen-

[10] K. Hofmann, S. W. Wood, C. C. Brinton, J. A. Montibeller, and F. M. Finn, *Proc. Natl. Acad. Sci. U.S.A.* **77,** 4666 (1980).

[11] E. A. Bayer, H. Ben-Hur, G. Gitlin, and M. Wilchek, *J. Biochem. Biophys. Methods* **13,** 103 (1986); E. A. Bayer, H. Ben-Hur, and M. Wilchek, this volume [8].

[12] K. Hofmann, D. B. Melville, and V. Du Vigneaud, *J. Biol. Chem.* **141,** 207 (1941).

[13] K. Hofmann and A. E. Axelrod, *J. Biol. Chem.* **187,** 29 (1950).

trated NH_4OH, 75 : 25 : 5, v/v) reveals the presence of a single iodine- and ninhydrin-positive spot of R_f 0.6.

2-Iminobiotin. The diaminocarboxylic acid sulfate (2.5 g) is treated with barium carbonate (5 g) and dissolved in 30 ml of hot water, and the solution is immediately filtered through a prewarmed fritted glass funnel. The precipitate is washed with an additional 10 ml of hot water. Cyanogen bromide (1.9 g) is added to the filtrate, and crystals of 2-iminobiotin (free base form) start to form within 5 min. The reaction mixture is left at 4° for 12 hr. The yield of 2-iminobiotin is 1.8 g (mp 260°, decomposition). To convert the free base to the hydrobromide salt, 2-iminobiotin is suspended in water (40 ml), heated to approximately 45°, and 1% (v/v) HBr is added dropwise until all of the 2-iminobiotin dissolves. The solution is concentrated *in vacuo*, and the residue is recrystallized from hot 2-propanol : methanol (6 : 4, v/v). 2-Iminobiotin hydrobromide has a melting point of 222–223°. Thin-layer chromatography on silica gel eluting with $CHCl_3$: H_2O : concentrated NH_4OH (70 : 25 : 5, v/v) reveals the presence of a single iodine-positive spot with an R_f of 0.38.

2-Iminobiotin N-Hydroxysuccinimide Ester. The *N*-hydroxysuccinimide ester of 2-iminobiotin is prepared by dissolving 2-iminobiotin hydrobromide (324 mg, 1 mmol) and *N*-hydroxysuccinimide (115 mg, 1 mmol) in dry *N,N*-dimethylformamide (3 ml) at 4°. Dicyclohexylcarbodiimide (206 mg, 1 mmol) is added, and the reaction is allowed to proceed for 1 hr at 4° and then overnight at room temperature. The dicyclohexylurea is removed by filtration and the organic solvent taken off *in vacuo*. The residue is recrystallized from hot 2-propanol to yield the *N*-hydroxysuccinimide ester (58% yield, mp 160–161°). Analysis, calculated for $C_{14}H_{21}N_4O_4SBr$: C 39.91, H 5.02, N 13.30, S 7.61%; found, C 40.00, H 5.20, N 13.20, S 7.34%.

Preparation of 2-Iminobiotin Affinity Columns

2-Iminobiotin–6-aminohexyl-Sepharose 4B via CNBr Activation of Sepharose. 2-Iminobiotin (400 mg, 1.23 mmol) is added to 100 ml of 6-aminohexyl-Sepharose 4B (40 ml of packed resin in water), prepared by the method of Porath[14] or purchased from Pharmacia (Piscataway, NJ), and the pH is adjusted to 4.8 with HBr (1%, v/v). 1-Cyclohexyl-3-(2-morpholinoethyl)carbodiimide metho-*p*-toluene sulfonate (4.24 g, 10 mmol) is added portionwise over a period of 10 min. The pH is kept at 4.8 throughout the reaction by the addition of 1% HBr and is constant after 3–5 hr. The resin is washed with 1 *M* NaCl (2 liters), water (2 liters),

[14] J. Porath, this series, Vol. 18B, p. 13.

and packed into a column. The binding capacity of the affinity matrix, using these coupling conditions is 0.75 mg purified avidin/ml of swollen gel. The resin has a reported $t_{1/2}$ of 3 days at pH 11, 25°. It is available from several commercial sources.

2-Iminobiotin–6-aminohexyl-Sepharose CL via Chloroformate Activation of Sepharose. Bayer *et al.* have reported the preparation of an alkaline pH-stable 2-iminobiotin-containing resin.[11] 1,6-Diaminohexane is coupled to *p*-nitrophenyl chloroformate-activated Sepharose CL, prepared by the method of Wilchek and Miron.[15] Chloroformate activation is performed in acetone–dimethylaminopyridine. The amine-containing resin is reacted with 2-iminobiotin *N*-hydroxysuccinimide ester in dimethylformamide : dioxane (1 : 1, v/v) for 20 hr at 4°. The resin has a reported capacity of 12 mg avidin/ml resin and is stable to the alkaline pH conditions used in the purification of avidin/streptavidin.

Affinity Isolation of Avidin. Homogenized egg whites from 24 fresh chicken eggs are diluted with water (2 : 1, v/v), and the solution is brought to 70% saturation with ammonium sulfate (enzyme grade, Schwarz/Mann) at 4°. After stirring for 2 hr, the mixture is centrifuged (8000 *g*, 20 min), the supernatant is brought to 100% saturation, and the mixture is left stirring at 4° overnight. After centrifugation (8000 *g*, 30 min at 4°), the pellet is dissolved in water (40 ml) and dialyzed against water (3 times, 2 liters each at 4°). The dialyzate is adjusted to pH 11 with 1 *N* NaOH, and NaCl (1 *M*) is added. The crude avidin solution is applied to 2-iminobiotin–6-aminohexyl-Sepharose 4B (40 ml) which had previously been equilibrated with 50 m*M* sodium carbonate (pH 11) containing 1 *M* NaCl. The column is washed with equilibrating buffer (20 ml/hr) until the absorbance at 282 nm returns to baseline. Avidin is eluted from the column with 50 m*M* ammonium acetate (pH 4) containing 0.5 *M* NaCl. Protein content is measured by absorbance at 282 nm and avidin content by its ability to bind either 4-hydroxyazobenzene-2′-carboxylic acid or [^{14}C]biotin.[16] The appropriate fractions are pooled, dialyzed against water, and lyophilized.

As can be seen from Fig. 2, avidin is eluted as a sharp peak after application of the low-pH buffer. Greater than 90% of the crude avidin applied to the column is recovered in the specifically eluted fractions, and the yield of avidin from 24 eggs is in the range of 15–20 mg. The avidin obtained by this procedure is pure as judged by its ability to bind 14.4 μg of [^{14}C]biotin/mg of protein; literature values for pure avidin range from

[15] M. Wilchek and T. Miron, *Biochem. Int.* **4**, 629 (1982).

[16] N. M. Green, this series, Vol. 18A, p. 418.

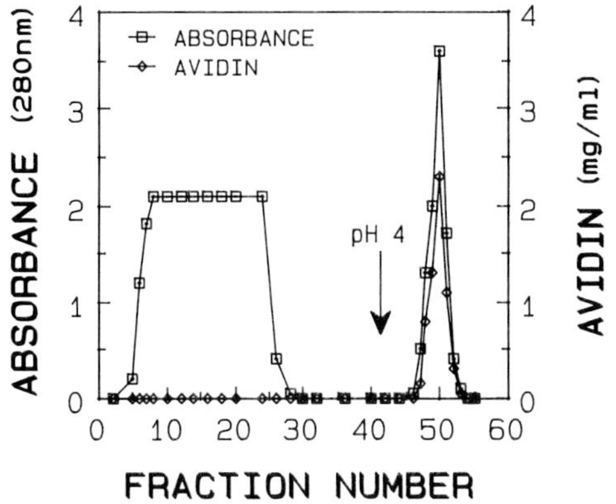

FIG. 2. Affinity purification of avidin on 2-iminobiotin–6-aminohexyl-Sepharose 4B. Crude avidin was loaded onto the column at pH 11 and specifically eluted by application of pH 4 buffer.

13.8 to 15.1.[1,4] Sodium dodecyl sulfate–polyacrylamide gel electrophoresis of the specifically eluted fractions reveals a single polypeptide with an apparent molecular weight slightly larger than the hemoglobin monomer (16,000). Native avidin (68,000) is composed of four identical subunits.[1] As a further indication of purity, if the specifically eluted protein is treated with biotin and rechromatographed on the affinity column, no protein is retained and eluted at pH 4.

Affinity Isolation of Streptavidin. Both Hofmann *et al.*[10] and Bayer *et al.*[11] have described the purification of streptavidin from cultures of *Streptomyces avidinii* on 2-iminobiotin affinity resins. In the method of Bayer *et al.*,[11] no preliminary purification of the streptavidin-containing culture medium is required. After adjusting the pH of the medium to pH 11, streptavidin is adsorbed onto the affinity resin, and, after washing the column with high-salt buffer (pH 11), homogeneous protein is obtained by elution with pH 4 buffer.

Affinity Purification of Avidin/Streptavidin Derivatives

^{125}I-Labeled Avidin and Streptavidin. Avidin (15 μg) in 50 μl of 40 m*M* borate (pH 8.0) is added to 100 μCi Bolton–Hunter reagent and left at 4° for 60 min. The reaction is terminated by the addition of 150 μl of 200 m*M* glycine in 40 m*M* borate (pH 8.0). After 20 min at 4°, 2 ml of 50 m*M* ammonium carbonate (pH 11) containing 0.5 *M* NaCl and 1 mg/ml bovine serum albumin is added. The radiolabeled avidin is purified as described for avidin except that all buffers contain bovine serum albumin. The

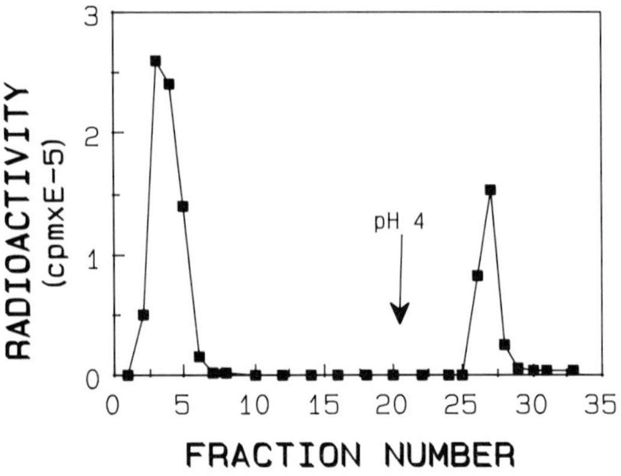

FIG. 3. Affinity purification of ^{125}I-labeled avidin on a 2-iminobiotin column.

column size is 4 ml, and 1.5-ml fractions are collected. ^{125}I-Labeled avidin is stored frozen at −20°.

Avidin is iodinated with ^{125}I-labeled Bolton–Hunter reagent rather than by the chloramine-T or lactoperoxidase methods, because (1) avidin loses biological activity rapidly in the presence of oxidizing agents, e.g., H_2O_2 and *N*-bromosuccinimide; (2) avidin contains only a single, deeply buried, tyrosine residue per subunit; and (3) up to 60% of primary amino groups in avidin can be blocked without loss of biotin-binding activity. Avidin is readily iodinated by ^{125}I-labeled Bolton–Hunter reagent and retains full biological activity as judged by its ability to bind to and be specifically eluted from a 2-iminobiotin affinity column (Fig. 3). Approximately 26% of the applied ^{125}I was found in the specifically eluted fractions, and greater than 95% of the ^{125}I in these fractions could be precipitated by 10% trichloroacetic acid. We have used the same procedure for the preparation and purification of ^{125}I-labeled streptavidin.

Rhodamine-Conjugated Avidin. To a solution of avidin (4 mg, 59 nmol) dissolved in 5 ml of 100 m*M* sodium bicarbonate (pH 8) at 4° is added 25 μl of rhodamine B isothiocyanate (0.25 mg, 47 μmol) dissolved in *N,N*-dimethylformamide. The reaction is left at 4° for 15 hr in the dark, and the conjugated avidin is purified as described above for avidin. The column size is 45 ml, and 6-ml fractions are collected. The avidin concentration is determined with [^{14}C]biotin, and the number of rhodamine groups introduced is calculated by the method of Schechter *et al.*[17] As can be seen from Fig. 4, rhodamine-conjugated avidin does not elute as a sharp peak but rather with a long trailing edge. If the ratio of rhodamine to

[17] Y. Schechter, J. Schlessinger, S. Jacobs, K.-J. Chang, and P. Cuatrecasas, *Proc. Natl. Acad. Sci. U.S.A.* **77**, 4833 (1978).

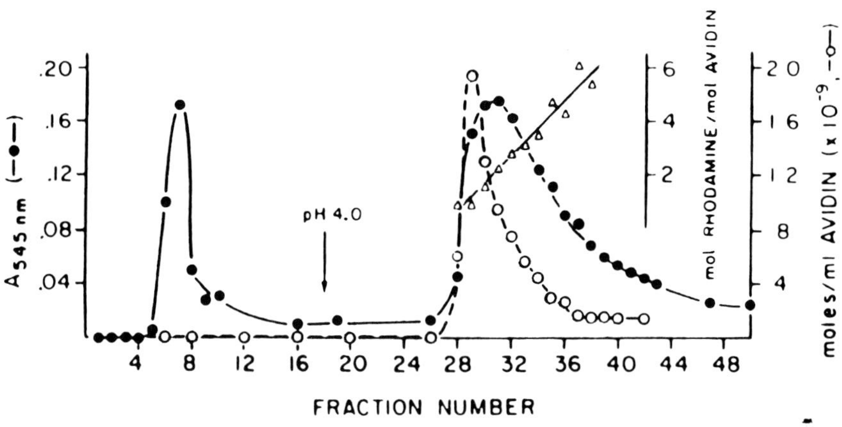

FIG. 4. Affinity purification of rhodamine-conjugated avidin on a 2-iminobiotin column.

avidin is calculated along the specifically eluted peak, the ratio increases. This retardation effect allows the separation of rhodamine–avidin conjugates with well-defined extents of dye incorporation. The 2-iminobiotin affinity column is also suitable for purifying rhodamine-conjugated streptavidin and fluorescein-conjugated avidin/streptavidin derivatives.

Enzyme-Conjugated Avidin/Streptavidin Derivatives. The horseradish peroxidase and the alkaline phosphatase conjugates of avidin/streptavidin can also be purified by chromatography on 2-iminobiotin columns. For these derivatives the affinity column is equilibrated at pH 10 rather than pH 11, and specific elution is achieved with pH 6 buffer.

Conclusion

The observation that 2-iminobiotin interacts with avidin/streptavidin in a pH-dependent manner has formed the basis for the development of a facile and efficient method for the affinity isolation of both proteins. This affinity isolation procedure is also applicable to a range of substituted avidin/streptavidin probes, allowing for the isolation of highly purified, biologically active reporter molecules free from damaged protein and unconjugated reporter molecules.

Acknowledgments

This work was supported in part by Grants Gm34029 and HD00577 from the National Institutes of Health. G.A.O. is the recipient of an Irma T. Hirschl Career Scientist Award and B.F.G. is supported by Training Grant GM07260 from the National Institutes of Health.

[18] Avidin- and Streptavidin-Containing Probes

By EDWARD A. BAYER and MEIR WILCHEK

The collection of avidin-containing probes (Table I) represents the second major "partner" of the avidin–biotin system. In this chapter, we provide protocols for the preparation of many types of these probes that are not included in the other chapters in the book. Hence, in addition to describing the synthesis of avidin-containing probes commonly in use in our laboratory, we include preparations by others which, we feel, fill a void.

As already mentioned in the introductory chapter [2], there are two general approaches in preparing a desired avidin-containing probe: (1) direct (covalent) coupling between the probe and avidin and (2) complex formation between native avidin and a biotinylated probe. Of course, the latter approach includes both the sequential (stepwise) approach and the preparation of preformed complexes.

Derivatized Avidins

Fluorescein-Derivatized Avidin

In theory, avidin is a better choice than streptavidin for derivatization via amino groups (e.g., for a fluorescent probe). The high number of free lysines provides a good base for extensive derivatization, and the p*I* of the resultant avidin derivative is reduced to an acceptable level vis-à-vis non-specific binding. Nonglycosylated avidin is an even better choice, since the lack of sugars removes the other major source of errant binding.

Reagents

Avidin (Belovo Soc. Coop., Bastogne, Belgium; or STC Laboratories, Winnipeg, Manitoba), 10 mg/ml, or nonglycosylated avidin,[1] dissolved in 10 m*M* phosphate buffer (pH 7.4)

Fluorescein isothiocyanate (FITC),[2] isomer I (Sigma Chemical Co., St. Louis, MO), 2.5 mg/ml in 0.5 *M* sodium bicarbonate buffer (pH 9.5)

[1] Y. Hiller, E. A. Bayer, and M. Wilchek, this volume [6].

[2] We use 25 μg fluorescein isothiocyanate per milligram of avidin, which is equivalent to about 20 mol FITC/mole avidin.

TABLE I
COMMONLY USED AVIDIN-CONTAINING PROBES[a]

Probes (conjugates)	Applications
Enzymes	Immunoassay, diagnostics, blotting, affinity cytochemistry [light microscopy (LM), electron microscopy (EM)], affinity perturbation
Radiolabels	Immunoassay, cytochemistry, cytological probe
Fluorescent agents	Affinity cytochemistry (fluorescence microscopy), flow cytometry, immunoassay, diagnostics
Chemiluminescent agents	Immunoassay, diagnostics
Chromaphores	Immunoassay, diagnostics
Heavy metals	Affinity cytochemistry (LM, EM), immunoassay, blotting
Colloidal gold	Affinity cytochemistry (LM, EM), immunoassay, blotting
Ferritin	Affinity cytochemistry (EM), macromolecular carrier
Hemocyanin	Affinity cytochemistry (EM), macromolecular carrier
Phages	Affinity cytochemistry (EM), affinity targeting, diagnostics
Macromolecular carriers	Cross-linking studies, signal amplification, affinity targeting, drug delivery, cytological probe, affinity fusion, affinity perturbation, affinity partitioning
Liposomes	Affinity fusion, drug delivery, affinity targeting, signal amplification
Solid supports	Affinity chromatography, immobilization, selective retrieval, selective elimination

[a] As described in [2] in this volume, the same probes can also be derivatized with biotin, and native (underivatized) avidin can be used (either by sequential application or in preformed complexes) to cross-link the biotinylated probe with a biotinylated binder.

Sephadex G-50 column (Pharmacia, Uppsala, Sweden), 10–20 ml bed volume
DEAE-cellulose column (DE-52, Whatman), 10 ml bed volume
10 m*M* Sodium phosphate buffer (pH 8.0) (PB)
0.1 *M* NaCl in PB
0.25 *M* NaCl in PB

Procedure. The avidin solution (1 ml) is mixed with 0.1 ml FITC, and the reaction is carried out overnight at 4°. The conjugate is applied to the Sephadex G-50 column (preequilibrated with PB), and the first peak is collected. The sample is loaded on a DEAE-cellulose column, preequilibrated with PB, and washed with the same buffer. Unlabeled avidin can be collected from the effluent. In order to elute FITC-labeled avidin, rinse the column with 0.1 *M* NaCl in PB. Collect the labeled peak, and then elute the column again, this time with 0.25 *M* NaCl in PB. The peak fractions are pooled separately and stored at −20°.

The average number of fluorescein (F) molecules coupled to avidin

(Av) can be calculated from the following equation:

$$F/Av = \frac{0.3(A_{495})}{A_{280} - 0.35(A_{495})}$$

The initial FITC-labeled avidin peak was found to contain an average of 6 fluorescein groups per avidin tetramer; the second labeled peak contained an average of about 9 fluorescein groups per avidin molecule.

Tritiated Avidin

Tritiated forms of the avidins can be prepared at relatively high specific radioactivity as a stable radiolabeled alternative to iodinated derivatives.

Reagents

Avidin (Belovo or STC Laboratories), nonglycosylated avidin,[1] or streptavidin,[3] 1 mg/ml in 50 m*M* $NaHCO_3$
N-Succinimidyl [2,3-^{3}H]propionate[4] (Radiochemical Center, Amersham, Buckinghamshire, England), 120 Ci/mmol, 1 mCi
Dimethylformamide
Sephadex G-50 column, 10 ml bed volume
Phosphate-buffered saline (pH 7.4) (PBS)

Procedure. The solution (100 μl) of the desired protein[5] is introduced to the vial containing the dimethylformamide solution of the tritiated reagent. After 1 hr at room temperature, the reaction mixture is applied to the Sephadex G-50 column (preequilibrated with PBS), and the first peak is collected. The peak fractions are pooled and stored in aliquots at −20°.

Comments. This procedure routinely results in stable tritiated derivatives of about 10^7 cpm/μg for avidin and about 3×10^6 cpm/μg for streptavidin. The ratio of egg-white avidin to reagent used in this protocol represents a compromise for achieving optimal levels of both biotin-binding activity and radiolabeling. The given ratio has the added advantage of reducing the p*I* of the radiolabeled derivative; if nonglycosylated avidin is used, the resultant conjugate is similar to streptavidin in its nonspecific binding properties.

[3] L. Chaiet and F. J. Wolf, *Arch. Biochem. Biophys.* **106,** 1 (1964).

[4] The reagent obtained from Amersham is dissolved in 1 ml of toluene. In order to carry out the radiolabeling of avidin, the toluene is evaporated in the original vial. The residue is then redissolved in 10 μl dimethylformamide.

[5] Final concentrations of 0.3 mg/ml for avidin or nonglycosylated avidin can be used in order to enhance the level of tritiation of the product; we usually perform the tritiation of streptavidin at 1 mg/ml.

Avidin-Conjugated Proteins

Ferritin–Avidin Conjugates

There are many methods available for the conjugation of two proteins. Several approaches have been taken to produce viable ferritin–avidin conjugates, including glutaraldehyde coupling,[6] cross-linking of bromoacetylated ferritin with thiolylated avidin,[7] and reductive alkylation.[8] The latter procedure consists of periodate-induced oxidation of the oligosaccharide residues on avidin; the aldehydes produced react with amino groups of the ferritin, and the resultant Schiff base is further reduced to a stable bond with borohydride.

Reagents

Avidin (Belovo or STC Laboratories),[9] 15 mg dissolved in 5 ml acetate-buffered saline (pH 4.5)
Ferritin, 2 times recrystallized (Sigma), 100 mg in 0.1 *M* NaCl 1 ml solution
Sodium metaperiodate (Merck, Darmstadt, Federal Republic of Germany)
Sodium borohydride (Merck)
Acetate-buffered saline [50 m*M* sodium acetate buffer (pH 4.5) containing 0.15 *M* NaCl]
0.1 *M* Borate-buffered saline (pH 8.5)
PBS (pH 7.4)
Iminobiotin–Sepharose column[10]

Procedure. The solutions of avidin and ferritin are mixed and periodate (0.66 ml, 0.1 *M* solution) is added to a final concentration of 10 m*M*. The solution is stirred for 30 min on ice, dialyzed against acetate-buffered saline for 6 hr at 4°, and immediately dialyzed overnight at 4° against borate-buffered saline. A fresh solution of sodium borohydride (10 mg/ml in 10 m*M* NaOH) is prepared, and 0.5 ml is added to the ferritin–avidin conjugates in an ice bath. After 1 hr, the solution is dialyzed against PBS. The conjugates are washed twice by centrifugation (100,000 *g*, 3 hr

[6] H. Heitzmann and F. M. Richards, *Proc. Natl. Acad. Sci. U.S.A.* **71,** 3537 (1974).

[7] L. Angerer, N. Davidson, W. Murphy, D. Lynch, and G. Attardi, *Cell* **9,** 81 (1976).

[8] E. A. Bayer, E. Skutelsky, D. Wynne, and M. Wilchek, *J. Histochem. Cytochem.* **24,** 933 (1976).

[9] STC avidin has been shown by us to contain the full complement of oligosaccharides and is thus similar in its molecular properties to that obtained from Sigma (the former is about 10 times cheaper). Belovo avidin is even cheaper than STC avidin, but its sugar residues are partially degraded.

[10] E. A. Bayer, H. Ben-Hur, and M. Wilchek, this volume [8].

at 4°) in order to remove free (unconjugated) avidin molecules, and the pellet is resuspended to 1 mg/ml ferritin. The avidin-containing species are separated from the free ferritin by affinity chromatography on an iminobiotin–Sepharose column.[10] The conjugates are passed through a sterile Millipore filter (HA 0.2 μm) and stored in aliquots under sterile conditions at 4°.

Comments. The procedure given here results in a high yield of unit-paired conjugates, which are particularly suitable for precise ultrastructural localization of biotinylated sites associated with subcellular structures. The procedure can, of course, be used to conjugate avidin to other proteins. In cases where the given protein is sensitive to periodate, avidin alone can be preoxidized, the periodate removed by dialysis or gel filtration, and the oxidized avidin reacted further with the desired protein. Owing to the directionality of the above-described chemistry, this procedure is applicable to cases where relatively small conjugates are required.

Avidin–Toxin Conjugates

Although we have not included in this volume the use of the avidin–biotin system for therapeutic purposes, the use of avidin–drug or avidin–toxin conjugates has recently been seriously considered in studies of cancer therapy. This basic approach is also being used for imaging purposes using conjugates of avidin with radioactive metals. In most of these cases, a biotinylated antibody is first delivered to the target (tissue or organ), and then the avidin-associated probe is introduced. The probe can be either covalently attached to avidin or complexed via a biotin bridge. In some cases, after the biotinylated antibody is interacted with the system, native (underivatized) avidin is delivered, followed sequentially by the biotinylated toxin or radioactive metal.

In this chapter, we describe only the preparation of an avidin–toxin conjugate (avidin–ricin A chain). In this case, an active disulfide-containing avidin derivative is interacted with the A chain of ricin (which contains a free sulfhydryl group), thus forming a mixed disulfide conjugate.[11]

Reagents

Avidin (Belovo or STC Laboratories), nonglycosylated avidin,[1] or streptavidin,[3,10] 10 mg in 4 ml of PBS
Ricin A chain (Sigma), 5 mg
Acetate-buffered saline (pH 4.5)
PBS (pH 7.4)

[11] N. Hashimoto, K. Takatsu, Y. Masuho, K. Kishida, T. Hara, and T. Hamaoka, *J. Immunol.* **132,** 129 (1984); K. A. Krolick, J. W. Uhr, and E. S. Vitetta, this series, Vol. 93 [21].

N-Succinimidyl 3-(2-pyridyldithio)propionate (SPDP, Pharmacia), 10 mg/ml in dimethylformamide

Sephacryl S-200 column (1.6 × 88 cm, Pharmacia), equilibrated with PBS

Procedure. An aliquot (30 μl) of SPDP is added to the solution of avidin, and the reaction is allowed to proceed at room temperature for 30 min. The reaction mixture is dialyzed against acetate-buffered saline for 3 hr in order to remove extraneous reagent.

Ricin A chain is dissolved in 2.5 ml of acetate-buffered saline, and the solution is dialyzed against the same buffer for 1 hr. The dialyzed solutions of SPDP-modified avidin and ricin A chain are mixed, and the reaction is allowed to proceed overnight at room temperature. The mixture is dialyzed against PBS for 3 hr, and the avidin–ricin A chain conjugates are separated from unreacted avidin and toxin by gel filtration on a Sephacryl S-200 column. The conjugates are passed through a sterile Millipore filter (HA 0.2 μm) and stored in aliquots under sterile conditions at 4°.

Comments. There are many other methods that can be used to couple a toxin to a given carrier. However, in order to obtain an active avidin–toxin conjugate, the bond connecting the two should be subject to disassociation. The preferred method is to form a disulfide linkage that would be susceptible to reducing conditions which characterize the intracellular matrix.

Internalization is very important for toxin action; the use of cocomplexes containing biotinylated antibody and biotinylated toxin connected via a native (underivatized) avidin bridge may enhance cross-linking on the cell surface and result in internalization of the complexed components into the target cells. The native avidin may also serve as a scavenger to remove rapidly from the circulation free residual toxin and antibody molecules that fail to react with target cells.

Avidin–Penicillinase Conjugates

One of the earliest examples of a reagent to be used for conjugating proteins is glutaraldehyde.[12] Although this is an example of a homobifunctional reagent and the directionality of the cross-linking cannot be controlled, glutaraldehyde is still a very popular reagent for covalent coupling of proteins. The reason for this (over and above historical precedence and its use out of habit) is simply that the procedure is easy and yields reasonably good results.

As an example of this mode of cross-linking, we detail below the coupling of avidin to penicillinase, an interesting enzyme which should find extensive use in immunoassays. One of the possible substrate for-

[12] S. Avrameas, *Immunochemistry* **6,** 43 (1969).

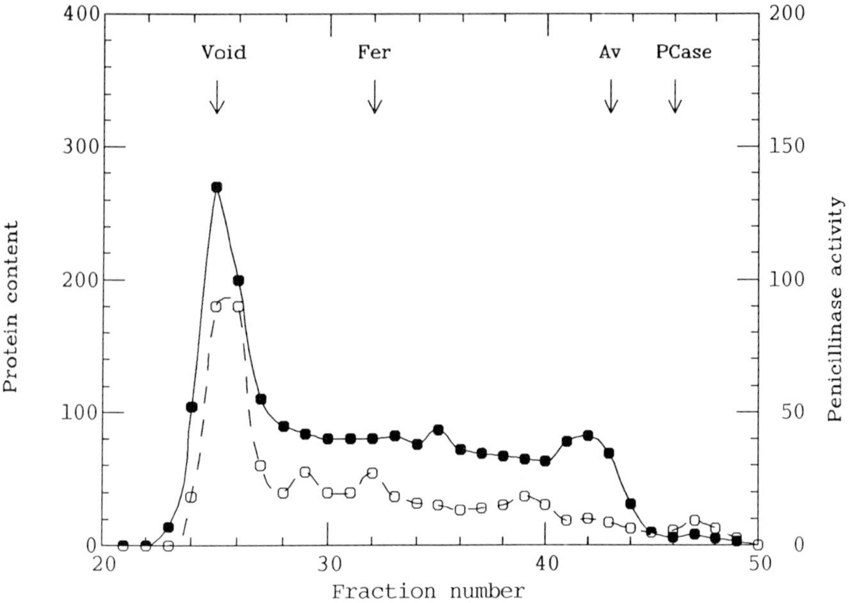

FIG. 1. Gel filtration of avidin–penicillinase conjugates. Following glutaraldehyde cross-linking of avidin and penicillinase, the reaction mixture was dialyzed and applied to a Sephacryl S-300 column (1.6 × 76 cm) preequilibrated with PBS. The resultant fractions (2 ml) were analyzed for protein content (●) and enzymatic activity (○). Protein content is given as μg/fraction, and penicillinase activity represents the reciprocal of time required for the disappearance of color in the substrate solution (hr^{-1}). Void, Fer, Av, and PCase represent the positions of eluted standards for the void volume, ferritin, avidin, and penicillinase, respectively.

mats (presented below) results in the disappearance (as opposed to production) of color, which can be particularly useful in hybridoma technology, pregnancy tests, or any assay where a yes/no answer is required.

Reagents

Avidin (Belovo or STC Laboratories), nonglycosylated avidin,[1] or streptavidin,[3,10] 2 mg (~30 nmol) dissolved in 0.4 ml PBS

Penicillinase from *Bacillus cereus* (Sigma, ~2000 units/mg), 0.5 mg protein (~20 nmol enzyme) dissolved in 0.1 ml PBS

PBS (pH 7.4)

Glutaraldehyde solution: 12.4 μl of an 8% (v/v) solution (Polysciences, Inc., Warrington, PA) is brought to 1.5 ml with PBS

Sephacryl S-300 column (1.6 × 76 cm)

Substrate solution[13]: 3 ml of 0.5% gelatin (w/v in distilled water), 0.5

[13] The substrate solution for penicillinase is prepared immediately before use.

ml of 25 mM iodine in 0.125 M potassium iodide, 0.1 ml of 1% (w/v) soluble starch, and 1 ml of benzylpenicillin [3 mg/ml in 0.1 M phosphate buffer (pH 7)]

Procedure. The solutions of avidin and penicillinase are mixed, and the glutaraldehyde solution is added (0.05% final concentration). After 2 hr at room temperature, the slight precipitate that forms is centrifuged, and the supernatant fluids are dialyzed against PBS. The contents of the dialysis tubing are subjected to gel filtration on a Sephacryl S-300 column, and the void-volume fractions are pooled, passed through a sterile Millipore filter (HA 0.2 μm), and stored in aliquots under sterile conditions at 4°.

Comments. Glutaraldehyde-mediated cross-linking of two protein species usually results in the formation of high molecular weight multimers of the two. This is indeed the case with avidin and penicillinase (Fig. 1). Using the conjugation procedure described above, very little free (unconjugated) avidin and penicillinase remain; the largest fraction appears in the void volume (>1,000,000), and a trail of lower molecular weight multimers can also be recovered. The void-volume fractions (25–27 in Fig. 1) are pooled, and the preparation was found to exhibit high levels of both biotin-binding and penicillinase activities. The preparation is appropriate for immunoassay studies.[14] Penicillinase can also be biotinylated and used in conjunction with underivatized avidin to form complexes, similar to those described below.

Avidin–Protein Complexes

Avidin and Streptavidin Complexes with Biotinylated Enzymes

Preformed complexes between avidin and biotinylated enzymes provide an alternative to classic covalent cross-linking of two proteins to form the corresponding conjugate. The major advantages in using conjugates is that the resultant signal is often enhanced. Moreover, the stock solutions of avidin and many of the biotinylated enzymes can be stored effectively with essentially no deleterious effect on their performance. The preparation of such complexes is convenient, reliable, and reproducible.

In forming complexes, a compromise must be reached between the amount of (biotinylated) enzyme which reacts with avidin and the average number of free (available) biotin-binding sites which remain on the avidin molecule. The former is needed to provide a good signal, and the latter is essential for the primary interaction with the desired biotinylated binder

[14] R. H. Yolken, this volume [61].

(e.g., antibody). The most effective ratio of avidin or streptavidin for complex formation with a given biotinylated enzyme is determined empirically by optimization experiments. In a convenient procedure we designed for this purpose, biotinylated bovine serum albumin is used as a model target system for determining the required conditions for efficient complex formation. Biotinylated alkaline phosphatase (B–AP) is the test enzyme in this example.

Reagents

Avidin (Belovo or STC Laboratories), nonglycosylated avidin,[1] or streptavidin,[3,10] 1 mg/ml in PBS[15]
B–AP,[16] 500 units/ml in 0.15 *M* NaCl[15]
PBS (pH 7.4)
Bovine serum albumin (BSA), 2% (w/v) in PBS
Lysozyme, 2% (w/v) in PBS
Biotinylated BSA,[16] 1 mg/ml in PBS
Substrate solution: 10 mg naphthol AS-MX phosphate (free acid) is dissolved in 200 μl of dimethylformamide, and the solution is mixed with a solution containing 30 mg Fast Red dissolved in 100 ml of 0.1 *M* Tris-HCl (pH 8.4)

Optimization Procedure. Complexes consisting of different ratios of avidin (or streptavidin) and B–AP are prepared in the following manner. The stock solution of avidin is diluted to 1, 3, 10, and 30 μg/ml (4 ml total solution each).[17] Similarly, the stock solution of B–AP is diluted (e.g., 1, 2, 5, and 10 units/ml, 4 ml total).[18] A 1-ml sample of each concentration of avidin is combined with an equivalent volume of one of the B–AP solutions in matrix fashion, yielding 16 separate solutions of varying ratios and quantities of avidin to B–AP. The solutions are allowed to stand for about 30 min at room temperature before being incubated with dot blots containing the target material.

The biotinylated target protein is applied to dot blots in the following manner. Samples (1 μl) containing serial dilutions of biotinylated BSA, are applied onto 16 dot blots.[19] The strips are rinsed with PBS, treated with quenching solution,[20] and rinsed again, and each blot is incubated in one of the above-described preformed complex preparations. After 30

[15] Stock solutions of avidin (streptavidin) and B–AP are stored in aliquots at −20°. The stock solutions are used no more than twice (thawed, refrozen, and rethawed).

[16] E. A. Bayer and M. Wilchek, this volume [14].

[17] Lysozyme solution is used as a diluent for avidin, and BSA is used for streptavidin.

[18] BSA is used as a diluent for B–AP.

[19] E. A. Bayer, H. Ben-Hur, and M. Wilchek, this volume [48].

[20] BSA is used as quencher for streptavidin-based complexes, and a 1 : 1 mixture of BSA and lysozyme is used for avidin-based complexes.

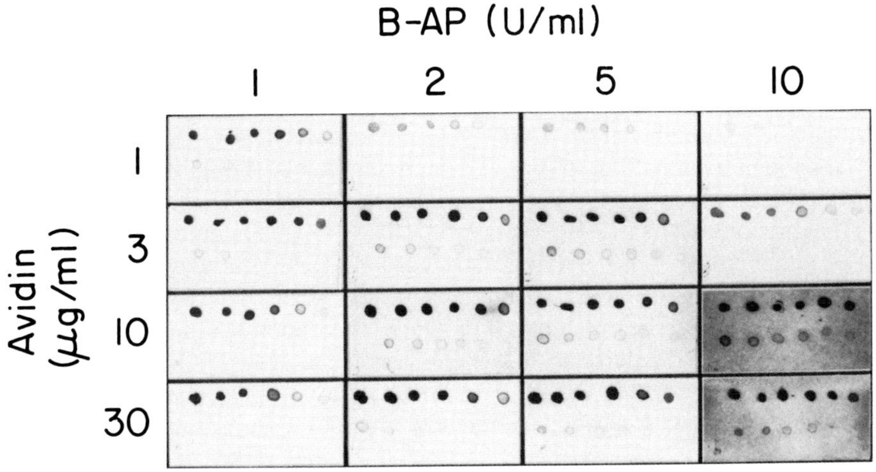

FIG. 2. Dot-blot optimization of complex formation between avidin and B–AP. Onto each nitrocellulose strip, 1-μl samples containing serial dilutions (top row: 1000, 300, 100, 30, 10, and 3 ng/μl; bottom row: 1, 0.3, 0.1, etc.) of biotinylated BSA were applied. The strips were rinsed, quenched, rinsed again, and incubated in solutions containing the different complexes. The designated amounts indicate the initial concentrations of either B–AP or avidin solutions that were combined 1 : 1 and applied to the corresponding nitrocellulose strip. Following incubation, the strips were rinsed, and substrate solution was added.

min, the strips are rinsed again with PBS, and substrate solution is added. Bands usually appear within 30 min; at the desired amplification, the reaction is terminated by washing the strips with tap water.

Standard Procedure. Once the system has been optimized for a given biotinylated enzyme preparation, the procedure for complex formation can be standardized using a given ratio of components. Using the B–AP prepared in our laboratory,[16] we found the following protocol to result in optimal levels of complexation of the enzyme with avidin or streptavidin. A 100-fold dilution[17,18] of the above stock solutions of both avidin (or streptavidin) and B–AP are combined and used after a 30-min incubation at room temperature.

Comments. The optimal ratio for complex formation corresponds to about 2 μg avidin per unit of B–AP (Fig. 2); initial concentrations (before mixing) of 10 μg/ml avidin and 5 units/ml B–AP are generally used. At relatively low avidin-to-B–AP ratios (i.e., 1 μg/ml and 10 units/ml, respectively), little labeling can be observed, presumably since most of the biotin-binding sites of avidin are involved in complex formation and few free sites are available to combine with the biotinylated target protein. Conversely, at high avidin-to-B–AP ratios (i.e., 30 μg/ml and 1 unit/ml, respectively), the labeling pattern is also impaired, apparently because extraneous uncomplexed avidin molecules compete with the complexes, thereby reducing the amount of enzyme available for detection. It should be noted that the concentration of complexes introduced into the system should be regulated in order to reduce the level of nonspecific binding. Specifically, high background levels are observed at B–AP concentrations higher than 10 units/ml, and such conditions should be avoided.

The optimal ratio for streptavidin-containing complexes is identical to that of avidin. The use of streptavidin usually results in much lower levels of either nonspecific binding or background, and it is generally recommended when available.

For new batches of biotinyl enzyme, such optimization experiments should always be performed. Stored samples of reagents should also be subjected periodically to such experiments; the complexes themselves are not stable for long periods of time and cannot be stored.

Avidin and Streptavidin Hydrazides

Hydrazide derivatives of avidin and streptavidin have been shown to be effective for selectively labeling glycoconjugates on blots.[19,21] The avidin hydrazide is combined with a biotinylated enzyme at an optimized ratio to form an amplified enzyme–hydrazide preparation, which can be used to label periodate-oxidized or galactose oxidase-treated saccharides.

Reagents

Avidin (Belovo or STC Laboratories), nonglycosylated avidin,[1] or streptavidin,[3,10] 50 mg/ml
Adipic acid dihydrazide (Sigma), 160 mg
Water-soluble carbodiimide,[22] 160 mg
Distilled water
5 *N* HCl
PBS (pH 7.4)

Procedure. The dihydrazide is dissolved by heating in 5 ml of distilled water, and the pH is adjusted to 5 with HCl. The solution is added to solid avidin or streptavidin (50 mg of the desired protein). The water-soluble carbodiimide is added in solid form, and the reaction is allowed to proceed for 3 hr at room temperature with periodic 30-min adjustments of the pH to 5. The avidin hydrazide is dialyzed exhaustively against PBS. The product is diluted to 50 ml with PBS, passed through a sterile Millipore filter (HA 0.2 μm), and stored in 1-ml aliquots under sterile conditions at 4°. Both avidin and streptavidin hydrazide are stable for years under these conditions.

[21] E. A. Bayer, H. Ben-Hur, and M. Wilchek, *Anal. Biochem.* **161,** 123 (1987).
[22] 1-Ethyl-3-(3-dimethylaminopropyl)carbodiimide hydrochloride (Sigma).

Complexes of Biotinylated Enzymes with Avidin (or Streptavidin) Hydrazides. To a solution of avidin hydrazide or streptavidin hydrazide (15 μg/ml, dissolved in 2% w/v lysozyme in PBS) is mixed an equal volume of biotinyl alkaline phosphatase[15,16] (7.5 μg/ml).[23]

Comments. The optimal ratio of hydrazide probe to biotinyl enzyme for complex formation is determined by dot-blot optimization experiments similar to that described above for avidin–enzyme complexes, except that the target molecule consists of a suitable glycoconjugate (e.g., fetuin) which is oxidized by enzymatic or chemical means.[19] The ratio of reagents is based on the results of such an experiment. As mentioned above, for new batches of biotinyl enzyme and avidin (or streptavidin) hydrazide, such optimization experiments should always be performed. Stored samples of reagents should also be subjected periodically to such experiments. The solution is applied to blots about 30 min after initiation of complex formation; the complexes themselves are not stable for long periods of time and cannot be stored.

Derivatization of a protein via its carboxyl group results in increased p*I* values; a neutral protein such as streptavidin becomes basic, and a basic protein such as avidin becomes even more basic. This phenomenon generates high levels of nonspecific binding which can easily be countered by quenching with appropriate reagents.[23] Thus, in lieu of using albumin (a negatively charged protein) as a quencher, we employ lysozyme (the p*I* of which is even higher than that of avidin).[21]

Immobilized Avidins

Avidin can be immobilized to solid matrices in many different ways. The classic method involving CNBr activation of Sepharose has been described explicitly in the past.[24] In this section, we present two alternative methods for immobilization of avidin to Sepharose.

Cyano-Transfer Activation of Sepharose

In the first method, the use of the extremely hazardous CNBr for activation is precluded by using a cyano-transfer derivative or intermediate.[25]

[23] For complexes containing avidin hydrazide, 2% lysozyme is used as a diluent; for complexes containing streptavidin hydrazide, the diluent consists of 2% BSA.

[24] E. A. Bayer, E. Skutelsky, and M. Wilchek, this series, Vol. 62 [55]; E. A. Bayer and M. Wilchek, *Methods Biochem. Anal.* **26,** 1 (1980); A. Bodanszky and M. Bodanszky, *Experientia* **26,** 237 (1970).

[25] M. Wilchek, T. Miron, and J. Kohn, this series, Vol. 104 [1].

Reagents

Avidin (Belovo or STC Laboratories), nonglycosylated avidin,[1] or streptavidin,[3,10] 200 mg in 100 ml of 0.1 *M* sodium bicarbonate
Sepharose 4B or CL-4B, 50 ml of swollen gel
N-Cyanotriethylammonium tetrafluoroborate (CTEA), 540 mg, or 1-cyano-4-dimethylaminopyridinium tetrafluoroborate (CDAP), 225 mg[26] (Sigma)
0.2 *M* Triethylamine (aqueous solution)
Acetone
Distilled water
Washing medium: acetone : 0.1 *N* HCl (1 : 1, v/v)
Storage medium: acetone : dioxane : water (60 : 35 : 5, v/v)
Coupling medium: 0.1 *M* sodium bicarbonate (pH 8.5)
PBS (plus 0.1% sodium azide, pH 7.4)

Procedure. The Sepharose is washed with water, 30% acetone, and 60% acetone, successively, and resuspended in 10 ml of 60% acetone. The suspension is cooled to 0°, and the desired cyano-transfer reagent is added. The triethylamine solution (5.4 ml for the CTEA-activated resin or 1.8 ml for the CDAP-activated resin) is then added dropwise with vigorous stirring. After 2 min, the reaction mixture is transferred to 100 ml of ice-cold washing medium. The resin can be maintained in this manner for about 1 hour without loss of active groups.[27]

The activated resin is washed with large volumes of cold water, followed by a rapid wash with coupling medium. The solution of avidin is mixed with the resin, and the interaction is carried out for 1 hr at room temperature and overnight at 4°. The resin is washed with coupling buffer and PBS,[28] and stored in PBS.

Comments. The above procedure results in about 80–90% coupling,[28] yielding about 3.5 mg avidin/ml swollen gel. The end products of the reaction are identical to those of the CNBr activation procedure, and a certain amount of leakage of avidin may thus be expected.[25] Nevertheless, these reagents are nonvolatile solid cyanoderivatives which can be stored and handled without use of a fume hood; they should therefore be used in place of CNBr since they are more convenient, more efficient and, of course, safer.

[26] Both reagents can be obtained from Sigma.

[27] For prolonged storage of cyano-transfer-activated Sepharose, the resin is washed extensively with storage medium and maintained at −20°. When required, the activated resin is reswollen for 5 min in cold washing medium and washed according to the procedure for immediate coupling.

[28] In the case of CTEA-activated resins, the A_{280} of the washes can be followed in order to determine the extent of coupling.

p-Nitrophenyl Chloroformate-Activated Sepharose

One of the major problems with CNBr (or cyano-transfer) activation of Sepharose is the observed leakage of ligand owing to the instability of some of the bonds formed. Alternative procedures have therefore been developed which provide improved chemical attachment of proteins and other ligands.[25] One of the most effective is the use of chloroformates that result in activated carbonate groups on the resin. Subsequent interaction with amines yields stable and uncharged carbamate (urethane) derivatives.

Reagents

Avidin (Belovo or STC Laboratories), nonglycosylated avidin,[1] or streptavidin,[3,10] 200 mg in 100 ml of 0.1 *M* sodium bicarbonate
Sepharose CL-4B (Pharmacia), 50 ml
p-Nitrophenyl chloroformate (Aldrich Chemical Company, Milwaukee, WI), 3 g
4-Dimethylaminopyridine (Sigma), 2.1 g
Acetone (absolute)
2-Propanol
Distilled water
Coupling medium: 0.1 *M* sodium bicarbonate (pH 8.5)
PBS (plus 0.1% sodium azide, pH 7.4)

Procedure. The activation procedure is carried out as described elsewhere in this volume.[10] The solution of avidin is then mixed with the washed, activated resin; the interaction is carried out for 1 hr at room temperature and overnight at 4°. The resin is washed successively with water, 0.2 *M* acetic acid, water, and (rapidly) with 10 m*M* NaOH. The resin is then washed exhaustively with distilled water, followed by coupling buffer and PBS. The immobilized avidin is stored in PBS.

Comments. Using the above procedure, about 3 mg avidin can be coupled per milliliter of Sepharose. The covalent bonds formed are very stable, and leakage is negligible. In other respects, the performance of such columns is equivalent to that of CNBr-activated resins.

[19] Phycobiliprotein–Avidin and Phycobiliprotein–Biotin Conjugates

By ALEXANDER N. GLAZER and LUBERT STRYER

The high sensitivity of fluorescence detection has encouraged the widespread use of fluorescent molecules or their biospecific conjugates for immunoassays, fluorescence-activated cell sorting, histochemistry, and numerous other applications in biology. In 1982, we showed that the unique properties of the phycobiliproteins could be exploited to generate novel fluorescent reagents that permit high-sensitivity multiparameter analyses of cells and molecules.[1]

The phycobiliproteins are found in all cyanobacteria (blue-green algae), as well as in two groups of eukaryotes, the red algae and the cryptomonads.[2] These brilliantly colored proteins are abundant cell constituents in all of these organisms and function as photosynthetic accessory pigments. Their strong colors arise from the presence of multiple covalently bound open-chain tetrapyrrole prosthetic groups (bilins). In intact cells, the phycobiliproteins are components of a macromolecular complex, the phycobilisome, which absorbs light over a large range of wavelengths and transfers the excitation energy to membrane-bound chlorophyll *a*-containing reaction centers. Upon breakage of cells, the phycobilisome dissociates, and the phycobiliproteins are released in water-soluble form. Previous contributions to this series should be consulted for information on preparation and characterization of phycobilisomes,[3] purification and properties of the individual phycobiliproteins,[4] and structures and modes of linkage of the bilins.[4,5] In this chapter, we focus on those properties of the phycobiliproteins directly relevant to their use as fluorescent reagents.

Phycobiliproteins purified to homogeneity are very stable and can be stored for long periods of time without detectable change in aggregation state or spectroscopic properties. The stability of these proteins impressed early investigators. In a 1910 paper that included a good review of

[1] V. T. Oi, A. N. Glazer, and L. Stryer, *J. Cell Biol.* **93,** 981 (1982).
[2] A. N. Glazer, *in* "The Biochemistry of Plants" (M. D. Hatch and N. K. Boardman, eds.), p. 51. Academic Press, London, 1981.
[3] A. N. Glazer, this series, Vol. 167, p. 304.
[4] A. N. Glazer, this series, Vol. 167, p. 291.
[5] A. N. Glazer, this series, Vol. 106, p. 359.

the earlier literature, Kylin[6] remarked, "Some of our phycoerythrin and phycocyanin crystal preparations have kept for twenty years in their mother liquor (about 10% ammonium sulfate solution), on slides under Canada balsam, without losing their shape or optical properties." Phycobiliproteins are highly soluble in suitable aqueous buffers, and solutions of 10–20 mg/ml can be prepared. In general, the proteins are stable from pH 5 to 9, although the aggregation state of certain of the phycobiliproteins is pH dependent over this range.[2] The isoelectric points of the phycobiliproteins lie between 4.5 and 5.5. Consequently, they are acidic at physiological pH values. Doubtless this property, coupled with the hydrophilic nature of the purified proteins, contributes to the absence of nonspecific binding of phycobiliproteins to cell surfaces.[1] The amino acid sequences of numerous phycobiliproteins have been determined both directly and from analyses of cloned genes.[7–9] On average, these proteins contain about 7 lysines per 100 residues[7] and can be coupled through ε-amino groups to a variety of small molecules or macromolecules without alteration in spectroscopic properties or stability of the phycobiliprotein.[1]

To date, the most widely used conjugates of the phycobiliproteins are those of R-phycoerythrin, B-phycoerythrin, and allophycocyanin. It should be emphasized that the stability and spectroscopic properties of these proteins vary with their organismal origin. The properties discussed here are those of *Gastroclonium coulteri* R-phycoerythrin,[10] *Porphyridium cruentum* B-phycoerythrin,[11] and *Anabaena variabilis* allophycocyanin.[12] *Gastroclonium coulteri* (Rhodymeniales) is a seaweed found along the Pacific coast from central to southern California.[13] *Porphyridium cruentum* (Bangiales) is a widely distributed unicellular red alga, and *Anabaena variabilis* is a common filamentous cyanobacterium.[14]

R-Phycoerythrin is a protein of 240,000 daltons with the subunit composition $(\alpha\beta)_6\gamma$ and carries 34 bilins.[10] The $(\alpha\beta)_6\gamma$ molecule is a stable assembly that does not dissociate even at 10^{-12} *M*. Two phycoerythrobilins are covalently attached to each α subunit, and two phycoerythrobilins

[6] H. Kylin, *Z. Physiol. Chem.* **69,** 169 (1910).

[7] H. Zuber, R. Brunisholz, and W. Sidler, *in* "Photosynthesis" (J. Amesz, ed.), p. 233. Elsevier, New York, 1987.

[8] A. N. Glazer, *Biochim. Biophys. Acta* **768,** 29 (1984).

[9] D. A. Bryant, *Can. Bull. Fish. Aquat. Sci.* **214,** 423 (1986).

[10] A. V. Klotz and A. N. Glazer, *J. Biol. Chem.* **260,** 4856 (1985).

[11] A. N. Glazer and C. S. Hixson, *J. Biol. Chem.* **252,** 32 (1977).

[12] D. A. Bryant, A. N. Glazer, and F. Eiserling, *Arch. Microbiol.* **110,** 61 (1976).

[13] G. M. Smith, "Marine Algae of the Monterey Peninsula," 2nd Ed. Stanford University Press, Stanford, 1969.

[14] R. Rippka, J. Deruelles, J. B. Waterbury, M. Herdman, and R. Y. Stanier, *J. Gen. Microbiol.* **111,** 1 (1979).

and one phycourobilin attached to each β subunit. The γ subunits carry three phycourobilins and one phycoerythrobilin.[10] The absorption maxima for the phycoerythrobilins and phycourobilins within native R-phycoerythrin lie at 530–565 nm and 495 nm, respectively.[2,4,5] R-Phycoerythrin has absorption maxima at 565 nm (ε_M 1.96 $\times$ 10^6 M^{-1} cm^{-1}), 539 nm (ε_M 1.62 $\times$ 10^6 M^{-1} cm^{-1}), and 498 nm (ε_M 1.53 $\times$ 10^6 M^{-1} cm^{-1}),[15] and a fluorescence emission maximum at 578 nm ($Q = 0.82$).[16] B-Phycoerythrin is likewise a stable $(\alpha\beta)_6\gamma$ 240,000 dalton complex containing 34 bilins in a disk-shaped molecule, 110 Å in diameter and 60 Å thick.[11] This complex contains 32 phycoerythrobilins and 2 phycourobilins, has absorption maxima at 545 nm (ε 2.41 $\times$ 10^6 M^{-1} cm^{-1}) and 563 nm (ε 2.33 $\times$ 10^6 M^{-1} cm^{-1}),[11,17] and a fluorescence emission maximum at 578 nm ($Q = 0.98$).[18]

Anabaena variabilis allophycocyanin is an $(\alpha\beta)_3$ complex of 100,000 daltons with an absorption maximum at 650 nm (ε 6.96 $\times$ 10^5 M^{-1} cm^{-1})[19] and a fluorescence emission maximum at 660 nm ($Q = 0.68$).[18] At low protein concentrations ($<10^{-9}$ M), allophycocyanin tends to dissociate to the monomer, $\alpha\beta$, with concomitant decrease in fluorescence quantum yield and shift in the absorption maximum to 620 nm.[20] In the context of the use of allophycocyanin in fluorescent conjugates, the instability of the trimer is clearly an undesirable property. This problem has been obviated by the preparation of a very stable allophycocyanin $(\alpha\beta)_3$ complex by cross-linking the trimer with the zero-length cross-linker 1-ethyl-3-(dimethylaminopropyl)carbodiimide.[20] This unusually heat-stable derivative can be readily conjugated to other molecules, and its spectroscopic properties have been extensively investigated.[21]

Phycobiliprotein conjugates make possible fluorescence-based analyses of far higher sensitivity than that achievable with other fluorophores. For example, at the argon ion laser line at 488 nm, ε is 1.28 $\times$ 10^6 (Q = 0.82) for R-phycoerythrin, whereas ε is 8 $\times$ 10^4 ($Q = 0.9$) for fluorescein, the most commonly used fluorescent chromophore for labeling. Consequently, a solution of R-phycoerythrin excited at 488 nm emits 14.5 times more fluorescence than that of an equimolar solution of fluorescein. The

[15] M.-H. Yu, A. N. Glazer, K. G. Spencer, and J. A. West, *Plant Physiol.* **68,** 482 (1981).

[16] D. J. W. Barber and J. T. Richards, *Photochem. Photobiol.* **25,** 565 (1977).

[17] D. J. Lundell, A. N. Glazer, R. J. DeLange, and D. M. Brown, *J. Biol. Chem.* **259,** 5472 (1984).

[18] J. Grabowski and E. Gantt, *Photochem. Photobiol.* **28,** 39 (1978).

[19] G. Cohen-Bazire, S. Béguin, S. Rimon, A. N. Glazer, and D. M. Brown, *Arch. Microbiol.* **111,** 225 (1977).

[20] L. J. Ong and A. N. Glazer, *Physiol. Veg.* **23,** 777 (1985).

[21] S. W. Yeh, L. J. Ong, J. H. Clark, and A. N. Glazer, *Cytometry* **8,** 91 (1987).

observed intensity ratio also depends on the wavelength dependence of the efficiency of the detection system, inasmuch as the emission maxima of these two fluorophores differ. A phycoerythrin/fluorescein intensity ratio of 10 was measured when equimolar solutions of the fluorophores were flowed through a cell sorter.[1] The sensitivity of detection can be increased further by exciting at the absorption maximum of the phycobiliprotein, e.g., 545 nm for B-phycoerythrin. This is particularly important for the detection of a cell surface antigen present at low density or for the detection of a minor population of cells. It is also important to note that the phycobiliproteins are more photostable than fluorescein. The photodestruction quantum yield (ϕ) for fluorescein is 2.7×10^{-5}, whereas the corresponding values for R-phycoerythrin (1.1×10^{-5}), B-phycoerythrin (6.6×10^{-6}), allophycocyanin (4.5×10^{-6}), and C-phycocyanin (2.5×10^{-6}) are significantly lower.[22,23] A few examples of the many immunological studies dependent on the intense fluorescence emission of the phycobiliproteins are cited below.

McHugh *et al.*[24] used a streptavidin–phycoerythrin conjugate to detect the Gerbich (Ge) antigen on the surface of human erythrocytes. In this study, the cells were exposed to anti-Ge antibody, followed by biotinylated goat $F(ab')_2$ anti-human IgG, and finally streptavidin–B-phycoerythrin. One-color immunofluorescence as detected by flow cytometry permitted detection of as few as 500 copies of antigen per cell. Erythrocytes from normal donors expressed an average of 7,800 Ge antigens per cell. In a parallel procedure, using fluoresceinated avidin, the signal-to-noise ratio was insufficiently high for satisfactory detection of this low-level antigen. Phycoerythrin has also been used as a probe for antigen-binding memory B cells.[25] Memory B cells in spleens from phycoerythrin-primed mice stain brightly with phycoerythrin and can be readily isolated by means of the fluorescence-activated cell sorter. Dinitrophenyl-B-phycoerythrin and -C-phycocyanin were used as fluorescent antigens to study the dynamic properties of antigen (antidinitrophenyl)–IgE–receptor complexes on rat mast cells.[26] A combination of flow cyto-

[22] R. A. Mathies and L. Stryer, *in* "Applications of Fluorescence in the Biomedical Sciences" (D. L. Taylor, A. S. Waggoner, F. Lanni, R. F. Murphy, and R. Birge, eds.), p. 129. Alan R. Liss, New York, 1986.

[23] J. C. White and L. Stryer, *Anal. Biochem.* **161,** 442 (1987).

[24] T. M. McHugh, M. E. Reid, D. P. Stites, E. S. Chase, and C. H. Casavant, *Vox Sang.* **53,** 231 (1987).

[25] K. Hayakawa, R. Ishii, K. Yamasaki, T. Kishimoto, and R. R. Hardy, *Proc. Natl. Acad. Sci. U.S.A.* **84,** 1379 (1987).

[26] J. C. Seagrave, G. G. Deanin, J. C. Martin, B. H. Davis, and J. M. Oliver, *Cytometry* **8,** 287 (1987).

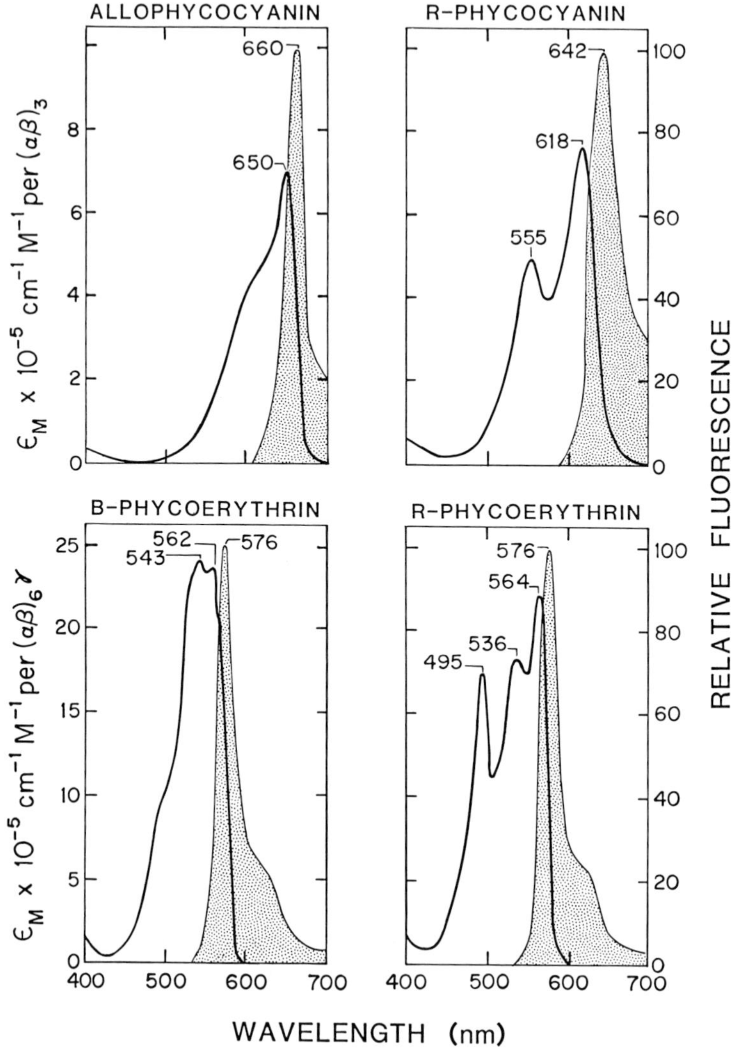

FIG. 1. Absorption and fluorescence emission (dotted) spectra of *Anabaena variabilis* allophycocyanin, *Porphyridium cruentum* R-phycocyanin and B-phycoerythrin, and *Gastroclonium coulteri* R-phycoerythrin.[27]

metry and fluorescence microscopy was used to quantify antigen binding and to follow the internalization and intracellular localization of the bound antigen.

The absorption maxima of the various phycobiliproteins span the wavelength range from 495 to 670 nm and the emission maxima the range from 565 to 680 nm. Moreover, the phycobiliproteins have broad absorption bands and large Stokes shifts (Fig. 1).[27] These properties can be exploited by the simultaneous use of several different phycobiliprotein conjugates and those of dye fluorophores in multiparameter flow cytometric analyses (reviewed in Refs. 27–29). For example, simultaneous measurements at four emission wavelengths, with two-laser excitation, have been made on lymphocytes stained with four different monoclonal antibodies conjugated to fluorescein, phycoerythrin, Texas red, and allophycocyanin, respectively.[30]

Biotin and avidin (or streptavidin) conjugates of the phycobiliproteins are convenient derivatives for many applications. Methods for the preparation of such conjugates are given below. Detailed instructions for the preparation of phycobiliprotein conjugates are also provided by Hardy.[31]

Preparation of Phycoerythrin–Biotin[1]

A 50-μl aliquot of *N*-hydroxysuccinimidobiotin (1 mg/ml) in dimethyl sulfoxide is added to 1 ml of 2.7 mg/ml R-phycoerythrin (or B-phycoerythrin) in 50 m*M* sodium phosphate (pH 7.5) to give a reagent-to-phycoerythrin molar ratio of 13. After 90 min at room temperature, the reaction is quenched by the addition of 10 μl of 100 m*M* glycylglycine (pH 7.5), and the solution is dialyzed for 3 days at 4° against 50 m*M* sodium phosphate (pH 6.8). An average of one biotin per phycoerythrin is incorporated under these conditions.

Preparation of Phycoerythrin–Avidin[1]

Biotinylated phycoerythrin (1 ml) obtained in the above manner is added slowly with stirring to 1 ml of 5 mg/ml avidin in the pH 6.8 phosphate buffer. The molar ratio of tetrameric avidin to phycoerythrin is 20. The mixture of phycoerythrin–biotin–avidin, biotin–phycoerythrin, unmodified phycoerythrin, and avidin is fractionated by high-performance liquid chromatography (HPLC) on a Varian G3000SW gel filtration column eluted with 200 m*M* sodium phosphate (pH 6.8). High molecular weight phycoerythrin–biotin–avidin complexes elute in the void volume. The elution volume of the second peak corresponds to that expected for a

[27] A. N. Glazer and L. Stryer, *Trends Biochem. Sci.* **9,** 423 (1984).

[28] D. R. Parks and L. A. Herzenberg, this series, Vol. 108, p. 197.

[29] M. N. Kronick, *J. Immunol. Methods* **92,** 1 (1986).

[30] J. A. Steinkamp, R. C. Habbersett, and C. C. Stewart, *Cytometry* **8,** 353 (1987).

[31] R. R. Hardy, *in* "Handbook of Experimental Immunology" (D. M. Weir, L. A. Herzenberg, C. C. Blackwell, and L. A. Herzenberg, eds.), 4th Ed., p. 31.1. Blackwell, Edinburgh, Scotland, 1986.

1 : 1 phycoerythrin–avidin conjugate. High molecular weight complexes (or aggregates) should always be removed because their presence leads to high background signals arising from nonspecific binding. Direct conjugation of phycobiliproteins to avidin is discussed by Hardy.[31]

Acknowledgments

Work in the authors' laboratories was supported by National Science Foundation Grant DMB 8816727 (A.N.G.), National Institutes of Health Grant GM 28994 (A.N.G.), and NIH Grant GM 24032 (L.S.).

[20] Monovalent Avidin Affinity Columns

By RONALD A. KOHANSKI and M. DANIEL LANE

Monovalent and tetravalent avidin affinity columns retain biotin-containing compounds. Originally described by Green and Toms,[1] avidin tetramers coupled to agarose retain the capacity for the nearly irreversible binding of biotin. Once bound, however, the biotin-containing compound is not easily released, even in the presence of high concentrations of free biotin or extremes of pH. However, denaturation followed by renaturation of the covalently linked avidin subunits lowers the affinity for biotin. The change in dissociation constant from approximately 10^{-15} *M* (tetravalent) to about 10^{-7} *M* (monovalent) means that biotinyl groups will be bound efficiently but can be displaced by millimolar concentrations of free biotin, simply by mass action. Because the affinity for biotin is lower with avidin monomers than with avidin tetramers,[2] columns containing these reagents became known as avidin monomer affinity columns. It is also characteristic of avidin monomers that biotin binding is very weak below about pH 2 (K_D ~1 m*M*), which allows monovalent avidin columns to be recycled.

Historically, monovalent avidin affinity columns have been used to purify biotin-dependent carboxylases,[3–5] which contain endogenous biotinyl groups linked through amide bonds to specific lysyl ε-amino groups

[1] N. M. Green and E. J. Toms, *Biochem. J.* **133,** 687 (1973).

[2] J. Moss and M. D. Lane, *Adv. Enzymol.* **35,** 321 (1971).

[3] K. Henrikson, S. H. G. Allen, and W. L. Maloy, *Anal. Biochem.* **94,** 366 (1979).

[4] R. A. Gravel, K. F. Lam, D. Mahuran, and A. Kronis, *Arch. Biochem. Biophys.* **201,** 669 (1980).

[5] N. B. Beaty and M. D. Lane, *J. Biol. Chem.* **257,** 924 (1982).

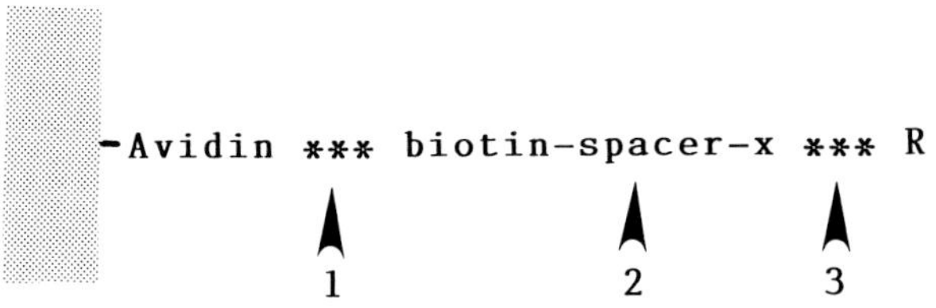

FIG. 1. Three possible methods to elute a biotin-containing compound adsorbed to an avidin affinity column. The most general method is displacement from a monovalent column with excess free biotin (method 1).[3-5,7] A second method involves the chemical structure of the spacer arm through which compound x is linked to biotin and therefore bound by the avidin column. Compound x can be released if the spacer arm contains a selectively cleavable group (method 2; e.g., a disulfide bond[9]). Method 3 applies if compound x is used to adsorb another molecule R (e.g., x is a hormone and R a receptor); then R may be recovered by weakening the interaction between R and x (cf. Refs. 6 and 8). The simplest and most general elution method, method 1, can be used only with monovalent avidin columns.

of the enzyme. Introduction of biotin through chemical coupling to a second molecule broadened the utility of avidin affinity columns for specific retrieval of receptors[6-8] (if the second molecule is a hormone) or recovery of a reporter group from enzyme-catalyzed reactions[9] (if the second molecule is a substrate). The reader is referred to two diverse applications of monomeric avidin columns: (1) purification of acetyl-CoA carboxylase,[5] an enzyme possessing a covalently bound biotinyl prosthetic group, and (2) purification of the insulin receptor[7] by use of the bifunctional affinity ligand $N^{\alpha,B1}$-biotinyl(ε-aminocaproyl)insulin. The isolation of any biotin-containing compound by avidin affinity chromatography is then reduced to the problem of its recovery from the affinity matrix, as illustrated in Fig. 1.

Preparation of Monovalent Avidin–Agarose

Tetrameric (native) avidin is first coupled to cyanogen bromide-activated agarose. Second, avidin subunits not covalently coupled to the solid phase are removed with the two-step denaturation protocol of Beaty and Lane.[5] Before use, the column is saturated with biotin, and then the "exchangeable" binding sites are freed at pH 1.5 leaving the nonexchangeable sites occupied by tightly bound biotin ($K_D \sim 10^{-15}$ M). Affinity chromatography can then be performed between pH 5.5 and 9.

[6] K. Hofmann and F. M. Finn, *Ann. N.Y. Acad. Sci.* **447,** 359 (1985); F. M. Finn, G. Titus, D. Horstman, and K. Hofmann, *Proc. Natl. Acad. Sci. U.S.A.* **81,** 7329 (1984).

[7] R. A. Kohanski and M. D. Lane, *J. Biol. Chem.* **260,** 5014 (1985).

[8] E. Hazum, I. Schvartz, Y. Waksman, and D. Keinan, *J. Biol. Chem.* **261,** 13043 (1986).

[9] B. Roseman, J. Lough, E. Houkom, and T. Herman, *Biochem. Biophys. Res. Commun.* **137,** 474 (1986).

Coupling Avidin to Agarose. Cyanogen bromide-activated agarose is either obtained commercially or is prepared with Sepharose CL-4B (Pharmacia, Piscataway, NJ) according to the activation procedure of March *et al.*[10] All of the following steps are performed at 4° or with ice-cold buffers. Washing of the agarose is done on a medium-frit sintered glass funnel. The activated agarose (40 ml packed volume) is washed successively with 800 ml each of 0.1 *M* sodium bicarbonate (pH 8.5), water, and 0.1 *M* sodium phosphate (pH 7). The activated and washed agarose is suspended in 30 ml of 0.1 *M* sodium phosphate (pH 7), and then 100 mg of avidin is added in 10 ml of 10 m*M* sodium phosphate (pH 7). Prior to adding the avidin, a 50-μl aliquot, diluted to 1 ml, is reserved to measure the absorbance at 280 nm. The slurry is rotated at 4° for 18–24 hr. Uncoupled avidin is removed by filtration, and the extent of coupling is determined by the loss of absorbance (280 nm) from the filtrate. The coupling efficiency is typically in the range 80–90% under these conditions.

The coupled agarose is washed with 200 ml of 10 m*M* sodium phosphate (pH 7), then 100 ml of 0.1 *M* 2-aminoethanol in 10 m*M* sodium phosphate (pH 7), and then resuspended in 80 ml of the latter buffer to block unreacted sites on the activated agarose. After rotating for 18–24 hr at 4°, the coupled and blocked agarose is washed with 1 liter of 10 m*M* sodium phosphate (pH 7). Tetrameric avidin–agarose may be stored at 4° for at least 2 years in 0.02% (w/v) sodium azide in a buffer at neutral pH.

Removing Uncoupled Avidin Subunits. Tetrameric avidin–agarose is packed from a 30% (w/v) slurry into a glass column. The solutions of denaturants are freshly prepared and filtered before use (cellulose acetate filters, 0.45-μm pores). Three bed volumes of 6 *M* guanidine hydrochloride in 0.2 *M* KCl (pH 1.5) is passed through the column. The column is washed with 10 m*M* sodium phosphate (pH 7) until the pH of the effluent is 6–7. Two bed volumes of 3 *M* guanidine thiocyanate in 0.2 *M* KCl (pH 1.5) is passed through the column, followed by reequilibration with 10 m*M* sodium phosphate buffer (pH 7). A flow rate of 2–3 bed volumes per hour is used with the denaturants, and the affinity matrix is never allowed to run dry. After the final equilibration with phosphate buffer, the monovalent avidin affinity matrix may be stored at 2–4° in any nondenaturing buffer at neutral pH with 0.02% sodium azide.

Characteristics of Monovalent Avidin–Agarose

There are, in fact, two classes of biotin-binding sites on monovalent avidin–agarose. These are distinguished by the ability of millimolar levels of unlabeled biotin to readily *displace* bound [^{14}C]biotin, i.e., exchange-

[10] S. C. March, I. Parikh, and P. Cuatrecasas, *Anal. Biochem.* **60,** 149 (1974).

able binding sites, versus [^{14}C]biotin that cannot be displaced, i.e., nonexchangeable binding sites. This definition is adopted to the elution of biotin-containing compounds. Biotin bound at the exchangeable sites is readily *dissociated* at pH 1.5–2.0, whereas biotin at the nonexchangeable sites remains tightly bound. This fact is used to generate (or regenerate) an avidin column with the nonexchangeable sites "blocked" and the exchangeable sites available for affinity chromatography.

Biotin-Binding Capacity and Characteristics of Monovalent Avidin Affinity Columns. To determine the total biotin-binding capacities in each class of sites, microcolumns are prepared using 0.2 ml of avidin–agarose packed in 1-ml plastic syringes, each fitted with a glass wool plug. The columns are characterized at 4° in 10 m*M* sodium phosphate buffer (pH 7). Each column is charged with 20 μM [^{14}C]biotin of known specific activity until the effluent radioactivity per microliter matches that of the stock solution. The columns are washed with buffer until the effluent radioactivity falls to 2% of the charge. The [^{14}C]biotin bound at the exchangeable sites is then eluted with 1 m*M* unlabeled biotin. The total [^{14}C]biotin-binding capacity of the column is calculated as the difference between applied radioactivity and that in the flowthrough from charging and washing the column. The exchangeable sites capacity is determined from [^{14}C]biotin eluted by unlabeled 1 m*M* biotin. The nonexchangeable sites capacity can be calculated as total *minus* exchangeable sites or by measuring radioactivity associated with the affinity column after the elution with unlabeled biotin. Results of this procedure are illustrated in Ref. 7.

The changes in [^{14}C]biotin binding from a typical conversion of tetravalent to monovalent avidin–Sepharose CL-4B are summarized in Table I. The tetravalent form exhibits greater than 99% irreversible binding of [^{14}C]biotin, whereas 87% is bound reversibly to the monovalent avidin column. The term "monovalent" avidin columns is appropriately

TABLE I
BIOTIN BINDING BY TETRAVALENT AND MONOVALENT AVIDIN–SEPHAROSE CL-4B[a]

	[^{14}C]Biotin bound (nmol/ml)			
Avidin–Sepharose	NES	EXS	Total	EXS/NES
Tetravalent	84.3	0.68	85	0.01
Monovalent	2.7	17.3	20	6.4

[a] NES, Nonexchangeable sites; EXS, exchangeable sites. Adapted from Kohanski and Lane.[7]

applied, therefore, to the affinity column where the higher affinity, nonexchangeable sites have been blocked with biotin and the lower affinity, exchangeable sites are open and available for affinity chromatography.

Two other facts concerning biotin binding have practical importance for protein purification or the characterization of biotinylated compounds using monovalent avidin columns. The on-rate constant[11] for biotin binding to tetravalent avidin–Sepharose is approximately $1 \times 10^4\ M^{-1}\ sec^{-1}$, which is almost 4 orders of magnitude slower than the association rate constant in solution.[12] Binding of biotin conjugates is slower by a factor of 3- to 30-fold, and as a consequence adsorption to avidin affinity columns should be done with a low flow rate. The half-time for dissociation of the tetravalent avidin–[^{14}C]biotin complex has been estimated to be between 4 and 6 months; this is another practical meaning of "irreversible" binding. On the other hand, the dissociation of [^{14}C]biotin from the exchangeable sites of monovalent avidin–Sepharose has an off-rate constant of $1.8 \times 10^{-3}\ sec^{-1}$. Therefore, to exchange bound [^{14}C]biotin with millimolar concentrations of unlabeled biotin requires about 35 min (a half-time for exchange of ~5 min); this time is longer for the displacement of larger biotin conjugates. Thus, smaller elution volumes result from slower flow rates when displacing biotin conjugates from monovalent avidin columns.

Stability of Monovalent Avidin Columns. Both tetravalent and monovalent avidin affinity matrices should be stored at neutral pH, and both deteriorate if kept under acidic conditions.[7] Tetravalent avidin columns that are used with a biotin conjugate tend to release small amounts of the conjugate on storage, but they can be regenerated with a fresh charge of the conjugate; this is because of the finite, albeit long, half-time for dissociation[13] and does not seem to be due to a loss of avidin subunits or to avidin denaturation. Tetravalent avidin–Sepharose stored for longer than 1 year can be converted to monovalent avidin–Sepharose, giving the same results as with the freshly prepared material. If only small amounts of monovalent avidin–Sepharose are to be used at any given time, storage in the tetravalent form is preferred.

[11] The on- and off-rate constants were measured by Kohanski and Lane (unpublished results); binding of micromolar concentrations of [^{14}C]biotin to tetravalent avidin–Sepharose was quenched with millimolar concentrations of unlabeled biotin. The dissociation from the exchangeable sites of monovalent avidin–Sepharose was measured by dilution of preequilibrated material into millimolar concentrations of unlabeled biotin. Calculation of the dissociation constant ($K_D = k_{off}/k_{on} = 0.18\ \mu M$), assuming k_{on} is the same for exchangeable and nonexchangeable sites, agrees with the dissociation constant measured by direct titration of monovalent avidin–Sepharose ($K_D = 0.2\ \mu M$, Ref. 7).

[12] N. M. Green, *Adv. Protein Chem.* **29,** 85 (1975).

[13] See K. Hofmann and F. M. Finn, *Ann. N.Y. Acad. Sci.* **447,** 359 (1985).

Monovalent avidin–Sepharose, stored in the unused state for 6 months at 4° (pH 7) will lose only 1.5% of its *total* biotin-binding capacity, and some reversion of exchangeable biotin-binding sites to nonexchangeable binding sites will occur.[7] The increase in nonexchangeable biotin binding can be reversed by retreatment of the affinity matrix with 3 *M* guanidine thiocyanate at pH 1.5; however, there will be a slight loss of capacity after this procedure.[7] In contrast, monovalent avidin columns that are in frequent use are stable for up to 1.5 years with minimal deterioration of exchangeable binding-site capacity, provided that they are kept at neutral pH between uses. Unless there is some apparent loss in exchangeable biotin binding with use or after storage, repeated exposure to guanidine hydrochloride or guanidine thiocyanate should be avoided.

Affinity Chromatography with Monovalent Avidin Columns

The use of monovalent avidin columns for affinity purification entails the following steps:

1. Block the nonexchangeable biotin-binding sites by charging the column with 20 μM to 2 m*M* biotin, at neutral pH.
2. Remove biotin from the exchangeable sites by washing the column with 0.1 *M* glycine in 0.2 *M* KCl (pH 1.5).
3. Ensure that only one biotin-containing compound is present in the solution that is the starting material to undergo purification.
4. Adsorb the biotin-containing compound to the monovalent avidin column at a flow rate of less than 0.3 column volumes/hour.
5. Elute the biotin-containing compound with free biotin.
6. Regenerate the exchangeable biotin-binding sites by washing the column with 0.1 *M* glycine in 0.2 *M* KCl (pH 1.5).

In practice, blocking the nonexchangeable biotin binding sites is accomplished the same way that the [^{14}C]biotin-binding capacity of the column is measured. The column is saturated with 20 μM biotin at pH 7, washed at pH 1.5 to remove the more weakly bound biotin, and then equilibrated with a buffer in which the protein or other compound of interest is stable (in the pH range 5.5–9). Over this pH range, nonexchangeably bound biotin remains bound, and binding to the exchangeable sites is unaffected.[3] It is recommended that trace amounts of [^{14}C]biotin be used to monitor the preparation of the column. This will ensure that the capacity and characteristics of the avidin column are known prior to the affinity purification step.

In order to recover a single biotin-containing compound using monovalent avidin columns, there must be only one biotin-containing compound in solution. If a biotinylated "reporter group" or hormone is to be added to recover selectively a second molecule, other biotin-containing compounds must be removed before that biotinylated compound is added. This is easily accomplished by first removing endogenous biotinyl groups on a tetravalent avidin column.

The amount of monovalent avidin–agarose employed should be 5- to 10-fold in excess over the amount of biotinyl groups present in solution. This rule of thumb is derived from empirical observations on the purification of acetyl-CoA carboxylase[5] at pH 7, where a 40-ml column with a theoretical capacity for 100 mg of the enzyme can, in fact, purify only 20 mg at saturation. This general rule also applies in cases where two dissociation constants must be considered to obtain efficient affinity purification (see Fig. 1). As an example,[7] saturation of the insulin receptor with $N^{\alpha,B1}$-biotinyl(ε-aminocaproyl)insulin requires a concentration of the bifunctional hormone of 0.1 μM, at pH 8.25. To recover the bifunctional hormone–receptor complex (40–60 μg of pure receptor) efficiently, a 10-fold molar excess of exchangeable biotin-binding sites is required (a 0.6-ml monovalent avidin column).

A major consideration in both examples is that, following the adsorption of the protein to monovalent avidin–Sepharose, washing the column is performed with the same buffer in which the protein is adsorbed. This is a direct consequence of the elution protocol, which requires the presence of only 0.8–2 mM biotin to displace specifically a pure protein from the monovalent avidin column. The low-stringency wash and mild elution conditions give a stepwise recovery of protein that is remarkably high for an affinity chromatography step (60–93%). As mentioned earlier, during the elution step, a low flow rate is required to obtain the product in a minimal volume; a typical flow rate is 0.1–0.4 column volumes per hour, with 50–80% of the total protein eluted in one column volume. Also, because the elution is achieved by displacement with free biotin, a gradient in biotin concentration is not required. The column is regenerated for use by washing extensively with any "column buffer" and then washing with 4–10 column volumes of glycine buffer at pH 1.5. The column is then stored at neutral pH and may be reused with uniform results.

The major advantage of monovalent avidin affinity columns is primarily the simple and mild elution by displacement with biotin, which results in very high yields. The ease with which many compounds can now be biotinylated means that a single affinity matrix may be used for the isolation of many different compounds of biochemical interest.

[21] Prevention of Nonspecific Binding of Avidin

By RAYMOND C. DUHAMEL and JAMES S. WHITEHEAD

For the purposes here, nonspecific binding is defined as any unwanted binding of biotin-containing proteins or avidin conjugates to the material under study. The unwanted binding may in some cases involve specific high-affinity binding to extraneous components, such as endogenous biotin-containing proteins or lectins. Unwanted binding of biotin or avidin conjugates may be classified into four categories: (1) unwanted binding of the avidin or biotin conjugates to endogenous biotin-containing proteins or endogenous avidin, respectively, (2) unwanted binding of avidin to endogenous lectins, (3) unwanted binding of avidin to macromolecules through nonspecific ionic or hydrophobic interactions, and (4) nonspecific binding arising from other components in the detection system besides biotin or avidin (such as the antibodies, lectins, or reporter molecules to which they are conjugated). Although the last category may not directly involve biotin or avidin, a few comments are included concerning frequently encountered causes of nonspecific binding to antibodies and nitrocellulose since these are often used with biotin or avidin reagents.

For each category of nonspecific binding, the cellular component or tissue type associated with the binding and the most appropriate method for blocking the nonspecific binding are discussed. For simplicity, the focus of the discussion is avidin, but much of what is said can be equally well applied to streptavidin reagents. The advantages and disadvantages of streptavidin relative to avidin are also discussed.

Unwanted Binding to Endogenous Biotin or Avidin

Site of Binding. Endogenous avidin-binding activity is due to the presence of biotin-containing proteins in certain mammalian tissues. Free, unconjugated biotin is also present in tissues, but it is easily washed from the material under study and not relevant to this discussion. Biotin-containing cytoplasmic binding sites for avidin seem to be restricted to kidney, liver, pancreas,[1,2] and brain.[3,4] These tissues contain elevated levels

[1] G. S. Wood and R. Warnke, *J. Histochem. Cytochem.* **29,** 1196 (1981).
[2] R. F. Rowley and G. S. Eisenbarth, *Diabetes* **31,** 107 (1982).
[3] W. Y. Naritoku and C. R. Taylor, *J. Histochem. Cytochem.* **30,** 253 (1982).
[4] S. M. Levine and W. B. Macklin, *Brain Res.* **444,** 199 (1988).

of biotin-dependent carboxylases,[5] presumably in mitochondria, and it has long been recognized that the biotin moiety of these carboxylases is accessible to avidin binding. In fact, inactivation of carboxylase activity by avidin is a criterion for demonstrating biotin dependence.[6] Avidin also binds to biotin-containing components of the cell surface of certain bacteria, including *Escherichia coli,* which may be present in cell extracts from bacterial cultures.[7] Other bacteria[8] as well as other microorganisms,[9] plants,[10] and insects[11] may also contain accessible bound biotin residues.

Endogenous biotin-binding activity is not of concern in mammalian tissues since avidin has not been found in mammals, but some tissue-staining protocols involve the use of egg white, which is a rich source of avidin, to aid in attaching sections to glass slides. Avidin is isolated from chicken eggs and is present in certain other tissues of oviparous vertebrates. Avidin is produced under hormonal control in the oviducts of birds, amphibians, and reptiles. It was long thought that avidin was found only in the reproductive tissues, but it has been recognized that avidin production can be induced by tissue injury in a variety of other tissues, primarily as an inflammatory response. Avidin is also produced by avian cells in tissue culture and can be induced by Rous sarcoma virus.[4]

Blocking Method. The method of choice for blocking endogenous avidin-binding activity is a two-step pretreatment performed prior to incubation with biotinylated antibody and avidin conjugates.[12–21] The two-step

[5] M. E. Bramwell, *J. Biochem. Biophys. Methods* **15,** 125 (1987).

[6] R. G. Duggleby, P. V. Atwood, J. C. Wallace, and D. B. Keech, *Biochem.* **21,** 3364 (1982).

[7] H. A. Elo and J. Korpela, *Comp. Biochem. Physiol.* **78B,** 15 (1984).

[8] B. R. Jennings, H. Mincer, J. Turner, V. Baselski, and R. T. Kelly, *J. Clin. Microbiol.* **18,** 1250 (1983).

[9] M. E. Collins, M. T. Moss, S. Wall, and J. W. Dale, *FEMS Microbiol. Lett.* **43,** 53 (1987).

[10] B. J. Nikolau, E. S. Wurtele, and P. K. Stumpf, *Anal. Biochem.* **149,** 448 (1985).

[11] T. F. Tsai, R. A. Bolin, M. Montoya, R. E. Bailey, D. B. Francy, M. Jozan, and J. T. Roehrig, *J. Clin. Microbiol.* **25,** 370 (1987).

[12] T. Letonja and C. Hammerberg, *J. Parasitol.* **73,** 962 (1987).

[13] S. Mori, T. Akiyama, Y. Morishita, S.-I. Shimizu, K. Sakai, K. Sudoh, K. Toyoshima, and T. Yamamoto, *Virchows Arch. B* **54,** 8 (1987).

[14] K. Sakai, M. Takiguchi, S. Mori, O. Kobori, Y. Morioka, H. Inoko, M. Sekiguchi, and K. Kano, *J. Natl. Cancer Inst.* **79,** 923 (1987).

[15] T. V. Tuazon, E. E. Schneeberger, A. K. Bhan, R. T. McCluskey, A. B. Cosimi, R. T. Schooley, R. H. Rubin, and R. B. Colvin, *Am. J. Pathol.* **129,** 119 (1987).

[16] R. Jonsson, L. Klareskog, K. Backman, and A. Tarkowski, *Clin. Immunol. Immunopathol.* **45,** 235 (1987).

[17] S. Mori, Y. Morishita, K. Sakai, S. Kurimoto, M. Okamoto, T. Kawamoto, and T. Kuroki, *Acta Pathol. Jpn.* **37,** 1909 (1987).

[18] C. J. Verdi, T. M. Grogan, R. Protell, L. Richter, and C. Rangel, *Hepatology* **6,** 6 (1986).

[19] V. Glezerov, *J. Histotechnol.* **9,** 15 (1986).

avidin–biotin block consists of incubation with excess unlabeled avidin followed by incubation with excess free biotin. Avidin in the first step binds to the endogenous biotin, and the excess biotin in the second step blocks residual biotin-binding sites in the immobilized avidin. For histological sections, Wood and Warnke recommend 1.0–0.1 mg/ml avidin and 0.1–0.01 mg/ml biotin.[1] An abbreviated procedure consists of mixing the avidin conjugate with normal blocking serum and diluting the primary antibody in a buffer containing biotin. The rare occurrence of endogenous avidin can be blocked by preincubation with free biotin alone.

Unwanted Binding of Avidin to Endogenous Lectins

Site of Occurrence. Many tissues contain membrane-associated lectins. Most of the endogenous lectins encountered in staining recognize terminal α-linked mannose or terminal β-linked galactose residues. The galactose-specific lectins generally do not present problems, because neither avidin nor most of the enzymes used in conjugates contain terminal galactose residues. Avidin consists of 10% carbohydrate, which includes terminal mannose in the side chains. The mannose-specific lectins reactive with avidin seem to be restricted to liver, kidney, pancreas, and, occasionally, brain. Lectin binding is generally seen only in frozen sections since mammalian lectins are destroyed by most fixatives and paraffin-embedding procedures.

Blocking Method. It is possible to block lectin binding specifically by diluting the avidin conjugate in 200 mM α-methyl-D-mannoside. The glycoside is used rather than the free sugar, because the former competes for the binding site on the lectin more effectively. It should be noted, however, that prevention of lectin binding may not completely eliminate nonspecific binding in a particular tissue, since many of the same tissues that exhibit lectin binding are also rich in biotin-containing proteins. The two-step avidin-biotin blocking procedure described above eliminates nonspecific binding whether it is due to endogenous biotin or to lectins.

Nonspecific Binding of Avidin Owing to Ionic and Hydrophobic Interactions

Site of Binding. Avidin in some circumstances will bind to the nucleus of cells. Avidin binding has been reported to be specific for condensed

[20] T. A. Van Dyke, C. Finlay, D. Miller, J. Marks, G. Lozano, and A. J. Levine, *J. Virol.* **61,** 2029 (1987).

[21] J. Bresser and M. J. Evinger-Hodges, *Gene. Anal. Techn.* **4,** 89 (1987).

chromatin rather than uncondensed chromatin,[22] but diffuse staining of interphase nuclei has also been observed.[1,23]

Nonspecific binding can also occur in the cytoplasm. For example, avidin conjugates can bind to the cytoplasmic granules of some mast cells.[24] These cells contain heparin, a polysaccharide containing many sulfate and carboxyl groups. Since heparin is highly negatively charged, it has the potential to bind to proteins of high isoelectric point such as avidin ($pI > 10$), antibodies, and some enzymes. Nonspecific avidin staining of mast cells appears to be highly dependent on the ionic strength and pH of the buffers used in staining, the tissue of origin, and the fixative used in preparing sections for histology. Tissues fixed in Methacarn or Carnoy's fixatives seem to have the greatest potential for nonspecific staining of mast cells.

Nonspecific avidin staining of the cell surface has also been reported. It has been suggested that avidin binds to acidic components of the cell capsule.

Blocking Method. Ionic binding to avidin conjugates can frequently be suppressed by raising the ionic strength of dilution buffers by the addition of 0.3–0.5 *M* sodium chloride.[25,26] Raising the pH to 9.4 has also proved effective in some instances.[24] High pH does not affect the avidin–biotin interaction, since it is effective across a pH range from 3 to 10, but high pH may be incompatible with the primary interaction under study and must therefore be used with caution. Many lectins, for example, become inactive at high pH.[24]

Nonspecific binding that cannot be blocked by raising the ionic strength or pH may be due to hydrophobic or other protein–protein interactions. Addition of protein to diluent solution has proved effective at blocking such interactions. Bovine serum albumin is effective at a concentration of 1–3%, but it should be a crystalline grade. Spurious background staining has been associated with the use of noncrystalline grades. With antibodies, it is common practice to include nonimmune serum in the diluent to serve as a carrier and to minimize nonspecific binding. Poly-L-lysine, a highly basic polypeptide, has been used to block nonspecific binding of streptavidin–peroxidase in a nitrocellulose system.[27] Gelatin has been used to prevent nonspecific interactions with avidin-stabilized

[22] M. H. Heggeness, *Stain Technol.* **52,** 165 (1977).

[23] R. C. Duhamel and D. A. Johnson, *J. Histochem. Cytochem.* **33,** 711 (1985).

[24] G. Bussolati and P. Gugliotta, *J. Histochem. Cytochem.* **31,** 1419 (1983).

[25] C. J. P. Jones, S. M. Mosley, I. J. M. Jeffrey, and R. W. Stoddart, *Histochem. J.* **19,** 264 (1987).

[26] R. K. Clark, Y. Tani, and I. Damjanov, *J. Histochem. Cytochem.* **34,** 1509 (1986).

[27] L. Scopsi, B.-L. Wang, and L.-I. Larsson, *J. Histochem. Cytochem.* **34,** 1469 (1986).

gold sols.[28] Nonfat dry milk at a concentration of 5–10% has been used to block nuclear and cytoplasmic nonspecific binding of avidin conjugates in histological sections.[23] However, it has been reported that, in a nitrocellulose blot procedure, although nonfat dry milk eliminated the vast majority of the background staining, several distinct bands exhibited spurious binding to avidin–peroxidase, but not to biotin–peroxidase. The use of 0.15 or 0.3 *M* NaCl instead of nonfat dry milk eliminated both general background staining and binding to the spurious bands.[26]

Presence of Biotin in Protein Solutions. The presence of biotin in protein carrier solutions, such as nonimmune serum and nonfat dry milk, may reduce the signal-to-noise ratio of the detection system. Avidin–biotin detection systems are often so sensitive that even small amounts of free biotin in buffers can significantly suppress the signal. This is especially true with ABC complexes of avidin–biotin–peroxidase. The usual response of the experimenter to a reduced signal is to increase the concentration of the primary reagents, but this can have the effect of increasing the nonspecific background significantly. A better solution is to reduce or eliminate the free biotin in the affected reagent. Biotin-containing solutions can be dialyzed to remove free biotin, if necessary.

Nonspecific Binding Unrelated to Biotin or Avidin

Antibody Procedures. Since antibodies are frequently conjugated with either biotin or avidin, some commonly encountered causes of nonspecific binding to antibodies deserve comment. Since most tissues contain some immunoglobulin, residual cross-reactivity of the secondary antibody for endogenous immunoglobulin in the target tissue can result in spurious binding of secondary antibodies. Unless the cross-reactivity is very strong, however, adding 2% serum from the species of the tissue to the diluent of the secondary antibody will block the cross-reactivity.

IgM antibodies in whole serum are a frequent source of nonspecific staining because IgM is notoriously "sticky." Since many IgM antibodies are cryoglobulins which tend to precipitate when stored in the cold for lengthy periods, they can frequently be eliminated by centrifuging the primary antibody prior to use. If necessary, ammonium sulfate fractionation or DEAE chromatography can be employed to eliminate IgM class immunoglobulins.

Many animals, especially rabbits, have constitutive antibodies against tissue elements such as cytokeratins and connective tissues. These are

[28] O. Behnke, T. Ammitzboll, H. Jessen, M. Klokker, K. Nilausen, J. Tranum-Jensen, and L. Olsson, *Eur. J. Cell Biol.* **41**, 326 (1986).

generally IgM antibodies of low affinity, however, and false staining may be eliminated by using a higher dilution of primary antibody. Ideally, it is sound practice to circumvent this problem by affinity purification of the primary antibody on an antigen column, but diluting the primary antibody as much as possible and longer incubation of the primary antibody at 4° will generally eliminate the binding problems.

If protein is used as a blocking agent or carrier it should be compatible with the specificity of the secondary antibody. For example, nonfat dry milk should not be used with primary antibodies made in goat or sheep, because secondary antibodies to goat or sheep IgG generally cross-react strongly with immunoglobulin present in bovine milk.

Nitrocellulose Procedures. Detection systems that involve the binding of target macromolecules to nitrocellulose or other membrane materials require that a blocking solution be used to inactivate all remaining binding sites on the membrane. In some protocols, protein solutions such as 1–3% albumin or 5% nonfat dry milk are used as irreversible blocking agents. Subsequent incubations with avidin–biotin reagents are subject to the same problems of nonspecific binding owing to ionic or hydrophobic interactions with the immobilized blocking protein as discussed above. Other protocols use Tween 20, a detergent, as a blocking agent. The latter does not block irreversibly and must be included in all subsequent diluent and wash buffers. RIA grade Tween 20 may be preferred. Since Tween 20 binds to proteins, the addition of excess carrier proteins may render the Tween 20 ineffective.

Advantages/Disadvantages of Streptavidin

The previous discussion has focused primarily on avidin, but many of the comments apply equally to streptavidin. Streptavidin is only slightly anionic, contains no carbohydrate, and is assumed to bind biotin with the same affinity as avidin.[29] It has been suggested that the lack of carbohydrate and the lower isoelectric point of streptavidin would eliminate many of the sources of nonspecific binding associated with avidin.[30] The presumption of an advantage to streptavidin does not necessarily hold in all circumstances. In many cases, the native isoelectric point of avidin is irrelevant, since it is altered by derivatization with fluorochromes or by conjugation to enzymes of low isoelectric point.

There are certain disadvantages to streptavidin. Streptavidin is usually more expensive than avidin. The biotin-binding cleft in streptavidin is different than that of avidin and, in some cases, may require a longer

[29] L. Chaiet and F. J. Wolf, *Arch. Biochem. Biophys.* **106,** 1 (1964).
[30] T. V. Updyke and G. L. Nicolson, *J. Immunol. Methods* **73,** 83 (1984).

spacer arm for biotinylation to achieve optimum binding with streptavidin.

The lack of carbohydrate potentially eliminates the problem of nonspecific binding to endogenous lectins, but if the reagent system includes other macromolecules containing terminal mannose residues, such as horseradish peroxidase, unwanted binding of the reagent to lectins may still occur. Furthermore, as discussed above, the same tissues that contain endogenous avidin-binding lectins may also contain endogenous biotin-containing proteins.

General Recommendations

Frequently, background staining that is initially ascribed to endogenous biotin or avidin binding proves to be due to other components of the system. To determine whether nonspecific binding occurs in a particular system, controls should be performed in which each reagent in the system is systematically omitted.

An almost universal method for preventing nonspecific binding in avidin–biotin systems is the two-step sequential pretreatment with avidin and biotin. This approach may add unnecessary time and expense to the staining process, however, and it is essential only when the unwanted binding is due to endogenous biotin-containing protein or avidin. In many applications, raising the ionic strength of dilution buffers by the inclusion of 0.3 *M* NaCl will be sufficient to eliminate nonspecific binding arising from ionic interaction. If the unwanted binding is due to endogenous lectins, the addition of 200 m*M* α-methyl-D-mannoside will usually eliminate it. High salt and the glycoside can be combined, if desired. If the appropriate controls indicate that nonspecific binding has not been fully suppressed, the addition of protein blocking agents such as albumin, nonfat dry milk, or nonimmune serum can be considered, subject to the caution that the biotin content should not diminish the intensity of the reaction.

Although the focus of this chapter has been nonspecific staining encountered with the avidin–biotin system, it should not be inferred that nonspecific binding will inevitably occur. In fact, if the proper procedures for general immunohistochemical staining are followed, nonspecific staining is rarely seen. When used in an appropriate way and with a full awareness of its principles, the avidin–biotin system offers the potential for maximum staining sensitivity with no significant nonspecific background.

[22] Enzymatic and Radioactive Assays for Biotin, Avidin, and Streptavidin

By ERNEST V. GROMAN, JEFFREY M. ROTHENBERG, EDWARD A. BAYER, and MEIR WILCHEK

Since the mid-1980s there has been published a variety of methods to assay avidin, streptavidin, and biotin (Table I). These methods claim a substantial increase in sensitivity over previously published assays. This report, together with the other chapters in this section, summarizes these recent publications and mentions syntheses of the various reporters (tracers) for these assays. Salient characteristics of each of these assays are summarized in Table I. Table II shows the structures of the various tracers. We conclude this report with an unpublished procedure developed in our laboratory to assay biotin, avidin, and streptavidin, using imidazole derivatives of biotin.

Assays for Biotin and Avidin Using Enzyme Labels

Niedbala *et al.*[1] report a procedure for measuring avidin and biotin using biotinylated glucose-6-phosphate dehydrogenase as the reporter group. The assay procedure is based on the observation that biotinylated glucose-6-phosphate dehydrogenase is inactivated following complexation with avidin.[2] In this assay the extent of inactivation of glucose-6-phosphate dehydrogenase is determined by the number of unbound binding sites on avidin after reaction with biotin. Thus, in a typical assay format for biotin, the concentration of avidin and biotinylated glucose 6-phosphate is held constant and the biotin concentration varied to generate a standard curve. Increasing concentrations of biotin result in fewer free binding sites on avidin which in turn lead to higher levels of enzyme activity. The assay has the advantage of being rapid and simple to run, but assay performance is sensitive to the quality of the preparation of biotinylated glucose-6-phosphate dehydrogenase; moreover, the assay lacks sensitivity.

Bayer *et al.*[3] have published a sensitive assay procedure for biotin and avidin using microtiter plates coated with biotinylated bovine serum al-

[1] R. S. Niedbala, F. Gergits, and K. J. Schray, *J. Biochem. Biophys. Methods* **3,** 205 (1986).
[2] T. T. Ngo, H. M. Lenhoff, and J. Ivy, *Appl. Biochem. Biotechnol.* **7,** 443 (1982).
[3] E. A. Bayer, H. Ben-Hur, and M. Wilchek, *Anal. Biochem.* **154,** 367 (1986).

TABLE I
METHODS FOR MEASURING BIOTIN[a]

Detection method	Tracer	Sensitivity (fmol/tube)	Sample size (μl)	Incubation period	Assay format	Refs.
Chemiluminescence	ABEI–biotin	25	25	2 hr	Homogeneous	[b]
Enzyme	Biotin–G6PDH	10,000	100	65 min	Homogeneous	1
Enzyme	Biotin–AP	1	100	24 hr	Sequential/microtiter plate	3
γ-Counting	^{125}I-Labeled avidin	10	100	17 hr	Sequential/microtiter plate	4
γ-Counting	^{125}I-Labeled tyramine biotinamide	5	200	25 min	Solution phase/charcoal	6,7
γ-Counting	^{125}I-Labeled tyramine biotinamide	3	100	3.5 hr	Solution phase/second antibody separation	6,7
γ-Counting	^{125}I-Labeled biotin analog[c]	25	Various	24 hr	Sequential/cellulose–avidin solid phase	8
γ-Counting	^{125}I-Labeled biotin glycyltyrosine	1	Various	60 min	Sequential/Sepharose–streptavidin solid phase	10
γ-Counting	^{125}I-Labeled histamine biotinamide	10	1000	65 min	Simultaneous/magnetic avidin solid phase	This report

[a] ABEI, Aminobutylethylisoluminol; G6PDH, glucose-6-phosphate dehydrogenase; AP, alkaline phosphatase.

[b] E. J. Williams and A. K. Campbell, *Anal. Biochem.* **155,** 249 (1986).

[c] See Table II.

bumin. The coated microtiter plate is first reacted with streptavidin, then biotin, and finally biotinylated alkaline phosphatase. The main drawback of this assay is the generation of the enzyme signal, which takes up to 20 hours to achieve a maximum.

Assays for Biotin and Avidin Using Radioactive Tracers

Mock and DuBois[4] have developed a solid-phase radioligand assay for measuring biotin. The assay is run by incubating fixed amounts of ^{125}I-

[4] D. M. Mock and D. B. DuBois, *Anal. Biochem.* **153,** 272 (1986).

TABLE II
STRUCTURES OF ^{125}I-LABELED BIOTIN DERIVATIVES

Compound	Structure	Ref.
N-[β-(4-Hydroxy-[3-^{125}I]iodophenyl)ethyl] biotinamide	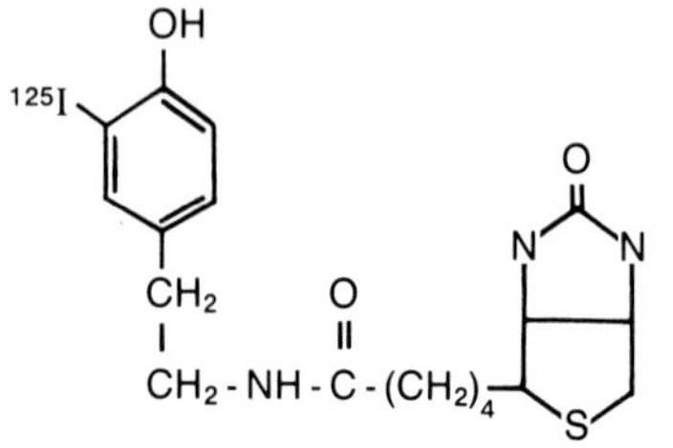	6
Bis-*N,N'*-[(6-aminohexyl)biotinamide]-[3-(4-hydroxy-[3-^{125}I]iodophenyl)propionic acid]	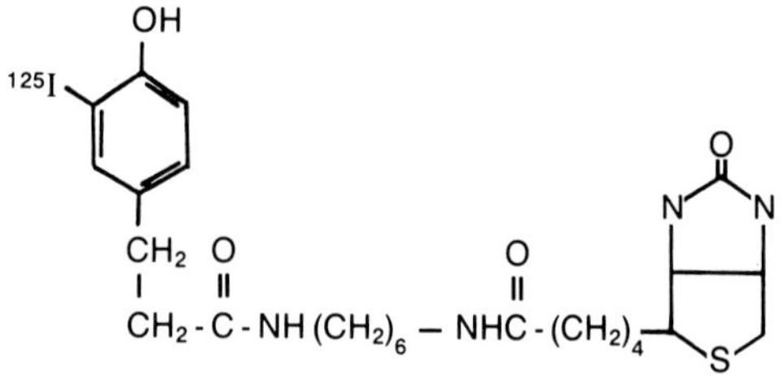	8
Biotinylglycine-(^{125}I-labeled tyrosine)	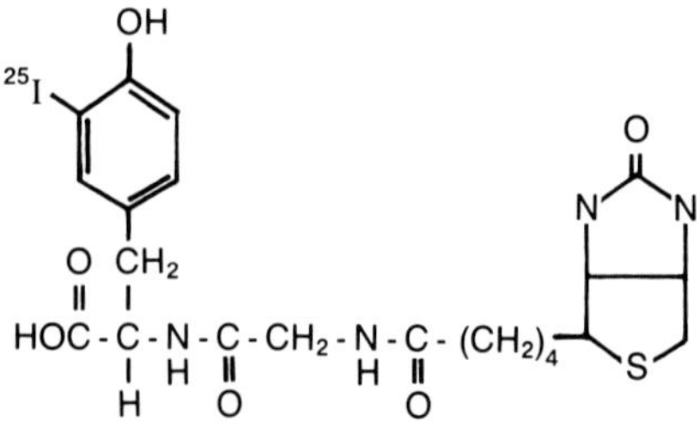	10
^{125}I-Labeled histamine biotinamide	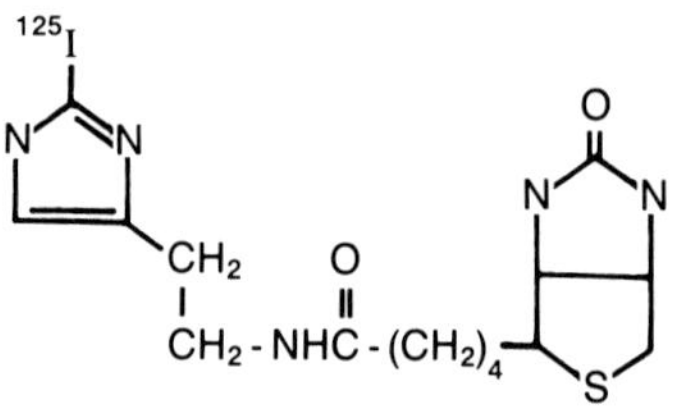	This rep

labeled avidin with varying amounts of biotin or unknown sample. A portion of the first incubation is then transferred to a microtiter plate coated with biotinylated bovine serum albumin. Increasing amounts of biotin in the standards occupy more of the biotin-binding sites on ^{125}I-labeled avidin, and this results in fewer counts bound to the plate. Procedures for iodinating avidin and coating biotinylated bovine serum albumin are detailed by Mock and DuBois.[4]

Ebrahim and Dakshinamurti[5] have presented an assay for biocytin. The procedure utilizes an anion-exchange resin to remove biotin during sample preparation (serum or urine). The remaining biocytin is measured by a competitive binding assay using avidin and [^{3}H]biotin as tracer.

Livaniou *et al.*[6,7] describe in detail the synthesis of two radioiodinated biotin derivatives in which the biotin ureido group remains intact. The synthesis is performed by coupling (at pH 8.5, 20–22°, for 90 min) biotinyl-*N*-hydroxysuccinimide ester to tyramine, which is radioiodinated prior to coupling using the chloramine-T method. Two ^{125}I-labeled derivatives are produced, *N*-[β-(4-hydroxy-3-iodophenyl)ethyl]biotinamide and *N*-[β-(4-hydroxy-3,5-diiodophenyl)enthyl]biotinamide. The final products are separated by thin-layer chromatography. The specific activity of the tracer was estimated to be about 350 Ci/mmol. The radioligand assay for biotin used one of these tracers, avidin as binding protein, and an antiavidin antibody as a separation reagent.

Chan and Bartlett[8] report the preparation of ^{125}I-labeled *N*-(6-aminohexyl)biotinamide. *N*-(6-Aminohexyl)biotinamide is synthesized and purified as described by Horsborough and Gompertz.[9] Iodination of *N*-(6-aminohexyl)biotinamide is accomplished with 3-(*p*-hydroxyphenyl)-propionic acid *N*-hydroxysuccinimide ester, which is iodinated using Iodogen. Assays for biotin are performed using avidin bound to cellulose or to carbonyl diimidazole-activated, glycerol-coated controlled-pore glass beads.

Smith *et al.*[10] report the synthesis of the biotin analog, biotinylglycyltyrosine, its iodination (Iodobeads), and its purification by HPLC to a specific activity of 2000 Ci/mmol. When this ^{125}I-labeled biotin analog is used in conjuction with immobilized streptavidin (CNBr-activated Sepharose 4B), an assay which detects 1 fmol of biotin can be obtained.

Synthesis of Iodinated Imidazole Derivatives of Biotin

The synthesis of the iodinated derivatives ^{125}I-labeled histamine biotinamide and ^{125}I-labeled histidine biotinamide and their use in the detection of biotin, streptavidin, and avidin are presented below. The

[5] H. Ebrahim and K. Dakshinamurti, *Anal. Biochem.* **162,** 319 (1987).
[6] E. Livaniou, G. P. Evangelatos, and D. S. Ithakissios, *J. Nucl. Med.* **28,** 1430 (1987).
[7] E. Livaniou, G. P. Evangelatos, and D. S. Ithakissios, *Clin. Chem.* (*N.Y.*) **33,** 1983 (1987).
[8] P. W. Chan and K. Bartlett, *Clin. Chem. Acta* **159,** 185 (1986).
[9] T. Horsborough and D. Gompertz, *Clin. Chem. Acta* **82,** 215 (1978).
[10] P. J. Smith, R. M. Warren, and C. von Holt, *FEBS Lett.* **215,** 305 (1987).

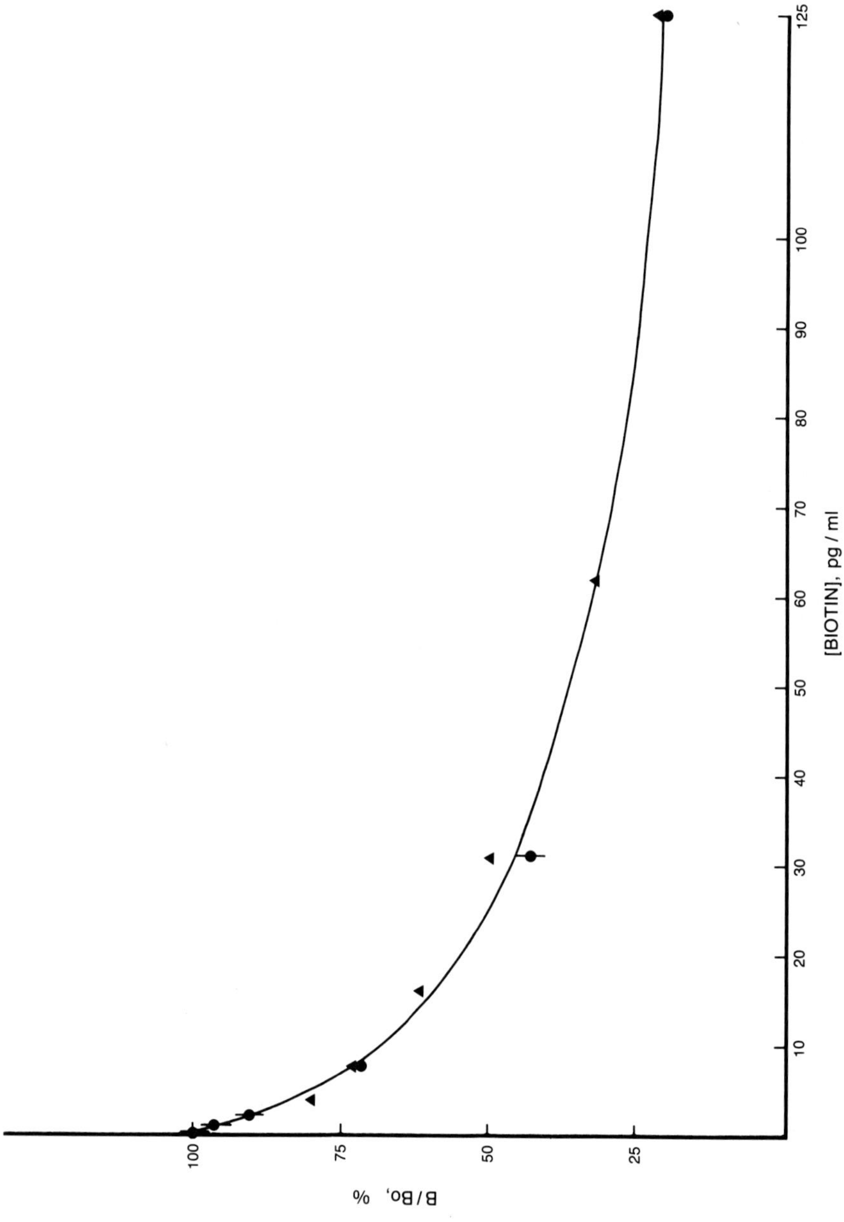
100
75
50
25
B/Bo, %
10
20
30
40
50
60
70
80
90
100
125
[BIOTIN], pg / ml

iodinated compounds are synthesized in a two-step process whereby histamine or histidine is initially iodinated and subsequently reacted with the *p*-nitrophenyl ester of biotin. The iodinated derivative is purified by passage over two ion-exchange columns to obtain radiochemically pure tracer (specific activity 100 Ci/mmol). Iodination, synthesis, and purification can be accomplished easily within a single day. The assay of biotin (Fig. 1), streptavidin (Fig. 2), and avidin (Fig. 3) is accomplished using a magnetically attractive solid phase to which avidin, rabbit antistreptavidin antiserum, or rabbit antiavidin antiserum is covalently attached.[11,12] All assays are performed in less than 75 min and have lower limits of detection of 10 fmol (biotin), 0.1 ng (streptavidin), and 0.1 ng (avidin).

Iodination of Histamine and Coupling to Biotin

The iodination of histamine and the coupling of ^{125}I-labeled histamine to the *p*-nitrophenyl ester of biotin are accomplished as follows. To 10 μl of histamine (20 μg/ml) in 0.5 *M* potassium phosphate (pH 8) are added in sequence 0.5 mCi of Na^{125}I (5 μl) (Amersham, Buckinghamshire, England) and 10 μl of chloramine-T (0.5 mg/ml dissolved in distilled water). The reaction is allowed to proceed for 3 min at room temperature. Next, 10 μl of sodium metabisulfite (1 mg/ml dissolved in distilled water) is added, and the reaction mixture is incubated for 1 min at room temperature. To the iodination mixture is added 50 μl of a freshly prepared solution of biotin *p*-nitrophenyl ester (0.8 mg/ml dissolved in dry dimethylformamide) at 0, 60, and 120 min. After 180 min the reaction mixture is diluted with 1 ml of 10 m*M* sodium acetate (pH 5), applied to a 4-ml QAE-Sephadex A-25 column equilibrated and eluted with the same buffer. The

[11] Magnetic avidin (Cat. No. M4620), carrier magnetic particles (M4100), and magnetic separator rack (4103) can be purchased from Advanced Magnetics, Inc., 61 Mooney St., Cambridge, MA 02138.

[12] The preparation of rabbit antiavidin and rabbit antistreptavidin was accomplished using standard procedures with Freund's complete adjuvant.

FIG. 1. Standard curve for biotin. The assay was performed using 10 μl of ^{125}I-labeled histamine biotinamide (total 20,562 cpm), 1 ml of standard dissolved in phosphate-buffered saline (pH 7.4)–0.1% bovine serum albumin, and 0.1 ml of magnetic avidin[11] [diluted 1/4000 into carrier magnetic particles[11] (0.5 mg/ml) in distilled water]. The resultant suspension was incubated for 60 min at room temperature, after which the magnetic avidin was separated magnetically for 5 min, using a magnetic separator,[11] and the clear supernatant was aspirated. The magnetic pellets were counted for 1 min. Circles indicate the averages of duplicate measurements, with the duplicate range indicated by a vertical line. Triangles indicate single points run on a separate assay. Maximum binding is 33% of total counts.

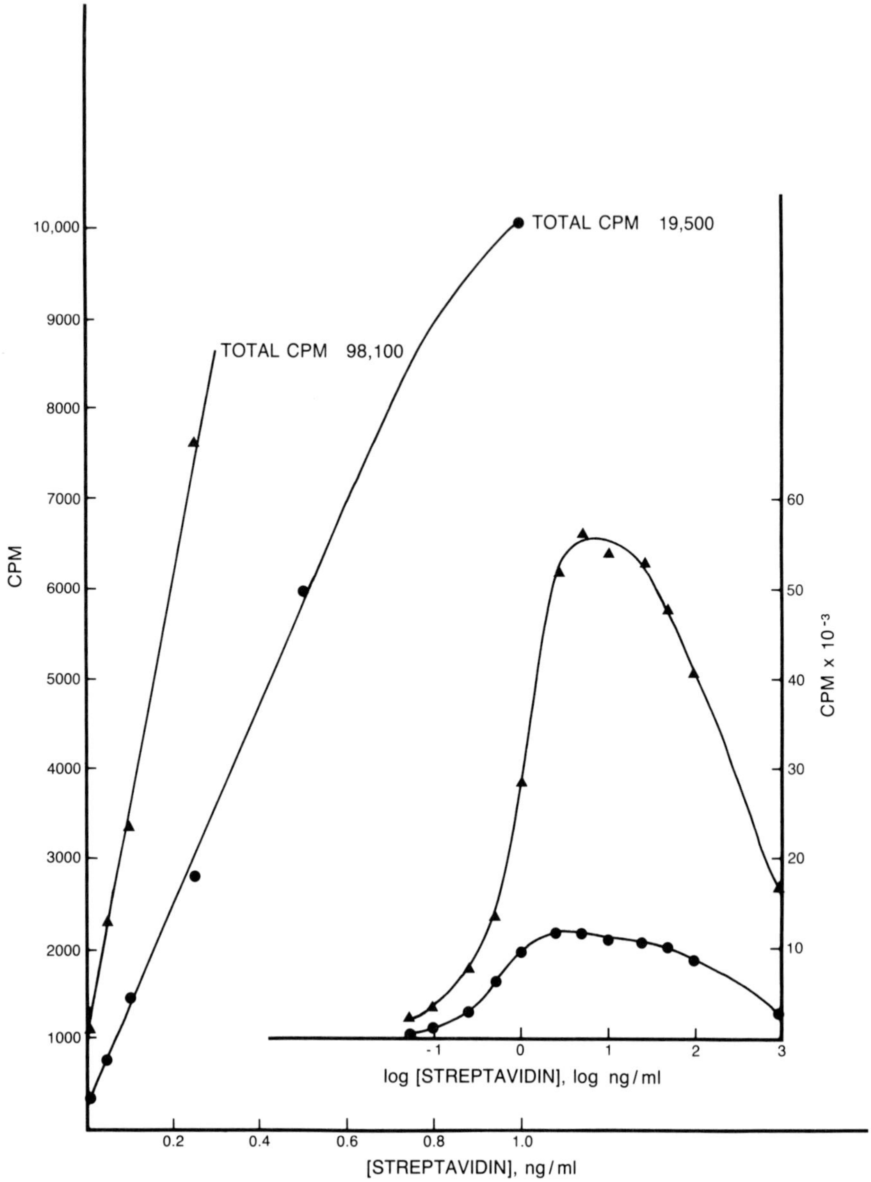

FIG. 2. Standard curve for streptavidin. The assay was performed using 50 μl of ^{125}I-labeled histamine biotinamide at the indicated count rates and 1 ml of standard dissolved in phosphate-buffered saline (pH 7.4)–0.1% bovine serum albumin which were incubated at room temperature for 1 hr. Following the incubation 0.1 ml of magnetic rabbit antistreptavidin[12] was added to each tube; the tubes were vortexed and then incubated for 15 min at room temperature. The magnetic particles were separated magnetically for 5 min using a magnetic separator, and the cleared supernatant was aspirated. The pellet containing the magnetic antistreptavidin–streptavidin–^{125}I-labeled histamine biotinamide complex was counted for 1 min. Assays were run using single-point determinations.

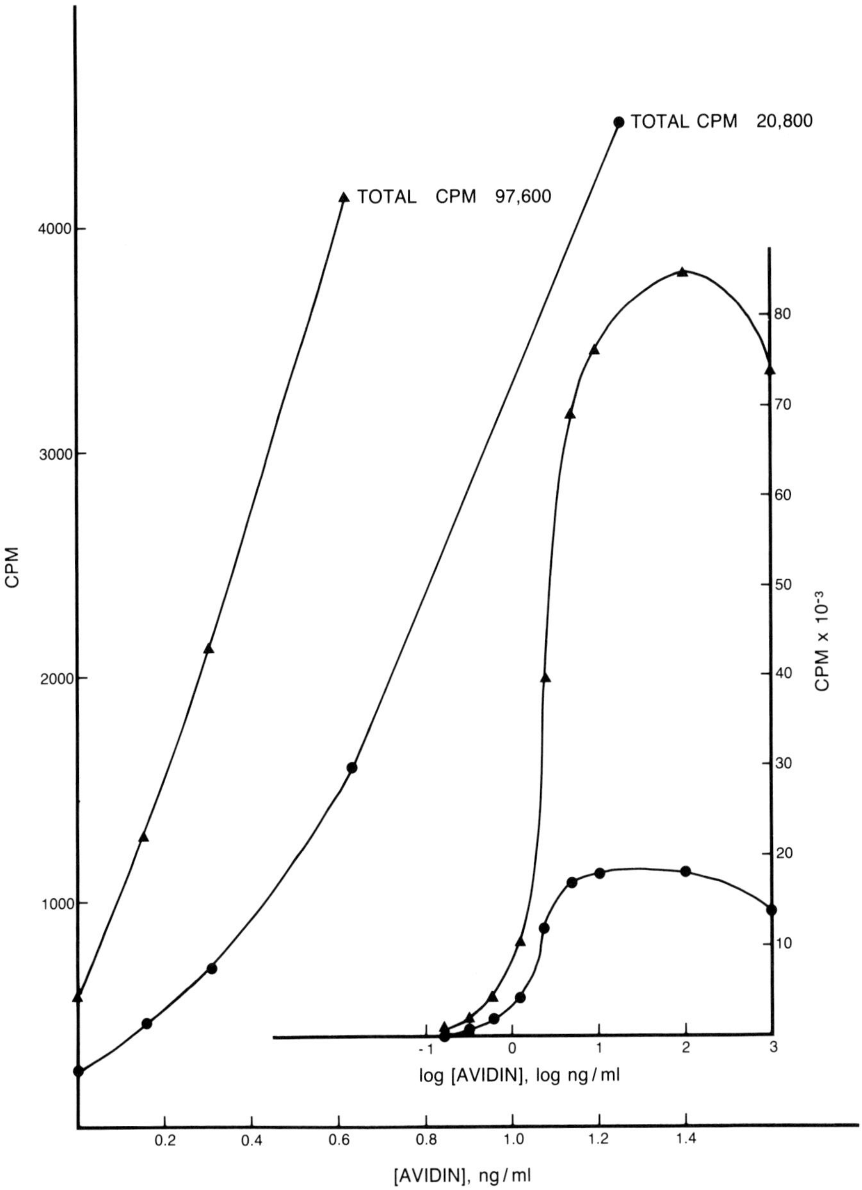

FIG. 3. Standard curve for avidin. The assay was performed using 50 μl of ^{125}I-labeled histamine biotinamide at the indicated count rates and 1 ml of standard dissolved in phosphate-buffered saline (pH 7.4)–0.1% bovine serum albumin which were incubated at room temperature for 1 hr. Following the incubation 0.1 ml of magnetic rabbit antiavidin[12] was added; the tube was vortexed and then incubated for 15 min at room temperature. The magnetic particles were separated magnetically for 5 min, using a magnetic separator, and the cleared supernatant was aspirated. The pellet containing the magnetic antiavidin–avidin–^{125}I-labeled histamine biotinamide complex was counted for 1 min. Assays were run using single-point determinations.

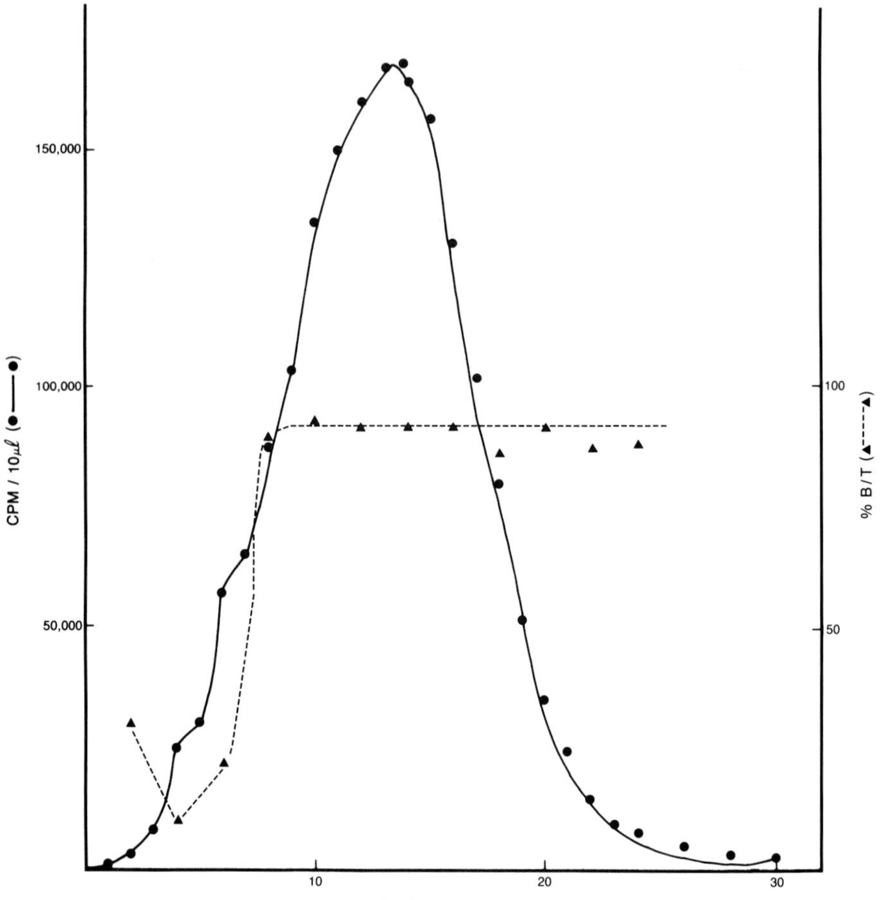

FIG. 4. Elution profile of ^{125}I-labeled histamine biotin derivative from a SP-Sephadex C-25 column. Fractions of 2.2 ml were collected. Aliquots of 10 μl were counted as indicated, and binding to magnetic avidin[11] was measured as follows. To 20 μl of tracer from each tube was added 1 ml of phosphate-buffered saline (pH 7.4)–0.1% bovine serum albumin and 10 μl of magnetic avidin.[11] The suspension was mixed and then incubated for 60 min at room temperature. The tubes were next placed in a magnetic separator rack[11] for 5 min, and the clear supernatants were aspirated. The magnetic pellets were counted and percent bound counts were calculated.

entire eluate (7 ml) is applied to a 4-ml SP-Sephadex C-25 column equilibrated and eluted with acetate buffer (Fig. 4). The imidazole-containing derivatives of biotin are very stable, and, unlike the tyrosine-based radiolabels, these tracers are exceptionally soluble in aqueous solutions.

Iodination of Histidine and Coupling to Biotin

The iodination of histidine and the coupling of ^{125}I-labeled histidine to the *p*-nitrophenyl ester of biotin are accomplished essentially following the procedure described above for histamine, except that histidine is substituted for histamine. The reaction mixture is diluted with 10 ml of 10 m*M* citric acid and applied to a 4-ml column of SP-Sephadex C-25 equilibrated with 10 m*M* citric acid. The column is eluted with 90 ml of 10 m*M* citric acid and then with 10 and 30 ml of 10 m*M* potassium phosphate, 0.15 *M* sodium chloride (pH 7.4) and 0.1 *M* potassium phosphate, 1.5 *M* sodium chloride (pH 7.4), respectively. All of the active tracer elutes between 95 and 110 ml of the column eluate.

Conclusions

Since the last publication of assay methods for biotin and its binding proteins in this series, substantial progress has been made in improving the quality, sensitivity, and convenience of assay procedures for these compounds. The research scientist now has a wide variety of methods and iodinated tracers to choose from for the use in the development of assays. Finally, the operation and development of these assays have been substantially simplified by the introduction of commercial reagents (e.g., magnetic solid-phase avidin) that can be used in the running of these assays.

[23] Colorimetric Enzyme Assays for Avidin and Biotin

By Edward A. Bayer, Haya Ben-Hur, and Meir Wilchek

A variety of assays are available for quantitative determination of avidin and biotin.[1] Most of these, however, are either cumbersome to perform or rely on the use of highly radioactive materials.[2] Others are either insensitive or not applicable for analysis of avidin or biotin in biological fluids owing to interfering substances.

For the assay of biotin, two major approaches have been used: microbiological assays, based on the growth of biotin-requiring strains, and the various avidin-based assays, which usually employ some form of isotope

[1] For a list of the assays for avidin and biotin which have been published previously in this series, see [2], Table II, in this volume.

[2] See chapters [22], [24], and [25] in this section.

dilution. For the assay of avidin (and streptavidin), the comparatively insensitive spectrophotometric and fluorometric assays are complemented by radioimmunoassay and by interaction of avidin with radioactive biotin.

In this chapter, we present convenient reciprocal assays for biotin, avidin, or streptavidin. The assays are based on the interaction of avidin with biotin-coated microtiter plates; residual free biotin-binding sites on the immobilized protein are interacted secondarily with a biotinylated enzyme that hydrolyzes a colored substrate. For the assay of biotin, the free vitamin is allowed to compete at selected concentrations with the biotinylated enzyme. The assays are extremely sensitive and convenient, and they circumvent the use of radiolabels.

Assay for Avidin and Streptavidin

Reagents

Biotinylated bovine serum albumin (B–BSA), 3 μg/ml, or biotin-ε-aminocaproyl–bovine serum albumin (Bcap–BSA), 1 μg/ml,[3,4] diluted in coating buffer

Avidin or streptavidin

Biotinylated alkaline phosphatase (B–AP) or biotin-ε-aminocaproyl–alkaline phosphatase (Bcap–AP),[3] 1 unit/ml

Coating buffer: 15 m*M* sodium carbonate buffer (pH 9.6)

Quenching solution: 0.3% (w/v) BSA

[3] E. A. Bayer and M. Wilchek, this volume [14].

[4] The exact concentration of B–BSA or Bcap–BSA optimal for coating the microtiter plates depends on the presence of the spacer group and on the amount of biotin moieties in the given preparation. In many of our preparations, 3 μg/ml has been found to be an optimal value for B–BSA and 1 μg/ml for Bcap–BSA. Nevertheless, this should be determined empirically for every preparation of the biotinylated protein. A typical experiment for doing so is carried out in our laboratory on microtiter plates. In such an experiment, B–BSA or Bcap–BSA, diluted in coating buffer, is applied to 11 of the columns on the plate at half-logarithmic incremented concentrations varying from 300 μg/ml to 3 ng/ml (the test concentrations can, of course, be further refined and varied, for example, linearly within a narrow range, say, from 30 μg/ml to 0.1 μg/ml). To the last column, coating buffer is added as a blank. After a suitable incubation period, the plates are rinsed 3 times with PBS and incubated with BSA solution (see text). To 7 of the rows of the microtiter plates, avidin or streptavidin is then added at half-logarithmic incremented concentrations varying from 10 μg/ml to 10 ng/ml. To the last row, PBS is added as a blank. After three PBS washes, B–AP or Bcap–AP solution is added, incubation is allowed to proceed for 30 min, the plates are washed, substrate solution is added, and the hydrolysis is checked at selected time intervals (e.g., 15 and 30 min, 1, 2, 3, 4, and 6 hr, and overnight). The optimum curves for the desired concentration ranges are drawn, and the appropriate concentration of the biotinylated carrier protein is established for future experiments.

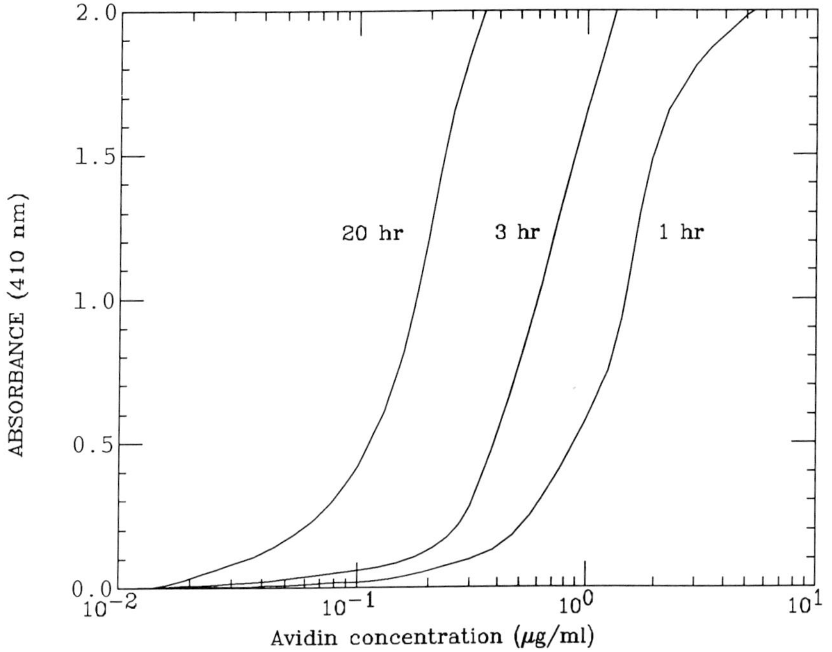

FIG. 1. Standard assay for avidin. B–BSA plates were treated successively with standard concentrations of avidin, B–AP, and substrate solution, and the plates were assayed after 1, 3, and 20 hr as shown.

Washing solution: 10 m*M* phosphate-buffered saline (pH 7.4, PBS)

Substrate solution: 10 mg *p*-nitrophenyl phosphate dissolved in 10 ml of 1 *M* diethanolamine buffer (pH 9.8) containing 0.5 m*M* $MgCl_2$

Procedure. A solution (0.1 ml) of B–BSA or Bcap–BSA is distributed in wells of microtiter plates and kept at 4° for 16–20 hr.[5] The wells are washed once, incubated for 1 hr with 0.2 ml quenching solution, and washed again.

Samples (0.1 ml) containing known concentrations of commercially prepared avidin or streptavidin (or serial dilutions of egg white, cell-free growth medium of *Streptomyces avidinii*, or any other biotin-binding protein) are added to the wells, and the plate is incubated for 30 min at room temperature. The plate is washed 3 times, B–AP or Bcap–AP solution

[5] B–BSA- or Bcap–BSA-coated microtiter plates are stored in hermetically sealed plastic bags and kept at 4° in PBS containing 0.1% sodium azide. Plates can be stored in this manner for at least 1 year without affecting their performance and are washed once before use.

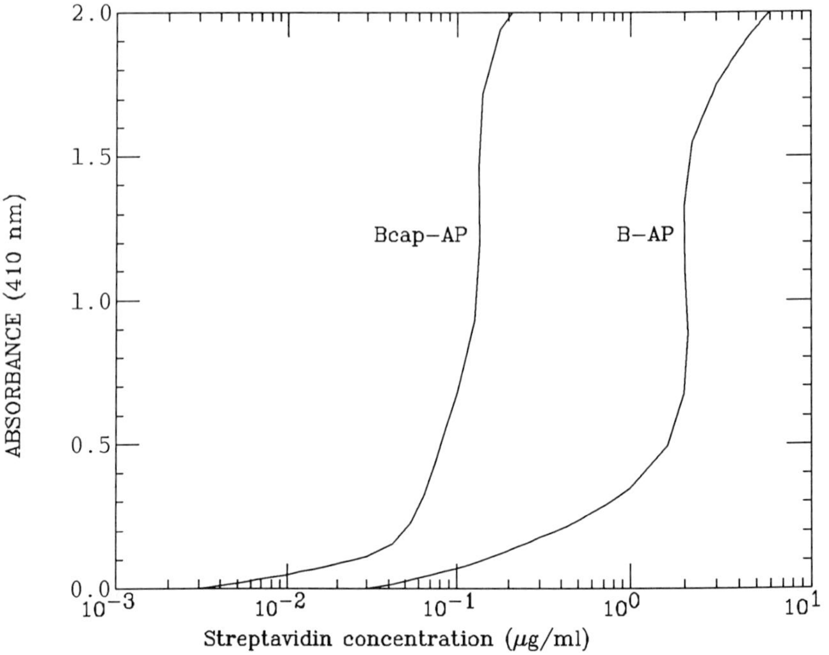

FIG. 2. The effect of spacer arm on the biotinylated enzyme. Streptavidin was assayed on B–BSA plates with either B–AP or Bcap–AP, and the plates were assayed after overnight incubation with substrate solution.

(0.1 ml) is then added to each well, and the plate is incubated for 30 min at room temperature. The plate is washed again, and substrate solution (0.1 ml) is added. The plate is read at various time intervals in an EL-310 EIA reader (Biotek Instruments, Inc., Burlington, VT), using a 410-nm filter.

Comments. A representative assay for avidin is presented in Fig. 1. One of the interesting and powerful characteristics of the assay is demonstrated: by extending the time of incubation, the assay becomes more sensitive. Thus, in a single experiment, the effective range of quantification can be varied over at least 2 orders of magnitude.

Another interesting feature of the system is the use of a biotinylated enzyme with an extended spacer arm, e.g., Bcap–AP instead of B–AP. This substitution results in an enhancement of the effective lower limit of sensitivity by a factor of at least 1 order of magnitude (Fig. 2). On the other hand, the substitution of Bcap–BSA for B–BSA does not significantly improve the assay system.

In using this approach, however, one should keep in mind that the signal is dependent on the capacity of the avidin sample to bridge between

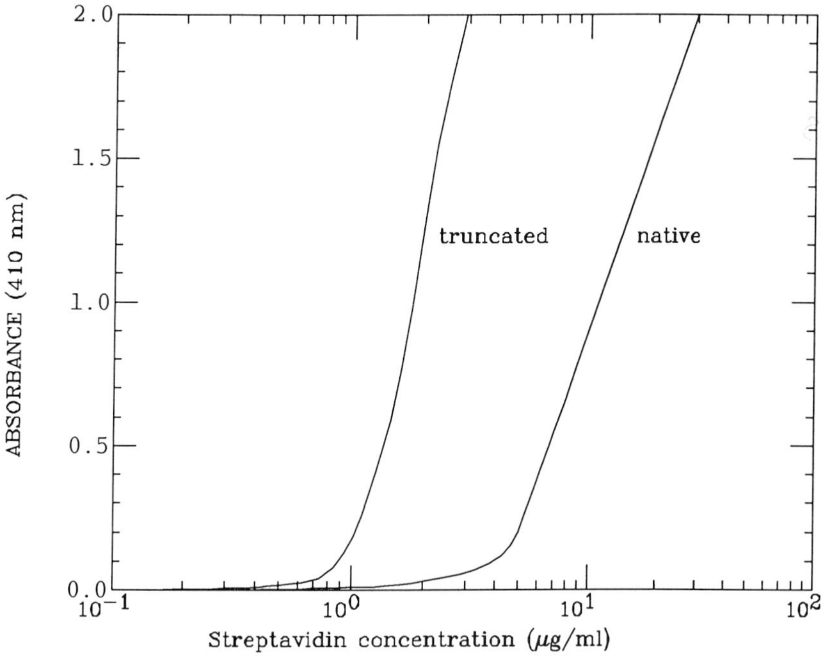

FIG. 3. Signal enhancement of truncated versus native streptavidin. Either the native or a truncated form of the protein was assayed on B–BSA plates with B–AP, and the plates were assayed after a 1-hr incubation with substrate solution.

two biotinylated molecules, i.e., the immobilized carrier protein and the enzyme. Thus, the integrity of the four biotin-binding sites (at least two in any given avidin molecule must be active) is imperative, and molecular alterations of avidin may hinder the assay. This effect has been used in the past for analyzing the biotin-binding capacity of avidin modified either by various group-specific reagents or by enzymatic means.[6] In some cases, surprising results may ensue. For example, in our studies on the bacterial biosynthesis of streptavidin, we were surprised to discover that a proteolysed truncated form of the molecule exhibited significantly *higher* biotin-binding activities than that of the native intact protein when using this assay system (Fig. 3).[7] The native molecule was found to comprise a

[6] G. Gitlin, E. A. Bayer, and M. Wilchek, *Biochem. J.* **242,** 923 (1987); G. Gitlin, E. A. Bayer, and M. Wilchek, *Biochem. J.* **250,** 291 (1988); G. Gitlin, E. A. Bayer, and M. Wilchek, *Biochem. J.* **256,** 279 (1988); Y. Hiller, J. M. Gershoni, E. A. Bayer, and M. Wilchek, *Biochem. J.* **248,** 167 (1987).

[7] E. A. Bayer, H. Ben-Hur, Y. Hiller, and M. Wilchek, *Biochem. J.* **259,** 369 (1989).

remarkably stable "core" with relatively short protease-sensitive appendages at both the N and C termini.[8,9] The data were interpreted in the following way: the terminal appendages of the native molecule sterically inhibit the interaction between the binding site and the biotinylated enzyme; their removal results in enhanced signals for the truncated (core) streptavidin that equal that of egg-white avidin. Indeed, when alternative assay systems are employed in which interaction with the free vitamin is examined, the biotin-binding activities of native and truncated streptavidin are essentially identical.

Assay for Biotin

Reagents

B–BSA, 20 μg/ml[10] in coating buffer
Streptavidin solutions,[11] containing 1, 5, or 25 μg/ml in PBS
Standard solutions of biotin (from 0.1 pg/ml to 10 ng/ml) or biotin unknown
B–AP, 1 unit/ml
Coating buffer: 15 m*M* sodium carbonate buffer (pH 9.6)
Quenching solution: 0.3% (w/v) BSA
Wash solution: PBS (pH 7.4)
Substrate solution: 10 mg *p*-nitrophenyl phosphate dissolved in 10 ml of 1 *M* diethanolamine buffer (pH 9.8) containing 0.5 m*M* $MgCl_2$

Procedure. Microtiter plates are coated with B–BSA.[5] The wells are washed once, incubated for 1 hr with 0.2 ml of quenching solution, and washed again.

Streptavidin solution (0.1 ml), at one of the above concentration levels, is applied to all of the wells of the plate, which is incubated for 30 min at room temperature. The plate is washed 3 times and incubated for another 30 min with standard concentrations of biotin (or serial dilutions of unknown samples). The plates are again washed briefly, B–AP solution

[8] E. A. Bayer, H. Ben-Hur, G. Gitlin, and M. Wilchek, *J. Biochem, Biophys. Methods* **13,** 103 (1986).

[9] C. E. Argarana, I. D. Kuntz, S. Birken, R. Axel, and C. R. Cantor, *Nucleic Acids Res.* **14,** 1871 (1986); A. Pähler, W. A. Hendrickson, M. A. Gawinowicz Kolks, C. E. Argarana, and C. R. Cantor, *J. Biol. Chem.* **262,** 13933 (1987).

[10] For biotin assay, the use of extended spacers, e.g., Bcap–BSA and Bcap–AP, has not been examined. Moreover, the use of B–BSA concentrations other than that reported above has not been investigated. There may thus be room for improvement in the assay of biotin through rigorous examination of the experimental conditions. In any case, such experiments should always be performed when using a new preparation of either the biotinylated carrier protein or the enzyme.

[11] E. A. Bayer, H. Ben-Hur, and M. Wilchek, this volume [8].

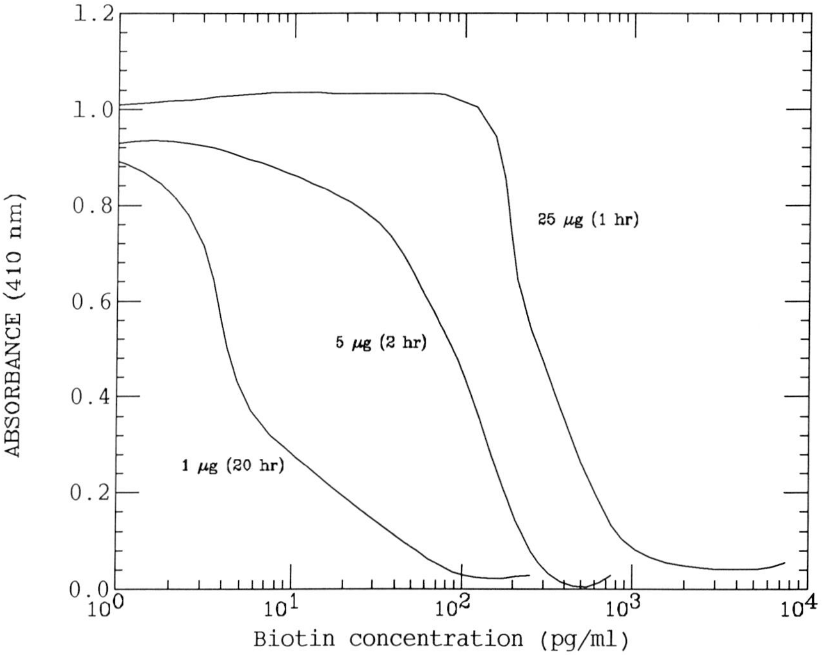

FIG. 4. Standard assay for biotin. B–BSA plates were coated with the designated concentrations (μg/ml) of streptavidin. Known concentrations of biotin were applied, followed by B–AP and substrate solution. The respective plates were assayed at the designated time intervals (in parentheses).

(0.1 ml) is added to each well, and the plate is incubated for 30 min at room temperature. The plate is washed again, and substrate solution (0.1 ml) is added. The plate is read at various time intervals in an EL-310 EIA reader using a 410-nm filter.

Comments. A representative assay for biotin is shown in Fig. 4. By altering the amount of streptavidin used to coat the B–BSA plates, a large range of biotin concentrations may be determined. Thus, plates prepared with 1, 5, or 25 μg/ml streptavidin are suitable for determining biotin concentrations of approximately 2–50, 20–300, and 100–800 pg/ml, respectively. Sensitivity of the assay is gained at the expense of time required for color formation of the yellow product: 30–60 min for the high range, several hours for the mid-range, and overnight for the low biotin concentration range.

[24] Sequential Solid-Phase Assay for Biotin Based on ^{125}I-Labeled Avidin

By DONALD M. MOCK

Interest in accurate measurement of biotin concentrations in plasma and urine has been stimulated by wider application of avidin–biotin techniques in biomedical research, by recent advances in the understanding of biotin-responsive inborn errors of metabolism, and by several reports describing biotin deficiency associated with parenteral nutrition. This chapter describes a biotin assay utilizing radiolabeled avidin in a sequential solid-phase method.[1] The assay has increased sensitivity compared to published methods based on avidin binding, correlates with expected trends in biotin concentrations in blood and urine in a rat model of biotin deficiency, and can utilize radiolabeled avidin available from a commercial source. The assay will detect several avidin-binding substances that have binding affinities equal to or not greatly decreased compared to that of biotin. Thus, the assay is not specific for biotin; when coupled with the HPLC separation described here, the method is a powerful tool for measuring biotin and biotin analogs in complex biological mixtures.

Assay Method

Principle. A fixed amount of ^{125}I-labeled avidin is incubated with varying concentrations of biotin in buffer (standard curve) and with several dilutions of the unknown samples (e.g., serum or urine). A portion of the first incubation is then transferred to microtiter plates previously coated with biotin covalently linked to bovine serum albumin (BSA). After a second incubation, the wells are washed with buffer and counted individually. Increasing amounts of biotin in the standards or unknowns occupy an increasing proportion of the biotin-binding sites on ^{125}I-labeled avidin in the first incubation, and progressively fewer counts are bound to the well in the second incubation. "Biotin" concentration (actually, total concentration of avidin-binding substances) in the unknowns is determined by comparing counts bound to the standard curve.

Safety Considerations. When the iodination procedure described here is properly performed, radioactive iodine gas should not be released. However, as a precaution against errors that result in production of $^{125}I_2$

[1] D. M. Mock and D. B. DuBois, *Anal. Biochem.* **153,** 272 (1986).

(e.g., acidification), a properly operating fume hood should be used for all steps through the final molecular sieve separation of $^{125}I^-$ from ^{125}I-labeled avidin. Use of a hood also provides an appropriate area for containment of spills, which are more likely to occur than release of $^{125}I_2$.

Materials and Reagents

U-bottomed microtiter plates (Cat. No. 001-010-2201), Immulon II Removawell strips (011-010-6302), and Removawell holders (011-010-6601), Dynatech Laboratories, Inc., Alexandria, VA

d-Biotin and BSA, RIA grade (Sigma Chemical Co., St. Louis, MO)

Avidin-D (Vector Laboratories, Burlingame, CA)

N-2-Hydroxyethylpiperazine-*N'*-2-ethanesulfonic acid (HEPES, Calbiochem, La Jolla, CA)

All other chemicals should be reagent grade or better. The water used is distilled and deionized in all buffers, washes, etc. Additional chemicals required are mono- and dibasic sodium phosphate, sodium chloride, sodium hydroxide, sodium carbonate and bicarbonate, and sodium azide. Additional materials will be needed if ^{125}I-labeled avidin is synthesized rather than purchased; these are mentioned in the description of the synthetic methods below.

Equipment

γ counter for ^{125}I quantitation

Twelve (or 96-)-channel pipet (e.g., Titertek, Flow Laboratories, Inc., McLean, VA), optional

Twelve (or 96-)-channel plate washer (e.g., Miniwasher, Skatron AS, Lier, Norway), optional

Preparation of ^{125}I-Labeled Avidin

The assay works well with ^{125}I-labeled avidin synthesized in any of three ways, namely, radioiodination with chloramine-T, Iodo-Beads, or the Bolton–Hunter reagent.

Preparation of HPP–Avidin. Avidin-D is derivatized with *N*-succinimidyl-3-(*p*-hydroxyphenyl)propionate (SHPP, Pierce Chemical Co., Rockford, IL) by the method of Finn *et al.*[2] The reagent is dissolved in 2-propanol : ethyl acetate (3 : 2) at a concentration of 5 mg/ml. A 100-μl aliquot of the SHPP solution is added to 2 ml of avidin [9.5 mg/ml in 0.2 *M* borate buffer (pH 9.0) prepared from sodium borate and pH adjusted with 0.2 *M* boric acid]. The reaction mixture is incubated in an ice bath with occasional shaking for 3 hr and then dialyzed overnight against 4 liters of

[2] F. M. Finn, G. Titus, J. A. Montibeller, and K. Hofmann, *J. Biol. Chem.* **225,** 5742 (1980).

0.5 *M* phosphate buffer (pH 7.5). The dialyzed HPP–avidin product is aliquoted into vessels (e.g., 10 or 50 μl in 1.5-ml microcentrifuge tubes), capped, and stored at −70° until used for iodination.

Chloramine-T Radioiodination of HPP–Avidin. Labeling of HPP–avidin is carried out by the method of Hunter and Greenwood.[3] A 40-μl volume of HPP–avidin (200 μg) in 50 m*M* phosphate buffer (pH 7.5, prepared from Na_2HPO_4 and NaH_2PO_4 mixed in a ratio of 84 to 16% by weight) is added to a reaction vial containing 1 mCi of carrier-free ^{125}I (100 mCi/ml, Amersham Corp., Arlington Heights, IL). A 12.5-μl aliquot of a chloramine-T solution (2 mg/ml in the same phosphate buffer) is added to the mixture, which is then allowed to incubate for 15 min at room temperature. The reaction is stopped by the addition of 50 μl of a solution of sodium metabisulfite (0.5 mg/ml in water). The entire reaction mixture is transferred to 1 ml of Tris–saline [20 m*M* Tris-HCl, 0.15 *M* saline (pH 7.4 at 25°)] containing 1 mg/ml BSA and 1 mg/ml sodium iodide.

Separation of Free from Bound ^{125}I. Free $^{125}I^-$ is separated from ^{125}I-labeled avidin by molecular sieve chromatography as follows. Apply the sample to the drained bed of a 0.9 × 25 cm Sephadex G-25 fine (Pharmacia, Uppsala, Sweden) column previously equilibrated with 50 m*M* phosphate buffer (pH 7.4) containing 0.1% BSA (w/v), which serves here as a carrier for the ^{125}I-labeled avidin to prevent loss by adsorption to the column. Drain the sample into the bed of the column and begin collecting 1.0-ml fractions into tubes which contain 1.0-ml aliquots of 0.1% BSA in phosphate buffer. After the sample is drained into the bed, wash the column walls with several milliliters of buffer and continue collecting fractions. Additional buffer is added as necessary to collect 30 fractions. Quantitate the radioactivity in 50 μl of each fraction using a γ counter; determine the position and separation between the two peaks. Pool the four or five fractions of the first peak that have the highest activity. The ^{125}I-labeled avidin runs in the exclusion volume and thus elutes first. The specific activity of the ^{125}I-labeled avidin should be approximately 5 μCi/μg. Specific activities up to 40 μCi/μg can be obtained by using less avidin in the radioiodination step. The pooled ^{125}I-labeled avidin is aliquoted (10 μl/tube); tubes are capped and frozen at −20°.

Radioiodination of HPP–Avidin by the Iodo-Beads Method. The Iodo-Beads method of iodination is less complex and less time-consuming than the chloramine-T method. The ^{125}I-labeled avidin synthesized by this method is more uniform in stability and specific activity.

Wash four Iodo-Beads (Pierce Chemical Co.) twice with 5 ml of

[3] W. M. Hunter and F. C. Greenwood, *Nature (London)* **194,** 495 (1962).

50 m*M* phosphate buffer (pH 7.4). Drain the beads on filter paper. Place 180 μl of the same phosphate buffer and the four Iodo-Beads in a vial containing 2 mCi of ^{125}I; the original vial from the supplier (e.g., Amersham) is suitable. Mix by thumping the closed vial, and allow the mixture to stand at room temperature for 5 min. The Iodo-Beads are now preloaded with ^{125}I.

Pipet 4 μl of HPP–avidin into the vial containing the preloaded Iodo-Beads; close and mix for 20 min by thumping every 5 min at room temperature. Next, separate the aqueous reaction mixture from the Iodo-Beads by transferring the liquid from the reaction vial to a 1.5-ml microcentrifuge tube using a pipet with a disposable tip. Wash the Iodo-Beads with 200 μl of 50 m*M* phosphate buffer (pH 7.4) and transfer the wash to the same tube. Separation of ^{125}I-labeled avidin from free $^{125}I^-$ is performed on a Sephadex G-25 column as described above. The final pool of ^{125}I-labeled avidin should have a specific activity of 25–50 μCi/μg, and the concentration should be about 200,000 cpm/μl. The pooled ^{125}I-labeled avidin is aliquoted (~25 μl/tube), and the tubes are capped and stored at 4° or frozen at −20°.

^{125}I-Labeled Avidin by Bolton–Hunter Reagent. ^{125}I-Labeled avidin synthesized by the Bolton–Hunter reagent method[4] can be obtained from New England Nuclear (Boston, MA).

Comments. The ^{125}I-labeled avidin synthesized by each of these methods will bind quantitatively (>95%) to an iminobiotin column[5] at pH 9 and will be released at pH 4. The ^{125}I-labeled avidin declines in specific activity and affinity for binding to the biotin–BSA plates more rapidly than would be predicted from simple radioactive decay. We routinely discard ^{125}I-labeled avidin that is more than 2 months old.

Biotinylation of BSA

BSA is biotinylated according to the method of Heitzmann and Richards.[6] A 500-mg amount of BSA is dissolved in 50 ml of ice-cold 0.1 *M* $NaHCO_3$ (pH 7.5). A 5-ml volume of biotin *N*-hydroxysuccinimide ester (BNHS, Pierce Chemical Co.) at a concentration of 12 mg/ml in *N,N*-dimethylformamide is added to the BSA solution, which is incubated with gentle stirring at 4° overnight. This mixture is then dialyzed exhaustively against distilled, deionized water to remove unreacted BNHS and finally buffered by dialyzing against HEPES buffer (see below). The biotin–BSA

[4] A. E. Bolton and W. M. Hunter, *Biochem. J.* **133,** 529 (1973).

[5] K. Hofmann, S. W. Wood, C. C. Brinton, J. A. Montibeller, and F. M. Finn, *Proc. Natl. Acad. Sci. U.S.A.* **77,** 4666 (1980).

[6] H. Heitzmann and F. M. Richards, *Proc. Natl. Acad. Sci. U.S.A.* **71,** 3537 (1974).

is aliquoted (200 μl/tube), and the microcentrifuge tubes are capped and frozen at −20°. The biotin–BSA preparation is stable for at least 1 year; each tube contains enough biotin–BSA to coat 10 plates (i.e., 96 wells: 8 strips of 12 wells each).

The degree of biotinylation of BSA is assessed by measuring the amount of ^{125}I-labeled avidin that will bind to a well which is coated with biotin–BSA as follows. Polystyrene wells (96 wells/plate, Immulon II) are coated with either biotin–BSA or native BSA as described below. Serial dilutions of ^{125}I-labeled avidin are incubated in the wells overnight at 4°; the wells are then washed, separated, and counted. Wells coated with biotin–BSA will reproducibly bind 50–100 times more ^{125}I-labeled avidin than wells coated with native BSA.

Performing the Assay

Buffers

HEPES buffer [0.2 *M* HEPES, 2 *M* NaCl (pH 7.0)]: Dissolve 190.64 g HEPES, 467.52 g NaCl, and 0.8 g sodium azide in about 3 liters of water; adjust the pH to 7.0 with 10 *M* NaOH and bring to 4.0 liters

Coating buffer [50 m*M* bicarbonate (pH 9.6)]: Dissolve 1.59 g sodium carbonate and 2.93 g sodium bicarbonate in 900 ml water and bring to 1.0 liter with water; this buffer can be stored at 4° for at least 2 weeks

Washing buffer [20 m*M* HEPES, 0.2 *M* NaCl (pH 7.0)]: Dilute HEPES buffer (above) 1/10 with water

Blocking buffer [0.2 *M* HEPES, 2 *M* NaCl (pH 7.0) with 0.01% BSA and 0.02% sodium azide]: Dissolve 0.1 g BSA in 1 liter of HEPES buffer; this buffer is stable at 4° for at least 2 weeks

Avidin buffer (HEPES buffer with 0.1% BSA): Dissolve 1 g BSA in 1 liter of HEPES buffer; stable at 4° for at least 2 weeks

Biotin Standards. A 100 n*M* biotin stock solution is prepared by dissolving 24.43 mg of *d*-biotin in 1 liter of distilled water. When aliquoted into 1 ml fractions and stored at −70°, this reference biotin stock is stable for at least 2 years.

A series of biotin standards is prepared from the reference stock solution by dilution in HEPES buffer. The dilutions are performed in disposable graduated polyethylene specimen containers having a volume greater than 200 ml. Twenty milliliters of the reference stock is diluted 1/10 to yield 200 ml of a 1.00×10^7 fmol/ml solution. Three subsequent 1/10 dilutions are performed in the same manner to generate 1.00×10^6, 1.00×10^5, and 1.00×10^4 fmol/ml standards. Fifty milliliters of the 1×10^4

fmol/ml solution is diluted to 200 ml to yield the 2500 fmol/l biotin standard. The final 16 dilutions of 1/1.5 in the series yield concentrations (in fmol/ml) of 1.67×10^3, 1.11×10^3, 741, 494, 330, 220, 146, 97.5, 65.0, 43.4, 28.9, 19.3, 12.8, 8.6, 5.7, and 3.8; the dilutions are prepared by diluting 120 ml of each stock to 180 ml. This scheme leaves 60 ml of the previous dilution to use as a standard.

The dilutions are stored at −70° after aliquoting 0.4-ml volumes. This dilution scheme yields 150 sets of tubes with 20 biotin concentrations ranging from 1×10^7 to 3.8 fmol/ml.

Preparing Coated Plates. The biotin–BSA-coated wells (plates of 96 wells) must be prepared at least 1 day before the assay. The biotin–BSA stock (prepared and stored as described above) is diluted 1/1000 in coating buffer. A 96-well plate requires 20 ml of coating buffer. Prepare the coating solution by adding 20 μl of biotin–BSA stock (10 mg/ml) to 19.98 ml of coating buffer to produce a final concentration of biotin–BSA of 10 μg/ml.[7] Add 200 μl of this biotin–BSA coating solution to each well of the Immulon II strips (8 strips per plate). Cover the wells and incubate at 4° at least overnight but not more than 5 days; 4 days has given the best results. After incubation, aspirate the coating solution and wash each Immulon II strip twice with washing buffer as follows: fill the well to the brim with washing buffer and then aspirate the washing buffer.[8] Next add 400 μl of blocking buffer to each well of the washed plates. Incubate for at least 2 hr at room temperature before using. Plates can be stored in blocking buffer for 1 month at 4°. At the time of assay, aspirate the blocking buffer and wash the plate twice with washing buffer immediately before transferring the contents from the first incubation microtiter wells. Do not allow drying of the biotin–BSA-coated wells for more than 1 minute.

Incubation 1. Add 100 μl of each biotin standard dilution to a well of the microtiter uncoated plate. Run triplicates of each standard. To determine maximum counts bound (i.e., zero concentration of biotin), use 100 μl of HEPES buffer; run in triplicate. These three wells are subsequently handled just as if they contained standards and yield a value for the maximum binding in the absence of biotin. To determine the number of counts transferred to Immulon II (and thus calculate a maximum percent binding which is equal to the maximum counts bound divided by the total counts transferred to Incubation 2), again add 100 μl HEPES buffer to three wells; after adding the ^{125}I-labeled avidin as described below,

[7] Biotin–BSA coating solution should be prepared in polypropylene tubes, *not* in polystyrene tubes, because the biotin–BSA will adsorb to the polystyrene.

[8] A multichannel well washer is a worthwhile timesaving device.

100 μl of the contents of each of these three wells is counted directly instead of being transferred to Incubation 2, thus yielding a mean value for the total counts in the final incubation. Also, add 100 μl of each dilution of each unknown sample to a well in the microtiter plate. Triplicates are run for each dilution of each unknown.

After all standards and unknowns have been added to the wells in the microtiter plate, add 50 μl of ^{125}I-labeled avidin solution to each well. The ^{125}I-labeled avidin solution is made by diluting the stock ^{125}I-labeled avidin with avidin buffer (see above) to a concentration of approximately 10,000 cpm/50 μl. As the labeled avidin degrades with age, more avidin will be required to maintain 10,000 cpm/50 μl. A 50-μl aliquot of diluted avidin should be counted to measure total cpm/μl before each assay. Mix the contents of Incubation 1 by alternately drawing up and expelling the contents of each well with a pipet. Incubate for 1 hr at room temperature (the length of this incubation is not critical).

Incubation 2. After the first incubation, transfer 100 μl from each well in the microtiter plate to a well of the Immulon II plate. Allow the samples to incubate in the coated wells for at least 4 hr at room temperature or up to 3 days at 4°. Then aspirate the contents of the Immulon II wells and wash twice with washing buffer to remove all unbound ^{125}I-labeled avidin. Quantitate the bound ^{125}I-labeled avidin by counting. The Removawell strips of Immulon II can be broken apart for placement into a suitable carrier for the γ counter.

Data Analysis and Interpretation

A standard curve (Fig. 1A) depicts uncorrected counts bound to the well versus the log of the biotin concentration. A log–logit transformation (Fig. 1B) does not consistently produce a linear dependence on the log of the biotin concentration. Competitive isotope dilution assays in liquid phase characteristically have linear log–logit plots; the fact that this assay is sequential rather than competitive and has a solid-phase binding step may be responsible for the more complex curve shape. Determination of biotin concentration graphically is acceptable, but tedious when assaying large numbers of unknowns. We use a polynomial curve-fitting program to fit the curve to $r^2 > 0.99$; typically a fourth or fifth degree polynomial is suitable.

The assay sensitivity has typically been about 100 fmol/ml (10 fmol/assay well) but has occasionally been 10 fmol/ml (1 fmol/assay well) with an avidin preparation of unusually high specific activity and unusually high affinity for the solid phase. The intraassay variability was determined on 10 replicates on rat urine (~1/400 dilution) in a single assay; the coeffi-

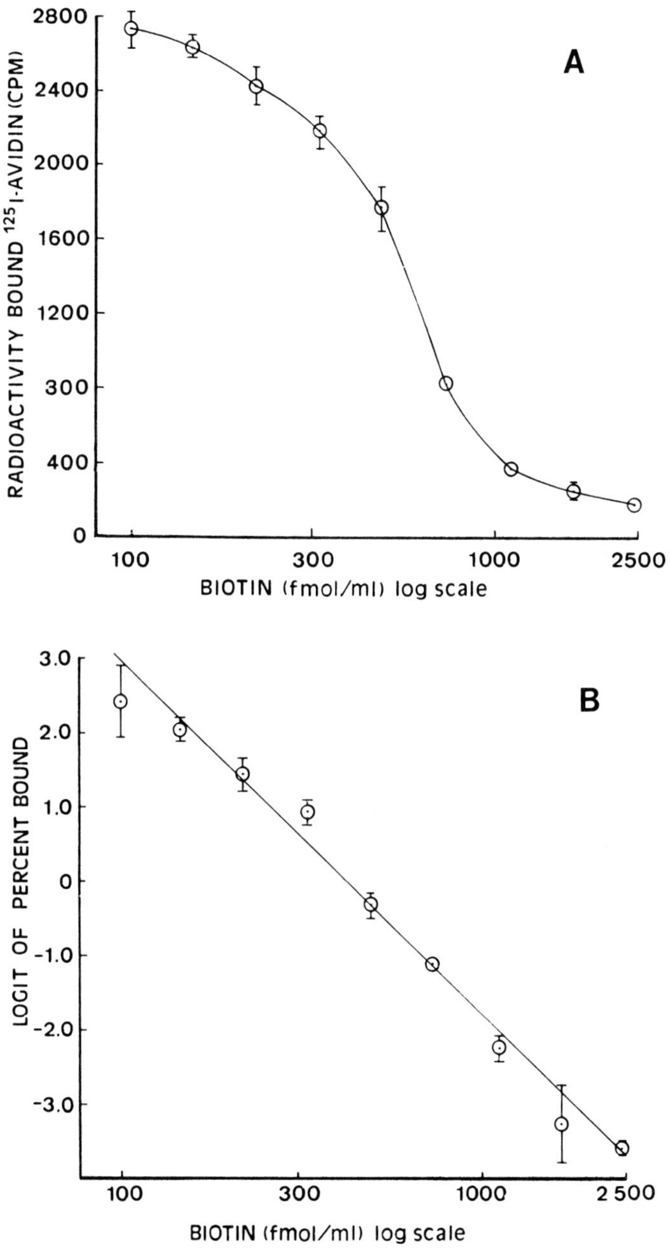

FIG. 1. (A) Plot of uncorrected counts bound to the coated wells versus log of biotin concentration in the series of biotin standards. (B) Plot of logit of percent binding versus log of biotin concentration.

cient of variation was 10%. The long-term precision (interassay variability) was estimated from results of triplicate determinations from assays of a sample by three different technicians on 6 different days; the coefficient of variation (calculated from the standard deviation of the means of the triplicates) was 6%, and the intraclass correlation coefficient on that data was −0.77.

We have investigated the use of an alternate coating protein (keyhole limpet hemocyanin), and the use of biotinylating agent with a longer spacer (NHS-LC-biotin, Pierce Chemical Co.). These modifications resulted in modestly higher maximum percent binding (85 versus 65%) but gave essentially the same value for biotin (i.e., total avidin-binding substances) in urine and ultrafiltrates of plasma.

We have also investigated the detectability of biotin analogs in this assay. For example, a series of standard concentrations of biocytin, biotin sulfoxide, and bisnorbiotin were assayed. The biocytin standard curve was identical to the biotin standard curve, within experimental error. For the others, the point of half-maximal binding (the inflection point of the sigmoid curve) was shifted toward greater concentrations of the analog, and the true concentration of the analog would have been underestimated by a factor of 4 to 5 for both analogs. The lower binding affinities of some biotin analogs[9] may allow the solid-phase biotin (biotinylated BSA) to compete for binding to ^{125}I-labeled avidin in the second step and thus modify the standard curve. This observation emphasizes two limitations of the assay that may well apply to other avidin-binding assays: (1) in a complex biological mixture of biotin, biotin analogs (e.g., metabolites), and even protein-bound biotin, the measured total avidin-binding substances may not equal the stoichiometric sum of all avidin-binding substances, and (2) a more accurate measure could be made if the avidin-binding substances are separated chromatographically and each avidin-binding substance identified and quantitated versus the standard curve for that compound.

A tacit assumption in previous measurements of biotin in human plasma has been that biotin is the only avidin-binding substance present (if using an avidin-binding assay) or that biotin is the only growth-promoting substance present (if using a bioassay). Using paper chromatography and a bioassay, Lee *et al.*[10] demonstrated the presence of substantial amounts of biotin metabolites in human urine. More recently this assumption concerning human plasma was examined by using reversed-phase HPLC to separate biotin analogs in ultrafiltrates of human plasma. Figure 2 depicts

[9] N. M. Green, *Adv. Protein Chem.* **34,** 1967 (1975).

[10] H. Lee, L. D. Wright, and D. B. McCormick, *J. Nutr.* **102,** 1453 (1972).

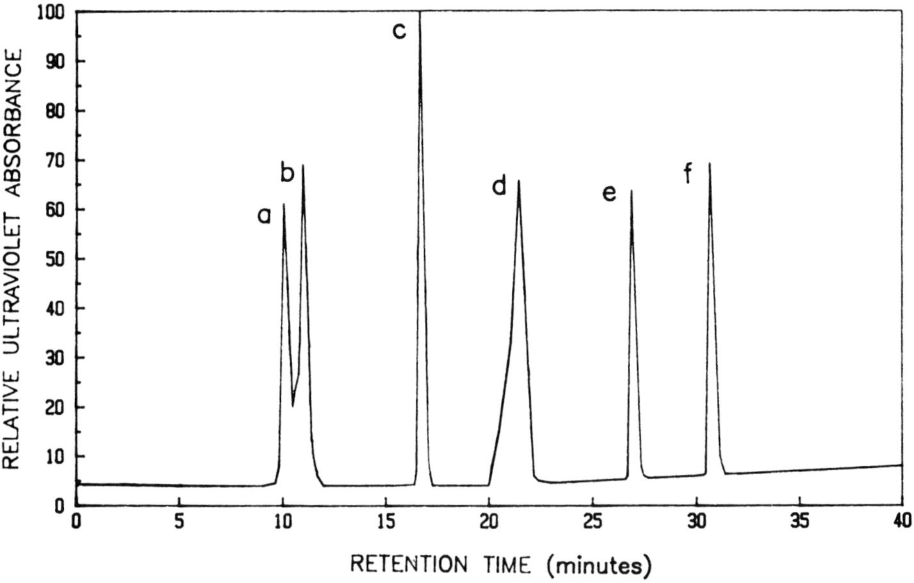

FIG. 2. Chromatography of a mixture of biotin analogs. The biotin analogs are biotin *l*-sulfoxide (a), biotin *d*-sulfoxide (b), biocytin (c), biotin (d), dethiobiotin (e), and biotin methyl ester (f). Chromatography parameters were a linear gradient from 0% A to 50% A at 1.0 ml/min for 40 min. [Solution A, methanol; solution B, 10 m*M* potassium phosphate (pH 4.5)].

the HPLC separation of a mixture of biotin analogs; in this mixture, concentrations were sufficient (e.g., micromolar range) for detection by ultraviolet absorption. In human plasma, however, biotin and biotin analogs are present in nanomolar concentrations. Using the ^{125}I-labeled avidin assay to measure the fractions collected after chromatography, a substantial amount of avidin-binding substances that are not biotin were detected.[11-13] Recovery of total avidin-binding substances was 94 ± 9%. Biotin accounted for only 60 ± 6% ($\bar{X}$ ± 1 SD) of total avidin-binding substances; four other peaks were detected that did not coelute with biocytin or biotin, the only recognized biotin vitamers which occur naturally.

[11] D. M. Mock, *Fed. Proc., Fed. Am. Soc. Exp. Biol.* **46,** 1159 (1987).

[12] D. M. Mock and G. L. Lankford, *FASEB J.* **3,** A1058 (1989).

[13] T. Suormala, E. R. Baumgartner, J. Bausch, W. Holick, and H. Wick, *Clin. Chim. Acta* **177,** 253 (1988).

[25] Fluorometric Assay for Avidin–Biotin Interaction

By DONALD M. MOCK and PAUL HOROWITZ

Absolute quantitation of biotin and avidin can be important in many applications of avidin–biotin technology and in studies of biotin nutriture. Many assays for biotin in physiological fluids such as blood and urine are based on the interaction of biotin with avidin and require standardization of the avidin to be used in the assay. Biotin can be quantitated by weight, and then avidin can be quantitated by measuring changes in the optical absorbance spectrum that occur when biotin displaces 4′-hydroxyazobenzene-2-carboxylic acid (HABA) from avidin.[1] An alternate approach is quantitation of avidin by absorbance at 280 nm or by changes in the absorbance spectrum that occur when HABA binds to avidin; biotin can then be quantitated by the optical absorbance changes that occur with displacement of HABA from avidin by biotin.

The interaction of avidin and biotin can also be quantitated by measuring the fluorescence changes that occur when the fluorescent probe 2-anilinonaphthalene-6-sulfonic acid (2,6-ANS) is displaced by biotin from the biotin-binding site on avidin.[2] The probe probably binds to a hydrophobic region at or near the biotin-binding site on avidin. The transition from an aqueous to a hydrophobic environment is associated with a greater than 6-fold increase in the fluorescence intensity, and the wavelength of maximum emission shifts from 463 to 422 nm. When biotin is added, it stoichiometrically displaces the 2,6-ANS (mole ratio 1 : 1) and reverses the enhancement of fluorescence intensity and the shift in wavelength of maximum emission. This fluorometric method is approximately an order of magnitude more sensitive than the HABA method and is less subject to interference from naturally occurring chromophores; the method is also relatively insensitive to naturally occurring fluorophores. Although this method can be used to standardize either avidin or biotin, for simplicity the standardization of avidin is described in this chapter. The differences when used to standardize biotin are briefly discussed at the end of the chapter.

[1] N. M. Green, *Biochem. J.* **94,** 23c (1965).

[2] D. M. Mock, G. Lankford, D. DuBois, N. Criscimagna, and P. Horowitz, *Anal. Biochem.* **151,** 178 (1985).

Assay Method

Materials. 2,6-ANS is available commercially from Molecular Probes (Junction City, OR). Avidin-D is affinity purified and available commercially from Vector Laboratories (Burlingame, CA). Biotin (*d*-biotin) is from Sigma Chemical Co. (St. Louis, MO). Because the fluorescence of the probe and perhaps even the interaction of the probe and avidin may be sensitive to pH, quantitation should be conducted in an appropriate buffer.[3]

Equipment. We have successfully used two spectrophotofluorometers for this standardization: (1) Farand Mark V spectrophotofluorometers and (2) Shimadzu RF-540 recording spectrophotofluorometers. The excitation wavelength is 328 nm, the emission wavelength 408 nm. A slit width of 5 nm for both excitation and emission beams is suitable.

The wavelength of maximum intensity depends on the solvent system in which the probe is dissolved. Because the equilibrium dissociation constant is about 200 μM, neither the bound nor the free probe concentrations are negligible with respect to the other at the concentrations typically used. Thus, the wavelength of maximum intensity of the probe–avidin complex has been determined by extrapolation to infinite avidin concentration (i.e., all probe bound). The wavelengths are chosen to maximize the difference in fluorescence intensity between bound and free 2,6-ANS based on published maxima for excitation and emission while avoiding scattered light from the excitation beam. In the buffer system used, the true maximum wavelength of emission of free 2,6-ANS is 463 nm and that of bound 2,6-ANS 422 nm. Sensitivity is still satisfactory when monitoring emission at 408 nm; thus, within limits, the choice of excitation and emission wavelengths is not critical.

Titration of 2,6-ANS–Avidin Complex by Biotin

Initially, avidin and 2,6-ANS are added to a quartz cuvette, mixed, and placed in the fluorometer. The maximum fluorescence enhancement is reached in the few seconds required for mixing and insertion of the cuvette; no extra incubation time is required at this point or during the titration. Biotin is then added in increments from a concentrated stock solution, the components are mixed, and the fluorescence intensity is measured after each addition. A typical starting volume is 2.00 ml, and a typical titration would be fifteen 10-μl additions of a 0.12 mM biotin

[3] We chose 0.2 M N-2-hydroxyethylpiperazine-N'-2-ethanesulfonic acid (HEPES Ultrol, Calbiochem, La Jolla, CA), pH 7.0, in 2 M NaCl because this buffer gives a satisfactory avidin–biotin interaction in a biotin assay (see D. M. Mock, this volume [24]).

solution in deionized, distilled water. Biotin does not dissolve easily in water at concentrations greater than 2 mg/ml (~8 mM). It may be necessary to make the solution slightly basic with $NaHCO_3$ or NaOH to facilitate dissolving the biotin and then correct back to pH 7.0 with HCl.

The choice of concentration of avidin (or amount of dilution of the unknown avidin source) depends on at least two factors: (1) the sensitivity of the fluorometer and (2) the optical path length of the cuvette. Most fluorescence cuvettes have an optical path length of 1 cm. For experiments requiring high concentrations of 2,6-ANS and thus producing high inner filter effects, we have successfully used a flow cell (Shimadzu RF-540 flow cell, 2040382504) with a path length of 2 mm.

When initially setting up the method, it is worthwhile to vary the concentrations of avidin and of 2,6-ANS in the cuvette to achieve the following result when a saturating amount of biotin is added: The decrease in fluorescence intensity should be great enough for accurate quantitation (e.g., ±2% reproducibility) on the available fluorometer. A "saturating" biotin concentration is at least 4-fold greater than the avidin concentration. Typical avidin concentrations are 1–200 μM. Typical 2,6-

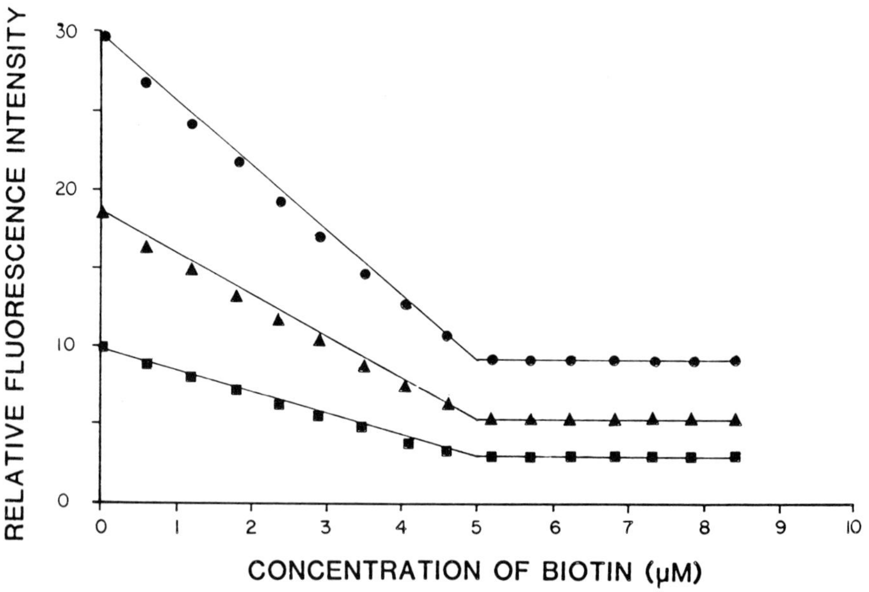

FIG. 1. Titrations of the complex between avidin and 2,6-ANS with biotin at several concentrations of 2,6-ANS. The corrected fluorescence intensity is plotted as a function of the concentration of biotin. The 2,6-ANS concentrations shown are 94.1 μM (top curve), 47.0 μM (middle curve), and 23.5 μM (bottom curve). The avidin concentration is 1.74 μM in all three titrations.

ANS concentrations are roughly equimolar; the point of equivalence (hence, the accuracy of the method) is not dependent on the 2,6-ANS concentration over a fairly wide range (Fig. 1), because the association constant for biotin and avidin is very high ($K_a = 10^{15}\ M^{-1}$). For the same reason, the accuracy of this method is not affected by the concentration of avidin in the cuvette (Fig. 2) unless the avidin concentration is too small for accurate quantitation of the fluorescence change or so great that the accompanying high concentrations of 2,6-ANS cause an inner filter effect so large that fluorescence increases cannot be accurately measured and corrected.

The point of equivalence is defined as the point at which all binding sites on avidin are occupied by biotin with no excess free biotin. Because of the high binding affinity, the effects due to equilibrium are negligible at the concentrations used for this method. However, nonlinearity of the

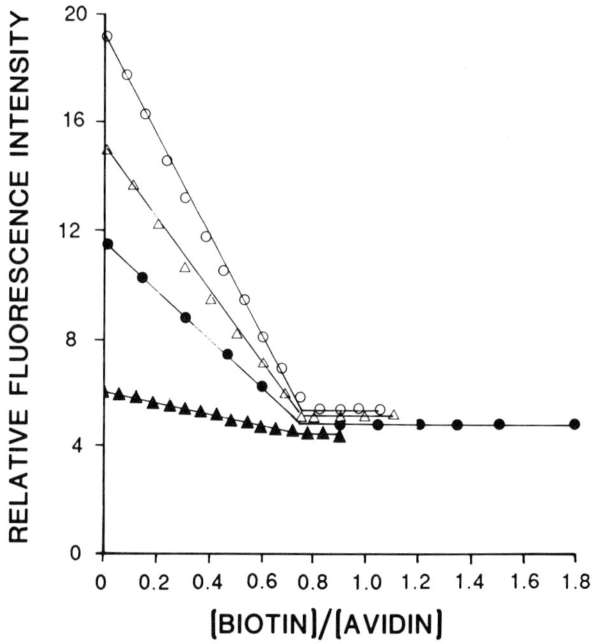

FIG. 2. Titration of the complex between avidin and 2,6-ANS with biotin at several avidin concentrations. The corrected fluorescence intensity is plotted as a function of the ratio of the biotin concentration to the monomeric avidin concentration. The avidin concentrations, determined independently by optical absorbance at 280 nm, are 1, 4, 6, and 8 μM from bottommost curve to topmost. The 2,6-ANS concentration is fixed at 50 μM. The point of equivalence should be 1.0, but this particular batch of avidin consistently demonstrates a value of about 0.75.

titration curve near the equivalence point is sometimes seen. Whether this represents heterogeneity in the avidin preparation or cooperativity in the avidin tetramer (i.e., K_a of one site depends on the number of sites already occupied in that avidin molecule) is not clear.

The point of equivalence can be determined graphically or by linear regression as the point of intersection of the two linear segments of the titration curve; if nonlinearity is seen near the equivalence point, these points should be excluded from the linear regression. The unknown avidin concentration at the point of equivalence $[A]_e$ is related to the biotin concentration at the point of equivalence $[B]_e$ by the following equation:

$$[A]_e = [B]_e/4$$

Keep in mind that the titration may have added significantly to the volume in the cuvette. An alternate equation for the *original* avidin concentration in the cuvette $[A]_o$ (in μmol/liter) is the following:

$$[A]_o = \text{biotin}/4V_I$$

where biotin is the total micromoles of biotin added at the equivalence point and is equal to the volume of biotin stock added (μl) times the concentration of biotin stock (in μmol/μl), and V_I is the *initial* volume in the cuvette (in liters; e.g., 2.00×10^{-3} liter).

The factor of 4 reflects the assumption that all four sites of tetrameric avidin are available for biotin binding. Using absorbance at 280 nm to standardize avidin independently,[2] we have occasionally found samples of affinity-purified avidin that had three rather than four biotin-binding sites available (Fig. 2). This unusual stoichiometry was confirmed by HABA titration.[2] This finding emphasizes the caveat that multimeric binding proteins need only a single functional binding site to be purified by affinity methods. Biotin contamination could result in essentially irreversible occupation of approximately 25% of the sites in a given lot of avidin.

Inner filter effects become important when the number of excitation photons reaching the fluorophores in the region of the cuvette monitored for emission is significantly reduced because of the absorption by fluorophores encountered earlier in the monitored region; at this point, the fluorescence intensity is no longer linear with concentration of fluorophore. Since changes in fluorescence intensity are linear with added biotin in the range of 2,6-ANS concentrations used, the same equivalence point will be determined whether or not the fluorescence intensity is corrected for the inner filter effect. For experiments measuring the binding characteristics of the probe, however, a correction factor according to the equation of McClure and Edelman[4] is necessary:

$$I_{corr} = I_{obs} \times 2.303A_{ex}/(1 - 10^{-A_{ex}})$$

where I_{corr} is the corrected fluorescence intensity, I_{obs} the observed fluorescence intensity, and A_{ex} the absorbance of the probe at the excitation wavelength. A_{ex} is calculated as the product of the extinction coefficient at the excitation wavelength times optical path length times concentration of probe.

This fluorescence method will give erroneously low values for avidin if significant amounts of biotin or biotin analogs are present in the avidin sample, provided that the biotin analogs have an intact heterocyclic ring structure and thus bind as tightly to avidin as biotin. Biocytin and biotin methyl ester are examples of such biotin analogs. The effect of biotin analogs with altered ring structures (e.g., dethiobiotin, biotin sulfoxide, diaminobiotin) on this method has not been studied in detail.

Titration of Biotin by Avidin–2,6-ANS Complex

To use the fluorometric method to measure biotin, the source of avidin must first be standardized. With a pure preparation, absorbance at 280 nm can be measured, or the avidin can be quantitated by weight. Alternatively, a primary biotin standard can be used to standardize the avidin, and the avidin, in turn, used to standardize samples of unknown biotin concentration. The titrations to determine the equivalence point are conducted as discussed above if the concentration of biotin in the unknown sample is sufficiently large. If not, a concentrated solution containing avidin and 2,6-ANS can be used to titrate the biotin solution. In this reverse titration, the fluorescence should first increase linearly at a modest slope with sequential additions of avidin–2,6-ANS solution because biotin will immediately displace all the probe into the aqueous phase; at the equivalence point, the slope will increase dramatically because no free biotin remains to displace the probe. Biotin concentration is calculated as follows:

$$[B]_e = [A]_e \times 4$$

This reverse titration (i.e., biotin titrated by avidin) also works if 2,6-ANS is added to the cuvette containing the unknown concentration of biotin rather than along with the avidin. Because the amount of free probe changes little with the avidin additions (except for dilution) until the point

[4] W. O. McClure and G. M. Edelman, *Anal. Biochem.* **6,** 559 (1967).

of equivalence is reached, the fluorescence intensity increases only after reaching the point of equivalence. Thus, the method of reverse titration may offer a minor advantage in precision compared to titrating with avidin–2,6-ANS solution.

Neither fluorescence enhancement nor reversal is seen with the interaction of 2,6-ANS, streptavidin, and biotin.

Section IV

Applications

A. Isolation and Purification
Articles 26 through 35

B. Localization
Articles 36 through 46

C. Protein Blotting
Articles 47 through 53

D. Immunoassays
Articles 54 through 64

E. Gene Probes
Articles 65 through 72

F. Composite and Special Applications
Articles 73 through 79

[26] Isolation of Biologically Active Compounds: A Universal Approach

By MEIR WILCHEK and EDWARD A. BAYER

The potential use of the avidin–biotin system for the isolation of biologically active compounds was considered at an early stage in the development of avidin–biotin technology. In this context, early studies were mainly directed toward the isolation of biotin-containing subunits of carboxylases[1] and their peptides[2] using an avidin column. The idea of using biotinylated ligands as a sandwich between an avidin column and a biologically active compound destined for isolation was the natural outgrowth of the perennial forerunner of this system, i.e., the use of haptenized ligands in the mediation between an antihapten antibody column and the desired ligand-binding protein.[3] Thus, it was shown that an antidinitrophenyl (anti-DNP) antibody column can bind DNP-labeled trypsin inhibitor which can be used to bind trypsin. At about the same time, trials with biotinylated trypsin inhibitors gave similar results using an avidin column.

The advantage of the antibody column was that the trypsin could be eluted either alone or in complexed form with the DNP-labeled inhibitor under mild conditions, whereas with an avidin column only trypsin could be eluted under mild conditions. Removal of the biotinylated inhibitor in the latter case required 6 *M* guanidine-HCl. Since the original idea in using the antibody or avidin column was to produce a universal reusable column (like an HPLC or ion-exchange column) that could be used for many different purification procedures, the avidin column was neglected by us until the demonstration of such a principle would become experimentally plausible. An approach for such a possibility was recently described by us using an iminobiotin-modified avidin column.[4]

In the interim, we continued to use the avidin column mainly as an analytical tool for the determination of efficiency of biotinylation of antibodies, lectins, etc.,[5,6] which were used mainly for localization purposes.

[1] M. Berger and H. G. Wood, *J. Biol. Chem.* **250,** 927 (1974).

[2] A. Bodanszky and M. Bodanszky, *Experientia* **26,** 327 (1970).

[3] M. Wilchek and M. Gorecki, *FEBS Lett.* **31,** 149 (1973).

[4] M. Wilchek and E. A. Bayer, *in* "Protein Recognition of Immobilized Ligands." (T. W. Hutchens, ed.), p. 83. Alan R. Liss, New York, 1989.

[5] E. A. Bayer, E. Skutelsky, D. Wynne, and M. Wilchek, *J. Histochem. Cytochem.* **24,** 933 (1976).

[6] E. A. Bayer, M. Wilchek, and E. Skutelsky, *FEBS Lett.* **68,** 240 (1976); E. A. Bayer and M. Wilchek, this volume [14].

From time to time, we tried to use avidin-containing columns for isolation purposes, e.g., for isolation of the insulin receptor (with Barry Ginsberg at the National Institutes of Health in 1974). In the meantime, the use of avidin columns flourished thanks to the work of Finn and Hofmann (see [27] in this volume).

The use of avidin columns prepared by direct coupling of avidin to the carrier or through immobilized biotin has now developed into a major tool in molecular biology for the binding of DNA and RNA, both for hybridization[7] and for purification purposes (e.g., purification of splicing components).[8] In many of the studies, the complex is dissociated by the use of cleavable biotin derivatives.[9]

In this section, we have included a variety of chapters that exemplify various approaches which have been used to date. Several of the chapters deal with the use of avidin–biotin technology for the isolation of both polypeptide and steroid hormone receptors. In this context, procedures are described for obtaining effective biotinylation of the desired hormone. In other chapters, procedures are presented for isolation of antigens, separation of cells using biotinylated antibody, isolation of glycoconjugates using biotinylated lectins, and isolation of spliceosomes and Z-DNA using biotinylated DNA.

[7] I. C. Gillam, *Trends Biotechnol.* **5,** 332 (1987).
[8] S. W. Ruby and J. Abelson, *Science* **242,** 1028 (1988); P. J. Grabowski, this volume [34].
[9] T. M. Herman, E. Lefever, and M. Shimkus, *Anal. Biochem.* **156,** 48 (1986).

[27] Isolation and Characterization of Hormone Receptors

By FRANCES M. FINN and KLAUS HOFMANN

Principles of Technique

Remarkable progress in the techniques of molecular biology have advanced the field of protein sequencing immeasurably. These advances coupled with a substantial increase in the sensitivity of peptide sequencing make it possible to sequence any protein that can be isolated even in picomolar amounts. Despite these successes, isolation of peptide or protein receptors poses a number of difficult problems owing to their scarcity, their location in a complex, lipid-rich environment, and the difficulty with which even starting material (plasma membranes) is prepared. A

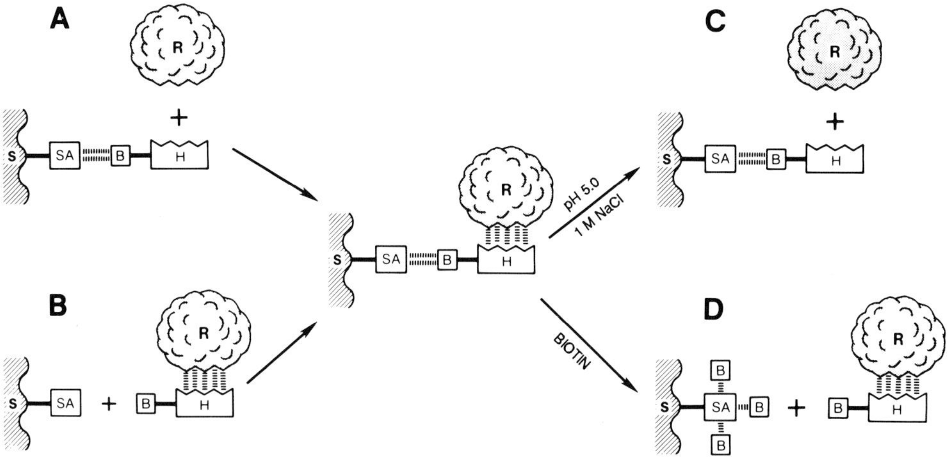

FIG. 1. Application of the avidin–biotin system to the isolation of hormone receptors. Dashed lines represent noncovalent bonds.

highly specific isolation method such as affinity chromatography is essential to the success of such an undertaking. The nonspecific attachment of peptide hormones to activated Sepharose resins can generate a number of different species with varying affinities for the desired receptor. For these reasons, we have devoted our efforts to developing affinity resins based on avidin–biotin technology where the hormone can be attached in a targeted fashion.

Bayer and Wilchek[1] have pioneered the application of avidin–biotin technology to the solution of biological problems. The uncommonly strong noncovalent interaction between avidin and biotin renders this pair ideally suited for detection and retrieval of molecules that function as components of a binding system. The interactions between hormones and their receptors constitute a particularly promising area for application of this technique. Feasible isolation protocols are illustrated in Fig. 1. In the method shown in Fig. 1A, soluble receptor (R) is percolated through an affinity column to form the complex shown in the center of Fig. 1. The column is then exhaustively washed to remove contaminating materials. Alternatively, the biotinylated hormone ligand (B–H) can be added to a solution of solubilized receptor to form the soluble complex BHR. Percolating a solution containing this complex through a column of immobilized succinylated avidin (suc-avidin or SA) as in Fig. 1B will result in the formation of the same complex obtained by method A. Both these

[1] E. A. Bayer and M. Wilchek, *Methods Biochem. Anal.* **26,** 1 (1980).

schemes have been used for the isolation of insulin receptors from human placenta.[2,2a] Removal of functional, purified insulin receptor from affinity resins has been achieved by eluting the column with pH 5.0 acetate buffer containing 1 *M* sodium chloride (Fig. 1C).[3] The replacement of biotin in the ligand BHR by a biotin analog such as dethiobiotin that exhibits weaker affinity for suc-avidin than biotin provides a route to the isolation of receptor–ligand complexes as illustrated in Fig. 1D. Such an approach would permit dissociation of the receptor–ligand complex from the column by biotin, which will occupy sites from which the dethiobiotinylated hormone dissociates. This method has not yet been used.

There are a number of advantages to the avidin–biotin technique. (1) Formation of the affinity column is highly specific, namely, through the avidin–biotin interaction, and is achieved by simply mixing the biotinylated ligand with the Sepharose-immobilized suc-avidin resin. (2) Hormone ligands can be prepared by well-defined solution methods that can be monitored accurately. This eliminates the need for performing chemical manipulations such as coupling the hormone to the resin or deprotecting the coupled product on the resin. (3) The amount of ligand on the column can be varied to achieve optimal operation. (4) Once the technique is worked out for a particular receptor, it can easily be scaled up for production of large amounts, and it is highly reproducible.

Tools

In order to exploit the avidin–biotin interaction for receptor purification, certain design principles must be observed in preparing the biotinylated hormone. Essentially three criteria must be met: (1) derivatization of the ligand with biotin must not substantially weaken its ability to interact with its receptor; (2) the biotinylated hormone must retain strong affinity for avidin; and, most importantly, (3) the biotinylated hormone must be able to interact simultaneously with avidin and its receptor. Recognition of the importance of these principles has evolved through development of a biotinylated derivative of insulin useful for insulin receptor isolation and can perhaps best be illustrated using this example.

[2] F. M. Finn, G. Titus, D. Horstman, and K. Hofmann, *Proc. Natl. Acad. Sci. U.S.A.* **81,** 7328 (1984).

[2a] F. M. Finn and K. Hofmann, unpublished results.

[3] Y. Fujita-Yamaguchi, S. Choi, Y. Sakamoto, and K. Itakura, *J. Biol. Chem.* **258,** 5045 (1983).

Insulin Receptor Isolation

Biotinylated Probe

The *N*-hydroxysuccinimide ester of biotin was chosen to introduce biotin into insulin. Three sites capable of reacting with the ester are present in the insulin molecule, namely, the primary amino groups on the amino-terminal ends of the A and B chains and the ε-amino group on the lysine in position 29 of the B chain. The results of a large body of structure–function studies with insulin indicated that modification of the N terminus of the A chain results in loss of biological activity. Reaction at the ε-amino group is less detrimental to activity, but the $N^{\varepsilon,B29}$-biotinylinsulin–avidin complex had very low biological activity,[4] indicating that criterion 3 had not been met.

Derivatization at the $N^{\alpha,B1}$ position has little effect on the biological activity of insulin.[5] Thus, a series of insulins derivatized at this position with biotin, iminobiotin, or dethiobiotin attached either directly or via a spacer arm were synthesized (Fig. 2), and the effect of derivatization on biological activity was assessed using stimulation of glucose oxidation in isolated rat adipocytes (Fig. 3). $N^{\alpha,B1}$-Biotinylated insulin and congeners containing biotin or dethiobiotin separated from insulin by a six-carbon spacer arm were all full agonists whose dose–response curves did not differ significantly from that of insulin. This was also true of $N^{\alpha,B1}$-(N-{3-[(3-{6-[6-(5-methyl-2-oxo-4-imidazolidinyl)hexanamido]hexanimido}-propyl)amino]propyl}succinamoyl)insulin (Fig. 2, structure **IX**). Derivatives with longer spacer arms such as **VII** and **VIII** were also full agonists, but their dose–response curves were shifted to the right (Fig. 3). Those derivatives whose biological activity is indistinguishable from insulin meet the requirements set forth in criterion 1 and are all potential affinity ligands.

Detailed descriptions of methods for the synthesis of biotinyl, iminobiotinyl, and dethiobiotinyl reagents have been published previously.[6]

The second consideration that must be met for a biotin probe to be functional is that of providing sufficient space between the biotin and the molecule derivatized to allow the avidin–biotin interaction to occur. $N^{\alpha,B1}$-Biotinylinsulin (**I**) forms a complex with suc-avidin that has a $t_{1/2}$ for

[4] J. M. May, R. H. Williams, and C. de Haen, *J. Biol. Chem.* **253,** 686 (1978).

[5] R. A. Pullen, D. G. Lindsay, S. P. Wood, I. J. Tickle, T. L. Blundell, A. Wollmer, G. Krail, D. Brandenburg, H. Zahn, J. Gliemann, and S. Gammeltoft, *Nature* (*London*) **259,** 369 (1976).

[6] F. M. Finn and K. Hofmann, this series, Vol. 109, p. 418.

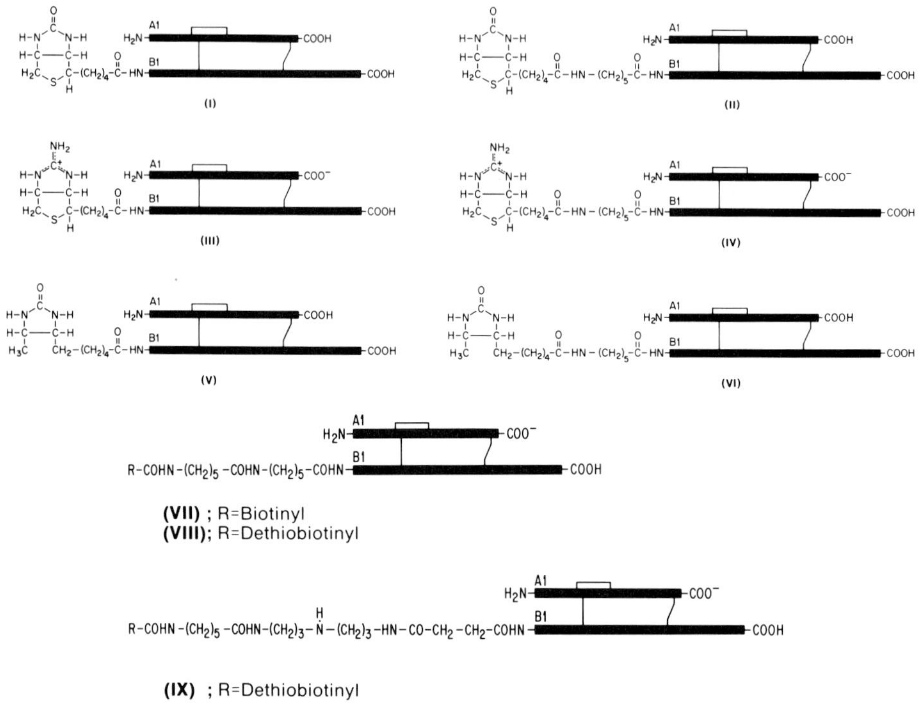

FIG. 2. Simplified structures of $N^{\alpha,B1}$-biotinyl, iminobiotinyl, and dethiobiotinyl derivatives of insulin: biotinylinsulin **(I)**, 6-(biotinylamido)hexylinsulin **(II)**, iminobiotinylinsulin **(III)**, 6-(iminobiotinylamido)hexylinsulin **(IV)**, dethiobiotinylinsulin **(V)**, 6-(dethiobiotinylamido)hexylinsulin **(VI)**, 6-(biotinylamidohexylamido)hexylinsulin **(VII)**, 6-dethiobiotinylamidohexylamido)hexylinsulin **(VIII)**, and (*N*-{3-[(3-{6-[6-(5-methyl-2-oxo-4-imidazolidinyl)hexanimido]hexanamido}propyl)amino]propyl}succinamoyl)insulin **(IX)**.

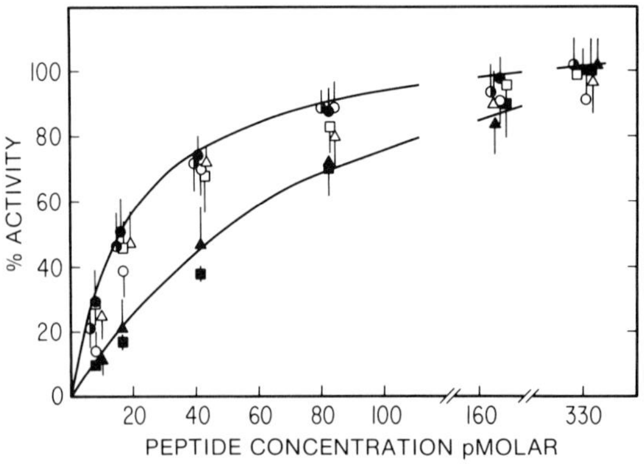

FIG. 3. Stimulation of glucose oxidation in isolated rat epididymal adipocytes by insulin and insulin derivatives: (●) insulin; (○) compound **I**; (□) compound **II**; (■) compound **VII**; (△) compound **VI**; (▲) compound **VIII**; (◑) compound **IX**. The top curve connects the data points for insulin activity. Vertical bars represent standard deviation (SD), $n = 9$.

TABLE I
DISSOCIATION RATES FOR SUC-AVIDIN COMPLEXES[a]

Ligand	$t_{1/2}$	K_{-1} (sec^{-1})
Biotin	127 days[b]	6.3×10^{-8}
Compound **I**	2.6 hr	7.4×10^{-5}
Compound **II**	76 days	1.1×10^{-7}
Compound **VII**	70 days	1.2×10^{-7}
Dethiobiotin	14 hr	1.4×10^{-5}
Compound **VI**	6 hr	3.2×10^{-5}
Compound **VIII**	7 hr	2.8×10^{-5}
Compound **IX**	14 hr	1.4×10^{-5}
[Bct^{25}]$ACTH_{1-25}$ amide	20 days	4.0×10^{-7}

[a] The dissociation kinetics for the complexes are biphasic. Green *et al.*[7] and Chignell *et al.*[8] have shown that avidin possesses two classes of biotin-binding sites. The kinetic constants given are those for the slower rate.

[b] Finn *et al.*[9]

dissociation of 2.6 hr (Table I)[7–9] compared with 200 days[10] for the avidin–biotin complex and 127 days for the suc–avidin–biotin complex. The most obvious explanation for this finding is that the attachment of insulin to biotin imposes a steric impediment to the interaction of the biotin residue with its avidin binding site. Green and colleagues have determined that bound biotin lies 8–9 Å beneath the van der Waals surface of the molecule.[7] Derivatization of insulin at the $N^{\alpha,B1}$ position with *N*-hydroxysuccinimido-6-(biotinylamido)hexanoate to give **II**, which provides a spacer arm of approximately 10 Å between biotin and insulin, overcame the problem of steric interference.

Avidin Molecule

Avidin is a highly basic glycoprotein composed of four identical subunits, each having a biotin-binding site.[11,12] The use of this protein for the construction of affinity columns has the serious disadvantage that avidin binds nonspecifically to many components, especially negatively charged

[7] N. M. Green, L. Konieczny, E. J. Toms, and R. C. Valentine, *Biochem. J.* **125,** 781 (1971).

[8] C. F. Chignell, D. K. Starkweather, and B. K. Sinha, *J. Biol. Chem.* **250,** 5622 (1975).

[9] F. M. Finn, G. Titus, J. A. Montibeller, and K. Hofmann, *J. Biol. Chem.* **255,** 5742 (1980).

[10] N. M. Green, *Biochem. J.* **101,** 774 (1966).

[11] N. M. Green, this volume [5].

[12] N. M. Green, *Adv. Protein Chem.* **29,** 85 (1975).

membrane proteins. This property can be eliminated by exhaustively succinoylating avidin with succinic anhydride to form suc-avidin. There are nine lysine and eight arginine residues per avidin subunit. Succinoylation changes the direction of migration of avidin from the cathode to the anode on electrophoresis without significantly altering its biotin-binding characteristics. The dissociation constant for the avidin–biotin complex is 4×10^{-8} sec^{-1} [12]; for suc-avidin–biotin, it is 6.3×10^{-8} sec^{-1}.[13]

The large avidin molecule (M_r 68,000) has the potential to interfere with the interaction between the biotinylated hormone and its receptor. The biologically inactive avidin–$N^{\varepsilon,B29}$-biotinylinsulin complex, referred to earlier, illustrates this problem. Once it has been established that the biotinylated hormone forms a strong complex with avidin, so that activity resulting from complex dissociation can be ruled out, the biological activity of the avidin–biotinylated hormone complex should be measured. If the avidin–biotinylated hormone complex is active, it is safe to assume that the biotinylated hormone interacts simultaneously with avidin and the receptor.

In the case of the biotinylated derivatives of insulin, the biological activity was measured using glucose oxidation in isolated adipocytes[13] (Fig. 4). To ensure complex formation, excess avidin or succinoylavidin was added to the derivatives prior to exposure to the cells. Most of the derivatives that had shown high biological activity were able to interact with avidin and receptor simultaneously. The exception was $N^{\alpha,B1}$-biotinylinsulin (Fig. 4A). When a 50-fold excess of suc-avidin was added to this biotinylated insulin, so that complex formation was highly favored, biological activity was eliminated. This suggests that $N^{\alpha,B1}$-biotinylinsulin cannot interact simultaneously with suc-avidin and receptor.

The biological activity of the other complexes was 20–30% that of the uncomplexed derivatives. Thus, a number of the insulin derivatives had met all three criteria and could be predicted to function successfully as ligands for insulin affinity chromatography. From the synthetic point of view, $N^{\alpha,B1}$-6-(biotinylamido)hexylinsulin (**II**) is the simplest to prepare, and it and the protected (diBoc) derivative were therefore chosen for constructing affinity resins.

Materials

Wheat Germ Agglutinin Sepharose 4B

CNBr-activated Sepharose 4B (15 g) (Pharmacia) is rapidly washed with 1 m*M* HCl (3 liters) at room temperature on a Büchner funel. The

[13] F. M. Finn, G. Titus, and K. Hofmann, *Biochemistry* **23**, 2554 (1984).

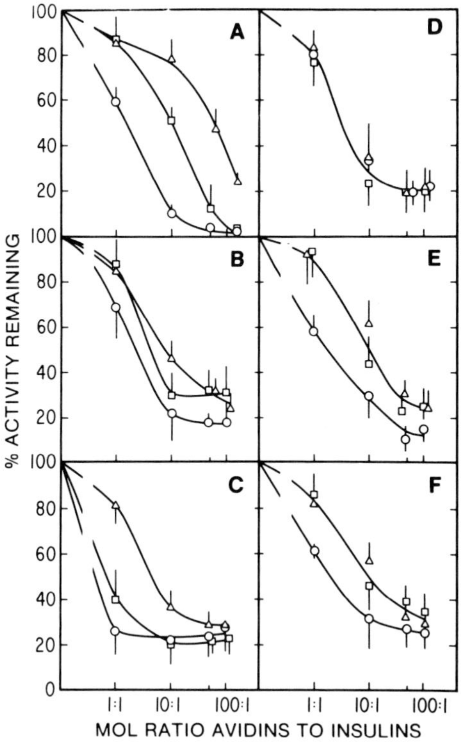

FIG. 4. Inhibition by (□) avidin, (△) suc-avidin, and (○) streptavidin of glucose oxidation stimulated by biotinyl and dethiobiotinyl derivatives of insulin. Glucose oxidation was stimulated by (A) compound **I,** (B) compound **II,** (C) compound **VII,** (D) compound **IX,** (E) compound **VI,** and (F) compound **VIII.** Vertical bars represent SD, $n = 9$.

resin is sucked dry, immediately transferred to a plastic bottle (250 ml), and a solution of wheat germ agglutinin (500 mg) (E-Y Laboratories, San Mateo, CA) in 0.1 *M* $NaHCO_3$ (100 ml) containing 0.5 *M* NaCl is added at 4°. The suspension is rotated at 4° for 24 hr and is filtered on a Büchner funnel; the filtrate normally contains negligible amounts of protein. The filter cake is resuspended in 50 ml of 1 *M* ethanolamine-HCl (pH 8.0), and the suspension is rotated for 3 hr at 4°. The resin is collected by filtration on a sintered glass filter and subjected to 3 washing cycles, each consisting of a wash with 0.1 *M* $NaHCO_3$, 1.0 *M* NaCl (pH 8.0) buffer (1 liter) and 0.1 *M* sodium acetate, 1.0 *M* NaCl (pH 4.0) buffer (1 liter). The washed resin is stored at 4° in pH 4.0 buffer to which sodium azide (0.05%, w/v) is added.

Sepharose 4B–Immobilized Suc-Avidin

CNBr-activated Sepharose 4B (10 g) is washed at room temperature with 1 m*M* HCl (2 liters), and the washed resin, in a plastic bottle, is rotated at 4° for 22 hr with a solution of avidin (100 mg) dissolved in 0.1 *M* $NaHCO_3$ (50 ml) containing 0.5 *M* NaCl. The resin is collected by filtration and resuspended in 1 *M* ethanolamine-HCl (25 ml, pH 8.0). The filtrate, which can be assayed for protein,[14] normally contains only negligible amounts of avidin. The suspension is rotated at 4° for 3 hr, and the resin is collected and washed with 3 cycles (600 ml each) of pH 8.0 and pH 4.0 buffers and then with 100 ml of 0.5 *M* Na_2CO_3 (pH 9.0). The washed resin is slurried in the sodium carbonate buffer (75 ml), the suspension is cooled in an ice bath, and succinic anhydride (59 mg) in dioxane (1.8 ml) is added with vigorous overhead stirring. Stirring is continued for 1 hr at ice-bath temperature and for 4 hr at room temperature. The resin is filtered, washed with 3 cycles (600 ml each) of pH 8.0 and pH 4.0 buffers, and stored in pH 4.0 buffer to which sodium azide (0.05%) is added.

The [^{14}C]biotin binding capacity of the resin can be determined in the following manner. An aliquot (0.2 ml) of a 1 : 1 slurry of the resin in pH 4.0 buffer is layered on 0.3 ml of Sephadex G-50 in a Pasteur pipet, and the resin is washed with 50 ml of 0.1 *M* NaCl. A solution of [^{14}C]biotin (Amersham, specific activity 50 mCi/mmol) is diluted with 0.1 *M* NaCl to 1 m*M*, applied to the resin, and allowed to diffuse into the resin for 30 min at room temperature. The biotin solution is recycled 3 times through the resin, and then fractions (2 ml each) are collected. The column is washed with 0.1 *M* NaCl (16 ml), and portions (0.1 ml) of the eluate fractions are counted. Binding capacity is the difference between applied and recovered biotin.

Insulin Affinity Resin (Human Placental Receptor)

Suc-avidin Sepharose 4B (5.0 ml of settled resin) is poured into a column for equilibration with buffer. The resin is washed with 15 volumes of 50 m*M* NH_4HCO_3. If the resin has been stored for several weeks or longer, it is first washed by cycling with the pH 8.0 and pH 4.0 buffers (15 volumes × 3 of each buffer) prior to equilibration with NH_4HCO_3 buffer. The equilibrated resin is suspended in 6 volumes of NH_4HCO_3 buffer and transferred to a beaker (100 ml). The probe [N^{α,B^1}-6-(biotinylamido)hexyl-N^{α,A^1},$N^{\varepsilon,B^{29}}$-(Boc)$_2$-insulin][4] (700μg) dissolved in 28 ml of the NH_4HCO_3 buffer is added dropwise with overhead stirring over a period of 1 hr at

[14] O. H. Lowry, N. J. Rosebrough, A. L. Farr, and R. J. Randall, *J. Biol. Chem.* **193,** 265 (1951).

room temperature. The resin is stirred for an additional 30 min, poured into a column (0.9 × 7 cm), and washed with 1 liter of 50 m*M* *N*-(2-hydroxyethyl)piperazine-*N*′-2-ethanesulfonic acid (HEPES, pH 7.6), 1 *M* NaCl, 0.1% Triton X-100 to remove any nonspecifically bound probe. The washed resin is stored in the HEPES buffer to which sodium azide (0.05%, w/v) is added.

Methods

Insulin Binding (Charcoal Method)

The charcoal method, described by Williams and Turtle,[15] is based on the ability of charcoal to bind free insulin but not receptor-bound insulin. It is difficult to measure nonspecific binding with this method because the charcoal becomes saturated when excesses of insulin are used. For this reason, it is used only as a screening assay for column eluates.

Solutions

TMA buffer: 50 m*M* Tris(hydroxymethyl)aminomethane (Tris), 10 m*M* $MgSO_4$, 1% (w/v) bovine serum albumin (BSA); adjust pH to 7.6 with HCl

Charcoal suspension: 2.5 g Norit A in 100 ml TMA buffer

Insulin: a solution containing 1 mg/ml in 0.01 *N* HCl is diluted to 1 nmol/ml with TMA

^{125}I-Labeled insulin: 0.1 pmol/250 μl TMA

Chase solution: 0.1 pmol ^{125}I-labeled insulin plus 33 pmol insulin/250 μl TMA

Assay. For each assay, one tube (12 × 75 mm, polystyrene) for measuring total binding (250 μl ^{125}I-labeled insulin) and one tube for measuring nonspecific binding (250 μl chase solution) are prepared. Receptor (10 or 20 μl) is added to each set of tubes, and the contents are mixed by vortexing. Reagent additions are made at ice-bath temperature, and the assay tubes are incubated for 50 min at 24°, after which they are returned to the ice-bath, and charcoal suspension (1 ml) is added to each. To keep the charcoal suspended, the suspension is stirred with a magnetic stirrer. The tubes are centrifuged for 15 min at 1900 *g*, and the supernatant is decanted into counting vials and counted. Standards (250 μl of ^{125}I-labeled insulin) are also counted.

[15] P. F. Williams and J. R. Turtle, *Biochim. Biophys. Acta* **579,** 367 (1979).

Insulin Binding (PEG Method)

In the PEG assay, used by Cuatrecasas for insulin receptor isolation studies,[16] the receptor-bound insulin is precipitated by poly(ethylene glycol) (PEG) while free insulin remains in solution. When crude receptor preparations such as the solubilized receptor are assayed, meaningful results can be obtained only if proteolytic activity is inhibited. For this purpose, *N*-ethylmaleimide, at 1 m*M*, inhibits the insulinase activity described by Duckworth and Kitabchi.[17] After wheat germ affinity chromatography, very little proteolytic activity is present.

Solutions

Tris–albumin: 50 m*M* Tris, 0.1% (w/v) BSA; adjust pH to 7.4 with HCl

Insulin: a stock solution of 1 mg/ml in 0.01 *N* HCl is diluted to 2 nmol/ml with Tris–albumin before use

^{125}I-Labeled insulin: specific activity 0.1 μCi/pmol; various dilutions of this stock solution are made with Tris–albumin to produce solutions containing 0.05, 0.1, 0.3, 0.6, 1.0, 1.5, 3.0, 5.0, and 10.0 pmol/150 μl

γ-Globulin: 0.1% (w/v) bovine γ-globulin (Miles Scientific) in 0.1 *M* sodium phosphate buffer (pH 7.4)

PEG (A): 25% (w/v) PEG (Sigma Chemical Co., M_r 8,000) in water

PEG (B): 8% (w/v) PEG in 0.1 *M* Tris-HCl (pH 7.4)

Assay. For each point, two tubes to measure total and two tubes to measure nonspecific binding are prepared. Microfuge tubes (1.5 ml) are placed in an ice bath. To each tube are added 100 μl insulin (nonspecific binding) or the same volume of Tris–albumin (total binding), 150 μl of ^{125}I-labeled insulin dilutions, receptor, and Tris–albumin to bring the volume to 300 μl. Receptor is added last, and the tubes are mixed by vortexing before and after the addition of receptor. The assay tubes are incubated at 4° for 16 hr. γ-Globulin (0.5 ml) and PEG A (0.5 ml) are added, and the contents of the tubes are mixed by vortexing. After 15 min at ice-bath temperature, the tubes are centrifuged in a microfuge (Beckman Instruments) for 5 min, and the supernatant is aspirated. The pellet is washed with 1 ml of PEG B, which is also aspirated. The tubes (without covers) are placed in counting vials and counted in a γ counter. Duplicate standards for each concentration of ^{125}I-labeled insulin are pipetted directly into counting vials.

[16] P. Cuatrecasas, *Proc. Natl. Acad. Sci. U.S.A.* **69,** 318 (1972).

[17] W. C. Duckworth and A. E. Kitabchi, *Endocr. Rev.* **2,** 210 (1981).

Insulin Receptor Isolation (Human Placenta)

Step 1: Preparation of 40,000 g Pellet. All operations are performed at 4° unless otherwise noted. A fresh term placenta is dissected to remove the chorionic and amnionic membranes. The tissue is cut into pieces (~5 × 5 cm), and these are washed several times (to remove blood) with 0.25 *M* sucrose adjusted to pH 7.5 with $NaHCO_3$. The sucrose is decanted, the tissue is weighed, and 0.25 *M* sucrose, 5 m*M* Tris-HCl (pH 7.4), 0.1 m*M* phenylmethanesulfonyl fluoride (PMSF) is added (2 ml/g). The suspension is homogenized using a Polytron homogenizer (Brinkmann) at a setting of 5 for 1 min. The suspension is centrifuged at 650 *g* for 10 min. The centrifugate often contains three layers, a spongy material, an infranatant liquid, and a pellet. The infranatant is saved and centrifuged for 10 min at 12,000 *g*. Sodium chloride and $MgSO_4$ are added to a final concentration of 0.1 *M* and 0.2 m*M*, respectively, the volume of the supernatant is adjusted to 1000 ml, and the suspension is centrifuged at 40,000 *g* for 40 min. The pellet is resuspended with the aid of a Dounce homogenizer in 50 m*M* Tris-HCl (pH 7.4), 0.1 m*M* PMSF, 2 m*M* EDTA (1200 ml) and washed by centrifugation at 40,000 *g* for 40 min. The pellet is resuspended in 50 m*M* HEPES (pH 7.6), 0.1 m*M* PMSF, 2 m*M* EGTA, 25 m*M* benzamidine-HCl to a final volume of 85.5 ml (protein concentration ~15 mg/ml). The suspension can be stored frozen (flash freeze in a dry ice–acetone bath and store at −70°) or solubilized immediately.

Step 2: Solubilization. To the suspension from Step 1 is added the same buffer containing 20% Triton X-100 (4.5 ml). The sample is stirred for 30 min at 4°. In earlier procedures, solubilization was performed at room temperature; however, the same degree of solubilization is achieved at 4°, and with less proteolytic destruction. Insoluble material is removed by centrifugation at 100,000 *g* for 2 hr. The supernatant from this step can either be flash frozen and stored at −70° or immediately applied to a wheat germ affinity column. A sample (1.0 ml) is removed for determinations of protein[18] and binding activity.

Step 3: Wheat Germ Agglutinin Affinity Chromatography. The wheat germ agglutinin Sepharose 4B resin (20 ml) is washed in a column with 4 column volumes of 50 m*M* HEPES (pH 7.6), 10 m*M* $MgCl_2$, 0.1 m*M* PMSF, 0.1% Triton X-100 and 4 column volumes of 50 m*M* HEPES (pH 7.6), 10 m*M* $MgCl_2$, 0.1 m*M* PMSF, 0.1% Triton X-100, 2 m*M* EGTA, 25 m*M* benzamidine-HCl. Magnesium chloride is added to the receptor to a concentration of 10 m*M*. A portion (one-quarter) of the soluble receptor is

[18] S. Udenfriend, S. Stein, P. Bohlen, W. Dairman, W. Leimgruber, and M. Weigele, *Science* **178,** 871 (1972).

slurried with the washed resin, the slurry is transferred quantitatively with the aid of the rest of the receptor solution to a 250-ml plastic bottle, and the slurry is rotated for 16 hr. The rotated mixture is transferred to a chromatography column (3 cm in diameter) that is washed with 1.0 liter of the HEPES–$MgCl_2$–PMSF–Triton buffer and eluted with 0.3 *M* N-acetylglucosamine (GlcNAc), 50 m*M* HEPES (pH 7.6), 0.1 m*M* PMSF, 0.1% Triton X-100 (~80 ml). A portion of the elution buffer (0.5 column volumes) is allowed to diffuse into the column, and the column is kept for 30 min before elution is started. Fractions (4 ml each, 15–20 fractions) are collected and assayed for binding activity by the charcoal assay. Fractions containing receptor are pooled, flash frozen, and stored at −70°. A sample (1.0 ml) is reserved for protein and binding activity determinations. For reuse, the resin is washed with 4 column volumes of 50 m*M* HEPES (pH 7.6), 0.1 m*M* PMSF, 0.1% Triton X-100, and it is stored in the same buffer to which sodium azide (0.05%) is added.

Step 4: Insulin Affinity Chromatography. The affinity column (5 ml, 0.9 × 7 cm) is washed with 500 ml of 50 m*M* HEPES (pH 7.6), 0.1 m*M* PMSF, 0.1% Triton X-100, 1 *M* NaCl. The wheat germ agglutinin eluate from Step 3 is adjusted to 1 *M* with NaCl. The receptor solution (800 pmol of binding activity/5 ml of resin) is cycled[19] over the column for 12 hr at a flow rate of 0.2 ml/min. This flow rate is sufficient to recycle the receptor approximately 8 times over the resin. The flowthrough is retained for assay and reapplication and can be stored frozen at −70° until used. The resin is washed with 1.2 liters of HEPES–salt–Triton buffer at a flow rate of 50 ml/hr. The column is eluted at 4° with 50 m*M* sodium acetate (pH 5.0), 1 *M* NaCl, 0.1% Triton, 0.1 m*M* PMSF.[3] Owing to the detrimental effects on binding of exposing the receptor to pH 5.0, the eluate fractions (2 ml) are collected into 0.5 *M* HEPES (pH 7.6), 0.1 m*M* PMSF (1 ml) to neutralize them as soon as possible. Although insulin binding is not affected when the receptor is eluted in this manner, autophosphorylation activity is significantly diminished by exposure to pH 5.0 even for short periods. Higher yields of receptor can be obtained by lowering the pH of the eluting buffer to 4.0,[2] but binding activity is rapidly destroyed at pH 4.0 even at 4°. The $t_{1/2}$ for the decay of binding is 67 min at this temperature.

We have developed an alternative procedure for eluting the insulin affinity column which preserves both binding and autophosphorylation activity.[20] Once the column has been washed with the Triton-containing buffer, it is washed with an additional 4 column volumes of the buffer to which inhibitors (bacitracin, 2 mg/ml; antipain, 1.5 μg/ml; and 0.1 m*M*

[19] K. Hofmann and F. M. Finn, *Ann. N.Y. Acad. Sci.* **447,** 359 (1985).
[20] K. Ridge, F. M. Finn, and K. Hofmann, *Proc. Natl. Acad. Sci. U.S.A.* **85,** 9489 (1988).

PMSF) are added. The Triton is then exchanged for octyl-β-glucoside by adding 0.5 column volumes of 50 m*M* HEPES (pH 7.6), 1 *M* NaCl, 0.6% octyl-β-glucoside buffer containing the inhibitors. The column is equilibrated to room temperature and is eluted with the same buffer. Fractions (20, 2 ml each) are collected and immediately chilled to ice-bath temperature.

The data presented in Fig. 5 show the elution profiles of a single insulin affinity column from which receptor was eluted using different conditions. Several conclusions can be drawn from these results. At 4° an acidic eluent is essential for removing receptor from the column; at room temperature, however, receptor can be retrieved if the column has been equilibrated with pH 7.6 octyl-β-glucoside-containing buffer. The binding between insulin and its receptor is apparently diminished in the presence of this detergent. For this reason, the affinity column is not washed with copious amounts of octyl-β-glucoside-containing buffers to remove contaminating proteins but rather is equilibrated briefly with the detergent prior to elution. Yields of receptor using the newer elution conditions are generally 60%, based on the amount of receptor bound. This is the same yield that has been obtained using the pH 5.0 sodium acetate–Triton eluent.

Another advantage of using octyl-β-glucoside instead of Triton is noted when receptor preparations are concentrated. Owing to the low

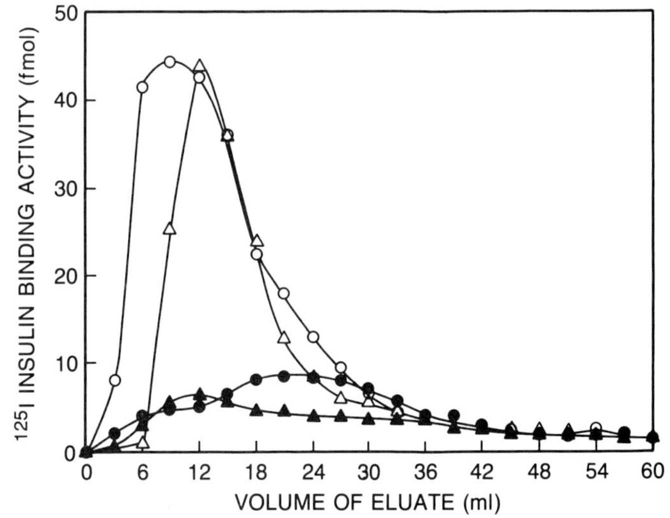

FIG. 5. Elution profiles for insulin receptor eluted from the insulin affinity column by various methods: (○) pH 7.6, 25°, using octyl-β-glucoside; (●) pH 7.6, 25°, using Triton X-100; (△) pH 5.0, 4°, using Triton X-100; (▲) pH 7.6, 4°, using octyl-β-glucoside.

critical micelle concentration (0.2–0.9 m*M*) and the high aggregation number (100–155) for Triton X-100, the detergent does not pass through the membranes used for concentrating the receptor and thus detergent concentration increases proportionately with decreasing volume. Octyl-β-glucoside has a critical micelle concentration of 19–25 m*M*[21] and an aggregation number of 27,[22] both of which favor its passage through membranes.

Properties of Insulin Affinity Column

Coupling avidin to CNBr-activated Sepharose is essentially quantitative. As 10 g of Sepharose produces 40 ml of settled resin, each milliliter should contain 147 nmol of biotin-binding sites, which is the value found when the resin is titrated with [^{14}C]biotin. In view of the amount of biotinylated insulin probe mixed with the resin (700 μg, 117 nmol/5 ml), only approximately one in six biotin-binding sites should be occupied by the biotinylated insulin probe. This ligand density was originally used so that there would be an excess of biotin-binding sites available to rebind ligands that might dissociate. Fortuitously, affinity columns having this ligand density have functioned well in the purification of human placental receptor. To ensure maximum binding of the wheat germ-purified receptor, amounts considerably in excess of the column binding capacity are applied, and the yield of receptor is calculated from the amount bound (applied − flowthrough). Routinely, 60% of the bound receptor can be eluted either at 4° using the pH 5.0 sodium acetate–Triton buffer or at 25° using the pH 7.6 HEPES–octyl-β-glucoside buffer. The question then arises as to the fate of the remaining 40%. Either it is nonspecifically bound and hence not released by conditions that weaken the insulin–receptor binding, or it is lost during the exhaustive washing to remove contaminants. Although nonspecific binding does occur, it is most apparent when a freshly prepared column is used or when a column that has been recently cycled through the pH 4 and pH 8 buffer wash is employed. In this case, nonspecific binding results in lower receptor yields. Once the column has been "seasoned" by repeated use, 60% yields are the average. Thus, it is likely that material is lost during the washing procedure.

When the same columns were employed to isolate rat liver insulin receptor, this receptor did not bind to the resin.[23] The affinity of the rat

[21] W. J. DeGrip and P. H. M. Bovee-Geurts, *Chem. Phys. Lipids* **23,** 321 (1979).

[22] P. Rosevear, T. VanAken, J. Baxter, and S. Ferguson-Miller, *Biochemistry* **19,** 4108 (1980).

[23] K. Hofmann, H. Romovacek, G. Titus, K. Ridge, J. A. Raffensperger, and F. M. Finn, *Biochemistry* **26,** 7384 (1987).

liver receptor for insulin is not as high as that of the human placental receptor, as judged by Scatchard[24] assays. Affinity chromatography was successful when the ligand density was increased 10-fold.

Properties of Purified Receptor

Molecular Weight. Intact receptor, composed of two α and two β subunits has an M_r, on SDS–PAGE, of 350,000, but other species of M_r 320,000 and 290,000[25] have been identified. The latter arise from proteolytic cleavage of the β subunit and represent $\alpha_2\beta\beta'$ and $\alpha_2\beta'_2$, respectively. Reduction of the intact receptor prior to SDS–PAGE produces a pattern showing the M_r 135,000 α subunit and the M_r 95,000 β subunit. In the case where proteolysis has occurred, a third component of M_r 45,000 (corresponding to the β' subunit) is found. Proteolytic cleavage can be minimized during the isolation of human placental receptor by performing Steps 1, 2, and 3, including applying the sample to the wheat germ affinity column, without interruption. The wheat germ column can then be washed overnight; the majority of proteolytic enzymes are removed during this procedure. The addition of proteolytic inhibitors is also advisable, but speed in receptor preparation is the critical factor in isolating a preparation containing a minimal amount of β' fragments.

Proteolytic cleavage of the β subunit of the rat liver receptor constitutes a major problem and occurs even during elution of the insulin affinity column.[23] With this receptor it was necessary to equilibrate the insulin affinity column with a buffer containing a "cocktail" of protease inhibitors containing aprotinin, pepstatin, antipain, leupeptin, benzamidine, and bacitracin. The receptor was eluted and stored in their presence as well. Including the inhibitors not only diminished proteolysis but noticeably enhanced the autophosphorylation capacity of the isolated receptor.

Claims[26] that benzoylarginine ethyl ester, a trypsin substrate, protects the human placental receptor from proteolysis have prompted the use of this compound and N^{α}-*p*-(tosyl-L-lysyl)chloromethane, a trypsin inhibitor, during rat liver receptor isolation. Neither agent, however, was effective.

^{125}I-Labeled Insulin Binding. The PEG precipitation assay is commonly used to determine the insulin-binding capacity of receptor preparations.[16] The most serious drawback of this method was discovered when highly purified insulin receptor became available and could be radiola-

[24] G. Scatchard, *Ann. N.Y. Acad. Sci.* **51,** 660 (1949).

[25] J. Massague, P. F. Pilch, and M. P. Czech, *Proc. Natl. Acad. Sci. U.S.A.* **77,** 7137 (1980).

[26] S. Kathuria, S. Hertmen, C. Grunfeld, J. Ramachandran, and Y. Fujita-Yamaguchi, *Proc. Natl. Acad. Sci. U.S.A.* **83,** 8570 (1986).

beled with ^{125}I. With this material it was shown[2] that PEG does not quantitatively precipitate the insulin–receptor complex; thus, insulin-binding activity is grossly underestimated by this assay. The insulin-binding activity of receptor preparations obtained by avidin–insulin affinity chromatography generally ranges from 16 to 32 μg insulin bound/mg receptor protein. This corresponds to 1–1.9 mole of insulin bound/mol of receptor, assuming a molecular weight of 350,000 for the receptor. The proteolytically cleaved species of the receptor bind insulin as well as the intact receptor. Both human placental and rat liver receptor preparations lacking intact β subunits are fully competent in insulin binding. Roth *et al.*[27] digested human placental insulin receptor with a collagenase preparation and found that insulin binding activity did not decrease.

Autophosphorylation. Activation of the tyrosine kinase activity of the insulin receptor occurs when insulin binds to the α subunit. This binding results in a conformational rearrangement in the β subunit that allows it to be phosphorylated by ATP. The reaction is intramolecular.[28] Furthermore, an intact β subunit is necessary for autophosphorylation to occur; incorporation of radioactive phosphate has not been detected in the M_r 45,000 proteolytic fragment. Most of the autophosphorylation studies have been performed with wheat germ-purified receptor. Although this is a relatively crude preparation, it is presumably preferred because the highly purified receptor eluted from the insulin affinity column with a pH 5.0 buffer incorporates relatively low amounts of phosphate. We have observed that only 0.1–0.2 pmol ^{32}P is incorporated per insulin-binding site into receptor prepared in this manner.[20] Human placental receptor preparations eluted at pH 7.6 with octyl-β-glucoside-containing buffer, however, incorporated as much as 1 pmol ^{32}P per insulin-binding site. When the receptor is assayed in the presence of octyl-β-glucoside, the dose–response curve for insulin stimulation is shifted to the right compared with that obtained with receptor in Triton X-100. This result is consistent with the assumption that octyl-β-glucoside weakens insulin binding. Octyl-β-glucoside can be exchanged for Triton X-100 during receptor concentration; when this is done the insulin dose–response autophosphorylation curves are restored to normal, and enhanced autophosphorylation is still evident.

The stoichiometry of receptor phosphate incorporation is not quantitative for at least two reasons. First, PEG precipitation of the insulin–receptor complex is not quantitative, as mentioned above. Measurements based on values obtained with this assay are obviously also not quantita-

[27] R. A. Roth, M. L. Mesirow, and D. J. Cassell, *J. Biol. Chem.* **258,** 14456 (1983).
[28] M. J. Shia, J. B. Rubin, and P. F. Pilch, *J. Biol. Chem.* **258,** 14450 (1983).

tive. The second reason has to do with the amount of phosphate already present on the isolated receptor. As phosphorylation is a physiological event, it is reasonable to suppose that the isolated receptor may have some phosphate groups incorporated as a result of normal metabolism. Hence, the amount of radiolabeled phosphate that can be introduced into the receptor will vary according to the degree to which it has already been phosphorylated.

Another Approach to Insulin Receptor Isolation Using Avidin–Biotin Technology

$N^{\alpha,B1}$-6-(Biotinylamido)hexylinsulin (Fig. 2, compound **II**) was employed by Kohanski and Lane[29] to isolate insulin receptor complexes from 3T3-L-1 cells. The purification scheme followed the usual sequence of steps, i.e., preparation of a crude membrane fraction, solubilization with Triton X-100, and wheat germ Sepharose chromatography. Biotin-containing proteins were removed from the wheat germ eluate by treatment with avidin–Sepharose, and compound **II** was added to the biotin-free extract. The biotinyl hormone–receptor complex was added to a denatured avidin–Sepharose CL-4B column. The column was then washed carefully with buffer so as not to remove the biotinyl hormone–receptor complex, and, finally, the complex was displaced by biotin.

The important difference between this scheme and the one described previously lies in the constitution of the affinity column. In this case, avidin, not suc-avidin, was coupled to Sepharose CL-4B, and the avidin–Sepharose resin was washed with guanidine hydrochloride and guanidinium thiocyanate at pH 1.5 (to dissociate avidin tetramers), leaving monomeric subunits of avidin covalently attached to the resin. The affinity of the avidin monomer for biotin is several orders of magnitude lower than that of the tetramer.[30] The reduced affinity of the avidin monomer facilitates displacement of the complex by biotin; however, the weaker binding of the biotinyl hormone–receptor complex precludes the exhaustive washing usually performed to remove nonspecifically bound contaminants. As biotin is used to elute the complex from this column, this may not be a serious consideration in so far as the purity of the product is concerned.

Attempts were made to remove the $N^{\alpha,B1}$-6-(biotinylamido)hexylinsulin from the purified receptor. The complex was reapplied to the wheat germ Sepharose resin, and the column was washed to remove the hor-

[29] R. A. Kohanski and M. D. Lane, *J. Biol. Chem.* **260,** 5014 (1985).

[30] J. Moss and M. D. Lane, *Adv. Enzymol.* **35,** 321 (1971).

mone ligand. The eluate from this column, however, was still heavily contaminated with ligand–receptor complex.

ACTH-Binding Protein

The adrenocorticotropic hormones (ACTHs) are 39-amino acid peptides elaborated by the adenohypophysis. They stimulate the adrenal cortex in vertebrate animals to produce steroid hormones (corticoids). The observation that ACTH stimulates steroid hormone production in isolated adrenal cells[31,32] provides a bioassay for the hormone.

The amino acid sequence covering positions 1–24 (Fig. 6) is identical in all known ACTHs. Species variations occur in the C-terminal portion of the molecule (positions 25–39), which has no known physiological function. A great number of truncated ACTH peptides have been synthesized and have been shown to possess substantial biological activity.[33] $ACTH_{1-24}$ (Synacthen, compound **I** in Fig. 6), which has full biological activity, is used clinically for measuring adrenal function in humans.

Strategy for Isolating ACTH-Binding Protein

The ACTH receptor has proved to be an elusive molecule. ACTH has been shown to stimulate corticosteroidogenesis in isolated adrenal cells as well as adenylate cyclase in adrenal membranes. Quantitative measurements of binding of ^{125}I-labeled ACTH analogs to adrenal cells and membranes have also been reported; however, solubilization and isolation of a functional ACTH receptor have not been achieved. The reasons for this are not clear, but the possibility must be considered that solubilization destroys receptor binding. For this reason, the alternative of cross-linking ACTH to the receptor in the plasma membrane and solubilizing the cross-linked complex was explored. Here we review the isolation and characterization of a protein that binds ACTH specifically and with high affinity (see below).[34] Although it is inviting to regard this protein as the cAMP-linked ACTH receptor, we prefer to refer to it as an ACTH-binding protein until its biological role can be definitively established.

The strategy is to attach [^{125}I][Phe^{2},Nle^{4},DTBct^{25}]$ACTH_{1-25}$ amide (**III**, Fig. 6) covalently to the binding protein. A schematic representation of a plausible structure for the cross-linked complex is shown in Fig. 7.

[31] P. W. C. Kloppenborg, D. P. Island, G. W. Liddle, A. M. Michelakis, and W. E. Nicholson, *Endocrinology* **82,** 1053 (1968).

[32] N. D. Giordano and G. Sayers, *Proc. Soc. Exp. Biol. Med.* **136,** 623 (1971).

[33] K. Hofmann, *in* "Handbook of Physiology" (E. Knobil and W. H. Sawyer, eds.), Sect. 7, Vol. 4, Part 2, p. 29. American Physiological Society, Washington, D.C., 1975.

[34] K. Hofmann, C. J. Stehle, and F. M. Finn, *Endocrinology* **123,** 1354 (1988).

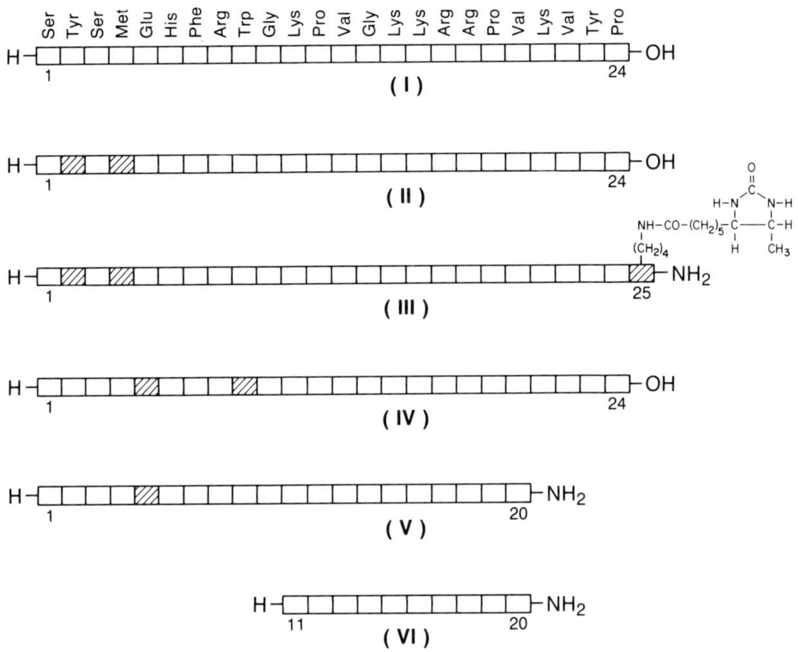

FIG. 6. Simplified amino acid sequences of truncated ACTH analogs: **I,** $ACTH_{1-24}$; **II,** $[Phe^{2},Nle^{4}]ACTH_{1-24}$; **III,** $[Phe^{2},Nle^{4},DTBct^{25}]ACTH_{1-25}$ amide; **IV,** $[Gln^{5},Phe^{9}]$-$ACTH_{1-24}$; **V,** $[Gln^{5}]ACTH_{1-20}$ amide; and **VI,** $ACTH_{11-20}$ amide.

The cross-linked binding protein, which contains dethiobiotin, can then be purified by affinity chromatography on suc-avidin Sepharose (Fig. 1). In practical terms, this means that the covalently linked ACTH–binding protein complex is prepared and solubilized, and the complex is applied to

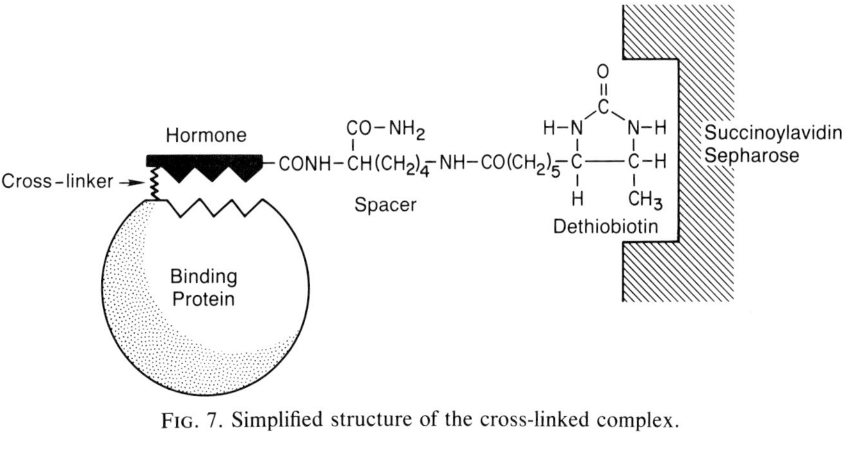

FIG. 7. Simplified structure of the cross-linked complex.

a suc-avidin column (Fig. 1B) to form the suc-avidin–biotinylated ACTH–binding protein complex. The method of Fig. 1C cannot be used to retrieve the binding protein from the affinity column as the binding protein is cross-linked to the hormone. Rather, the cross-linked complex is displaced from its binding site on the suc-avidin column. The yields of radioactive material obtained when elution is performed by the method shown in Fig. 1D, i.e., saturating the resin with biotin, have thus far not been satisfactory. Instead, the cross-linked complex is eluted from the column with 6 *M* guanidinium chloride (pH 1.5), 0.1% Triton X-100. Covalent attachment of the binding protein to the hormone precludes subsequent biological activity measurements (binding), and, thus, denaturing the binding protein is not an important consideration. Acidic guanidinium chloride was first employed by Cuatrecasas and Wilchek[35] to elute avidin from a Sepharose-immobilized biocytin column. In addition to weakening the binding between suc-avidin and the dethiobiocytin portion of the cross-linked hormone–binding protein complex, these conditions promote dissociation of the suc-avidin subunits.

Biotinylated Probe

Our attempts to develop ACTH analogs that can be employed for receptor identification and isolation began with the synthesis of $ACTH_{1-25}$ fragments containing biocytin (Bct) or dethiobiocytin (DTBct) amide (Fig. 6) in position 25. Attaching the biotin or dethiobiotin at residue 25 by way of a lysine seemed the best course of action since this area of the ACTH molecule is not necessary for physiological activity. When the $DTBct^{25}$ analog of $ACTH_{1-24}$ was tested for its ability to stimulate steroidogenesis, its activity was indistinguishable from that of $ACTH_{1-24}$.[36]

Because labeling ACTH or ACTH fragments by conventional iodination techniques destroys biological activity owing to oxidation of the methionine residue at position 4 and incorporation of iodine into the tyrosine residue at position 2, we prepared an analog of $ACTH_{1-24}$ containing phenylalanine in position 2 and norleucine in position 4, i.e., $[Phe^2, Nle^4]ACTH_{1-24}$ (**II**, Fig, 6).[37] This peptide is now commercially available. Iodination of $[Phe^2,Nle^4]ACTH_{1-24}$, using Iodogen as the oxidizing agent, has been accomplished without any detectable loss of biological activity. The monoiodo and diiodo derivatives have been separated by HPLC, and the dose–response curves for steroidogenesis and

[35] P. Cuatrecasas and M. Wilchek, *Biochem. Biophys. Res. Commun.* **33,** 235 (1968).

[36] H. Romovacek, F. M. Finn, and K. Hofmann, *Biochemistry* **22,** 904 (1983).

[37] K. Hofmann, H. Romovacek, C. J. Stehle, F. M. Finn, A. A. Bothner-By, and P. K. Mishra, *Biochemistry* **25,** 1339 (1986).

cAMP formation in bovine adrenal cortical cells of the iodinated molecules are superimposable on those for $ACTH_{1-24}$.

With these analogs we had potential tools for receptor retrieval that had met the first criterion for applicability as affinity ligands, namely, they possessed full biological activity.

Suc-Avidin–Biotinylated ACTH Complexes

The rates of dissociation of $[Bct^{25}]ACTH_{1-25}$ amide and $[DTBct^{25}]$-$ACTH_{1-25}$ amide from their suc-avidin complexes show the same order as the corresponding underivatized biotins, i.e., the dethiobiotin-containing compound dissociates much faster than the biotin-containing derivative (Table I). Although the rate of dissociation of $[Bct^{25}]ACTH_{1-25}$ amide is greater than that for biotin, the DTBct derivative has the same dissociation rate constant as dethiobiotin. The space provided by attaching the biotins to the ε-amino group of Lys^{25} is apparently sufficient to allow the biotins access to their binding site on the suc-avidin molecule. Unlike insulin, where a well-defined three-dimensional conformation has been established, ACTH, which seems to possess little detectable conformation, is probably a more flexible molecule, and this, too, may contribute to the accessibility.

By far the most important consideration in designing potential ligands is whether the suc-avidin complexes of the biotinylated ACTH analogs can bind to the ACTH binding protein. This question was answered by measuring the biological activity of the more stable complex suc-avidin–$[Bct^{25}]ACTH_{1-25}$ amide to eliminate the possibility that any biological activity detected could be ascribed to uncomplexed probe. When the ability of the suc-avidin$[Bct^{25}]ACTH_{1-25}$ amide 1 : 1 complex to stimulate steroidogenesis or cAMP formation in isolated adrenal cortical cells was compared with that of the unbound peptide, the complex was capable of fully stimulating the cells (Fig. 8).[38] The log dose–response curves were shifted to the right by an order of magnitude, indicating that the complex was 10% as active as the unbound peptide on a molar basis. Activity measured in the presence of a 10-fold excess of suc-avidin was the same as for the 1 : 1 complex, demonstrating that the activity was indeed due to complex and not dissociated $[Bct^{25}]ACTH_{1-25}$ amide. The results of these assays indicated that $[Bct^{25}]$ and $[DTBct^{25}]$ derivatives had the properties required to function successfully as ligands for the isolation of ACTH-binding protein. In order to have probes that could be iodinated, the $[Phe^{2},Nle^{4}]$ analogs of the biocytin and dethiobiocytin-containing molecules were prepared.

[38] F. M. Finn, C. J. Stehle, H. Romovacek, and K. Hofmann, *Biochemistry* **24,** 1960 (1985).

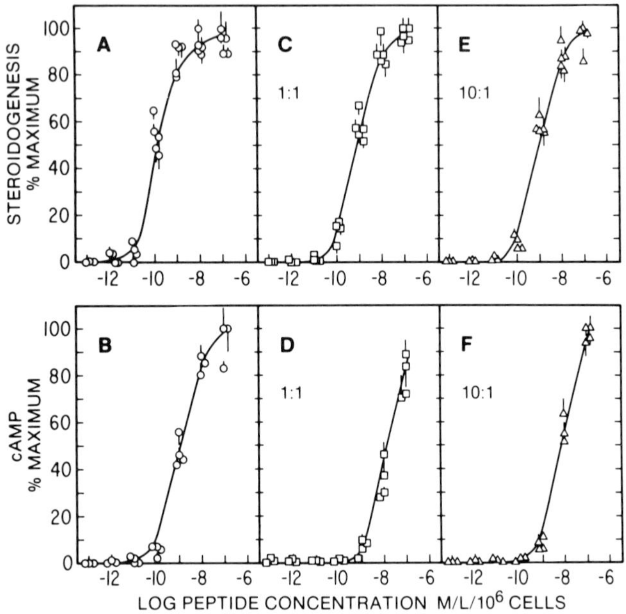

FIG. 8. Effect of suc-avidin on the ability of [Bct[25]]$ACTH_{1-25}$ amide to stimulate steroidogenesis and cAMP production in calf adrenal cortical cells. (A, B) [Bct[25]]$ACTH_{1-25}$ amide; (C, D) 1 : 1 ratio of suc-avidin and [Bct[25]]$ACTH_{1-25}$ amide; (E, F) 10 : 1 ratio of suc-avidin and [Bct[25]]$ACTH_{1-25}$ amide.

Materials

Buffers

Vehicle: 0.1 m*M* HCl, 0.9% NaCl (w/v)

KRBG: Krebs–Ringer bicarbonate buffer containing 0.2% glucose (w/v)

KRBGAC: KRBG containing 0.5% BSA (w/v) and additional $CaCl_2$ to make the solution 8 m*M* in calcium

KRBGACI: KRBGAC containing (per ml) 1.5 μg pepstatin, 2 mg bacitracin, and 15 μg benzamidine

Buffer concentrate: 0.3 *M* HEPES (pH 7.6) containing (per ml) 7.5 μg pepstatin, 10 mg bacitracin, 75 μg benzamidine, and 2.5 mg sodium azide

Dialysis buffer: 50 m*M* HEPES (pH 7.6) containing (per ml) 1.5 μg pepstatin, 2 mg bacitracin, 15 μg benzamidine, and 0.5 mg sodium azide

[Phe^2, Nle^4, $DTBct^{25}$]$ACTH_{1-25}$ Amide

A description of the synthesis of [Phe^2,Nle^4,$DTBct^{25}$]$ACTH_{1-25}$ amide is beyond the scope of this chapter. A detailed procedure can be found in Refs. 36 and 38.

[^{125}I][Phe^2, Nle^4, $DTBct^{25}$]$ACTH_{1-25}$ Amide

The procedure for iodinating [Phe^2,Nle^4,$DTBct^{25}$]$ACTH_{1-25}$ amide, adapted from a general procedure for iodinating proteins and Iodogen,[39] has been shown to produce products with full biological activity.[37] To a 1.5-ml microfuge tube (Beckmann) coated with Iodogen (Pierce) (16 nmol) are added 0.3 *M* phosphate buffer (pH 7.4, 45 μl), $Na^{125}I$ (1.6 nmol, 3.5 mCi), and [Phe^2,Nle^4,$DTBct^{25}$]$ACTH_{1-25}$ amide (10 nmol) in vehicle (30 μl). After 10 min at room temperature, the reaction is terminated by addition of 0.3 *M* phosphate buffer (500 μl), and the diluted reaction mixture is kept at room temperature for 15 min. A portion of the reaction mixture (5 μl) is diluted to 1 ml with KRBGACI, and aliquots (5 μl each) are counted for determination of the total radioactivity in the reaction solution. Two aliquots of the diluted sample (5 μl each) are applied to paper strips (1 $\times$ 20 cm) for ascending chromatography in 10% trichloroacetic acid. The radiolabeled ACTH derivative remains at the point of application while the unincorporated iodine moves with the solvent front. The ratio of radioactivity in these two areas is used to compute the percent iodine incorporation. The reaction mixture is applied to a Sephadex G-10 column (0.9 $\times$ 10 cm) equilibrated and eluted with KRBGACI. Fractions (0.5 ml each) are collected, and 5-μl portions are counted. The ^{125}I-labeled hormone elutes in fraction 7.

Methods

ACTH-Binding Protein Assays

To [^{125}I][Phe^2,Nle^4]$ACTH_{1-25}$ amide (0.025–5.5 pmol in 50 μl of KRBGACI) and $ACTH_{1-24}$ (10 nmol in 50 μl vehicle) or vehicle alone are added adrenal membranes (200 μg protein) suspended in KRBGACI (150 μl). The suspensions are incubated with shaking for 30 min at 23° and centrifuged at 40,000 *g* for 20 min at 4°. The supernatants are aspirated, the pellets washed with ice-cold KRBGAC, and the tips of the tubes excised and counted. The binding data are analyzed by the radioligand

[39] P. R. P. Salacinski, C. McLean, J. E. C. Sykes, V. V. Clement-Jones, and P. J. Lowry, *Anal. Biochem.* **117,** 136 (1981).

analysis computer programs of McPherson (Elsevier-Biosoft) based on the method of Munson and Rodbard.[40] Competitive inhibition assays are performed in a similar manner using 1.7 pmol of [^{125}I][Phe^{2},Nle^{4}, DTBct^{25}]ACTH$_{1-25}$ amide and various amounts of unlabeled competitors.

Isolation of ACTH-Binding Protein

Step 1: Preparation of Bovine Adrenal Cortical Membrane Fraction. Unless otherwise noted, all operations are performed at ice-bath temperature. Steer adrenals (~10) are collected and transported in 0.25 *M* sucrose to the laboratory where they are defatted and cortical slices made with a Stadie–Riggs tissue slicer after discarding capsular material. Approximately 20 g of cortical tissue is collected. Sucrose is added to a concentration of 9 ml/g of tissue, and the suspension is homogenized for 1 min with a Polytron at a setting of 4.

The homogenate is diluted to 500 ml with 1 m*M* $NaHCO_3$ (pH 7.5), stirred for 5 min, and filtered through 4 layers of cheesecloth. The filtrate is centrifuged at 10,400 *g* for 20 min. The supernatant is decanted and centrifuged at 50,000 *g* for 45 min, and the pellet from this step is resuspended with the aid of a Dounce homogenizer in 0.25 *M* sucrose to a volume of 84 ml. At this point, the material can be flash frozen and stored at −70° for several days. Alternatively, the homogenate can be subjected to sucrose density-gradient centrifugation using a discontinuous gradient composed of 54% sucrose (w/w) (7 ml) and 33.7% sucrose (w/w) (14 ml). Both sucrose solutions contain 0.1 m*M* PMSF. The homogenate (14 ml) is layered on the lighter sucrose solution, and the suspension is centrifuged for 2 hr at 40,000 *g*. The material between the application solution and the 33.7% sucrose layer is harvested with a Pasteur pipet. A portion (0.5 ml) is removed for determination of protein,[14] and the remainder of the sample can be stored for several weeks at −70°. Immediately before the membranes are used they are washed with 10 volumes of 1 m*M* $NaHCO_3$ by centrifugation for 30 min at 40,000 *g* and resuspended in an appropriate solution.

Step 2: Cross-Linking Probe to Membranes. Membranes (24 mg protein) are suspended in KRBGACI (18 ml), and [^{125}I][Phe^{2},Nle^{4}, DTBct^{25}]ACTH$_{1-25}$ amide (120 pmol) in KRBGACI (6 ml) and ACTH$_{1-24}$ (1.2 μmol) in vehicle (6 ml) or vehicle alone (6 ml) are added. The suspensions are incubated, with shaking, for 30 min at 23° and are centrifuged for 20 min at 20,000 *g*. The pellets are resuspended in 1 m*M* $NaHCO_3$

[40] P. J. Munson and D. Rodbard, *Anal. Biochem.* **107,** 220 (1980).

(pH 7.5, 30 ml), and a solution of disuccinimidylsuberate [4.48 mg in dimethyl sulfoxide (0.4 ml)] is added to a final concentration of 0.4 m*M*. The suspension is incubated, with shaking, for 15 min at 23°, the tubes are transferred to an ice bath, and Tris-HCl (pH 7.5, 34.5 m*M*, 4.0 ml) is added to a final concentration of 4 m*M*. The suspensions are kept at ice-bath temperature for 5 min and centrifuged as above. The pellets are resuspended in 30 ml of 0.1 *M* acetic acid, incubated, with shaking, for 30 min at 23°, and washed by centrifugation for 20 min at 20,000 *g*. Aliquots of the supernatants from the bicarbonate and acetic acid washes are counted to compute cross-linking efficiency.

Step 3: Solubilization. Membranes (10 mg protein) cross-linked to the radioactive probe in the presence or absence of added $ACTH_{1-24}$ are suspended in 50 m*M* HEPES (pH 7.6, 4% SDS, w/v, 1.0 ml), and the suspension is stirred at room temperature for 30 min, after which it is diluted with 50 m*M* HEPES buffer to 0.4% SDS and protease inhibitors (pepstatin, 15 μg/10 μl dimethyl sulfoxide, bacitracin, 20 mg/100 μl KRBG, and benzamidine, 150 μg/10 μl KRBG) added. The suspension is centrifuged at 100,000 *g* for 1 hr at 23° to remove insoluble material. Aliquots of the 0.4% SDS solution are counted prior to and following centrifugation to determine the degree of solubilization.

Step 4: Detergent Exchange. Sephadex G-100 [80 ml settled resin in 50 m*M* HEPES (pH 7.6)] is equilibrated with HEPES (pH 7.6, 0.1% Triton X-100) at room temperature, and solubilized, cross-linked membranes (4 ml) are added. The column is developed with the same buffer, fractions (2 ml each) are collected and counted, and fractions containing radioactivity are pooled.

Step 5: Suc-Avidin Affinity Chromatography. Sepharose 4B immobilized suc-avidin resin (0.5 ml) is equilibrated with the HEPES–Triton buffer, and the pooled, radioactive fractions from Step 4 are cycled twice over the resin. The resin is cooled to 4°, washed with the HEPES–Triton buffer (15 column volumes), and washed with 3 cycles of pH 8 and pH 4 buffers (15 column volumes each). The column is eluted with 6 *M* guanidinium chloride (pH 1.5%, 0.1% Triton X-100). Fractions (0.5 ml each) are collected into counting vials containing 5 μl of 0.4 *M* Tris to neutralize them, and buffer concentrate (100 μl) is added to each. Fractions containing radioactivity are pooled and concentrated to a small volume (250 μl) in a Micro Confilt cell (Bio Molecular Dynamics). The cell is refilled to the original volume twice with dialysis buffer. A control column in which biotin-binding sites are saturated with biotin before application of the cross-linked complex is used to determine the conditions necessary for removal of nonspecifically bound material.

Problems Encountered

The major difficulty when working with $ACTH_{1-24}$ and its derivatives is the property of these analogs to bind nonspecifically to a large number of materials such as glass, plastics, and biological membranes. These peptides have a net positive charge as a result of the presence of three arginine and four lysine residues whose charge is counterbalanced by only one acidic residue (glutamic acid). In addition, the high content of arginine and lysine makes these peptides attractive substrates for proteolytic enzymes with tryptic specificity. Methods had to be devised to inhibit proteolysis and lower nonspecific binding in order to perform meaningful binding experiments. The combination of pepstatin, bacitracin, and benzamidine was chosen after a systematic study of the ability of a number of proteolytic inhibitors to protect $[^{125}I][Phe^2,Nle^4]ACTH_{1-24}$ from destruction. Unexpectedly, addition of these inhibitors suppressed nonspecific binding as well. Despite the presence of the inhibitors, we were unable to generate meaningful binding data with bovine adrenal cells, and, therefore, we focused our attention on bovine adrenal cortical membranes. These membranes contain an ACTH-sensitive adenylate cyclase system[41] and can be prepared in large scale and stored for weeks at $-70°$ without losing binding ability.

Sodium dodecyl sulfate (SDS) was selected for solubilizing the $[^{125}I]$-$[Phe^2,Nle^4,DTBct^{25}]ACTH_{1-25}$ amide cross-linked complex after comparing the solubilizing efficiency of a number of detergents; SDS solubilized 70% of the radioactivity on the average. Unfortunately, SDS could not be used with a suc-avidin column because it dissociates the suc-avidin subunits (Table II). This difficulty was overcome by exchanging the SDS for Triton X-100 on a Sephadex column. Once the cross-linked complex has been solubilized, solubility can be maintained in Triton X-100.

Properties of ACTH-Binding Protein

In the presence of the protease inhibitors, $[^{125}I][Phe^2,Nle^4,DTBct^{25}]$-$ACTH_{1-25}$ amide bound to the cortical membranes (Fig. 9), and nonspecific binding was only 21%. Scatchard[24] analysis of the binding data indicated that a single class of binding sites with a K_D of 2.7×10^{-9} *M* was present on the membranes. Exposing the membranes to trypsin destroyed binding, indicating that the responsible entity is a protein.

As a result of the large number of structure–activity studies done on ACTH over the years, a wealth of information exists on which analogs should and should not be expected to bind to a receptor. Competitive

[41] F. M. Finn, C. C. Widnell, and K. Hofmann, *J. Biol. Chem.* **247,** 5695 (1972).

TABLE II
EFFECT OF DETERGENTS ON AVIDIN AND SUC-AVIDIN SUBUNIT DISSOCIATION[a]

Molecule	Detergent		
	2% SDS	0.1% Triton	0.1% Lubrol
Avidin	Monomer	Tetramer	Tetramer
Avidin–biotin	Tetramer	Tetramer	Tetramer
Suc-avidin	Monomer	Tetramer	Tetramer
Suc-avidin–biotin	Monomer	—	—

[a] Molecular sizes of the various avidins and their biotin complexes were determined by gel filtration on Sephadex G-100 columns (0.9 × 150 cm) in buffered solutions of the detergents indicated.

binding studies using various nonradioactive ACTH analogs were performed to characterize the binding specificity of the ACTH-binding protein (Fig. 10). The ability of the analogs to compete for binding correlated well with their ability to stimulate cAMP formation in bovine adrenal cortical cells, i.e., $ACTH_{1-24}$ **(I)** > [Gln^5]$ACTH_{1-20}$ amide **(V)** >

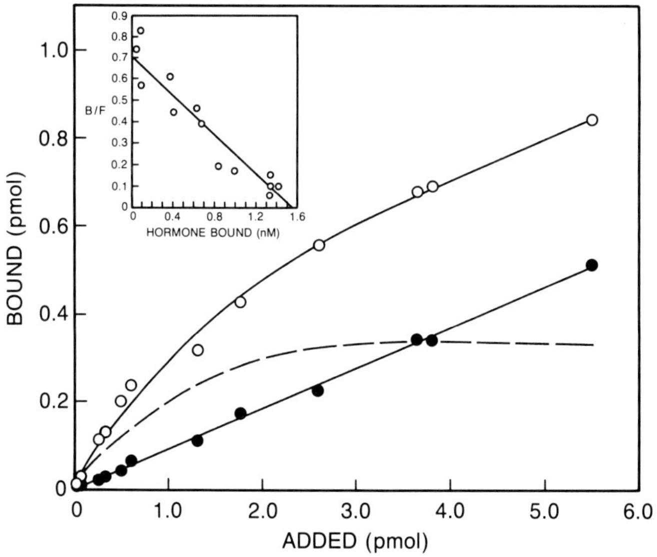

FIG. 9. Binding of [^{125}I][Phe^2,Nle^4]$ACTH_{1-25}$ amide to adrenal particulates: (○) total binding; (●) nonspecific binding; (---) calculated specific binding. Insert is a Scatchard plot of the results.

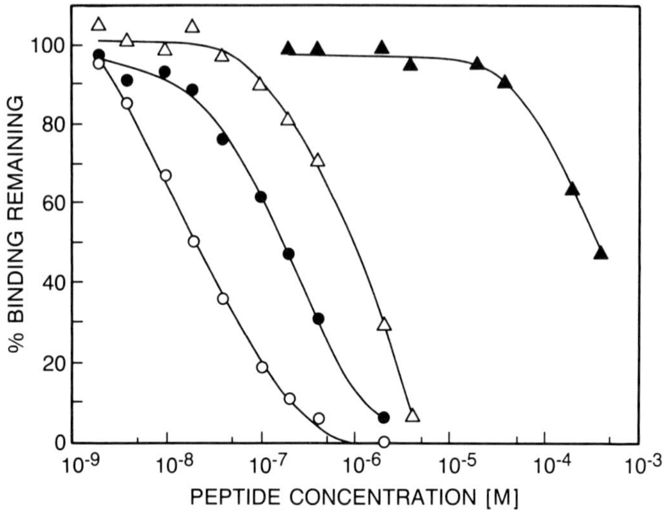

FIG. 10. Competitive inhibition of binding of [^{125}I][Phe^2,Nle^4,DTBct^{25}]ACTH$_{1-25}$ amide to adrenal particulates by analogs: (○) ACTH$_{1-24}$; (●) ACTH$_{1-20}$ amide; (△) [Gln^5,-Phe^9]ACTH$_{1-24}$; (▲) ACTH$_{11-20}$ amide.

[Gln^5,Phe^9]ACTH$_{1-24}$ **(IV)**. ACTH$_{11-20}$ amide **(VI)** is a biologically inactive analog that inhibits ACTH$_{1-24}$-stimulated adenylate cyclase with a 50% inhibition ratio of 400 : 1,[42] i.e., this peptide should be expected to be a poor competitor. Angiotensin II, which binds to the same membranes, failed to compete with the probe.

Cross-linking the ^{125}I-labeled probe to the membrane protein was performed in the presence or absence of a large excess of unlabeled ACTH$_{1-24}$ to demonstrate the specificity of the process. The labeled components were identified by subjecting the membranes to SDS–PAGE[43] followed by autoradiography. The bulk of the radioactivity was present in a single sharp band of M_r 45,000. This band was absent from the samples containing excess unlabeled ACTH$_{1-24}$. Some radioactivity was also present in the high molecular weight region of the gels. The same pattern was observed under reducing and nonreducing conditions. SDS–PAGE of the suc-avidin-purified cross-linked material gave a pattern identical to that observed with the crude membranes.

As has been stated earlier, it is tempting to speculate that the ACTH-binding protein represents the cAMP-coupled ACTH receptor; however,

[42] F. M. Finn, P. A. Johns, N. Nishi, and K. Hofmann, *J. Biol. Chem.* **251,** 3576 (1976).

[43] U. K. Laemmli, *Nature (London)* **227,** 680 (1970).

it is premature to draw this conclusion. The protein labeled here could be only the binding subunit of a larger complex, or cross-linking could have occurred between the radiolabeled probe and a neighboring protein rather than to the receptor itself. For these reasons we reserve judgment on whether the material identified here corresponds to the receptor until its dynamic function in a reconstituted adenylate cyclase system can be established unequivocally.

Estrogen Receptor Purification

Nontransformed estrogen receptor, i.e., the 8 S form of the receptor, has been purified by avidin–biotin affinity chromatography.[44] A number of biotinyl derivatives of estradiol were synthesized by coupling the 7α-carboxylic acid of estradiol to the carboxyl group of biotin using diamines as spacer groups. One of these, estradiol 7α-$(CH_2)_{10}CONH(CH_2)_2$-$O(CH_2)_2O(CH_2)_2NHCO(CH_2)_3NH$–biotin, was used to purify estrogen receptor from unfractionated cytosol.

The avidin–biotin technique was chosen because it presented certain advantages not available with the previously used methods. For example, receptor in a crude tissue extract can be complexed with the biotinyl-estradiol probe, and the resulting complex binds rapidly to the avidin–Sepharose column. Speed is essential to decrease the risk of inactivation of the receptor. Furthermore, the high affinity of the avidin–Sepharose for biotinyl derivatives permits the purification of receptor from samples in which it is present in very low concentrations.

The approach that was used was patterned after the method of Fig. 1B. The biotinylated estradiol derivative was mixed with crude receptor, unbound biotinylestradiol probe was removed by charcoal dextran adsorption, and the complex was applied to avidin–Sepharose. The column was washed, and the estradiol–receptor complex was removed by a variation of the method shown in Fig. 1C; receptor was displaced by [^{3}H]estradiol.

Avidin and not suc-avidin was used for construction of the affinity column in this instance. As previously discussed, avidin binds nonspecifically to many negatively charged molecules. In this case, however, nonspecific binding to avidin would not be expected to contribute to receptor inhomogeneity as the receptor is eluted in a highly specific fashion, namely, by addition of estradiol.

Each of the three criteria that must be met for a biotin-containing hormone ligand to be employed effectively were examined before a ligand

[44] G. Redeuilh, C. Secco, and E.-E. Baulieu, *J. Biol. Chem.* **260,** 3996 (1985).

was chosen for affinity chromatography. The binding of the biotinylestradiols to avidin was demonstrated by their ability to displace the dye 4-hydroxyazobenzene-2′-carboxylic acid from avidin and by gel filtration of mixtures of radiolabeled biotinylestradiols and avidin in the presence or absence of biotin. The binding of the various derivatives to estrogen receptor was assessed by measuring their ability to compete with [^{3}H]estradiol. The ability of the biotinylestradiols to bind simultaneously to receptor and avidin was determined by comparing the effectiveness of the free ligands with their avidin complexes in competitive binding assays. On the basis of these assays it was determined that only ligands having spacer arms of 42 Å were able to bind simultaneously. Using avidin–biotin technology, the authors were able to achieve a 500- to 1500-fold purification of estrogen receptor in a single step without the receptor undergoing transformation. Once most of the contaminants had been removed, conventional chromatographic techniques could be employed without inactivation of the receptor occurring.

Concluding Remarks

In this review of avidin–biotin technology, two techniques for receptor isolation have been discussed. Receptors have been purified by binding biotinyl-containing hormones noncovalently to them, forming an affinity resin with Sepharose-immobilized suc-avidin, and eluting either the unbound receptor or the hormone–receptor complex by one of several strategies. More recently, we have begun to apply the technique to isolation of a cross-linked biotinyl hormone–receptor complex. This usage expands the applicability of affinity chromatography using the hormone as a ligand for those cases in which hormone–receptor binding is lost on solubilization. The fact that the affinity column can be formed by simply mixing the biotinyl hormone–receptor complex permits cross-linking the biotin-containing hormone to a membrane-bound receptor before solubilization. In many cases, it may be possible to employ reversible cross-linkers so that the hormone can be removed from the receptor once purification has been achieved, simplifying subsequent sequence analysis. Thus, the potential of this method is only beginning to be explored.

Acknowledgments

The studies performed in the authors' laboratories have been funded by U.S. Public Health Service Grants DK21292 and DK01128 and by National Science Foundation Grant DCB-8602134.

[28] Biotinylation of Peptide Hormones: Structural Analysis and Application to Flow Cytometry

By WALTER NEWMAN, L. DAWSON BEALL, and ZAFAR I. RANDHAWA

The detection of cell surface structures by flow cytometry has become an established technique. The use of biotinylated monoclonal antibodies plus fluorochrome–avidin is a common approach to the labeling of cell surfaces, and the biotinylation of antibodies is a routine procedure that rarely results in the loss of binding activity. In contrast, the biotinylation of peptide hormones often results in loss of biological activity,[1] making their use in receptor detection problematic.

However, the development of monoclonal antibodies to peptide hormone receptors on cell surfaces usually lags behind the biochemical characterization of the receptor, retarding studies of receptor expression. It is often quite difficult to generate monoclonal antibodies which by their binding pattern and/or effects on cell physiology can be attributed to interaction with a peptide hormone receptor. This is especially true where one is dealing with peptide hormones, and by inference their receptors, which are highly conserved in evolution. To circumvent this problem we have developed a procedure for the biotinylation of a parathyroid hormone analog and transforming growth factor β (TGF-β) type 1 that allows for full retention of activity. We have chosen these peptides in part because they are highly conserved. The active 34-amino acid amino-terminal region of human parathyroid hormone (PTH) is 91% homologous to the bovine peptide (bPTH) and 94% homologous to the porcine peptide.[2] Likewise, TGF-β type 1, a homodimer of 25 kDa, shows only a single amino acid substitution in the murine compared to the human protein.[3] Biotinyl-bPTH and biotinyl-TGF-β are shown here to be useful for flow cytometric detection of receptors. This opens up the possibility of sorting cells, looking for heterogeneity in receptor expression, and multiparameter analysis of receptors in conjunction with other cell surface, cytoplasmic, or nuclear structures. Of course, biotinylated peptide hormones may also be efficient substitutes for monoclonal antibodies for affinity isolation

[1] A. Niendorf, M. Dietel, H. Arps, J. Lloyd, and G. V. Childs, *J. Histochem. Cytochem.* **34,** 357 (1986).

[2] J. F. Habener, M. Rosenblatt, and J. T. Potts, Jr., *Physiol. Rev.* **64,** 985 (1984).

[3] R. V. Derynck, J. A. Jarrett, E. Y. Chen, and D. V. Goeddel, *J. Biol Chem.* **261,** 4377 (1986).

of receptors[4,5] and visualization of receptors by microscopy.[1] The lessons learned with these two peptide hormones should be applicable to the biotinylation of other proteins and will facilitate structure–function studies of the peptide hormones themselves, as has been shown by Lobel *et al.* for curarimimetic toxin,[6] and Hochhaus *et al.* for β-endorphin.[7]

Structure–function studies, however, require knowledge of the number and location of biotin residues within a protein. The detection of biotin by the spectrophotometric method of Green[8] has the disadvantage that it can only detect nanomoles of biotin, and no information regarding the location of the residue is obtained. In addition, steric constraints on the binding of avidin to closely spaced biotin molecules in a protein may give erroneous results. An alternative method to determine which lysine residues are biotinylated is the use of tryptic digestion. This technique, as used by Lobel *et al.*,[6] takes advantage of the loss of a tryptic cleavage site where lysine is derivatized but also requires nanomoles of peptide. The procedure presented here, however, takes advantage of the fact that biotinylated lysine can be sequenced directly at picomolar levels.

Reagents

Bovine parathyroid hormone analog [Nle^8,Nle^{18},Tyr^{34}]PTH(1–34) amide (bPTH)[9]
Transforming growth factor β type 1 (TGF-β) purified from human platelets[10]
Biotinyl-ε-aminocaproic acid-*N*-hydroxysuccinimide (NHS-biotin)
Dimethyl sulfoxide (DMSO), silylation grade
Bovine serum albumin (BSA), RIA grade
FITC-avidin
Biotinyllysine (biocytin)

Buffers

Buffer 1: 0.5 *M* borate (pH 8.0)
Buffer 2: 0.5 *M* borate (pH 6.0)

[4] F. M. Finn, G. Titus, D. Horstman, and K. Hofmann, *Proc. Natl. Acad. Sci. U.S.A.* **81,** 7328 (1984).
[5] D. P. Brennan and M. A. Levine, *J. Biol. Chem.* **262,** 14795 (1987).
[6] P. Lobel, P. N. Kao, S. Birken, and A. Karlin, *J. Biol. Chem.* **260,** 10605 (1985).
[7] G. Hochhaus, B. W. Gibson, and W. Sadee, *J. Biol. Chem.* **263,** 92 (1988).
[8] N. M. Green, this series, Vol. 18, p. 418.
[9] M. Rosenblatt, M. Coltrera, G. L. Shephard, and J. T. Potts, Jr., *Biochem. J.* **20,** 7246 (1981).
[10] J. L. Cone, D. R. Brown, and J. E. DeLarco, *Anal. Biochem.* **168,** 71 (1988).

Buffer 3: 1% (w/v) BSA and 1 m*M* $MgCl_2$ in phosphate-buffered saline (PBS), 0.02% (w/v) NaN_3 (pH 7.2)
Buffer 4: Dulbecco's PBS, 1% BSA, 0.02% NaN_3

Materials and Columns

Reversed-phase Nova-Pak C_{18}, 5 μm, 3.9 × 150 mm
Reversed-phase C_4, 5 μm, 4.6 × 250 mm
Sulfoethylaspartamide (SEA), 4.6 × 200 mm
Pierce reactivials with magnetic spin bar, siliconized

Biotinylation of Parathyroid Hormone

NHS-biotin (0.132 micromole), freshly dissolved in DMSO, is combined with bPTH (0.12 μmol) in 1 ml of buffer 1. The reaction is allowed to proceed for 45 min with constant stirring at 23°. NHS-biotin (0.132 μmol) in 3 μl DMSO is added a second time, and the reaction is continued with stirring for an additional 45 min. This gives a final molar ratio in the reaction mixture of 0.7 mol biotin per lysine residue. This two-step reaction is necessary to avoid precipitation of the protein. The formation of a faint precipitate may be ignored. These reaction conditions are designed to result in some degree of biotinylation of bPTH without the extensive derivatization that can lead to denaturation. The bPTH analog we use has three lysine residues available for biotinylation on the ε-amino groups of lysine at positions 13, 26, and 27.

It is very important with bPTH that all plastic and glassware with which the material comes into contact be siliconized. It is our experience that commercially siliconized glass tubes need to be resiliconized. In addition, it is advisable to add 100 μl of PBS containing 0.1% BSA to the siliconized tubes in the fraction collector to prevent loss of protein. If protein determinations are necessary, this step can be omitted. However, it is possible to arrive at a working concentration of biotinyl-bPTH by performing titrations.

Chromatography

C_{18} Reversed-Phase HPLC. Separation of the reaction products is performed by reversed-phase C_{18} HPLC. Buffer A consists of 90% water, 10% acetonitrile, and 0.1% trifluoroacetic acid (TFA). Buffer B consists of 10% w/v water, 90% acetonitrile, and 0.1% TFA. The reaction is stopped by application to the column, which begins with 100% buffer A for 10 min. At 10 min the gradient is adjusted to reach 39% buffer B by 22 min, followed by readjustment to reach 47% buffer B by 56 min. The flow rate throughout is 2.5 ml/min; UV detection is at 226 nm. From this

gradient, five peaks are obtained that are manually collected, lyophilized, and rechromatographed on C_{18} under the same gradient conditions. Individual fractions are collected and lyophilized for further fractionation, as necessary (see below). DMSO and the unreacted biotin elute from the C_{18} column in the first 5 min. This separation technique takes advantage of the increased hydrophobicity that the biotin molecule confers on the peptide. In general, the higher the degree of substitution, the later the retention time. Amino acid sequence analysis of the individual peaks shows that there are mixtures of biotinylated bPTH peptides in some peaks; ion-exchange chromatography allows for additional separation of components.

Ion Exchange. Sulfoethylaspartamide is the best column for rechromatography of the biotinylated bPTH derivatives. The column conditions for separation are as follows: buffer A consists of 5 m*M* KH_2PO_4 plus 25% acetonitrile; buffer B is the same as buffer A, plus 400 m*M* KCl. The pH of both buffers is 5.25. The lyophilized sample is brought up in 500 μl of 10 m*M* acetic acid plus 500 μl of buffer A and applied to the column directly. The first 10 min of the run is 100% buffer A. From that point a linear gradient of buffer B to 100% is applied over a period of 30 min at a flow rate of 1 ml/min.

Table I summarizes the retention times of five of the biotinylated bPTH derivatives from C_{18} reversed-phase and SEA ion-exchange chromatography as well as the number of biotin residues per bPTH molecule

TABLE I
CHROMATOGRAPHIC PROPERTIES OF BIOTINYLATED bPTH DERIVATIVES

Retention time on C_{18} (min)	Retention time on sulfoethylaspartamide (min)					
	24.6	26.4	27.3[a]	28.3	29.4	32.3
22.5						Native bPTH
23.8				1344	391	
24.1				132	1125	
24.4		1076		707	2127	
25.0		1086	709		577	
25.4	460[b]	183	198			
Number of biotin residues determined by microsequencing	3		2		1	0

[a] Blocked at N terminus.
[b] Values represent picomoles of peptide recovered from sulfoethylaspartamide; 245 nmol of bPTH was applied to the C_{18} column.

as determined from sequencing analysis (see below). The yield of the individual fractions of biotinyl-bPTH after ion exchange is in the range of 1% (Table I). We have pursued these additional purification steps in order to facilitate a structure–function analysis.[11] In practice, a mixture of the dibiotinylated derivatives, even with some monobiotinyl-bPTH, is satisfactory for most binding studies. The tribiotinylated bPTH has lost much of its activity. For routine flow cytometric applications we use the material that elutes from C_{18} at 25.0 min. This material can be obtained in a final yield of 20%.

Flow Cytometry

Data analysis is performed by a Coulter Flow Cytometer Model 541 equipped with an argon laser and a quartz flow cell. Data reduction is performed on an EPICS Easy 88 computer workstation. ROS 17/2.8 osteoblastic osteosarcoma cell line (ROS) of rat origin was used for receptor binding studies.[12] Cells were detached by incubation with PBS containing 2 m*M* EDTA and washed 3 times in buffer 3. For sample preparation, 50 μl of a cell suspension in buffer 3 at 1.5×10^6/ml is added to wells of a round-bottomed 96-well plate which contain 10 ng of biotinylated bPTH in 50 μl. Cells are incubated for 45 min at 4°, washed 3 times with 150 μl of buffer 3 by centrifugation, and incubated in 100 μl of buffer 3 containing 3 μg/ml fluorescein isothiocyanate (FITC)–avidin. Cells are incubated at 4° for 30 min, washed 3 times, and resuspended in buffer 3 for flow cytometric analysis. Control incubations are performed with FITC–avidin only. Table II summarizes representative data obtained with a C_{18} fraction 25.0 biotinylated bPTH derivative on the flow cytometer.

Biotinylation of TGF-β

Our experience with TGF-β is shown here to give some idea of what may be expected from the biotinylation of larger proteins.[13] The optimum reaction conditions are as follows: 4 nmol of TGF-β in 200 μl of buffer 2 is mixed with 176 nmol of NHS-biotin in 3 μl of DMSO. This results in a molar ratio of 2.8 biotin molecules per lysine residue. The reaction proceeds at 23° with constant stirring for 1 hr. As with bPTH, the reaction is stopped by immediate application to the reversed-phase column.

[11] W. Newman, L. D. Beall, M. A. Levine, J. L. Cone, Z. I. Randhawa, and D. R. Bertolini, *J. Biol. Chem.* **264,** 16359 (1989).

[12] R. J. Majeska, S. B. Rodan, and G. A. Rodan, *Endocrinology* **107,** 1494 (1980).

[13] W. Newman, L. D. Beall, D. R. Bertolini, and J. L. Cone, *J. Cell. Physiol.* **141,** 170 (1989).

TABLE II
FLOW CYTOMETRIC DETECTION OF bPTH AND TGF-β RECEPTORS

Cell pretreatment[a]	Mean channel fluorescence intensity
I. No addition	58
FITC–avidin	62
Biotin–bPTH + FITC–avidin	95
Excess bPTH + Biotin–bPTH + FITC–avidin	65
II. No addition	53
FITC–avidin	53
Biotin–TGF-β + FITC-avidin	125
Excess TGF-β + Biotin–TGF-β + FITC-avidin	70

[a] In procedure I, ROS cells are used for detection of bPTH receptors; in procedure II, human endothelial cells are used for detection of TGF-β receptors. There are 256 channels in the fluorescence intensity scale, which represents a 3-log range of fluorescence intensities. Every 25 channels represents an approximate doubling of fluorescence intensity. Biotin–bPTH is fraction 25.0 from C_{18} reversed-phase chromatography. Biotin–TGF-β is the major peak from C_4 reversed-phase at 26.5 min.

Chromatography

A C_4 reversed-phase column works best for repurification of biotinylated TGF-β. Buffer A consists of 90% water, 10% acetonitrile, and 0.1% TFA. Buffer B is 10% water, 90% acetonitrile, and 0.1% TFA. The sample is mixed with 800 μl of buffer A and applied to the column, which maintains 100% buffer A for 15 min. This is followed by an increase to 35% buffer B in 1 min, at which point the gradient is programmed to reach 48% buffer B in 26 additional min. We routinely find the major peak of protein eluting with a retention time of 26.5 min, 3 min after the elution of native TGF-β. As with bPTH, these reaction conditions are designed to give some biotinylation of all the TGF-β in the reaction without causing overbiotinylation and possible denaturation. Minor peaks of biotinylated TGF-β, amounting to no more than 5% of the end products, may be found to elute at 25.1 and 27.2 min. The major peak of material, obtained in a yield of approximately 50%, is fully active in standard assays of TGF-β function. This biotinylated TGF-β is probably a mixture of TGF-β with biotin on different lysine residues. Because there are a total of 16 lysines in TGF-β, and because the reaction product shows full biological activity, further fractionation would be necessary only for structure–function studies.

Flow Cytometry

Human umbilical vein endothelial cells are detached in phosphate-buffered saline containing 2 m*M* EDTA for 2 min at room temperature. The cells are centrifuged and resuspended in buffer 4 to 1.5×10^6/ml. Fifty microliters of the cell suspension is added per well to a round-bottomed 96-well plate and mixed with 10 ng of biotinylated TGF-β in a volume of 50 μl. Maximum binding occurs after 45 min at 4°. The cells are washed 3 times with buffer 4 and incubated for 30 min in FITC–avidin, 2 μg/ml in buffer 4. After 3 additional washes in buffer 4, cells are ready for flow cytometry. As with biotinylated bPTH, background fluorescence levels are determined with FITC–avidin only. Greater than 80% of the fluorescence signal generated by biotinylated TGF-β (and biotinylated bPTH) can be blocked by a 100-fold excess of the unmodified peptide hormone premixed with the biotinylated species in the first incubation step. Data from a representative experiment are shown in Table II.

Structural Analysis of Biotinylated Peptides

Amino Acid Sequencing

Biotin is covalently attached to the peptide through an amide bond that links the carboxyl group of the valeric acid side chain of biotin to the ε-amino group of lysine. This mechanism has been described previously in this series.[14]

Analysis of Biocytin (Biotinyllysine). The phenylthiohydantoin derivative of biocytin elutes near the phenylthiohydantoin derivative of tyrosine and is quantified using the peak height of the lysine residue. The lysine derivative is stable under the conditions of automated Edman degradation. About 8% is debiotinylated (converted to lysine) as shown in Table III.

Analysis of Biotinylated bPTH. The sequence analysis is performed according to the experimental conditions described in detail elsewhere in this series.[15] Both unmodified and biotinylated bPTH samples are subjected to automated Edman degradation. The unique position of the elution of the biocytin derivative makes the sequence assignment unambigu-

[14] E. Bayer and M. Wilchek, this series, Vol. 34, p. 265.

[15] M. W. Hunkapillar, R. M. Hewick, W. J. Dreyer, and L. E. Hood, this series, Vol. 91, p. 399.

TABLE III
DEBIOTINYLATION[a] OF BIOCYTIN[b] AND TRIBIOTINYL-bPTH

Sample	Sequence position	Percentage of total	
		BK[c]	K
Biocytin	—	92	8
Tribiotinyl–bPTH	13	92	8
	26	88	12
	27	91	9

[a] Percentage of total converted to lysine as determined by automated Edman degradation.
[b] HPLC purified on a C_{18} column.
[c] The value for biotinyllysine (BK) was calculated by using the peak height of standard lysine (K).

ous. The sequences of native bPTH and the biotinylated bPTH eluting at 24.6 min from the ion-exchange column (Table I) are shown in Table IV.

The results of amino acid sequence analyses demonstrate that the lysine residues at positions 13, 26, and 27 are occupied by biotinyllysine in the sequence of biotinyl-bPTH. On-line debiotinylation is approximately 10%. Hence, sequence analysis facilitates a structure–function study of biotinylated peptides.[11]

For proteins greater than 40 amino acids in length, fragmentation may be necessary to obtain information regarding the location of the biotin residue(s). It is important, therefore, to establish that biocytin is stable under the conditions utilized for chemical degradation with cyanogen bromide. The data in Table V show that biocytin is stable when degraded with a 50-fold molar excess of cyanogen bromide in 70% v/v formic acid for 22 hours at room temperature. This approach should assist in a structure–function analysis of TGF-β.

Conclusions

A general rule can be made as to the fractionation of biotinylated peptide hormones, as illustrated in Table I. First, reversed-phase chromatography provides for the separation of biotinylated species at later retention times than the unmodified peptide owing to the increased hydrophobicity of the biotin molecule. Conversely, ion-exchange chromatography of biotinylated peptides takes advantage of the loss of a charged residue (lysine). The more highly biotinylated species will elute at lower salt

TABLE IV
COMPARISON BETWEEN UNMODIFIED AND BIOTINYLATED bPTH SAMPLES

Sequence position	One-letter sequence	bPTH	
		Unmodified	Biotinylated
1	A	100[a] (500)	100[a] (428)
2	V	97	88
3	S	47	57
4	E	61	42
5	I	76	67
6	Q	60	62
7	F	71	58
8	NL[b]	80	74
9	H	19	16
10	N	54	51
11	L	59	55
12	G	53	48
13	K/BK[b]	50 (K)	47 (BK)[b]
14	H	19	15
15	L	46	43
16	S	24	24
17	S	19	20
18	NL[b]	33	33
19	E	23	16
20	R	28	18
21	V	22	22
22	E	13	10
23	W	8	6
24	L	14	16
25	R	16	12
26	K/BK[b]	13 (K)	11 (BK)[b]
27	K/BK[b]	13 (K)	12 (BK)[b]
28	L	14	16
29	Q	13	16
30	D	14	10
31	V	11	11
32	H	6	4
33	N	10	10
34	Y	7	7

[a] Values of sequencer yield at each cycle of Edman degradation are compared when the yield of amino-terminal alanine is equal to 100. Picomole yield of amino-terminal alanine residue is given in parentheses.

[b] The values for norleucine (NL) and biotinyllysine (BK) were calculated by using the peak height of standard leucine (L) and lysine (K), respectively.

TABLE V
STABILITY OF BIOCYTIN

Biocytin reaction conditions	On-line debiotinylation[a] (% of total biocytin)	Amount of biocytin analyzed (pmol)
Control[b]	4.0	58
70% Formic acid[c]	3.8	305
70% Formic acid and 50-fold excess of CNBr[b]	3.6	239

[a] Percent of total converted to lysine as determined by automated Edman degradation.

[b] Biocytin is incubated in 10 m*M* acetic acid for 22 hr at room temperature.

[c] Biocytin is incubated for 22 hr at room temperature.

concentrations than the less biotinylated species, and the underivatized material will elute last. Nevertheless, there may be exceptions to these rules depending on the effects of biotinylation on the secondary structure of the protein in question. Also, the SEA column has not yet been shown to be useful for biotinylated peptides other than bPTH.

Application of these techniques to other peptide hormones will have to proceed on a trial and error basis. However, where flow cytometry is the intended application, certain generalizations can be made. First, unlike the demands of affinity chromatography, flow cytometry will likely require that at least two well-spaced biotin molecules be available for cross-linking the receptor on the cell surface to FITC–avidin. For this reason, the selective monobiotinylation of peptide hormones[16] is not a practical approach. Unless there are greater than approximately 25,000 receptors/cell, monobiotinyl derivatives will probably not give a sufficient signal-to-noise ratio. Knowledge of the amino acid sequence or the amino acid composition of the peptide will allow one to determine in a preliminary way whether lysine residues are the best candidates for derivatization. Alternatively, tyrosine, histidine, cysteine, or carbohydrate side chains may also be utilized as attachment sites for biotin (for review, see Ref. 17). Site-directed mutagenesis studies, if they have been performed on the peptide in question, may also give some clues as to which residues are not critical for function. Second, one should have in hand a sensitive bioassay with which to measure the activity of the biotinylated derivatives. Third,

[16] F. M. Finn and K. H. Hofmann, this series, Vol. 109, p. 418.

[17] M. Wilchek and E. A. Bayer, *Anal. Biochem.* **171,** 1 (1988).

a cell line that has a large number of receptors and preferably a tolerable level of autofluorescence for testing derivatives as they are synthesized is essential. It is preferable for the cell line to grow in suspension culture, but adherent cells have also been used. Trypsinized cells can be incubated for several hours (if the receptor is trypsin sensitive) in growth medium in polypropylene tubes or non-tissue culture coated plasticware and then processed for flow cytometry as described. Fourth, unlike flow cytometry with monoclonal antibodies, biotinylated peptide hormones must be carefully evaluated at the cell-binding phase for the optimum time, temperature, requirement for divalent cations, and specificity of binding as determined by competition binding with underivatized ligands.

Assuming the decision to modify ε-amino groups of lysine with NHS–biotin has been made, our experience suggests that the following conditions are a reasonable starting point: a pH range of 6.0–8.0, a molar ratio of biotin to lysine of 2, a protein concentration of approximately 0.5 mg/ml, and a reaction time of 1 hr at 23°. The precise conditions for optimum biotinylation may vary.

A thorough characterization of the biotinylated peptide hormone will be necessary if structure–function studies are to be performed. Since peptide hormones are often not available in large quantities, and fractionation of active biotinylated species may also reduce the available amounts of peptide, it can be a considerable advantage to use the microsequencing techniques outlined here for determination of biotin location and the purity of individual fractions. Small peptides can be sequenced directly, and larger proteins can be characterized after fragmentation by chemical or enzymatic means.

Acknowledgments

We thank James Cone for expert technical assistance, Cindy Utley for excellent secretarial assistance, and Michael A. Levine for helpful discussions.

[29] Purification of Gonadotropin-Releasing Hormone Receptors

By ELI HAZUM

The primary regulator of the reproductive cycle is the hypothalamic decapeptide gonadotropin-releasing hormone (GnRH), which stimulates gonadotropin release from the anterior pituitary. The GnRH receptor is the initial site of action of the hormone in the mammalian pituitary and represents a family of hormone receptors that has not been investigated at the molecular level. Therefore, elucidation of the GnRH receptor structure has theoretical and practical implications both in reproductive biology and cancer. For most peptide hormones, such studies are greatly facilitated by developing techniques for solubilization and purification of the hormone-binding protein from the membrane under conditions that preserve its hormone-binding activity, characteristic affinity, and specificity.

The high affinity constant (10^{15} M^{-1}) between the glycoprotein avidin and the vitamin biotin provides an important experimental advantage for a wide variety of biological applications. Avidin–biotin complexes have been used as mediators in localization, isolation, and immunological studies (reviewed in Refs. 1–6). Recently, biotinylated peptide hormones have been used for the localization and isolation of receptors on cell surfaces. This chapter summarizes the synthesis and application of biotinylated GnRH to highlight the potential use of this method for receptor purification.

Preparation of Biotinylated GnRH

It is important to design the synthesis of biotinylated peptide hormones in such a way that they retain high binding affinity and biological activity. To achieve this goal, [D-Lys^6]GnRH has been selected as the starting material for derivatization since (1) substitution of D-amino acids in position 6 of GnRH results in more potent and metabolically stable derivatives, and (2) the ε-amino group of lysine serves as a spacer for substitution reactions and thus the GnRH conformation is less likely to be disturbed.

Synthesis of [Biotinyl-D-Lys^6]GnRH[7,8]

pGlu-His-Trp-Ser-Tyr-(biotinyl-D-Lys^6)-Leu-Arg-Pro-Gly-NH_2 (Fig. 1) is prepared by reaction of [D-Lys^6]GnRH (0.6 mg, 0.45 mmol; Penin-

[1] E. A. Bayer and M. Wilchek, *Methods Biochem. Anal.* **26,** 1 (1980).

[2] E. A. Bayer, E. Skutelsky, and M. Wilchek, this series, Vol. 83, p. 195.

[3] G. V. Childs, in "Immunocytochemistry" (P. Petrusz and G. Bullock, eds.), p. 85. Academic Press, New York, 1983.

[4] M. Wilchek and E. A. Bayer, *Immunol. Today* **5,** 39 (1984).

[5] F. M. Finn and K. H. Hofmann, this series, Vol. 109, p. 418.

[6] E. Hazum, this series, Vol. 124, p. 47.

[7] G. V. Childs, Z. Naor, E. Hazum, R. Tibolt, K. N. Westlund, and M. B. Hancock, *Peptides* **4,** 549 (1983).

[8] G. V. Childs, Z. Naor, E. Hazum, R. Tibolt, K. N. Westlund, and M. B. Hancock, *J. Histochem. Cytochem.* **31,** 1422 (1983).

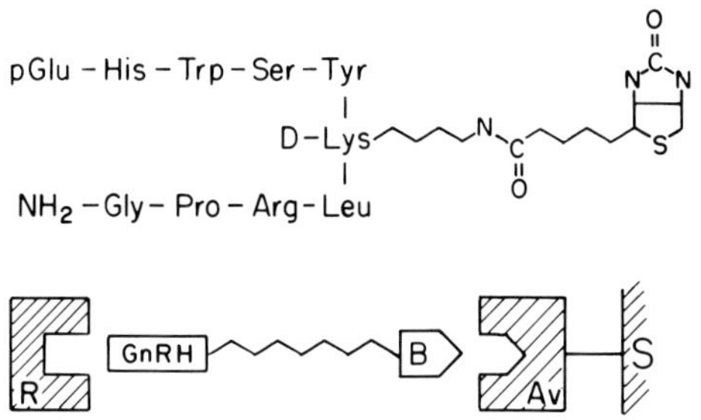

FIG. 1. Schematic representation (bottom) illustrating the bifunctional ligand [biotinyl-D-Lys[6]]GnRH (top) as the mediator for absorption of GnRH receptor to avidin–Sepharose. Av, Avidin; B, biotin; S, Sepharose; R, receptor.

sula Laboratories, Inc., Belmont, CA) with 2 molar equivalents of *d*-biotin *p*-nitrophenyl ester in methanol–dimethylformamide (10 : 1, v/v) in the presence of 1.2 equivalents of triethylamine. After standing at 24° for 3 hr, the product is precipitated by the addition of ether and washed 3 times with ethyl acetate in order to remove unreacted *d*-biotin *p*-nitrophenyl ester. The product is purified by preparative high-voltage paper electrophoresis (Whatman No. 3 paper, 60 min at 60 V/cm) in pyridine–acetate buffer (pH 3.5; electrophoretic mobility, 0.62). The pure product reveals a negative test with ninhydrin reagent, indicating the absence of free amino groups. [Biotinyl-D-Lys[6]]GnRH exhibits a higher binding affinity and biological potency than GnRH and [D-Lys[6]]GnRH.

Isolation of Receptors Using Avidin–Biotin Complexes

Immobilized forms of avidin can be used for the isolation of receptors. This can be accomplished either by binding the biotinylated hormone to avidin columns followed by subsequent interaction with the solubilized receptor or by prior incubation of the biotinylated hormone with the solubilized receptor and then immobilization on avidin columns (shown schematically in Fig. 1). The receptor can be eluted directly from the column, or the biotinylated hormone–receptor complex can be eluted with biotin and the receptor subsequently dissociated from the hormone. In the latter case, it may be advisable to use succinylated avidin, which exhibits a reduced affinity constant. The advantages of the method are (1) the hormone can be attached to the support via a single defined site that is not involved in its biological function; (2) the anchoring of the hormone to the support is unequivocal and proceeds in high yield; (3) the chemical manipulations are performed with the free hormone and thus its effect on binding and biological activity can be readily assessed; and (4) the technique is highly reproducible and can be readily scaled up for production of larger quantities of receptors.

This novel procedure has been applied recently to the purification of GnRH receptors from other solubilized membrane proteins.[9,10] The following steps are involved: solubilization of rat pituitary GnRH receptor with 3-[(3-cholamidopropyl)dimethylammonio]-1-propane sulfonate (CHAPS) without alteration of binding affinity; immobilization of [biotinyl-D-Lys^6]GnRH on avidin–agarose; equilibration of the solubilized GnRH receptor with the affinity resin; and elution of the receptors with high salt, acidic pH, or GnRH analogs. Following 2 cycles of affinity chromatography, the GnRH receptor has been purified to homogeneity. The overall recovery of the purified receptor is 4–10% of the initial activity in the CHAPS extract, and the calculated purification is approximately 10,000- to 15,000-fold.

Solubilization and Purification of GnRH Receptors[9,10]

Iodination and Pituitary Membrane Preparations

[D-Ser(*t*-Bu)^6,de-Gly^{10},ethylamide]GnRH (buserelin) is iodinated by the lactoperoxidase method.[11] The specific activity of the labeled peptide is approximately 1.0 mCi/μg, as measured by self-displacement in the pituitary radioligand receptor assay. Pituitary membranes are prepared from 25- to 28-day-old Wistar-derived female rats. Briefly, the glands are homogenized gently with a tight-fitting Dounce homogenizer at 4° in 10 m*M* Tris-HCl (pH 7.4, Tris buffer) and centrifuged for 10 min at 1000 *g*. The supernatant is then centrifuged at 4° for 20 min at 20,000 *g*. The pellet is resuspended in Tris buffer and centrifuged at 20,000 *g* for 20 min, and the pellet stored at −20°.

Solubilization of GnRH Receptors

Pituitary membrane preparations are washed with Tris buffer at 4° by centrifugation (20 min at 20,000 *g*). The pellet is resuspended in Tris buffer containing 5 m*M* CHAPS (Sigma Chemical Co., St. Louis, MO), shaken for 60 min at 4° and centrifuged (60 min at 100,000 *g*). This procedure is repeated, and the supernatants are combined and used to measure binding. Usually, the solubilized receptor is kept in 1 m*M* CHAPS, 10 m*M* Tris buffer containing 10% glycerol (CHAPS–Tris–glycerol) and 1

[9] E. Hazum, I. Schvartz, Y. Waksman, and D. Keinan, *J. Biol. Chem.* **261,** 13043 (1986).
[10] E. Hazum, I. Schvartz, and M. Popliker, *J. Biol. Chem.* **262,** 531 (1987).
[11] E. Hazum, *Endocrinology* **109,** 1281 (1981).

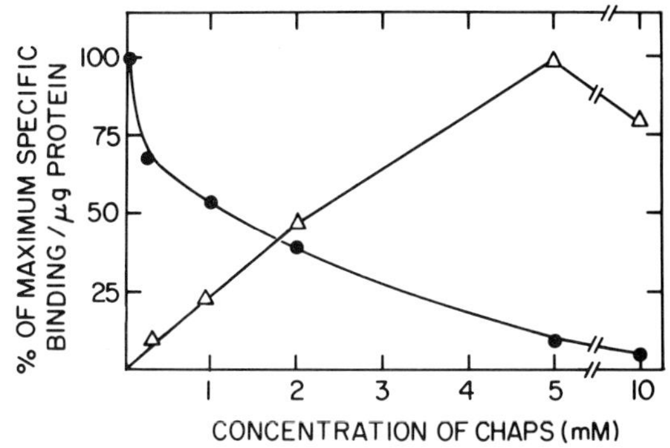

FIG. 2. Solubilization of GnRH receptors by different concentrations of CHAPS. Pituitary membranes (1.7 mg) are washed twice in 10 m*M* Tris-HCl (pH 7.4). The pellet is incubated with different concentrations of CHAPS for 1 hr at 4° while stirring. At the end of the incubation, the tubes are centrifuged at 4° for 1 hr at 100,000 *g*, and binding is conducted for both the membranal (●) and solubilized (△) receptor. The concentration of CHAPS in all solubilized samples is adjusted to 1 m*M* during the binding incubation.

m*M* phenylmethylsulfonyl fluoride. This crude preparation can be stored at −20° for extended periods of time under these conditions.

Binding Assays to Membranal and Solubilized Receptors

The binding of radioiodinated buserelin to membranal receptors is assessed as previously described.[12] Briefly, the labeled buserelin (50,000 cpm) is incubated with pituitary membranes (10–20 μg protein) in a total volume of 0.5 ml Tris buffer containing 0.1% (w/v) bovine serum albumin (BSA) for 90 min at 4° (equilibrium conditions). The binding is measured by filtration under vacuum through Whatman GF/C filters. The solubilized receptors (25–50 μg protein) are incubated with the labeled buserelin (50,000 cpm) in 0.5 ml of 10 m*M* Tris, 0.1% BSA containing 1 m*M* CHAPS for 2.5 hr at 4°. The reaction is stopped by the addition of 0.3 ml ice-cold dextran-coated charcoal [0.5 g dextran T-70 (Pharmacia, Piscataway, NJ) and 5.0 g activated charcoal (Norit A, Fisher) dissolved in 1000 ml of phosphate-buffered saline (PBS)]. The tubes are left on ice for 10 min and then centrifuged for 20 min at 2000 *g* at 4°. The supernatant fluids are collected and counted in a γ counter. Specific binding represents the bound radioactivity in the presence of 10^{-7} *M* unlabeled buserelin subtracted from the total bound radioactivity. Figure 2 shows the effect of increasing concentrations of CHAPS (0.25–10 m*M*) on the specific binding activity of both solubilized and membrane-associated receptors.

[12] D. Keinan and E. Hazum, *Biochemistry* **24,** 7728 (1986).

Affinity Chromatography on GnRH Resin

Avidin–agarose (Sigma, 1.5 ml containing 2.5 mg of avidin) is incubated with 10^{-6} *M* [biotinyl-D-Lys^6]GnRH in Tris buffer for 5 hr at 24°. The resin is washed extensively with Tris buffer (10 times, 20 ml each wash) and subsequently equilibrated with CHAPS–Tris–glycerol. The resin is incubated with the solubilized receptors (1–2 mg) in 5–10 ml of CHAPS–Tris–glycerol for 12 hr at 4° and poured into a column (0.5 × 4

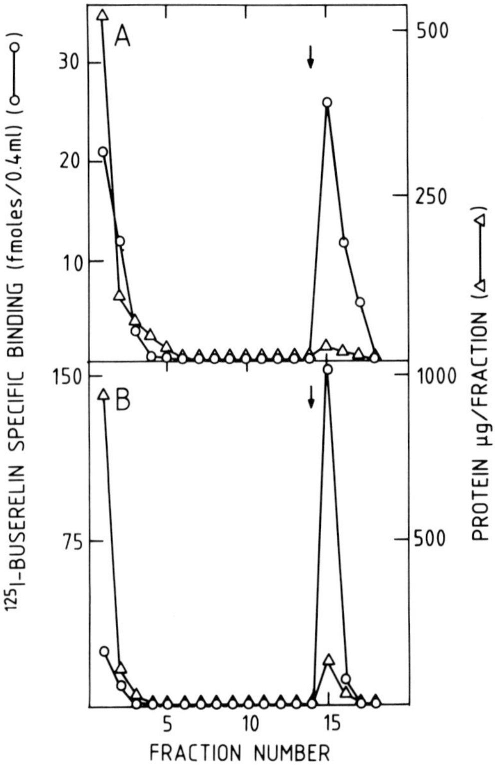

FIG. 3. Typical elution of rat GnRH receptor from the affinity column with acidic pH or high salt. [Biotinyl-D-Lys^6]GnRH immobilized on avidin–agarose resin (1.5 ml) is incubated with the solubilized receptor (A, 1 mg/4 ml; B, 1.8 mg/6.5 ml) in 1 m*M* CHAPS, 10 m*M* Tris, 10% glycerol for 12 hr at 4° and poured into a column (0.4 × 5 cm). The supernatant is collected by centrifugation, and the column is washed with 13 ml of the same buffer. (A) The resin is incubated (30 min at 4°) with the above buffer adjusted to pH 5.5; fractions (1 ml each) are collected and neutralized immediately to pH 7.4. (B) The resin is incubated (60 min at 4°) with 2 ml of 0.5 *M* NaCl in the chromatography buffer. The supernatant is collected by centrifugation, and eluate fractions (1 ml) are dialyzed. All fractions are sampled for protein and GnRH-binding activity. Arrows indicate the starting point of the elution buffer.

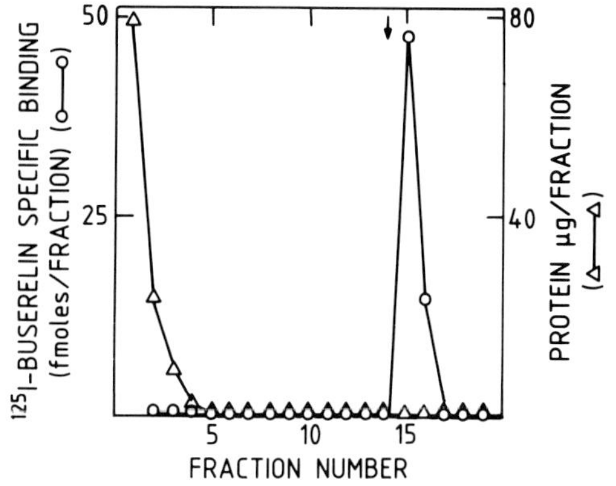

FIG. 4. Elution of rat GnRH receptor from a second affinity column with buserelin. The fractions with the highest receptor activity from the NaCl-eluted affinity column are dialyzed, pooled (0.3 mg/1.3 ml), and rechromatographed in the second affinity-purification step. The experimental details are as in the legend to Fig. 3, except that elution of the receptor is carried out by incubating the resin (3 hr at 4°) with 5×10^{-9} *M* buserelin (arrow). The fractions (0.3 ml) are diluted (1 : 5), dialyzed, and sampled for protein and binding activity. The total protein content in fraction 15 is 0.1 μg.

cm). The eluate is collected by centrifugation, and the column (kept at 4°) is washed with 13 ml of the above buffer. The solubilized receptor is eluted using three procedures: (1) with CHAPS–Tris–glycerol (pH 5.5), the eluate fractions being neutralized immediately to pH 7.4 with 1 *N* NaOH (Fig. 3A); (2) with 0.5 *M* NaCl in the same buffer (pH 7.4), the samples being dialyzed against CHAPS–Tris–glycerol to remove excess NaCl (Fig. 3B); or (3) with 5×10^{-9} *M* buserelin, the samples being diluted (1 : 5) and dialyzed before assays (Fig. 4). Further experimental details are presented in the legends to Figs. 3 and 4. GnRH-binding activity is detected by the charcoal assay.

Comments

The avidin–biotin complex system has been used widely to study biological interactions that involve the specific binding between a protein and a ligand. In the present chapter, this strategy has been applied to the purification of GnRH receptors. The development of a two-step affinity chromatography procedure permits preparation of large quantities of the pure receptor that can be used for detailed studies on the structure and function of the receptor, for the development of monospecific antibodies, as well as for partial amino acid sequencing. These studies will advance our understanding of the molecular basis for the action of GnRH, which has a pivotal role in the regulation of reproduction.

Acknowledgments

I am grateful to Mrs. M. Kopelowitz for typing the manuscript and to Dr. C. Webb for useful suggestions. This work was supported by the U.S.–Israel Binational Science Foundation, the Fund for Basic Research administered by the Israel Academy of Sciences and Humanities, and the Minerva Foundation, Federal Republic of Germany.

[30] Biotinylestradiol for Purification of Estrogen Receptor

By G. Redeuilh, C. Secco, and E. E. Baulieu

Affinity resins suitable for purification of different steroid hormone receptors have been synthesized. They were necessarily used empirically, in contrast to soluble ligands which could be utilized under controlled conditions in terms of affinity and concentration. Therefore, soluble biotinylated hormones, potential ligands of both receptors and avidin–agarose columns, represent an attractive approach to the purification of receptors. Purification of the estrogen receptor by the avidin–biotin system is based on the high affinity of both avidin and the receptor for biotinylestradiol derivatives.[1] Estradiol 7α derivatives were selected on the basis of their binding affinity and specificity for the estrogen receptor.[2] It is known that chemical modification of the carboxyl group of biotin does not affect the binding of biotin to avidin and that avidin can be linked to solid supports without losing its binding capacity.[3,4]

Preparation of Biotinylestradiol Conjugates

Synthetic routes to biotin estradiol derivatives are illustrated in Scheme I. Estradiol-7α-carboxylic acid derivatives (estradiol-7α-butanoic acid and estradiol-7α-undecanoic acid prepared as described by Bucourt *et al.*[2]) are coupled to biotin by using linear diamines. Longer linkers are obtained by replacing biotin by 4-(*N*-biotinyl)aminobutanoic acid

[1] G. Redeuilh, C. Secco, and E. E. Baulieu, *J. Biol. Chem.* **260,** 3996 (1985).

[2] R. Bucourt, M. Vignau, V. Torelli, H. Richard-Foy, C. Geynet, C. Secco-Millet, G. Redeuilh, and E. E. Baulieu, *J. Biol. Chem.* **253,** 8221 (1978).

[3] L. D. Wright, H. R. Skeggs, and E. L. Cresson, *J. Am. Chem. Soc.* **73,** 4144 (1951).

[4] N. M. Green and E. J. Toms, *J. Biochem.* **133,** 687 (1973).

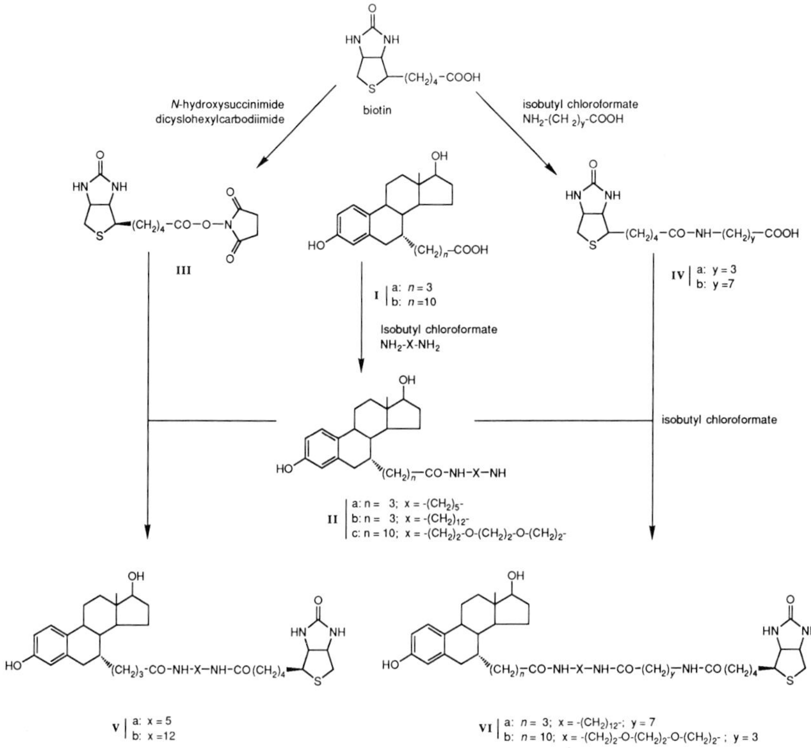

SCHEME I

or 8-(*N*-biotinyl)aminooctanoic acid. Amide bonds are obtained by the mixed anhydride procedure using isobutyl chloroformate in tri-*n*-butylamine-containing medium. This procedure is more efficient than the most commonly used techniques using carbodiimide or *N*-hydroxybenzotriazole in the presence of dicyclohexylcarbodiimide and those using *N*-ethoxycarbonyl-2-ethoxy-1,2-dihydroquinoline as coupling agent. High yield and satisfactory purity lead to an easier purification by column chromatography. All manipulations are carried out at low temperature (<25°), and the intermediary estradiolamine derivatives have to be purified by chromatography on a silicon dioxide column before use in order to avoid problems arising from the complex composition of reaction products. Purity of biotinylestradiol compounds is checked by thin-layer chromatography (TLC) on silicon oxide plates. In two solvent systems, a single spot appears which is 4-dimethylaminocinnamaldehyde positive, indicating the presence of biotin, and H_2SO_4–methanol or phosphomolybdic acid positive, indicating the presence of estradiol.

Synthesis of Biotinylestradiol (Scheme I)

N-(5-Aminopentyl)-4-[3,17-dihydroxyestra-1,3,5(10)-triene]-7α-butanamide **(IIa).** To a solution of estradiol-7α-butanoic acid **(Ia)** (100 mg) in dioxane (10 ml) containing tri-*n*-butylamine (80 μl), cooled at 10–15°, is added freshly distilled isobutyl chloroformate (45 μl). After an additional 1-hr incubation period with gentle stirring, the mixture is added to a solution of pentamethylenediamine (5-fold excess) in dioxane cooled at 10–15°. The mixture is stirred for an additional 2 hr at room temperature, and the solvent is then removed to dryness under vacuum. The residue is washed successively with 0.1 *M* $NaHCO_3$ and water and chromatographed through a column of silicon dioxide. The chromatographically homogeneous fractions are pooled to give 81 mg product (72%). TLC: R_f 0.42 ($CHCl_3$–methanol–NH_4OH, 80 : 20 : 3).

N-(12-Aminododecanyl)-4-[3,17-dihydroxyestra-1,3,5(10)-triene]-7α-butanamide **(IIb).** Compound **IIb** is prepared from the corresponding mixed anhydride and dodecamethylenediamine, as described above for compound **IIa.** The crude product is purified by chromatography on silicon dioxide (yield 69%). TLC: R_f 0.71 ($CHCl_3$–methanol–NH_4OH, 80 : 20 : 3).

N-(8-Amino-3,6-dioxaoctanyl)-11-[3,17-dihydroxyestra-1,3,5(10)-triene]-7α-undecanamide **(IIc).** To a solution of estradiol-7α-undecanoic acid **(Ib)** (100 mg) and tri-*n*-butylamine (70 μl) in dioxane (15 ml), cooled at 10–15°, is added isobutyl chloroformate (35 μl). The solution is stirred for 40 min at room temperature. The mixture obtained is added dropwise to a solution of 1,8-diamino-3,6-dioxaoctane in 30 ml of dioxane, and the reaction is maintained at room temperature for 2 hr. After distillation of the solvent under vacuum, the desired product is chromatographed on a column of silicon dioxide (yield 100 mg, 78%). TLC: R_f 0.37 ($CHCl_3$–methanol–NH_4OH, 85 : 15 : 1.5).

d-Biotinyl-N-hydroxysuccinimide Ester **(III).** Compound **III** is prepared from biotin and *N*-hydroxysuccinimide by coupling with dicyclohexylcarbodiimide according to Bayer and Wilchek.[5]

4-(N-d-Biotinyl)aminobutanoic Acid **(IVa).** Compound **IVa** is already known.[6] Isobutyl chloroformate (160 μl) is added to a solution of biotin (250 mg) in dimethylformamide (20 ml) containing tri-*n*-butylamine (320 μl). After 10 min at room temperature, the mixture is added to a suspension of 4-aminobutyric acid in dimethylformamide (20 ml) at 5° and

[5] E. Bayer and M. Wilchek, this series, Vol. 34, p. 265.

[6] N. M. Green, L. Konieczny, E. J. Toms, and R. C. Valentine, *Biochem. J.* **125,** 781 (1971).

stirred for 2 hr. The solvent is distilled under vacuum, and the crude precipitate is dissolved in warm aqueous ethanol. After filtration and cooling, the product is precipitated by acidification and purified by recrystallization from water (yield 50%). TLC: R_f 0.22 (toluene–methanol–acetic acid–acetone, 14 : 4 : 1 : 1).

8-(N-d-Biotinyl)aminooctanoic Acid **(IVb).** Compound **IVb** is prepared from 8-aminooctanoic acid in dimethylformamide–water (v/v), and the mixed anhydride is obtained from biotin and isobutyl chloroformate. The product is recrystallized from ethanol (yield 66%). TLC: R_f 0.45 (toluene–methanol–acetic acid–acetone, 14 : 4 : 1 : 1).

Biotinylestradiol **Va.** A solution of amine **IIa** (Scheme I) in ethanol is stirred at room temperature for 2 hr, in the presence of a 1.5-fold excess of *d*-biotinyl-*N*-hydroxysuccinimide ester **(III).** After removal of the solvent, the residue is chromatographed through a column of silicon dioxide. Elution by toluene–methanol–acetic acid–acetone (14 : 4 : 1 : 1) yields 83.5 mg (72%). TLC: R_f 0.3 (toluene–methanol–acetic acid–acetone, 14 : 4 : 1 : 1); 0.46 ($CHCl_3$–methanol, 80 : 20).

Biotinylestradiol **Vb.** Compound **Vb** is obtained by the same procedure as for **Va** and purified by column chromatography (yield 60%). TLC: R_f 0.43 (toluene–methanol–acetic acid–actone, 14 : 4 : 1 : 1); 0.61 ($CHCl_3$–methanol, 80 : 20).

Biotinylestradiol **VIa.** To a solution of 66 mg of **IIb** in dimethylformamide (10 ml) are added, at 15°, the mixed anhydride obtained from biotinylaminooctanoic acid (112 mmol) and isobutyl chloroformate (18 μl) in dimethylformamide containing tri-*n*-butylamine. The temperature is allowed to rise, and the mixture is stirred for 2 hr at room temperature. The solvent is distilled under vacuum, and the residue is then dissolved in $CHCl_3$–methanol (90 : 10) and chromatographed through a column of silicon dioxide. Elution is performed by $CHCl_3$–methanol (85 : 15) (yield 64 mg, 65%). TLC: R_f 0.26 (toluene–methanol–acetic acid–acetone, 14 : 4 : 1 : 1); 0.23 ($CHCl_3$–methanol, 80 : 20).

Biotinylestradiol **VIb.** Compound **VIb** is obtained by the same procedure as described above for **VIa** and is purified by chromatography on silicon dioxide ($CHCl_3$–methanol, 80 : 20) (yield, 34%). TLC: R_f 0.38 (toluene–methanol–acetic acid–acetone, 14 : 4 : 1 : 1); 0.58 ($CHCl_3$–methanol, 80 : 20).

Biotinyl[^{3}H]estradiols **Va** *and* **VIa.** Compounds **Va** and **VIa** are prepared from the corresponding [^{3}H]estradiol-7α-carboxylic acid obtained by reduction of the respective estrone-7α-carboxylic acid by NaB_3H_4 (20 Ci/mmol). They are obtained by the same procedure as described above for the nonradioactive derivatives. Chemical and radiochemical purity is verified by TLC.

Binding of Biotinylestradiol to Estrogen Receptor and Avidin

The biotinylestradiol derivatives must bind with high affinity simultaneously to both avidin and the receptor. The binding affinity of biotinyl estradiol derivatives for the molybdate-stabilized (9 S; M_r ~310,000) nontransformed receptor is determined by a competitive binding assay against 2 nM [^{3}H]estradiol (Fig. 1). The concentrations of competitors necessary to inhibit the specific binding by 50% are as follows: estradiol, 3.6×10^{-9} M (100%); **Va,** 3.5×10^{-7} M (1%); **Vb,** 5×10^{-8} M (7%); **VIa,** 8.5×10^{-8} M (4%); and **VIb,** 3.5×10^{-8} M (10%).

When avidin is added to a solution containing the **Va** or **Vb** derivative (short spacer arms), the relative binding activity is reduced dramatically. In contrast, avidin causes only a moderate decrease in the relative binding activity of the two long-chain compounds **VIa** and **VIb.** This difference in behavior of the biotinylestradiol derivatives may be due to the likely possibility that both the estradiol-binding site on the estrogen receptor and the biotin-binding site on avidin are buried in deep clefts. It has been reported[6] that the biotin-binding site of avidin could be a depression of the surface of the protein and that the carbonyl group of biotin is buried about 9–10 Å beneath the van der Waals surface of the molecule. On the other

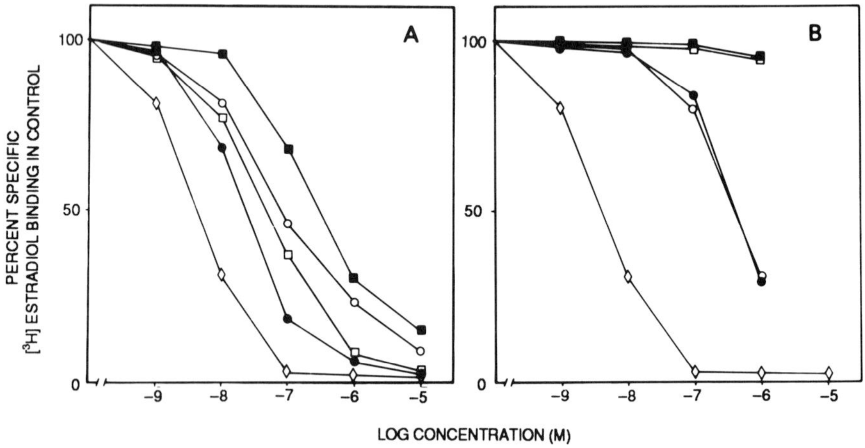

FIG. 1. Competitive inhibition against [^{3}H]estradiol for binding of biotinylestradiol to the receptor. Cytosol was incubated with 2 nM [^{3}H]estradiol either in the absence or in the presence of the indicated concentration of competitors (A) or with the same concentration of competitors preincubated for 30 min at room temperature with avidin (2 μM final concentration). (B). ◇, Nonradioactive estradiol; ■, biotinylestradiol **Va;** □, biotinylestradiol **Vb;** ○, biotinylestradiol **VIa;** ●, biotinylestradiol **VIb.** Specific binding is expressed as percentage of control activity (without competitors).

hand, the binding of the receptor to specific adsorbents is maximal when the ligand is coupled to the matrix by a long spacer arm containing more than 14 atoms.[2] Therefore, for biotinylestradiol–avidin complexes wherein biotinylestradiol derivatives bear short spacer chains, the estradiol residue is not available for receptor-binding sites. Only when both estradiol and biotin moieties are separated by a long spacer arm, (~42 Å, assuming an average length of 1.5 Å per bond) will they be available to bind simultaneously to the respective proteins.

Synthesis of Avidin–Agarose

Avidin (30 mg), dissolved in 20 ml of coupling buffer [0.1 *M* sodium bicarbonate, 0.5 *M* sodium chloride (pH 7.9)], is added to a slurry of CNBr-activated Sepharose CL-4B (5 g), and the suspension is stirred end-over-end overnight at 4°. The resin is collected by filtration, washed with coupling buffer, and quenched with additional incubation in a solution of 1 *M* 2-aminoethanol (pH 9.0) for 1 hr at 20°. The avidin–Sepharose is poured into a column and successively washed with coupling buffer (200 ml), 3 *M* guanidine hydrochloride (100 ml), and exhaustively with 10 m*M* phosphate buffer (pH 7.4), in which it is stored. The binding capacity of avidin–agarose for biotinylestradiol is determined either by the dye-binding assay[7] or by direct measurement of radioactive compounds. In all cases, the binding capacity is approximately 0.10 μmol/ml of packed gel.

Measurement of Receptor–Ligand Complexes

Measurement of Receptor-Binding Activity in Absence of Nonradioactive Ligand in Sample

Samples containing estrogen receptor are incubated with saturating concentrations (5–30 n*M*) of [^{3}H]estradiol at 0–4° until equilibrium is reached. Excess unbound steroid is removed by incubation with a charcoal (0.5%)–dextran T 80 (0.05%) suspension in phosphate buffer (1 : 1, v/v) for 4–17 hr at 4°. After centrifugation at 1,500 *g* for 10 min, the radioactivity in the supernatant is determined. The background of nonspecific binding is determined by either isotopic dilution or the double-concentration measurement method.[8]

[7] N. M. Green, this series, Vol. 18, p. 418.

[8] H. Richard-Foy, G. Redeuilh, and R. Richard-Foy, *Anal. Biochem.* **88,** 367 (1978).

Measurement of Receptor-Binding Activity in Presence of Nonradioactive Ligand in Sample

When nonradioactive ligand is present in the extract, the total number of binding sites is determined using a liquid-phase exchange. Samples are incubated with 1 μM nonradioactive ligand until equilibrium is reached. After removal of the unbound ligand by charcoal treatment, incubation for 2–3 hr at room temperature in the presence of a large excess of [^{3}H]estradiol (30 nM) leads to partial exchange of the ligand bound to the receptor by the [^{3}H]estradiol. The [^{3}H]estradiol–receptor complexes are then measured as above. The concentration of receptor complexes is determined with reference to a standard cytosol preparation treated in the same way.

Purification of Nontransformed Estrogen Receptor

The estrogen receptor is an intracellular protein. In crude cytosol prepared from unstimulated target tissue, the receptor is present in the "nontransformed" state as an "8–9 S" complex (M_r ~310,000) and releases the hormone-binding subunits (4–5 S) when incubated in media of increased ionic strength.[9] It has been reported[9] that the nontransformed form of the receptor can be stabilized by molybdate. We recently demonstrated[10] that the nontransformed form of the receptor is a heterooligomer complex which includes two hormone-binding subunits (M_r ~66,000) associated with two molecules of a heat-shock protein (M_r ~90,000). The procedure described here is used for the purification of calf uterus cytosol estrogen receptor stabilized in a nontransformed form by sodium molybdate.

The methodological principle includes three steps: (1) cytosol is incubated with saturating amounts of biotinylestradiol in order to obtain biotinylestradiol–receptor complexes, (2) the biotinylestradiol–receptor complexes are adsorbed on avidin–immobilized agarose, and (3) after appropriate washing the release of receptors is performed by exchange with an excess of free [^{3}H]estradiol.

Step 1. Calf uteri are homogenized in 3 volumes of phosphate buffer [10 mM potassium phosphate, 10 mM thioglycerol, 20 mM sodium molybdate, 10% glycerol (pH 7.5)]. The homogenate is centrifuged at 4° for 1 hr at 100,000 g, and the supernatant (referred to as "cytosol") is incubated for 2–3 hr with compound **VIb** (10^{-6} M final concentration). The excess of

[9] G. Shyamala and L. Leonard, *J. Biol. Chem.* **255,** 6028 (1980).

[10] G. Redeuilh, B. Moncharmont, C. Secco, and E. E. Baulieu, *J. Biol. Chem.* **262,** 6969 (1987).

free biotinylestradiol is removed by adding 0.1 volume of a charcoal (5%)–dextran (0.5%) suspension.

Step 2. Cytosol, labeled with biotinylestradiol, is loaded onto a column containing avidin–Sepharose or incubated in batch with avidin–Sepharose. The total volume of cytosol is 16 ml/ml of gel. Before elution of the receptor, the gel is washed (until the A_{280} falls to zero) successively with phosphate buffer and phosphate buffer containing 2.5 *M* urea. The gel is then reequilibrated in phosphate buffer. A critical point in the purification procedure is the washing of the gel before elution, which maintains the receptor in its nontransformed form. Therefore, it is necessary to avoid high ionic strength buffer (>0.3 *M* KCl) that dissociates the receptor even in the presence of 20 m*M* molybdate, a constraint which does not favor the efficiency of purification.

Step 3: For elution of the estrogen receptor, the gel is removed from the column and suspended in phosphate buffer containing 20 μM [^{3}H]estradiol (specific activity 4 Ci/mmol) for 30 min at 28–30°. The gel is removed by filtration and washed with buffer, and the filtrate containing receptor is immediately cooled at 0–4°. After 1 hr, the binding activity of the filtrate is determined by filtration of a sample (1 ml) through an Ultrogel AcA 34 column (50–70 ml). The activity of the eluate is determined by the measurement of the radioactivity of fractions containing macromolecular bound [^{3}H]estradiol.[11] About 40% of the receptor bound to the adsorbent is recovered in the eluate. Starting from cytosol, the purification of the receptor is 500- to 1500-fold, depending on the specific activity of the original cytosol. Table I summarizes a typical result of this type of purification procedure.

General Discussion

The biotinylestradiol–avidin system presents some advantages in comparison with more conventional bioaffinity chromatography obtained by immobilization of a ligand either on an insoluble resin or on a water-soluble polymer. The receptor present in the extract can be immediately complexed with the biotinylestradiol ligand that in turn binds rapidly to the immobilized avidin. These quick steps decrease the risk of inactivation of the receptor. Biotinylestradiol derivatives might also be useful for the purification of receptor complexed with endogenous hormone, since exchange of the endogenous hormone by biotinylestradiol can be per-

[11] G. Redeuilh, R. Richard-Foy, C. Secco, V. Torelli, R. Bucourt, E. E. Baulieu, and H. Richard-Foy, *Eur. J. Biochem.* **106,** 481 (1980).

TABLE I
PURIFICATION OF NONTRANSFORMED ESTROGEN RECEPTOR FROM CALF UTERUS[a]

Component	Protein (mg)	Receptor (pmol)	Specific activity (pmol/mg of protein)	Yield (%)
Cytosol	2250	1483	0.659	100
Eluate	0.360	377	1046	25.4

[a] The experiments are performed as described in the text. Cytosol (150 ml), containing receptor labeled with biotinylestradiol **VIb,** was loaded onto a 10-ml avidin–Sepharose column. In this experiment, 85% of the complex was bound to the column. The eluate was filtered through an Ultrogel AcA 34 column in order to eliminate excess free steroid.[11] Protein concentrations were measured according to W. Schaffner and C. Weissman [*Anal. Biochem.* **56,** 502 (1973)].

formed directly in solution, avoiding incubation at increased temperatures in the presence of a charcoal–dextran suspension in order to empty the binding sites before using conventional bioaffinity chromatography. The high affinity of avidin for biotinyl compounds makes this procedure applicable for the purification of the receptors present in biological samples at very low concentrations. This method also avoids the dilution effect of the water-soluble biospecific polymer procedure.[12] In addition, biotinylestradiol compounds may be used histochemically to detect receptors and measure them, using avidin labeled with ferritin, fluorescein, or rhodamine. There is no reason not to extend this method to other steroidal or nonsteroidal ligands, provided that appropriate biotinyl derivatives are found which exhibit sufficient affinity both for the corresponding receptor and for avidin.

[12] P. Hubert, J. Mester, E. Dellacherie, J. Neel, and E. E. Baulieu, *Proc. Natl. Acad. Sci. U.S.A.* **75,** 3143 (1978).

[31] One-Step Immunoaffinity Purification of Transferrin

By EDWARD A. BAYER and MEIR WILCHEK

We have previously promoted the use of avidin–biotin technology for improving the versatility of affinity chromatography.[1] For this purpose, the affinity column would consist of an avidin- or streptavidin-containing resin to which various biotinylated binders can be bound. The immobilized binder would then be used for the purification of its molecular counterpart. A major advantage of this approach would be that following the preparation of the initial avidin-containing affinity matrix, additional chemical manipulations would not be necessary. In this chapter, we present procedures for the immobilization of a biotinylated antibody specific for transferrin. The column can be used for the one-step isolation of transferrin directly from plasma.

Immobilization of Biotinylated Antibody on Avidin-Containing Column

Reagents

Avidin-containing affinity resin,[2] ~1.5 mg protein/ml Sepharose
Biotinylated rabbit anti-(human)-transferrin antibody (polyclonal)[3]
Phosphate-buffered saline (pH 7.4, PBS)
0.1 *M* Acetic acid

Procedure. A solution (1 ml) containing 1.7 mg biotinylated antitransferrin antibody is applied to a 1-ml avidin column. The column is washed with 5 column volumes of PBS or until the A_{280} reaches a minimum value (<0.01). The column is then washed with 5 column volumes of acetic acid, and the PBS washing procedure is repeated. During the above washes, 1-ml fractions are collected in order to monitor the efficiency of binding of the biotinylated antibody.

[1] E. A. Bayer and M. Wilchek, *Trends Biochem. Sci.* **3,** N257 (1978); E. A. Bayer and M. Wilchek, *Methods Biochem. Anal.* **26,** 1 (1980); M. Wilchek and E. A. Bayer, *Immunol. Today* **5,** 39 (1984).

[2] E. A. Bayer and M. Wilchek, this volume [18].

[3] Antitransferrin antiserum was obtained from Sigma Chemical Co. (St. Louis, MO). The relevant antibodies were affinity purified by conventional affinity chromatography on a transferrin–Sepharose (CNBr activated) column (M. Wilchek, T. Miron, and J. Kohn, this series, Vol. 104, [1]). A 20-ml column was used to isolate 13.7 mg of antibody from 7 ml of rabbit antiserum. In later experiments (after the validity of the method was demonstrated), biotinylated transferrin was combined with an avidin column to isolate the antibodies.

Comments. Using this procedure, almost quantitative binding of the biotinylated antibody can be achieved. The acetic acid wash removes about 15% of the bound antibody, but the remaining fraction (about 1.3 mg/ml resin) is stable.

We have also attached biotin to the antibody via a long-chain (caproylbutyryl) spacer with essentially identical results. Biotinylation via the sugar residues of the antibody (periodate-induced alkylation using biotin hydrazide) was less efficient, and only about one-third of the antibody could be attached to the affinity column using this approach.

Isolation of Transferrin

Reagents

Immobilized antitransferrin column, ~2 mg biotinylated antibody/ml avidin–Sepharose (~1.7 mg avidin/ml Sepharose)
Whole plasma (human)
PBS (pH 7.4)
0.1 *M* Acetic acid

Procedure. A 1-ml affinity column is washed successively with PBS, acetic acid, water, and PBS again (5-ml volumes each). The plasma (1 ml) is diluted to 5 ml with PBS and applied to the column. Following passage of the effluent fraction, the column is washed with 25 ml of PBS.[4] The absorbed material is released with acetic acid (2-ml fractions), and the eluate is immediately dialyzed (against distilled water) and lyophilized.

Comments. The bulk of the eluted material (0.9 mg) was eluted in the first two fractions. A second run yielded about 0.8 mg material. SDS–PAGE of the eluted fractions indicated that both preparations were equally pure (Fig. 1). Only low levels of other contaminating plasma proteins could be detected. The purified transferrin contained even lower levels of contaminating antibody than that of the commercially available (Sigma) protein (against which the antitransferrin antibodies used in this study were prepared). We have obtained comparable results using a similar process for isolating glycoproteins on biotinylated lectin/avidin–Sepharose columns.

In the experiments described here, avidin–Sepharose (CNBr activated) was used as the affinity matrix, although later research demonstrated that other types of resins or modes of activation are suitable for this purpose. Of course, columns containing nonglycosylated avidin[5] or

[4] The first wash consists of 5 ml of solution. The column is washed further with 2-ml fractions until the A_{280} reaches a minimal value. The effluent fraction is saved for further analysis.

[5] Y. Hiller, E. A. Bayer, and M. Wilchek, this volume [6].

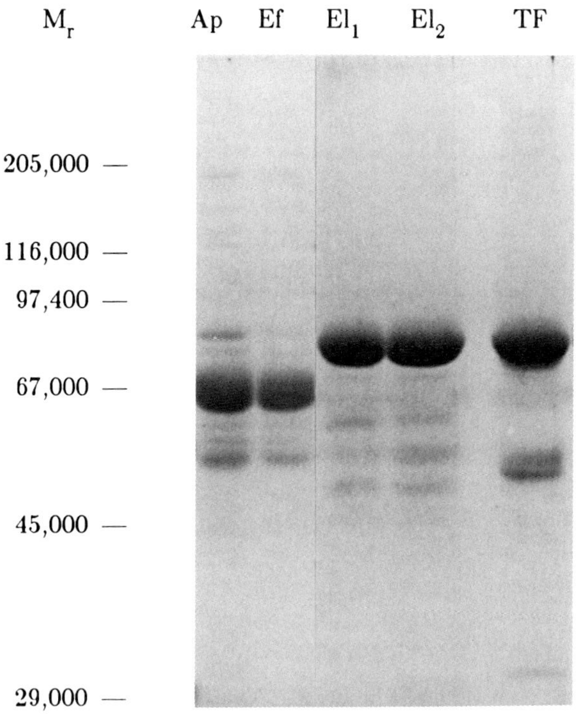

FIG. 1. Purification of transferrin directly from human plasma using a biotinylated transferrin/avidin–Sepharose affinity column. The gel shows the composition of the plasma samples upon application to the column (Ap), the effluent fraction (Ef), the eluent fractions of both the first and second trials (El_1 and El_2, respectively), and a commercial sample of transferrin (TF). To each lane, 20 μg protein was applied. Note the disappearance of the transferrin band in Ef compared to that of Ap; the other bands were retained.

streptavidin[6] can be employed to offset the nonspecific binding properties of the native egg-white protein. Nevertheless, the data presented in this chapter demonstrate that even columns containing native avidin can be used (together with an appropriate biotinylated binder) to achieve remarkable levels of purification in a single step.

The results of this study indicate that the use of an avidin-containing affinity column for the immobilization of biotinylated binders is a very efficient alternative for conventional affinity chromatography. The process can easily be scaled up, and it is expected that such technology will be used for the immunoaffinity isolation of large amounts of precious antigens.

[6] E. A. Bayer, H. Ben-Hur, and M. Wilchek, this volume [8].

[32] Lectin-Mediated Isolation of Cell Surface Glycoproteins

By GEOFFREY M. W. COOK and J. WILLIAM BUCKIE

Introduction and Principles

Complex carbohydrates (glycoproteins, glycolipids, and proteoglycans) are ubiquitous components of animal cell surfaces, where, because of the diversity of structures possible within their glycan moieties, they are considered to play a pivotal role in determining the specificity of many biological recognition phenomena.[1] Of this group of compounds, cell surface glycoproteins have been the subject of much study, it being suggested that alterations in their carbohydrate groups may be responsible for the aberrant social behavior of cancer cells.[2,3]

In order to investigate the specific roles played by surface glycoproteins, it is usually necessary to isolate species of interest from the cell surface to allow biochemical characterization. This usually involves preparing plasma membrane fractions, from which the constituent glycoproteins can be isolated following appropriate solubilization. Alternatively, if interest is centered on changes in glycan structure, then the release of surface glycopeptides from intact cells by proteolytic enzyme digestion can provide material for biochemical analysis. Neither approach is without its attendant drawbacks. The isolation of plasma membranes, especially as vesicular preparations, may be greatly affected by the physiological state of the cell, making comparative studies with such material very difficult.[4] On the other hand, although the nature of the carbohydrate groups of membrane glycoproteins is frequently inferred from the nature of the glycopeptides obtained by proteolytic enzyme treatment of intact cells, it has been shown that such treatments can result in accelerated release of materials from the cytoplasm.[5]

In an effort to overcome these problems we have devised an indirect affinity chromatography procedure, utilizing biotinylated lectins, for the specific isolation of surface glycoproteins.[6] Intact cells, whose polypep-

[1] G. M. W. Cook, *J. Cell Sci. Suppl.* **4,** 45 (1986).

[2] P. H. Atkinson and J. Hakimi, *in* "The Biochemistry of Glycoproteins and Proteoglycans" (W. J. Lennarz, ed.), p. 191. Plenum, New York, 1980.

[3] L. A. Smets and W. P. Van Beck, *Biochim. Biophys. Acta* **738,** 237 (1984).

[4] D. M. Neville, *in* "Biochemical Analysis of Membranes" (A. H. Maddy, ed.), p. 27. Chapman & Hall, London, 1976.

[5] M. Jett and G. A. Jamieson, *Biochem. Biophys. Res. Commun.* **55,** 1225 (1973).

[6] J. W. Buckie and G. M. W. Cook, *Anal. Biochem.* **156,** 463 (1986).

tides have been metabolically labeled with [^{35}S]methionine, are treated on ice with an excess of biotinylated concanavalin A (Con A). The biotinylated Con A reacts only with glycoproteins at the cell periphery.[7] After solubilization in nonionic detergent, the biotinylated Con A–glycoprotein complexes are retrieved on immobilized streptavidin, from which the radiolabeled glycoproteins can be quantitatively recovered by treating the affinity medium with 2% (w/v) sodium dodecyl sulfate. The eluted material is analyzed by standard two-dimensional gel electrophoretic techniques. This procedure may be adapted to examine alterations taking place in the oligosaccharide groups of surface glycoproteins. In this case, the surface carbohydrate residues of intact cells are metabolically labeled with radioactive sugars. The cells are then treated with biotinylated lectin and solubilized, and biotinylated lectin–glycoprotein complexes are then retrieved on immobilized streptavidin. The affinity matrix is in turn treated with pronase to release the labeled glycans as glycopeptides.

In this chapter, the methodology for the recovery of surface glycoproteins is described by reference to results obtained with acute lymphoblastic leukemic cells (ALL cells) of AKR mice as a model system. The adaption of the method for examining changes in surface glycans is illustrated with data obtained with quiescent and Con A-activated T lymphocytes.

Indirect Affinity Chromatography of Surface Glycoproteins

Preparation of Immobilized Streptavidin. Streptavidin is commercially available from a number of suppliers. Alternatively, it can be prepared inexpensively from the culture broth of *Streptomyces avidinii* (ATCC 27419). The strain is grown for 72 hr at 30° in medium A[8]; the culture fluid harvested by filtration, dialyzed against deionized water, and concentrated by lyophilization may be used as source material. The streptavidin is readily retrieved using the procedure described in detail by Hofmann *et al.*[9]; the affinity medium used by these authors is available

[7] The presence of biotinylated lectin at the surface of intact cells may be readily detected by indirect fluorescence microscopy using fluorescein isothiocyanate (FITC)–streptavidin. By maintaining cells on ice during incubation with biotinylated lectin and FITC–streptavidin, there is no evidence that either internalization of lectin or "patching" of surface material occurs.

[8] E. O. Stapley, J. M. Mata, I. M. Miller, T. C. Denny, and M. B. Woodruff, *Antimicrob. Agents Chemother.* **3,** 20 (1963).

[9] K. Hofmann, S. W. Wood, C. C. Brinton, J. A. Montibeller, and F. M. Finn, *Proc. Natl. Acad. Sci. U.S.A.* **77,** 4666 (1980). See also E. A. Bayer, H. Ben-Hur, and M. Wilchek, this volume [8].

commercially (Sigma). Routinely, 0.5 g CNBr-activated Sepharose 4B (Pharmacia Fine Chemicals) is swollen in 10 ml of 1 m*M* HCl for 1 hr then washed on a sintered glass funnel with 50 ml of coupling buffer consisting of 0.5 ml NaCl containing 0.1 *M* $NaHCO_3$ (pH 8.3). Streptavidin (10 mg) is dissolved in coupling buffer (2 ml), combined with the washed and activated gel, and mixed end-over-end at 4° for 18 hr. The conjugated gel (~1.5 ml swollen volume) is collected by centrifugation (1,800 *g* for 5 min) at room temperature and then washed with 10 ml of coupling buffer. Any unreacted groups on the gel are blocked by incubation with 5 ml of 1 *M* ethanolamine (pH 8.4) for 4 hr at room temperature. The gel is then washed with 10 ml of 0.5 *M* NaCl containing 0.1 *M* acetate (pH 4.0) followed by 10 ml of phosphate-buffered saline (pH 7.2); streptavidin–Sepharose 4B may be stored for several months at 4° in the latter buffer containing 0.02% (w/v) sodium azide.

The biotin-binding capacity of the gel is readily determined using a colorimetric assay based on 4-hydroxyazobenzene-2′-carboxylic acid (HABA) (Fluka AG).[10] Samples of the immobilized streptavidin (75 μl packed volume) are mixed with 0.1 m*M* HABA and then washed by centrifugation (500 *g* for 5 min) with 1 ml phosphate-buffered saline. The biotin-binding capacity is measured by adding 25-μl aliquots of *d*-biotin (10 mg/ml in phosphate-buffered saline) to the pink-colored HABA–streptavidin–gel complex and observing the gel particles under a magnifying glass; the end point is reached when the pink color disappears. Gels prepared by this procedure have a biotin binding capacity of 80 μg biotin/ml of gel.

Preparation of Biotinylated Lectin. Con A (crystallized 3 times and lyophilized; Miles Scientific, U.K.) is derivatized with biotin-ε-aminocaproic acid *N*-hydroxysuccinimide ester (Calbiochem-Behring) essentially as described previously.[11] A solution (20 μl) containing biotin-ε-aminocaproic acid *N*-hydroxysuccinimide (17 mg/ml) in dimethylformamide is added to a solution of Con A (1 mg/ml) in phosphate-buffered saline containing 10 m*M* α-methyl-D-mannopyranoside (Sigma) in a 1 : 50 (v/v) ratio, 750 nmol of biotin derivative being required per milligram of lectin. The solution is kept at room temperature for 4 hr and dialyzed overnight at 4° against the same buffer with one buffer change. The extent of biotinylation is determined by the HABA procedure[10] after extensive digestion with pronase. Samples of derivatized lectin (1 mg) are dialyzed against 50 m*M* Tris-HCl containing 10 m*M* $CaCl_2$ (pH 7.5) and then treated with pronase (100 μg; 83,500 units/mg) at 37° for 24 hr. Using these conditions, Con A is found to contain approximately 60 μg of

[10] N. M. Green, this series, Vol. 18 [74].

[11] E. A. Bayer, E. Skutelsky, and M. Wilchek, this series, Vol. 83 [12].

biotin/mg of lectin (i.e., about 240 nmol of biotin/mg), corresponding to an average content of 27 biotin residues per tetrameric Con A molecule.

Metabolic Labeling of Cells. In order to aid the detection of surface glycoproteins, cells are subjected to short-term labeling with [^{35}S]methionine in Eagle's minimal essential medium without methionine but containing L-glutamine (0.3 mg/ml) (Flow Laboratories). ALL cells (5 $\times$ 10^7/ml) are incubated at 37° in the above medium containing [^{35}S]methionine (71 μCi/ml, 1300 Ci/mmol) (Amersham International) for 2 hr. Following the pulse period, the cells are washed twice by centrifugation (500 *g* for 5 min) in 50 ml of phosphate-buffered saline and then incubated at a concentration of 1 $\times$ 10^7 cells/ml at 37° for 6 hr in RPMI 1640 medium containing penicillin (50 units/ml), streptomycin (50 μg/ml), and L-glutamine (0.3 mg/ml).

Indirect Affinity Isolation of Surface Glycoproteins. Radiolabeled ALL cells are treated with ice-cold phosphate-buffered saline containing biotinylated Con A (1 mg/ml) for 20 min, at a concentration of 100 μg of Con A per 1 $\times$ 10^6 cells.[12] Cells are then washed twice by centrifugation (500 *g* for 5 min) in 50 ml of ice-cold phosphate-buffered saline, followed by a final wash in 10 ml of the same buffer. The cell suspension is solubilized by a 30-min incubation in a solution (1 ml per 10^7 cells) of ice-cold solubilization buffer containing 0.1% (v/v) Nonidet P-40, 50 m*M* borate (pH 8.6), phenylmethylsulfonyl fluoride (0.2 m*M*), leupeptin (0.5 mg/liter), pepstatin A (0.7 mg/liter), and EDTA (1 m*M*).

Following solubilization the preparation is centrifuged at 100,000 g_{av} for 30 min, after which the supernatant is passed through streptavidin–Sepharose 4B (0.2–0.3 ml of gel packed in a Pasteur pipet fitted with a glass wool plug and coupled to an LKB Microperpex peristaltic pump) with 5 ml of solubilization buffer. Fractions (0.5 ml) are collected and monitored for radioactivity in a toluene-Triton X-100 (2 : 1, v/v)-based scintillant containing 0.6% (v/v) 2-(4'-*tert*-butylphenyl)-5-(4"biphenyl)-1,3,4-oxadiazole. Radioactivity remaining attached to the immobilized streptavidin is eluted by treating the affinity medium with 2% (w/v) sodium dodecyl sulfate (SDS, 200 μl) at 100° for 5 min. The SDS is removed by extensive dialysis against 0.1% Nonidet P-40 overnight, and the material retained in the dialysis sac is concentrated by freeze drying. Lyophilized material may then be solubilized in the lysis buffer described by O'Farrell[13] and subjected to two-dimensional (2D) polyacrylamide gel electrophoresis.

[12] Binding studies with [^{3}H]Con A [K. D. Noonan, *in* "Concanavalin A as a Tool" (H. Bittiger and H. P. Schnebli, eds.), p. 191. London, 1976] indicated that this amount is sufficient to represent a 4- to 5-fold excess over that needed to saturate the surface of lymphocytes. It is recommended that a similar excess be used with other cell types.

[13] P. H. O'Farrell. *J. Biol. Chem.* **250,** 4007 (1975).

Discussion of Results. An example of results obtained using this procedure is illustrated in Fig. 1A. More than 50 Con A-binding glycoproteins can be identified in ALL cells by this technique; the major species are designated 1–19. For comparison, a fraction containing sheets of isolated plasma membranes[14] from [^{35}S]methionine-labeled ALL cells and subjected to 2D gel electrophoresis is also shown (Fig. 1B). Surface glycoproteins common to both preparations are apparent; however, glycoproteins 11, 13, 17, and 19 are unique to preparations obtained using intact cells. The same result is obtained when isolated plasma membranes from ALL cells[14] are treated with biotinylated Con A and solubilized and the biotinylated lectin/glycoprotein complexes retrieved on immobilized streptavidin; again, glycoproteins 11, 13, 17, and 19 are absent. This suggests that certain surface glycoproteins are lost during the isolation of plasma membrane fractions. Interestingly, spot 20 is found only when isolated membranes are used which suggests that this component is not available for lectin binding in intact cells. By comparison with the results of fractionating detergent-solubilized [^{35}S]methionine-labeled plasma membranes on immobilized Con A,[6] we have confirmed that the polypeptide components retrieved with biotinylated Con A do indeed represent the major Con A-binding glycoproteins at the surface of ALL cells.

In using the indirect affinity procedure described here, it is essential that various experimental conditions are observed. It has been demonstrated that a high level of biotinylation is required for the retention of biotin–Con A on immobilized avidin,[15] and we find that the same criterion applies to immobilized streptavidin. In agreement with the work of others[16] the inclusion of a 6-aminohexanoic acid spacer between the biotin residue and the lectin results in the formation of a more stable complex with the immobilized streptavidin.[17] We have made a direct comparison between Con A derivatized with biotin-*N*-hydroxysuccinimide ester (31–35 biotin residues per tetrameric Con A molecule) and find that when equivalent amounts of both types of derivatized Con A were examined, 72% of the biotin–Con A added adhered to immobilized streptavidin. This figure was increased to greater than 93% when the derivative containing the spacer arm was used.

[14] A. Warley and G. M. W. Cook, *Biochim. Biophys. Acta* **323,** 55 (1973).

[15] E. A. Bayer, M. Wilchek, and E. Skutelsky, *FEBS Lett.* **68,** 240 (1976).

[16] K. Hofmann, G. Titus, J. A. Montibeller, and F. M. Finn, *Biochemistry* **21,** 978 (1982).

[17] As iminobiotin–avidin dissociation is sensitive to changes in pH (see Ref. 9) we have attempted the derivatization of Con A with 2-iminobiotin-*N*-hydroxysuccinimide, it being envisaged that this approach might provide a route to retrieval of affinity-bound material without having to resort to the use of denaturing agents. Unfortunately, in our hands the use of this reagent results in the precipitation of the lectin.

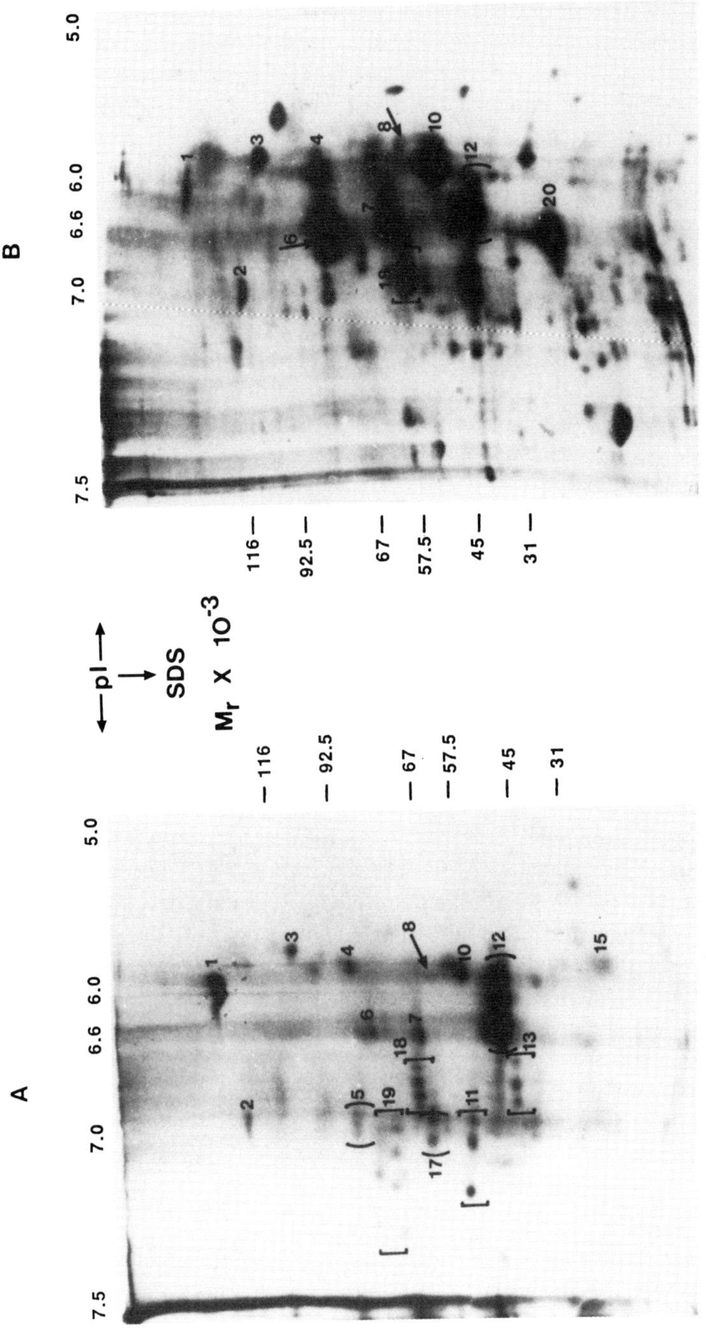

FIG. 1. Autoradiographs from two-dimensional gel electrophoresis of Con A-binding glycoproteins retrieved from (A) intact [^{35}S]methionine-labeled ALL cells by indirect affinity chromatography (1×10^6 cpm were examined in this gel) and (B) plasma membrane proteins derived from the same cells (2×10^6 cpm). Major radiolabeled species are designated 1–20; it should be noted that Con A-binding glycoproteins 11, 13, 17, and 19 are not detected in isolated sheets of plasma membranes. [Adapted from J. W. Buckie and G. M. W. Cook, *Anal. Biochem.* **156**, 463 (1986).]

It is essential to keep the cells at 4° when treating them with biotinylated lectin; when this condition is met no internalization of lectin is observed. Seventy-seven percent of the radiolabeled glycoproteins retrieved on immobilized streptavidin resist elution with a 0–0.5 *M* α-methylmannoside gradient, indicating that they bind with high affinity to the lectin and that the biotinylated lectin–glycoprotein complexes which have been observed[7] to form only at the cell periphery are not easily dissociated. When detergent-solubilized cellular proteins are subjected to chromatography on immobilized streptavidin, the binding of SDS-elutable radiolabeled species is reduced to 2.5% of that achieved in affinity chromatography with biotinylated Con A. When intact cells are incubated with biotinylated Con A, the addition of 200 m*M* α-methylmannoside reduces the binding of radiolabeled polypeptides to streptavidin–Sepharose 4B to 5% of that achieved with biotinylated Con A alone.

Adaptation of Method for Monitoring Changes in Glycan Structure

The use of intact viable cells, as described above, may more closely approximate the availability of surface receptors for ligand binding *in vivo* (glycoprotein–lectin interactions in this instance) and therefore may be particularly appropriate for comparative studies. Particular attention is currently being paid to the function of carbohydrates at the cell periphery, and we have adapted the above methodology to monitor changes in surface glycans. The procedure is illustrated with reference to the surface of murine T lymphocytes activated by Con A.

Preparation of Quiescent and Con A-Stimulated T Lymphocytes. Cell suspensions from the spleens of adult 'R' strain mice, a substrain of AKR mice,[18] sacrificed by cervical dislocation, are prepared by gently pressing the dissected tissue through a sterile stainless steel gauze (0.5 mm mesh) into RPMI 1640 medium. Cells are then washed by centrifugation (500 *g* for 10 min) twice in the same medium and resuspended at a concentration of 2.5×10^6 cells/ml in RPMI 1640 containing penicillin (50 units/ml), streptomycin (50 μg/ml), L-glutamine (0.3 mg/ml), and 3% (v/v) fetal bovine serum. Cultures to be activated are incubated at 37° in the presence of 2.5 μg/ml Con A[19] for 48 hr. This concentration of Con A over a 48-hr period is sufficient to stimulate T cells, as shown by the 38-fold increase in [*methyl*-^{3}H]thymidine incorporation. T lymphocytes are then

[18] G. M. W. Cook and W. Jacobson, *Biochem. J.* **107,** 549 (1968).

[19] This optimum concentration was found by testing a range of 0.1–10 μg/ml Con A for [*methyl*-^{3}H]thymidine incorporation.

enriched from both the quiescent and activated splenocyte cultures using the soybean agglutination procedure.[20]

Metabolic Labeling of T Lymphocytes with [2-^{3}H]Mannose. Quiescent and activated T lymphocytes are suspended in Eagle's minimal essential medium supplemented with 10% fetal calf serum and containing penicillin (50 units/ml), streptomycin (50 μg/ml), and L-glutamine (0.3 mg/ml) to a concentration of 5×10^6 cells/ml, to which is added [2-^{3}H]mannose (1 μCi/ml, 13.8 Ci/mmol; Amersham International). Routinely, we incubate lymphoid cells for 8 hr with [2-^{3}H]mannose, it having been deduced experimentally in our laboratory that at the end of this period the specific activity of radiolabeled sugar in isolated plasma membranes[14] is optimal. When using this procedure with other cell types, however, investigators should optimize their own conditions.

Isolation of Cell Surface Glycopeptides. T lymphocytes (1×10^7 cells/ml) metabolically labeled with radioactive sugar are treated at 0° with biotinylated Con A (1 mg/ml) and processed by indirect affinity chromatography as described above. However, instead of treating the radiolabeled glycoproteins retrieved on immobilized streptavidin with 2% SDS, the affinity medium (0.2 ml of packed gel) is digested in 1 ml of 50 m*M* Tris-HCl (pH 7.5) containing 10 m*M* $CaCl_2$ with pronase (1 mg/ml) at 37° for 48 hr; the beads are removed by centrifugation (1000 *g* for 5 min), and the clear supernatant fluid, in which the radioactivity is recovered quantitatively as glycopeptides, is inactivated (100° for 5 min) prior to analysis.

Application of Gel Filtration to Analysis of Glycopeptides

Fifty times as much radioactivity, in the form of [2-^{3}H]mannose-labeled glycopeptides, is recovered using the indirect procedure from Con A-stimulated T lymphocytes compared to quiescent T lymphocytes. Insufficient material labeled with radioactive sugar is available from the latter cells for analysis. However, material recovered from the activated T cells is readily available for gel filtration analysis (see Fig. 2). The major glycopeptide (see Fig. 2A) is susceptible to various glycosidase treatments. That it is readily cleaved by endoglycosidases H (EC 3.2.1.96) and D (EC 3.2.1.96) and by α-mannosidase (EC 3.2.1.24) but is resistant to *N*-acetyl-β-glucosaminidase (EC 3.2.1.30) indicates the presence of an *N*-linked, high-mannose Con A-binding structure which is free of terminal β-linked *N*-acetylglucosamine residues. Reduction of the endoglycosidase-released oligosaccharide with NaB^3H_4 and careful sizing by gel

[20] Y. Reisner, A. Ravid, and N. Sharon, *Biochem. Biophys. Res. Commun.* **72,** 1585 (1976).

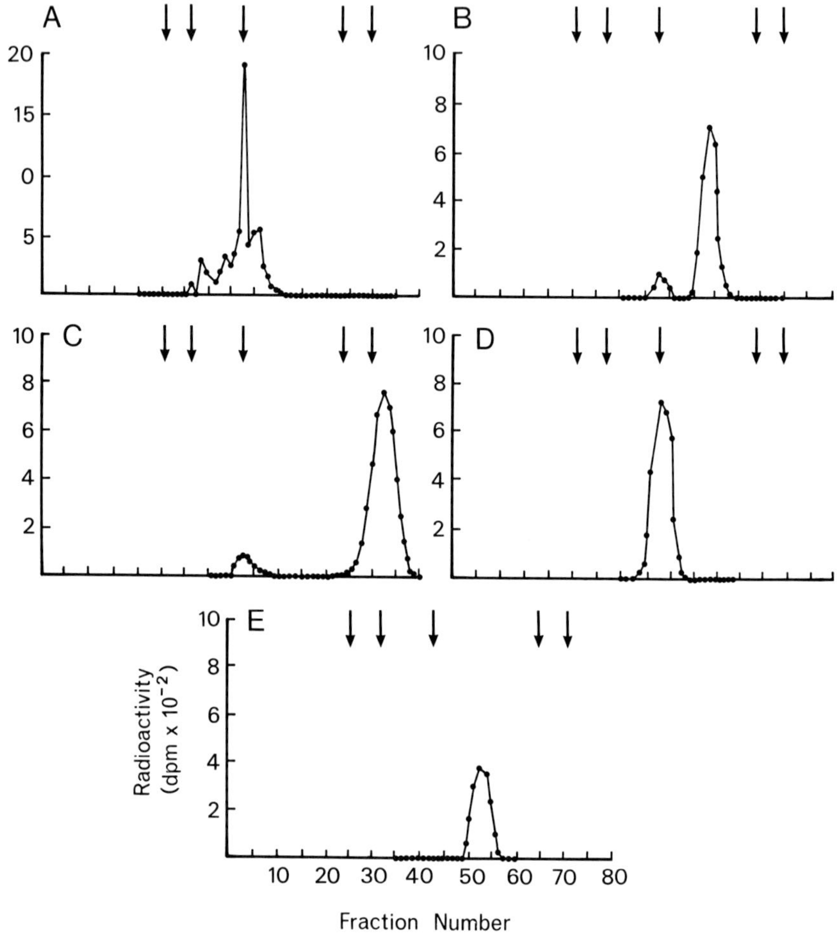

FIG. 2. Analysis of [2-^{3}H]mannose-labeled glycopeptides by gel filtration and glycosidase digestion. Radiolabeled glycopeptides isolated from Con A-activated T-lymphocytes by indirect affinity chromatography on immobilized streptavidin were analyzed by gel filtration on a BioGel P-6 column (1.5 × 60 cm) (A). Fractions (1.75 ml) were assayed for radioactivity, and the elution position of the various glycopeptides was determined relative to standard marker carbohydrates. As indicated by the arrows left to right, the latter consisted of Blue dextran, fetuin large oligosaccharide, desialylated fetuin oligosaccharide, stachyose, and mannose. The susceptibility of the glycopeptides in fractions 41–46 to endoglycosidase H (B), α-mannosidase (C), N-acetyl-β-glucosaminidase (D), and endoglycosidase D (E) treatments was the basis for partial characterization of the oligosaccharide by gel filtration analysis, performed under identical conditions to those used in (A).

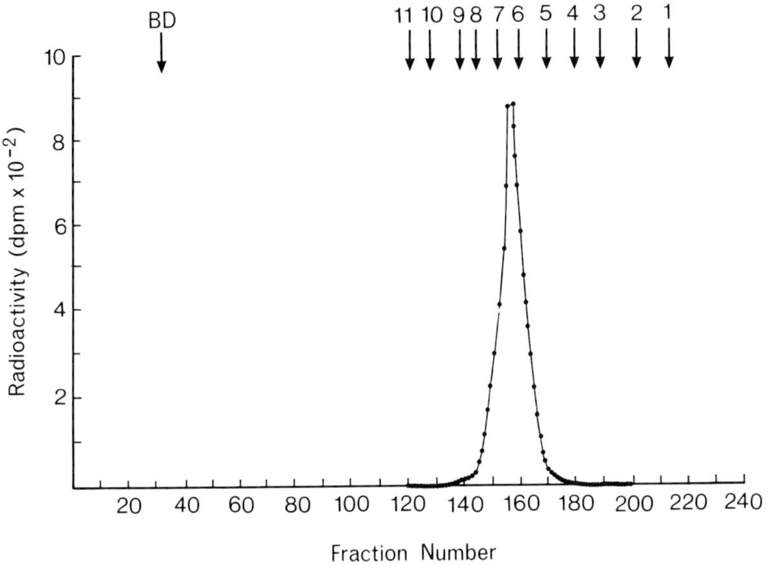

FIG. 3. BioGel P-4 column chromatography of the oligosaccharide liberated from the surface glycopeptide of T lymphocytes by treatment with endoglycosidase D (see Fig. 2E). Prior to gel filtration the oligosaccharide was reduced with NaB^3H_4.[21]. Arrows indicate the eluting positions of glucose oligomers (numbers indicate the glucose units), and BD indicates the elution position of Blue dextran.

filtration[21] shows an oligosaccharide eluting at a glucose unit value of 6.65 (Fig. 3), characteristic of either $Man_5GlcNAcOH$ or Fucα-1→2Galβ1→3GlcNAcβ1→3Galβ1→4GlcOH; in view of the susceptibility of the oligosaccharide to α-mannosidase and endoglycosidase treatments, this latter structure is unlikely to represent the structure described here. Upon digestion with endoglycosidases, ovalbumin glycopeptide V, known to bear a $Man_5GlcNAc_2Asn$ residue,[22] produces an oligosaccharide eluting in an identical position to the lymphocyte material. The methodology described here enables one to detect an N-linked $Man_5GlcNAc_2$ structure at the surface of activated T lymphocytes, but not at the periphery of quiescent cells.

Comments

The methodology detailed above differs from the routine usage of avidin–biotin technology in that it uses intact cells and exploits the prop-

[21] K. Yamashita, T. Mizuochi, and A. Kobata, this series, Vol. 83 [6].
[22] C. C. Huang, H. E. Mayer, and R. Montgomery, *Carbohydr. Res.* **13,** 127 (1970).

erties of streptavidin. As avidin has an isoelectric point of 10.5, there is the possibility that net negatively charged glycoproteins will bind nonspecifically to the net positively charged avidin molecule. In addition, avidin is a glycoprotein containing a high-mannose asparagine-linked oligosaccharide moiety that is likely to interact with the carbohydrate-binding site of appropriate lectins, such as Con A. We have evidence that complexes formed between biotinylated lectin and membrane glycoproteins are dissociated by passage through immobilized avidin, probably as a result of the oligosaccharide residues on the avidin competing for the carbohydrate-binding sites on the lectins. These problems are obviated by using streptavidin, which is devoid of covalently linked carbohydrate.

Acknowledgment

This work was supported by a project grant from the Medical Research Council, London. G.M.W.C. is a member of the Council's External Scientific Staff, and J.W.B. was in receipt of a Research Studentship from the Medical Research Council, held at the Department of Pharmacology, University of Cambridge.

[33] Immunoselective Cell Separation

By JAN WORMMEESTER, FRANK STIEKEMA, and CORNELIS DE GROOT

Introduction

Identification of cell populations (especially lymphocyte subpopulations) using highly purified polyclonal or monoclonal antibodies labeled with fluorochromes against cell surface antigens has been well documented. Immunolabeling of cells with biotinylated antibodies and fluorochrome-conjugated avidin (or streptavidin), thus using the specific and high-affinity binding of avidin to biotin,[1] has served to extend the applicability and sensitivity of this technique. The cells labeled with fluorochromes can be obtained at high purity using a fluorescence-activated cell sorter, although this procedure is time-consuming and therefore not the method of choice for the separation of large quantities of cells.

[1] M. Wilchek and E. A. Bayer, *Immunol. Today*. **5,** 39 (1984).

Indirect rosetting techniques, using the specific antibody-mediated binding of erythrocytes to the cells to be selected, provide detection equal or superior to fluorescence techniques and are easily exploited in both positive and negative separation of cell populations on the basis of the differential density of the rosetted cells.[2] This method can be applied in an easy-to-perform and versatile cell separation which can be extended to large quantities of cells with concomitant maintenance of sterile conditions if desired without the use of expensive equipment. To this end, erythrocytes can be coupled to avidin[3–5] or biotin.[6] This allows the possibility of forming indirect rosettes with cell populations tagged with biotin-conjugated (monoclonal) antibodies. Rosetting with avidin-coupled erythrocytes is obtained upon mixing with the tagged cells; with biotin-coupled erythrocytes, the avidin–streptavidin bridge method can be used.[6] Compared to the indirect rosette technique with antibody-coupled erythrocytes, the use of avidin-coupled erythrocytes is not complicated by unwanted binding, e.g., with Fc receptors, or (cross-) reactivity to surface immunoglobulins of B lymphocytes.

Coating avidin to sheep red blood cells (SRBC) has been reported by Fürfang and Thierfelder[5] using the pyridyl sulfide technique, which is reliable but time-consuming compared to the chromic chloride method used by others.[3,4] The basic principle of chromic chloride coupling of proteins to erythrocytes, as well as applications of the coated erythrocytes, was reviewed recently by Coombs *et al.*[7] When applying indirect rosetting of human lymphocytes, it should be remembered that human T lymphocytes have CD2 molecules at their surface, i.e., binding sites for sheep erythrocytes (SRBC). Therefore, with human lymphocytes, SRBC should be avoided as rosetting cells. The use of ox red blood cells (OxRBC) can overcome this problem, but these cells give a less reliable coupling of protein using the chromic chloride method.[7] Therefore, the avidin bridge method[6] using biotinylated OxRBC could be the method of choice for human lymphocyte separation.

Here, we review optimal conditions for biotinylation of antibodies and procedures for coupling of avidin to erythrocytes, rosette formation, and separation of rosetted cells.

[2] K. Mills, this series, Vol. 121, p. 726.

[3] J. Wormmeester, F. Stiekema, and K. De Groot, *J. Immunol. Methods* **67,** 389 (1984).

[4] D. Levitt and R. Danen, *J. Immunol. Methods* **89,** 207 (1986).

[5] J. Fürfang and S. Thierfelder, *J. Immunol. Methods* **91,** 123 (1986).

[6] P. J. Lucas and R. E. Gress, *J. Immunol. Methods* **99,** 185 (1987).

[7] R. R. A. Coombs, M. L. Scott, and M. P. Cranage, *J. Immunol. Methods* **101,** 1 (1987).

Materials and Methods

Biotinylation of Antibodies

Reagents

23 m*M* Biotinyl-*N*-hydroxysuccinimide (8 mg/ml, BNHS; Sigma Chemical Co., St. Louis, MO) in *N,N*-dimethylformamide

Polyclonal or monoclonal antibody (purified at least to a γ-globulin fraction by ammonium sulfate precipitation), 1–5 mg/ml in 0.1 *M* $NaHCO_3$ (pH 8.0)

Procedure (According to Guesdon et al.[8]*).* Add 25 μl of the BNHS solution per milligram of antibody in solution and mix well. After incubation for 2 hr at room temperature, the reaction mixture is dialyzed overnight at 4° against Dulbecco's phosphate-buffered saline solution (DPBS, pH 7.4; Gibco BRL, Paisley, Scotland). After testing for performance, the biotinylated antibody can be stored in small aliquots at −20°. Repeated freezing and thawing should be avoided.

The water-soluble reagent sulfosuccinimidobiotin (sulfo-NHS-biotin; Pierce Chemical Co., Rockford, IL) or its spacer-arm analog sulfosuccinimidyl 6-(biotinamido)hexanoate (NHS-LC-biotin; Pierce) can be used in essentially the same protocol. This eliminates the exposure of antibodies to organic solvent.

Cell Suspensions

Human and mouse lymphocytes can be obtained by density-gradient centrifugation, using standard protocols provided by the manufacturers. Human lymphocytes are isolated using a density-gradient medium with a density of 1.077 g/ml (e.g., Lymphoprep, Nycomed AS, Oslo, Norway).[9] Mouse lymphocytes are isolated using a density medium with a density of 1.086 g/ml (e.g., Lympho-paque, Nycomed AS).

Red Blood Cells

Sheep and ox red blood cells should be kept in Alsever's solution. They have a maximum shelf-life of 3 weeks at 4°.

[8] J. Guesdon, T. Ternynck, and S. Avrameas, *J. Histochem. Cytochem.* **27,** 1131 (1979).

[9] A. Bøyum, this series, Vol. 108, p. 88.

Avidin Coupling: Chromic Chloride Method[3]

Reagents. A 1% "aged" chromic chloride solution should be prepared according to Goding.[10] Briefly, 1% (w/v) $CrCl_3 \cdot 6H_2O$ is dissolved in 0.15 *M* NaCl, adjusted to pH 5.0 with 1 *M* NaOH, and left at room temperature for 3 weeks with weekly readjustments to pH 5.0 with 0.1 *M* NaOH. Just before use, a 100-fold dilution of the 1% stock solution is prepared in 0.15 *M* NaCl.

Procedure. SRBC (1×10^9) are washed at least 5 times with 0.15 *M* NaCl and made up to a volume of 7 ml with 0.15 *M* NaCl. Twenty microliters of avidin–fluorescein isothiocyanate (FITC) (5 mg/ml; Vector Laboratories Inc., Burlingame, CA) is added with thorough mixing. Then 3 ml of the 0.01% chromic chloride solution is added with constant stirring. After incubation for 30 min at room temperature or overnight at 4°,[11] the reaction is stopped by mixing with an equal volume of DPBS. After 2 washings in DPBS or biotin-free tissue culture medium, the avidin-coupled erythrocytes are ready for use. The coupling efficiency can be examined directly by fluorescence microscopy.

Comments. Special attention should be paid that no phosphate ions are present in the reagents, as they will inhibit the coupling completely.

Avidin Coupling: Avidin Bridge Method (Modified from Lucas and Gress)[6]

Reagents

2 m*M* Sulfosuccinimidyl 6-(biotinamido)hexanoate in DPBS
5 mg/ml Avidin (Sigma) in DPBS

Procedure. Red blood cells (sheep or ox) are washed twice with DPBS and adjusted to 1×10^9/ml. One milliliter of the RBC suspension is mixed with 100 μl of the NHS-LC-biotin solution and incubated for 1 hr at 4° with occasional stirring. The cells are washed at least twice with cold DPBS. After reconstituting the cells to 1 ml, 200 μl of the avidin solution is added with constant stirring. After incubation for 30 min at 4° with occasional stirring, the avidin-coupled cells are washed twice with DPBS or biotin-free tissue culture medium and are ready for use. The coupling efficiency can easily be examined by the incubation of a small aliquot of the cells with a fluorochrome-conjugated and biotinylated protein (e.g., FITC–bovine serum albumin–biotin) using fluorescence microscopy.

[10] J. W. Goding, *J. Immunol. Methods* **10,** 61 (1976).
[11] N. R. Ling, S. Bishop, and R. Jefferis, *J. Immunol. Methods* **15,** 279 (1977).

Comments. Avidin-coupled erythrocyte suspensions can be stored in tissue culture medium for a maximum of 1 week at 4° when sterile procedures are used. After storage, they should be washed prior to use.

Rosette Formation

Lymphocytes (5×10^7/ml) are directly or indirectly labeled with optimized dilutions of biotinylated antibodies, washed twice with cold Hanks' balanced salt solution (HBSS, pH 7.4; Gibco BRL) and adjusted to a concentration of 5×10^7/ml HBSS. One milliliter of the labeled cell suspension is mixed with an equal volume of the avidin-coupled erythrocyte suspension (1×10^9/ml) in a 10-ml test tube and centrifuged for 5 min at 200 *g* and 4°. After incubation for 20 min at 4°, the cells are resuspended using a Vortex mixer at medium speed. Rosette formation can be determined by mixing a small aliquot with an equal volume of brilliant cresyl blue solution [0.1% (w/v) in DPBS] and by counting in a hemocytometer. Lymphocytes binding four or more erythrocytes are considered positive. Controls using nonbiotinylated antibody should be negative.

Separation of Rosetted Cells

Reagents. A stock solution of 25 % (w/v) Nycodenz (Nycomed AS) in distilled water (density 1.132 g/ml) is diluted with HBSS to a final density of 1.090 g/ml.

Procedure. The rosetted suspension is layered onto 10 ml of Nycodenz solution of 1.090 g/ml in a 40-ml test tube and centrifuged for 30 min at 1000 *g* and 4°. The nonrosetted cells are collected from the interface and washed twice with cold HBSS. The erythrocytes in the pellet are lysed using NH_4Cl and washed twice with cold HBSS.

Limitations of Procedure

As described here for lymphocytes, the method is applicable to numerous cell populations provided that the starting cell suspension has a narrow range of specific gravity well below 1.090 g/ml. For example, the method failed to isolate CD1a-positive cells (Langerhans cells) from a crude human epidermal cell suspension owing to the high density of contaminating keratinocytes.[12] Moreover, only a one-parameter (positive or negative) separation is possible; if a two-parameter separation, e.g., size

[12] M. B. M. Teunissen, J. Wormmeester, M. L. Kapsenberg, and J. D. Bos, *J. Invest. Dermatol.* **91,** 358 (1988).

as well as immunophenotype, is desired, the fluorescence-activated cell sorter is recommended.

Conclusion

The avidin–biotin method used for negative and positive selection shares the advantages (fast, applicable to large quantities of cells) and disadvantages (one-parameter separation, positively selected cells are labeled with antibodies) as the indirect rosetting technique using antibody-coupled erythrocytes. However, the former technique is preferable when used for the separation of mouse and rat cells, because unwanted binding to surface immunoglobulins of B lymphocytes (which occurs when applying the latter technique to murine or rat monoclonal antibodies) is avoided.

In addition to the method described here, the avidin–biotin system for cell selection has been used in immunoadsorption with avidin immobilized onto columns.[13] However, in using an avidin-containing column, the yields of positively selected cells can be low because of the strong binding of cells to the immobile substrate.[14] Such a method would therefore be more valuable for negative selection. In using such a procedure, special care should also be taken to avoid nonspecific binding of cell subpopulations (e.g., B lymphocytes and monocytes) to the matrix used.[15]

[13] R. J. Berenson, W. I. Bensinger, and D. Kalamasz, *J. Immunol. Methods* **91,** 11 (1986).
[14] R. S. Basch, J. W. Berman, and E. Lakow, *J. Immunol. Methods* **56,** 269 (1983).
[15] R. W. Braun and G. Kümel, this series, Vol. 121, p. 737.

[34] Isolation and Analysis of Splicing Complexes

By PAULA J. GRABOWSKI

The splicing of messenger RNA precursors (pre-mRNAs) is a remarkably accurate process which requires recognition of the precise boundaries of an intervening sequence. This recognition process is far from understood but is believed to involve the stepwise association of small nuclear ribonucleoprotein particles (snRNPs) and accessory factors with the pre-mRNA molecule.[1–3] Essential for the first cleavage–ligation step of splicing is the assembly of a complete spliceosome structure.

The spliceosome was first identified as a multicomponent complex that can be resolved by sedimentation through glycerol gradients.[4–7] The spliceosome was found to be distinct from other splicing complexes in that it contains the RNA intermediate of splicing, the 5′ exon, and the lariat IVS-3′ exon RNA species. The integral components of the spliceosome and their arrangement are of vital interest since the process by which cellular factors associate with the pre-mRNA is likely to result in the recognition of splice sites and in the juxtaposition of the correct splice site pairs. Spliceosome assembly may, in addition, reflect changes in splice site recognition leading to alternative splicing. Finally, the structural entity responsible for catalysis of the nuclear pre-mRNA splicing event has not yet been determined and may be related to the structure of the spliceosome.

Identification of the snRNP components of the spliceosome has recently been achieved using biotin–streptavidin affinity chromatography.[8,9] The snRNPs U2, U4, U5, and U6 were found to be stable constituents of the spliceosome, present at approximately unit stoichiometry.[8] These snRNA constituents of the spliceosome have been confirmed by Northern analysis of splicing complexes separated by gel electrophoresis.[10] At present, three additional splicing complexes have been identified. A presplicing complex, which kinetically precedes formation of the spliceosome, was found to contain only U2 snRNP.[11] More recently, an altered form of the spliceosome, lacking U4 snRNA, has been identified.[12] The U4-depleted spliceosome is formed simultaneously with the first cleavage–ligation step of splicing, underscoring the importance of this complex in the reaction process. Later in the reaction, a postsplicing complex can be resolved, containing the excised intron lariat RNA and

[1] R. A. Padgett, P. J. Grabowski, M. M. Konarska, S. Seiler, and P. A. Sharp, *Annu. Rev. Biochem.* **55,** 1119 (1986).

[2] T. Maniatis and R. Reed, *Nature* (*London*) **325,** 673 (1987).

[3] M. R. Green, *Annu. Rev. Genet.* **20,** 671 (1986).

[4] P. J. Grabowski, S. R. Seiler, and P. A. Sharp, *Cell* **42,** 345 (1985).

[5] D. Frendewey and W. Keller, *Cell* **42,** 355 (1985).

[6] E. Brody and J. Abelson, *Science* **228,** 963 (1985).

[7] A. Bindereif and M. R. Green, *Mol. Cell. Biol.* **6,** 2582 (1986).

[8] P. J. Grabowski and P. A. Sharp, *Science* **233,** 1294 (1986).

[9] A. Bindereif and M. R. Green, *EMBO J.* **6,** 2415 (1987).

[10] M. M. Konarska and P. A. Sharp, *Cell* **49,** 763 (1987).

[11] M. M. Konarska and P. A. Sharp, *Cell* **46,** 845 (1986).

[12] A. I. Lamond, M. M. Konarska, P. J. Grabowski, and P. A. Sharp, *Proc. Natl. Acad. Sci. U.S.A.* **85,** 411 (1988).

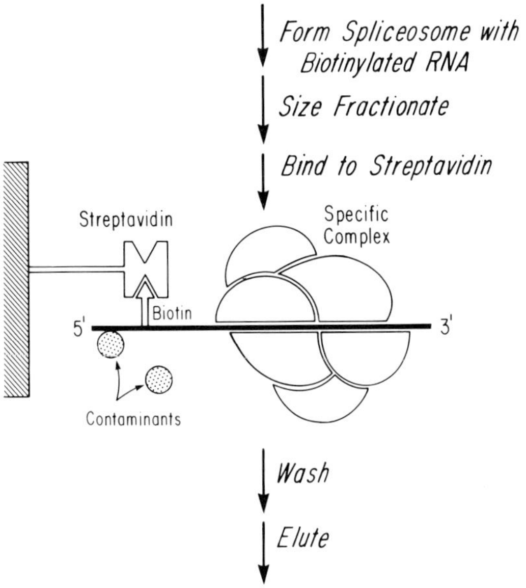

FIG. 1. Scheme for affinity purification of splicing complexes.

the snRNA species U2, U5, and U6 but not U4.[10] These data suggest a sequential pathway of spliceosome assembly and dismantlement that is intimately involved in all aspects of the splicing process, including specific recognition of splice sites and catalysis of phosphodiester bond cleavage and formation, followed by the release of reaction products in distinct forms of the ribonucleoprotein complex, and, finally, the recycling of spliceosome subunits.

The principal difficulty in identifying genuine components of splicing complexes is due to contamination by the abundant snRNPs present in nuclear extracts of HeLa cells. This problem can be circumvented by affinity selection of splicing complexes using the high-affinity interaction of biotin and streptavidin.

Methods

The scheme for purification of splicing complexes by biotin–streptavidin affinity chromatography is shown in Fig. 1. In the first step, splicing complexes are assembled on a biotinylated pre-mRNA molecule followed by treatment with heparin. Stable complexes are then separated by centrifugation through glycerol gradients, and gradient fractions containing distinct complexes are incubated with streptavidin–agarose beads. Specific complexes bind tightly to the column; unbound and loosely bound contaminants are removed by washing. Bound components are then eluted and analyzed.

Synthesis of Biotinylated Pre-mRNA

RNA is prepared using SP6 RNA polymerase, essentially as described by Melton *et al.*,[13] in the presence of low levels of biotin-UTP.[8] It is desirable to include the cap dinucleotide G(5′)ppp(5′)G in the transcription reaction in order to produce pre-mRNA substrates with authentic 5′ ends for splicing reactions.[14] T7 or T3 RNA polymerases can also be used to incorporate biotin-UTP. A plasmid cleaved by a restriction endonuclease at a site to give a desired runoff transcription product is used as the template DNA. A typical reaction is performed as follows.

1. For each RNA to be synthesized, a 100-μl transcription reaction is prepared at room temperature, containing 85 μl of Transcription Mix, 1 μl of a 2.5 m*M* stock of biotin-11-UTP (Bethesda Research Laboratories, Gaithersburg, MD), 10 μl of 1 mg/ml template DNA, and 20 units SP6 RNA polymerase. The Transcription Mix contains 4 μl of 2.5 mg/ml bovine serum albumin, 10 μl of 10× SP6 buffer [400 m*M* Tris-HCl (pH 7.9), 60 m*M* $MgCl_2$, 20 m*M* spermidine], 5 μl of 0.2 *M* dithiothreitol, 5 μl of 10 m*M* G(5′)ppp(5′)G cap dinucleotide, 10 μl of a solution containing 5 m*M* each of ATP, CTP, GTP and UTP, 10 μl of ^{32}P-labeled UTP (10 μCi), and water to give a final volume of 100 μl.
2. The reaction mixture is incubated at 37° for 30 min to allow synthesis of RNA transcripts.
3. Ten units of RNase-free DNase I is added, and the mixture is incubated at room temperature for 10 min. These conditions result in complete degradation of template DNA.
4. The reaction mixture is then eluted through a Sephadex G-50 spin column in 20 m*M* HEPES (pH 7.5), 0.25 *M* sodium acetate, 0.1 m*M* EDTA to remove unincorporated nucleotides. The column flowthrough is then precipitated with ethanol.

The amount of biotin-UTP in the transcription reaction is critical. A titration of biotin-UTP in the transcription reaction showed that full-length RNA chains 450–1400 nucleotides in length are synthesized only in the presence of low levels of the biotin-UTP analog.[8] Reaction conditions

[13] D. A. Melton, P. A. Krieg, M. R. Rebagliati, T. Maniatis, K. Zinn, and M. R. Green, *Nucleic Acids Res.* **12,** 7035 (1984).

[14] M. M. Konarska, R. A. Padgett, and P. A. Sharp, *Cell* **38,** 731 (1984).

are optimized for full-length RNA chains and for the maximum level of RNA transcripts which can be produced in the presence of the analog.[8] Optimal conditions were found to be 25 μM biotin-UTP and 0.5 mM UTP in the transcription reaction. These conditions allow production of 80% of RNA transcript compared to control reactions lacking biotin-UTP. Higher levels of biotin-UTP result in two effects: a decrease in the total level of transcription and a decrease in the average chain length of the RNA produced.

Purification of RNA transcripts for splicing is performed as described by Grabowski *et al.*[16] It is important to have a highly purified pre-mRNA substrate for splicing in order to achieve efficient splicing complex formation. Although gel purification of the RNA is not strictly required, the cleanest results are obtained with RNA purified in this manner. The RNA prepared from one 100-μl transcription reaction is loaded onto two lanes of a 10% (w/v) polyacrylamide–8 M urea gel of lane dimensions 8 mm width by 1.5 mm thickness. RNA is eluted for 18 hr at 4° in buffer containing 500 mM ammonium acetate, 10 mM magnesium acetate, 0.1 mM EDTA, and 0.1% (w/v) sodium dodecyl sulfate (SDS). Eluted RNA is subsequently purified by binding to DEAE-cellulose as described.[16]

Preparation of Splicing Complexes

Splicing complexes are assembled on an adenovirus pre-mRNA substrate in a HeLa cell nuclear extract. Saturating levels of substrate RNA are used in order to optimize the level of splicing complexes to be selected. The amount of substrate RNA that gives saturation is determined by titrating substrate RNA against a constant level of nuclear extract in the splicing reaction. Preparative splicing reactions, 300 μl in volume, are prepared by combining 252 μl of Reaction Mix (see below), 24 μl of a 0.4 mg/ml stock of ^{32}P-labeled pre-mRNA, and 24 μl of 18.75 mM ATP. Control reactions are prepared as above but without pre-mRNA.

The Reaction Mix is prepared by combining the following components for each 0.3-ml splicing reaction: 3 μl of creatine phosphate, 4 μl of 100 mM $MgCl_2$, 6 μl of 1 M HEPES (pH 7.5), 132 μl of nuclear extract,[17] and 120 μl of water. The Reaction Mix is preincubated at 30° for 10 min prior to the addition of substrate RNA. For control reactions incubated in the absence of ATP, the Reaction Mix is prepared without creatine phosphate and preincubated as indicated.

Complete splicing reactions are incubated at 30° for 15 min, then stopped by the addition of 6 μl of 200 mg/ml heparin. Incubation is contin-

[16] P. J. Grabowski, R. A. Padgett, and P. A. Sharp, *Cell* **37,** 415 (1984).

[17] J. D. Dignam, R. M. Lebowitz, and R. G. Roeder, *Nucleic Acids Res.* **11,** 1475 (1983).

ued for 8 min, after which the mixture is immediately loaded onto glycerol gradients. Reactions are incubated with heparin, a polyanion, to remove contaminants loosely bound to splicing complexes by electrostatic interactions. An alternative approach is treatment with 300 m*M* potassium chloride.[9]

Biotin–Streptavidin Selection

Each heparin-treated, 300-μl splicing reaction is loaded immediately onto an 11.5-ml, 10–30% (v/v) glycerol gradient, preequilibrated at 4°. Gradients contain 10–30% glycerol, 25 m*M* potassium chloride, 20 m*M* HEPES (pH 7.5), 1.0 m*M* magnesium chloride. Centrifugation is carried out in an SW41 rotor for 14 hr at 25,000 rpm, 4°. The following steps were performed at 15°.

Fractionation of Gradients. Each gradient is fractionated into 20 fractions, and a profile of the gradient is obtained by scintillation counting of a 5-μl sample per fraction. The gradient profile of a 15-min splicing reaction containing both the presplicing complex and the spliceosome is shown in Fig. 2.

Sample Preparation. Streptavidin–agarose beads are prepared in Eppendorf tubes in the following way. For each complex to be analyzed, a 100-μl sample of streptavidin–agarose beads (Bethesda Research Laboratories) is added to one tube and washed once with 1 ml of Binding Buffer containing 1.0 *M* potassium chloride and 2 times with 1 ml Binding Buffer [0.1 *M* potassium chloride, 20 m*M* HEPES (pH 7.5), 10% glycerol, 1.0 m*M* magnesium chloride, 0.1 m*M* EDTA]. Two gradient fractions (~500 μl each) for each peak are pooled and mixed with 0.1 mg/ml glycogen and 0.1 mg/ml bovine serum albumin (BSA) (Boehringer-Mannheim). The samples are then added to one tube of washed streptavidin–agarose. Vortexing of streptavidin–agarose beads is strictly avoided.

Binding. Binding is allowed to occur on a rotating device for 2–16 hr. Complete binding occurs after 2 hr, but the stability of splicing complexes for at least 18 hr at 4° permits overnight incubations.

Washing. Each column is washed 5 times with 1 ml of Binding Buffer.

Elution. To elute the bound components, 450 μl of a solution containing 1% SDS, 1 m*M* EDTA, 50 μg/ml BSA, and 50 μg/ml glycogen is added to each sample and incubated at 90° for 5 min. Eluted material is extracted with phenol and precipitated with ethanol.

Comments. The efficiency of selection was determined to be approximately 75%, by titrating increasing amounts of biotinylated RNA against a constant amount of streptavidin–agarose. Splicing complexes containing biotinylated RNA were selected at approximately the same efficiency as the naked biotinylated RNA. Complexes treated with heparin prior to sedimentation in glycerol gradients were found to give significantly higher efficiency of selection compared to complexes sedimented in gradients without heparin treatment.

What limits the observed binding saturation to 75%? First, even though the level of biotin-UTP in the transcription reaction should yield on average five biotin-UTP residues for each 450-nucleotide RNA chain, it is likely that some RNA chains are synthesized without any biotin residues. Second, because binding is carried out under native conditions, the secondary and tertiary structure of RNA could potentially mask biotinylated residues. Both of these effects are likely to contribute to the less than quantitative selection of the biotinylated RNA.

Analysis of Selected Components

To detect RNA components purified by streptavidin–agarose selection, ethanol-precipitated RNA species are labeled with ^{32}PCp and T4 RNA ligase.[18] Labeled RNA molecules are separated in denaturing polyacrylamide gels in parallel with markers for known HeLa cell snRNAs. The result of a typical experiment is shown in Fig. 2. The 25 S presplicing complex (lane a) shows specific selection of U2 snRNA compared to the control reaction (lane b). The 35 S spliceosome region shows the selection of snRNA species U2, U4, U5, and U6 (duplicate reactions, lanes c and d), compared to the control reaction (lane e). The identity of these RNA species is based on comigration with known snRNA species and partial RNase T1 digestion patterns which match those of the known snRNAs.

Perspectives and Critique

Biotin–streptavidin affinity selection is a valuable tool permitting analysis of the RNA components of splicing complexes assembled on a wide variety of pre-mRNA substrates. Analysis of one adenovirus and two β-globin model substrates strongly suggests a common set of snRNPs stably associated in the spliceosome: U2, U5, and U4 plus U6.[8,9,12] In addition, preliminary evidence suggests that an additional RNA species is a constituent of an adenovirus but not a β-globin substrate.[12] Of great interest is the identity of components and their arrangement in splicing complexes formed as a consequence of cell-type-specific or developmental stage-specific splicing events. In these cases, alternative splice sites are activated, presumably owing to recognition by cellular factors, unique in

[18] T. E. Englund, A. G. Bruce, and O. C. Uhlenbeck, this series, Vol. 65, p. 65.

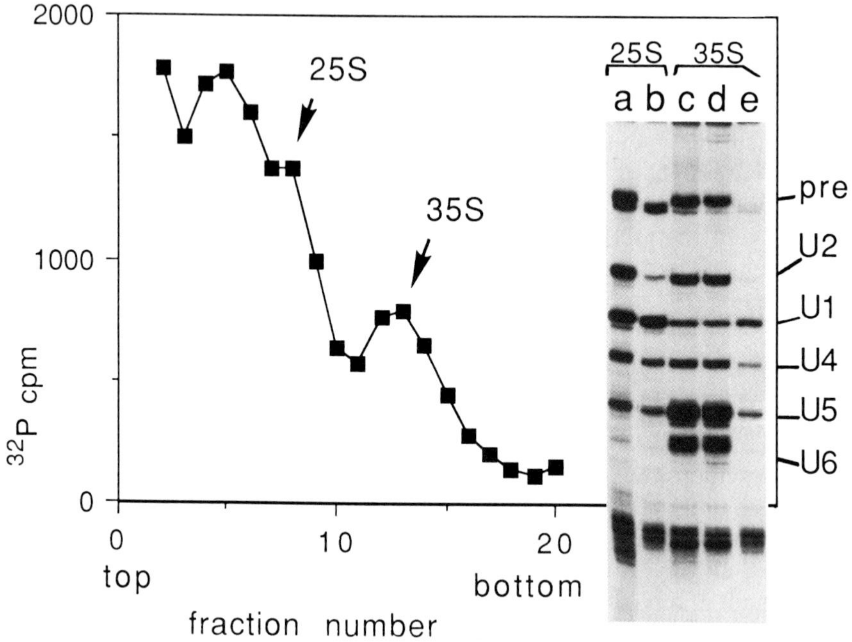

FIG. 2. Analysis of spliceosome components isolated by biotin–streptavidin selection. Complete splicing reactions containing ^{32}P-labeled pre-mRNA were size fractionated by sedimentation through 10–30% (v/v) glycerol gradients. Control splicing reactions, in which pre-mRNA was omitted, were sedimented in parallel gradients. (Inset) Fractions corresponding to the 25 S presplicing complex (lane a) and the 35 S spliceosome (duplicate, lanes c and d) were selected on streptavidin–agarose beads. The eluted RNA components were 3′-end-labeled with ^{32}PCp and T4 RNA ligase, then resolved on a 10% (w/v) polyacrylamide–8 *M* urea gel. Control reactions containing the 25 S region but lacking pre-mRNA (lane b) and containing the 35 S region but lacking pre-mRNA (lane e) were analyzed in parallel.

form or level in a particular cell type. Such factors could in principle be readily isolated by the approach presented here.

The identification of unknown RNA constituents of splicing complexes for which sequences or probes are unavailable signifies one of the greatest utilities of the method. Unknown RNA species that specifically associate with pre-mRNA substrates can be identified initially by molecular weight, using denaturing gel electrophoresis. The 3′-end-labeled RNA molecules can then be gel purified and sequenced using chemical or enzymatic means.

Limitations inherent in the selection of splicing complexes using biotin–streptavidin affinity selection are 2-fold. First, only stable complexes can be studied by this method. Factors that transiently bind to pre-mRNA molecules would not be detected. U1 snRNP is an example of a cellular factor shown definitively to bind 5′ splice sites,[19] and yet U1 RNA is not found to be a stable component of splicing complexes analyzed either by the method described here or by Northern analysis of electrophoretically separated splicing complexes.[8,10–12] Loosely bound components of splicing complexes will require other methods for their identification.

The second limitation of the method is the significant level of background RNA and protein that is eluted from streptavidin–agarose columns in the absence of biotinylated RNA substrates. The background is primarily due to nonspecific binding of nuclear extract components to the streptavidin–agarose. Three conditions were found to be essential in order to optimize signal over background: the use of saturating pre-mRNA substrate levels in splicing reactions, treatment of splicing reactions with heparin, and sedimentation in glycerol gradients to enrich for distinct complexes. In order to investigate in greater detail the components and structures of splicing complexes using this method, it will be necessary to achieve reduced background levels. Different selection matrixes should be considered as a way of reducing background levels. A matrix containing biotin covalently linked to cellulose has recently been shown to provide efficient, specific binding and very low background levels arising from nonspecific binding of cellular components.[20] In this case, purified streptavidin is used as a linker to select biotinylated substrates.

Acknowledgment

I thank Phillip Sharp for critical comments on the manuscript.

[19] Y. Zhuang and A. M. Weiner, *Cell* **46,** 827 (1986).
[20] M. S. Kasher, D. Pintel, D. C. Ward, *Mol. Cell. Biol,* **6,** 3117 (1986).

[35] Z-DNA Affinity Chromatography

By RICHARD FISHEL, PAUL ANZIANO, and ALEXANDER RICH

Introduction

The adsorption of nucleic acids to a column matrix has greatly improved our ability to identify and purify RNA- and DNA-specific metabolic proteins.[1,2] In the case of DNA, both duplex and single-stranded as well as specific nucleotide sequences have been added to cellulose, agarose, or Sephacryl matrix materials.[3,4] In many of the methods used for linking the DNA to the matrix material, the cross-link is poorly understood, although it generally involves multiple and random sites along the length of the attached substrate DNA. It has been suggested that the efficiency of binding and chromatography of appropriate proteins would increase if the cross-link were specific and situated at the end of the DNA.[5]

Recent interest in alternate DNA configurations and their attendant binding proteins has led us to develop a method for quantitative addition of left-handed Z-DNA to a chromatography matrix.[6] The technique utilizes the stable interactions inherent in the avidin–biotin bond as a point of attachment and is similar to previously reported methods.[7,8] We have recently used this material to identify and purify Z-DNA-binding proteins.[9]

Z-DNA is a left-handed conformation of the DNA double helix. Although both right-handed and left-handed conformations have antiparallel sugar phosphate backbones, they differ considerably in their external form (Fig. 1). Z-DNA has a dinucleotide repeat, and the base pairs are "flipped over" relative to the right-handed B-DNA conformation. Considerable work has been carried out on the interconversion between these

[1] H. Shaller, C. Nusslein, F. Bonhoeffer, C. Kurz, and I. Nietzschmann, *Eur. J. Biochem.* **26,** 474 (1972).
[2] L. Yarbrough and I. Hurwitz, *J. Biol. Chem.* **249,** 5394 (1974).
[3] G. Herrick and B. Alberts, this series, Vol. 21D, p. 198.
[4] S. H. Hall and E. A. Smuckler, *Biochemistry* **13,** 3795 (1974).
[5] P. T. Gilham, this series, Vol. 21D, p. 191.
[6] C. J. Harris, S. P. Moore, and R. Fishel, submitted for publication.
[7] P. R. Langer, A. A. Waldrop, and D. C. Ward, *Proc. Natl. Acad. Sci. U.S.A.* **78,** 6633 (1981).
[8] H. Delius, *Nucleic Acids Res.* **13,** 5457 (1985).
[9] R. A. Fishel, K. Detmer, and A. Rich, *Proc. Natl. Acad. Sci. U.S.A.* **85,** 36 (1988).

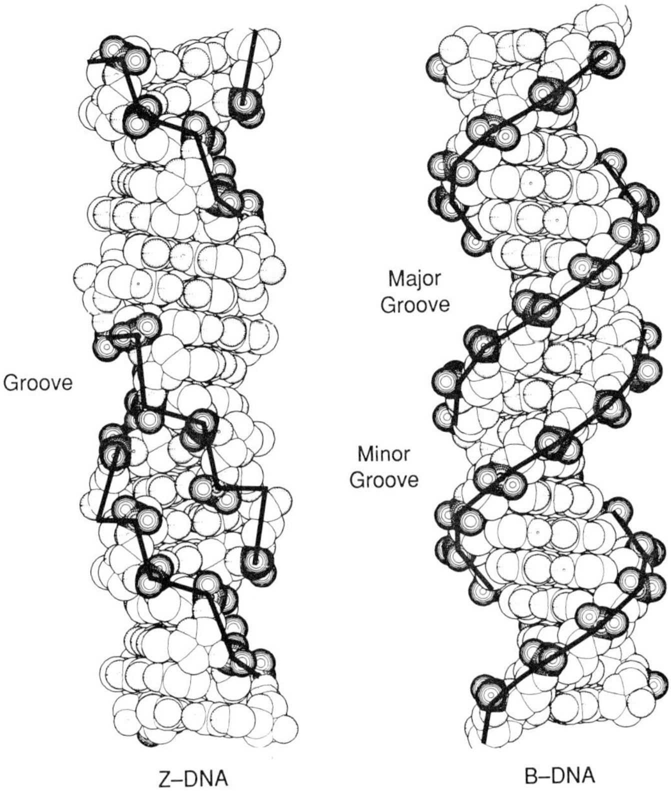

FIG. 1. van der Waals diagrams of Z-DNA and B-DNA. Heavy solid lines are drawn from phosphate to phosphate and illustrate the irregularity of the Z-DNA backbone.

two forms (reviewed in Refs. 10 and 11). There is an equilibrium between B- and Z-DNA, but Z-DNA is a higher energy form that requires something to stabilize it. The DNA polymer poly(dC-dG) forms Z-DNA readily in 4 *M* NaCl solution,[12,13] and it can be stabilized in the left-handed form by bromination of cytosine at the C-5 position or by bromination of guanine at C-8.[14,15] Several such chemical modifications have been used to stabilize poly(dC-dG) in the Z-DNA conformation in a low-salt solution.

[10] A. Rich, A. Nordheim, and A. H. J. Wang, *Annu. Rev. Biochem.* **53,** 791 (1984).
[11] T. M. Jovin and D. M. Soumpasis, *Annu. Rev. Phys. Chem.* **38,** 521 (1987).
[12] F. M. Pohl and T. M. Jovin, *J. Mol. Biol.* **67,** 375 (1972).
[13] T. J. Thamann, R. C. Lord, A. H. J. Wang, and A. Rich, *Nucleic Acids Res.* **9,** 5443 (1981).
[14] L. P. McIntosh, I. Greiger, F. Eckstein, D. A. Zarling, J. H. van de Sande, and T. M. Jovin, *Nature (London)* **294,** 83 (1983).
[15] R. D. Hotchkiss, *J. Biol. Chem.* **175,** 315 (1948).

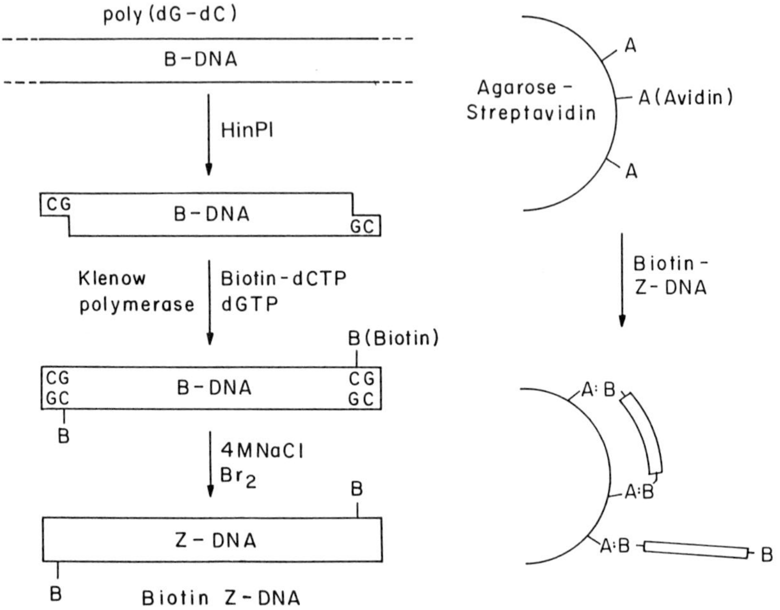

FIG. 2. Flow diagram for the preparation of Z-DNA affinity columns.

In this chapter we describe in detail several methods for preparing Z-DNA that is stable under physiological conditions. This material is then cross-linked to an agarose column matrix at its ends via biotin and streptavidin for use as a chromatography medium. A simple reproducible method for attaching biotin to the ends of virtually any DNA molecule is described. A generalized flow diagram for the construction of a Z-DNA affinity column is shown in Fig. 2.

Reagents

Ultrapure Tris-HCl, Tris-OH, sodium chloride, and EDTA are purchased from American Research Products Company (Solon, OH). Tris-buffered solutions are prepared by mixing appropriate volumes of 1 *M* Tris-HCl and 1 *M* Tris-OH solutions that result in the desired pH (see Buffer Tables in Sigma Catalog).

Sodium citrate is purchased from Mallinckrodt (St. Louis, MO) and a stock solution of 0.2 *M* is prepared at pH 7.2 with sodium hydroxide (Mallinckrodt). A 10-fold dilution of this stock yields a solution with a final pH of 6.8.

Liquid bromine is purchased from Aldrich Chemical Co. (Milwaukee, WI), and a stock solution is prepared by overlaying 25 ml of the bromine with 50 ml glass-distilled water in a sintered glass container. The water phase is saturated with bromine by repeated inversion followed by settling.

Poly(dG-dC), 100 OD units, and poly(dG-m^{5}dC), 25 OD units, from Pharmacia (Piscataway, NJ) are dissolved in 10 mM Tris (pH 7.5), 1 mM EDTA (TE buffer) and biotinylated as described below. The DNA is generally dissolved at a concentration of 100 A_{280} units/ml.

A bubbler attachment for a nitrogen gas source is constructed by heat sealing the end of a 10-cm piece of PE240 tubing (Fisher, Pittsburgh, PA) with flamed forceps. The end of the sealed tube is then pincushioned with a 22-gauge needle.

Biotinylated-dCTP is purchased from ENZO Biochemicals (New York, NY) as a 0.3 mM solution. dGTP is purchased from Pharmacia Biochemicals and dissolved in 0.1 mM EDTA (pH 8.0). The nucleotide is neutralized to pH 7.0 with 2 M NaOH and the concentration checked spectrophotometrically.

*Hin*PI and *Escherichia coli* DNA polymerase I Klenow fragment are purchased from New England Biolabs (Beverly, MA). Biotin–agarose and streptavidin are purchased from Sigma Chemical Co. (St. Louis, MO).

Biotinylation of DNA

An enzymatic DNA synthesis using the Klenow fragment[16] of *E. coli* polymerase I is the basis for the biotinylation reaction. Virtually any DNA substrate can be used as a substrate for biotinylation if it contains a G or an A on the template strand, since both biotinylated dCTP and dUTP are commercially available. To biotinylate poly(dG-dC), the polymer solution with an average length of 4 kb and varying between 0.5 and 25 kb is first partially restricted with *Hin*PI. A 2 mg/ml poly(dG-dC) solution containing 100 units/ml of *Hin*PI is reduced to an average length of 1.5 kb (varying between 0.1 and 9 kb) after incubation for 6 hr at 37° under the recommended buffer conditions. This digestion leaves a 2-nucleotide 5′ overhang that is a template for fill-in DNA synthesis.

Poly(dG-m^{5}dC), however, is resistant to digestion by the three known restriction enzymes that recognize alternating d(GC)$_n$ sequences. Thus, an alternate method for generating single-stranded tails that are substrates for DNA synthesis must be devised. A convenient method appears to be sonication, although the energy required to shear DNA must be determined empirically for each instrument. We have found that three short bursts on ice (30 sec, duty cycle 3 at 60%) with a Bronson microtip

[16] H. Jacobsen, H. Klenow, and K. Overgaard-Hansen, *Eur. J. Biochem.* **45,** 623 (1974).

TABLE I
EFFICIENCY OF BIOTINYLATION POLYMERASE REACTION[a]

Enzyme	Incorporation[b] (%)
Escherichia coli polymerase I	4.1
Escherichia coli Klenow fragment	8.8
Micrococcus luteus polymerase	1.3
T4 DNA polymerase	6.0
T7 DNA polymerase	0.8
Avian myeloblastosis virus (AMV) reverse transcriptase	9.2

[a] The template for polymerase was partially *Hin*PI-restricted poly(dG-dC).

[b] Using [^{3}H]dGTP and biotin-dCTP. The second nucleotide added in this reaction is guanine. The calculated maximum incorporation for polynucleotides of average length 2500 bp is approximately 9%. Reaction conditions were as described by the manufacturer at 37° for 1 hr with 100 U/ml of enzyme and 2 mg/ml of poly(dG-dC) DNA.

sonicator reduces the average length of the polymer to 1–2 kb (varying between 0.1 and 7 kb). Many of these reduced-length fragments appear to be templates for DNA synthesis in the biotinylation reaction as shown below.

DNA Synthesis

We have tested several polymerase enzymes for efficiency of biotinylation using biotin-dCTP and [^{3}H]dGTP. The results (Table I) suggest that the Klenow fragment of DNA polymerase I is the most efficient fill-in enzyme for this reaction. We use the following conditions: 40 m*M* potassium phosphate (pH 7.5), 6.6 m*M* $MgCl_2$, 1.0 m*M* 2-mercaptoethanol, 0.5 m*M* dGTP, 6 μM biotin-dCTP, 100 U/ml Klenow fragment, and 20 A_{260} units/ml DNA. Incubation at 25° for 16 hr is sufficient to label over 95% of the DNA molecules on at least one end, based on their ability to be retained on a column matrix. Unincorporated biotinylated nucleotides must be removed from the DNA in order to maximize binding to the column matrix. This is accomplished during the bromination step for the poly(dG-dC) polymer by ultrafiltration. Unincorporated nucleotides can be removed from the poly(dG-m^{5}dC) polymer by dialysis, ultrafiltration, or molecular sieve chromatography.

Preparation of Z-DNA

The most convenient commercially available substrate for use in the production of large quantities of stable Z-DNA is poly(dG-dC). The con-

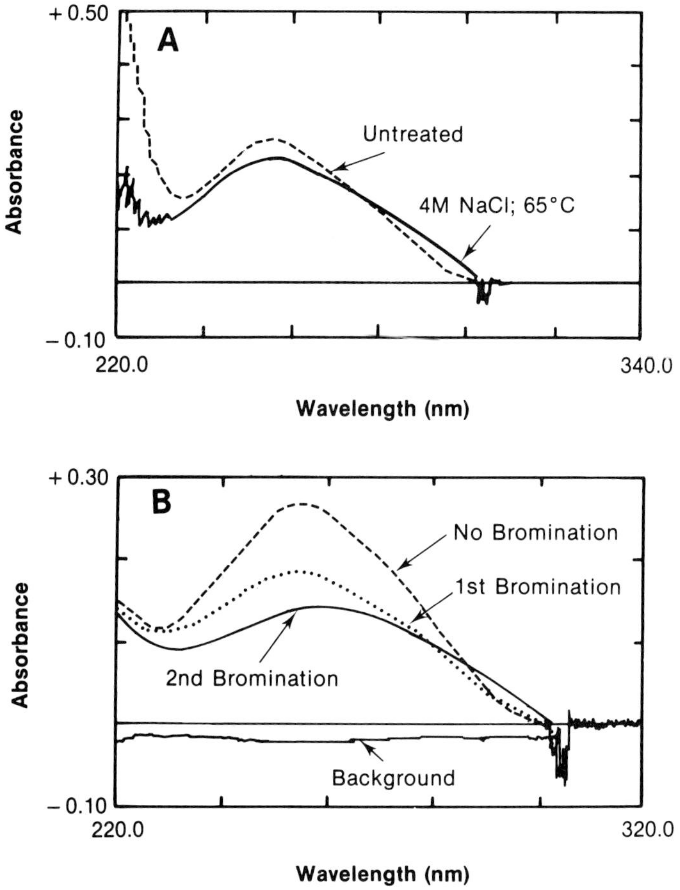

FIG. 3. Bromination of poly(dG-dC). (A) Absorption spectra of the high-salt-induced red shift indicative of the B-DNA to Z-DNA transition of poly(dG-dC). (B) Stabilization of the B-DNA to Z-DNA transition by bromination as shown by the change in absorption spectra. Chemical bromination was carried out as described in the text, and a 1/60 dilution in TNE (10 mM Tris, 50 mM NaCl, 1 mM EDTA) was done to determine if the Z-DNA configuration has been stabilized in low salt.

version of this polymer from the B-DNA form to the Z form can be monitored either by changes in absorption spectroscopy (Fig. 3) or by changes in the circular dichroism (Fig. 4).

Poly(dG-dC)

Bromination of biotinylated poly(dG-dC) is performed in a bromination buffer that consists of 20 mM sodium citrate (pH 6.8), 4 M NaCl,

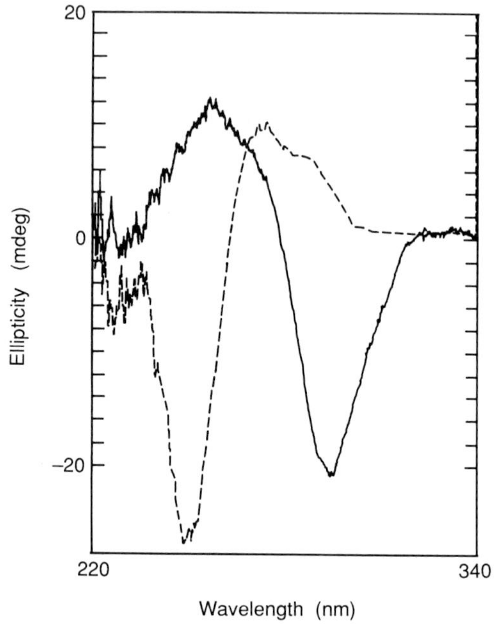

FIG. 4. Circular dichroism spectra of poly(dG-dC), illustrating the B form (– – –) and brominated poly(dG-dC) in the Z form (——) under low-salt (TNE) ionic conditions.

1 m*M* EDTA, and 7 OD units/ml DNA (generally 21 units in 3 ml), and the procedure is similar to that of Moller *et al.*[17] The DNA is flipped to the Z-DNA configuration by incubation at 65° for 10 min in this bromination buffer (Fig. 3A). A 10% (v/v) saturated bromine solution is prepared by mixing 0.1 ml of the bromine-saturated aqueous phase (described in the reagents section) with a concentrated DNA-free bromination buffer solution such that the final buffer concentration in the 1 ml solution matches that of the DNA-containing solution. The 10% bromine solution should be prepared immediately before addition to the DNA.

Bromination is carried out in a *suitably ventilated area* and in two steps to minimize overbromination and potential precipitation of the DNA substrate. The extent of bromination is monitored by scanning a 1/60 diluted sample in 10 m*M* Tris (pH 7.5), 50 m*M* NaCl, 1 m*M* EDTA (TNE buffer) from 220 to 340 nm (see Fig. 3B). The spectral changes for the conversion of poly(dC-dG) to the Z-DNA conformation are considerably greater when measured using circular dichroism. An example is shown in

[17] A. Moller, A. Nordheim, S. A. Kozlowski, D. Patel, and A. Rich, *Biochemistry* **23,** 54 (1984).

Fig. 4, where the circular dichroism spectra of both poly(dC-dG) and brominated poly(dC-dG) are measured in the low-salt TNE buffer. Using circular dichroism spectra, it is possible to determine the degree of transition (θ) with reasonable sensitivity.[17]

We have routinely used the A_{295}/A_{260} ratio of 0.38 as an indication of stably brominated Z-DNA (see Fig. 5). At time zero, 0.2 ml of the freshly prepared bromine solution is added to 3 ml of DNA solution, and the DNA is incubated at 25° for 15 min. After incubation, the unreacted bromine is removed by carefully bubbling nitrogen through the solution for 5 min. The extent of bromination is then tested spectrophotometrically

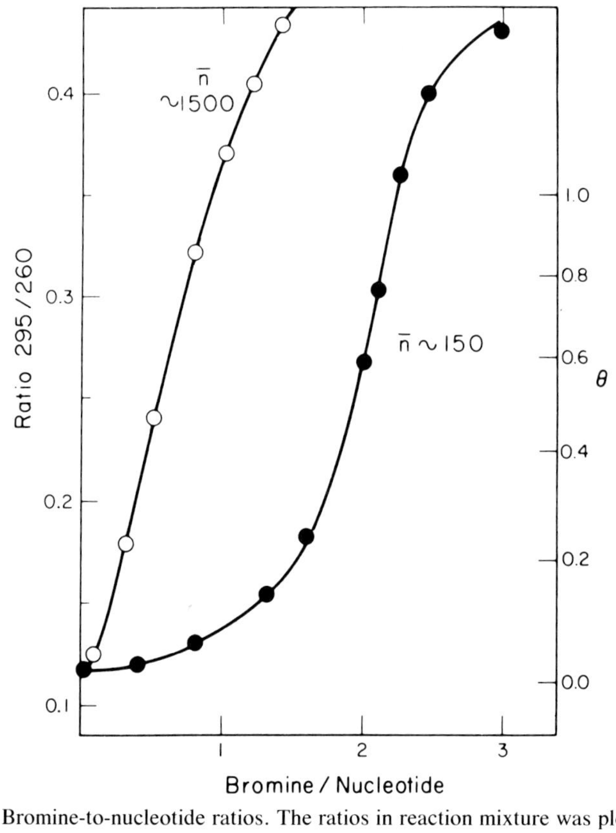

FIG. 5. Bromine-to-nucleotide ratios. The ratios in reaction mixture was plotted against the A_{295}/A_{260} ratios and the corresponding degree of transition (θ) values were measured by circular dichroism spectra in low-salt buffer.[11] Two different polymers were used with chain lengths of n = 1500 bp (○) and n = 150 bp (●). Note that the degree of transition is not linear, and the absorbance continues to change with added bromine even after the transition is complete (θ = 1.0).

(Fig. 3A).[18] A single bromination step, however, does not stabilize the polymer in the Z-DNA configuration; a second identical bromination step is required to stabilize the Z-DNA configuration in low salt (Fig. 3B). In practice, we have found that strict adherence to the two bromination steps described produces DNA with an A_{295}/A_{260} ratio of 0.38 in better than 95% of the experiments.

Quantitative recovery of the brominated DNA in TNE buffer is carried out by ultrafiltration. The DNA in the high-salt bromination solution is diluted 4-fold and applied to an Amicon 8MC ultrafiltration unit containing a YM10 filtration membrane. The filtrate is successively diluted and washed 3 times with 10 ml of TNE buffer and concentrated to a volume of 1 ml. Recommended procedures are followed to reduce nonspecific binding of the DNA to the membrane.[19] Following filtration the sample is divided into several Eppendorf tubes, 2 volumes of 95% ethanol is added, and precipitation of the DNA is carried out at −80°. The dried precipitate is resuspended in 0.5 ml per 20 OD units starting material. After confirming that the brominated poly(dG-dC) is in the Z-DNA configuration by circular dichroism (see Fig. 4), the polymer is linked to the column material.

Poly(dG-m⁵dC)

The pyrimidine–purine sequence (dG-m^{5}dC) occurs naturally in mammalian DNAs.[20] Furthermore, in the presence of Mg^{2+} or Mn^{2+}, polynucleotides with this sequence form Z-DNA at physiological salt concentrations.[21] Because brominated DNA is unlikely to occur *in vivo*, a column that consists of poly(dG-m^{5}dC) may be more physiologically relevant for the isolation of Z-DNA-binding proteins. Figure 6 illustrates conditions in which poly(dG-m^{5}dC) is stabilized in the Z-DNA configuration in low salt. A circular dichroism spectrum of this material is essentially identical to that shown in Fig. 4. In practice we have found that the Z-DNA configuration of poly(dG-m^{5}dC) is maintained in the presence of 10 m*M* $MgCl_2$ up to approximately 0.3 *M* KCl. However, above this salt concentration the polymer abruptly flips to the B-DNA configuration, which is maintained up to a salt concentration of approximately 2.5 *M* KCl. Above 2.5 *M* KCl

[18] A scanning spectrophotometer is required to unambiguously observe the shifts inherent in the B- to Z-DNA transition reaction.

[19] Amicon recommends stirring the sample without pressure for 30 min prior to removal from the apparatus. We generally stir for 45 min followed by a second wash step with 0.5 volume of TNE and stir an additional 45 min.

[20] M. Behe and G. Felsenfeld, *Proc. Natl. Acad. Sci. U.S.A.* **78,** 1619 (1981).

[21] J. Chaires, *Biochemistry* **25,** 8436 (1986).

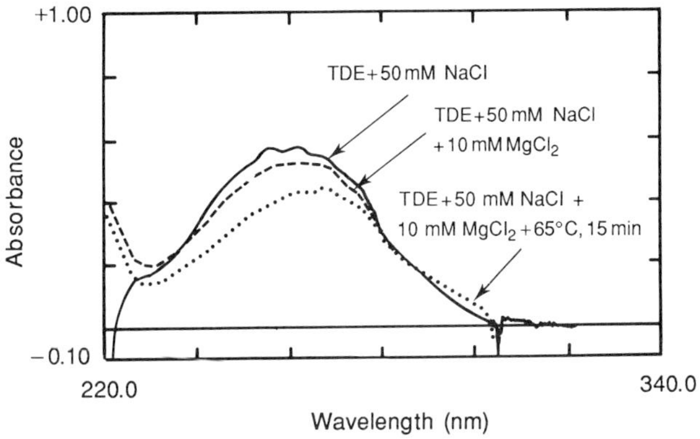

FIG. 6. B-DNA to Z-DNA transition of poly(dG-m^{5}dC) under low-salt conditions. Poly(dG-m^{5}dC) was resuspended in 20 m*M* Tris (pH 7.5), 1 m*M* dithiothreitol, 1 m*M* EDTA (TDE buffer) containing 50 m*M* NaCl. The absorption spectrum was measured without added $MgCl_2$, in the presence of 10 m*M* $MgCl_2$, and in the presence of 10 m*M* $MgCl_2$ after incubation at 65° for 15 min to thermally induce the B to Z transition.

the polymer flips back to the Z-DNA configuration. For proteins that elute below 0.3 *M* salt, this property of the polymer is not an issue. For proteins that bind to Z-DNA and not to B-DNA, this conversion of poly(dG-m^{5}dC) can be used as an additional purification step. This property of poly(dG-m^{5}dC) has been observed and characterized by others.[21,22]

Addition of Biotinylated DNA to Solid Matrix

There are presently three commercially available avidin–biotin-based solid matrix materials that can be used in the preparation of a chromatography column (available from Sigma Chemical Co.). The least expensive is biotin cross-linked to agarose generally via an amide or ether bond and containing a 12-carbon spacer.[23] We have found that linkage of biotinylated DNA to the biotin–agarose matrix via streptavidin in TNE buffer results in retention of virtually all of the DNA substrate. However, this quantitative linkage is dependent on the absence of unincorporated biotinylated nucleotides and a highly washed biotin-agarose that is resuspended in buffered low salt (TNE). The quantitative results of retention experiments are shown in Table II.

Before linking the DNA to the biotin–agarose, it must be washed with

[22] J. B. Chaires, personal communication.

[23] See Sigma Catalog.

TABLE II
ADDITION OF BIOTINYLATED Z-DNA POLYNUCLEOTIDE TO COLUMN MATRIX[a]

Step	Brominated poly(dG-dC)[b]	Poly(dG-m^{5}dC)
Starting amount (mg)	2.2	1.6
1 *M* KCl wash[c]	0.02	0.63
2.5 *M* KCl wash[c]	0.06	0.38
Total bound Z-DNA (%)	96.4	36.9

[a] Quantification of Z-DNA polynucleotide bound to 2 ml of biotin–agarose column matrix. An appropriate quantity of streptavidin was first added to the DNA, and this solution was then mixed with 2 ml of washed biotin–agarose (see text).

[b] Stably brominated Z-DNA prepared according to the procedure described.

[c] Salt wash of streptavidin–biotin–DNA column complex constructed according to the described procedure.

high salt followed by washing and resuspension in low salt. This is accomplished by multiple low-speed centrifugation in a 50-ml polyethylene conical tube. Vacuum aspiration is the most convenient method for removing the supernatant from the wash solution. Preparation of the column material begins with a low-speed centrifugation for 3 min, followed by aspiration of the overlying buffer. We then routinely wash 3 times with 3 volumes of TE buffer plus 2.5 *M* KCl followed by 5 times with TNE buffer.

Linkage of the DNA to the washed biotin–agarose is accomplished by first attaching streptavidin to the DNA, then linking the streptavidin–DNA complex to the biotin agarose. We use a 5-fold molar excess of streptavidin to the DNA ends, calculating the number of DNA ends based on the average nucleotide length of the substrate. For an average length of 2 kb, approximately 1 μg streptavidin should be added per microgram of Z-DNA. We use the extinction coefficient equivalent of 14.4 A_{260} units/mg of Z-DNA. It is important not to confuse moles of streptavidin with moles of binding sites, since there are four biotin-binding sites per molecule of streptavidin. The initial streptavidin addition is designed to place one streptavidin molecule per end of the substrate DNA. Incubation of the DNA–streptavidin solution at 25° for 60 min is sufficient to complete the initial binding reaction. This solution is then added to the washed biotin–agarose and placed on a rotator wheel at room temperature overnight.

The experiments shown in Table II indicate that greater than 96% of the biotinylated Br-poly(dG-dC) is linked to the column material while approximately 37% of the biotinylated poly(dG-m^{5}dC) is linked to the

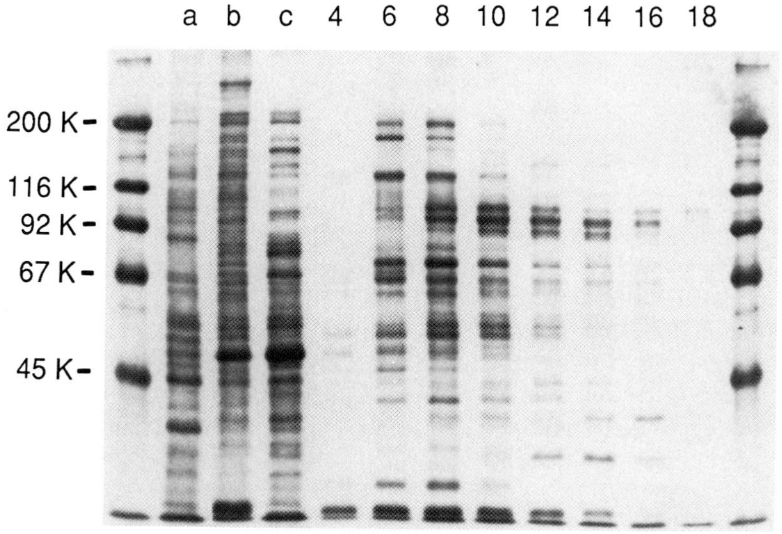

FIG. 7. Protein elution pattern following Z-DNA column chromatography. Protein fractions from a gradient elution of a brominated poly(dG-dC) Z-DNA column were analyzed by SDS–PAGE according to established procedures. The protein extract was prepared from a human tumor cell line (HBP-ALL) by isolation of the cell nuclei followed by salt extraction of the nuclear proteins.[9] Lane a, cytoplasmic fraction; lane b, nuclear extract; lane c, effluent fraction from Z-DNA column; lanes 4–18, increasing salt elution fractions from the Z-DNA column. The end lanes contain marker molecular weight proteins.

biotin–agarose. Both these materials have been used to isolate Z-DNA-binding proteins with essentially the same elution profiles through 0.3 *M* salt. As discussed above, the poly(dG-m^{5}dC) changes configuration above 0.3 *M* salt, and its elution pattern becomes divergent with that from the Br-poly(dG-dC) column.

In using the affinity column, a cell extract is passed over the column in a low-salt buffer, for example, with 50 m*M* KCl. By gradually raising the salt concentration, proteins are eluted in various fractions. These columns have been employed using a continuous-gradient elution[6,9] or using block elution (e.g., 0.1 *M* KCl, 0.2 *M* KCl, etc.). In many cases it is desirable to add competing B-DNA, which can be derived from *E. coli* or a eukaryotic source. In some cases, polymers such as poly(dI-dC) may also be used for this purpose. This helps to remove nonspecific and B-DNA-binding proteins.

An example of the fractionation obtained using the Z-DNA affinity column is shown in Fig. 7. An extract was made from a human lymphoblastic cell line and passed over the column starting with 50 m*M* KCl.[9]

As the KCl gradient increased, various protein components were eluted. This column has been used to concentrate proteins that have a strand transferase activity, enzymes that may play a role in genetic recombination.

Summary

In this chapter we have detailed a method that can be generalized to link virtually any DNA substrate to a chromatography matrix at its ends via an avidin–biotin linkage. We have used this technique to construct a left-handed Z-DNA column for the purpose of identification and purification of Z-DNA-binding proteins. This technique for the linkage of DNA to a column matrix by avidin–biotin technology can be modified, however, to produce linked multimeric sequences specific for regulatory or other DNA-binding proteins.

Acknowledgments

The authors thank Dr. Sharon Moore for helpful discussion and critical review of the manuscript, Candice Harris and Greg Beal for technical assistance, and Julie Ratliff for preparation of the manuscript. Research was sponsored in part by the National Cancer Institute, U.S. Department of Health and Human Services, under Contract NO1-CO-74101 with BRI (R.F.), and grants from the National Institutes of Health and the Office of Naval Research (A.R.).

[36] Affinity Cytochemistry

By MEIR WILCHEK, EDWARD A. BAYER, and EHUD SKUTELSKY

Although in recent years the use of avidin–biotin technology has progressed rapidly in a variety of areas, such as immunoassay and gene probes, the broadest application remains in the localization of target molecules, both on cell surfaces and in the intracellular matrix.

It seems that in the mid-1970s the use of the avidin–biotin complex was perfectly suited for the requirements of localization studies. Indeed, the then-current use of protein–protein conjugates (namely, binders such as antibodies or lectins covalently cross-linked to appropriate markers such as ferritin or enzymes) was fraught with difficulty; the conjugation procedure and/or the covalent attachment of large macromolecules to the binding protein often deleteriously affected both the physicochemical properties and the biological activity of the resultant conjugate. There were also problems in purifying the active conjugates. A high degree of nonspecific or errant binding was commonly obtained.

The intercession of the high-affinity avidin–biotin complex was shown to circumvent many of these problems, and its adaptation to localization studies was rapid and extensive, as shown in Tables III, IV, and V in [3] in this volume. As in other areas, the use of avidin–biotin technology was preceded by a study in which hapten antihapten antibodies were used for a similar purpose.[1] Nevertheless, the applicative advantages of the avidin–biotin system were immediately clear (see discussion in [2] in this volume).

Since localization usually involves a cell or tissue which can be considered a solid-phase system, the application of avidin–biotin technology to other fields (e.g., immunoassays) was clearly an outgrowth of the initial development of the technique for localization studies. In fact, the mediation of the system in histocytochemical studies is rapidly becoming the basis for current progress in diagnostic pathology, thereby underscoring the interrelationship between localization and assay.

Avidin–biotin technology can be applied for two basic purposes in this field: direct localization and binder-mediated visualization of cell-associated components. The direct approach consists of group-specific biotinylation of amino acids or sugars using one of the reactive biotin-containing reagents described in [13] in this volume. This approach was introduced by Heitzmann and Richards[2] for labeling and localizing lysines using biotinyl-*N*-hydroxysuccinimide ester (BNHS) and sialic acids using periodate oxidation and biotin hydrazide on isolated membranes.

This approach was later developed further by us and applied to the localization of amino acid and sugar residues on cell membranes of intact cells.[3] We showed that it is possible to label and localize sialyl residues on whole erythrocytes and lymphocytes using periodate oxidation of the corresponding cells followed by biotin hydrazide and avidin-conjugated ferritin.[4] This enabled, for the first time, the direct localization of sialic acids per se (and not negative charges via electrostatic interactions) on membranes. Using the combination of neuraminidase and galactose oxidase, it was also possible to localize the penultimate galactose and to measure the distance of the sialic acid from the cell membrane. Use of galactose oxidase alone (in the presence of biotin hydrazide) revealed the

[1] M. E. Lamm, G. C. Koo, C. W. Stackpole, and U. Hämmerling, *Proc. Natl. Acad. Sci. U.S.A.* **69,** 3732 (1972).

[2] H. Heitzmann and F. M. Richards, *Proc. Natl. Acad. Sci. U.S.A.* **71,** 3537 (1974).

[3] E. A. Bayer, E. Skutelsky, D. Wynne, and M. Wilchek, *J. Histochem. Cytochem.* **24,** 933 (1976).

[4] E. Skutelsky, D. Danon, M. Wilchek, and E. A. Bayer, *J. Ultrastruct. Res.* **61,** 325 (1977).

distribution of terminal galactosyl residues on appropriate cell types.[5] At about the same time, we tried to use the avidin–biotin system for localizing the affinity-labeled biotin transport protein on yeast protoplasts.[6] These studies have all been summarized in earlier volumes.

The localization of specific amino acids on cell surface membranes, using the group-specific reagents described in [13] in this volume, has been given little attention; the general distribution of amino acids on cell membranes renders such studies in nature much less informative. Much more information could be derived from a system wherein affinity labels are prepared in which one part of the reagent contains a biotin moiety that can serve as a handle for isolation and localization of the corresponding receptor. An example of such a study is the localization of the ATPase site of myosin.[7]

The binder-mediated approach, introduced by us,[8] is of much broader application and is the most widely accepted approach for localization. Virtually any biologically active molecule (antibodies, lectins, hormones, lipids, etc.) can be (and a great many have been) employed to localize the respective cell-based receptor (see [2] in this volume for a more extensive list of binders). It is no wonder that the complementary tables in [3] in this volume for this strategy are the longest, owing to the different procedures for localization which have been introduced. These comprise the various immunocytochemical approaches, which include the use of biotinylated derivatives of primary antibody, second antibody, or protein A. In all cases, enhanced signals can be obtained with little or no background.

Chapters [37]–[46] include representative works in which cellular components are localized. Selected examples of various target molecules, including lectin receptors, antigens, hormone receptors, and low molecular weight membrane constituents are discussed as well as the supramolecular localization of ATPase sites on myosin. Procedures for labeling intracellular components using postembedding techniques are also given, as well as those for double-labeling techniques, the use of avidin-containing complexes and avidin-containing conjugates for light, fluorescent, and electron microscopic visualization.

[5] E. Skutelsky and E. A. Bayer, *J. Cell Biol.* **96,** 184 (1983).

[6] E. A. Bayer, E. Skutelsky, T. Viswanatha, and M. Wilchek, *Mol. Cell. Biochem.* **19,** 23 (1978).

[7] K. Sutoh, this volume [46].

[8] E. A. Bayer, M. Wilchek, and E. Skutelsky, *FEBS Lett.* **68,** 240 (1976).

[37] Double Fluorescence Immunolabeling Studies

By ANNE H. DUTTON, MARGIE ADAMS, and S. J. SINGER

Introduction

Double fluorescence immunolabeling experiments for the simultaneous localization of two cell or tissue components provide a powerful means of structural analysis in cell biology. Two primary antibodies can be distinguished in principle by directly conjugating each with a different fluorophore. In practice, however, such double fluorescent labeling procedures usually result in an inadequate signal-to-noise ratio with current conventional epifluorescence instruments (although they may prove more useful if video enhancement techniques are utilized). It is therefore generally necessary to employ secondary fluorescent antibody reagents on top of the primary antibodies in order to obtain an adequate signal. Thus, in many double fluorescent immunolabeling experiments (cf. Ref. 1), two different species of primary antibodies (often rabbit and mouse, or rabbit and guinea pig antibodies) prepared against the two antigens in question were used, followed by two fluorescently conjugated secondary antibodies, suitably affinity purified and cross-absorbed, that discriminated between the two primary antibodies. However, the very rapidly expanding production and availability of mouse monoclonal antibodies (MAbs) to a wide range of antigens make it increasingly necessary to use methods whereby secondary reagents can discriminate between two closely similar primary antibodies. The avidin–biotin system can provide one component of such a general technology, in conjunction with a hapten-sandwich method[2] for the second.

Our objectives have been to doubly label intracellular as well as surface antigens using two mouse MAbs. Intracellular antigens, for a variety of reasons including steric hindrance, are often more difficult to immunolabel with a satisfactory signal-to-noise ratio than are surface antigens. As a result, previously published methods of biotinylation[3–5] with the biotin

[1] B. Geiger and S. J. Singer, *Proc. Natl. Acad. Sci. U.S.A.* **77,** 4769 (1980).
[2] L. Wofsy, this series, Vol. 92, p. 472.
[3] H. Heitzmann and F. M. Richards, *Proc. Natl. Acad. Sci. U.S.A.* **71,** 3537 (1974).
[4] M. H. Heggeness and J. F. Ash, *J. Cell Biol.* **73,** 783 (1977).
[5] E. A. Bayer and M. Wilchek, *Methods Biochem. Anal.* **26,** 1 (1980).

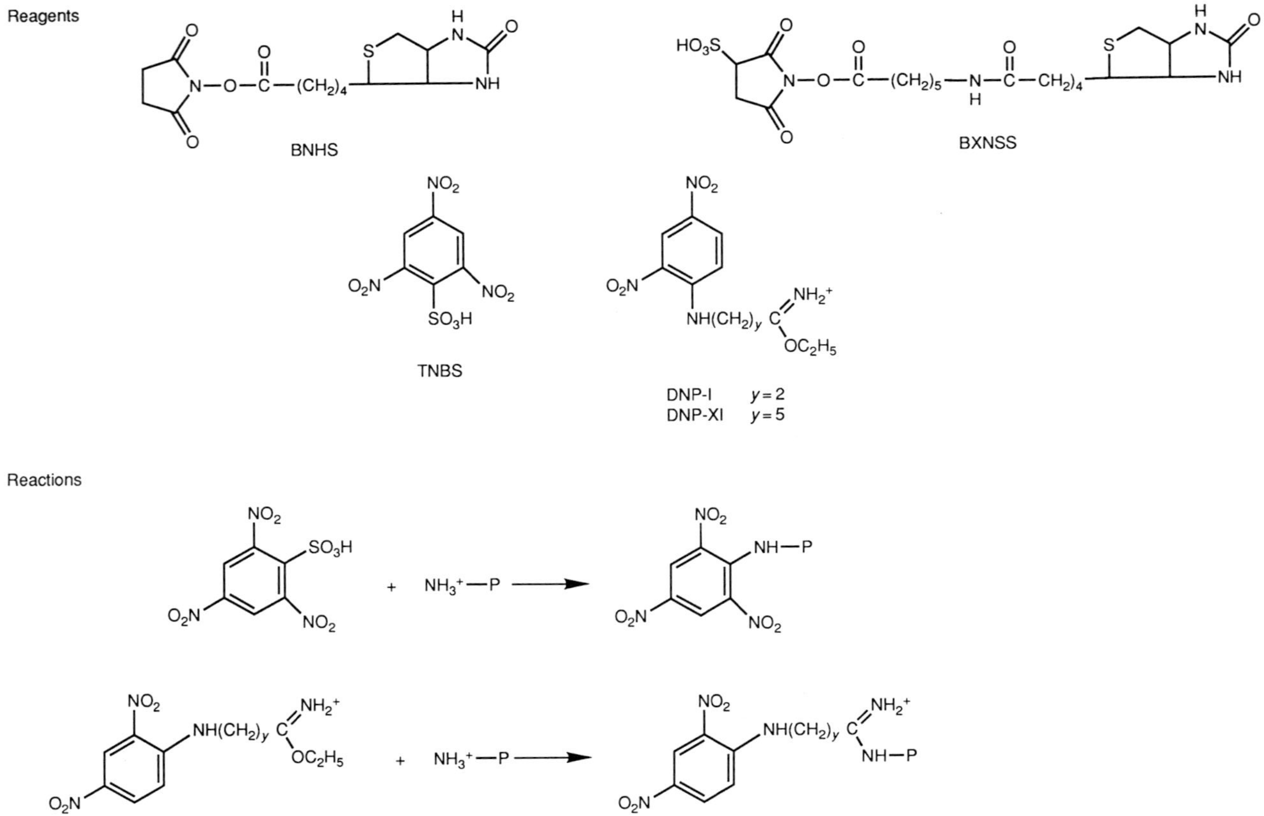

FIG. 1. Reagents used to modify the monoclonal antibodies, together with the chemical reactions involved in the trinitrophenylation and dinitrophenylation of the ε-amino groups of antibodies.

TABLE I
MOUSE MONOCLONAL ANTIBODIES

Designation[a]	Antigen[b]	Class
XIV B4	Vinculin	IgG_1
1019	Talin	IgG_1
20	Zeugmatin[6]	IgG_{2a}
108	Enactin[7]	IgG_1
30B6	Integrin[8]	IgG_{2a}

[a] These MAbs were prepared by different investigators in our laboratories.
[b] Antigens from chicken tissues and cells.

ester of *N*-hydroxysuccinimide (BNHS) (Fig. 1) for one of the two MAbs did not in our hands provide satisfactory immunolabeling for several intracellular antigens, whereas they had been useful with surface antigens.[2] The ultimate criterion for successful labeling which we employed was that the results obtained had to be at least as satisfactory (with respect to signal-to-noise ratio and specificity) as those obtained for the same two antigens using the more conventional double immunofluorescent labeling techniques mentioned above (i.e., two different species of primary polyclonal antibodies followed by two discriminating fluorescently tagged secondary antibodies).[1]

A problem we encountered is that each MAb has distinctive properties, a fact which makes it difficult to prescribe a single procedure that will be generally optimal. Five different MAbs have been examined in this study (Table I),[6-8] and they exhibited differences in the extent to which they could be modified and remain soluble or give low background labeling. The caution must therefore be emphasized at the outset that with each different MAb it may be necessary to try variations in the procedures such as those described below in order to optimize the double immunolabeling. In addition, other double-labeling methods involving avidin–biotin have been described[9,10] for use with immunoenzyme systems. Another method has been reported for double fluorescent labeling

[6] P. A. Maher, G. F. Cox, and S. J. Singer, *J. Cell Biol.* **101,** 1871 (1985).
[7] G. F. Cox, P. A. Maher, and S. J. Singer, unpublished.
[8] A. A. Rogalski and S. J. Singer, *J. Cell Biol.* **101,** 785 (1985).
[9] D. Y. Mason, Z. Abdulaziz, B. Falini, and H. Stein, *Ann. N.Y. Acad. Sci.* **420,** 127 (1983).
[10] D. M. Boorsma, J. Van Bommel, and E. M. H. Vanderhaaij-Helmer, *J. Microsc.* (*Oxford*) **143,** 197 (1986).

involving two biotinylated MAbs added sequentially,[11] but this method seems to us to have serious theoretical shortcomings.

The procedures that we have adopted differ from those of Wofsy[2] in two principal respects: (1) for biotinylation of one MAb the reagent BXNSS[12] (Fig. 1) is used instead of BNHS, and (2) the haptenic group 2,4-dinitrophenol (DNP) is introduced onto the second MAb instead of azobenzenearsonate, by reaction with one of several reagents shown in Fig. 1. DNP is used in conjunction with affinity-purified polyclonal rabbit anti-DNP antibodies and fluorescently labeled goat anti-rabbit IgG antibodies. BXNSS introduces a biotinyl group that is spaced further away from the protein surface than in the case of BNHS, and the sulfonic acid group improves solubility; we believe these factors improve the streptavidin labeling. The DNP hapten[13] is a powerful immunogen that elicits polyclonal rabbit anti-DNP antibodies having an unusually large average affinity for the hapten; this system has therefore been widely useful in immunochemistry. In addition, anti-DNP antibodies are commercially available, whereas antibenzenearsonate antibodies are not. Three different DNP reagents were tried (Fig. 1): TNBS[14] (the 2,4,6-trinitrophenyl group is recognized by anti-DNP antibodies); DNP-PI[15]; and DNP-XI.[16] These reagents react with ammonium groups on protein molecules under mild conditions. The imidates DNP-PI and DNP-XI have the additional feature that their reaction with ammonium groups converts the latter into amidines (Fig. 1) which preserves the positive charge of the ammonium residue. Consequently, reaction with imidates produces minimal effects on protein structure and activity.[16] DNP-XI presumably positions the DNP group at a greater distance from the protein surface than does DNP-PI.

Coupling Reactions with Monoclonal Antibodies

Biotinylation with BXNSS[4]

The advantage of BXNSS over BNHS for labeling studies is the greater accessibility of the biotin group after coupling to the protein surface.

[11] G. Colucci, M. Colombo, E. D. Ninno, and F. Paronetto, *Gastroenterology* **85,** 1138 (1983).

[12] K. Hofmann, G. Titus, J. A. Montibeller, and F. M. Finn, *Biochemistry* **21,** 987 (1982).

[13] H. N. Eisen and G. W. Siskind, *Biochemistry* **3,** 996 (1964).

[14] A. F. S. A. Habeeb, *Anal. Biochem.* **14,** 328 (1966).

[15] A. M. Efros, *Zh. Obsch. Khim.* **30,** 3565 (1960).

[16] M. J. Hunter and M. L. Ludwig, *J. Am. Chem. Soc.* **84,** 3491 (1962).

Reagents

BXNSS (referred to as NHS-LC-biotin, Pierce Chemical Co., Rockford, IL)

Dimethyl sulfoxide (DMSO)

Buffer: phosphate-buffered saline (PBS): 10 m*M* phosphate, 150 m*M* NaCl (pH 7.4)

Procedure. A solution of BXNSS is freshly prepared containing 1 mg in 0.2 ml of DMSO. The DEAE-purified MAb is dissolved in PBS (0.258 mg in 0.5 ml of PBS). This solution is mixed with 14 μl of the BXNSS solution and incubated for 1 hr at room temperature. The mixture is then dialyzed against several changes of PBS.

The number of biotin groups coupled per MAb molecule was not measured. With different MAbs, larger volumes of the BXNSS solution (up to 20 μl) could be used without encountering subsequent nonspecific labeling, but all conjugates made with 14 μl gave satisfactory specific labeling results.

Trinitrophenyl and Dinitrophenyl Coupling

Three reagents were tested, but we did not carry out a systematic study of all combinations of reagents and MAbs. TNBS was useful only in connection with cell surface labeling with MAb 30B6. With the other MAbs that were directed to intracellular antigens, TNBS-modified MAbs generally gave too large a nonspecific background labeling. The two DNP-imidates, DNP-PI and DNP-XI, yielded more satisfactory results. However, these are not commercially available as yet. The extent of hapten coupling to the MAb was determined spectrophotometrically, using extinction coefficients of 1.40×10^4 for TNP groups and 1.74×10^4 for DNP groups.

Trinitrophenylation with TNBS[14]

Reagents

TNBS (Pierce)

Buffer: PBS (pH 7.4)

Procedure. TNBS solution is freshly prepared, at 1.0 mg/ml in water. A solution containing 0.258 mg MAb in 0.44 ml PBS is mixed with 0.2 ml of the TNBS solution, and the mixture is stirred at room temperature for 2 hr. After this time, the mixture is dialyzed against several changes of PBS. Under these conditions, 10 TNP groups were conjugated per molecule of MAb 30B6.

Dinitrophenylation with DNP-PI and DNP-XI

Preparation of DNP-Nitriles

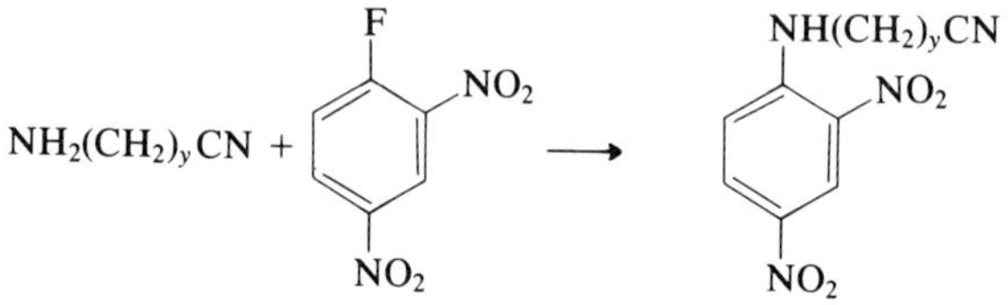

Ia: DNP-propionitrile, $y = 2$
Ib: DNP-capronitrile, $y = 5$

Reagents

3-Aminopropionitrile (ICN-K&K, Plainview, NY)
6-Aminocapronitrile (Aldrich Chemical Co., Milwaukee, WI)
2,4-Dinitrofluorobenzene (Aldrich)

Procedure for preparation of DNP-propionitrile.[15] To 7.0 g (0.1 mol) of 3-aminopropionitrile in 100 ml of dioxane is added 37.2 g (0.2 mol) of 2,4-dinitrofluorobenzene in 100 ml of dioxane. After stirring the reaction mixture at room temperature for 2 hr, a heavy yellow precipitate forms. After filtration, the precipitate is recrystallized twice from ethanol–ethyl acetate (1 : 1 by volume) (mp 134–135°).

Procedure for preparation of DNP-capronitrile.[16] To 5.6 g (50 mmol) of 6-aminocapronitrile dissolved in 50 ml of ethanol is suspended with stirring 4.2 g (50 mmol) of $NaHCO_3$. To this suspension is added 12.3 g (66 mmol) of dinitrofluorobenzene in 50 ml of ethanol. Upon stirring at room temperature for 30 min, a gum forms. The gum is collected and recrystallized from acetone–ether–petroleum ether (1 : 2 : 2 by volume) (mp 78–80°).

Preparation of DNP-Imidates[16]

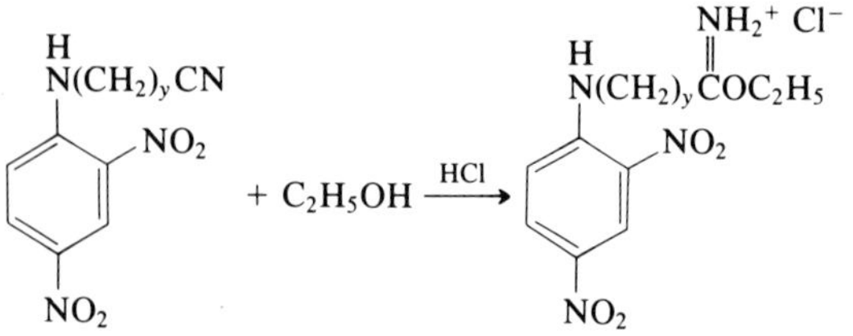

I

IIa: DNP-PI, $y = 2$
IIb: DNP-XI, $y = 5$

Reagents

DNP-nitriles (described above)
HCl gas (dry)
Ethanol, absolute
Ethyl acetate (stored over molecular sieve)

Procedure. Either of the two DNP-nitriles (5 mmol) is dissolved in 50 ml of dry ethyl acetate and 50 ml of anhydrous ethanol. The mixture, cooled in ice water, is saturated with dry HCl and stored at 4° for 24 hr. Ether is then added until the mixture becomes cloudy. Yellow crystals precipitate out after 4 days. After filtration, the crystals are washed with ether, and the solvent is removed under vacuum (mp 126–129° for DNP-PI and 114–115° for DNP-XI).

Coupling Reaction for DNP-Imidates with MAbs

Reagents

DNP-PI and DNP-XI (described above)
Buffers: triethanolamine-HCl, 0.3 *M* (pH 9, TEA buffer); PBS (pH 7.4)

Procedure. Saturated solutions of either DNP-PI or DNP-XI are freshly prepared in TEA buffer. To a solution of DEAE-purified MAb (0.258 mg in 0.2 ml PBS) is added 0.2 ml of either imidate solution, and the mixture is kept for 2 hr at 4°. The mixture is then dialyzed against several changes of PBS. With MAbs 1019 and 20, this procedure attached 3.0 and 2.0 DNP groups per molecule of protein, respectively. With some MAbs, such as XIV B4, larger volumes of imidate solutions could be added up to 0.5 ml, thereby increasing the extent of coupling to 5.0 groups per molecule without loss of specificity of immunolabeling.

Double Fluorescence Labeling Procedures

Several pairs of antigens were examined by double fluorescence labeling using biotinylated and DNP- (or TNP-) conjugated MAbs prepared as described above. The procedures and results of two such experiments with cultured cells are described next.

Reagents

Fluorescein-conjugated streptavidin (Amersham Corporation, Arlington Heights, IL)
Rhodamine-conjugated streptavidin

Fluorescein- and rhodamine-conjugates of F(ab′)$_2$ fragments of goat anti-rabbit IgG antibodies (all from Jackson Immunoresearch Labs, Avondale, PA)

Rabbit anti-DNP antiserum (Miles Scientific, Naperville, IL), or affinity-purified rabbit anti-DNP antibodies

Chick Embryonic Heart Cells: Zeugmatin and Integrin Double Labeling

The cells from 8-day embryonic hearts[17] are cultured on fibronectin-coated coverslips for 24 hr. After rinsing in PBS, the cells are fixed for 5 min at 37° in 3% (w/v) paraformaldehyde in PBS and then permeabilized by treatment for 5 min at room temperature with 0.5% (v/v) Triton X-100 in cytoskeletal buffer[18] [60 m*M* PIPES, 25 m*M* HEPES, 10 m*M* EGTA, 1 m*M* $MgCl_2$ (pH 6.9)]. The permeabilized cells are then incubated for 30 min at room temperature with a solution containing both modified MAbs: In the example presented here, BXNSS-modified MAb 20 to zeugmatin and TNBS-modified MAb 30B6 to integrin, both at a concentration of 50–100 μg/ml. After thorough rinsing with PBS, the coverslips are treated for another 30 min with a solution of affinity-purified rabbit anti-DNP antibodies[13] at a concentration of 10 μg/ml. Following rinsing with PBS, the cells are incubated 30 min with a solution containing fluorescein-conjugated streptavidin at a concentration of 5–10 μg/ml and rhodamine-conjugated F(ab′)$_2$ fragments of goat anti-rabbit IgG antibodies at 5–10 μg/ml. After rinsing with PBS, the coverslips are mounted in 50% (v/v) glycerol in PBS (pH 8.5) and examined at 63× magnification in a Zeiss Photoscope III epifluorescence instrument.

The results reproduced in Fig. 2 show that zeugmatin and integrin display distinguishable distributions in the embryonic heart cell. Zeugmatin, in accord with previous results,[6] is associated with the Z bands (Fig. 2a, bracket) of the myofibrils inside the cell. Integrin, a cell surface antigen,[8] is strongly labeled on the plasma membrane (Fig. 2b, arrowhead) and appears in an interestingly nonuniform surface distribution, probably at sites where the myofibrils are attached to the cell membrane (corresponding to the so-called costameres observed to be present in skeletal muscle).[19] For the purposes of this chapter on labeling meth-

[17] P. J. Kronebusch and S. J. Singer, *J. Cell Sci.* **88,** 25 (1987).

[18] M. Schliwa, U. Euteneuer, J. C. Bulinski, and J. G. Izant, *Proc. Natl. Acad. Sci. U.S.A.* **78,** 1037 (1981).

[19] J. V. Pardo, J. D. Siliciano, and S. W. Craig, *Proc. Natl. Acad. Sci. U.S.A.* **80,** 1008 (1983).

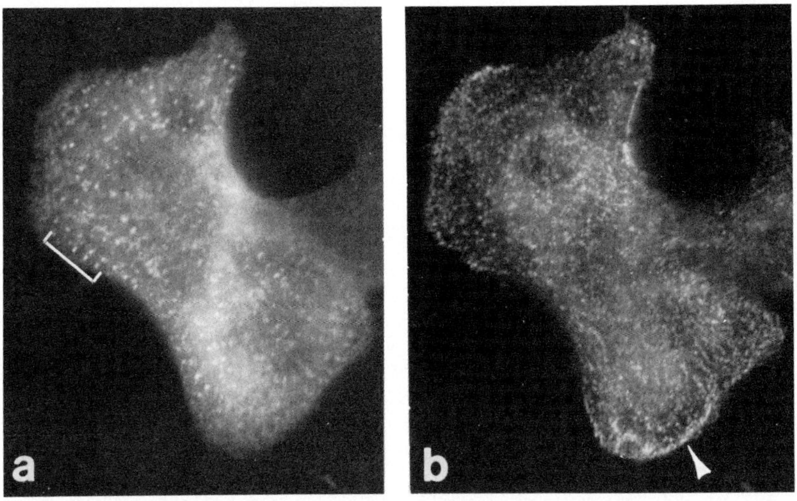

FIG. 2. Double fluorescence immunolabeling of chick embryonic heart cells for (a) zeugmatin and (b) integrin, using the detailed procedures described in the text. The bracket in (a) designates the periodic array of zeugmatin labeling on the Z bands of the myofibrils. The arrowhead in (b) points to a region of the plasma membrane that is intensely labeled for integrin. ×1000.

ods, the main point is to appreciate the discrete double fluorescence labeling results obtainable with the two MAbs using the procedures we have described.

Chick Embryonic Fibroblasts: Vinculin and Talin Double Labeling

The procedures used for double labeling of chick embryonic fibroblasts are closely similar to those described above. MAb 1019 directed to talin is modified with DNP-PI, and MAb XIV B4 specific for vinculin is conjugated with BXNSS. After labeling the fixed, permeabilized fibroblasts with the two modified primary antibodies at a concentration of 50–100 μg/ml and subsequent treatment with the rabbit anti-DNP antibodies, the final stage involves treatment with a mixture of fluorescein-conjugated $F(ab')_2$ fragments of goat anti-rabbit IgG antibodies and rhodamine-conjugated streptavidin. The double labeling results are shown in Fig. 3. As expected from previous results, both talin and vinculin labelings are superimposed in the region of the focal adhesions at the cell periphery, but, in addition, talin labeling is widely dispersed in association with the stress fibers whereas vinculin labeling is less dispersed.

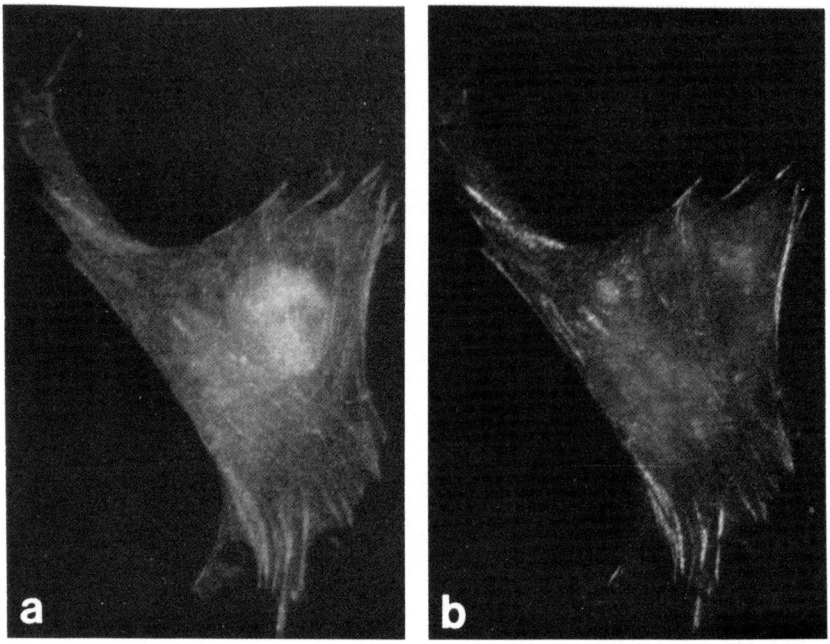

FIG. 3. Double fluorescence immunolabeling of chick embryonic fibroblasts for (a) talin and (b) vinculin, using the detailed procedures described in the text. The intensely labeled streaks at the cell periphery are associated with the focal adhesions made by the ventral cell surface with the substratum. ×1000.

Comments

Note that these procedures, because they involve simultaneous labeling with the two primary antibody reagents, allow equal access to the two antigens under study. On the other hand, double labeling procedures that involve two consecutive stages of primary antibody labeling[11] may result in limited accessibility to the second antigen by virtue of the labeling reagents attached to the first antigen. This may be an important consideration for two antigens such as vinculin and talin which are at least in part in ultrastructural proximity to one another.

Other experiments similar to those illustrated by Figs. 2 and 3 were carried out in which the MAbs involved were alternatively modified by biotinylation and by some of the different reagents described for DNP conjugation. DNP-XI-modified MAb XIV B4 to vinculin yielded sharper labeling results than did the DNP-PI-modified MAb, but we did not extensively investigate this comparison of the two reagents with the other

MAb. With the procedures and experience described in this chapter, we are confident that many pairs of MAbs can be successfully used in double fluorescence labeling, especially if, as discussed above, the effort is made to optimize the modification reactions with individual MAbs.

In addition, although our emphasis in this chapter has been on the use of two MAbs, the same double labeling procedures can be successfully employed with two polyclonal antibodies, as in cases where both antibodies have been elicited in rabbits. Polyclonal antibodies after modification generally exhibit greater solubility and less background labeling than MAbs similarly treated, and they should therefore present even fewer problems in implementing our procedures.

Acknowledgments

The original studies described in this chapter were supported by U.S. Public Health Service Grants GM-15971 and AI-06659. We are grateful to Dr. Abraham Kupfer who prepared MAb 1019 directed to chicken talin. S.J.S. is an American Cancer Society Research Professor.

[38] Detection of Cell Surface Components Using Fluorescent Microspheres

By ARTHUR J. SYTKOWSKI

The use of fluorescent probes and flow cytometry has proved of great value in the identification and physical separation of cells bearing surface determinants of interest.[1-4] Although the introduction of the avidin–biotin system has both enhanced the sensitivity and reduced nonspecific background, the limits of detection of the instrumentation remain in the range of 2000–10,000 surface determinants per cell.

Fluorescent microspheres offer the potential for enhanced sensitivity, since they can be manufactured to possess the fluorescence equivalent of several thousand fluorescein molecules per sphere. Moreover, they can be derivatized, thus providing a means to couple antibodies or other

[1] H. M. Shapiro, *Cytometry* **3,** 227 (1983).
[2] J. C. Cambier and J. G. Monroe, this series, Vol. 103, p. 227.
[3] M. R. Loken and A. M. Stall, *J. Immunol. Methods 50,* R85 (1982).
[4] L. Voet, M. Krefft, H. Mairhofer, and K. L. Williams, *Cytometry* **5,** 26 (1984).

probes to their surfaces covalently. Indeed, the detection of cell surface determinants with antibody-coated microspheres has been reported.[5,6] However, the density of the antigens detected was within the limits obtainable without the use of microsphere amplification.

Unpublished studies in our laboratory have indicated that the number of antibody molecules that can be immobilized directly on derivatized microspheres is limited to 1000–2000 molecules. Moreover, only 10–20% of these immobilized antibodies retain their capacity to bind antigen, presumably because of steric factors. We hypothesized that streptavidin-mediated coating of *biotinylated* fluorescent microspheres (BFM) with either biotinylated antibody or another biotinylated cell surface probe would increase the number of probe molecules immobilized, thus increasing their effective concentration and the probability of binding to the surface determinant. In addition, the biotin–streptavidin bridge should reduce steric hindrance and permit greater retention of biological activity. This approach has proved extremely successful.[7] The procedures described below contain several useful modifications of our original methodology.

Biotinylation of Fluorescent Microspheres

One hundred microliters of fluorescent microspheres (FX Covaspheres, Duke Scientific, Palo Alto, CA), which possess surface primary amino groups, are washed twice with Dulbecco's phosphate-buffered saline (PBS) by centrifugation at 10,000 *g* for 6 min in a 500-μl Eppendorf tube and are sonicated for 5 min. Examination with a fluorescence microscope should reveal discrete microspheres with no aggregates. NHS-LC-Biotin [sulfosuccinimidyl 6-(biotinylamido)hexanoate (Pierce Chemical Co., Rockford, IL), 750 μl of a 1 m*M* solution in PBS] is added to the spheres, and the mixture is rocked gently for 60 min at room temperature. The biotinylated microspheres are washed 3 times with PBS and dialyzed against PBS to remove excess NHS-LC-biotin. The BFM stock preparation ($\sim 1 \times 10^9$ BFM/ml) can be stored for several weeks at 4°. Using ^{125}I-labeled streptavidin binding, we have found that this method yields approximately 100,000 active biotin molecules per sphere. The investigator is encouraged to verify the coupling of NHS-LC-biotin to the amino mi-

[5] J. E. Cupp, J. F. Leary, E. Cernichiari, J. C. S. Wood, and R. A. Doherty, *Cytometry* **5,** 138 (1984).

[6] D. R. Parks, V. M. Bryan, V. T. Oi, and L. A. Herzenberg, *Proc. Natl. Acad. Sci. U.S.A.* **76,** 1962 (1979).

[7] D. M. Wojchowski and A. J. Sytkowski, *Biochim. Biophys. Acta* **857,** 61 (1986).

crospheres in this way. BFM are also available commercially (Diversified Biotech, Newton Centre, MA).

Streptavidin Coating of BFM

The BFM stock is sonicated for 10 min. A convenient volume (100 μl) of BFM is added to a small Eppendorf tube. Streptavidin (10 μl of a 1 mg/ml solution in PBS) is added, and the suspension is rocked gently for 30 min at room temperature. The streptavidin–BFM complexes are separated from unbound streptavidin by gel filtration through Sepharose 4B (0.7 $\times$ 6.0 cm column, Bio-Rad Econocolumn). Glass columns fitted with porous polyethylene disks will permit the passage of particles of up to 30 μm. The spheres are readily visible as 2–3 turbid fractions at the void volume. The stoichiometry of streptavidin molecules per BFM can be estimated by [^{3}H]biotin binding.

Coating of Streptavidin–BFM with Biotinylated Probe

The method of coating streptavidin–BFM with biotinylated probe (antibody, hormone, etc.) is analogous to the procedure for coating BFM with streptavidin described above. For the purpose of this discussion, it is assumed that the investigator has demonstrated that the *biotinylated* probe, when coupled to avidin or streptavidin, retains its capacity to bind to the cell surface determinant of interest. Ideally, the concentration of the biotinylated probe should be severalfold greater than the effective concentration of streptavidin coated on the BFM. However, this cannot always be accomplished. At a density of 100,000 streptavidin molecules per BFM and a BFM concentration of 1 $\times$ 10^9 spheres/ml, the concentration of the bound streptavidin is approximately 1.7 $\times$ 10^{-7} M. The number of BFM per milliliter can be determined by microscopic examination using a hemacytometer. BFM can be concentrated easily by centrifugation and resuspension of the pellet in the desired volume. After any such centrifugation procedure, sonication is recommended.

The biotinylated probe is incubated with streptavidin–BFM with gentle rocking for 30 min at room temperature. Separation of the complete BFM–streptavidin–biotinylated probe construct (BFM probe) from excess unbound biotinylated probe is achieved by gel filtration as described above. The preassembled BFM probes can be stored at 4° in the presence of sodium azide and one or more protease inhibitors (e.g., EDTA, leupeptin). The shelf-life of each unique BFM probe must be determined empirically by the investigator.

Reacting BFM Probes with Target Cells

The target cells and the BFM probes each are resuspended in PBS containing 1% normal rabbit serum (or other suitable source of carrier protein). BFM probes (1×10^8) are mixed with $5–50 \times 10^5$ target cells in one well of a 96-well flat-bottomed tissue culture plate (250 μl final volume). The number of target cells used is inversely proportional to their size. The dish is centrifuged at 150 *g* for 10 min to achieve a monolayer of BFM probes and cells. The monolayer is incubated for 60 min at 4°. Then it is resuspended gently, layered on 3 ml of 40% (v/v) Percoll (in PBS–serum) in a 15-ml conical polystyrene test tube, and centrifuged for 10 min at 250 *g* in a swinging-bucket rotor. The supernatant containing unbound BFM probes at the interface is removed carefully, and the cell pellet is resuspended gently in PBS–serum. This cell suspension is now ready for analysis and/or sorting by flow cytometry. Examination by a combination of light and fluorescence microscopy may also be employed.

Comments

Avidin–biotin technology has gained widespread use in several areas of cell biology and immunology, including the study of cell surface determinants.[8] Although fluorescent microspheres have been available for several years, they have been employed relatively infrequently in such experiments. The biotinylation of fluorescent microspheres has resulted in a versatile reagent. The concentration of probe can approach micromolar, thereby increasing the probability of binding to low-density cell surface molecules (<200 per cell).[7] The signal afforded by the fluorescent microspheres themselves is very high, and background is easily gated out. Biotinylated microspheres can be extremely useful in studies of cell surface antigens and receptors, including the cloning of genes for these proteins by expression.[9–11]

[8] E. A. Bayer and M. Wilchek, *Methods Biochem. Anal.* **26,** 1 (1980).

[9] M. E. Kamarck, J. A. Barbosa, L. Kuhn, P. G. Messer-Peters, L. Shulman, and F. H. Ruddle, *Cytometry* **4,** 99 (1983).

[10] C. P. Stanners, T. Lam, J. W. Chamberlain, S. S. Stewart, and G. B. Price, *Cell* **27,** 211 (1981).

[11] P. Kavathas, V. P. Sukhatme, L. A. Herzenberg, and J. R. Parnes, *Proc. Natl. Acad. Sci. U.S.A.* **81,** 7688 (1984).

[39] Immunohistochemistry

By Su-Ming Hsu

Any substance tagged with biotin can be visualized with an avidin-conjugated or an avidin–biotin-complexed substance. The substance can be an enzyme, a fluorescent dye, electron-dense particles (ferritin, colloidal gold), or a radioactive isotope. Therefore, the avidin–biotin detection technique can be applied to the following: (1) antibody–antigen reactions (immunohistochemistry, immunofluorescence, immunoelectron microscopy, or Western blotting), (2) lectin histochemistry, (3) receptor–ligand reactions, (4) *in situ* hybridizations with biotin-labeled nucleic acid probes, (5) enzyme-linked immunoassays (ELISA), and (6) flow cytometry. The method of avidin–biotin interaction employed is very sensitive because of the extraordinary affinity between avidin and biotin. This chapter focuses on the use of avidin–biotin reagents for histochemical or immunohistochemical staining.

Principles of Avidin–Biotin Techniques

Basically, three avidin–biotin techniques can be used: (1) the labeled avidin–biotin (LAB) technique, (2) the bridged avidin–biotin (BRAB) technique, and (3) the avidin–biotin complex (ABC) technique. In the LAB method, the biotin-labeled substance is visualized directly by means of an avidin conjugate, such as avidin–fluorescein or avidin–peroxidase conjugates. Although the use of an avidin–peroxidase conjugate yields reasonable sensitivity, one of the major drawbacks is that the avidin conjugates usually cause a high background reaction as well as nonspecific staining of some tissue elements, such as the kidney.[1–3]

In the BRAB method, avidin is used as a link between two or more biotin-conjugated molecules, such as biotin-labeled antibody and biotin-labeled peroxidase. This technique is based on the biochemical characteristics of avidin, a substance which has four binding sites for biotin. Theoretically, the BRAB technique may produce a higher sensitivity than the LAB method because of the amplification effected by the biotin–avidin–

[1] S.-M. Hsu, L. Raine, and H. Fanger, *Am. J. Clin Pathol.* **75,** 734 (1981).
[2] S.-M. Hsu, L. Raine, and H. Fanger, *Am. J. Clin. Pathol.* **75,** 816 (1981).
[3] S.-M. Hsu, L. Raine, and H. Fanger, *J. Histochem. Cytochem.* **29,** 577 (1981).

biotin interaction. The avidin molecules that bind to biotin-labeled antibody can potentially react with an additional 2–3 biotin–peroxidase molecules. In our experience, however, the BRAB technique does not produce the maximal sensitivity that is expected on theoretical grounds, perhaps because of the steric hindrance between molecules.[3]

To enhance the reaction sensitivity further, we have developed a technique in which we use an ABC.[1-3] The same idea of enhancing sensitivity has also been recently applied to the formation of avidin–biotin–alkaline phosphatase and avidin–biotin–glucose oxidase complexes. The formation of an ABC is based on the following biochemical characteristics of avidin and of biotin conjugates: (1) As mentioned above, avidin has four binding sites for biotin conjugates. Thus, avidin can form a link with two or more biotin conjugates (e.g., biotin-labeled peroxidase). (2) Several biotin molecules can be coupled to one peroxidase molecule. Thus, each biotin-peroxidase conjugate can react with several avidin molecules. (3) With an optimal ratio of avidin to biotin–peroxidase conjugate, the formation of a complex containing multiple peroxidase molecules is possible. (4) The sensitivity of the detecting system is increased when the number of peroxidase molecules present is increased.

The optimal ratio of avidin to biotin–peroxidase conjugate is determined empirically. The sensitivity is also affected greatly by the degree of biotinylation of peroxidase. A reliable ABC can be obtained from a commercial source (Vector Laboratories, Burlingame, CA). Alternatively, the ABC can be prepared in the laboratory of the user according to a previously published method.[1-3] In general, the optimal amounts of avidin and biotin–peroxidase conjugate are about 10 and 2.5 μg/ml, respectively. It should be noted that the availability of free biotin-binding sites in the complex is created by incubation of excessive amounts of avidin with respect to the biotin–peroxidase molecules.

ABC Immunohistochemical Staining Method

The following is a step-by-step description of the method for detecting antigen in tissue sections with our ABC technique.

1. Deparaffinize and rehydrate sections routinely, if paraffin sections are used. Fix the sections in cold acetone for 5 min, if frozen sections are used.
2. Wash the slides in Tris-buffered saline (TBS) or phosphate-buffered saline for 5 min.
3. Incubate with 1% nonimmune or preimmune serum in TBS. The serum used should preferably be derived from the same animal species as that used to prepare the secondary antibody.
4. Incubate with primary antibody (e.g., mouse monoclonal antibody

against human IgG) at a proper dilution (usually 1–5 μg/ml) for 30 min–2 hr, depending on the affinity of the antibody.
5. Wash in TBS for 5 min.
6. Incubate with biotin-labeled secondary antibody (e.g., biotin-labeled horse anti-mouse Ig) at a dilution of 1 to 200 for 30 min.
7. Wash in TBS for 5 min.
8. Incubate the tissue sections with the ABC reagent for 30 min. The ABC reagent can be obtained from Vector Laboratories and prepared as instructed by the manufacturer.
9. Wash in TBS for 5 min.
10. Develop in a substrate such as diaminobenzidine–hydrogen peroxide. The substrate is prepared as follows: dissolve 100 mg of diaminobenzidine-HCl (Sigma Chemical Co., St. Louis, MO) in 200 ml of TBS; stir and filter; add 1 ml of 8% nickel chloride (aqueous solution) and 25 μl of 3% hydrogen peroxide.[4]
11. Terminate the reaction by washing the sections in TBS.

Advantages of Avidin–Biotin Systems

The conjugation of biotin to a protein (e.g., immunoglobulin) is a relatively mild chemical reaction; thus, biotinylation usually does not affect the biological or physiological affinity of the protein. Since biotin is a small molecule, the biotin-labeled substance maintains a size comparable to original one. Because the molecular size of the antibody is very important for its good penetration into tissues (e.g., in immunoelectron microscopy), it may be best to use the biotin-labeled Fab fragments of an antibody.

Any substance, when labeled with biotin, can be detected with an avidin-conjugated or -complexed substance, such as avidin–fluorescein, avidin–ferritin, or ABC. In other systems, such as methods using fluorescein-labeled antibody or peroxidase-labeled antibody, the application is restricted. The avidin–biotin system offers a greater flexibility than do other enzyme-conjugated detecting systems.

Biotin-labeled immunoglobulins and lectins are very stable.[5,6] They maintain the same reactivities for up to 5 years when stored at 4°. The same is true for avidin and biotin-conjugated peroxidase.

Selection of Tissue Sections

Although the ABC method is very sensitive for the detection of antigens in tissue sections or cytospin smears, it may be less satisfactory if

[4] S.-M. Hsu and E. Soban, *J. Histochem. Cytochem.* **30,** 1079 (1982).
[5] S.-M. Hsu and L. Raine, *J. Histochem. Cytochem.* **29,** 1349 (1981).
[6] S.-M. Hsu and L. Raine, *J. Histochem. Cytochem.* **30,** 157 (1982).

antigenic epitopes are destroyed during tissue processing. Precautions should be taken to ensure optimal staining according to the following criteria: (1) the tissue sections should provide sufficient cytological detail, allowing the determination of the type of cells stained as well as the distribution and relationship of these cells to nonstained cells; and (2) the antigenic determinants should be maximally preserved to guarantee sensitivity of detection. These two aspects are generally not compatible; the tissue-processing technique that is suitable for morphological study is generally detrimental to antigen preservation, and vice versa. None of the available tissue preparation methods meet both criteria.

Tissue fixed in B5, Bouin, or formalin and embedded in paraffin provides sufficient cytological and histological detail; however, antigens tend to be destroyed or partially inhibited, and this results in a high rate of false-negative staining.[7–11] Nevertheless, paraffin sections are usually suitable for the detection of abundant cytoplasmic antigens or carbohydrate antigens. Examples include immunoglobulin in plasma cells, lysosomal enzymes in macrophages, and lacto-*N*-fucopentoase (heptan X) in granulocytes.[12] An additional advantage of the paraffin-embedding technique is that the paraffin blocks can be stored at room temperature, thus eliminating their costly maintenance in a low-temperature (liquid nitrogen or $-70°$) freezer. Very few staining methods in which monoclonal antibodies (MAbs) are used can be applied to formalin-fixed paraffin sections, because the single or few antigenic epitopes (except for carbohydrate moieties) are likely to be destroyed.[9]

Frozen sections fixed in acetone are the most suitable for staining with MAbs.[7,13] Most antigen epitopes are very fragile and can be easily destroyed by improper tissue handling or by temperature changes, tissue fixation, or storage. Fixatives such as methanol and ethanol should be avoided. The frozen tissue blocks can be stored in a liquid nitrogen freezer for an indefinite period. After sectioning, they can be stored at 4° for up to 2 weeks, but they should be stained as soon as possible (within 2–3 days). Whereas virtually all MAbs can be applied successfully to frozen sections, very few MAbs directed against carbohydrate epitopes can be used in such paraffin sections.[12]

Cytospin smears prepared from cell suspensions (pleural fluid, tissue

[7] S.-M. Hsu, J. Cossman, and E. S. Jaffe, *Am. J. Clin. Pathol.* **80,** 21 (1983).
[8] S.-M. Hsu, J. Cossman, and E. S. Jaffe, *Am. J. Clin. Pathol.* **80,** 429 (1983).
[9] S.-M. Hsu, H.-Z. Zhang, and E. S. Jaffe, *Am. J. Clin. Pathol.* **80,** 415 (1983).
[10] S.-M. Hsu, *J. Histochem. Cytochem.* **31,** 258 (1983).
[11] S.-M. Hsu, H.-Z. Zhang, and E. S. Jaffe, *Hybridoma* **2,** 403 (1983).
[12] S.-M. Hsu and E. S. Jaffe, *Am. J. Clin. Pathol.* **82,** 29 (1984).
[13] S.-M. Hsu and E. S. Jaffe, *Am. J. Pathol.* **114,** 387 (1984).

culture medium, bone marrow aspirates, etc.) can be immunostained as frozen sections. Acetone is the fixative of choice.

Use of Biotin-Labeled Primary Antibody

For practical purposes, the method of choice for immunohistological staining is the three-step ABC method in which a biotin-labeled secondary antibody is applied, followed by the ABC. The three-step method does not require direct conjugation of primary antibodies, which sometimes is not feasible because of the small quantities of antibody available. The sensitivity of the three-step ABC method is enhanced by amplification through primary antibody–secondary antibody reactions. However, the secondary antibody may introduce some degree of nonspecificity or an unwanted reaction.[8] For example, the three-step ABC method is not suitable for staining of mouse tissue with mouse monoclonal antibodies because the secondary antibody (e.g., biotin-labeled rabbit anti-mouse Ig) binds not only to the added monoclonal antibody but also to the interstitial or cytoplasmic Ig in tissue sections.

One can overcome this problem by using the biotin-labeled MAb and then ABC, eliminating the need for a secondary antibody. This two-step ABC method is most useful in the detection of immunoglobulin in tissue sections, as well as in a system in which the primary antibody and the tissue sections have the same source (e.g., mouse MAb in mouse tissues, human autoimmune antibody in human tissues). The two-step ABC method is less sensitive than the three-step method but yields much lower background staining.[8]

Lectin Histochemistry

By using a biotin-labeled lectin, one can apply the ABC method to the detection of the lectin receptor (carbohydrate moiety) associated with glycoprotein on the cell membrane or in the cytoplasm. Since carbohydrates are generally resistant to destruction by fixation, ABC–lectin histochemical staining can be applied to formalin- or B5-fixed, paraffin-embedded tissue sections. However, staining results achieved with paraffin sections may be different from those with frozen sections.[10] The staining procedure for the ABC–lectin histochemical method is the same as that for the two-step ABC method.[14]

Biotin-labeled lectin can be injected *in vivo* or applied to cell cultures. The fate of biotin-labeled lectin can be used, for example, for studies of retrograde axonal transport or of modulation of cell-surface glycoprotein.

[14] S.-M. Hsu and H. J. Ree, *J. Histochem. Cytochem.* **31,** 538 (1983).

Background and Nonspecific Staining

Biotin is a coenzyme of decarboxylase. It is present in many tissues, such as liver, pancreas, kidney, and intestine. In fixed tissues, the biotin-containing proteins are likely to be destroyed by the formalin-fixation or paraffin-embedding procedures; thus, the use of ABC does not produce unwanted binding to tissues or cells. In frozen sections of liver, kidney, or pancreas, the sections must be pretreated with avidin (25 μg/ml) for 15 min before biotin is added at a saturated concentration for 15 min, with an interval for washing in buffer. This pretreatment with avidin can successfully block endogenous biotin molecules. Since avidin is a tetramer with four binding sites, the sections require an additional biotin treatment for saturation of all free biotin-binding sites associated with avidin, so that the added avidin will not react with biotin-labeled antibody or ABC. Biotin, when bound to avidin, will not react with a second avidin molecule.[15]

Avidin–biotin complexes can react with mast cells in both fixed and frozen sections. The reaction cannot be prevented by avidin–biotin pretreatment. One can prepare the ABC in a high-pH (8.5) solution or in high-ionic strength buffer such as 0.5 *M* NaCl. High-pH or high-ionic strength solutions will not affect the ABC–biotin reaction but can prevent or minimize unwanted binding to mast cells.

In test systems such as ELISAs or Western blots, the background staining of ABC can be minimized effectively by preincubation of the cellulose paper or plastic wells with 0.01% Triton X.

Several types of cells, including red blood cells, neutrophils, and eosinophils, contain peroxidase. One has to distinguish the staining due to added peroxidase (i.e., in the ABC) from that due to endogenous peroxidase. If the presence of an endogenous peroxidase reaction does not affect the accuracy of interpretation, blocking of endogenous peroxidase is not necessary. In paraffin sections, the endogenous peroxidase activity can be blocked by pretreatment of the sections with 3% hydrogen peroxide in absolute methanol for 30 min. This treatment does not affect the preservation of antigens, if such antigens remain intact throughout the fixation and embedding processes.

In frozen sections, which are most commonly used for MAbs, blocking with H_2O_2–methanol can inhibit the antigen and thus decrease the sensitivity of detection. Nevertheless, the endogenous peroxidase reaction in eosinophils or granulocytes does not affect the interpretation of the results in most antibody–antigen reactions. Thus, blocking is unnecessary. If the study involves the detection of antigen in neutrophils/eosino-

[15] S.-M. Hsu and L. Raine, *Am. J. Clin. Pathol.* **77,** 775 (1982).

phils, the use of avidin–fluorescein or avidin–rhodamine conjugates is recommended.

Summary

The use of the ABC technique, in which an avidin–biotin interaction is used for immunohistochemical studies, is a simple and straightforward procedure. All required reagents are commercially available. The main advantages of this technique are its sensitivity, specificity, and flexibility. It is crucial for the success of this staining procedure that the antibodies, the type of tissues, and the fixatives that are used be selected properly. The method can easily be adapted for electron microscopy, flow cytometry, *in situ* hybridization, Western blotting, and ELISA.[16]

[16] P.-L. Hsu, S.-M. Hsu, and E. Appella, *Gene Anal. Techn.* **2,** 30 (1985).

[40] Immunocytochemical Detection of Human T- and B-Cell Antigens

By ROGER A. WARNKE and LAWRENCE M. WEISS

Background

In the past, the identification of human T cells and B cells relied on sheep red blood cell rosette assays applied to cell suspensions and frozen tissue sections for T cells as well as on the identification of surface immunoglobulins on B cells in suspension through the use of polyclonal antibodies. When investigators became interested in identifying B cells in normal and neoplastic lymphoid tissue sections, the application of polyclonal antibodies to immunoglobulins was complicated not only by their lack of specificity in some instances but by the large amounts of immunoglobulin that may be present in tissue fluids and bound to connective tissue. In addition, it proved very difficult to generate polyclonal antibodies monospecific to T cells in part because of the large number of nonlineage-specific cell surface antigens present on immunizing cell populations.

The somatic cell hybridization technique, which results in stable hybridomas each secreting pure antibody of identical specificity and affin-

TABLE I
SELECTED MONOCLONAL ANTIBODIES REACTIVE WITH HUMAN T- AND B-CELL ANTIGENS GROUPED BY PREDOMINANT REACTIVITY PATTERN

Antibody	Cluster designation	Source
Pan T cell		
Leu 1	5	Becton-Dickinson (BD)
Leu 4	3	BD
Leu 5	2	BD
Leu 9	7	BD
BF1	—	M. B. Brenner *et al.*,[a] T-Cell Sciences
T-cell subset		
Leu 2	8	BD
Leu 3	4	BD
Leu 6	1	BD
Pan B cell		
Leu 12	19	BD
B1	20	Coulter
T015	22	D. Y. Mason, DAKO
MB1	37	M. P. Link *et al.*[b]
Immunoglobulins		
κ, λ	—	BD
μ, δ	—	R. Levy, Stanford
γ	—	Coulter
α	—	Bethesda Research Laboratories

[a] M. B. Brenner, J. McLean, H. Scheft, R. A. Warnke, N. Jones, and J. L. Strominger, *J. Immunol.* **138,** 1502 (1987).

[b] M. P. Link, J. Bindl, T. C. Meeker, C. Carswell, C. A. Doss, R. A. Warnke, and R. Levy, *J. Immunol.* **137,** 3013 (1986).

ity,[1] has led to the generation of numerous highly specific probes that can be used to identify a wide variety of differentiation antigens, not only on B cells and T cells (Table I) but on a wide range of other cell types. For example, use of monoclonal antibodies that recognize sheep erythrocyte receptors on T cells and different types of complement receptors on B cells and other cell types has largely replaced the rosette assays. Thus, the exquisite sensitivity and specificity of monoclonal antibodies make them ideally suited for localizing T cells, B cells, and other cell types in frozen tissue sections of hyperplastic and neoplastic disorders. Panels of monoclonal antibodies can now be used to investigate cellular interactions in response to antigenic stimulation, to investigate the immunoarchi-

[1] G. Kohler and C. Milstein *Nature* (*London*) **256,** 495 (1975).

tecture of reactions to viruses such as the human immunodeficiency virus (HIV), and to provide immunophenotypic criteria for the diagnosis of B-cell and T-cell lymphomas.[2,3]

However, the cell surface and cytoplasmic antigens which now can be detected by a variety of monoclonal antibodies are sometimes present in normal or tumor cell populations in insufficient amounts for detection with traditional indirect antibody methods. In addition, small amounts of antigen are generally easier to detect with fluorescent methods and flow cytometry analysis than with tissue section methods employing enzyme labels and light microscopic analysis. Thus, increasing the sensitivity for detecting monoclonal antibody binding has become increasingly important. The exquisite sensitivity of avidin–biotin systems has proved extremely useful in this regard. We describe a method using monoclonal antibodies and frozen tissue sections or cytospin preparations with a detection system employing anti-mouse antibodies coupled to biotin and peroxidase coupled to avidin.[4]

Method

Equipment

- Cryostat and chucks
- Freezer (−70°)
- Standard light microscope
- Coplin jars

Materials

- Plastic capsules (BEEM) (Ted Pella, Inc., Redding, CA)
- Embedding medium (OCT) (Miles Scientific, Naperville, IL)
- Freezing bath: isopentane and dry ice in insulated flask or thermos
- Glass slides
- Acetone, methanol, ethanol, xylene
- Desiccant (Drierite, W. A. Hammon Drierite Co., Xenia, OH)
- First-stage antibodies (Table I)
- Second-stage antibodies coupled to biotin
  - Biotinylated horse anti-mouse (Vector Laboratories, Burlingame, CA)

[2] G. S. Wood, *in* "AIDS and Other Manifestations of HTLV III Infection" (G. P. Wormser, ed.), p. 867. Noyes, Park Ridge, New Jersey, 1987.

[3] L. J. Picker, L. M. Weiss, L. J. Medeiros, G. S. Wood, and R. A. Warnke, *Am. J. Pathol.* **128,** 181 (1987).

[4] R. Warnke and R. Levy, *J. Histochem. Cytochem.* **28,** 771 (1980).

Biotinylated goat anti-mouse SP (Jackson Laboratories, West Grove, PA)

Avidin coupled to peroxidase

Avidin–peroxidase (Vector)

Streptavidin–peroxidase (Jackson)

3,3-Diaminobenzidene tetrahydrochloride (DAB, Sigma Chemical Co., St. Louis, MO)

H_2O_2

$CuSO_4$ (0.5% in normal saline)

Mounting medium, xylene-based

Methylene blue (2% in distilled water)

Phosphate-buffered saline (PBS, 25× stock solution): dissolve 188 g K_2HPO_4, 33 g NaH_2Po_4, and 180 g NaCl in 1 liter of distilled water

PBS working solution: to 960 ml of distilled water, add 40 ml of 25× PBS and 20 mg of thimerosal (Sigma)

Tissues. Human tissue is obtained at the time of surgical biopsy or resection. The tissue is kept moist with normal saline at all times to avoid drying of surfaces. If immediate freezing of tissue is not feasible, a delay in freezing may be acceptable since many human B- and T-lymphoid antigens are easily detectable in tissue shipped or stored in cold saline for up to 48 hr after removal from the patient.[5] Such delays in the transport of unfrozen specimens may be preferable to improper freezing or partial thawing in transport. We do not use ammonium sulfate-based transport media for lymphoid organs as it is our experience that some antigens are not well preserved and the morphology may be poorly preserved.[6] Tissue slices (3–5 mm thick, 5–10 mm on a side), are generally frozen in plastic capsules surrounded by OCT medium by immersion in an isopentane–dry ice bath for at least 1 min. Capsules are then stored frozen at −70° until used. We use plastic capsules for ease in storage and because they avoid desiccation of the specimens, but tissues can also be frozen directly on a chuck surrounded by OCT or frozen with OCT on foil and wrapped in foil or kept in a plastic container to avoid desiccation.

Cutting and Fixation of Frozen Sections[7]

When cutting the plastic capsule to transfer the frozen tissue to a cryostat chuck and when securing it to the chuck, it is critical that the tissue remains frozen. Four- to five-micron sections are cut on a cryostat

[5] R. A. Warnke, J. Bindl, and R. Doggett, *Histochem. J.* **15,** 637 (1983).

[6] M. J. Borowitz, B. P. Croker, and J. Burchette, *J. Histochem. Cytochem.* **30,** 171 (1982).

[7] R. V. Rouse and R. A. Warnke, *in* "Handbook of Experimental Immunology" (D. M. Weir, ed.), Vol. 4, p. 116.1. Blackwell, Oxford, 1986.

and lifted from the blade with room temperature glass slides. Sections are air–dried for 1–5 min or more, prior to prefixation in acetone (5 to 10 dips). After drying, the sections are stored at $-20°$ for up to 2 weeks. Sections can be stored for longer periods in the presence of desiccant. Immediately before staining, sections are transferred from the freezer to 4° acetone for 10 min and allowed to air dry prior to staining. To preserve the tissue block for reuse, the cut surface may be covered with OCT and immediately immersed in the freezing bath. The block can then be removed from the chuck by briefly immersing the shank in hot water until the frozen block slides off. The block of tissue can then be wrapped in foil and stored frozen.

In an alternative method of fixation and storage, the frozen sections are cut, air dried, immersed in acetone (at 4°) for 10 min, and stored in a container with desiccant. After storage of such slides at $-20°$, they can be brought to room temperature for staining without further fixation in acetone.

Avidin–Biotin–Peroxidase Staining Procedure

Frozen sections or cytospins, which have been stored at $-20°$, are removed and fixed immediately in cold acetone at 4° for 10 min prior to drying at room temperature. First-stage mouse monoclonal antibody is applied at appropriate dilution (usually 0.5–10 μg/ml) in a volume sufficient to cover the slide without any drying out (usually 20–50 μl) by means of a pipet or dropper bottle (Wheaton). Incubate for 30 min in a humidified chamber at room temperature, and wash with PBS.

Biotinylated second-stage anti-mouse antibodies are then applied at appropriate dilution (usually 10–20 μg/ml) followed by incubation for 30 min as described for the first-stage antibody. The preparation is washed again with PBS, and incubated with avidin–peroxidase for 30 min. The slide is washed twice with PBS, incubated with DAB (1–3 mg/ml) in 0.3% H_2O_2 in PBS (freshly made) for 5 min, washed twice with PBS, and rinsed by dipping 10 times in distilled water. The slide is then incubated with $CuSO_4$ solution for 5 min, rinsed with PBS, dipped in distilled water several times, and counterstained for 10 min with methylene blue. The stained preparation is rinsed with 100% alcohol, then xylene, and a coverslip with xylene-based mounting medium is placed on the tissue sample.

Variations

Paraffin- or Plastic-Embedded Tissue

Monoclonal and polyclonal antibodies that detect fixation- and processing (denaturation)-resistant antigenic determinants have been described. The previously described method of staining frozen sections can be utilized for paraffin-embedded tissue after removal of the paraffin in xylene and rehydration in decreasing concentrations of ethanol. In addition, fixation and processing protocols that preserve certain T- and B-cell antigens in plastic-embedded tissue allow the detection of such antigens after enzyme pretreatment.[8]

Biotin and Avidin Reagents in Coplin Jars[9]

In order to stain a large number of sections simultaneously in the minimal amount of technical time, the second- and third-stage reagents can be applied in Coplin jars. Biotinylated antibodies and avidin-conjugated horseradish peroxidase are diluted in appropriate volumes of PBS with the addition of 5% (by volume) normal human serum and 0.2% thimerosal as a preservative. Both staining and storage of the Coplin jars occur at 4°. Incubation times are determined by a control and range from 30 to 50 min. A major application of Coplin jar detection is for use in intensive screening of hybridomas from one or more fusions.

Avidin–Biotin Complexes

The avidin–biotin complex technique or multiple applications of biotin and avidin detection reagents may be used to enhance the sensitivity of antigen detection.[10] Further modifications requiring availability of the antigen in a pure form have also been devised.[10,11] Some studies claiming superior sensitivity of a particular detection system have not controlled for all the variables in the comparisons. In practice, a variety of three-layer detection methods are suitable for most applications, and the more time-consuming and technically difficult multilayered approaches are seldom necessary.

Advantages and Disadvantages of Method

The detection of T cells and B cells in tissue sections allows simple collection and storage of frozen samples with little initial processing. Such stored tissue samples can be used multiple times. More importantly, rare cells or cells defined by location, such as T cells within B follicles, can be identified and studied (Fig. 1). Their relation to nonsuspendable cell types such as dendritic reticulum cells can also be studied. Further, in contrast

[8] J. H. Beckstead, *J. Histochem. Cytochem.* **33,** 954 (1985).

[9] J. M. Bindl and R. A. Warnke, *Am. J. Clin. Pathol.* **85,** 490 (1986).

[10] W. W. Hancock and R. C. Atkins, this series, Vol. 121, p. 828.

[11] S. M. Hsu and H. J. Ree, *Am. J. Clin. Pathol.* **74,** 32 (1980).

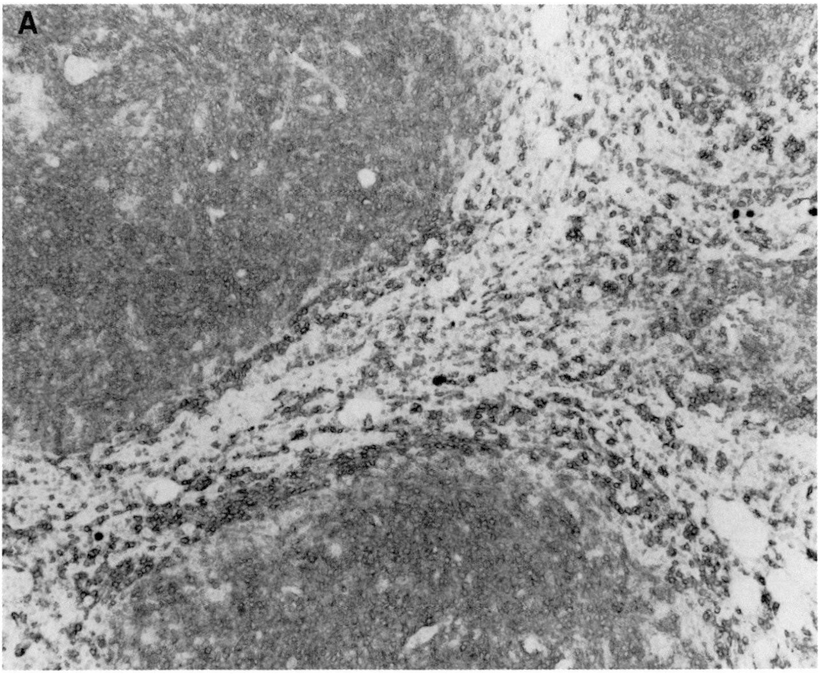

FIG. 1. Follicular lymphoma. (A) The T015 antibody stain labels the neoplastic B cells located predominantly within follicles. (B) The anti-Leu 1 antibody stain labels T cells located predominantly between follicles but also those scattered within the neoplastic follicles. Avidin–biotin–peroxidase detection, frozen sections. Magnification: ×119.

to cell suspension studies, the cytoplasm and nucleus of cells, as well as the cell membranes, are available for staining. Standard histological or more specialized enzymatic counterstains may be employed to identify the cell types stained by immunological means.

Quantification of stained cells or of antigen density is more readily accomplished by cell suspension staining methods generally in combination with automated cell sorters. Quantification in tissue sections may be affected by variations in the thickness of sections and may be complicated by nonspecific staining of connective tissue. In addition, endogenous peroxidase in eosinophils and mast cells or pseudoperoxidase in erythrocytes may complicate quantification. Nevertheless, reproducible quantification methods have been developed for stained cells in frozen tissue sections.[12] Although endogenous peroxidase may be blocked in tissue

[12] C. F. Garcia, L. M. Weiss, J. Lowder, C. Komoroske, M. P. Link, R. Levy, and R. A. Warnke, *Am. J. Clin. Pathol.* **87,** 470 (1987).

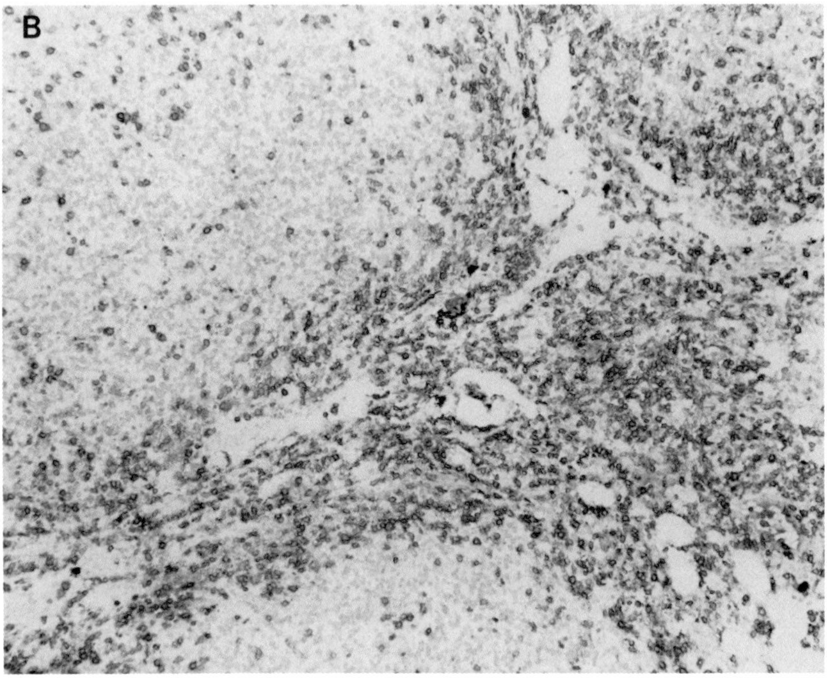

FIG. 1. (*continued*)

sections, some T- and B-cell antigens are labile to such blocking procedures. Furthermore, it is not possible to study capping or synthesis of antigens or to perform functional studies on cells in tissue sections.

[41] Ultrastructural Localization of Retinal Photoreceptor Proteins

By DAVID S. PAPERMASTER, BARBARA G. SCHNEIDER, and IZHAK NIR

Since their introduction in the mid-1970s,[1-3] avidin–ferritin (AvF) conjugates have proved to be extremely useful in immunocytochemical stud-

[1] H. Heitzmann and F. M. Richards, *Proc. Natl. Acad. Sci. U.S.A.* **71,** 3537 (1974).
[2] E. A. Bayer, E. Skutelsky, D. Wynne, and M. Wilchek, *J. Histochem. Cytochem.* **24,** 922 (1976).
[3] E. A. Bayer, M. Wilchek, and E. Skutelsky, *FEBS Lett.* **68,** 240 (1976).

ies of the distribution and biosynthesis of photoreceptor proteins. More recently, preparations of streptavidin–gold[4] have demonstrated certain advantages, especially in the capacity to select freely the diameter of the dense particle for localization of biotinyl conjugates by electron microscopy.[5] We have previously described procedures for localization of antibodies to opsin in tissues embedded in glutaraldehyde cross-linked bovine serum albumin (BSA). These procedures included descriptions of the use of avidin–ferritin with a first-stage biotinyl antibody or a three-stage technique consisting of a primary antibody followed by biotinyl secondary antibody and by AvF.[6] Since then, new embedding media suitable for electron microscopic immunocytochemistry have been introduced that are more readily sectioned and provide greater tissue contrast. The development of streptavidin–gold (SaG) has greatly simplified the techniques of immunocytochemical localization at high resolution.[7] In this chapter, therefore, we describe some of these advances that are of practical importance.

The AvF and SaG conjugates can be used in several ways. Either one is suitable for direct visualization of bound biotinyl antibodies. Amplification of the signal from the binding of an antibody or detection of an antibody whose reactivity is destroyed by biotinylation is readily achieved by using a series of multiplier stages. For example, the second-stage reagent can be a biotinyl antiantibody followed by a third stage of AvF[8,9] or SaG. A simple four-stage technique is also possible: (1) first-stage antibody, (2) biotinyl antiantibody, (3) avidin or streptavidin, and (4) biotinylferritin. Although multistage techniques would appear to be cumbersome, the rapidity of the reaction of streptavidin or avidin with biotinyl conjugates and the absence of nonspecific background binding make such multistage labeling procedures simple. These reagents can be used for surface preembedding immunocytochemistry of antigens exposed on the exterior domains of cells, followed by embedding in conventional media (Fig. 1A), or they may be used in postembedding protocols for detection of antigens revealed by thin sectioning of tissues embedded in appropriate hydrophilic media. Furthermore, streptavidin–fluorescein

[4] A. S. Polans, L. G. Altman, and D. S. Papermaster, *J. Histochem. Cytochem.* **34,** 659 (1986).

[5] In contrast, avidin–gold conjugates are apparently unsuitable unless they are used at high pH because of their highly cationic charge and nonspecific binding.

[6] B. G. Schneider and D. S. Papermaster, this Series, Vol. 96, p. 485.

[7] C. Bonnard, D. S. Papermaster, and J. P. Kraehenbuhl, *in* "Immunolabelling for Electron Microscopy" (J. M. Polak and I. M. Varndell, eds.), p. 95. Elsevier, New York, 1984.

[8] F. J. Roll, J. A. Madri, J. Albert, and H. Furthmayr, *J. Cell Biol.* **85,** 597 (1980).

[9] D. S. Papermaster, P. Reilly, and B. G. Schneider, *Vision Res.* **22,** 1417 (1982).

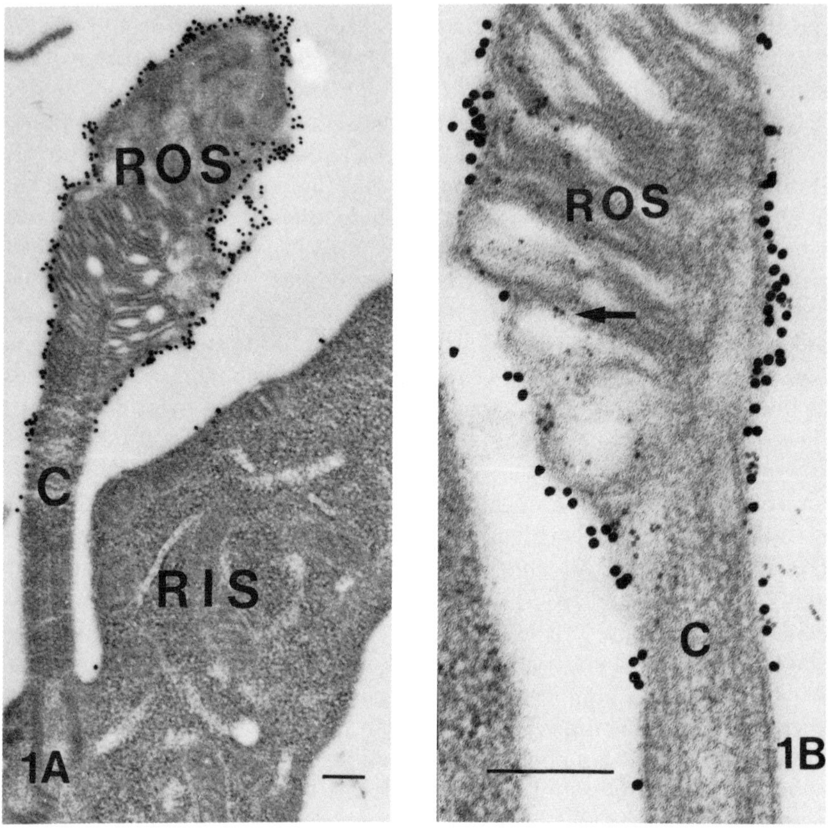

FIG. 1. Developing photoreceptors of a 10-day-old C57/B16 mouse labeled with biotinyl sheep antiopsin followed by streptavidin–gold (15 nm) using the preembedding protocol. (A) Bound antibody is detected on the developing rod outer segment (ROS) plasma membrane. The interior of the ROS is unlabeled. Only scant labeling is detected on the rod inner segment (RIS) plasma membrane. Tissue was embedded in LR Gold after preembedding labeling. Magnification: ×29,000; bar, 0.2 μm. C, Cilium. (B) Preembedding and postembedding reactions may be combined to illustrate penetrability of cells to biotinyl probes. During early stages of mammalian photoreceptor development, some disks remain open to the extracellular compartment rather than fully enclosed by the plasma membrane of the outer segment. After preembedding, detection of bound antibody with streptavidin–gold (15 nm) was performed as in (A): tissue was embedded in LR Gold, and thin sections were labeled with a second probe to detect biotinyl antibodies that had penetrated inside the open disks. In this case we used avidin–ferritin conjugates (arrow), which are readily distinguished from the larger SaG (15 nm) particles. The postembedding reaction revealed the biotinyl antibodies inside open disks. This result illustrates the usefulness of biotinyl antibodies and postembedding detection as a way to investigate domains accessible only to small probes such as the biotinyl antibody. Such approaches may also be useful in studies of endocytosis using other biotinyl ligands. Magnification: ×87,500; bar, 0.2 μm.

or –rhodamine conjugates are available for fluorescence microscopy or cell sorting to evaluate reactions prior to electron microscopy.

Finally, some biological problems may be readily studied by preincubation of tissues with biotinyl conjugates followed by detection of the bound or internalized biotinyl conjugate by a postembedding protocol on thin section surfaces. This latter technique[10] is highly dependent on the choice of embedding media. To date, only aldehyde-cross-linked BSA or methacrylate-derived hydrophilic plastic resins (e.g., LR Gold) have proved useful for this approach (Fig. 1B). This approach should have potential for studies of endocytosis, since biotinyl conjugation of a protein to be endocytosed may trivially alter uptake by its receptor and internalization. Such studies, of course, should be conducted in the presence of controls, including the presence of excess free biotin to avoid inadvertent binding through biotin receptors.

In general, many antigens of the retina have been readily studied by electron microscopy (EM) immunocytochemistry.[6,11] For example, antibodies to the N-terminal domain of opsin reacted in a surprising fashion. Ten different polyclonal and monoclonal antibodies reacted readily with exodomains of the N terminus of opsin exposed on the plasma membrane of fixed or detergent-treated photoreceptors. Native, unfixed rod membranes were unreactive, however, despite the accessibility of the oligosaccharides of opsin to lectins (these are located at positions 2 and 15 in the N-terminal sequence of rhodopsin). These results suggest that the polypeptide backbone of this portion of the molecule is sterically inaccessible unless the membrane is perturbed chemically.[4] Finally, antibodies to opsin have proved useful for studies of retinas from rats and mice with inherited retinal degenerations. These studies have revealed the altered distribution of opsin during early stages of photoreceptor development and subsequent stages of photoreceptor degeneration.[12–14]

Preparation of Reagents

Details of the preparation of antibodies and uses of AvF have been described previously.[6] The length of the spacer arm linking biotin to the conjugate is often an important variable. Studies of the contribution of the

[10] I. Nir, B. G. Schneider, and D. S. Papermaster, *J. Histochem. Cytochem.* **32,** 643 (1984).

[11] B. G. Schneider, D. S. Papermaster, G. I. Liou, S.-L. Fong, and C. D. Bridges, *Invest. Ophthalmol. Visual Sci.* **27,** 679 (1986).

[12] I. Nir and D. S. Papermaster, *Invest. Ophthalmol. Visual Sci.* **27,** 836 (1986).

[13] I. Nir, G. Sagie, and D. S. Papermaster, *Invest. Ophthalmol. Visual Sci.* **28,** 62 (1987).

[14] I. Nir, D. Cohen, and D. S. Papermaster, *J. Cell Biol.* **98,** 1788 (1984).

length of the biotinyl linker have indicated that the longer spacer arm (biotinyl-ε-aminocaproyl-*N*-hydroxysuccinimide) has a greatly increased rate of reaction when compared to the shorter, conventional biotinyl-*N*-hydroxysuccinimide.[7]

Preparation of streptavidin–gold has also been described in detail.[7] For preparation of 10-nm streptavidin–gold conjugates, colloidal gold particles are prepared according to the procedure of Frens.[15] All glassware is siliconized to eliminate adsorption of gold colloid to the walls. To 100 ml of 0.01% $HAuCl_4$, 3 ml of 1% sodium citrate solution is added; after being boiled for 30 min, the burgundy-colored solution is cooled to 4°. Streptavidin–gold complexes are prepared using principles described by Geoghegan and Ackerman[16] to evaluate the appropriate pH and concentration of streptavidin for optimal binding and stabilization of the gold colloid. To 20 ml of gold colloid solution, 200 μl of 1.0 *M* $NaHCO_3$ and 0.5 ml streptavidin (1 mg/ml) in 1 m*M* phosphate buffer (pH 7.4) are added, and the solution is stirred for 10 min at room temperature. A small aliquot is tested for stability by addition of an equal volume of 1 *M* NaCl. Unstable particles form a dark blue precipitate. A solution (200 μl) containing 2% polyethylene glycol 6000 (Carbowax) is added, and the mixture is centrifuged at 40,000 rpm (SW 40 Beckman rotor, 30 min, 4°) through a 37.5% (w/w) sucrose cushion in 0.1 *M* phosphate buffer (pH 7.4) containing 0.02% polyethylene glycol. The resulting pellet is resuspended in 0.1 *M* phosphate buffer (pH 7.4) containing 0.02% polyethylene glycol and 0.05% sodium azide and stored at 4° until use. Preparation of larger diameter streptavidin–gold complexes is similar except that the first centrifugation is performed at 3000 rpm (JA21 Beckman rotor) for 30 min at 4° and a second centrifugation is at 11,000 rpm for 30 min using the same rotor.

Streptavidin–gold conjugates are now readily available commercially (Janssen Life Sciences Products, Piscataway, NJ; Amersham, Arlington Heights, IL; Bethesda Research Laboratories, Gaithersburg, MD), and fluorescent conjugates of streptavidin are also available commercially from a variety of sources (Amersham, BRL). Labeling densities are not constant when the size of the gold probe is varied. In general, 5-nm conjugates label at higher density than larger probes. Bound 5-nm gold conjugates can be enlarged *in situ* on postembedded tissue section surfaces by a process termed silver latensification (Janssen). Not all 10-nm probes label at comparable density, however, so that other, as yet undefined, variables also contribute to labeling density. Thus, in quantitative studies, we usually include a positive control of antiopsin and label a

[15] G. Frens, *Nature* (*London*) **241,** 20 (1973).
[16] W. D. Geoghegan and F. A. Ackerman, *J. Histochem. Cytochem.* **25,** 1187 (1977).

section of photoreceptors to determine if the outer segment labeling density is appropriate for the dilution of reagents we have chosen.

Preembedding Labeling

Preembedding labeling is primarily used for the detection of antigens exposed on the external surfaces of cells. Although, superficially, this appears to be a simple approach to immunocytochemistry, several factors can affect the successful application of the technique. Adhesion of extracellular molecules on cell surfaces may obscure antigenic determinants. These substances may or may not respond to simple washing procedures. In complex tissues such as the retina, washing steps may affect superficial domains (e.g., the tips of rod outer segments, ROS) more than deeper areas of the cell such as paraciliary or inner segment domains. Careful observation and quantification of labeling density along the length of the cell often reveal a gradient of labeling from distal to proximal regions of the photoreceptor when such washing is incomplete.

Dark-adapted frogs, mice, or rats are sacrificed, and the retina is isolated under dim red light so that separation from the pigment epithelium is accomplished with little adhesion. The initial rinsing step and all subsequent rinses and incubations are carried out under continuous gentle shaking in order to facilitate the diffusion of solutes between the photoreceptor cells and to wash out easily extracted interphotoreceptor matrix molecules. Following an initial rinse in phosphate-buffered saline (PBS) containing 1 m*M* $CaCl_2$ and 0.5 m*M* $MgSO_4$, for 15 min at 4°, the retina is fixed in freshly prepared 1% (v/v) glutaraldehyde in 0.15 *M* phosphate buffer (pH 7.0). After fixation for 30 min at room temperature, the retina is sliced into 1 × 2–3 mm strips and left in the fixative for an additional 30 min. The fixed tissue is washed in PBS for 5 min, in 50 m*M* glycine (in PBS) for 10 min, in 2% bovine serum albumin (BSA) (in PBS) for 30 min, and in PBS for 5 min. Retina slices are transferred to solutions of 0.1 mg/ml biotinyl antiopsin for 90 min at room temperature. Following a 60-min rinse in PBS, the tissues are incubated with AvF at 0.2 mg/ml in 1% BSA, or with 10- to 20-nm SaG (BRL) diluted 1 : 3 for 90 min at room temperature, rinsed in PBS for 60 min, and treated with 1% OsO_4, in 0.15 *M* phosphate buffer (pH 7.0) for 30 min, dehydrated with ethanol and propylene oxide, and embedded in epoxy resins. Controls include substitution of biotinyl-IgG solutions for the antibody (Fig. 1).

Postembedding Labeling

Major advances in techniques for immunocytochemical localization of antigens by postembedding techniques have been recently described.

Epon etched with sodium ethoxide or sodium metaperiodate, aldehyde-cross-linked bovine serum albumin,[17] and frozen sucrose[18] were originally employed as embedding media, and the latter is still in active use for antigens that are sensitive to extensive fixation or organic solvents. The most important change has been the introduction of a variety of methacrylate-derived hydrophilic plastics which provide easily sectioned tissue blocks that are suitable for labeling with antibodies to numerous antigens. These new media include Lowicryl K4M, Lowicryl K11, LR White, and LR Gold (Polysciences, Warrington, PA). There is no single embedding medium that is suitable for all antigens, and we have often embedded tissues in numerous media until the one which provides the best combination of tissue and antigen preservation is determined empirically.

These new plastics are usually polymerized by ultraviolet light or chemical initiators. Since the polymerization reactions are exothermic and many antigens and organelles are thermolabile, the polymerization steps are usually conducted at reduced temperature. Roth and colleagues[19] have emphasized polymerization of Lowicryl K4M at −35° for preservation of intracellular structures, especially the Golgi apparatus; our experience parallels theirs. We have developed, however, a rapid procedure for embedding in Lowicryl K4M at 4° that is useful when answers are needed quickly, e.g., for surgical pathology.[20] The specific changes we have introduced in the use of Lowicryl K4M[20,21] include rapid transfer of tissues at room temperature through the dehydration and infiltration steps and illumination of the tissue block at 4° with UV light from below at a short lamp–tissue distance. Slower polymerization may be important under some conditions, however, and tissues vary in their transparency to light so that blocks may require additional circumferential light exposure to become sufficiently firm to section easily.

Some individuals are easily skin-sensitized to these plastics. Latex gloves do not provide adequate protection, but polyethylene-coated gloves may offer superior protection from a resin-induced dermatitis. Care against exposure to unpolymerized resin and the execution of all embedding steps in a well-vented hood are also important. Comparative

[17] J. D. McLean and S. J. Singer, *Proc. Natl. Acad. Sci. U.S.A.* **65,** 122 (1970).

[18] K. T. Tokuyasu and S. J. Singer, *J. Cell Biol.* **71,** 894 (1976).

[19] J. Roth, M. Bendayan, E. Carlemalm, W. Villiger, and M. Garavito, *J. Histochem. Cytochem.* **29,** 663 (1981).

[20] L. G. Altman, B. G. Schneider, and D. S. Papermaster, *J. Histochem. Cytochem.* **32,** 1217 (1984).

[21] L. G. Altman, B. G. Schneider, and D. S. Papermaster, *in* "Proceedings of the 43rd Meeting of the Electron Microscopy Society of America" (G. W. Bailey, ed.), p. 430. San Francisco Press, San Francisco, California, 1985.

studies of labeling density with antibodies to opsin have demonstrated that this antigen is detected with comparable labeling densities in tissues embedded in glutaraldehyde-cross-linked BSA, Lowicryl K4M polymerized by the manufacturer's procedures, and Lowicryl K4M prepared by our rapid protocol.[21] Antibodies to the α chain of Na^+K^+-ATPase also behaved similarly in all three embedding protocols.

Recent studies of the biosynthesis and transport of newly synthesized opsin have revealed a vesicular post-Golgi pathway in amphibian retinal rod photoreceptors. The vesicles, bearing opsin in their membranes, cluster beneath a unique plasmalemmal domain, the periciliary ridge complex (PRC), which is organized at the base of the connecting cilium that joins the inner and outer segments (Fig. 2).[22] Vesicles appear to fuse with the membranes of the groove of the PRC.[23,24] How rhodopsin is constrained from rapidly diffusing over the surface of the rod inner segment is unclear. In normal cells, only the plasmalemma of the outer segment and its disks contain opsin at high density. In photoreceptors that are degenerating as a consequence of an inherited retinal dystrophy, however, opsin is distributed over inner segment plasma membrane domains,[12,13] a distribution that is also found in developing photoreceptors prior to the elaboration of the outer segment.[14] Most of these studies have used AvF or 10-nm SaG to detect bound biotinyl antibodies. Recent experience demonstrates greater labeling density with 5 nm SaG particles.

Localization of membrane antigens by postembedding techniques requires consideration of the geometric sources of the antigen to avoid misinterpretation of EM immunocytochemical studies. When investigating labeling of vesicles whose diameters approach the thickness of the thin section, we observed distribution of label at sites that do not immediately lie over the vesicular membrane. We recognized that this appearance arises from the potential of tangentially sectioned vesicle membranes. Since only the section surface is labeled, antigenic sites might be presented on the top surface of the tangentially sectioned vesicle and appear to reside within the center of a pale, poorly demarcated oval area while the bulk of the vesicle membrane appears to be several nanometers away because it lies deeper in the section (see Fig. 2, arrow). Alternatively, cross-sectioned vesicles whose plane of section is below the equator might generate a circle of label overlying an area not pale enough to

[22] D. S. Papermaster, this series, Vol. 81, p. 240.

[23] D. S. Papermaster, B. G. Schneider, and J. C. Besharse, *Invest. Ophthalmol. Visual Sci.* **26,** 1386 (1985).

[24] D. S. Papermaster, B. G. Schneider, D. Defoe, and J. C. Besharse, *J. Histochem. Cytochem.* **34,** 5 (1986).

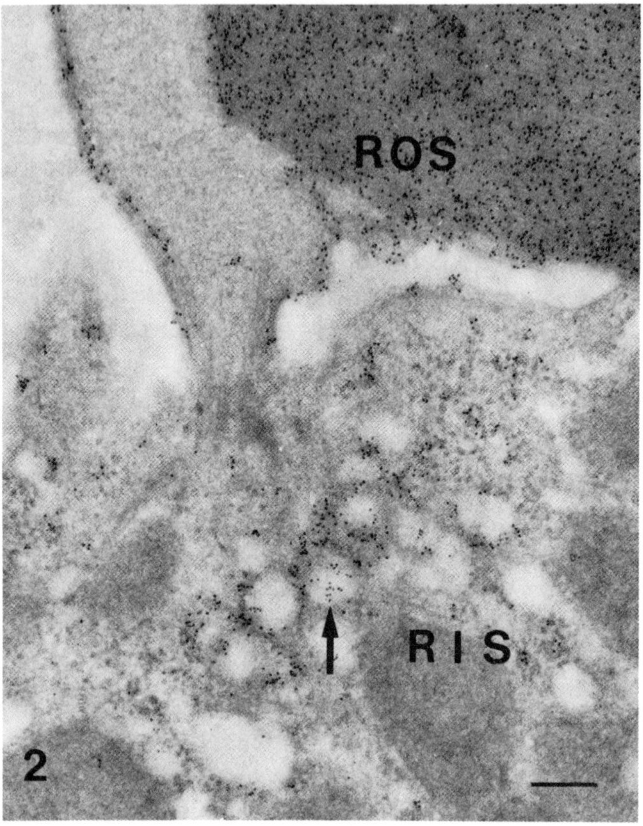

FIG. 2. Longitudinally sectioned *Xenopus laevis* rod photoreceptor embedded in Lowicryl K4M and labeled sequentially with affinity-purified rabbit antiopsin,[22] biotinyl sheep anti-rabbit $(Fab)_2$, and avidin–ferritin. Antibody labels the rod outer segment (ROS) disks and plasma membrane confluently. In addition, immunolabeled vesicles are clustered beneath the connecting cilium that joins the rod inner segment (RIS) to the ROS. These vesicles were shown, by a quantitative autoradiographic and immunocytochemical study, to bear newly synthesized opsin from the Golgi apparatus to this juxtaciliary domain of the cell.[23,24] Magnification: ×46,500; bar, 0.2 μm. A tangentially sectioned vesicle is labeled in its center because the label does not penetrate significantly into the section (arrow).

indicate the origin of the antigen from within a vesicular membrane. This issue is discussed thoroughly in Ref. 24. With these considerations in mind, it is possible to evaluate intracellular pathways of synthesis and transport of opsin and to generate results which correlate highly with earlier biochemical evidence of vesicular post-Golgi transport of opsin.

Thus, avidin–ferritin and streptavidin–gold conjugates have been stable, versatile, and reliable reagents for detection of numerous antibodies to retinal antigens and antigens of other tissues. Their use in diagnostic pathology is expanding new areas of investigation.

[42] Visualization of Intracellular Trafficking of Proteins

By RANDAL E. MORRIS and CATHARINE B. SAELINGER

The intracellular trafficking of proteins following internalization by eukaryotic cells is a subject of intense scientific interest. Unfortunately, few proteins have sufficient electron density to be visualized at the ultrastructural level. To visualize intracellular trafficking, therefore, the protein of interest must be directly conjugated to an electron-dense marker. Alternatively, the protein can be localized by various immunological methods in which the antibody molecules are labeled with an electron-dense marker. Because of their superb electron density, gold colloids are the preferred marker. However, direct conjugation of certain proteins to gold colloids results in an aberrant intracellular routing.[1–3] In order to circumvent this problem, we and others have found that the use of gold-labeled antibody techniques is most effective when used in conjunction with postembedding methods.[4] The disadvantage of the postembedding method is that, in most instances, fixation with osmium tetroxide must be omitted. This results in reduced specimen contrast, which makes interpretation difficult.

In this chapter we describe a method in which the intracellular routing of protein ligands can be followed by conventional electron microscopy. The technique described permits the intracellular visualization of biotinylated proteins when used in conjunction with avidin–gold colloids. We have successfully used this technique to follow the intracellular routing of

[1] M. C. Willingham, J. A. Hanover, R. B. Dickson, and I. Pastan, *Proc. Natl. Acad. Sci. U.S.A.* **81,** 75 (1984).

[2] M. R. Neutra, A. Ciechanover, L. S. Owen, and H. F. Lodish, *J. Histochem. Cytochem.* **33,** 1134 (1985).

[3] B. van Deurs, T. I. Tonnessen, O. W. Petersen, K. Sandvig, and S. Olsnes, *J. Cell Biol.* **102,** 37 (1986).

[4] M. C. Willingham, *J. Histochem. Cytochem.* **31,** 791 (1983).

several proteins. We suggest that this technique can be tailored to a variety of proteins and cell types.

Biotinylation of Proteins

The biotinylation of proteins is a mild procedure involving the nucleophilic attack on proteins by the *N*-hydroxysuccinimide ester group of biotinyl-*N*-hydroxysuccinimide ester (BNHS; Pierce Chemical Co., Rockford, IL). The procedure described below results in very stable products that retain good biological activity.

1. Dialyze the protein overnight at 4° against 50 m*M* borate buffer (pH 9.1; 19.06 g/liter $Na_2B_4O_7 \cdot 10H_2O$). We usually use 1 mg of protein in a volume of 1 ml and dialyze against 1 liter of buffer.
2. The biotinylation is done at a 5 : 1 molar ratio (BNHS/protein) and a 1 : 50 volume ratio (BNHS/protein) as described by Bayer *et al.*[5] Calculate the required amount of BNHS and dissolve in dimethylformamide (DMF). Since only 20 μl will be required for a 1 ml protein solution, we typically prepare 0.5 ml of DMF containing BNHS.
3. Place the dialyzed protein solution in a glass test tube and clarify by centrifugation if necessary. To the solution add the appropriate volume of DMF containing BNHS.
4. Incubate at 23° for 4 hr.
5. Dialyze the resultant solution against phosphate-buffered saline (pH 7.4, PBS; 8.01 g/liter NaCl, 1.18 g/liter Na_2HPO_4, 0.22 g/liter KCl, 0.26 g/liter $K_2HPO_4 \cdot H_2O$).
6. Store biotinyl proteins at 4° without azide or other anti-microbials. For long-term storage, they can be stored at −20° without deleterious effects.

This procedure results in biotinylation preferentially via ε-amino groups of lysyl residues. However, this can result in loss of biological activity if the lysine groups are required for specific activity. In such an event, the procedure can be modified to direct the nucleophilic attack to the α-amino groups. In this instance, the reaction is done between pH 5.0 and 6.0.[5,6]

Using the procedure described we have found that an average of 2.5 mol of biotin is incorporated for every mole of protein, i.e., *Pseudomonas*

[5] E. A. Bayer, E. Skutelsky, and M. Wilchek, this series, Vol. 62, p. 308.

[6] P. Cuatrecasas and I. Parikh, *Biochemistry* **11,** 2291 (1972).

exotoxin A.[7] (This is estimated by the method of Means and Feany[8] using 2,4,6-trinitrobenzenesulfonic acid.) Higher molar ratios (BNHS/protein) can be used to increase the level of incorporation. It has been our experience, however, that the higher the molar ratio, the greater the reduction in biological activity. We therefore rarely exceed a 5 : 1 (initial) molar ratio.

Preparation of Avidin–Gold Colloids

The interaction between egg-white avidin and biotin is among the most avid interactions known (K_d 10^{-15} M).[9] The binding of avidin to biotin is several orders of magnitude greater than that found for most antigen–antibody interactions and can be considered essentially an irreversible noncovalent interaction. This has important ramifications in the use of biotinyl protein–avidin–gold to study intracellular trafficking. Following internalization, ligands are routed through various acidic compartments, e.g., endosomes and lysosomes (pH 4.5–5.5), where certain receptors and ligands are known to dissociate.[10] The interaction between avidin and biotin is stable at these pH values,[9] and therefore one is assured of following the labeled ligand and not dissociated gold colloids.

Gold colloids of varying sizes can be easily prepared by the reduction of gold chloride. Colloids with average diameters ranging from 5 to 20 nm are routinely used for electron microscopy. It has been our experience that only 5- to 7-nm avidin–gold colloids can be used in the biotinyl ligand–avidin–gold method; larger size colloids, i.e., 12 and 18 nm, do not bind biotinyl ligands. The reasons for this are not known but possibly result from steric hindrance. Described below is a method used for the production of 5-nm gold sols by reduction of gold chloride with white phosphorus.[11]

1. To a 500-ml round-bottomed flask add 240 ml of doubly distilled water, 5.4 ml of 0.1 M K_2CO_3 and 6 ml of 0.5% gold chloride solution (Fisher Scientific, Fair Lawn, NJ).
2. Prepare a solution of white phosphorus-saturated diethyl ether. Saturation is complete when the solution has a distinctively gray color. *Note:* Unused white phosphorus-saturated ether is neutralized by the addition of an equal volume of 1% $CuSO_4$.

[7] R. E. Morris and C. B. Saelinger, *J. Histochem. Cytochem.* **32,** 124 (1984).

[8] G. E. Means and R. E. Feany, *in* "Chemical Modification of Proteins." Holden-Day, San Francisco, California, 1971.

[9] N. M. Green, *Adv. Protein Chem.* **29,** 85 (1975).

[10] H. J. Geuze, J. W. Slot, G. J. A. M. Strous, H. F. Lodish, and A. L. Schwartz, *Cell* **32,** 277 (1983).

[11] M. Horisberger, *Biol. Cell.* **36,** 253 (1979).

3. Initiate the reduction of the gold chloride solution by adding of 2 ml of a 1 : 5 white phosphorus–ether solution (1.6 ml of diethyl ether and 0.4 ml of white phosphorus-saturated ether).
4. Incubate the solution at 23° for 30 min, during which time it becomes a rust-brown color.
5. The reaction is driven to completion by refluxing. The end point of the reaction is noted when the solution becomes deep red in color. This usually requires 5–20 min of refluxing. The gold sols are cooled to 23°, pH adjusted (see below), and stabilized with avidin, i.e., converted to gold colloids.

Five-nanometer gold sols can also be produced by reduction of gold chloride with sodium citrate and tannic acid.[12,13] We advocate preparation with white phosphorus because we have experienced some reduction in the recognition of biotinyl ligand when using avidin–gold colloids prepared by the sodium citrate–tannic acid method. This may result because tannic acid is a mordant, and residual acid may denature the avidin.[14]

After the gold sols are prepared they are stabilized, i.e., converted to gold colloids, by the adsorption of avidin onto the sols. Stabilization occurs as the result of electrostatic interactions between the gold sols and the protein being adsorbed. Handley *et al.*[15] have shown that this interaction is very stable under a variety of conditions. We have used two sources of avidin with equal success, egg-white avidin and streptavidin. Egg-white avidin is a basic protein with a p*I* of 10.5.[9] At physiological pH, the protein carries a net positive charge and nonspecifically binds to cell membranes. For this reason egg-white avidin should be succinylated prior to use (see below). Streptavidin, on the other hand, has a neutral p*I* and requires no modification prior to use. Streptavidin is a protein secreted from *Streptomyces avidinii* that has essentially the same molecular weight, subunit structure, and binding affinity as the egg-white protein.[16] Because gold sols are susceptible to flocculation by the presence of ions, all proteins to be labeled with gold must be exhaustively dialyzed against low-molarity buffer prior to adsorption. Furthermore, the greatest stability of the colloids occurs when adsorption is carried out at a pH value slightly above the p*I* of the protein being adsorbed.[17] Accordingly, avidin

[12] H. Muhlpfordt, *Experientia* **38,** 1127 (1982).

[13] J. Slot and H. J. Geuze, *Eur. J. Cell Biol.* **38,** 87 (1985).

[14] K. Aoki, S. Kajiwara, R. Shinke, and H. Nishira, *Anal. Biochem.* **95,** 575 (1979).

[15] D. A. Handley, C. M. Arbeeny, L. D. Witte, and S. Chien, *Proc. Natl. Acad. Sci. U.S.A.* **78,** 368 (1981).

[16] L. Chaiet and F. J. Wolf, *Arch. Biochem. Biophys.* **106,** 1 (1964).

[17] W. D. Geoghegan and G. A. Ackerman, *J. Histochem. Cytochem.* **11,** 1187 (1977).

is dialyzed against repeated changes of 5 m*M* phosphate buffer (pH 7.5; 0.11 g/liter $NaHPO_4 \cdot H_2O$, 1.13 g/liter $Na_2HPO_4 \cdot 7H_2O$). Typically we dialyze against 4 changes of buffer for 40 hr at 4° prior to adsorption.

Succinylation of Egg-White Avidin

1. Dissolve 10 mg of egg-white avidin (Sigma Chemical Co., St. Louis, MO) in 10 ml of a 3 *M* sodium acetate solution. Add 10 mg of succinic anhydride; incubate overnight at 4°.
2. Incubate for 1 additional hour at 23°. Place the solution into a dialysis bag; allow sufficient space for the volume to double.
3. Dialyze against 5 m*M* phosphate buffer (pH 7.5) as described above.

The amount of avidin required to stabilize a known volume of gold sols is determined by the method of Geoghegan and Ackerman.[17]

1. Adjust the pH of the gold sols to 7.5 using either 0.1 *M* K_2CO_3 or 1% acetic acid. (*Note*: Submersion of the pH electrode into the gold sol solution will result in short-circuiting of the electrode. To avoid this, take a 5-ml sample of the sols and add 10 drops of a 1% polyethylene glycol solution (MW ≥20,000) prior to immersion of the electrode. This will stabilize the sols and prevent damaging the electrode.)
2. To 10 plastic test tubes add 1 ml of the 5 m*M* phosphate buffer (pH 7.5). Leave the first tube as a control tube; starting with the second tube, serially dilute the avidin solution 1 : 2, 1 : 4, 1 : 8, etc.
3. Add 5 ml of the pH-adjusted gold sols to each tube, mix well. Allow all tubes to stand for at least 1 min at 23°.
4. Add 1 ml of a 10% NaCl solution to each tube. This will cause flocculation, detected by a change in color from red to blue, in those samples which have not been stabilized by the protein. We typically allow 30 min for the reaction to go to completion.
5. To determine the end point of the reaction, that is, the greatest dilution which stabilizes the gold against flocculation, visually identify the tube with the greatest dilution that retains a clear red color. To guard against possible errors in dilution, we use twice the amount of protein indicated by the test. If, for example, the dilution assay indicated that stabilization occurs at 1 : 16, then we would use 1 : 8 as the stabilizing dilution. If we want to prepare 100 ml of avidin-stabilized gold, we would add 2.5 ml of undiluted protein [volume desired/(dilution factor × volume of gold stabilized in the test, i.e., 5)].

6. The protein is then added (concentration ≤1 mg/ml in 5 m*M* phosphate buffer) to a rapidly stirring solution of pH-adjusted gold sols.
7. After stirring for 15 min, a 5-ml sample is taken and tested for stability as in step 4.
8. The avidin–gold colloids are washed by centrifugation at 18,000 *g* for 3 hr at 4°. This results in the formation of a loose pellet and a tight pellet. The supernatant fluid is discarded, and the loose pellet is resuspended to the original volume in 5 m*M* phosphate buffer (pH 7.5) and transferred to another centrifuge tube. The remaining tight pellet is discarded.
9. The gold colloids are repelleted by a second centrifugation as above. The resulting loose pellet is resuspended in sufficient 5 m*M* phosphate buffer (pH 7.5) such that its color approximates that of the initial solution, i.e., step 7 (above). The resultant avidin–gold colloidal suspension is transferred to a 100-ml beaker and stirred. While stirring, an equal volume of double-strength stabilizing buffer [18.0 g/liter NaCl, 12.1 g/liter Tris-HCl, 400 mg/liter polyethylene glycol (MW ≥20,000) (pH 7.5)] is added.
10. After 15 min of stirring, the avidin–gold colloids are placed in plastic tubes and stored at 4° until needed.[18]

The procedure described varies slightly from previous published methods.[7,19] It is important to note that we use the avidin–gold colloids within 1 month after production. Avidin–gold colloids older than 1 month have an increased tendency to bind nonspecifically to cells.

Just before use, the avidin–gold colloids are absorbed against the cell line being studied. For example, 5 ml of avidin–gold colloids is absorbed against nearly confluent monolayers of mouse LM fibroblasts. Prior to absorption the LM cells are exhausted of exogenous biotin and cooled to 4°. Absorption is repeated 3 times for 30 min at 4°. Large aggregates and debris are removed by centrifugation at 18,000 *g* for 20 min at 4°.

Biotinyl Ligand–Avidin–Gold Method

The procedure described is limited to *in vitro* conditions. This restriction is due to the significant concentrations of biotin present in the sera of

[18] Since we usually use the avidin–gold colloids within 1 month of production, we do not add azide or similar agents. In any case, for internalization studies (as well as for other studies in which viable cells are used), it is, of course, not recommended to use any such toxic antimicrobial agents.

[19] R. E. Morris and C. B. Saelinger, *Infect. Immun.* **52,** 445 (1986).

animals.[20] This fact also dictates that, during the experimentation procedures, medium containing biotin or animal sera must not be used. Typically, we grow cells in complete medium with 10% sera. For 2 hr prior to the initiation of the experiment we exhaust the cells of exogenous biotin by incubation at 37° in HEPES-buffered Hanks' balanced salt solution (HBSS). Failure to include this step results in high levels of nonspecific binding by the avidin–gold colloids. The steps in the procedure are given below.

1. Wash all samples 3 times with cold HBSS; incubate at 4° for 15 min.
2. Add the biotinylated protein to the samples in a minimal volume of HBSS; incubate at 4° for 30 min.[21]
3. Wash all samples 3 times with cold HBSS.
4. Add the avidin–gold conjugates in a minimal volume of HBSS to each sample; incubate for 30 min at 4°.
5. To those samples which are to be warmed, add prewarmed HBSS and incubate on a warming plate. After appropriate incubation, wash samples 3 times with cold HBSS.
6. Wash all samples twice with cold 0.1 *M* sodium cacodylate buffer (pH 7.4) containing 0.05% $CaCl_2$ and 5.0% sucrose (SCB).
7. Fix all samples with SCB containing 2.0% paraformaldehyde and 2.5% (w/v) glutaraldehyde; fix for 30 min at 4°.
8. Repeat step 6.
9. Post fix all samples with SCB containing 1% osmium tetroxide reduced with 1% potassium ferrocyanide[22]; fix for 1 hr at 4°.
10. Repeat step 6.
11. Wash all samples twice with distilled water at 23°.
12. Begin dehydration by 3 washes with 70% (v/v) ethanol.
13. Stain *en bloc* for 10 min with 0.5% uranyl acetate in 70% (v/v) ethanol.
14. Continue dehydration by 2 washes with 70% ethanol followed by 3 washes with absolute ethanol.
15. Infiltrate and embed by standard procedures using epoxy resin.

[20] R. E. Morris and C. B. Saelinger, *Immunol. Today* **5,** 127 (1984).

[21] Concentrations and volumes of reagents used in these experiments vary from experiment to experiment. For example, in one instance we may add 1 ml of a given biotinyl protein to a 16 × 85 mm Leighton tube; in another instance we may add 2.5 ml to a 100 mm petri dish. Concentrations vary depending on the biotinyl ligand and the cell type. For most experiments we use LM cells grown on 16 × 85 mm glass Leighton tubes, using a 1-ml volume of biotinyl toxin at concentrations ranging from 1 ng/ml to 1 μg/ml.

[22] M. J. Karnovsky, *J. Cell Biol.* **51,** 146a (1971).

A control sample in which the biotinyl ligand is omitted should be included. This will measure nonspecific (background) binding of the gold colloid to cells. One of the advantages of the biotinyl ligand–avidin–gold system is that specific binding can be demonstrated by a competition assay. This is done by including a sample with a 100- to 200-fold excess of native ligand (not biotinylated) during the primary incubation step with the biotinyl ligand.

The procedure described above is done on glass Leighton tubes.[23] Following polymerization at 60°, the resin containing the monolayer is freed from the glass by submersion in an ice bath. In preparation for sectioning, small areas of the plastic are drilled out with an electric cork borer (E. H. Sargent and Company, Birmingham, AL) and mounted on 3/4 × 1/4 inch wooden dowels.[24] Alternatively, the cells can be grown on plastic petri dishes. At the conclusion of step 11, the cells are scraped from the plastic surface, placed in 1% low-temperature gelling agarose (Sigma), and pelleted (in a conical centrifuge tube) by low-speed centrifugation at 23°. Following overnight solidification at 4°, the cell-rich regions of the agarose are diced into pieces of about 1 mm^3 and treated as described above. The samples are ultimately embedded in epoxy resin in BEEM capsules.

The samples are viewed after preparation of ultrathin sections. The above procedure results in sufficient specimen contrast so that the samples do not require staining.

Comments

We have used the biotinyl ligand–avidin–gold method to follow the intracellular trafficking of several different proteins. The majority of our experience has been with the intracellular trafficking of *Pseudomonas* exotoxin A (PE) by mouse LM fibroblasts. We have shown that LM cells, a cell line exquisitely sensitive to PE, internalize biotinyl-PE–avidin–gold by receptor-mediated endocytosis, route the complex to the Golgi region in endosomes, and ultimately deliver the complex to the lysosomal compartment. The entire process takes 20–30 min at 37°.[19,25,26] We have also followed the internalization and intracellular trafficking of biotinyl diphtheria toxin by Vero cells, a cell line exquisitely sensitive to this toxin.

[23] J. S. Sutton, *Stain Technol.* **40,** 151 (1965).

[24] R. E. Morris, G. M. Ciraolo, D. A. Cohen, and H. C. Bubel, *In Vitro* **16,** 136 (1980).

[25] R. E. Morris, M. D. Manhart, and C. B. Saelinger, *Infect. Immun.* **40,** 806 (1983).

[26] R. E. Morris, *in* "Microbiology 1985" (L. Leive, ed.), p. 91. American Society for Microbiology, Washington, D.C., 1985.

The routing of diphtheria toxin in Vero cells is similar to that of PE in mouse LM cells.[27]

Others have shown that ligands bound to gold are routed differently in cells than are native ligands. For example, van Deurs *et al.*[3] showed that Vero and MCF-7 cells internalize native ricin by receptor-mediated endocytosis and that the ricin is routed by the endosomal system to the Golgi cisternae. In contrast, ricin–gold conjugates, although internalized by receptor-mediated endocytosis, are not observed in the Golgi. They estimated that each ricin–gold conjugate contains 2 to 4 ricin molecules per gold particle[28] and is therefore polyvalent. They further showed that monovalent conjugates, formed between ricin and horseradish peroxidase, are routed intracellularly as is native ricin. The authors concluded that the valency of the ligand conjugate dictates the intracellular trafficking pattern. Others have reported similar observations with transferrin–gold conjugates.[1,2]

Two experimental observations lead us to believe that the biotinyl ligand–avidin–gold technique circumvents this problem. First, biotinyl diphtheria toxin–avidin–gold complexes are routed differently in mouse LM cells than are biotinyl *Pseudomonas* toxin–avidin–gold complexes. Mouse LM cells are resistant to diphtheria toxin. The diphtheria toxin–gold conjugates are not internalized by receptor-mediated endocytosis and are not routed to the Golgi region.[29] This occurs in spite of the fact that LM cells have receptors for diphtheria toxin[30] and that diphtheria toxin and *Pseudomonas* toxin inhibit mammalian cell protein synthesis by identical mechanisms. We feel that this observation supports our hypothesis that the mechanism of entry and intracellular trafficking dictates the susceptibility of a cell to toxin. Second, our morphological observations are corroborated by biochemical data using native toxin.[31,32] We feel that the biotinyl ligand–avidin–gold method allows for the ligand to be processed normally because the biotinyl ligand is permitted to interact with its receptor prior to addition of gold. Thus, the initial binding of ligand is not influenced by the gold, i.e., the valency is not altered.

In conclusion, we advocate the use of the biotinyl ligand–avidin–gold technique as a tool to visualize intracellular trafficking. There are three

[27] R. E. Morris, A. S. Gerstein, P. F. Bonventre, and C. B. Saelinger, *Infect. Immun.* **50,** 721 (1985).

[28] B. van Deurs, L. R. Pedersen, A. Sundan, S. Olsnes, and K. Sandvig, *Exp. Cell Res.* **159,** 287 (1985).

[29] R. E. Morris and C. B. Saelinger, *Infect. Immun.* **42,** 812 (1983).

[30] J. R. Didsbury, J. M. Moehring, and T. J. Moehring, *Mol. Cell. Biol.* **3,** 1283 (1983).

[31] C. B. Saelinger, R. E. Morris, and G. Foertsch, *Eur. J. Clin. Microbiol.* **4,** 170 (1985).

[32] C. B. Saelinger and R. E. Morris, *Antibiot. Chemother.* (*Basel*) **39,** 149 (1987).

advantages of this technique over other methods for following intracellular trafficking at the ultrastructural level. First, because of the affinity of avidin for biotin, dissociation of the ligand from the electron-dense marker is negligible. Second, competition of native ligand and biotinyl ligand for a receptor on the cell surface can be measured. Third, the technique permits the use of fixation and embedding protocols which result in excellent ultrastructural preservation. This allows for more precise definition of intracellular locations.

[43] Localization of Lectin Receptors

By RICHARD M. PINO

Introduction

The location and chemical composition of cell surface components have often been examined by ultrastructural cytochemical methods.[1-5] A basic approach is the identification of monosaccharides using lectins.[6] With direct labeling, cells or tissues are treated with lectins that are coupled to ferritin,[7] which has inherent electron density. Lectins can also be coupled to horseradish peroxidase (HRP),[8] which can generate an electron-dense reaction product after incubation in a cytochemical medium. HRP methods are also applicable to light microscopic cytochemistry analysis.

The avidin–biotin method[9] is a reliable indirect method that can be used for lectin affinity[10,11] and immunocytochemical[4,10,12] localizations on

[1] D. Danon, L. Goldstein, Y. Marikovsky, and E. Skutelsky, *J. Ultrastruct. Res.* **38,** 500 (1972).

[2] P. P. H. DeBruyn and S. Michelson, *J. Cell Biol.* **82,** 708 (1979).

[3] E. Essner, R. M. Pino, and R. A. Griewski, *Curr. Eye Res.* **1,** 381 (1981).

[4] R. M. Pino, *Invest. Ophthalmol. Visual Sci.* **27,** 840 (1986).

[5] R. M. Pino, *Cell Tissue Res.* **250,** 257 (1987).

[6] N. Sharon and H. Lis, *Science* **177,** 949 (1972).

[7] G. L. Nicolson and S. J. Singer, *J. Cell Biol.* **60,** 236 (1974).

[8] W. Bernhard and S. Avrameas, *Exp. Cell Res.* **64,** 232 (1971).

[9] H. Heitzmann and F. M. Richards, *Proc. Natl. Acad. Sci. U.S.A.* **71,** 3537 (1974).

[10] E. A. Bayer, M. Wilchek, and E. Skutelsky, *FEBS Lett.* **68,** 240 (1976).

[11] S. M. Hsu and L. Raine, *J. Histochem. Cytochem.* **30,** 157 (1982).

[12] S. M. Hsu, L. Raine, and H. Fanger, *J. Histochem. Cytochem.* **29,** 577 (1981).

METHODS IN ENZYMOLOGY, VOL. 184

the cell surface. With this technique, carbohydrates on the cell surface are bathed in biotinylated lectins. The biotinylated lectins in turn are localized either with avidin–ferritin conjugates or with avidin followed by biotinylated HRP. By virtue of the multiple high-avidity binding capacity of avidin for biotin molecules,[13] the localization of the molecule in question is identified in an amplified manner. This chapter describes a reliable method for the localization of cell surface components using the avidin–biotin method.

Materials and Methods

Reagents. Biotinylated lectins, avidin, biotinylated HRP, and avidin–ferritin conjugates are available from a number of commercial sources. As with any commercial preparation, reagents may vary in concentration, binding affinity, biotinylation, etc. Because of this, we routinely use reagents from one source (Vector Laboratories, Burlingame, CA). For noncommercially available compounds, it is possible to perform biotinylation by a simple method.[14]

Appropriate hapten sugars are obtained from Sigma Chemical Co. (St. Louis, MO). Enzymes with specificity for carbohydrate domains of glycoproteins (Table I) are obtained from Miles Laboratories (Elkhart, IN), Calbiochem-Behring (San Diego, CA), or Sigma.

Tissue Preparation

In our studies, the endothelia of the bone marrow sinusoids,[15] of the choriocapillaris[16] and retinal capillaries,[17] and the apical surface of the retinal pigment epithelium[18] of the rat have been examined. For the endothelium, the vasculature is perfused with buffer to remove plasma components that might bind to the cell surface and affect the localizations.[15–17] To obtain the retinal pigment epithelium, the eyes are enucleated and dissected at the limbus, and the neural retinas are removed.[18] The resulting eye cups consisting of the sclera, choroid, and retinal pigment epithelium are fixed by immersion. We have used 1.25% glutaraldehdye–1% formaldehyde in either phosphate or cacodylate buffer with equal success for lectins. After fixation, the tissues are rinsed 3 times in 0.2 *M* buffer and

[13] N. M. Green, *Biochem. J.* **89,** 585 (1963).

[14] E. A. Bayer, E. Skutelsky, and M. Wilchek, this series, Vol. 62, p. 308.

[15] R. M. Pino, *Am. J. Anat.* **169,** 259 (1984).

[16] R. M. Pino, *Cell Tissue Res.* **243,** 145 (1986).

[17] R. M. Pino and C. L. Thouron, *Curr. Eye Res.* **5,** 625 (1986).

[18] R. M. Pino, *J. Histochem. Cytochem.* **32,** 862 (1984).

TABLE I
ENZYMES USEFUL IN LECTIN STUDIES

Enzyme	Action	Reaction conditions[a,b]
Endoglycosidase D[c] (E.C.3.2.1.96: mannosyl-glycoproteinendo-β-*N*-acetylglucosaminidase)	GlcNAcβ1 → 4 GlcNAc[d]	50 mU/ml in 50 m*M* citrate phosphate buffer (pH 6.0) plus 8% sucrose
N-Acetylhexosaminidase[e] (E.C.3.2.1.52)	β-GlcNAc	25 mU/ml in 50 m*M* citrate phosphate buffer (pH 5.0) plus 8% sucrose
Heparitinase[f] (E.C.4.2.2.8.:Heparitin-sulfate lyase)	Heparan sulfate	0.5 U/ml in 0.1 *M* acetate buffer (pH 7.0) plus 10% sucrose
Neuraminidase[g] (influenza virus) (E.C.3.2.1.18)	Sialic acid	1.5 U/ml in 0.1 *M* Tris maleate buffer (pH 6.5) plus 5 m*M* $CaCl_2$ and 5% sucrose
Pronase E[h] (E.C.3.4.21.14)	Bonds adjacent to hydrophobic amino acid residues	5–50 U/ml HBSS[i] (pH 7.2)
Proteinase K[j] (E.C.3.4.21.14)	Proteins at carboxyl, aromatic, or hydrophobic amino acid residues	10 U/ml HBSS (pH 7.2)
Trypsin[k] (E.C.3.4.21.4)	Arg-, Lys-	1 mg/ml HBSS (pH 7.2)

[a] At 37°; reaction conditions from referenced works.
[b] Addition of sucrose to bring buffer to approximately 325 mos*M*.
[c] N. Koide and T. Muramatsu, *J. Biol. Chem.* **245,** 4879 (1974).
[d] *N*-Acetylglucosamine.
[e] P. Kornfeld and S. Kornfeld, *J. Biol. Chem.* **245,** 2536 (1970).
[f] A. Linker and P. Hovingh, this series, Vol. 28, p. 902.
[g] M. E. Rafelson, S. Gold, and I. Priede, this series, Vol. 8, p. 677.
[h] R. M. Pino, E. Essner, and L. C. Pino, *J. Histochem. Cytochem.* **30,** 245 (1982).
[i] Hanks' balanced salt solution.
[j] W. Ebling, N. Hennrich, M. Klockow, H. Metz, H. D. Orth, and H. Lang, *Eur. J. Biochem.* **47,** 91 (1974).
[k] M. Simionescu, N. Simionescu, J. E. Silbert, and G. E. Palade, *J. Cell Biol.* **90,** 614 (1981).

kept in buffer overnight at 4°. This step serves to remove any remaining fixative that may affect the cytochemical reagents.

The application of cytochemical reagents requires adequate penetration into the tissue. Inadequate penetration may yield false-negative results. Althought a perfused vasculature forms an anastomosing network of "tunnels" from one side of the tissue to another, thinner sections assure the best penetration. Our laboratory used 40-μm nonfrozen sections obtained with a TC-2 (Smith-Farquhar) tissue sectioner. Vibratome or frozen sections are alternatives. With these sections, reagents must travel only 20 μm to reach the center of the tissue. In order to further enhance penetration, the vials (15 × 45 mm) containing the tissue are placed on a rotator during the incubation steps.

Cytochemical Incubation

The localization of lectin–receptor monosaccharides with avidin–ferritin is done as follows:[15–18] (1) incubation with biotinylated lectin (0.1 mg/ml PBS) for 2 hr, (2) PBS wash, 4 times for 15 min each, (3) incubation with avidin–ferritin conjugates (0.1 mg/ml PBS) for 2–18 hr, and (4) PBS wash, 4 times for 15 min each. Incubation in reagents for longer than 2 hr is usually not needed. However, for some experiments, tissue penetration may be a problem even if 40-μm sections are used. An overnight incubation at 4° will not compromise the morphology of the tissue and will provide adequate incubation of tissue. The treatment with avidin–ferritin, both in time and concentration, is the most critical step.

For experiments using HRP as a marker, the following protocol can be used: (1) incubation with biotinylated lectin (0.1 mg/ml PBS) for 2 hr, (2) PBS wash, 4 times for 15 min each, (3) Avidin treatment (0.1 mg/ml PBS) for 30–60 min, (4) PBS wash, 4 times for 15 min each, (5) incubation with biotinylated HRP (0.05–0.1 mg/ml PBS) for 2 hr, (6) PBS wash, 4 times for 15 min each, and (7) cytochemical incubation (37° for 30–60 min) with 15 mg 3,3′-diaminobenzidine in 50 m*M* Tris-HCl (adjusted to pH 7.0 after addition of the former) containing 0.01% H_2O_2 (final concentration), followed by filtering through Whatman #1 paper (2 ml reaction medium per vial).

The concentrations of the above reagents have been determined as optimal via experimentation. With both methods, as a control for carbohydrate specificity, the biotinylated lectin should be incubated with tissue in the presence of 0.1 *M* of the appropriate competing hapten sugar both prior to and during the treatment of tissue. Another important control is the omission of the biotinylated lectins in the incubation sequence. For the HRP marker, in addition to the above controls, avidin and biotinylated HRP should also be omitted in the incubation sequence. For initial experiments, it is also advisable to use tissue or cells with known carbohydrates on the cell surface as a positive control.

Enzyme Studies

To more fully characterize the sugars on the cell surface, prior to perfusion of fixative, specific enzymes can be perfused (Table I). Although enzyme studies may not be necessary, they can provide information concerning the nature of the cell surface that will augment the localization of lectin–receptor carbohydrates.

Final Processing

Following the above labeling procedures, the tissues are postfixed in 2% OsO_4 and washed in distilled water. We routinely use acidified 2,2-dimethoxypropane as a dehydrating agent as an alternative to ethanol.[19] This chemical, when added to water, will form acetone and methanol. In order for the reaction to work and maintain tissue morphology, residual traces of buffer should be removed, and overacidification of the 2,2-dimethoxypropane must be prevented. After dehydration, the tissues are rinsed in propylene oxide, infiltrated, and embedded in epoxy resin. Sections are stained with uranyl acetate and lead citrate or lead citrate alone.

Comments

The avidin–biotin method for the localization of lectins using biotinylated HRP as a marker produces a diffuse reaction product on the cell surface (Fig. 1). A major advantage of this approach is that the biotinylated HRP is small enough (MW ~40,000) to gain access to the abluminal front and some cytoplasmic vesicles of endothelial cells. In a few instances, we have been able to localize lectin–receptor monosaccharides in the stacks of the Golgi apparatus.[20] This HRP reaction product is visible at low magnification and is easily viewed by light microscopy. However, a major disadvantage is that the probe is not very discrete with respect to localizations on the cell surface. Diffusion artifacts associated with HRP and diaminobenzidine reactions are well known.[21]

With avidin–ferritin as a marker, high-resolution localizations are possible. The multiple binding capacity allows clusters of ferritin particles to mark specific sites (Fig. 2). Alternatively, discrete, more uniform localizations are possible (Fig. 3), depending on the composition of the cell sur-

[19] L. I. Muller and T. J. Jacks, *J. Histochem. Cytochem.* **23**, 107 (1975).
[20] R. M. Pino, unpublished observations.
[21] W.-L. Lin and E. Essner, *J. Histochem. Cytochem.* **34**, 1320 (1986).

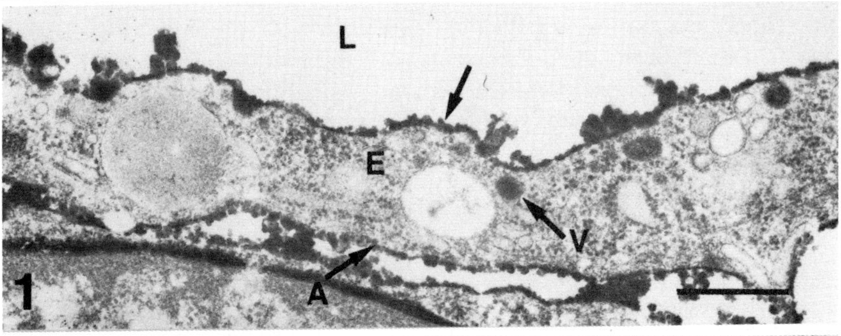

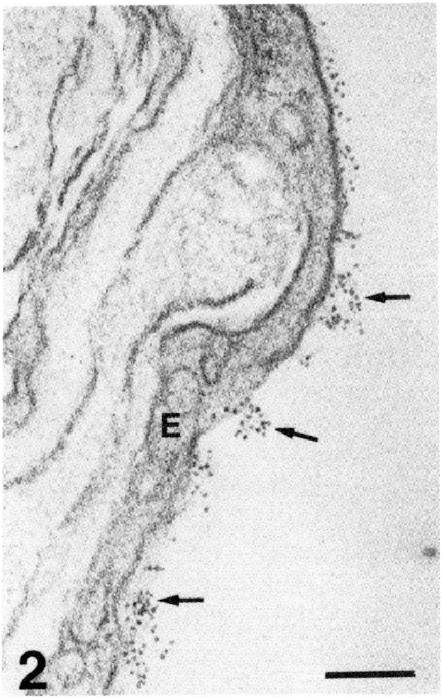

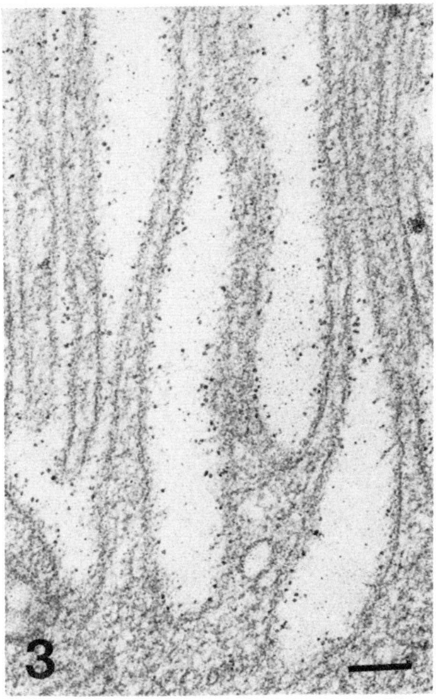

FIG. 1. Biotinylated concanavalin A–avidin–biotinylated HRP treatment of the bone marrow sinusoidal endothelium. Reaction product indicative of lectin binding is present on the luminal (arrow) and abluminal (A) fronts of the endothelium (E). Reaction product is also localized to vesicles (V). L, Lumen. Magnification: ×29,000; bar, 1 μm.

FIG. 2. Sequential treatment of retinal capillary with biotinylated *Ricinus communis* agglutinin and avidin-conjugated ferritin. Aggregates of avidin–ferritin particles (arrows) indicate areas containing a high degree of β-D-galactose residues on the endothelium (E). Magnification: ×108,000; bar, 0.1 μm.

FIG. 3. Treatment of the apical surface of the retinal pigment epithelium with biotinylated wheat germ aggutinin–avidin–ferritin. Numerous avidin–ferritin particles mark the location of the lectin binding sites. Magnification: ×84,000; bar, 1 μm.

face. This information is not obtainable with HRP methods (Fig. 1). We also feel that the localizations seen with the avidin–ferritin method are often better than those obtained with lectin–ferritin conjugates. This may be due to the more gentle conjugation used to prepare avidin- and biotin-

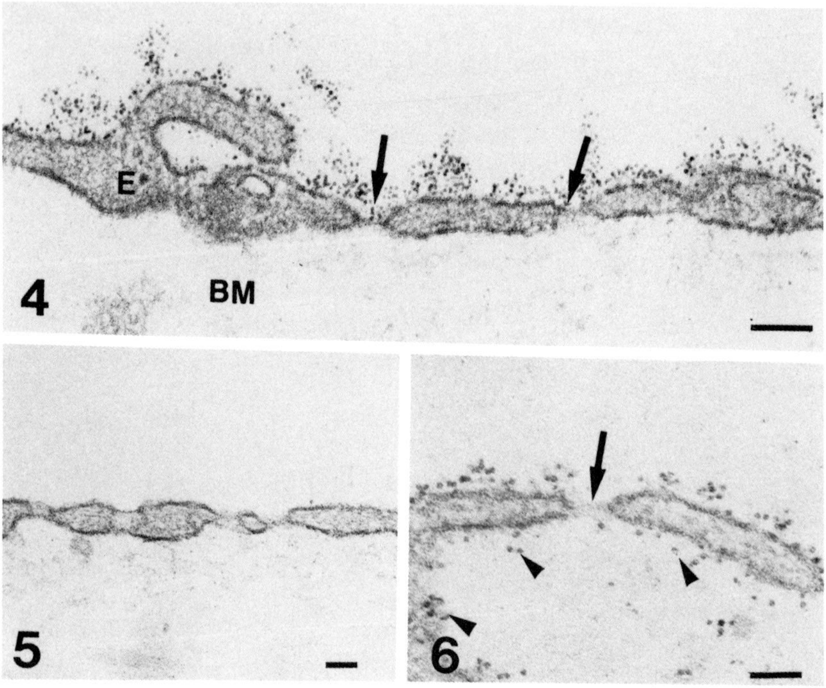

FIGS. 4–6. Treatment of the endothelium of the choriocapillaris with biotinylated wheat germ agglutinin–avidin–ferritin.

FIG. 4. A high degree of labeling is present at the luminal surface of the endothelium (E) including some diaphragmed fenestrae (arrows). The labeling here has been shown to be due to *N*-acetylglucosamine [R. M. Pino, *Cell Tissue Res.* **243**, 145 (1986)]. Note the absence of particles in Bruch's membrane (BM). Magnifications: ×82,000; bar, 1 μm.

FIG. 5. Cytochemical treatment following the perfusion of endoglycosidase D. In contrast to Fig. 4, there is now no labeling of the endothelium. This indicates that the *N*-acetylglucosamine of the glycocalyx is part of low-mannose oligosaccharides. Magnification: ×84,000; bar, 0.5 μm.

FIG. 6. Cytochemical treatment following proteinase K perfusion. Compared to Fig. 4, there is an absence of labeling on a diaphragmed fenestra (arrow). This indicates that the *N*-acetylglucosamine moieties of the fenestrae are components of glycoproteins with large amounts of hydrophobic amino acid residues. Note that following this digestion step, avidin–ferritin particles (arrowheads) have gained entrance into Bruch's membrane. Magnification: ×129,000; bar, 0.05 μm.

containing cytochemical reagents with less destruction of biological activities.[14]

The use of enzymes in conjunction with lectin-based localizations can add new dimensions in studying the structure of cell surface carbohydrates *in situ*. For example, we have shown that wheat germ agglutinin binds to the surface of the endothelium of the choriocapillaris.[16] This included a localization of the label to diaphragmed fenestrae (Fig. 4). The failure of neuraminidase to affect the pattern of localization suggested that this lectin marked *N*-acetylglucosamine residues. This contention was supported by the absence of labeling following treatment with *N*-acetylhexosaminidase.[16] This approach was taken one step further by the use of additional enzymes. The absense of localization after endoglycosidase D digestion (Fig. 5) showed that the *N*-acetylglucosamine moieties were on low-mannose carbohydrates. Furthermore, the loss of labeling with proteinase K (Fig. 6) treatment indicated that these moieties were components of proteins with high contents of hydrophobic amino acids.

The location and nature of some cells, such as the vascular endothelium of fenestrated capillary beds, often prevent precise biochemical analysis of the cell surface plasmalemma and of specializations such as diaphragmed fenestratae. This chapter has demonstrated that with the use of the avidin–biotin method in concert with digestive enzymes, a more detailed chemical analysis of the cell surface of these and similar cells is possible.

Acknowledgments

This research has been supported by Grant EY 03776 from the National Institutes of Health and a grant from the Cancer Association of Greater New Orleans. The author thanks Ms. Carol L. Thouron for expert technical assistance.

[44] Localization of Gonadotropin-Releasing Hormone Receptors

By GWEN V. CHILDS

Gonadotropin-releasing hormone (GnRH) is a decapeptide that binds to receptors on the pituitary gonadotrope and specific cells in the gonad. Whereas its function in the pituitary is stimulatory, less is known about its role in the gonads. Studies of its binding and receptor-mediated endocytosis in the different target cells may therefore be important to aid our understanding of its function and the factors that modulate this function. To this end, a number of investigators have attached labels to GnRH to morphologically trace its pathway into the target cells.[1–9]

Labels such as colloidal gold or ferritin provide discrete localization but may compromise the potency of the hormone. Radioactive labels allow full potency; however, the resolution of the signal after autoradiography does not allow accurate marking and identification of the smallest organelles in the pathway. Therefore, we developed an avidin–biotin labeling system that has the resolution of the gold or ferritin protocols and also maintains the full potency of the GnRH analog.[8–10] The purpose of this chapter is to describe the protocol for use and application of this detection system.

Description of Biotinylated GnRH Analog

The GnRH analog used for this protocol is one in which lysine is substituted for the glycine in the sixth position, thereby making it available for conjugation to a derivative of biotin that reacts with amino groups. The conjugate is synthesized by reacting [D-Lys^6]GnRH (0.6 mg) with D-biotin-*p*-nitrophenyl ester.[11] The potency of the [biotinyl-D-Lys^6]GnRH is tested by both binding assays and bioassays. The former is performed as described by Hazum.[11] The bioassays are performed by incubating 10^{-12} to 10^{-8} *M* of either the biotinylated or the unlabeled [D-Lys^6]GnRH analog with 3-day monolayer cultures of pituitary cells from female rats (mixed cycles) for 4 hr at 37°.[8,9] The amount of luteinizing hormone (LH) or follicle-stimulating hormone (FSH) released from the

[1] Z. Naor and E. Yavin, *Endocrinology* **111,** 1615 (1982).
[2] E. Hazum and A. Nimrod, *Proc. Natl. Acad. Sci. U.S.A.* **79,** 1747 (1982).
[3] C. Seguin, G. Pelletier, D. Dube, and F. Labrie, *Regul. Pept.* **4,** 183 (1982).
[4] C. R. Hopkins and H. Gregory, *J. Cell Biol.* **75,** 528 (1977).
[5] G. Pelletier, D. Dube, J. Guy, C. Seguin, and F. A. Lefebvre, *Endocrinology* **111,** 1068 (1982).
[6] T. M. Duello, T. M. Nett, and M. G. Farquhar, *Endocrinology* **112,** 1 (1982).
[7] L. Jennes, W. E. Stumpf, and P. M. Conn, *Endocrinology* **113,** 1683 (1983).
[8] G. V. Childs, Z. Naor, E. Hazum, R. Tibolt, and K. N. Westlund, *J. Histochem. Cytochem.* **31,** 1422 (1983).
[9] G. V. Childs, Z. Naor, E. Hazum, R. Tibolt, K. N. Westlund, and M. B. Hancock, *Peptides* **4,** 549 (1983).
[10] G. V. Childs, E. Hazum, A. Amsterdam, R. Limor, and Z. Naor, *Endocrinology* **119,** 1329 (1986).
[11] E. Hazum, this volume [29],

cultures in response to the stimulus is assayed by radioimmunoassays.[8,9,12]

Such tests show that there is no loss in binding affinity or biological activity of the biotinylated GnRH analog. In fact, the ED_{50} values for the assays (Table I) demonstrate that there is at least a 3-fold increase in potency when biotin is attached to the GnRH analog. Therefore, target cells can be stimulated with concentrations of the biotinylated analog that match those in portal blood.

Detection of Biotinylated GnRH at Light Microscopic Level

The labeled avidin detection system used in initial studies is the avidin–biotin–peroxidase complex (ABC, Vectastain Kit, Vector Laboratories, Burlingame, CA), which is obtained in kit form for use in immunocytochemical protocols. The detection system is applied to fixed pituitary cells from female rats (mixed cycles) plated in monolayer cultures and exposed to 10^{-12} or 10^{-10} *M* of biotinylated [D-Lys^6]GnRH for 30 sec–30 min at 37°. The biotinylated analog is diluted in minimum essential medium (MEM) containing 0.3% bovine serum albumin (crystalline, Sigma Chemical Co., St. Louis, MO). To some of the samples, a 10- to 1000-fold excess of unlabeled GnRH analog is added to compete with the biotinylated GnRH for binding sites. The cells are then fixed with glutaraldehyde [2.5% in 0.1 *M* phosphate buffer (pH 7.4)] for 30 min at 24° and washed 3 times in the same buffer containing 4.5% sucrose (w/v). Then the cells are prepared for the avidin detection system by treatment with 3–5% normal goat serum diluted in 50 m*M* phosphate buffer containing 0.3% crystalline bovine serum albumin. This solution is applied for 15 min at 24° to block nonspecific reactive sites and also is the diluent buffer for the avidin–biotin–peroxidase complex (ABC) detection system.

The ABC detection system requires the use of matched components of the Vectastain Kit. The complex is made 30 min prior to its use as follows: 25 μl of avidin is added to 25 μl of biotinylated peroxidase in 4 ml of the above diluent buffer. After 30 min, the ABC solution is added to the above groups of cells, and the incubation is continued at 24° for 60 min. Then the cells are washed 3 times in phosphate buffer and once in 50 m*M* acetate buffer (pH 6).

The peroxidase substrate used in the detection system for the biotinylated analog is nickel-intensified diaminobenzidine (DAB), made as follows: 0.45 g of nickel ammonium sulfate is added to 30 ml of 50 m*M* acetate buffer (pH 6) and dissolved while stirring. Then, 6 mg of DAB is

[12] R. E. Tibolt and G. V. Childs, *Endocrinology* **117,** 396 (1985).

TABLE I
POTENCY OF BIOTINYLATED ANALOGS OF GONADOTROPIN-RELEASING HORMONE

Parameter	Potency	
	[Biotinyl-D-Lys^6]GnRH	[D-Lys^6]GnRH
Binding affinity (ED_{50}) (radioreceptor assay)	0.7 n*M*	2.0 n*M*
Biological activity		
ED_{50} LH radioimmunoassay	0.075 n*M*	0.5 n*M*
ED_{50} FSH radioimmunoassay	0.02 n*M*	0.2 n*M*

added to the solution. Following its solubilization, 15 μl of 30% hydrogen peroxide (no more than 2 months old) is added. The solution is filtered immediately (Whatman No. 1 filter paper) and applied to the cells for 6 min. The solution is usually a light clear green at first and may turn cloudy during reaction. Since DAB is easily dissolved in water-soluble mounting media such as glycerol, the stained cells are then dehydrated in 50, 70, and 95% ethanol (v/v), followed by 2 changes of absolute ethanol (10 min in each solution). The samples are then dried, dipped in xylene, and mounted with permount.

The cell populations are then analyzed for the percentage of reactive cells as well as the density of the stain.[8-10,12,13] Quantification of the percentage of labeled cells in populations from diestrous female rats is shown in Table II. Table II also shows that when unlabeled [D-Lys^6]GnRH is added with the biotinylated analog, a 92% decrease in the percentage of labeled cells is observed. This indicates that the unlabeled analog is competing for GnRH receptors and supports the specificity of the reaction. Another control can be performed by adding a 1000-fold excess of another releasing hormone, e.g., corticotropin-releasing hormone, which should fail to block binding by the biotinylated GnRH.[8,9] Finally, the data in Table II also show that exposure to solutions containing no biotinylated GnRH results in a 97% decrease in the percentage of labeled cells. This indicates that nonspecific reactions with ABC or DAB solutions do not contribute significantly to the signal.

Thus, the biotinylated GnRH analog is a potent, specific probe for the GnRH receptor when used with ABC peroxidase detection systems. The black reaction for the biotinylated analog is sufficiently dense so that it can be followed by an immunocytochemical stain for LH or FSH anti-

[13] G. V. Childs, J. L. Morell, A. Niendorf, and G. Aguilera, *Endocrinology* **119,** 2129 (1986).

TABLE II
PERCENTAGE OF CELLS LABELED FOR BLOTINYLATED GnRH (FEMALE RATS)[a]

Conditions	Detection system: ABC–peroxidase (after fixation)	Detection system: Avidin–fluorescein (living cells)
Time (min) of exposure to biotinylated GnRH		
0	0.4 ± 0.3	1.6 ± 0.5
1	11.8 ± 1.6	10.0 ± 0.5
5	11.6 ± 3.5	11.5 ± 0.9
15	12.7 ± 2.0	8.9 ± 2.0
30	10.0 ± 1.6	6.0 ± 2.0
Plus 10^{-6} *M* [D-Lys^6]GnRH (5 min)	1.2 ± 0.2	1.38 ± 0.5
No biotinylated GnRH but with 100 μg/ml avidin–fluorescein plus 10^{-6} *M* [D-Lys^6]GnRH (15 min)	—	1.0 ± 0.4

[a] Values give the percentage of labeled cells ± SE.

gens.[9] Figure 1 (see color plate opposite p. 400) shows a double-labeled cell, exposed to biotinylated GnRH for 3 min, fixed, and then stained with the ABC complex and the black peroxidase substrate. This reaction was then followed by an immunocytochemical stain for LH with a red peroxidase substrate (see Ref. 9). The gonadotrope has extended a process as a result of the stimulation which is filled with LH stores (red). The biotinylated GnRH is seen in black patches in or on the cellular process. Quantitative analyses showed that 90% of all gonadotropes (containing LH or FSH antigens) exhibited a reaction for biotinylated GnRH.

In the next phase, biotinylated GnRH is used in combination with avidin-derivatized fluorescein to learn if the labeled analog can be localized on living cells. This avidin detection system was developed especially for use with cells in fluorescence-activated cell sorters (Avidin–fluorescein, Cell Sorter Grade, Vector Laboratories). The protocol for exposure of the cells to biotinylated GnRH is identical to that used for the ABC detection system, except that the cells are not fixed. Rather, the reaction is stopped by washing 3 times in cold phosphate-buffered saline [(PBS) 50 mM phosphate buffer (pH 7.4) containing 0.9% NaCl] maintained at ice-bath temperatures. Then cold avidin–fluorescein is added, and the incubation is allowed to proceed in the dark at 2–4° for 15–20 min. The avidin–fluorescein is diluted to a concentration between 10 and 25 μg/ml in PBS. After washing in cold PBS, the cells are examined immediately by placing the coverslip in a shallow well filled with cold PBS and covering the well with another coverslip. The remaining sets of cells are kept cold in the refrigerator as each coverslip is examined with the fluorescence microscope (Leitz, Inc. Optometric, Houston, TX).

With the avidin–fluorescein protocol, no more than 12 coverslips are stained at a time. Counting is done rapidly, because the living cells internalize the GnRH–receptor–avidin–fluorescein complex when warmed to 24°. Such internalization quenches the fluorescence as the complex is brought into the acidic environment of a receptosome, resulting in loss of signal and lowered counts of labeled cells. Furthermore, Table II shows that fewer cells are detected with this system if applied after 5 min of exposure to the biotinylated GnRH. This is undoubtedly the result of the rapid uptake of the GnRH–receptor complex, making it unavailable for reaction with avidin–fluorescein (in the living state).

Tests for specific binding are also conducted with the avidin–fluorescein detection system. These tests involve omission of the biotinylated GnRH or addition of unlabeled [D-Lys^6]GnRH with the biotinylated GnRH. Additional tests are conducted, however, to determine if any of the labeling is due to nonspecific uptake of avidin–fluorescein during stimulation. The latter tests are conducted at 0–2° and involve the addition of unlabeled [D-Lys^6]GnRH with as much as 100 μg/ml avidin–fluorescein for 15 min. Table II shows that after all of these controls, there is greater than 90% loss in the percentage of labeled cells.

The detection systems allows study of the events leading up to receptor-mediated endocytosis by target cells.[8,9] During the first minute of exposure to the biotinylated GnRH, the label is distributed diffusely over the cell surface. After 3 min, the label is seen in patches, usually at one pole of the cell (Figs. 1 and 2, see color plate opposite p. 400). The patches are larger and less frequent after 5 min of exposure, and by 10–30 min granular deposits are seen scattered at one pole of the target cell. Often the gonadotropes respond to the stimulus by sending out processes that appear to concentrate the biotinylated ligand. This dynamic labeling

FIG. 1. Double label for biotinylated GnRH (black) followed by immunocytochemical stain for LH (red). The gonadotrope was stimulated for 3 min with the biotinylated GnRH and has responded by extending a cellular process that resembles a ruffle (arrow). The process is filled with LH antigen, and the black reaction for the GnRH shows that GnRH binds to receptors that are concentrated on the cellular processes. (From Ref. 9, with permission.) Magnification: ×1010; bar, 20 μm.

FIG. 2. Avidin–fluorescein label for biotinylated GnRH (arrows) applied to living cells at 0–2° after stimulation with biotinylated hormone for 3 min. Cellular processes extend in response to the stimulation and are the sites of highest concentration of the GnRH receptor. Magnification: ×1010; bar, 20 μm.

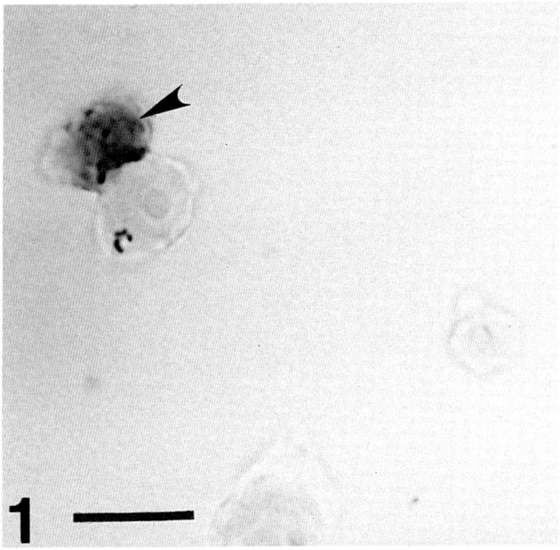
1

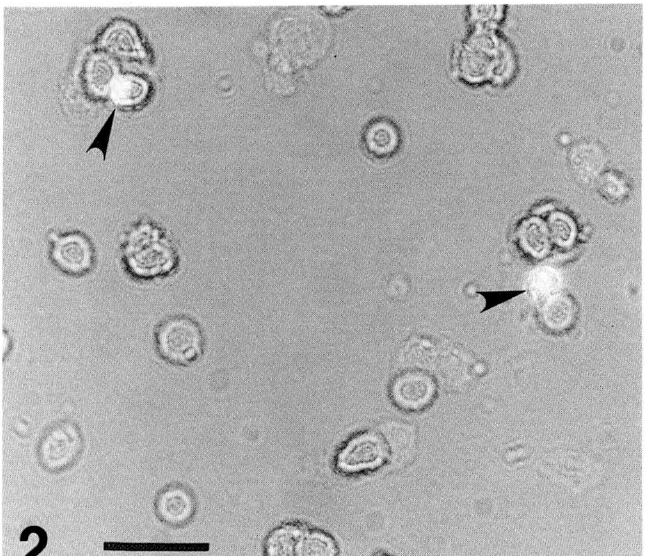
2

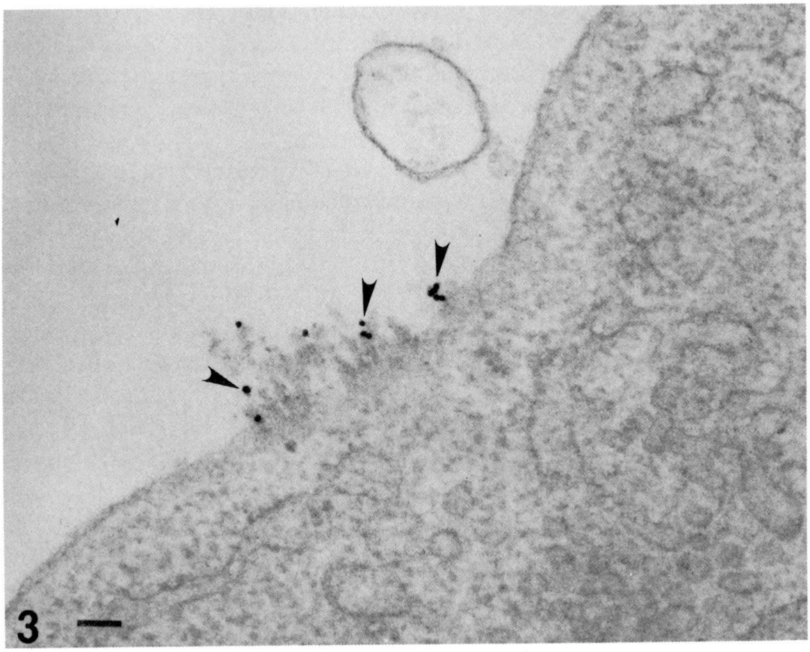

FIG. 3. Avidin–gold label for biotinylated GnRH on an ovarian granulosa cell exposed to GnRH for 3 min. The label has clustered on the surface and appears associated with fine microvilli (arrows). Original magnification: ×58,900; bar, 0.1 μm.

pattern is seen most dramatically in the avidin–fluorescein-labeled living cells (Fig. 2).

Localization of Biotinylated GnRH at Electron Microscopic Level

After the specificity of the stain is established and the target cells are identified by immunocytochemistry, the protocols are applied to cells in suspension in order to investigate the pathway of internalization of the GnRH–receptor complex at the electron microscopic level.[10] Initially, three different avidin detection systems have been applied to localize biotinylated GnRH: ABC peroxidase, avidin–gold, and avidin–ferritin (Vector Laboratories). The protocol for exposure to biotinylated GnRH is similar to that described for the light microscopic studies, except that the living cells are stimulated in suspension and then fixed by adding an equal volume of 4% glutaraldehyde directly to the cell suspension. After 30 min at 24°, the cells are centrifuged at 900 rpm and then washed in phosphate buffer containing 4.5% sucrose 4 times over a period of 1 hr. After each wash or exposure to the staining solutions (at room temperature), the cells are centrifuged at 900 rpm for 5 min and then resuspended in the next solution in the protocol.

After a 15-min exposure to the same blocking buffer used for the light microscopic studies, the cells are incubated in one of the avidin detection systems for 60 min. The ABC solution is made as described for the light microscopic studies; the avidin–gold (10–15 nM) is diluted 1 : 10, and the avidin–ferritin is diluted to 10 μg/ml prior to application in the protocol.

After exposure to the labeled avidin solutions, the cells are washed by 3 changes of phosphate buffer. The cell suspensions (treated with ABC peroxidase) are then exposed to nickel-intensified DAB (made as in the light microscopic studies). After a 5- to 6-min incubation in DAB, the suspension is centrifuged at 900 rpm for 5 min and washed 3 times in acetate buffer.

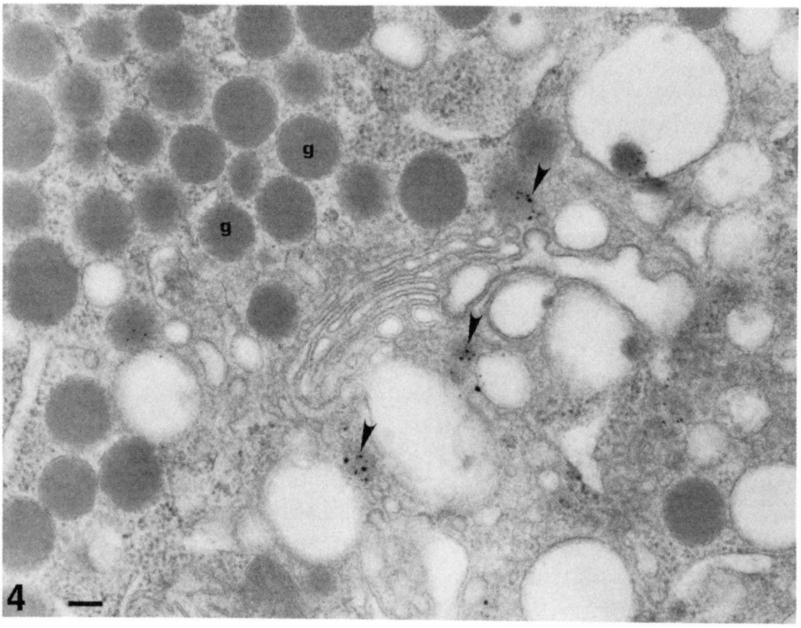

FIG. 4. Avidin–biotin–peroxidase complex label for biotinylated GnRH in the Golgi complex region of a pituitary gonadotrope. Dense deposits of peroxidase reaction product (arrows) are found on immature granules in the Golgi complex. Nearby storage granules (g) are unlabeled at this time after stimulation (3 min). Original magnification: ×47,120; bar, 0.1 μm. (From Ref. 10, with permission.)

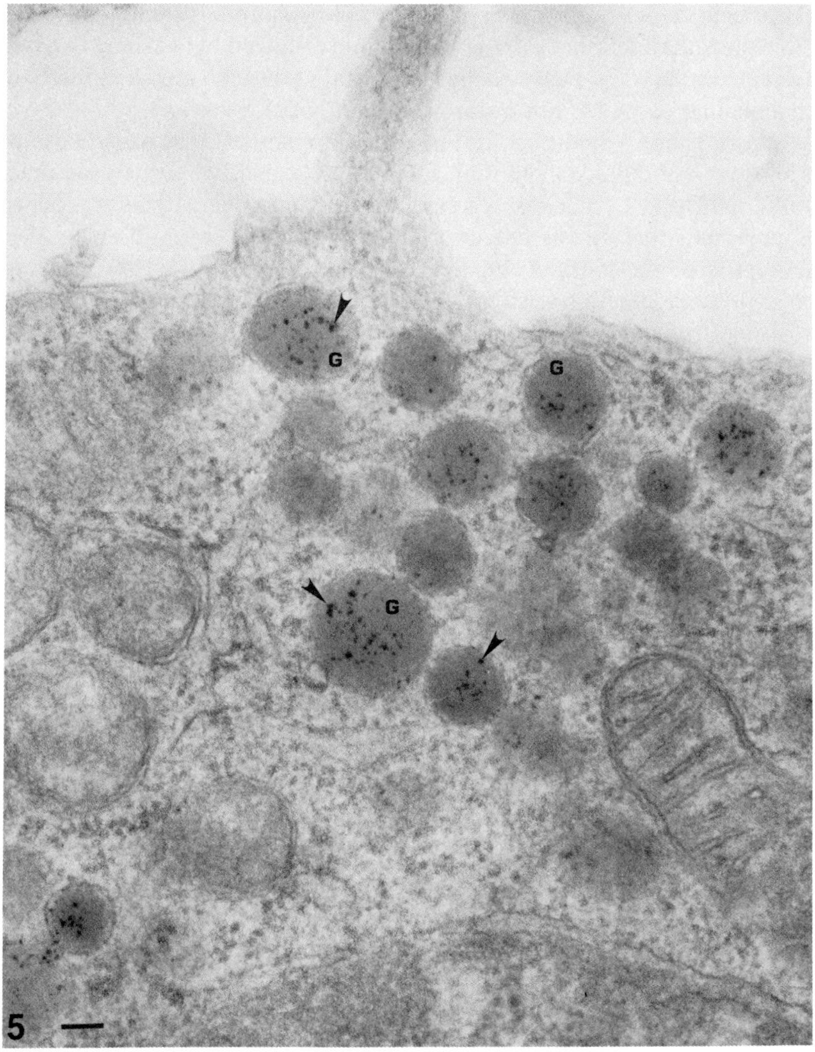

FIG. 5. Avidin–biotin complex label for biotinylated GnRH after 5 min of stimulation shows that the GnRH receptor, or biotinylated GnRH, is found in a subpopulation of secretory granules (G) in addition to secondary lysosomes (see Ref. 10). The region depicts a cluster of labeled granules (arrows) near the surface, and the working hypothesis is that the granule membrane may represent one route for receptor recycling. The label inside the granules may in fact be a biotinylated metabolite of the GnRH analog. Original magnification: ×55,200; bar, 0.1 μm.

All cell suspensions are exposed to 1% osmium tetroxide [diluted in 0.1 *M* phosphate buffer (pH 7.4)] for 30 min followed by washing twice in the same buffer. The cells are then dehydrated through a graded series of ethanols and embedded in Epon or Araldite 6005.[10]

Figure 3 shows avidin–gold stain for biotinylated GnRH on the surface of an ovarian granulosa cell, and Fig. 4 shows an ABC peroxidase reaction in the Golgi complex of a gonadotrope. The ABC method appears to be more sensitive for the detection of the ligand–receptor complex (or a metabolite of the biotinylated ligand) inside the cells than the gold- or ferritin-based detection systems. Indeed, the ABC detection system (Fig. 5) is able to define a subpopulation of granules as a respository for the ligand–receptor (or a biotinylated metabolite). Gold partcles larger than 5 nm penetrate cells poorly and fail to label intracellular compartments such as granules. They are, however, useful for quantitative studies of the changes in surface labeling.[10]

Advantages of Avidin Labeling Systems

These studies have shown that the biotinylated analog of GnRH is a sensitive, potent probe for GnRH receptors that can be detected in or on target cells with several different avidin detection systems. Changes in receptivity can be detected cytochemically following modulation of secretion.[12] More recently, the same protocols have been applied to localize a synthetic biotinylated analog of corticotropin-releasing hormone (CRH).[14–16] The affinity cytochemistry protocol for the biotinylated ligands can be followed by immunocytochemistry with a contrasting colored substrate or a different electron-dense marker to further identify the antigens inside the receptive cells.[10,14,16] It is quantifiable both by microdensitometry and by counting defined markers such as the gold particles. Finally, it can be applied to living cells and detected with preparations of labeled forms of avidin that are designed for use with such systems, allowing one to conduct studies of stimulus–secretion coupling in labeled, living cell populations.[15]

[14] K. N. Westlund, P. J. Wynn, S. Chmielowiec, T. J. Collins, and G. V. Childs, *Peptides* **5,** 634 (1984)

[15] G. V. Childs, G. Unabia, J. A. Burke, and C. Marchetti, *Am. J. Physiol.* **252,** E347 (1987).

[16] G. V. Childs, C. Marchetti, and A. M. Brown, *Endocrinology* **120,** 2059 (1987).

[45] Localization of Ganglioside G_{M1} with Biotinylated Choleragen

By HIROAKI ASOU

Affinity cytochemistry utilizing the avidin–biotin complex system for the visualization of reactive sites has several advantages over immunocytochemical procedures.[1,2] The avidin–biotin–peroxidase method is much more sensitive and yields a staining reaction with lower background than classic immunocytochemical techniques.[3,4] The very high affinity of avidin for biotin is almost irreversible. In addition, the availability of numerous biotinylation reagents for coupling as well as different avidin conjugates (e.g., fluorescein, peroxidase, and ferritin) provides the means of using a method which is more simple, versatile, and specific than most of the immunocytochemical methods.

In this chapter, I describe the use of the avidin–biotin complex system (with biotinylated choleragen and avidin–peroxidase) for the visualization of ganglioside G_{M1} in primary cell cultures from rat brain. Cholera toxin binds specifically and with high affinity to ganglioside G_{M1}, thus providing a means for demonstrating the localization of this ganglioside.[5–7]

Biotin Labeling of Choleragen

Biotinyl-*N*-hydroxysuccinimide ester is used for labeling the cholera toxin B subunit.[8,9] Briefly, 30 μl of biotinyl-*N*-hydroxysuccinimide ester solution (5.5 mg/ml) in dimethylformamide is added to 2 mg of cholera toxin B subunit (Campbell, CA) in 1 ml of Tris–EDTA buffer [50 m*M* Tris-HCl, 0.2 *M* NaCl, 1 m*M* Na_2EDTA (pH 7.5)]. The mixture is incubated at room temperature for 2 hr, kept at 4° overnight, and then dialyzed against Dulbecco's phosphate-buffered saline (PBS) for 3 days at 4° with three buffer changes. Biotinylated choleragen is stored at −20° until used.

[1] E. A. Bayer, M. Wilchek, and E. Skutelsky, *FEBS Lett.* **68,** 240 (1976).
[2] J. L. Berman and R. S. Basch, *J. Immunol. Methods* **36,** 335 (1980).
[3] S. M. Hsu, L. Raine, and H. Fanger, *J. Histochem. Cytochem.* **29,** 557 (1981).
[4] J. L. Guesdon, T. Ternyck, and S. Avrameas, *J. Histochem. Cytochem.* **27,** 1131 (1979).
[5] P. Cuatrecasas, *Biochemistry* **12,** 3547 (1973).
[6] H. A. Hansson, J. Holmgren, and L. Svennerholm, *Proc. Natl. Acad. Sci. U.S.A.* **74,** 3782 (1977).
[7] P. H. Fishman, *J. Membr. Biol.* **69,** 85 (1982).
[8] E. A. Bayer, E. Skutelsky, and M. Wilchek, this series Vol. 62, p. 308.
[9] H. Asou, E. G. Brunngraber, and I. Jeng, *J. Histochem. Cytochem.* **31,** 1375 (1983).

For the determination of biotin–binding activity to cholera toxin, ^{125}I-labeled cholera toxin (5×10^4 cpm) is included. The recovery of the biotinylated ^{125}I-labeled choleragen was 65–70% under the above conditions.

Cell Culture Conditions[9,10]

The preparation of primary cultures from rat embryonic cerebral hemispheres is carried out as follows. Ten to twelve fetuses (Sprague-Dawley strain) are removed at a gestational age of 18 days. The cerebral hemispheres are separated from the rest of the brain. The meninges and blood vessels are carefully removed using a microscope. The pooled hemispheres are then rinsed 3 to 4 times with Eagle's minimum essential medium (MEM) containing 10% fetal calf serum (FCS) but no sodium bicarbonate. The tissue is placed on a screen and gently smoothed into a paste using a glass pestle. The paste is then extruded through a stainless steel mesh (140 μm pore size, 100 mesh screen) into an MEM-supplemented tissue culture dish.

The cells are collected by centrifugation at 800 rpm for 5 min at room temperature, resuspended in culture medium consisting of MEM, 10% FCS, 0.6% glucose, and 0.05% sodium bicarbonate, and washed 3 times by centrifugation under the same conditions. The cells are counted and plated in MEM containing 10% FCS, 0.05% sodium bicarbonate, 0.6% glucose, 292 μg/ml L-glutamine, and 100 μg/ml kanamycin at a density of 1×10^6 cells per 2 ml medium in a glass Leighton tube (Matsunami, Tokyo). The sample is covered with a poly(L-lysine)-coated cover glass (12×32 mm). Cell viability is determined by the Trypan blue exclusion test. The cells are grown in a CO_2 incubator (IF-41, Yamato, Tokyo) at 37° in an atmosphere of 97% air and 3% CO_2. The medium is renewed every third day of culture.

Assay Procedure

In order to detect ganglioside G_{M1} cytochemically, the following procedure which employs biotinylated choleragen and avidin–peroxidase is used.[9,11,12] The cultured cells are fixed with 2% paraformaldehyde in PBS (pH 7.4) for 30 min at 4°. After rinsing 5 times with PBS containing 2 mg/ml bovine serum albumin (BSA), the cells are treated overnight with

[10] H. Asou, N. Iwasaki, S. Hirano, and D. Dahl, *Int. J. Dev. Neurosci.* **4,** 477 (1986).

[11] H. Asou, N. Mutou, S. Hirano, T. Turumizu, and Y. Horibe, *Acta Histochem. Cytochem.* **18,** 383 (1985).

[12] H. Asou and E. G. Brunngraber, *Neurosci. Lett.* **46,** 115 (1984).

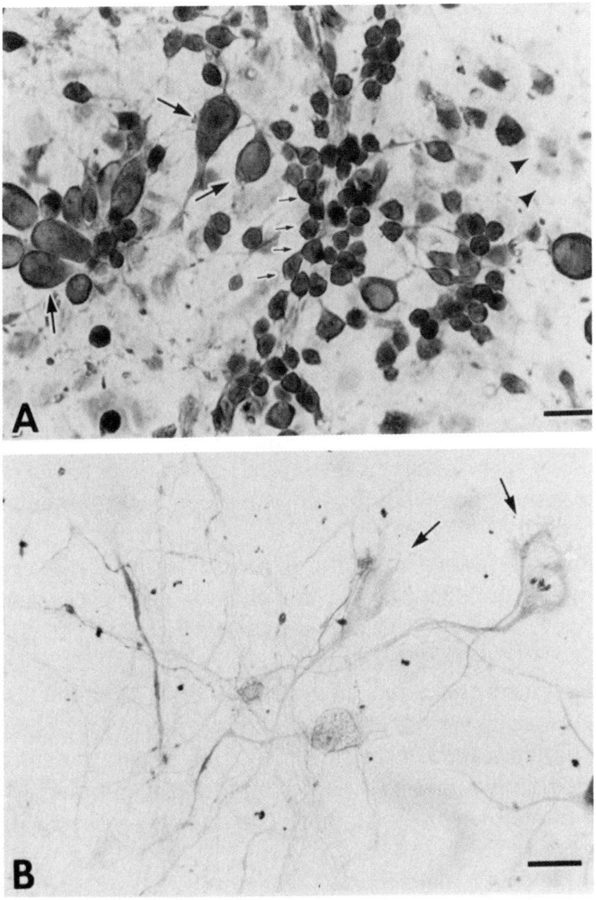

FIG. 1. (A) Staining for ganglioside G_{M1} utilizing biotinylated choleragen and avidin–peroxidase conjugate on 10-day-old cell cultures from rat embryonic cerebral hemispheres. The darkly stained cells are neurons (large arrows) and oligodendrocytes (small arrows). Numerous nuclei of unstained astroglial cells can be seen (arrowheads). Bar, 16 μm. Note that neurons were identified using antibodies to the neurofilament protein as a cell marker.[13] Oligodendrocytes were identified by using antibodies to hyaluronectin and cerebroside as cell markers.[14] Antiglial fibrillary acidic protein antibody was used to identify astrocytes.[15] Antibodies to ganglioside G_{M1} and the avidin–biotin complex method were used to identify G_{M1} in neurons and oligodendrocytes and the absence of this glycolipid in astrocytes.[9,12,16] (B) Staining for ganglioside G_{M1} with the avidin–biotin complex method on 6-day-old primary cultured neurons and their growth cones. A veillike lamellipodia bearing numerous filopodia were stained for ganglioside G_{M1} (arrows). Bar, 16 μm.

biotinylated choleragen (5 μg/ml) in PBS at 4°. Controls are treated with underivatized choleragen. After washing 3 times for 5 min with PBS–BSA solution, the cells are treated with avidin–proxidase (Sigma Chemical Co., St. Louis; 5 μg/ml) in PBS for 30–45 min at 37°. The cell preparation is washed several times with PBS–BSA solution, and the reaction is then developed with hydrogen peroxide as a substrate and 3, 3′-diaminobenzidine tetrahydrochloride (DAB) as an electron donor. A brown insoluble product is obtained. The DAB solution is removed by washing with distilled water. Hematoxylin staining is used to make the nuclei visible. The coverslips are then directly examined under a Nikon (Tokyo) microscope.

Cellular Localization of Ganglioside G_{M1}

Ganglioside G_{M1} serves as a membrane receptor for cholera toxin and is thought to be involved in specific cell receptor interactions on the oligodendrocyte and neuron membrane surface. The localization of ganglioside G_{M1} in primary cultured cells from embryonic rat cerebral hemispheres visualized by the avidin–biotin complex method is shown in Fig. 1.

The staining of 10-day-old cultures for G_{M1} showed that cells exhibiting a neuron and oligodendrocyte morphology gave a positive reaction. Astrocytes were not stained by this method (Fig. 1A).[13–16] G_{M1} was also found on the nerve growth cones (Fig. 1B). Although light microscopy is these studies indicates that ganglioside G_{M1} is probably localized on the membrane surface of growth cones, electron microscopic examination in required to obtain a more precise localization of this ganglioside. Since avidin has four active binding sites for biotin,[3,17] membrane receptors for biotinylated choleragen may become clustered upon cross-linking with avidin.

The present study suggests that other toxins, known to bind gangliosides specifically, may also be biotinylated and visualized using suitable markers conjugated to avidin.

[13] H. Asou, N. Iwasaki, S. Hirano, and D. Dahl, *Brain Res.* **332,** 355 (1985).
[14] H. Asou, E. G. Brunngraber, and B. Delpech, *J. Neurochem.* **40,** 589 (1983).
[15] H. Asou, S. Hirano, and E. G. Brunngraber, *Neurosci. Res.* **3,** 364 (1986).
[16] H. Asou, and E. G. Brunngraber, *Neurochem. Res.* **8,** 1045 (1983).
[17] E. Holtzmann, O. Wise, J. Wall, and A. Karlin, *Proc. Natl. Acad. Sci. U.S.A.* **74,** 3782 (1977).

[46] Visualization of Adenosinetriphosphatase Site of Myosin Using Photoreactive Adenosine 5′-Diphosphate Analog

By KAZUO SUTOH

Myosin has ATPase activity. Energy released from myosin-catalyzed ATP hydrolysis is converted to mechanical energy through the myosin–actin interaction. Thus, ATP hydrolysis is the most essential process of muscle contraction. In order to elucidate the molecular mechanism of the contraction, it is important to know where the ATPase site is located on the myosin molecule. This chapter describes a method for labeling the ATPase site with an avidin–biotin system.[1] First, the myosin ATPase site is covalently labeled with a photoreactive biotinylated ADP analog. Then avidin is attached to the biotin moiety of the ADP analog cross-linked to the ATPase site. The bound avidin, which can be visualized by electron microscopy, is a marker for the ATPase site.

Synthesis of Photoreactive Biotinylated ADP Analog

Synthesis of Biotinylated ADP Analog

N^6-[(6-Aminohexyl)carbamoylmethyl]-ADP (ACM-ADP, compound **I**, Fig. 1) is synthesized as previously described.[2] The *N*-hydroxysuccinimide ester of biotin (BNHS, **II**) is synthesized as described.[3] BNHS (1.2 mmol), freshly dissolved in 10 ml of dimethylformamide, is added dropwise to ACM-ADP (0.6 mmol) in 10 ml of water. The concentration of ACM-ADP is estimated by absorption at 267 nm (ε 17,300 M^{-1}).[2] The pH of the reaction mixture is maintained at 9.0 during the reaction (1–2 hr). The extent of biotinylation is followed by the ninhydrin reaction on TLC plates.

After completion of the reaction, 5 volumes of ice-cold acetone is added to the mixture to precipitate ADP derivatives. The precipitate is collected by low-speed centrifugation and dissolved in 50 ml of water. The pH is then adjusted to 3.2. The resulting solution is loaded onto a Dowex-1 column (2 × 10 cm) equilibrated with 0.5 *M* LiCl (pH 2.0). The biotinyl-

[1] K. Sutoh, K. Yamamoto, and T. Wakabayashi, *Proc. Natl. Acad. Sci. U.S.A.* **83,** 212 (1986).

[2] M. Lindberg and K. Mosbach, *Eur. J. Biochem.* **53,** 481 (1975).

[3] E. A. Bayer, E. Skutelsky, and M. Wilchek, this series, Vol. 62, p. 308.

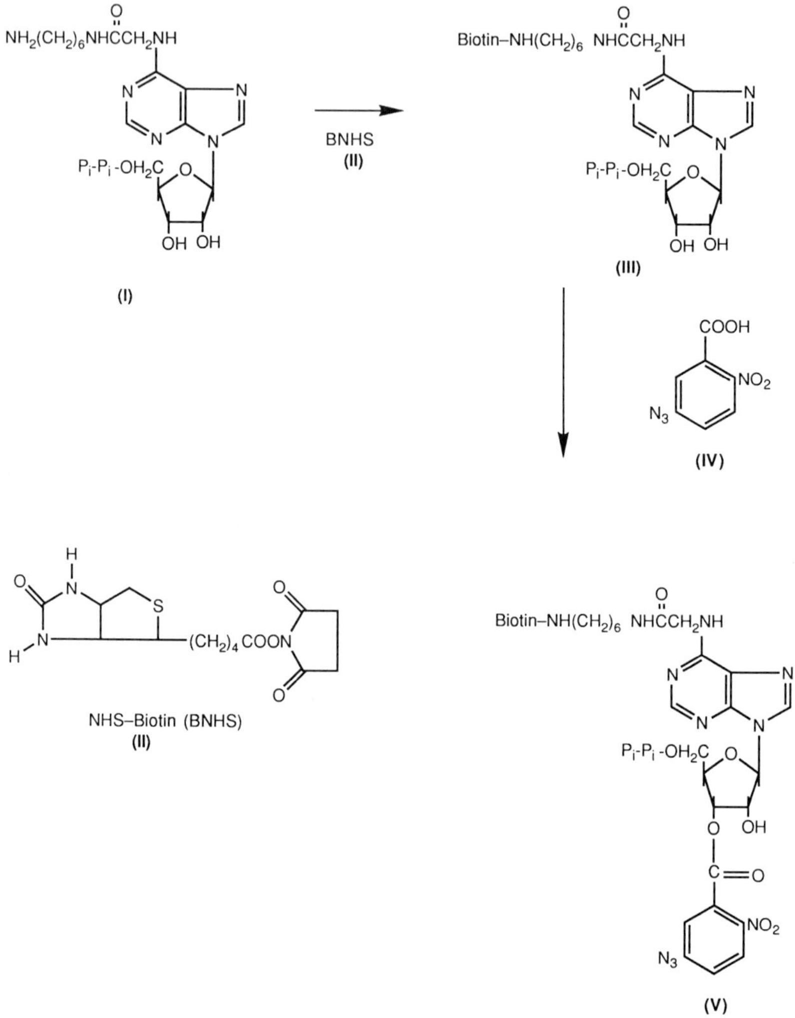

FIG. 1. Synthetic pathway to the biotinylated photoreactive ADP analog (**V**).

ated ADP analog (**III**) is eluted from the column by a linear gradient of LiCl from 0.5 to 1.0 *M* at pH 2.0 (total 1 liter). Elution is monitored by absorption at 267 nm. Fractions from the major peak are collected and neutralized. Pooled fractions are then concentrated to 50 ml by rotary evaporator at 40°. To the concentrated solution is added ice-cold acetone–ethanol (1 : 1, v/v) to precipitate the biotinylated ADP analog. The resulting precipitate is collected by low-speed centrifugation and dissolved in 50 ml of water. A purity check by HPLC with a DEAE-5PW column (Tosoh, Tokyo) shows a single elution peak; neither the starting material (ACM-ADP) nor any by-product is present in the final preparation. The concentration of the biotinylated ADP analog is estimated by absorption at 267 nm. Yield, about 50%.

Synthesis of Photoreactive Biotinylated ADP Analog

The biotinylated ADP analog (**III**, Fig. 1) is coupled with a photoreactive derivative, 5-azido-2-nitrobenzoic acid (**IV**),[4] which is synthesized according to the method of Lewis *et al.*[5] 5-Azido-2-nitrobenzoic acid (150 μmol) in 1 ml of dimethylformamide is first activated by the addition of solid carbonyldiimidazole (500 μmol). After incubation at 25° for 30 min, the mixture is poured into a solution (1 ml) of the biotinylated ADP analog (15 μmol in water). The coupling reaction is allowed to proceed at 35° for 4 hr. Then ADP derivatives are precipitated by the addition of 5 volumes of ice-cold acetone. The precipitate is collected by centrifugation and then dissolved in 5 ml of water. Excess 5-azido-2-nitrobenzoic acid is removed at this step.

The product (**V**) is finally purified by HPLC with the DEAE-5PW column. An aliquot (0.5 ml) is applied to the column, and ADP derivatives are eluted with a linear gradient of ammonium acetate from 0.1 to 0.65 *M*. Elution is monitored by simultaneous absorption at 267 and 320 nm. Since the 5-azido-2-nitrobenzoyl group is linked to 3′-OH group of ribose ring, the product (**V**) shows strong absorption at 320 nm. When monitored at this wavelength, only two peaks, corresponding to the product (**V**) and contaminating 5-azidonitrobenzoic acid, are detected. The product is eluted immediately after the starting material (**III**), which fails to react with 5-azido-2-nitrobenzoic acid.

Peak fractions are collected and lyophilized. The concentration of **V** is estimated by using the absorption of the 5-azido-2-nitrobenzoyl group at 320 nm (ε 9,000 M^{-1}). The yield of **V** from **III** is about 10%. The low yield is expected since modification at the 3′-OH of the ribose ring is a very inefficient reaction.[4] The final product is stable in the lyophilized state or in water when stored at −25° in the dark. Even after 1 year of storage, no degradation of **V** is observed when checked by HPLC with the DEAE-5PW column. After introduction of 5-azido-2-nitrobenzoic acid, all procedures are carried out under red safety light to protect the photoreactive group.

[4] R. J. Guillory and S. J. Jeng, this series, Vol. 46, p. 259 (1977).

[5] R. V. Lewis, M. F. Roberts, E. A. Dennis, and W. S. Allison, *Biochemistry* **16,** 5650 (1974).

Biotinylation of Myosin ATPase Site

Incorporation of Photoreactive Biotinylated ADP Analog into ATPase Site

The ADP analog synthesized as above is used to label the ATPase site of myosin by exploiting the fact that ADP is tightly trapped into the ATPase site in the presence of vanadate ions.[6] Heavy meromyosin (HMM) (1.8 mg/ml, 5 μM), a proteolytic fragment of myosin (M_r 3.5 × 10^5) carrying two ATPase sites, in 0.1 *M* NaCl, 20 m*M* imidazole, and 2 m*M* $MgCl_2$ (pH 7.0) is incubated with the ADP analog (**V**) (20 μM) in the presence of vanadate ions (0.7 m*M*). Since one HMM molecule has two equivalent ATPase sites, the molar ratio of the ATPase site to the ADP analog (**V**) is 1 : 2 under the above conditions. After incubation at 0° for 16 hr, the mixture is passed through a Dowex-1 column (0.5 × 5 cm) prewashed with 0.1 *M* NaCl and 20 m*M* imidazole (pH 7.0). The ADP analog (**V**) and vanadate ions that are not trapped in the ATPase site are retained in the column while those trapped are eluted in the flowthrough fraction as the HMM–ADP analog–vanadate ion complex (1 mol of ADP analog per 1 mol of ATPase site).

The isolated HMM–ADP analog–vanadate ion complex is very stable. After incubation for 24 hr at 4°, only a small amount of the trapped ADP analog is released from the complex. The number of molecules of the ADP analog trapped in the ATPase site can be counted by using the strong absorption of the 5-azido-2-nitrobenzoyl group at 320 nm.[1]

Covalent Cross-Linking of ADP Analog to ATPase Site

The ADP analog trapped in the ATPase site is covalently cross-linked to the site by activating the 5-azido-2-nitrobenzoyl group with UV light to form the corresponding nitrene.[7] This step is essential for attaching the avidin molecule to the ATPase site since the association of avidin and HMM through the biotinylated ADP analog that is noncovalently trapped in the ATPase site is not stable enough. Only the biotinylated ADP analog covalently cross-linked to HMM can hold avidin tightly.

After isolation, the HMM–ADP analog–vanadate ion complex is immediately irradiated for 10 min with a UV lamp (16 W, 356 nm) at a distance of 2 cm. The solution is placed on a parafilm sheet as drops (200 μl/drop). The Parafilm sheet is kept on ice to avoid thermal denaturation of HMM. After irradiation, drops are collected and dialyzed against 0.5 *M*

[6] C. C. Goodno, *Proc. Natl. Acad. Sci. U.S.A.* **76,** 2620 (1979).

[7] H. Bayley and J. R. Knowles, this series, Vol. 46, p. 69.

NaCl and 20 mM imidazole (pH 7.0). The yield of the covalent cross-linking is about 15%. The low yield is expected for photoreaction of the azide group, since the activated nitrene reacts not only with protein but also with water.[7] Until the photoreactive group is photolyzed, all steps are carried out under red safety light.

Avidin as Electron Microscopic Probe

Avidin Oligomer

As a first choice, a linear oligomer of avidin[8] is very useful as an electron microscopic probe of a biotinylated site because its characteristic elongated shape is easily discernible.[8,9] After obtaining some information by using the avidin oligomer, avidin monomer can be used for visualizing the biotinylated site at higher resolution,[9] although sometimes it is very difficult to discern globular avidin monomer bound on a protein (unfortunately, this is the case for HMM biotinylated at its ATPase site).

The avidin oligomer is prepared by exploiting the fact that divalent biotin induces polymerization of avidin since avidin has four biotin–binding sites.[8,9] Avidin (2 mg/ml) in 0.5 M NaCl and 20 mM imidazole (pH 7.0) is mixed with the divalent biotin (**IV**) (2 mM in dimethyl sulfoxide) in a molar ratio of 2 : 1 (divalent biotin/avidin). After incubation at 25° for 2 hr, 0.5 ml (1 mg) of the mixture is loaded on a G3000SW HPLC gel permeation column (Tosoh) and eluted by 0.5 M NaCl–20 mM imidazole (pH 7.0) at a flow rate of 1.0 ml/min. Under these conditions, monomer, dimer, trimer, tetramer, and higher-order oligomer forms are resolved. Peak fractions of the trimer (0.3–0.5 ml) are collected. The concentration of the collected fraction is 0.4 mg/ml, which is high enough for further use.

Attachment of Avidin to Biotinylated Site of HMM

The trimer of avidin (0.4 mg/ml) prepared as above is mixed with the biotinylated HMM (1.5 mg/ml) in a molar ratio of 1 : 6 (HMM/avidin trimer) in 0.5 M NaCl–20 mM imidazole (pH 7.0). The mixture is incubated at 4° for 4 hr to form stable HMM–avidin oligomer complexes bridged through the biotinylated ADP analog incorporated into the ATPase site. Since unbound avidin oligomer must be removed from electron microscopic samples of the HMM–avidin oligomer complex, the mixture (0.5 ml) is passed through a G4000SW HPLC column (Tosoh).

[8] N. M. Green, L. Koieczny, E. J. Toms, and R. C. Valentine, *Biochem. J.* **125,** 781 (1971).
[9] K. Sutoh, K. Yamamoto, and T. Wakabayashi, *J. Mol. Biol.* **178,** 323 (1984).

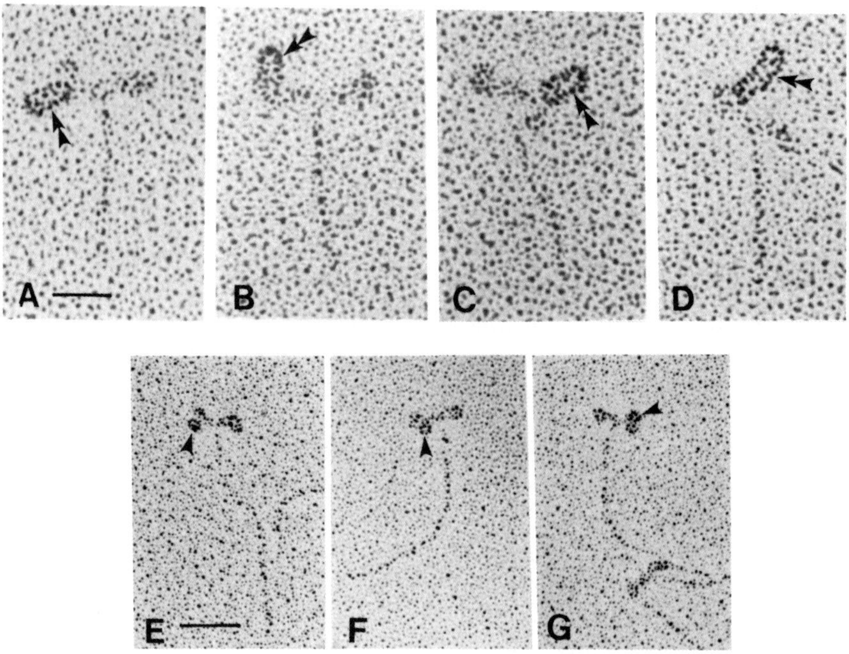

FIG. 2. Electron micrographs of HMM–avidin oligomer complexes (A–D)[1] (bar, 20 nm). the ATPase site of HMM is covalently modified with the biotinylated photoreactive ADP analog. Avidin oligomer attaches to the biotin moiety incorporated into the ATPase site. Avidin oligomers with characteristic elongated shapes are marked by double arrowheads. They are discernible from pear-shaped myosin heads. For comparison, electron microscopic images of myosin–avidin monomer complexes are shown (E–G)[9] (bar, 50 nm). In these complexes, the biotin moiety is covalently incorporated into one of the most reactive thiol groups of myosin (SH_1). Avidin monomers are indicated by arrowheads. The bound avidin monomers are discernible from the myosin heads.

Elution is carried out with 0.65 *M* ammonium acetate and monitored at 280 nm. A peak corresponding to the complex is collected. The protein concentration of the collected peak is more than 0.1 mg/ml.

The complex is then diluted to 5–10 μg/ml with 50% glycerol–0.65 *M* ammonium acetate (v/v) just before rotary shadowing. The shadowing is carried out with platinum–carbon at an angle of 10° on the rotary stage of a freeze-etching apparatus (JFD-7000, JEOL).[1,9] Examinations of shadowed images of HMM–avidin oligomer complexes revealed that avidin oligomers with characteristic elongated shapes bind onto the heads of HMM (Fig. 2A–D).[1,9] Since the avidin oligomers bind to HMM heads through biotin moieties that are covalently incorporated into the ATPase site, the site is located at the region where the avidin oligomer attaches. In place of avidin oligomer, avidin monomer can be used as an electron microscopic probe (Fig. 2E–G),[9] though it is discernible under favorable conditions only.

Advantages of Avidin–Biotin System as Electron Microscopic Probe

To visualize a specific site on a protein by electron microscopy, antibody has been used in many cases. What, then, are the advantages of the avidin–biotin system? Of course, the strong association of avidin and biotin ($K_D = 10^{-15}\ M$) is a significant advantage of the avidin–biotin system. Since electron microscopic examination is carried out on diluted solutions, strong binding of an electron microscopic probe to a protein is essential to get good electron microscopic images. Another advantage is that covalent labeling with biotin can be directed to specific sites such as the ATPase site as described in this chapter. These advantages of the avidin–biotin system have been exploited in order to attain an extremely high degree of resolution using the electron microscopic technique, i.e., visualization of the three-dimensional location of the ATPase site.[10] It must be mentioned here that the ADP analog (**V**) can be used not only to identify the three-dimensional location of the ATPase site as described here but also to identify the peptide segments located close to the site.[11] After some modification, the method would be applicable to other ATPase.

[10] M. Tokunaga, K. Sutoh, C. Toyoshima, and T. Wakabayashi, *Nature (London)* **329,** 635 (1987).
[11] K. Sutoh, *Biochemistry* **26,** 7648 (1987).

[47] Analysis of Proteins and Glycoproteins on Blots

By EDWARD A. BAYER, HAYA BEN-HUR, and MEIR WILCHEK

Blotting of proteins and other macromolecules onto nitrocellulose (or other suitable matrices such as positively charged nylon membranes) has become a standard technique in biological studies.[1,2] Using this approach, a suitable stain such as Amido black or Coomassie blue can be used for

[1] H. Towbin, T. Staehelin, and J. Gordon, *Proc. Natl. Acad. Sci. U.S.A.* **76,** 4350 (1979).
[2] J. M. Gershoni and G. E. Palade, *Anal. Biochem.* **131,** 1 (1983); J. M. Gershoni, *Methods Biochem. Anal.* **33,** 1 (1988).

the general identification of the proteins. General staining of DNA is usually accomplished with ethidium bromide. The utility of the technique is greatly expanded by applying specific binders (e.g., antibodies, lectins, ligands, or complementary nucleic acid fragments), associated via complexation or conjugation to a chromogenic or radiolabeled probe, for the identification of specific proteins or nucleic acids.

The avidin–biotin system can be used to identify proteins on blots according to the accessibility of relevant functional groups on the target proteins. For this purpose, we have used the biotin reagents described earlier in this volume (see [13]) and have developed optimal conditions for labeling lysines, cysteines or cystines, tyrosines and histidines, carboxyls, and carbohydrates following chemical or enzymatic oxidation. The final identification step after labeling with biotin is performed with either avidin or streptavidin complexes with, or conjugates of, the desired probe. The probe can either be an enzyme (e.g., alkaline phosphatase, β-galactosidase, peroxidase) or a radiolabel. The labeling of the protein with the biotin-containing reagents can be performed either before or after the blot transfer (Fig. 1).

Owing to its sensitivity, the avidin–biotin system can also be used to

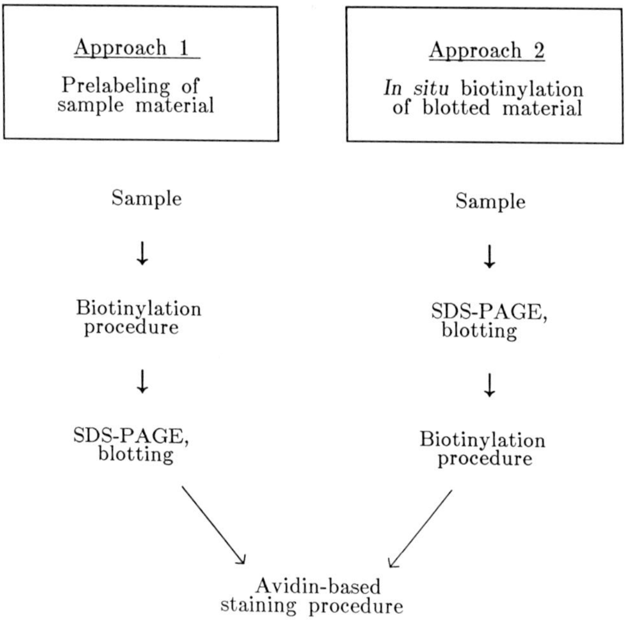

FIG. 1. Schematic of the two major labeling strategies for blot analysis of proteins and glycoproteins.

increase the signal obtained between specific proteins and their counterparts by the use of biotinylated antibodies or other specific interacting molecules followed by avidin probes. Thus, avidin–biotin technology can serve as a versatile and efficient means for group-specific detection, for affinity- and lectin-mediated staining, as well as for immunolabeling of blotted proteins and glycoproteins.

In this chapter, we present a series of procedures we have used in our laboratory to label proteins and glycoproteins on blots. The methodology can be used to analyze both purified and electrophoretically separated protein samples.

General Techniques

Although the primary consideration of this chapter is the mode of biotinylation, we include a brief description of the methodology we use for dot blotting, SDS–PAGE, and electrophoretic blot transfer of sample material. Related protocols in general use in other laboratories are equally suitable. The samples may include either prebiotinylated material or unbiotinylated material. In the latter case, the blotted material is subjected to one of the *in situ* biotinylation procedures (detailed below) before being stained by one of the avidin-based methods found in the final section of this chapter. In the former case, the blots are immediately ready for staining by the avidin-associated probe of choice.

Dot Blotting

Serial dilutions of protein solutions are applied in 1-μl aliquots to precut (2 × 3.5 cm) nitrocellulose strips and allowed to dry at room temperature. Additional treatments (biotinylation, washing, staining, etc.) are carried out in minimal volume (1–2 ml) in suitably sized polystyrene compartmented boxes (Althor Products, Wilton, CT).

SDS–PAGE

We usually perform the SDS–PAGE step overnight at low constant current (7 mA). This system is convenient, and the bands obtained are sharp. We employ a discontinuous buffer system using 3% stacking gels, and, depending on the nature of the sample material, the separating gel contains from 6 up to 15% acrylamide. The running buffer consists of 0.1% SDS in 25 m*M* Tris–glycine buffer (pH 8.9); samples are diluted with one-half volume of sample buffer which contains 9% (w/v) SDS, 30% (v/v) glycerol, 2% (v/v) 2-mercaptoethanol, and 0.2% bromphenol blue in 62.5 m*M* Tris-HCl buffer (pH 6.8). Electrophoresis is continued until the bromphenol blue marker migrates to the end of the gel. After electrophoresis, the gel is removed from the mold, and portions are either stained with Coomassie brilliant blue or subjected to blot transfer.

Blot Transfer

Electrophoretic transfer is performed for 2 hr in 15.6 m*M* Tris–glycine buffer (pH 8.3), using a gradient field.[2] A constant voltage power supply is used, set at 45 V. In all subsequent treatments (following the blotting of electrophoretically separated proteins), the volume of the appropriate reaction mixtures is kept to a minimum such that the solution uniformly covers the blot. In our laboratory, this is usually carried out using 0.5–1 ml/cm^2 in suitable-sized polystyrene compartmented boxes (Althor Products).

Prelabeling of Sample Material

The group-specific biotinylation of proteins in solution can be performed *in vitro* on a sample containing either a purified protein or a mixture of proteins (Fig. 1, Approach 1). Membrane and intact cells or tissues can also serve as sample material. Following the biotinylation step, the sample can be applied to dot blots or can be subjected to a suitable separation method (e.g., SDS–PAGE, isoelectric focusing, gel filtration) prior to blot transfer. In many cases, excess biotin reagent does not have to be removed, and the sample can be applied immediately to the gel. This does, however, lead to smearing of the samples in some instances (especially when relatively high concentrations of proteins are applied to the gel), and dialysis (or centrifugation of solid-phase material such as cells) is recommended when possible. With the exception of disulfide and periodate-induced staining, all of the techniques involve a single-step incubation of the sample with the appropriate biotinylating reagent. The signal level is usually superior to that achieved by direct labeling of the blots.

Prelabeling Amino Groups (Lysines). Lysine groups in proteins are labeled in solution using biotin-*N*-hydroxysuccinimide ester (BNHS) prior to SDS–PAGE, blot transfer, and staining. The protein sample (50 μg/ml to 50 mg/ml), dissolved in an appropriate buffer,[3] is treated with an

[3] Since BNHS reacts with amines, amine-containing buffers (Tris, glycine, etc.) or solutions containing proteins added as "stabilizers" should be avoided. Phosphate and bicarbonate buffers are commonly used, and the pH can be neutral or, preferably, alkaline (up to pH 9). Once the reaction has been completed the sample can be diluted with any type of sample buffer (e.g., containing Tris).

appropriate amount[4] of BNHS.[5] For example, a sample containing 1 mg/ml protein is treated with 0.1 ml of a BNHS solution (1 mg/ml in dimethylformamide). The reaction is carried out at room temperature for 1 hr, after which the sample is dialyzed[6,7] against an appropriate buffer and applied either to dot blots or to lanes of a gel prior to electrophoresis and blot transfer.

Prelabeling Phenols and Imidazoles (Tyrosines and Histidines). *p*-Diazobenzoylbiocytin is freshly prepared from the *p*-aminobenzoyl precursor,[8] and 0.3 mg[9] is added per milligram of protein.[10] The reaction is carried out at room temperature for 2 hr, and the sample is dialyzed[6,7] and applied either to dot blots or to lanes of a gel prior to electrophoresis and blot transfer.

Prelabeling Sulfhydryls (Cysteines). 3-(*N*-Maleimidopropionyl)biocytin (MPB)[11] is added to a protein sample[12] (30 μg/mg).[13] The reaction is

[4] The amount of BNHS per sample is best determined empirically. The optimum is dependent on many factors, including the number and accessibility of lysines in a given protein and the level of biotinylation desired by the investigator. We commonly use about 0.1 mg reagent/ml sample when the protein concentration is 1 mg/ml or less. At higher concentrations of proteins, we keep the ratio of BNHS to protein at about 10% (w/w).

[5] In addition to BNHS, various analogs (e.g., biotinyl-*N*-hydroxysulfosuccinimide ester and biotinylaminocaproyl-*N*-hydroxysuccinimide ester) may be used. The *N*-hydroxysuccinimide derivatives can be synthesized according to the procedures described by M. Wilchek and E. A. Bayer (this volume [13]), or they may be purchased commercially from Sigma or Calbiochem. Stock solutions of BNHS and biotinylaminocaproyl-*N*-hydroxysuccinimide ester can be stored at −20° in dry dimethylformamide, although we generally recommend fresh preparation of stock solutions.

[6] The dialysis step can be omitted in cases where time is a factor or where such treatment may be detrimental to the sample (e.g., limited amount of material). The biotinylated material can be stored at −20° prior to blotting.

[7] The biotinylation reaction can also be performed on membrane or cell suspensions, and, where necessary, the target material is washed by centrifugation (instead of dialysis as for proteins in solution). For cells, all steps should be performed in isotonic buffers, and organic solvents should be kept to a strict minimum or, preferably, omitted altogether. The use of water-soluble biotinylating reagents is highly recommended in this case. After biotinylation, the cells are washed and then lysed using a hypotonic buffered solution (e.g., 5–20 m*M* phosphate or Tris buffer). The resultant subcellular particles are washed by centrifugation, subjected to electrophoretic separation (if desired), blotted, and stained.

[8] M. Wilchek and E. A. Bayer, this volume [13].

[9] The optimum amount of DBB per sample is dependent on the distribution and accessibility of tyrosines and histidines in the sample.

[10] The concentration of protein is usually adjusted to approximately 1 mg/ml in 0.1 *M* sodium borate buffer (pH 8.4).

[11] Maleimide-containing biotin derivatives, other than MPB, can also be used (see M. Wilchek and E. A. Bayer, this volume [13] for details). Aqueous solutions of MPB and analogous maleimides can be stored for long periods of time at −20°.

carried out at room temperature for 1 hr, and then the sample is dialyzed[6,7] and applied either to dot blots or to lanes of a gel prior to electrophoresis and blot transfer.

Prelabeling Disulfides (Cystines). Protein samples[12] are treated with *N*-ethylmaleimide (10 mg/ml final concentration) for 1 hr in order to block endogenous SH groups. The solutions are dialyzed for 3 hr against phosphate-buffered saline (PBS, pH 7.4), and mercaptoethanol is then added to a final concentration of 2% (v/v) in order to reduce intrinsic cystinyl disulfide bonds of the target protein(s). After a 30-min incubation period, excess mercaptoethanol is removed by dialyzing the protein sample[7] overnight with 3 or 4 buffer changes. The maleimide-blocked, mercaptoethanol-reduced protein is then treated with MPB[11] (30 μg/mg protein)[13] at room temperature for 1 hr. The sample is dialyzed again[6,7] and applied either to dot blots or to lanes of a gel prior to electrophoresis and blot transfer.

Prelabeling Carboxyl Groups (Aspartic and Glutamic Acids). A protein sample (~1 mg/ml) is brought to 2.5 mg/ml with a solution (adjusted to pH 5) of biotin hydrazide or biocytin hydrazide,[14,15] and water-soluble carbodiimide[16] is added to a final concentration of 0.5 mg/ml.[17] After the reaction is carried out at room temperature for 1 hr, the reaction mixture is dialyzed[7] overnight to remove the excess of reagents, and the sample is applied either to dot blots or to lanes of a gel prior to electrophoresis and blot transfer.

Prelabeling Glycoconjugates

Glycoconjugates, in solution, on purified membrane fractions, or on intact cell surfaces, can be biotinylated by chemical or enzymatic oxidation combined with interaction with biocytin hydrazide.[14,18]

---

[12] In our original experiments, the protein solutions were adjusted to concentrations within the range of 0.5–1.0 mg/ml. MPB-induced biotinylation involving higher and lower protein concentrations have not been investigated.

[13] We have not examined the specificity of higher concentrations of MPB. In any experiment, positive and negative controls should always be included. The controls should consist of freshly prepared solutions of a representative "sulfhydryl" protein and a "disulfied" protein, respectively (e.g., aldolase, β-galactosidase, or ovalbumin as SH-containing proteins and chymotrypsinogen as an S—S-containing protein).

[14] See M. Wilchek and E. A. Bayer, this volume [13] for synthesis of these reagents. Aqueous solutions of biotin hydrazide and biocytin hydrazide can be stored for long periods of time at $-20°$.

[15] Other hydrazide-containing derivatives of biotin can also be used. Biotin hydrazide and the extended caproyl analog are available commercially from Sigma and Calbiochem. Biocytin hydrazide is not yet commercially available.

Periodate-Induced Biotinylation. For selective biotinylation of terminal sialyl residues, an ice-cold glycoprotein solution (0.5–3 mg/ml) is brought to 1.0 m*M* with sodium metaperiodate (or to 10 m*M* for general labeling of sugar residues). The reaction is carried out for 30 min at 4°, and the solution is dialyzed for 4 hr against PBS. The contents of the dialysis sac are brought to 2 mg/ml with biocytin hydrazide and incubated for 2 hr at room temperature. The sample is dialyzed again[6,7] and applied either to dot blots or to lanes of a gel prior to electrophoresis and blot transfer.

Enzyme-Induced Biotinylation of Glycoproteins in Solution. For selective biotinylation of terminal sialyl residues, the glycoprotein (0.5–3 mg/ml) is dissolved in PBS containing 2 mg/ml biocytin hydrazide, 1 m*M* $CaCl_2$, and 1 m*M* $MgCl_2$. The solution is treated with a mixture of immobilized forms of neuraminidase (0.03 units/ml glycoprotein solution) and galactose oxidase (3 units/ml glycoprotein solution).[19,20] For labeling of terminal galactose or *N*-acetylgalactosamine residues, neuraminidase-containing resin is omitted from the reaction mixture. The reaction is carried out for 2 hr at 37°, and the suspension is removed by centrifugation. The supernatant is dialyzed overnight and applied either to dot blots or to lanes of a gel prior to electrophoresis and blot transfer.

Enzyme-Induced Biotinylation of Membrane or Cell Preparations. For selective biotinylation of terminal sialyl residues, the cell or membrane preparation, suspended in PBS containing 2 mg/ml biocytin hydrazide, 1 m*M* $CaCl_2$, and 1 m*M* $MgCl_2$, is brought to 0.03 units/ml with neuraminidase and 3 units/ml with galactose oxidase[20] (for labeling of terminal galactose or *N*-acetylgalactose residues, neuraminidase is omitted from the reaction mixture). The reaction is carried out for 2 hr at 37°, and the suspension is washed by centrifugation. The cells are then lysed using a hypotonic buffered solution (e.g., 5–20 m*M* phosphate or Tris buffer), the resultant subcellular particles are washed by centrifugation, and the sample is applied either to dot blots or to lanes of a gel prior to electrophoresis and blot transfer.

---

[16] 1-Ethyl-3-(3-dimethylaminopropyl)carbodiimide-HCl (Sigma).

[17] The optimal concentrations of the reagents have not been rigorously examined for this particular reaction.

[18] M. Wilchek and E. A. Bayer, this series, Vol. 138 [35].

[19] Both neuraminidase (from *Vibrio cholerae,* Behringwerke AF, Marburg, FRG) and galactose oxidase (from *Dactylium dendroides,* Sigma) can be attached to appropriate matrices by a variety of means (see K. F. O'Driscoll, this series, Vol. 44 [12]; W. H. Scouten, this series, Vol. 135 [2]; and M. Wilchek, T. Miron, and J. Kohn, this series, Vol. 104 [1]).

[20] Stock solutions of galactose oxidase (100 units/ml in PBS containing 1 m*M* $CaCl_2$ and 1 m*M* $MgCl_2$) are stored in aliquots at −20°. Neuraminidase (1 unit/ml solution) is stored under sterile conditions at 4°.

In Situ Biotinylation of Blotted Material

Using the procedure for prelabeling of sample material (above), it is generally observed that increasing the amount of a given reagent causes a similar increase in the resultant signal. The only restriction is that using excess reagent may jeopardize the specificity of the reaction. The situation is more complicated for direct biotinylation of blotted material (Fig. 1, Approach 2). For most reagents (with the exception of MPB), the reagent concentration may be increased only to some optimal value, above which there is a decline in the amplitude of the resultant signal. There is still no definitive explanation for this phenomenon. In this section we describe procedures developed in out laboratory for the biotinylation of sample material directly on nitrocellulose blots.

Labeling Blotted Amino Groups (Lysines). From a stock solution of BNHS[5] in dimethylformamide (30 μg/ml), 0.1 ml is added per 0.9 ml of 0.1 *M* sodium bicarbonate solution. The latter is immediately applied to the blotted proteins, and the reaction is allowed to take place for 1 hr. The blots are rinsed and stained by one of the procedures detailed below.

The above-described final concentration of 3 μg/ml BNHS represents an optimum value.[21] In this case, the signal is exceptionally weak, and the direct BNHS-induced staining of blots is not recommended if an alternative approach is available.

Labeling Blotted Phenols and Imidazoles (Tyrosines and Histidines). The blotted proteins are treated with a *p*-diazobenzoylbiocytin[8] solution at a concentration of 10 μg/ml.[21] The blots are rinsed and stained by one of the procedures given below.

Labeling Blotted Sulfhydryls (Cysteines). Blotted proteins are treated for 1 hr with an aqueous solution of MPB[11] (30 μg/ml). The blots are rinsed and stained by one of the procedures detailed below.

Labeling Blotted Disulfides (Cystines). For dot blots, the blotted proteins are treated with a solution of 5 mg/ml *N*-ethylmaleimide for 1 hr in order to block free cysteinyl SH groups. The strips are rinsed well with PBS, and intrinsic cystinyl S—S groups are reduced for 1 hr with a 2% solution of 2-mercaptoethanol. The strips are rinsed well with PBS and treated for 1 hr with a solution of MPB[11] (30 μg/ml). The blots are rinsed and stained by one of the procedures given below.

For mixtures of proteins destined for electrophoretic separation, the samples are brought to 5 mg/ml *N*-ethylmaleimide *before* addition of mercaptoethanol-containing sample buffer and SDS–PAGE. The conditions

[21] The signal intensity is dependent on the concentration of the reagent. The value stated represents an optimum, below which there is apparently insufficient reagent for extensive signal production and above which an undefined inhibition of the signal occurs.

for electrophoresis and blot transfer are the same as for the other reagents. The strips are then treated for 1 hr with a solution of MPB[11] (30 μg/ml). The strips are rinsed and stained by one of the procedures detailed below.

Labeling Blotted Carboxyl Groups (Aspartic and Glutamic Acids). A viable procedure for biotinylation of carboxyl groups on blots has yet to be experimentally tested; the following represents an unverified approach for determining optimal conditions. Various aqueous solutions of biocytin hydrazide[14] are prepared ranging from 3 μg/ml to 1 mg/ml at half-logarithmic concentration increments. A portion of each solution is brought to a designated final concentration with water-soluble carbodiimide[16] (suggested range 0.1–3 mg/ml at half-logarithmic increments), and the resultant solutions are applied to dot blots containing varying concentrations of standard protein samples. The reaction is carried out at room temperature for 1 hr, after which the blots are rinsed and stained by one of the procedures given below. After determining the optimum reagent concentrations, the experimentally obtained values are used for future application, e.g., for blot transfers.

Labeling Glycoconjugates on Blots

Blotted glycoconjugates can also be biotinylated directly in a selective manner. In this case, the glycoconjugates are first oxidized either chemically (by periodate treatment) or enzymatically (by galactose oxidase treatment alone or in combination with neuraminidase). The oxidized sugars react with biocytin hydrazide.[22] The labeling pattern is almost identical to that observed using the avidin hydrazide procedure.[23] The latter procedure, however, is about 10-fold more sensitive than that described below, and avidin hydrazide is therefore recommended when periodate oxidation is desired. Nevertheless, unlike avidin hydrazide, biocytin hydrazide can be used for the direct enzyme-mediated labeling of sugars (although such labeling requires relatively high amounts of sample material).

Periodate-Induced Biotinylation. The blotted glycoconjugates are quenched using the appropriate solution.[24] The blot is rinsed 3 times with PBS, and a freshly prepared solution containing 1 m*M* sodium periodate

[22] The long-chain reagent is required in this case, since we have been unable to generate detectable signals using biotin hydrazide.

[23] E. A. Bayer, H. Ben-Hur, and M. Wilchek, this volume [48].

[24] For streptavidin-containing complexes, a solution of 2% BSA (in PBS) is used as quencher and diluent. For avidin-containing complexes, a 1 : 1 mixture of solutions containing 2% lysozyme (dissolved in distilled water) and 2% BSA is used.

(for labeling of terminal sialyl moieties) or 10 m*M* periodate (for labeling vicinal hydroxyls of sugars) is added. After 30 min, the blots are again rinsed 3 times with PBS. A solution of biocytin hydrazide[14] (3 μg/ml)[21] is applied to the strips, which are incubated for 1 hr. The strips are rinsed again with PBS and stained with radiolabeled or enzyme-complexed avidin as described below.

Enzyme-Induced Biotinylation. The strips are quenched with 2% bovine serum albumin (BSA) for 1 hr at room temperature, rinsed 3 times with PBS, and treated with a solution containing neuraminidase (0.03 units/ml) and galactose oxidase (3 μg/ml)[20] in PBS containing 3 μg/ml biocytin hydrazide,[14,21] 1 m*M* $CaCl_2$, and 1 m*M* $MgCl_2$. For labeling terminal galactosyl residues, neuraminidase is excluded from the latter solution. The strips are incubated for 2 hr at 37°, rinsed 3 times with PBS, and stained as outlined below.

General Method for Labeling Blotted Antigens and Receptors

The avidin–biotin system provides a simple, versatile method for labeling target macromolecules on blots using a specific biotinylated binder. Biotinylated antibodies, lectins, hormones, and other ligands (see [2] in this volume) can be used for this purpose.

Reagents

Biotinylated binding protein,[25] 3 μg/ml[26] in quenching solution
Quenching solution: 2% BSA in PBS
PBS[27] (pH 7.4)

Procedure. The strips are quenched for 2 hr at 37°, rinsed 3 times with PBS, and treated with the solution containing the biotinylated binding protein. The strips are incubated for 2 hr at 37°, rinsed 3 times with PBS, and stained as outlined below.

[25] Binding proteins and other biologically active macromolecules can be biotinylated by one of the procedures detailed in [14] in this volume.

[26] The above-listed concentration of biotinylated binder has been found to be effective for a variety of antibodies and lectins in a variety of studies. In any given system, of course, a range of concentrations should be tried and the results assessed for intensity of signal, level of background binding, and extent of nonspecific interactions.

[27] For certain binders, various additives (e.g., metal ions), or alternative buffer systems may be necessary. The buffer should be revised accordingly to ensure facile interaction of the binder with the blotted target molecule.

Staining of Blots

Numerous methodologies are available for staining nitrocellulose blots. The blots can be stained directly using an appropriate radioactive derivative of avidin or streptavidin. Blots can also be stained enzymatically using an avidin–enzyme conjugate.[28] Sequential application of native avidin and biotinylated enzyme or direct treatment with the respective preformed complex can also be used. In this section, we provide two simple and effective procedures by which we routinely stain blots containing biotinylated proteins.

Enzyme-Mediated Staining of Blots

Native (underivatized) avidin and biotinylated alkaline phosphatase (B–AP) are mixed in a predetermined optimized ratio. The complexes that form are subsequently interacted with the blot-immobilized target material.

Reagents

Preformed complexes[29]: solutions containing 10 μg/ml avidin or streptavidin[24] and 5 units/ml B–AP[30] (in 2% BSA) are combined 30 min before application[31]

Quenching solution[24]

PBS (pH 7.4)

Substrate solution: 10 mg naphthol AS-MX phosphate (Sigma, Chemical Co., St. Louis, MO; free acid dissolved in 200 μl of dimethylformamide) is mixed with a solution containing 30 mg Fast Red TR salt (Sigma) dissolved in 100 ml of 0.1 *M* Tris-HCl (pH 8.4)

Alternative substrate solution: 5-bromo-4-chloro-3-indolyl phosphate (50 mg/ml in dimethylformamide, 33 μl) and nitroblue tetrazolium (75 mg/ml in 70% dimethylformamide, 40 μl) in 10 ml of 0.1 *M* Tris-HCl buffer (pH 9.5) containing 0.1 *M* NaCl and 50 m*M* $MgCl_2$

Procedure. The blot containing the biotinylated sample material (either prelabeled in solution or via direct biotinylation of the blots) is quenched for 1 hr at room temperature, rinsed 3 times with PBS, and treated with the preformed complexes for 30 min at room temperature.

[28] Many appropriate avidin–enzyme conjugates or complexes are commercially available.

[29] E. A. Bayer and M. Wilchek, this volume [18].

[30] E. A. Bayer and M. Wilchek, this volume [14].

[31] Stock solutions of avidin or streptavidin (1 mg/ml) and B–AP (500 units/ml) can be stored in aliquots at −20°.

The blots are rinsed again, and either of the above substrate solutions is added. When the desired level of staining is reached, the blots are rinsed in tap water.[32]

Radiolabeling of Blotted Bands

Radiolabeling is still the most sensitive method of labeling blotted material. The use of radioactive avidin or streptavidin is particularly attractive since the researcher comes into contact with the radioactive material only in the final step of the procedure. The procedure is exceptionally sensitive, and the autoradiographic process can be significantly reduced (hours are usually sufficient).

Reagents

Radiolabeled avidin or streptavidin: 10^6 cpm diluted in quenching solution (see relevant preparation protocol in [17], [18], and [24] in this volume)[33]
Quenching solution[24]
Washing solution: 0.1% Tween 20 in PBS (pH 7.4)

Procedure. The blot containing the biotinylated sample material (prelabeled in solution or labeled via direct biotinylation of the blots) is quenched for 1 hr at room temperature, rinsed 3 times with PBS, and treated with the radiolabeled avidin for 1 hr at room temperature. The blots are rinsed extensively (overnight incubation) with wash solution, air dried, and exposed for 24 hr to Kodak X-OMAT film at −70° with an intensifying screen.

General Comments

The avidin–biotin system is usually more sensitive than Coomassie blue for general staining of gels, particularly if the sample material is prelabeled using biotin-containing reagents. In order to selectively label specific antigens or receptors on the cell surface, it is possible to use biotinylated antibody, lectin, or hormone followed by the avidin probe. By using biotinylated molecular weight standards in conjunction with selective (binder-mediated) biotinylation of specific target proteins, the relative positions of the biotinylated bands can be unequivocally assigned. Additional information can be obtained using the group-specific biotinylation of complex mixtures of macromolecular sample material (e.g., in cell

[32] The intensity of the bands is usually reduced when the blots are dried. The stained blots can be stored in the dried state, but it is recommended to photograph the blots when wet.
[33] Radioiodinated streptavidin is also commercially available (from Amersham).

membranes or extracts). Thus, only one blot and a single staining step (with the avidin probe) are required for comparative studies.

Owing to the basicity of avidin, the avidin-associated probe alone (without a biotinylated binder) can be used as a general stain for DNA and other highly acidic macromolecular target molecules.

[48] Direct Labeling of Blotted Glycoconjugates

By EDWARD A. BAYER, HAYA BEN-HUR, and MEIR WILCHEK

Glycoconjugates are a very important class of macromolecules that form a part of every complex biological system (e.g., cells, membrane preparations, extracellular fluids). When analyzing such a system, one of the most common characteristics of interest is the amount and distribution of glycosylated macromolecules. For example, when studying erythrocyte membranes, it would be advantageous to know the composition of glycoproteins in the membrane. If these are identified and cataloged in normal (healthy) cells, one can then examine their possible alteration in diseased cells.

Many approaches have appeared in the literature for the labeling of glycoconjugates in complex mixtures of biological materials. One of the earliest approaches, which is still in use, is the periodic acid–Schiff's (PAS) stain of polyacrylamide gels. In fact, most methods are still based on the chemical oxidation (using periodate) of vicinal hydroxyl groups of sugars for the identification of glycoconjugates. In this chapter, we describe a simple, sensitive, and effective procedure for selective labeling of glycoconjugates in electrophoretically separated material which has been blotted onto nitrocellulose membranes. The method is based on periodate-induced oxidation followed by secondary interaction with preformed complexes consisting of avidin hydrazide and biotinylated enzyme.

Reagents

Sodium metaperiodate stock: dissolve 22 mg $NaIO_4$ into 1 ml distilled water

1 m*M* Periodate solution: dilute stock solution 1 : 100 in phosphate-buffered saline (PBS, pH 7.4)

10 m*M* Periodate solution: dilute stock solution 1 : 10 in PBS

Bovine serum albumin (BSA), 2% in 0.15 *M* NaCl (saline)

Lysozyme,[1] 2% in 0.15 *M* NaCl
Avidin hydrazide,[2] 15 μg/ml in 2% lysozyme
Streptavidin hydrazide,[2] 15 μg/ml in 2% lysozyme
Biotinylated alkaline phosphatase (B–AP),[3] 7.5 units/ml (in BSA for streptavidin; in lysozyme for avidin)
ABC-HZ: mix equal volumes of avidin hydrazide and B–AP solutions at room temperature 30 min before use
StABC-HZ: mix equal volumes of streptavidin hydrazide and B–AP solutions at room temperature 30 min before use
Alkaline phosphatase substrate solution[4]: naphthol AS-MX phosphate (10 mg) is dissolved in 200 μl dimethylformamide; to this solution is combined 100 ml of a solution containing Fast Red TR salt (30 mg) dissolved in 0.1 *M* Tris-HCl buffer (pH 8.4)

Procedure. A nitrocellulose sheet, containing electrophoretically separated or dot-blotted material,[5] is placed in the appropriate quenching solution.[6] The treatment is carried out with shaking for 1 to 2 hr at room temperature. The sheet is rinsed 3 times with PBS. Sodium metaperiodate is added at the desired concentration.[7] The sheet is incubated with shaking for 30 min and again rinsed 3 times with PBS. Either ABC-HZ or StABC-HZ is then added, the nitrocellulose sheet is incubated for 1 hr at room temperature and then washed 5 times with PBS, and substrate solution is added until colored bands appear.[8] The sheet is rinsed with tap water to stop the reaction.

[1] Lysozyme (2 g) is mixed with 100 ml of 0.15 *M* NaCl. The suspension is stirred for 2 hr at room temperature, then centrifuged at 2000 *g*, and the supernatant fluids are stored in 10-ml aliquots at −20°.

[2] E. A. Bayer and M. Wilchek, this volume [18].

[3] E. A. Bayer and M. Wilchek, this volume [14].

[4] Prepared freshly prior to use. Other suitable substrate systems can also be used, notably 5-bromo-4-chloro-3-indolyl phosphate (50 mg/ml in dimethylformamide, 33 μl) and nitroblue tetrazolium (75 mg/ml in 70% dimethylformamide, 40 μl) in 10 ml of 0.1 *M* Tris-HCl buffer (pH 9.5) containing 0.1 *M* NaCl and 50 m*M* $MgCl_2$.

[5] The amount of sample material subjected to this staining procedure depends on the percentage of saccharides in the target material. For membrane samples, about 10–30 μg material per lane in SDS–PAGE gels is typically analyzed. For purified glycoproteins, 0.5–5 μg/lane is usually sufficient. For dot blots, as little as 1–3 ng can be detected. These values, however, are only initial estimates; for any experimental system, the optimal amount of target material is determined empirically.

[6] For blots stained with ABC-HZ, 2% lysozyme is used as a quencher; for StABC-HZ, the quenching solution contains a mixture of 1% BSA and 1% lysozyme.

[7] Periodate at a concentration of 1 m*M* is selective for sialic acids, whereas 10 m*M* periodate results in the general labeling of vicinal hydroxyl groups.

[8] Bands usually appear within 30 min. However, for low levels of label, the color reaction can be carried out overnight.

Comments. The choice of quencher is critical to the successful application of the method. A nonglycosylated macromolecule or mixture should be used. In the event that low levels of glycosylation appear in the quenching solution, this can be overcome by subjecting the quencher to successive periodate and borohydride pretreatments.

The use of avidin hydrazide complexes with B–AP is more sensitive for labeling blotted glycoconjugates than the biocytin hydrazide procedure.[9] On the other hand, avidin hydrazide cannot be used for direct labeling of glycoconjugates in solution phase or for the direct labeling of glycoconjugates on membranes or cells. In addition, the avidin hydrazide method is limited to periodate-induced oxidation, since its employment with galactose oxidase (either alone or combined with neuraminidase) leads to high levels of nonspecific binding which we have as yet been unable to reduce.

[9] E. A. Bayer, H. Ben-Hur, and M. Wilchek, this volume [47].

[49] Identification of Leukocyte Surface Proteins

By WALTER L. HURLEY AND EVE FINKELSTEIN

Leukocyte surface proteins include the major histocompatibility complex proteins, cell-specific antigens, and receptors for complement, immunoglobulins, chemotactic factors, enzymes, hormones, and others. A common approach to the study of leukocyte surface proteins is the use of immunological methods that identify a specific surface antigen. A limitation of such an approach is the requirement for antibodies specific to individual surface components. In addition, the immunological methods do not show the relationship of specific antigens with the entire spectrum of proteins on the cell surface. Study of leukocyte surface proteins can be hampered by contamination of sample preparations with intracellular proteins, and methods used in selectively labeling surface proteins must minimize the labeling of internal proteins.

The biotin–avidin complex provides a nonradioactive method for labeling and identifying surface proteins on leukocytes.[1] The method uses a reactive biotin compound to label exposed extracellular portions of surface proteins on intact leukocytes. These biotin-labeled proteins may be separated by gel electrophoresis. Protein blots are made of the electro-

[1] W. L. Hurley, E. Finkelstein, and B. D. Holst, *J. Immunol. Methods* **85,** 195 (1985).

phoretic gels, and biotin-labeled proteins are detected with an avidin–enzyme conjugate. The result is a protein blot representing labeled surface proteins of the leukocytes.

This approach to labeling surface proteins provides a means for establishing comparisons of leukocyte cell types and analyzing differences in leukocyte populations of body fluids.[1,2] This method may also be used to analyze general changes in surface protein profiles after exposure of leukocytes to cell mediators. Because the method provides an overview of surface proteins, it may be of value in conjunction with other approaches that identify specific surface proteins, such as identification of leukocyte antigens by immunoblotting.

Leukocyte Preparation

Any method of leukocyte collection and preparation may be used for biotin labeling provided that the cells are intact prior to labeling. The efficacy of maintaining intact cells during cell preparation should be determined by comparing biotinylation of intact cells and disrupted cells (disrupted by sonication or osmotic lysis). The biotin labeling procedure has been successfully used on leukocyte populations obtained by distilled water lysis of peripheral red blood cells,[1] by Ficoll–Hypaque separation of neutrophils and mononuclear leukocytes,[1,2] by pelleting milk leukocytes via low-speed centrifugation,[2] and from cultured lymphocyte cell lines. Cells should be washed 2 or 3 times with phosphate-buffered saline [PBS; 10 m*M* potassium phosphate, 150 m*M* sodium chloride (pH 7.4)] or Hanks' balanced salt solution (HBSS),[3] prior to biotin labeling, to remove loosely bound proteins from the cell surface. Although cells are generally maintained at 4° during washing, final protein blot patterns are not substantially altered by washing cells at room temperature. Cells should be diluted to 5×10^7/ml in PBS or HBSS prior to labeling.

Biotin Labeling

Generally, 100 μg of the water-soluble biotin derivative sulfo-*N*-hydroxysuccinimide (sulfo-NHS-biotin) is used to label 5×10^6 cells in 0.5-ml reaction volumes. This level of biotin reagent provides optimal labeling while minimizing background color development during the detection procedure. Fifty micrograms of sulfo-NHS-biotin provides satisfactory labeling, but 10 μg of reagent gives insufficient labeling. Amounts of labeling reagent required may depend on the specific biotin derivative

[2] W. L. Hurley and E. Finkelstein, *Am. J. Vet. Res.* **47,** 2418 (1986).

[3] J. H. Hanks, *Tissue Cult. Assoc. Man.* **1,** 3 (1975).

used. One hundred micrograms of *p*-diazobenzoylbiocytin[4] (DBB) also has been effective in labeling 5 × 10⁶ cells. Biotin reagent is solubilized in dimethyl sulfoxide (DMSO) and maintained frozen as a 10 mg/ml stock solution. This stock solution remains active for at least 6 months.

One hundred microliters of cell suspension (5 × 10⁶ cells) is diluted with 0.4 ml PBS (or HBSS). Biotin reagent (10 μl of stock solution in DMSO) is added to the cells, which are then incubated 10 min at 22° with shaking. Cells are centrifuged at 400 *g* for 2–5 min at 22°, followed by one washing with PBS and centrifugation as before. Low-speed centrifugations are used in these early washing steps to minimize cell disruption that may result in labeling of intracellular proteins by residual biotin reagent.

Pelleted cells are resuspended in 1 ml of PBS, disrupted by brief sonication (10 sec, Tekmar Sonic Disruptor, Tekmar Co., Cincinnati, OH), and centrifuged at 10,000 *g* for 2–5 min at 22°. Pelleted cellular material is resuspended in 1 ml of PBS and centrifuged at 10,000 *g* for 2–5 min at 22°. This wash procedure is repeated once. The final pellet is prepared for electrophoresis by resuspension in 100 μl of distilled water (volume may be varied to maximize the efficiency of detection method), one-fifth volume of gel sample buffer [25% 2-mercaptoethanol (v/v), 14.5% sodium dodecyl sulfate (SDS), and 0.28 *M* Tris (pH 6.8)], and one-fifth volume of dye solution (70% glycerol with 0.12% bromphenol blue). Samples are briefly sonicated (10 sec) and heated to 90° for 5–10 min prior to loading on the acrylamide gel.

Gel Electrophoresis

Samples are separated by SDS–polyacrylamide gel electrophoresis (SDS–PAGE).[5] Typically, 12.5 or 15% polyacrylamide slab gels are used (8 × 10 cm, 0.8 mm thick).[1] About 10 μl of sample, equivalent to about 3.5 × 10⁵ cells, is loaded per gel lane.

Protein Blotting and Avidin–HRP Detection

Proteins are transfered from gels to nitrocellulose by electroblotting (Transblot, Bio-Rad, Richmond, CA), at 0.06 A for 12–14 hr at 22°, in 25 m*M* Tris (pH 8.3), 192 m*M* glycine, and 20% methanol (v/v). Nitrocellulose blots are blocked with 2% bovine serum albumin (BSA), 0.1% Triton X-100 in PBS for 30–60 min at 37°. Blots are then incubated for 30–60 min

[4] M. Wilchek, H. Ben-Hur, and E. A. Bayer, *Biochem. Biophys. Res. Commun.* **138,** 872 (1986).

[5] U. K. Laemmli, *Nature (London)* **227,** 680 (1970).

at 37° in avidin–horseradish peroxidase (HRP) conjugate at 3.5 μg avidin–HRP/ml PBS containing 0.1% BSA. Generally, 4 ml of avidin–HRP is sufficient to cover an 8 × 10 cm blot. Blots may be placed on inverted covers from 96-well microtiter plates and avidin–HRP applied directly to the blot. The microtiter plate cover is placed on top of several wet paper towels in a covered transparent container (plastic container or glass dish). The covered dish acts as a wet-box to prevent dehydration of the avidin–HRP solution. The clear sides of the wet-box container allow repeated viewing of the blot so that exposure of the complete blot to avidin–HRP may be maintained. Larger volumes of avidin–HRP may also be used. Blots in this wet-box arrangement may be gently shaken.

After incubation with avidin–HRP, the blot is washed 3 times with 25 ml of 0.1% BSA, 0.05% Tween 20 in PBS, for 5–10 min per wash, at 37°. Bound avidin–HRP may be detected with the chromogen diaminobenzidine tetrahydrochloride (DAB is oncogenic). A freshly prepared DAB solution is made by adding 100 μl of 1% cobalt chloride to 5 ml of 10 m*M* Tris (pH 7.5) plus 2.5 mg DAB. The solution is mixed to dissolve the DAB and incubated for 10 min on ice in the dark. Hydrogen peroxide, 7.5 μl of a 30% solution, is added to the DAB solution, mixed, and immediately applied to the washed blot in a flat-bottomed dish. The DAB–hydrogen peroxide solution is washed over the blot for 0.5–1 min or until the desired color development is obtained. The blot is rinsed thoroughly in tap water and blotted dry with heavy blotting paper. The blot then may be dried at 75° for 10–15 min. This tends to clear some of the background color that develops on the nitrocellulose. To preserve the fragile blot in a durable and clearly visible form, the blot may be placed facedown on an acetate-sheet transparency that is precut to a size larger than the blot. The blot then is carefully sealed onto the transparency with transparent adhesive plastic.

Comments

The purity of a cell population will influence the protein blot profiles obtained with this method. While cell-type specific proteins may be identified, many leukocyte cell types have proteins of similar molecular weights when analyzed by this method.[1,2] Care should be taken when evaluating cell types from protein blot profiles in the absence of corresponding differential cell count data.

Major surface proteins, as determined by discrete bands on protein blots, are easily and reproducibly detected by this method. Minor bands may or may not be distinguishable, depending on the efficiency of labeling and on the clarity of protein separation on SDS–PAGE gels.

The sulfo-NHS-biotin labels free amine groups, primarily lysine residues exposed on proteins.[6] The intensity of biotin labeling of specific proteins will be influenced by the number of exposed lysines within the protein. Other water-soluble biotin derivatives may be used including DBB,[4] which labels tyrosines and histidines, and maleimidobutyrylbiocytin, which labels reduced cysteines.[7]

This method has also been used to identify surface proteins of a protozoan parasite by one- and two-dimensional polyacrylamide gel electrophoresis.[8] A similar method has been described to characterize surface antigens of human leukemic cells.[9] The latter study demonstrated that biotinylation of surface proteins does not interfere with their recognition by antibodies. A separate study on transblotting biotin-labeled proteins has shown that the increase in apparent molecular weight of biotinylated proteins is generally less than 10%, and that less than 1 ng biotinylated protein may be detected.[10]

[6] E. A. Bayer and M. Wilchek, *Methods Biochem. Anal.* **26,** 1 (1980).
[7] E. A. Bayer, M. G. Zalis, and M. Wilchek, *Anal. Biochem.* **149,** 529 (1985).
[8] P. R. Gardiner, T. W. Pearson, M. W. Clarke, and L. M. Mutharia, *Science* **235,** 774 (1987).
[9] S. R. Cole, L. K. Ashman, and P. L. Ey, *Mol. Immunol.* **24,** 699 (1987).
[10] M. Neumaier, V. Fenger, and C. Wagener, *Anal. Biochem.* **156,** 76 (1986).

[50] Staining of Proteins on Nitrocellulose Replicas

By WILLIAM J. LAROCHELLE and STANLEY C. FROEHNER

Principle

The high-affinity interactions of biotin with either avidin[1] or streptavidin[2] have been used extensively in signal-generating systems for immunofluorescence and enzyme-linked immunoassays. The biochemical derivatization of proteins in solution with biotin has also been well characterized.[3] Here we describe a chemically defined technique whereby amino groups of proteins are covalently derivatized with sulfosuccinimidobiotin after transfer to nitrocellulose paper. Depending on the sensitivity required, either of two techniques can be used to stain the proteins.

[1] P. Gyorgy, C. S. Rose, R. E. Eakin, E. E. Snell, and R. J. Williams, *J. Biol. Chem.* **140,** 535 (1940).
[2] L. Chaiet and F. J. Wolf, *Arch. Biochem. Biophys.* **106,** 1 (1964).
[3] E. A. Bayer, M. Wilchek, and E. Skutelsky, *FEBS Lett.* **68,** 240 (1976).

Both methods detect biotinylated proteins as dark bands against an essentially white background. The first method utilizes avidin or streptavidin conjugated to horseradish peroxidase. This procedure is rapid and detects less than 25 ng of protein in a single band. The second technique requires sequential incubations with streptavidin, rabbit antistreptavidin polyclonal antibody, and goat anti-rabbit IgG conjugated to horseradish peroxidase. The enhanced procedure, although slightly more lengthy, detects less than 5 ng of protein per band.

The methods described here permit direct comparison of stained replicas with a duplicate blot that has been probed with antibody or ligand. Problems associated with gel shrinkage on drying or altered electrophoretic mobility caused by biotinylation of proteins prior to gel electrophoresis are avoided. In principle, the procedure can be used in double-label experiments in which all proteins on the replica are biotinylated, and the same blot is then probed with radioactive antibody.

Methodology

General

All procedures are performed at ambient temperature. Proteins separated by SDS–PAGE are electrophoretically transferred to nitrocellulose membranes as described.[4] The procedure may be used with any gel system that is compatible with transfer of proteins to nitrocellulose paper.

Biotinylation of Proteins on Nitrocellulose Replicas

After electrophoretic transfer of the proteins to nitrocellulose membranes (0.45 μm, Bio-Rad, Richmond, CA), the replicas are rinsed briefly in 0.1 *M* sodium bicarbonate (pH 8.0; 0.25–0.50 ml/cm^2 nitrocellulose) and then soaked in the same buffer for 5 min. The replicas are then transferred to the same volume of bicarbonate buffer containing 10 μM sulfosuccinimidobiotin (Pierce Chemical Co., Rockford, IL).[5] After incubation for 45 min, the reaction is quenched by the addition of 1 *M* glycine (pH 6.5) to a final concentration of 1 m*M*. The replicas are then washed 2 or 3 times with 10 m*M* sodium phosphate, 0.15 *M* NaCl (pH 7.4; PBS) and blocked by incubation with PBS containing 5% newborn calf serum and 3% w/v bovine serum albumin (BSA) for 30 min. The replica is then stained with either detection system described below.

We have noted that the staining intensity is highly dependent on the

[4] W. J. LaRochelle and S. C. Froehner, *J. Immunol. Methods* **92,** 65 (1986).

[5] Sulfosuccinimidobiotin solutions should be prepared fresh each time.

concentration of sulfosuccinimidobiotin used to label the proteins. Concentrations greater than 10 μM result in a dramatic decrease in the staining intensity of *Torpedo* postsynaptic membrane proteins.[4] This may be due to a loss of proteins from the replica as the extent of biotinylation increases. In some cases, it may be necessary to determine empirically the optimal concentration of sulfosuccinimidobiotin for staining particular proteins.

Detection of Biotinylated Proteins with Horseradish Peroxidase-Conjugated Avidin or Streptavidin

After biotinylation, the replicas are incubated with either horseradish peroxidase conjugated to avidin (Cappel Laboratories, Malvern, PA; 5 μg/ml) or horseradish peroxidase conjugated to streptavidin (Bethesda Research Laboratories, Gaithersburg, MD; 1 μg/ml) for 1 hr. The enzyme conjugates are diluted in PBS containing 1% BSA and 0.05% Tween 80, and 10 ml is typically used for each replica. The replicas are then washed 3 times for 15 min each time with the same volume of PBS–0.05% Tween 80. Protein bands are then visualized by immersing the replicas in α-chloronaphthol (0.6 mg/ml in PBS containing 0.01% hydrogen peroxide). After allowing sufficient time for color development (usually 30 min), the replicas are rinsed with distilled water and dried between two sheets of dialysis membrane.

Streptavidin Immunochemical Detection of Biotinylated Proteins

The nitrocellulose replicas are biotinylated and subsequently blocked as described above. The blots are then incubated sequentially with the following reagents in PBS containing 1% BSA and 0.05% w/v Tween 80: streptavidin (1 μg/ml) for 1 hr, affinity-purified antistreptavidin IgG (0.5 μg/ml) overnight (this reagent can be prepared as described below), and horseradish peroxidase conjugated to goat anti-rabbit immunoglobulin (4 μg/ml; Cappel Laboratories) for 4 hr. Following each incubation, the replicas are washed 3 times for 15 min each with PBS–0.05% Tween 80. Color is then developed as described above. The concentrations of reagents and incubation times are determined empirically and are chosen to give maximum staining sensitivity.

Preparation of Antibodies to Streptavidin

Affinity-purified antibodies to complexes of streptavidin and biotin are prepared in rabbits according to the following procedures. Sulfosuccinimidobiotin is reacted with an excess of glycine to inactivate the succinimido group, and the product is then incubated in a 2-fold molar excess with streptavidin (0.4 mg/ml). An emulsion is prepared with equal volumes of the streptavidin–biotin complex and Freund's complete adjuvant. Each rabbit is injected subcutaneously and intradermally at several spots on the back with 1 ml of the emulsion containing 100 μg of streptavidin. Three and five weeks after the initial injection, the rabbits are injected with the same amount of antigen emulsified with Freund's incomplete adjuvant. One week after the third injection, the rabbits are injected intravenously with 50 μg of the streptavidin–biotin complex.

Antibody activity to streptavidin is monitored with a solid-phase assay. Except for the washing steps, all volumes are 50 μl. Wells of microtiter plates are coated with 300 ng of biotinylated BSA and blocked with PBS containing 0.02% sodium azide and 4% BSA. Streptavidin (350 ng) is then added to each well, and the plates are incubated for 2 hr and washed with PBS, 0.02% sodium azide, 0.05% Tween 80. Antiserum (serial dilutions beginning at 1/200 made in PBS, 0.02% sodium azide, 0.05% Tween 80, 1% BSA) is added to the wells, which, after incubation for 4 hr are washed. Bound antibodies are detected by incubation for 2 hr with alkaline phosphatase conjugated to goat anti-rabbit antibody (diluted 1/100; Cappel Laboratories). After the wells are washed, 65 μl of *p*-nitrophenyl phosphate [1 mg/ml in 50 m*M* bicarbonate buffer (pH 9.8), 1 m*M* magnesium chloride] is added. The reaction is terminated after approximately 30 min with 100 μl of 1.0 *M* sodium hydroxide. The absorbance is read at 405 nm with a Dynatech Minireader II.

Prior to use in staining replicas, antistreptavidin antibodies are purified from the antiserum. A total IgG fraction is isolated via chromatography on protein A–Sepharose by the method of Ey *et al.*[6] For affinity purification, 2 ml of the IgG fraction (8.6 mg dissolved in PBS) is incubated batchwise overnight at 4° with an Affi-Gel 10–streptavidin column (2 ml of Affi-Gel containing 5 mg of streptavidin). The column is then washed with PBS and eluted with 100 m*M* glycine-HCl (pH 2.5). Fractions (1 ml) are collected into tubes containing 100 μl of 1.5 *M* Tris-HCl (pH 8.8), dialyzed against PBS, and stored in aliquots at −70°.

[6] P. L. Ey, S. J. Prowse, and C. R. Jenkin, *Immunochemistry* **15,** 429 (1978).

[51] Streptavidin–Enzyme Complexes in Detection of Antigens on Western Blots

By CHRISTINE L. BRAKEL, MARK S. BROWER, AND KIMBERLY GARRY

The Western blot has become an important analytical tool for studies on protein structure and synthesis in addition to its recent clinical application in confirmation of serum antibodies to the AIDS (HTLV III) virus antigens.[1-3] The development of this procedure [which includes electrophoretic separation of proteins on a gel, electrophoretic transfer of the separated proteins to a suitable membrane, and detection of specific proteins (antigens) on the membrane by immunoabsorption of specific antibodies] has mirrored the development of the procedure from which it received its name, the Southern blot.[4] In most early applications, the development of the Western blot was accomplished by use of radiolabeled (^{125}I) antibodies or protein A,[5,6] although the application of nonradioactive detection systems to the development of Western blots was demonstrated in the initial report by Towbin *et al.*[1] In continuing use of Western blot procedures it was often the case that the immunoenzymatic assays were found to be inadequate in terms of sensitivity of detection when compared to the ^{125}I labeling and detection strategy.[7-12] While the use of enzyme-linked antibodies was commonly applied to investigations of proteins, it was rarely applied to the study of nucleic acids. However, just as the Southern blot led to the development of the Western blot, the techniques of the nonradioactive detection were soon applied to Southern blots. The most successful approach in nonradioactive hybridization and

[1] H. Towbin, T. Staehelin, and J. Gordon, *Proc. Natl. Acad. Sci. U.S.A.* **76,** 4350 (1979).
[2] W. N. Burnette, *Anal. Biochem.* **112,** 195 (1981).
[3] V. C. W. Tsang, J. M. Peralta, and A. R. Simons, this series, Vol. 92, p. 377.
[4] E. M. Southern, *J. Mol. Biol.* **90,** 809 (1975).
[5] C.-Y. Gregory Lee, Y.-S. Huang, P.-C. Hu, V. Gomel, and A. C. Menge, *Anal. Biochem.* **123,** 14 (1982).
[6] R. L. Dimond and W. F. Loomis, *J. Biol. Chem.* **251,** 2680 (1976).
[7] C. G. O'Connor and L. K. Ashman, *J. Immunol. Methods* **54,** 267 (1982).
[8] M. S. Blake, K. H. Johnston, G. J. Russell-Jones, and E. C. Gotschlich, *Anal. Biochem.* **136,** 175 (1984).
[9] D. A. Knecht and R. L. Dimond, *Anal. Biochem.* **136,** 180 (1984).
[10] A. L. DeBlas and H. M. Cherwinski, *Anal. Biochem.* **133,** 214 (1983).
[11] Z. Wojtkowiak, R. C. Briggs, and L. S. Hnilica, *Anal. Biochem.* **129,** 486 (1983).
[12] W. F. Glass, R. C. Briggs, and L. S. Hnilica, *Science* **211,** 70 (1981).

detection has been the avidin (streptavidin)–biotin methodology originally introduced for use with nucleic acids by Ward and co-workers.[13] The rising interest in this tool has hastened the development of biotin detection systems. As the sensitivity and utility of these biotin detection systems improved, the utility of the avidin–biotin system for development of Western blots was soon apparent.

Methodology

General Comments

The basic preparative procedures of Western blotting (gel preparation and running methods and transfer blotting methods) have been described in detail previously in this series,[3] and our concern here is the application of avidin–biotin systems to the development of the blot itself. The development of the Western blot by these techniques is carried out in three or four steps. First, the nitrocellulose membrane is blocked (quenched) so that unoccupied membrane sites are fully covered by a noninterfering protein. In the next step, the primary antibody is incubated with the blot to recognize and bind to the antigen of interest.[14] In the third step, i.e., labeling, secondary antibody, a biotinylated antiantibody directed against the primary antibody, is incubated with the filter. In the final step, the blot is developed by binding of the biotin detection complex or conjugate and development of the color reaction specific for the enzyme of the complex. Thus, the development of a Western blot is reduced to the detection of biotin. The use of biotinylated rather than enzyme-labeled secondary antibodies appears to be advantageous for several reasons. It is much simpler to prepare biotinylated antibodies than enzyme-linked antibodies. Furthermore, the use of the biotin label and subsequent detection appears to yield a greater sensitivity of detection owing to the layering of the detection systems.

Equipment

Following preparation of the Western blot there is little, if any, requirement for sophisticated equipment. Plastic boxes, trays, and disposable laboratory culture dishes are often used for the development steps of

[13] P. R. Langer, A. A. Waldrop, and D. C. Ward, *Proc. Natl. Acad. Sci. U.S.A.* **78,** 6633 (1981).

[14] While not the usual practice, the primary antibody can be biotinylated, and this incubation would be followed by the addition of the avidin (streptavidin)–enzyme conjugate or complex and the development of color, thus giving a three-step procedure.

the Western blot procedure. Some manufacturers provide slotted trays for the multiple strip developments as suggested by Tsang *et al.*[3] in an earlier volume of this series. These tools, while extremely convenient for the clinical laboratory, may consitute an unnecessary expense for the research laboratory, in which a refrigerator box may serve a similar function.

Buffers and Solutions

Phosphate-buffered saline (PBS): 7 m*M* Na_2HPO_4, 3 m*M* NaH_2PO_4, 130 m*M* NaCl (pH 7.4)
PBS–BSA: PBS containing 1.0% bovine serum albumin (BSA)
PBS–Tween: PBS containing 0.5% (v/v) Tween 20
Blocking buffers
- A: PBS containing 5% (w/v) BSA
- B: PBS containing 2% (w/v) BSA and 0.05% (v/v) Triton X-100

Washing buffers
- A: 10 m*M* potassium phosphate buffer (pH 6.5), 0.5 *M* NaCl, 1.0 m*M* EDTA, 2% (w/v) BSA, 0.5% Triton X-100
- B: 10 m*M* sodium phosphate buffer (pH 6.5), 0.5 *M* NaCl, 0.1% BSA, 0.5% Tween 20
- C: 0.3 M NaCl, 30 m*M* disodium citrate

Color development reagents
- Acid phosphatase reaction: 20 ml of 0.2 *M* acetate buffer (pH 5.8) is mixed with 5 ml of 5 m*M* naphthol AS-MX phosphate (Sigma Chemical Co., St. Louis, MO; prepared in acetate buffer) and 0.5 ml of a 4 mg/ml solution (in acetate buffer) of Fast Violet B salt (Sigma)
- Horseradish peroxidase reaction: 5 mg of diaminobenzidine (Polysciences, Warrington, PA) is dissolved in 10 ml of 10 m*M* Tris-HCl (pH 7.5), and 0.2 ml of 1% $CoCl_2$ and 0.2 ml of 1% H_2O_2 are added

Blocking the Membrane

Most commonly, the transfer membrane is blocked with solutions containing bovine serum albumin. For most applications, the use of BSA from 2 to 5% (w/v) in PBS with or without NaN_3 is adequate. Other reports have suggested that these procedures are not adequate and indicated that the addition of Tween 20 (0.1–1%, v/v) to the blocking solution provides reduced background staining of the developed membrane.[3] More recently, the use of the blocking solution nicknamed BLOTTO has found wide application for Western and Southern transfers, and even some application in immunohistochemistry.[15] BLOTTO usually is prepared from nonfat dry milk at concentrations of 2–5% and is used in buffered solutions with and without detergents such as Tween 20 and Triton X-100. The developed blots are, for the most part, considerably freer of background staining following this blocking procedure.

Blocking is carried out by incubating the membrane for 30 min to overnight at temperatures ranging from 37° for the shorter blocking procedures to 2° for the extended incubations. After blocking, the membrane is generally rinsed briefly with a solution similar to the diluent used for antibody dilution.

Addition of Antibodies

Primary antibodies are diluted in an appropriate buffer, usually PBS–Tween, to a concentration previously found to be adequate by experimentation. The diluted primary antibody solution is incubated with the membrane with gentle agitation for a period of 30–60 min at either 37° or at room temperature. The membrane is then washed with a suitable washing solution, e.g., PBS–Tween, to remove unbound primary antibody. The secondary, biotinylated antibody is then diluted and incubated with the membrane for 30–60 min at 20–37°. The membrane is washed several times to remove unbound secondary antibody before addition of the detection complex.

Addition of Detection Complex

Prior to addition of the detection complex, some researchers find it necessary to reblock the membrane with solutions such as blocking buffer B. Complexes of streptavidin–biotinylated acid phosphatase[16] are diluted into PBS containing 5 m*M* EDTA, and complexes of streptavidin–biotinylated horseradish peroxidase are diluted into PBS containing 1% BSA. The complexes are added to the membrane and incubated at room temperature or 37° for 30–60 min. Following a series of washes, generally three 5- to 10-min washes in washing buffer A followed by 2 washes in washing buffer C for the acid phosphatase complex or three 5- to 10-min washes in washing buffer B followed by two washes in washing buffer C

[15] D. A. Johnson, J. W. Gautsch, J. R. Sportsman, and J. H. Elder, *Gene Anal. Tech.* **1,** 3 (1984); R. C. Duhamel and D. A. Johnson, *J. Histochem. Cytochem.* **33,** 711 (1985); see also R. C. Duhamel and J.S. Whitehead, this volume [21].

[16] C. L. Brakel and D. L. Engelhardt, *in* "Symposium on Rapid Detection and Identification of Infectious Agents" (D. T. Kingsbury and S. Falkow, eds.), p. 235. Academic Press, New York, 1985.

for the horseradish peroxidase complex, the membranes are ready for color development.

Color Development

Acid Phosphatase Complex. The acid phosphatase complex is developed by incubation with naphthol AS-MX phosphate and Fast Violet B in acetate buffer (pH 5.8). The reaction is allowed to develop for 60 min–18 hr. The development of color can be accelerated by incubation at 37° or slowed by incubation at 2–8°.

Horseradish Peroxidase Complex. Development of color using the horseradish peroxidase complex is accomplished by incubation in a solution containing H_2O_2, diaminobenzidine, and $CoCl_2$. The color reaction with these reagents is generally rapid and is carried out for 30–60 min at room temperature. Longer development times for the horseradish peroxidase color reaction generally do not result in increased levels of sensitivity but do tend to generate increased background staining of the membrane.

Results

Application to Detection of α_1-Antitrypsin

The sensitivity of detection of the plasma protein α_1-antitrypsin was compared using ^{125}I-labeled and biotin-labeled secondary antibodies.[17] The results were comparable in sensitivity when the biotin-labeled secondary antibody was detected with the streptavidin–horseradish peroxidase complex. However, when the biotinylated antibody was detected with the acid phosphatase complex, a 5- to 10-fold increase in the level of detection was obtained. This was presumably due to the longer times for which the acid phosphatase remains active in the production of reaction products.

New Methods and Tools for Western Blots

While considerations of sensitivity include, of course, the ability of the primary antibody to recognize the antigen following electrophoresis under denaturing conditions, transfer, and partial to complete renaturation on the membrane, this capability of the primary antibody can be selected. The absolute sensitivity of Western blot detections depends greatly on the ability of the membrane to bind the antigen of interest efficiently. For

[17] M. S. Brower, C. L. Brakel, and K. Garry, *Anal. Biochem.* **147,** 382 (1985).

these purposes, many commercial enterprises now provide membranes with increased capacities that researchers can test and select for retention of the proteins of interest.

The use of streptavidin (or avidin)–alkaline phosphatase complexes in place of the acid phosphatase complex for Western and Southern blot detections yields similar levels of sensitivity.[16,18] The development of color using the alkaline phosphatase complexes is generally more rapid than with the acid phosphatase, but for some membranes, notably nylon-based membranes, the substrate and chromogen for the alkaline phosphatase color reaction may generate more background staining. Additionally, some of the recent enhancement techniques in the assays for horseradish peroxidase[19] may be applicable to the use of horseradish peroxidase–avidin/streptavidin in Western blot applications that demand high levels of sensitivity.

[18] J. J. Leary, D. J. Brigati, and D. C. Ward, *Proc. Natl. Acad. Sci. U.S.A.* **80,** 4045 (1980).

[19] J. Burns, V. T. W. Chan, J. A. Jonosson, K. A. Fleming, S. Taylor, and J. O'D. McGee, *J. Clin. Pathol.* **38,** 1085 (1985).

[52] Luminescent Detection of Immunodot and Western Blots

By MERLIN M. L. LEONG and GLYN R. FOX

The visualization and characterization of protein and DNA antigens immobilized on nitrocellulose filters after immunodot binding or Western blotting assays can be easily accomplished by the use of antibody–enzyme conjugates and chromogenic substrates.[1–6] Alternatively, antigens in tissue sections or on nitrocellulose filters can be detected with the aid of biotin–avidin complexes.[7–11] In the simplest assay, avidin, an egg-white

[1] J. S. Hanker, P. S. Yates, C. B. Metz, and A. Rustioni, *Histochem. J.* **9,** 789 (1977).

[2] S. Avrameas and T. Ternynck, *Immunochemistry* **8,** 1175 (1971).

[3] P. K. Nakane, *J. Histochem. Cytochem.* **16,** 557 (1968).

[4] P. K. Nakane and A. Kawaoi, *J. Histochem. Cytochem.* **22,** 1084 (1974).

[5] L. A. Sternberger, P. H. Hardy, Jr., J. J. Cuculis, and H. G. Meyer, *J. Histochem. Cytochem.* **18,** 315 (1970).

[6] R. Hawkes, E. Niday, and J. Gordon, *Anal. Biochem.* **119,** 142 (1982).

[7] J. L. Guesdon, T. Ternynck, and S. Avrameas, *J. Histochem. Cytochem.* **27,** 1131 (1979).

[8] S. M. Hsu, L. Raine, and H. Fanger, *J. Histochem. Cytochem.* **29,** 577 (1981).

[9] M. Wilchek and E. A. Bayer, *Immunol. Today* **5,** 39 (1984).

protein with an extremely high natural affinity for biotin, is covalently conjugated to an enzyme reporter molecule such as peroxidase. The avidin–peroxidase conjugate in conjunction with chromogenic substrates such 4-chloronaphthol is used to visualize the presence of antigens on fixed tissue sections or nitrocellulose filters. Permutations of the avidin–biotin interaction have been developed for the sensitive detection of antigens.[12–16]

We have discovered that the use of a light-emitting substrate, luminol, in combination with antibody–peroxidase or avidin–peroxidase conjugates yielded at least 2 to 4-fold increases in detection sensitivity over that attained with the usual chromogenic substrate, 4-chloronaphthol.[17,18] This light emission-based system can be easily modified to fit the technical format requirements of any immunodot binding assay, ELISA, or Western blotting assay. Furthermore, the required light-emitting substrate, luminol, is readily available, inexpensive, nontoxic, and noncarcinogenic, making it a reagent of choice in both the research and clinical laboratory. In view of the number of laboratories that may benefit from the use of this safer and more sensitive detection system, the purpose of this chapter is to describe the necessary protocols involved in the detection of antigens using avidin–peroxidase, streptavidin–peroxidase, and antibody–peroxidase conjugates and luminol.

Reagents

Nitrocellulose (BA-85; Bio-Rad Ltd., Mississauga, ON)
Pacific skim milk powder (Dairyland Foods, Vancouver, BC)
Tween 20 (Baker Chemical Co., Phillipsburg, NJ)
Avidin–peroxidase conjugate (ICN Immunobiologicals, Lisle, IL)
Streptavidin–peroxidase conjugate (ICN)
Luminol (Sigma Chemical Co., St. Louis, MO)

[10] R. H. Yolken, F. J. Leister, L. S. Whitecomb, and M. Santasham, *J. Immunol. Methods* **56,** 319 (1983).
[11] S.-M. Hsu and L. Raine, *J. Histochem. Cytochem.* **29,** 1349 (1981).
[12] C. Kendall, I. Ionescu-Matiu, and G. R. Dressman, *J. Immunol. Methods* **56,** 329 (1983).
[13] P. V. S. Rao, N. L. McCartney-Francis, and D. D. Melcalfe, *J. Immunol. Methods* **57,** 71 (1983).
[14] M. A. Shamsuddin and C. C. Harris, *Arch. Pathol. Lab. Med.* **107,** 514 (1983).
[15] C. Bonnard, D. S. Papermaster, and J.-P. Kraehenbuhl, *in* "Immunolabeling for Electron Microscopy" (J. M. Polak, and I. M. Varndell, eds.), p. 95. Elsevier, Amsterdam, 1984.
[16] K. Ogata, M. Arakawa, T. Kasahara, K. Shioriri-Nakano, and K. Hiraoka, *J. Immunol. Methods* **65,** 75 (1983).
[17] M. M. L. Leong, C. Milstein, and R. Pannell, *J. Histochem. Cytochem.* **34,** 1645 (1986).
[18] M. M. L. Leong, G. R. Fox, and J. S. Hayward, *Anal. Biochem.* **168,** 107 (1988).

Biotinyl-*N*-hydroxysuccinimide ester (Sigma)
Tris(hydroxymethyl)aminomethane (Trizma base; Sigma)
Hydrogen peroxide (30% stock solution; Anachemia Canada Inc., Montreal, QE)
DMSO (dimethyl sulfoxide; Aldrich Chemical Co., Milwaukee, WI)
Polaroid film (Type 667; Polaroid Corporation, Cambridge, MA)
X-Ray film (X-Omat AR; Eastman Kodak Co., Rochester, NY)
Rabbit serum (ICN)
Goat anti-rabbit IgG Fc fragment (ICN or Sigma)
Phosphate-buffered saline (PBS, pH 7.2): 8.00 g NaCl, 0.20 g KCl, 1.15 g Na_2HPO_4, 0.20 g K_2HPO_4 and 900 ml distilled water; adjust to pH 7.2 and add distilled water to a final volume of 1.0 liter
Blocking solution: to 50.00 g Pacific skim milk powder add PBS buffer to give a final volume of 1.0 liter; prior to use, centrifuge the solution at 25,000 *g* in a Sorvall preparative centrifuge (RC-2B; Beckman Instruments Inc., Irvine, CA) for 2 hr at 4° to pellet out insoluble particulate matter
Washing solution: to 1.0 liter of blocking solution (precentrifuged) add 0.50 ml Tween 20 and mix thoroughly
10 m*M* Tris-HCl buffer (pH 8.0): 1.21 g tris(hydroxymethyl)aminomethane and 900 ml distilled water; adjust the pH to 8.0 with HCl and add more distilled water to a final volume of 1.0 liter
Luminol solution: 20 ml of 10 m*M* Tris-HCl buffer (pH 8.0), 100 μl luminol (100 mg/ml in DMSO), and 50 μl 30% hydrogen peroxide; add luminol to the Tris-HCl buffer and mix throughly prior to addition of hydrogen peroxide
150 m*M* NaCl–10 m*M* phosphate buffer (pH 7.5): dissolve 8.77 g NaCl and 1.42 g Na_2HPO_4 in 800.0 ml of distilled water, adjust the pH to 7.5 with a solution of NaH_2PO_4 (4.14 g/liter), and then add more distilled water to yield a final volume of 1.0 liter

Biotinylation of Proteins

Proteins to be labeled with biotin molecules are biotinylated essentially according to the protocol of Della-Penna *et al.*[19] with minor modifications as follows: Dialyze proteins (1 mg in 2.5 ml of PBS) against 1 liter of 0.1 *M* $NaHCO_3$ (pH 8.0) overnight at 4°. Add 200 μl of biotinyl-*N*-hydroxysuccinimide ester (1 mg/ml in dimethylformamide) to the 2.5 ml of dialyzed proteins, and allow the mixture to react at room temperature for 6 hr with gentle stirring. Finally, dialyze the biotinylated proteins over-

[19] D. Della-Penna, R. E. Christoffersen, and A. B. Bennett, *Anal. Biochem.* **152,** 329 (1986).

night at 4° against 4 liters of 150 mM NaCl in 10 mM phosphate buffer (PBS, pH 7.5) with at least one buffer change. The biotinylated proteins are kept frozen at −20° until needed.

Detection of Biotinylated Proteins

The nitrocellulose strip containing the spotted or Western transferred biotinylated proteins is incubated in 100 ml of blocking solution for at least 1 hr at room temperature. After blocking, the nitrocellulose strip is sealed inside a polyethylene bag containing at least 500 μl of an appropriate dilution (e.g., a 1/5 dilution in PBS) of avidin–peroxidase (2.3 mg/ml stock solution) or streptavidin–peroxidase (0.25 mg/ml stock solution) conjugate and incubated for at least 1 hr at room temperature. After the incubation, the nitrocellulose strip is washed 2 times for 10 min each with agitation in 100-ml volumes of washing buffer. Last, the biotinylated proteins are visualized by placing the nitrocellulose strip in the light-emitting luminol solution for 3–5 min, after which the strip is sealed in a polyethylene bag or simply wrapped. The resulting light signals are then recorded on X-ray or Polaroid film (Type 667) by contact exposures (typically 5–60 sec).

Immunodot Binding Assay

Spot the test antigen in 2-μl volumes in PBS directly onto a strip of nitrocellulose (BA-85) with a micropipettor. Let the spotted antigen samples air dry at room temperature for at least 30 min. Block the nitrocellulose strip by placing it in a tray containing 50 ml of blocking solution for at least 60 min at room temperature. Remove the nitrocellulose strip from the blocking solution, transfer it to a new plastic tray, and spot 2-μl volumes of biotinylated antibody (0.1 mg/ml) over each antigen site. Allow the nitrocellulose strip to incubate in a covered plastic tray at room temperature for at least 30 min. Wash the nitrocellulose strip 2 times for 10 min each with agitation in 100-ml volumes of washing solution. Spot 2-μl volumes of avidin–peroxidase conjugate (0.23 mg/ml) over each antigen site and incubate again at room temperature for at least 30 min. Wash the nitrocellulose strip 2 times for 10 min each with agitation in 100-ml volumes of washing buffer.

To visualize the immunodot binding assay results, place the nitrocellulose strip in the light-emitting luminol substrate solution. After 3–5 min of soaking in the above luminol solution, the nitrocellulose strip is sealed inside a waterproof polyethylene bag, or simply wrapped with plastic wrap, and placed facedown on a sheet of X-ray film or Polaroid film (Type 667) in a darkroom in order to record the light signals emitting from the antigen sites. Permanent records of the immunodot binding results can be routinely obtained by using appropriate contact exposure times (typically 1–60 sec).

Western Blotting

Prepare a sodium dodecyl sulfate (SDS) polyacrylamide minigel of desired percentage in accordance to the technique of Laemmli.[20] Using a Hamilton microsyringe, load an appropriate volume of the protein antigen mixture to be electrophoretically separated (e.g., 10 μl/lane, 50–100 μg/ml). Perform electrophoretic separation under appropriate conditions (e.g., 15 mA at room temperature for 3 hr). After the electrophoretic run, the acrylamide gel is carefully removed from the apparatus, and the separated protein antigens on the gel are transferred onto a sheet of nitrocellulose according to the Western blotting technique of Burnette[21] using relevant conditions (e.g., a transfer time of 2 hr at 150 mA).

The nitrocellulose sheet containing the transferred protein antigens is then blocked for at least 1 hr at room temperature in the blocking solution. Next, the nitrocellulose sheet is placed inside a polyethylene bag containing at least 500 μl of an appropriate dilution of biotinylated antibody. The bag is completely sealed with a plastic sealer, and the nitrocellulose sheet is allowed to incubate for at least 30 min with the biotinylated antibody (0.2 mg/ml) undisturbed at room temperature. After the incubation, the nitrocellulose sheet is removed from the bag and washed 2 times for 15 min each with agitation in 100-ml volumes of washing buffer. The nitrocellulose sheet is sealed inside a new polyethylene bag containing at least 500 μl of an appropriate dilution of avidin–peroxidase conjugate (e.g., 0.5 mg/ml in PBS). The nitrocellulose sheet is allowed to incubate for at least 60 min at room temperature, after which it is again removed and washed 2 times for 15 min each with agitation in 100-ml volumes of washing buffer. The results are visualized and recorded as described above for the immunodot procedure.

Results

For demonstrating direct detection of biotinylated proteins, bovine serum albumin (BSA) can be used as a model system. For this purpose, BSA is biotinylated and spotted in decreasing concentrations on nitrocellulose filters. In initial experiments, the sensitivity and specificity of avidin–peroxidase and streptavidin–peroxidase conjugates can be com-

[20] U. K. Laemmli, *Nature (London)* **227,** 680 (1970).
[21] W. N. Burnette, *Anal. Biochem.* **112,** 195 (1981).

pared. Results of such experiments have shown that the use of avidin–peroxidase gives rise to a generally higher level of nonspecific binding of this conjugate to nonrelevant sites, e.g., binding to pencil marks and nonbiotinylated BSA control sites. On the other hand, the use of streptavidin–peroxidase conjugates yields the same detection sensitivity without the nonspecific background signals observed with the avidin–peroxidase conjugate.

In the next stage, biotinylated BSA in increasing concentrations can be separated on 12% polyacrylamide gels in the presence of SDS and transferred onto nitrocellulose filters by Western blotting. This is per-

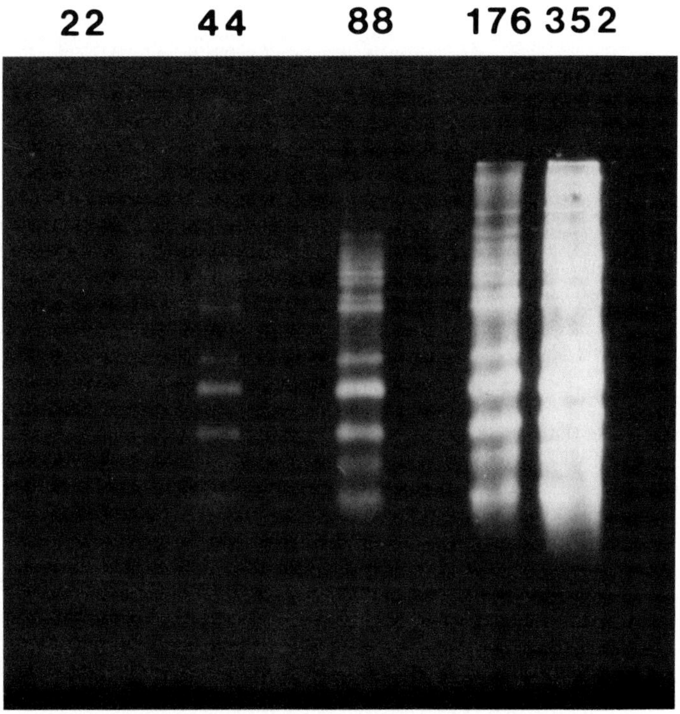

FIG. 1. Sensitive detection of biotinylated *Pinus nigra* proteins after electrophoretic separation and Western blotting. Increasing concentrations of biotinylated *Pinus nigra* proteins (22–352 ng/lane) were loaded onto a 15% polyacrylamide minigel and electrophoretically separated in the presence of SDS using a current of 30 mA for 4 hr. The separated proteins were transferred by Western blotting onto a nitrocellulose strip and blocked for 1 hr. After blocking, the strip was incubated with an avidin–peroxidase conjugate (0.23 mg/ml in PBS) for 30 min. The results were visualized by incubating the washed strip in luminol solution, sealing in a polyethylene bag, and, finally, making a contact exposure of 15-sec duration on Polaroid film (Type 665).

formed in order to determine whether the results for the dot-blot system is also applicable to SDS–PAGE. As before, it has been found that the use of avidin–peroxidase conjugate yields greater nonspecific background signal than streptavidin–peroxidase.

After establishing the conditions for the model protein, the methodology can be applied to a more complex system. For example, the direct detection of biotinylated proteins from *Pinus nigra* long shoot terminal buds can be carried out after electrophoretic separation on an SDS–polyacrylamide minigel and Western blotting. Results indicate that the direct detection with avidin–peroxidase and light-emitting luminol is capable of identifying and resolving a mixture of biotinylated pine proteins at a sample loading as low as 44 ng (Fig. 1).

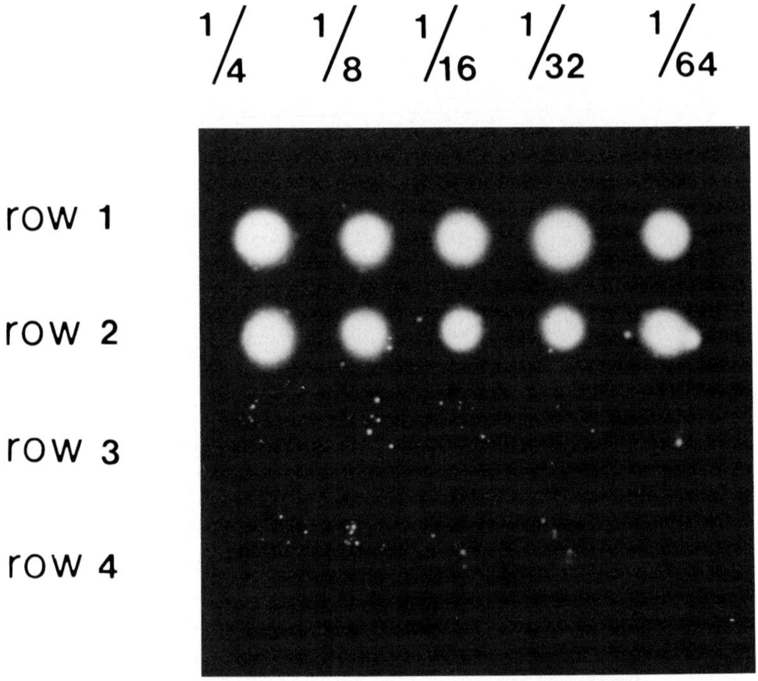

FIG. 2. Immunodot binding assay for the presence of IgE in samples of nonimmune human serum. Serial dilutions (1/4–1/64) of human sera and BSA in PBS were spotted in 2-μl volumes on a strip of nitrocellulose as follows: row 1, dilutions of nonimmune serum from male donor 1; row 2, dilutions of nonimmune serum from male donor 2; row 3, blank control (i.e., pencil marks only; no antigen spotted); and row 4, BSA (5 mg/ml in PBS) as a negative control. After blocking, the nitrocellulose strip was incubated with neat (1 mg/ml in PBS) biotinylated goat anti-human IgE, washed, and subsequently incubated with avidin–peroxidase (0.23 mg/ml) in PBS. The immunodot binding results were visualized by incubation in luminol solution followed by contact exposure of 60 sec using Type 667 Polaroid film.

The results of a typical immunodot binding assay using a biotinylated antibody and avidin–peroxidase conjugate are shown in Fig. 2. The antigen detected in this particular case was low levels of human IgE in two nonimmune human serum samples. A commercial goat anti-human IgE (ICN) was biotinylated and employed in this immunodot binding assay for IgE. An avidin–peroxidase conjugate and the light-emitting substrate luminol were used to visualize the immunodot binding results (Fig. 2). The great intensity and specificity of the emitted signals from the antigen sites adequately illustrate the sensitivity and simplicity of this light emission-based immunodot binding technique.

Figure 3 shows typical results of a Western or immunoblot using rabbit serum as a source of IgG Fc fragments. In this particular example, the detecting goat anti-rabbit IgG Fc fraction was first biotinylated and subsequently used to detect the presence of rabbit IgG Fc fragment on the Western blot. The bound, biotinylated goat anti-rabbit IgG Fc on the

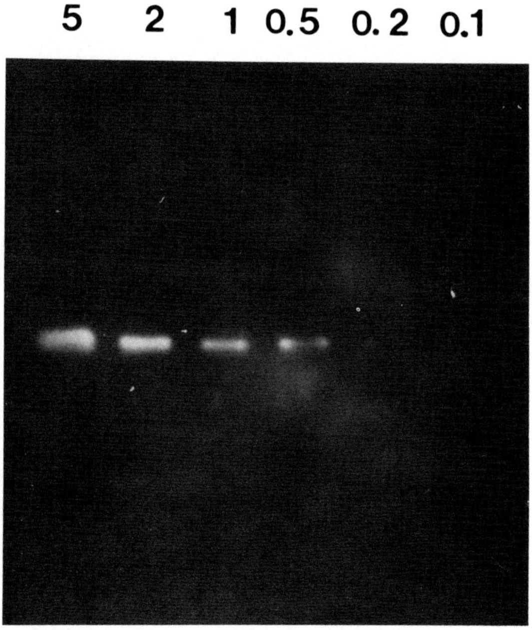

FIG. 3. Western blotting detection of rabbit IgG Fc. Rabbit serum in decreasing concentrations (5.0–0.1 μg/lane) was separated on a 10% polyacrylamide minigel in the presence of SDS using a current of 15 mA for 2.5 hr. After transfer to a sheet of nitrocellulose by Western blotting, the protein was incubated for 30 min with 500 μl of biotinylated goat anti-rabbit IgG Fc (0.2 mg/ml) in PBS, washed, and subsequently incubated for 30 min with avidin–peroxidase conjugate (0.125 mg/ml) in PBS. The results were visualized as described in the text.

Western blot was visualized by the use of luminol and avidin–peroxidase. The use of avidin–peroxidase conjugate yielded intense signals, revealing the presence of rabbit IgG Fc fragment in rabbit serum (Fig. 3).

Comments

We would like to point out that although the protocols described in this chapter utilize avidin–peroxidase conjugates for the final visualization of immunodots and immunoblots, such conjugates can be replaced by specific antibody–peroxidase conjugates. Indeed, in most of our routine immunodot binding and Western blotting assays, the first antigen-detecting antibody is used without modifications (i.e., nonbiotinylated), following which an antibody–peroxidase conjugate with binding specificity for the first antibody is employed to detect the bound first antibody. For instance, if we used a rat antibiotin antibody to detect the presence of biotin, a goat anti-rat IgG–peroxidase would be employed to visualize the presence of the bound rat antibiotin antibody.

The emission of light makes the final visualization of immunodot binding and Western blotting assays a relatively simple task. The light signals can be permanently recorded on X-ray or Polaroid films of varying ASA speeds or sensitivity. In this regard, we have recently developed a photodetection device that enables one to make multiple permanent records of immunodot, ELISA, Southern, and Western blotting results directly on Polaroid film without the need for darkroom and film processing facilities.[18] Furthermore, it is conceivable that modified, sensitive luminometers meeting the format requirements of immunodot binding, enzyme-linked, Western, and Southern blotting assays can be used to quantify the emitted light signals.

This light emission-based detection technique can potentially be extended to detection of any analyte provided that the reporter molecule, peroxidase (e.g., avidin–peroxidase or antibody–peroxidase conjugate), and the light-emitting substrate, luminol, can both be incorporated into the detection system. We have already shown that this light-emission technique can be used in immunodot binding, Western, and Southern blotting assays.[17] Further refinements in nonisotopic labeling of DNA probes by biotin-dUTP[22] or photoreactive biotin[23] and subsequent detection will permit sensitive visualization of hybridization results for single-copy or low-copy-number genes on DNA dot, Northern, and Southern blots. In this regard, preliminary data from our experiments involving

[22] J. J. Leary, D. J. Brigati, and D. C. Ward, *Proc. Natl. Acad. Sci. U.S.A.* **80,** 4045 (1983).

[23] A. C. Forster, J. L. McInnes, D. C. Skingle, and R. H. Symons, *Nucleic Acids Res.* **13,** 745 (1985).

photoreactive biotin labeling of DNA probes and observed signal enhancement by iodophenol[24] suggest that the above objective will be reached in the near future.

[24] M. M. L. Leong and G. R. Fox, *Anal. Biochem.* **172,** 145 (1988).

[53] Identification of Calmodulin-Binding Proteins

By MELVIN L. BILLINGSLEY, JOSEPH W. POLLI, KEITH R. PENNYPACKER, and RANDALL L. KINCAID

Introduction

The use of chemically modified proteins has proved indispensible for the elucidation of structural and functional characteristics of a wide range of proteins. Chemical perturbations of primary amino acid structure have been used (1) to investigate functional changes brought about by the perturbation and (2) to "map" the structural location of functional contacts between two interacting proteins.[1] Experimentally, there are three possible consequences of such a modification. First, the functional interaction can be totally destroyed by the chemical modification. Second, the functional interaction can be unperturbed by the chemical modification. Third, the functional interaction can be partially affected by the structural modification. When designing probes for detection of protein–protein interactions, the goal is to use a chemical modification that does not affect the functional interaction; for site-specific chemical modifications for determining functional contact points, the goal is to produce a range of perturbations that block, partially block, or leave unimpaired the ability of the modified protein to interact with its putative target.

Calmodulin is the major intracellular receptor for calcium; the primary and tertiary structures of the molecule have been well described.[2,3] On binding up to 4 mol of calcium/mol protein, the molecule undergoes a marked conformational change, exposing an hydrophobic α-helical region of the protein. The specificity of the Ca^{2+}-regulated event is determined by the Ca^{2+}-dependent interaction of this domain on calmodulin with a target binding protein. Thus, many of the varied Ca^{2+}-regulated cellular

[1] G. K. Ackers and F. R. Smith, *Annu. Rev. Biochem.* **54,** 597 (1985)

[2] J. C. Stoclet, D. Gerard, M.-C. Kilhoffer, C. Lugnier, R. Miller, and P. Schaffer, *Prog. Neurobiol.* **29,** 321 (1987).

[3] Y. S. Babu, C. E. Bugg, and W. J. Cook, this series, Vol. 139, p. 632.

events are ultimately mediated through the actions of calmodulin-binding proteins.[4]

One approach to characterize cellular calmodulin-binding proteins is to chemically modify calmodulin with fluorescent, radioactive, or other chemically reactive congeners.[5,6] Such modified calmodulins have been exploited to study conformational changes on Ca^{2+}-dependent binding, to monitor the distribution of calmodulin in the cell, and to identify calmodulin-binding proteins following sodium dodecyl sulfate–polyacrylamide gel electrophoresis (SDS–PAGE).[7–9] The use of an iodinated calmodulin gel overlay for the purpose of detecting calmodulin-binding proteins was first described by Carlin *et al.*[10] and Glenny *et al.*[11] In this procedure, putative calmodulin-binding proteins were subjected to SDS–PAGE and renatured *in situ*. The gel was incubated with trace amounts of ^{125}I-labeled calmodulin, washed extensively in Ca^{2+}-containing buffers (2–3 days), and dried and subjected to autoradiography. This procedure demonstrated that calmodulin-binding proteins could be resolved on SDS–PAGE, renatured, and that the ^{125}I-labeled calmodulin could interact with specific calmodulin-binding domains in a Ca^{2+}-dependent manner. The gel-overlay technique has been refined to permit detection and quantification of microgram quantities of calmodulin-binding proteins.[12]

There are several drawbacks to the gel-overlay technique, particularly regarding the overall time of the procedure (7–14 days of sample preparation, washing, and autoradiography) and the amount (liters) of low-level radioactive waste generated during washing. Flanagan and Yost described a modified procedure for detection of calmodulin-binding proteins using nitrocellulose blots (Western blots) of SDS–PAGE-resolved calmodulin-binding proteins.[13] By using Tween 20 to minimize the nonspecific binding on the blot, they were able to identify calmodulin-binding proteins from both one- and two-dimensional SDS–PAGE. Furthermore, they reduced the lengthy washing times needed to reduce background in the gel-overlay protocol. This method still required the use of ^{125}I-labeled

[4] C. B. Klee and T. C. Vanaman, *Adv. Protein Chem.* **35,** 213 (1982).

[5] R. L. Kincaid, M. L. Billingsley, and M. Vaughan, this series, Vol. 159, p. 605.

[6] T. J. Andreason, C. H. Keller, D. C. LaPorte, A. M. Edelmon, and D. R. Storm, *Proc. Natl. Acad. Sci. U.S.A.* **78,** 2782 (1981).

[7] R. L. Kincaid, M. C. Vaughan, J. C. Osborne, Jr., and V. A. Tkachuk, *J. Biol. Chem.* **257,** 10638 (1982).

[8] H. W. Jarrett, *J. Biol. Chem.* **261,** 4967 (1986).

[9] K. Luby-Phelps, F. Lanni, and D. L. Taylor, *J. Cell Biol.* **101,** 1245 (1985).

[10] R. K. Carlin, D. J. Corab, and P. Siekevitz, *J. Cell Biol.* **89,** 449 (1981).

[11] J. R. Glenny and K. Weber, *J. Biol. Chem.* **255,** 10551 (1980).

[12] G. R. Slaughter and A. R. Means, this series, Vol. 139, p. 433.

[13] S. D. Flanagan and B. Yost, *Anal. Biochem.* **140,** 510 (1984).

calmodulin and subsequent autoradiography for detection of the calmodulin-binding proteins.

In order to facilitate the rapid, nonradioactive detection of immobilized calmodulin-binding proteins, we devised a method for incorporating biotin into lysine residues of calmodulin.[14] By using biotin as a reporter ligand, avidin-linked enzyme systems could be used to detect biotinylated calmodulin (Bio–CaM) bound to renatured calmodulin-binding proteins. Of note is the caveat that not all proteins may "renature" following SDS–PAGE and blotting; hence, care should be exercised when interpreting negative results. In addition, biotinylation can be used to explore what portion(s) of the calmodulin molecule is important for interactions with target proteins. By observing how the modification of a specific residue affects subsequent calmodulin–enzyme interactions, one can determine whether biotinylation perturbs focal points of protein–protein interaction. If biotinylation of a specific residue does not affect calmodulin interaction directly, one can preincubate biotinylated calmodulin derivatives with avidin, and determine whether complexing biotinylated calmodulin with avidin can block subsequent binding or enzymatic activation.[15] However, the avidin–biotinylated calmodulin complexes may sterically or allosterically interfere with the calmodulin-binding protein interaction and, thus, cannot be used to directly prove that a particular site is necessary for interaction.

Biotinylated calmodulin can be used to study differential expression of calmodulin-binding proteins in tissues or cell populations. We have exploited this technique to elucidate the major calmodulin-binding proteins of lymphocytes. In addition, biotinylated calmodulin can also be used to identify specific calmodulin-binding peptides following proteolytic degradation of the intact protein; comparison of such "fingerprints" may indicate whether they are structural features common to a series of calmodulin-binding proteins. Finally, Bio–CaM may prove useful for microinjection studies, since the biotin reporter group can be easily localized in cells using avidin-based histochemistry. However, careful titration of injected Bio–CaM would be needed in order to avoid detection of unbound Bio–CaM.

One key point is that in order to detect the interaction between calmodulin and specific calmodulin-binding proteins, one must generate a chemically modified probe that is functionally competent. Owing to the high methionine content (8 residues) of calmodulin, oxidative approaches to

[14] M. L. Billingsley, K. R. Pennypacker, C. G. Hoover, D. J. Brigati, and R. L. Kincaid, *Proc. Natl. Acad. Sci. U.S.A.* **82,** 7585 (1985).

[15] D. Mann and T. C. Vanaman, this series, Vol. 139, p. 417.

modification often alter biological activity.[14,15] In addition, there are no cysteine residues, further narrowing the choice of sites for modification. Thus, numerous studies have focused on amine-directed agents for modification. However, as several studies have indicated, the 7 lysyl residues in calmodulin differ with respect to their relative reactivities; depending on the target protein, several of the lysines may be critical for calmodulin–target interactions.[16–19] Since many lysines may be modified during reaction, such "derivatives" are not homogeneous populations; thus, caution must be used when interpreting the results obtained with modified molecules.

One empirical approach is to produce a series of biotinylated calmodulins modified at lysyl residues (using *N*-hydroxysuccinimide esters), at tyrosyl or histidyl residues (using *p*-diazobenzoylbiocytin), at acidic amino acid residues (using biotin hydrazide and carbodiimide coupling), or at nondirected, accessible nucleophilic sites (using "photobiotin" acetate). Modification of acidic amino acids in calmodulin with biotin hydrazide has proved to be of limited utility in our hands and is not described in detail.

Preparation of Biotinylated Calmodulin Derivatives

We have described the preparation and uses of calmodulin modified using biotinyl-ε-aminocaproic acid *N*-hydroxysuccinimide ester in several prior publications, and readers are referred to Refs. 5, 14, and 20 for specific modification. In this chapter, we focus on the reagents needed for producing other forms of biotinylated calmodulin. All modifications are carried out in the presence of 1 m*M* $CaCl_2$ in order to ensure modification of a homogeneous population of fully liganded proteins, i.e., to avoid partially or nonliganded forms.

Reagents

Purified bovine brain or testis calmodulin, 2 mg/ml, in coupling buffer [0.1 *M* sodium phosphate buffer (pH 7.4) or 0.1 *M* HEPES (pH 7.0)][21]

[16] D. P. Giedroc, S. K. Sinha, K. Brew, and D. Puett, *J. Biol. Chem.* **260,** 13406 (1985).
[17] D. P. Giedroc, D. Puett, S. K. Sinha, and K. Brew, *Arch. Biochem. Biophys.* **252,** 136 (1987).
[18] F. M. Faust, M. Slisz, and H. W. Jarret, *J. Biol. Chem.* **262,** 1938 (1987).
[19] M. A. Winkler, V. A. Fried, D. L. Merat, and W. Y. Cheung, *J. Biol. Chem.* **262,** 15466 (1987).
[20] M. L. Billingsley, K. R. Pennypacker, C. G. Hoover, and R. L. Kincaid, *BioTechniques* **5,** 22 (1987).
[21] R. L. Kincaid, this series, Vol. 139, 3 (1987).

$0.1\ M$ $CaCl_2$
$0.1\ M$ EGTA
$1.0\ M$ NaCl
NHS-LC-biotin (Pierce Chemical Co., Rockford, IL), 26 mg/ml in dry *N,N*-dimethylformamide (10 : 1 molar ratio of derivative to calmodulin)
Photobiotin acetate (Research Organics, Cleveland, OH), 2 mg/ml in coupling buffer (10 : 1 molar ratio of derivative to calmodulin)
Sulfo-NHS-biotin (Pierce), 21 mg/ml in dry *N,N*-dimethylformamide (10 : 1 molar ratio of derivative to calmodulin)
Activated diazobenzoylbiocytin (Calbiochem, La Jolla, CA), see below for method of activation
Nitrocellulose membrane (BA-85, Schleicher & Schuell, Keene, NH)
Blocking solution [50 m*M* Tris-HCl (pH 7.4) containing 150 m*M* NaCl, 1 m*M* $CaCl_2$, 0.1% antifoam A (Sigma Chemical Co., St. Louis, MO), and 5% nonfat dry milk]
Wash buffer [50 m*M* Tris-HCl (pH 7.4) containing 1 m*M* $CaCl_2$ and 150 m*M* NaCl]
Avidin–alkaline phosphatase or avidin–horseradish peroxidase detection systems (Vector Laboratories, Burlingame, CA; BioMeda, Inc.)
Hydrogen peroxide, 30% stock solution (v/v)
p-Chloronaphthol (Sigma), 1 mg/ml, in wash buffer containing 20% methanol (v/v)
5-Bromo-4-chloro-3-indoyl phosphate *p*-toluidine (BCIP; Amresco), 50 mg/ml, in *N,N*-dimethylformamide
Nitroblue tetrazolium chloride (NBT; Amresco), 50 mg/ml, in 50% *N,N*-dimethylformamide
Phosphatase buffer [0.1 *M* Tris-HCl (pH 9.5) containing 100 m*M* NaCl and 50 m*M* $MgCl_2$]
Avidin (Sigma or other source)
d-Biotin (Sigma or Pierce)
Dialysis membrane (Spectrapor; 6,000 MW cutoff)

Preparation of Sulfo-NHS-Biotin- and NHS-LC-Biotin-Derivatized Calmodulin

Purified calmodulin, prepared using melittin–Sepharose,[21] is dialyzed against coupling buffer, and the concentration is adjusted to 1.0 mg/ml ($E^{1\%}_{280}$ = 1.8). A 2-mg aliquot of calmodulin solution is adjusted to 1 m*M* $CaCl_2$ and 0.15 *M* NaCl. An aliquot (25 μl) of either NHS-LC-biotin (0.66 mg/25 μl) or sulfo-NHS- biotin (0.53 mg/25 μl) is added at a 10-fold molar excess with respect to calmodulin. Previous studies have indicated that the 10:1 molar excess of NHS derivative gives optimal labeling with minimal loss of biological function.[20] The derivatization is carried out for 2 hr at 25°, and the mixture is then either desalted on a precalibrated Sephadex G-25 column (PD-10, Pharmacia) or dialyzed exhaustively against coupling buffer to remove unreacted reagent. The calmodulin derivatives are stored in coupling buffer with 10% glycerol at −20°.

Preparation of Photobiotin Acetate-Derivatized Calmodulin

Purified calmodulin (in 1 m*M* $CaCl_2$ and 150 m*M* NaCl) is dialyzed as described above. In subdued light, photobiotin acetate (0.71 mg/335 μl) is dissolved in water and added to the calmodulin solution (2 mg total) in a flat-bottomed tube kept at 4° using the precautions and procedures previously outlined.[22] The reaction mixture is then placed 10 cm from a broad-spectrum GE sunlamp and irradiated for 10 min at 4°. The reaction mixture is either dialyzed against coupling buffer or chromatographed over a PD-10 Sephadex column to remove unreacted photobiotin. The derivatized calmodulin is stored in coupling buffer containing 10% glycerol at −20°.

Preparation of p-Diazobenzoylbiocytin-Derivatized Calmodulin

p-Diazobenzoylbiocytin (DBB) reagent is activated as previously described[23]: 2 mg of DBB is dissolved in 15 μl of dimethyl sulfoxide. Eighty-five microliters of 1 *M* HCl is added, and the mixture is chilled on ice. To this mixture, 40 μl of 0.112 *M* sodium nitrate is added and allowed to react for 5 min. This activation is quenched by adding 100 μl of 1 *M* NaOH, and the reaction is diluted with 1.75 ml of coupling buffer. In order to start the modification, 1 ml of purified, dialyzed calmodulin (2 mg/ml coupling buffer, brought to 1 m*M* $CaCl_2$ and 150 m*M* NaCl) is reacted with 500 μl of the activated DBB mixture for 1.5 hr at 25°. Following incubation, the reactants are removed by chromatography over PD-10 Sephadex G-25 columns. The derivative is stored in coupling buffer containing 10% glycerol at −20°.

Characterization of Biotinylated Calmodulin Derivatives

A useful method for documenting the production of the various biotinylated species is to determine the ultraviolet spectrum of the calmo-

[22] E. Lacey and W. N. Grant, *Anal. Biochem.* **163,** 151 (1987).

[23] M. Wilchek, H. Ben-Hur, and E. A. Bayer, *Biochem. Biophys. Res. Commun.* **138,** 872 (1986).

dulin before and after derivatization. As can be seen in Fig. 1A–C, all of the modified derivatives result in spectral shifts relative to native calmodulin. Sulfo-NHS-biotin and NHS-LC-biotin derivatives (data not shown) give nearly identical spectra (Fig. 1A), with absorbance maxima shifted from 277 nm (native calmodulin) to 266 nm (NHS-modified calmodulin). Thus, there is no apparent difference in the extent or amount of labeling of calmodulin produced by the more water-soluble sulfo-NHS-biotin reagent or by the more hydrophobic NHS-LC-biotin derivative. Photobiotin (Fig. 1B) and DBB (Fig. 1C) produce a different type of spectral shift, with DBB-modified biotin exhibiting considerable absorbance between 320 and 260 nm. Thus, different chemistries of derivatization result in characteristic spectral patterns of the modified calmodulins.

A second method for demonstrating biotinylation is direct visualization of biotinylated calmodulins following transfer to nitrocellulose. Although calmodulin has been reported to bind poorly to nitrocellulose,[24] we have found that calmodulin easily and rapidly transfers from 20% polyacrylamide gels to nitrocellulose, provided that the blot is transferred for

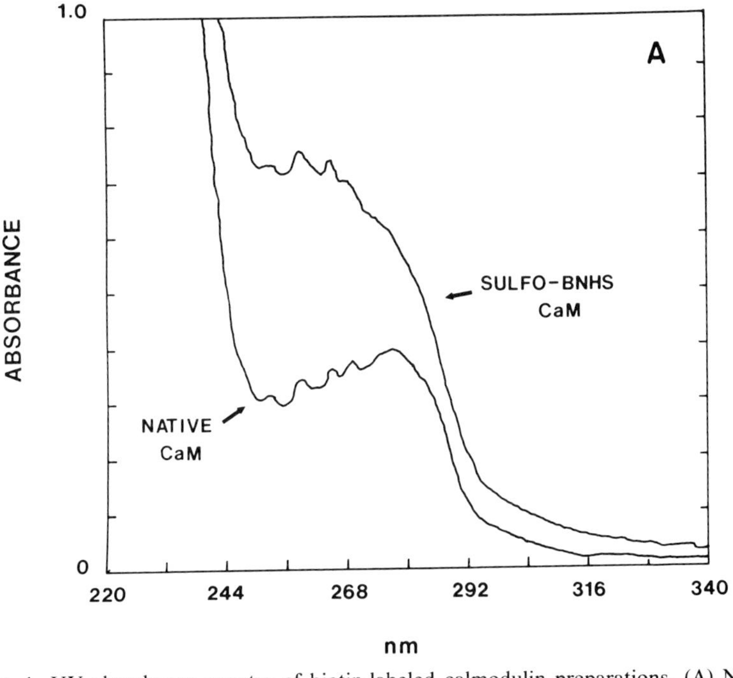

FIG. 1. UV absorbance spectra of biotin-labeled calmodulin preparations. (A) Native calmodulin and NHS-LC-biotin-derivatized calmodulin; (B) photobiotin-modified calmodulin; and (C) DBB-modified calmodulin. All solutions were approximately 2 mg/ml calmodulin in coupling buffer (pH 7.0) and were scanned using a Beckman DU-7 scanning spectrophotometer.

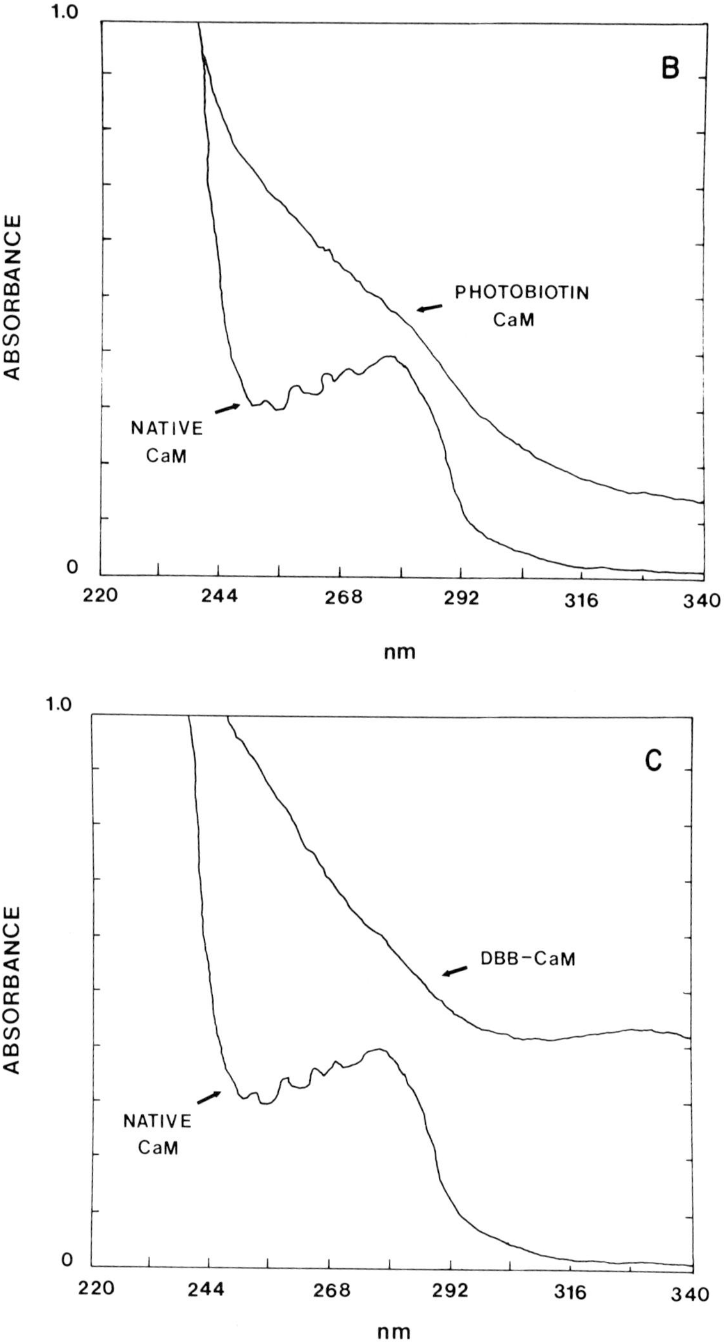

FIG. 1. (*continued*)

only 20 min at 200 mA; normal blotting times of 2–4 hr result in calmodulin passing through the nitrocellulose and into the transfer buffer. In order to further stabilize adsorption of calmodulin on nitrocellulose blots, the freshly transferred blot is immediately fixed in a solution of 10% acetic acid–25% (v/v) 2-propanol for 30 min prior to subsequent processing. Although others have shown that calmodulin will bind to nylon-based membranes,[24] we have found that most nylon membranes result in an unacceptable level of color background when incubated with avidin-based chromogenic enzymes. By reducing the time and current of cell transfer and by fixing the nitrocellulose prior to incubation, satisfactory binding and detection of biotinylated calmodulin on nitrocellulose can be achieved.

NHS-LC-Biotinylated calmodulin can be easily detected following transfer to nitrocellulose. For most modifications as little as 20 mg biotinylated calmodulin can be detected. Following dialysis, either 10 or 20 μg of native or biotinylated calmodulin is subjected to SDS–PAGE (20% gel), transferred, and fixed as described above, and nonspecific binding sites are blocked using blocking buffer for 30 min. Following 3 washes for 15 min each, the blot is incubated with avidin–peroxidase (1 μg/ml in wash buffer) for 30 min, washed 3 additional times in wash buffer, and reacted with substrate solution, consisting of 0.03% H_2O_2 and *p*-chloronaphthol (1 mg/ml) in wash buffer containing 20% methanol. Biotinylated calmodulin is easily visualized following this procedure, while native calmodulin does not generate a signal. Similar results can be obtained with the other calmodulin derivatives. On occasion, particularly if excess molar ratios of derivatization agent are used, several calmodulin peptides of altered electrophoretic mobility can also be visualized on blots; however, these preparations are not effective as probes for blot overlays.[20] The more slowly moving peptides most likely represent calmodulins containing multiple biotinylated amino acid residues and, possibly, having impaired Ca^{2+} binding, which also decreases electrophoretic mobility.

In order to determine whether the biotinylated calmodulin derivatives can recognize calmodulin-binding proteins immobilized on nitrocellulose (Fig. 2), fractions of rat cerebral cortex synaptosomes (lane 1; P2 fraction), cerebral cortical cytoplasm (lane 2; S2 fraction), and partially purified rat cytosolic calmodulin-binding proteins (primarily calcineurin at 60K and caldesmon at 150K and 105K) were subjected to SDS–PAGE and electroblotted. Identical blots are incubated with the biotinylated calmodulin derivatives as follows. Following a 30-min incubation in blocking buffer, blots are washed 3 times for 10 min in wash buffer and

[24] C. R. Egly and D. Daviaud, *Electrophoresis* **6,** 325 (1985).

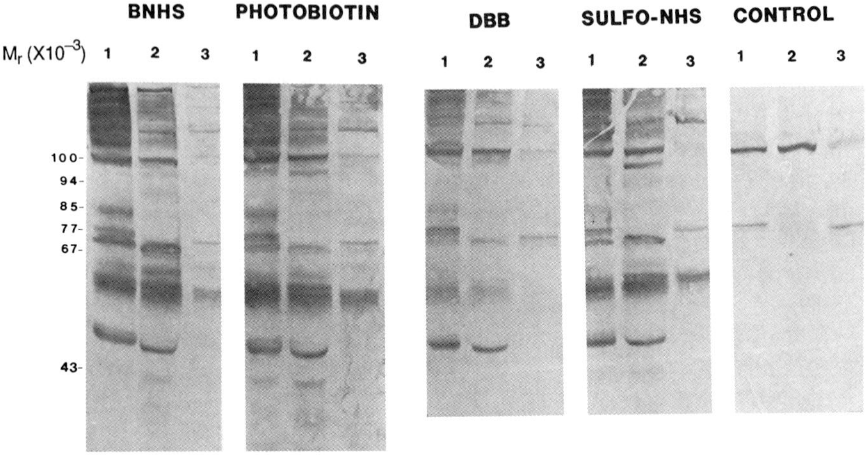

FIG. 2. Relative reactivity of several biotinylated calmodulin derivatives. Calmodulin was modified with NHS-LC-biotin (BNHS), sulfo-NHS-biotin, photobiotin, or DBB and incubated with blots containing cerebral cortex membranes (lane 1, 100 μg), cerebral cortical cytosol (lane 2, 100 μg), or partially purified calmodulin-binding proteins (lane 3, 10 μg). Following incubation with biotinylated calmodulins, the blots were incubated with avidin–alkaline phosphatase and developed with appropriate chromogens. An identical blot (Control) was reacted with avidin–alkaline phosphatase alone, revealing endogenous biotin-containing peptides having molecular weights of 75,000 and 100,000.

incubated with calmodulin biotinylated using NHS-LC-biotin (BNHS), photobiotin, DBB, or sulfo-NHS-biotin. Each congener is incubated at a concentration of 5 μg/ml in blocking buffer for 30 min, followed by 3 10-min washes in wash buffer. Avidin–alkaline phosphatase complexes (1 μg/ml wash buffer) are then incubated with the blot for 20 min, followed by 3 additional 10-min washes. A control consists of blotted samples incubated with avidin–alkaline phosphatase alone. The avidin detection system step should not be carried out in the usual blocking buffer because of occasional contamination of the nonfat dry milk with biotin. Each blot is incubated with BCIP–NBT (100 μl BCIP stock, 200 μl NBT stock in 30 ml of phosphatase buffer) until colored bands appear. The color development is carried out under subdued light and can proceed for as long as 12 hr without excessive background staining.

As shown in Fig. 2, all biotinylated calmodulin derivatives bind to a number of immobilized calmodulin-binding proteins from rat brain. Some differences in binding were noted, however. NHS-LC-biotin (BNHS) and sulfo-NHS-biotin derivatives of calmodulin recognize several brain calmodulin-binding proteins, with prominent staining of peptides at M_r 52,000, 60,000–62,000, 75,000, 85,000, and 105,000–240,000; this pattern is reminiscent of Bio–CaM used in previous studies.[14] Control incubations have demonstrated that the peptides at M_r 75,000 and 100,000 bind avidin–alkaline phosphatase alone, suggesting that these peptides contain biotin. DBB–calmodulin gives a less intense pattern of labeling (see staining of the 52,000 peptide) and fails to recognize the 60K subunit of calcineurin (Lane 3). Thus, modification of histidyl and/or tyrosyl residues with DBB appears to alter the recognition of specific binding domains. This may suggest that a specific histidine or tyrosine residue is important for interaction of DBB–calmodulin with immobilized calcineurin. Studies are in progress to investigate this possibility.

Additional methods for characterizing the calmodulin derivatives include tests of their ability to stimulate calmodulin-dependent enzymes, limited proteolysis of biotin–calmodulin, and direct sequencing of the modified molecule(s). Past studies have indicated that calmodulin modified with biotin-NHS derivatives retains the ability to stimulate calmodulin-dependent phosphatase activity[14] and calmodulin-dependent phosphodiesterase.[15] Interestingly, Mann and Vanaman observed that although BNHS-calmodulin could stimulate phosphodiesterase, preincubation with avidin inhibited calmodulin stimulation of phosphodiesterase activity.[15] We have observed that phosphodiesterase is poorly detected by BNHS–calmodulin on blot overlays, in agreement with the idea that the biotinylated lysine is obscured following calmodulin interaction with phosphodiesterase. Alternatively, the interaction of avidin may alter the calmodulin conformation such that interaction with phosphodiesterase cannot take place. Mann and Vanaman have sequenced BNHS–calmodulin and found that the biotin label is primarily localized on lysine-94. Thus, biotinylated calmodulin derivatives may prove quite useful in mapping points of interaction between calmodulin and its targets.

Biotinylated calmodulin can be purified using a variety of chromatographic techniques. Hydrophobic interaction chromatography can be used to purify calmodulin; however, this method does not resolve biotinylated from native calmodulin.[25] HPLC using DEAE resins has been used to separate modified from native calmodulin.[15] We have used avidin–agarose to purify biotinylated calmodulin; however, harsh elution conditions (pH 2.5; 6 *M* guanidine) must be used (M. Billingsley, unpublished observation).

[25] R. Gopalakrishna and W. B. Anderson, *Biochem. Biophys. Res. Commun.* **104,** 830 (1982).

Uses of Biotinylated Calmodulin

Identification of Tissue-Specific Calmodulin-Binding Proteins

The effect of Ca^{2+}-calmodulin in a given tissue is determined by the expression of particular calmodulin-binding proteins in the various cells of the tissue. We have used BNHS–calmodulin to determine the pattern of calmodulin-binding protein expression in various rat tissues. Membrane preparations, derived from heart, kidney, testis, lung, and pancreas, and equal amounts of tissue protein (100 μg/lane) were subjected to SDS–PAGE and transferred to nitrocellulose. As shown in Fig. 3, each tissue exhibits a distinct pattern of calmodulin-binding proteins. Although some calmodulin-binding proteins are common among tissues, others are specific to a given tissue.

Two controls are essential in such studies. One is to determine the Ca^{2+} specificity of the biotinylated calmodulin binding by including EGTA in an identical blot incubation (data not shown). Another complementary control is to incubate an identical blot with avidin–alkaline phosphatase alone in order to determine the extent of endogenous biotin-containing proteins in a tissue. It is also possible to block the reactivity of endogenous biotin-containing proteins by incubating the preblocked blot with avidin (0.1 mg/ml in wash buffer containing 1% bovine serum albumin) followed by a wash with excess biotin (1 mg/ml). We have found it easier to locate and "subtract" the biotin containing proteins from the calmodulin-binding proteins, in part because avidin "blocking" can give equivocal results (M. Billingsley and R. Kincaid, unpublished observation).

Detection of Reginal and Subcellular Differences in Calmodulin-Binding Protein Expression

Biotinylated calmodulin overlays can also be used to explore the differences in calmodulin-binding proteins in subcellular fractions from brain regions. An example of such a separation is shown in Fig. 4. Rat brain is dissected into cerebral cortex, cerebellum, hippocampus, and striatum. Each tissue is homogenized and separated into synaptosomal (P2) and cytosolic (S2) fractions. Individual lanes of a 10% polyacrylamide gel are loaded with 75 μg total protein, and the resolved proteins are transferred to nitrocellulose. Following blocking as described above, the blot is reacted with NHS-LC-biotinylated calmodulin (5 μg/ml blocking solution) and detected using avidin–alkaline phosphatase and BCIP–NBT chromagen systems.

As shown in Fig. 4, there are marked regional and subcellular differences in calmodulin-binding proteins. Most notable is the relative absence

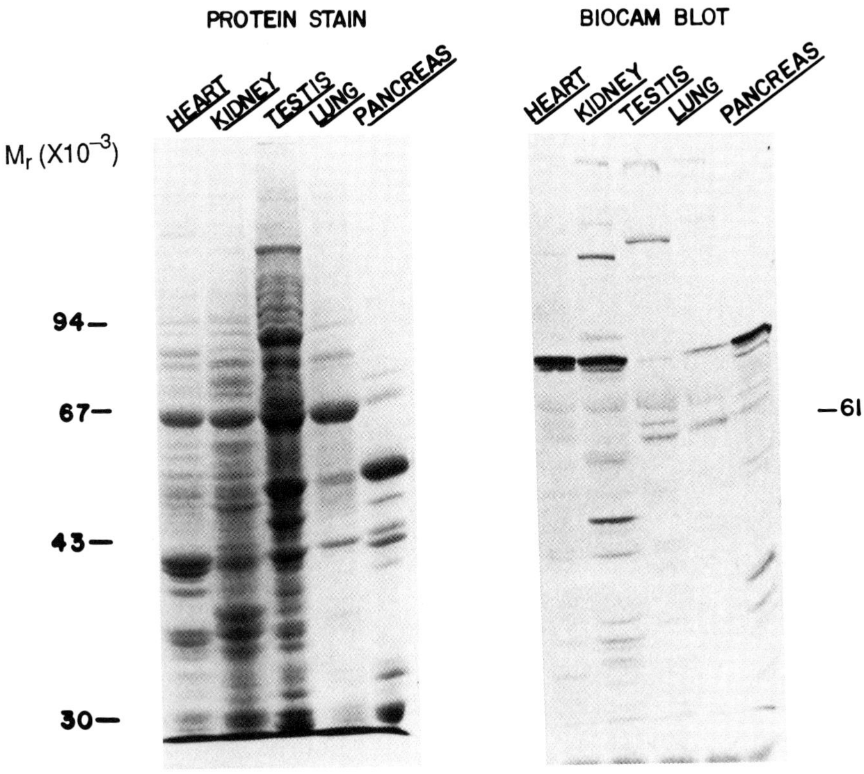

FIG. 3. Tissue distribution of membrane-associated rat calmodulin-binding proteins. Homogenates of rat heart, kidney, testis, lung, and pancreas were centrifuged at 20,000 *g* and subjected to SDS–PAGE and blotting as described. The blot was incubated with NHS-LC-biotinylated calmodulin and detected using avidin–alkaline phosphatase systems. The peptide at 75K bound avidin–alkaline phosphatase alone. The 61K marker corresponds to the approximate migration of purified calcineurin.

of the 52K subunit of calmodulin-dependent protein kinase II in cerebellum (lanes 2, 6); this finding has been noted by several investigators.[26,27] In addition, there are regional differences in calmodulin-binding peptides between 65K and 87K, with cerebellum exhibiting the greatest difference. The S2 fraction is relatively devoid of high molecular weight calmodulin-binding proteins (i.e., spectrin, caldesmon, and proteolytic fragments thereof), whereas the P2 fraction contains a 36K calmodulin-binding pro-

[26] S. D. Flanagan, B. Yost, and G. Crawford, *J. Cell Biol.* **94,** 743 (1982).
[27] T. L. McGuinness, Y. Lai, and P. Greengard, *J. Biol. Chem.* **260,** 1696 (1985).

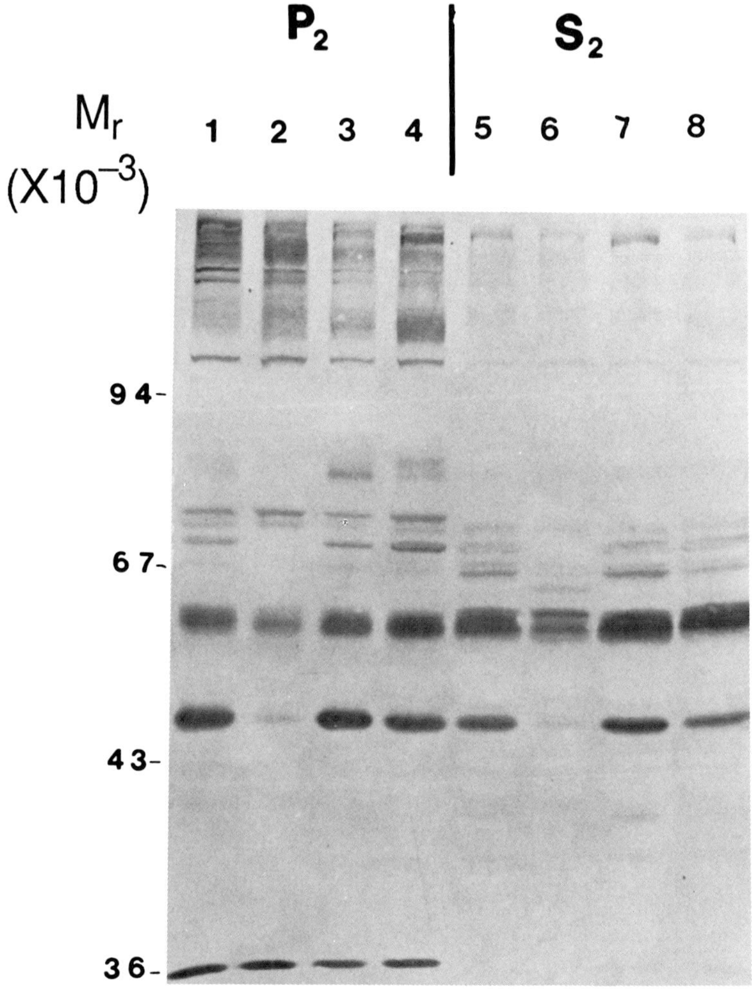

FIG 4. Regional and subcellular distribution of calmodulin-binding proteins in rat cortex (lanes 1, 5), cerebellum (lanes 2, 6), hippocampus (lanes 3, 7), and striatum (lanes 4, 8). Regions were separated into crude synaptosomal (P2) and cytosolic (S2) fractions as described. The blot was incubated with NHS-LC-biotinylated calmodulin and detected using alkaline phosphatase and chromogen systems.

tein termed cytosynalin[28] not seen in the cytosolic fraction. Thus, the biotinylated calmodulin overlay can be used to study regional and subcellular differences in expression of target enzymes. Of note is that, following electrophoresis and blotting, these results are obtained within 3 hr of

[28] K. Sobue, T. Okabe, K. Kadowaki, K. Itoh, T. Tanaka, and Y. Fujio, *Proc. Natl. Acad. Sci. U.S.A.* **84,** 1916 (1987).

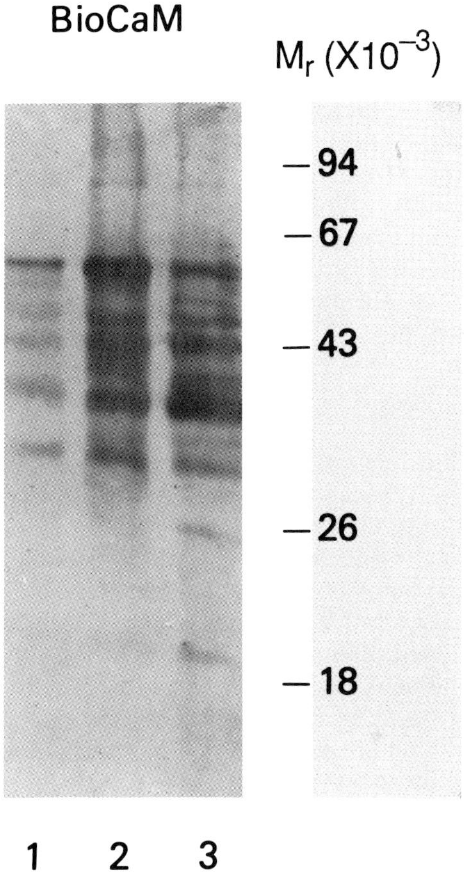

FIG. 5. Identification of calmodulin-binding peptides after limited proteolysis, SDS–PAGE, and transfer to nitrocellulose. Extracts of calmodulin–Sepharose eluates of T lymphocytes (lane 1), splenocytes (lane 2), and thymocytes (lane 3) were incubated for 30 min at 37° with staphylococcal V8 protease, electrophoresed on a 15% polyacrylamide gel in the presence of SDS, and transferred to nitrocellulose. The blot was incubated with biotinylated calmodulin and detected using avidin–alkaline phosphatase systems. T lymphocytes and splenocytes contain a 59K isoform of calcineurin, whereas thymocytes contain a 65K form; digestions indicate nearly identical calmodulin-binding peptides (lanes 2 and 3) from both the 59K and 65K forms. (Adapted, with permission, from Ref. 29.)

incubation; similar studies with iodinated calmodulin would take considerably longer.

Proteolytic Mapping of Calmodulin-Binding Peptides

We have also used biotinylated calmodulin overlays to obtain "fingerprints" of calmodulin-binding peptides following limited proteolytic digestion. In a study on the characterization of calmodulin-binding proteins from murine T and B lymphocytes and thymocytes, we have observed that thymocytes contain a major 65,000 calmodulin-binding protein which reacts weakly with antibodies against calcineurin.[29] Since the 59K calmodulin-binding subunit of calcineurin is the major calmodulin-binding peptide in B and mature T lymphocytes, we have used limited proteolysis of both the 65,000 and 59,000 peptides to show substantial relatedness between the calmodulin-binding peptides of these two proteins (Fig. 5). Although this approach may not be applicable to all calmodulin-binding proteins, the results indicate an important potential use of the calmodulin blot-overlay technique.

Summary

We have outlined and partially characterized a series of biotinylated calmodulin derivatives that may be useful in the study of calmodulin-binding protein expression, physical points of calmodulin–target interaction, and proteolytic maping of related calmodulin-binding proteins. Biotinylated calmodulins offer several advantages as probes of protein–protein interactions. First, biotinylation can be directed to different amino acid residues. Second, biotinylation can be carried out under mild, near-physiological conditions, reducing the likelihood that conditions of protein modification would destroy biological function. Third, biotinylated proteins are stable, and reagents needed for their preparation and detection are relatively inexpensive. Fourth, the sensitivity of avidin–chromogenic enzyme systems is approaching that of radioactivity, with the added advantage that chromogens can be visualized in a relatively short time with respect to autoradiography.

However, as with any protein modification procedure, one must be cautious when interpreting the results obtained with biotinylated proteins. For calmodulin-binding proteins, some interactions are impaired by modification of specific lysyl residues. On the other hand, interaction of biotinylated calmodulin with phosphodiesterase occurs, but this interaction may obscure recognition of the biotin residue by avidin.[15] One approach to circumvent this problem is to have a series of site-directed biotinylated proteins available for use as outlined in this chapter.

The choice of which agent to use is determined by the primary sequence of the protein of interest and whether any information is available concerning the effects of chemical modification on structure (i.e., acetyla-

[29] R. L. Kincaid, H. Takayama, M. L. Billingsley, and M. V. Sitkovsky, *Nature* (*London*) **330,** 176 (1987).

tion experiments, modification of free sulfhydryls). In the absence of such information, an empirical approach can be taken. Photobiotin affords an easy means for biotinylation of proteins; however, the sites of modification are not always predictable. NHS-biotin derivatives are readily available and are relatively easy to use. Finally, one may wish to biotinylate the protein while liganded to its normal interacting molecule, in the case of calmodulin, calcium ion is the obvious choice. However, calmodulin could also be biotinylated while bound to a specific binding protein such as calcineurin.[30] The latter method may be of use in determination of changes in reactivities of specific amino acid residues subsequent to binding. Finally, it may prove advantageous to biotinylate genetically engineered calmodulin, yeast calmodulin, or plant calmodulin to further define calmodulin–target protein interactions. Thus, the use of biotinylated calmodulin derivatives may offer insights into a range of structural and functional questions relevant to regulation of specific calmodulin-binding proteins.

Acknowledgments

This research was supported by a research grant from The International Life Sciences Institute Research Foundation and by U.S. Public Health Service Grant R01-AG06337 to M.L.B. The authors thank Ms. Doris Lineweaver for manuscript preparation.

[30] A. S. Manalan and C. B. Klee, *Biochemistry* **26,** 1382 (1987).

[54] Avidin–Biotin Mediated Immunoassays: Overview

By MEIR WILCHEK and EDWARD A. BAYER

Progress in the use of avidin–biotin technology in immunoassays has developed together with major advances of the immunodiagnostics field in general. Thus, whenever a particular improvement in a given step was desired, the applicability of the avidin–biotin complex was rapidly demonstrated.

At the beginning, in the late 1960s and early 1970s, when the need for improved diagnostics was established and different approaches (e.g., radioimmunoassay[1] and bacteriophage[2] assay systems for the quantification

[1] R. S. Yalow and S. A. Berson, *Nature (London)* **184,** 1648 (1959); D. S. Skelley, L. P. Brown, and P. K. Besch, *Clin. Chem.* **19,** 146 (1973).

[2] O. Mäkelä, *Immunology* **10,** 81 (1966).

of antigens) were suggested, the mediation of the avidin–biotin system was also examined.[3] Later, when enzyme and solid-phase immunoassay systems were introduced, it was immediately clear that the avidin–biotin system would also be incorporated somehow into such approaches.

The basis for further work in this direction was established when it was shown that biotin-modified cells or biotin-modified antibodies and lectins can interact simultaneously with their respective antigens or receptors and with an avidin-containing probe.[4,5] In fact, the binding of biotin or biotinylated molecules to cells is, in essence, a solid-phase assay system, where the cells are representative of an immobilizing matrix.

The main application of avidin–biotin technology in immunoassays is enhancement of the signal and/or speed of the assay. The signal is enhanced owing to the fact that many biotin residues can be introduced chemically to the antibody molecule. Thus, multiple avidin-containing probes (including enzymes and radiolabels) can be incorporated into the final step of the detection system. In addition, the four biotin-binding sites on the avidin molecule allow further signal enhancement by using a biotinylated detection probe either applied sequentially or complexed with avidin in a predetermined ratio. An alternative approach would be to use the avidin–biotin complex in the capture system. For this purpose, biotinylated antibody or biotinylated protein A can be immobilized to an avidin-containing matrix.

The introduction of the avidin–biotin system to enzyme immunoassay can mainly be attributed to Avrameas and co-workers[6] (see also [55] in this volume).

Our own efforts in this area have mainly been directed to the "least attractive" immunodiagnostic approach, namely, radioimmunoassay. The inherent disadvantages of radioiodination are further compounded by the presence of only one tyrosine residue on egg-white avidin and the apparent requirement of this residue for biotin binding. Nonetheless, there are various ways to circumvent this apparent "fatal flaw." One possibility is to use bacterial streptavidin, which has 6 tyrosines per subunit.[7] Another possibility is to increase the number of iodinatable groups on egg-white avidin by chemically incorporating phenols. This can be accomplished by various methods. For example, the Bolton–Hunter reagent will modify amino groups of lysine with phenols; the reagent itself

[3] J. M. Becker and M. Wilchek, *Biochim. Biophys. Acta* **264,** 165 (1972).

[4] H. Heitzmann and F. M. Richards, *Proc. Natl. Acad. Sci. U.S.A.* **71,** 3537 (1974); E. A. Bayer, E. Skutelsky, D. Wynne, and M. Wilchek, *J. Histochem. Cytochem.* **24,** 933 (1976).

[5] E. A. Bayer, M. Wilchek, and E. Skutelsky, *FEBS Lett.* **68,** 240 (1976).

[6] J.-L. Guesdon, T. Ternynck, and S. Avrameas, *J. Histochem. Cytochem.* **27,** 1131 (1979).

[7] L. Chaiet and F. J. Wolf, *Arch. Biochem. Biophys.* **106,** 1 (1964).

can be radiolabeled, or the modified protein can be iodinated when desired. In this case (and for most amine-specific modifications), egg-white avidin is preferred over streptavidin because of its large number of lysines per subunit. Extensive modification of the lysines would also serve to reduce the high p*I* of the native protein, thereby reducing one potential source of nonspecific binding. Nonglycosylated avidin[8] is perhaps the protein of choice for such modifications since its employment precludes a second major contributor (the oligosaccharide component of the native avidin molecule) to nonspecific interactions. Another method of increasing the number of phenolic groups in avidin is to prepare conjugates with either polytyrosine or tyrosine-containing copolymers.[9]

Chapters [56]–[64] in this volume illustrate the versatility of the avidin–biotin system in immunoassays. A more extensive literature summary providing details of the application of avidin–biotin technology in immunoassays and diagnostics is presented in [3] in this volume.

[8] Y. Hiller, E. A. Bayer, and M. Wilchek, this volume [6].
[9] M. Wilchek, *J. Solid-Phase Biochem.* **5,** 193 (1980).

[55] Avidin–Biotin System in Enzyme Immunoassays

By THÉRÈSE TERNYNCK and STRATIS AVRAMEAS

Avidin is a 66,000-Da glycoprotein present in egg white. One of its properties is the ability to bind with a high affinity (dissociation constant 10^{-15} *M*) to the small molecule biotin (or vitamin H).[1,2] Biotin is easily linked to proteins, which modifies little if any of their biological activities, and, when bound, retains its high affinity for avidin. The avidin–biotin system has been used to develop various procedures.[3] In immunoassays, in order to detect or quantify a constituent, biotinylated antibodies are allowed to react with the constituent, and then avidin (linked to a marker substance) is added. A procedure based on this principle was initially developed for the immunocytochemical localization of erythrocyte surface antigens; in this case, the biotin-labeled antibody was revealed by ferritin-labeled avidin.[3-5]

[1] N. M. Green, *Biochem J.* **89,** 585 (1963).
[2] N. M. Green, *Adv. Protein Chem.* **29,** 85 (1975).
[3] M. Wilchek and E. A. Bayer, *Immunol. Today* **5,** 39 (1984).
[4] H. Heitzmann and F. M. Richards, *Proc. Natl. Acad. Sci. U.S.A.* **71,** 3537 (1974).
[5] E. A. Bayer, M. Wilchek, and S. Skutelsky, *FEBS Lett.* **68,** 240 (1976).

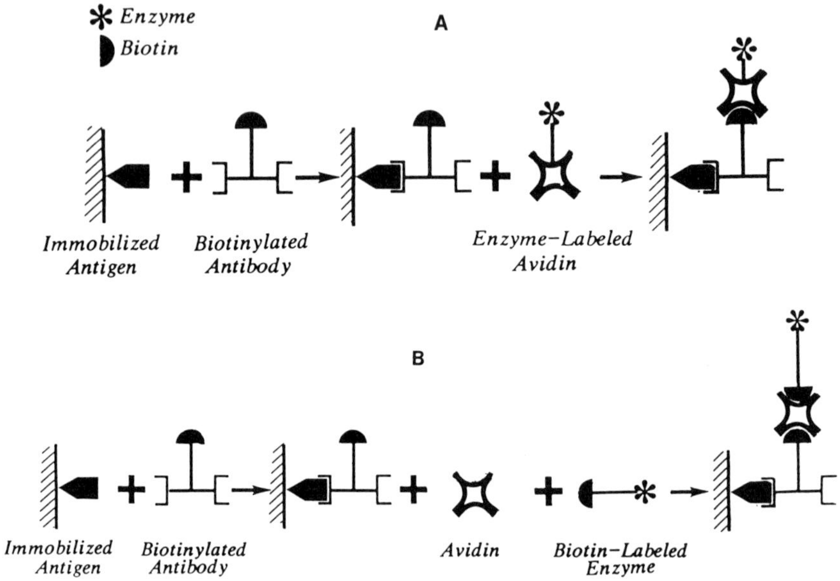

FIG. 1. Detection of antigens by the labeled avidin–biotin (LAB) system (A) or by the bridged avidin–biotin (BRAB) system (B).

The same principle was subsequently employed in immunofluorescence[6] and immunoenzymatic techniques.[7] In these latter procedures, two protocols for linking enzymes to proteins were described. In the first, the enzyme is covalently linked to avidin. The labeled antibody (or antigen) is allowed to react with the antigen (or the antibody), and, after washing, enzyme-labeled avidin is added. After further incubation and washing, the enzyme-associated antigen is stained histochemically or measured spectrophotometrically (Fig. 1A). In the following, this procedure is referred to as the labeled avidin–biotin (LAB) technique.

In the second procedure, biotin-labeled antibody (or antigen), biotin-labeled enzyme, and native unlabeled avidin are used. To quantify or to detect an antigen, the biotin-labeled antibody is allowed to react with the antigen. After washing to eliminate excess labeled antibody, avidin is added. After incubation and washing, the biotin-labeled enzyme is added. This step, after an additional incubation and washing, is followed by histochemical staining or measurement of the enzyme associated with the antigen by conventional procedures. This procedure, referred to as the bridged avidin–biotin (BRAB) technique (Fig. 1B), is based on the princi-

[6] M. H. Heggeness and J. F. Ash, *J. Cell Biol.* **73,** 783 (1977).

[7] J.-L. Guesdon, T. Ternynck, and S. Avrameas, *J. Histochem. Cytochem.* **27,** 1131 (1979).

ple that avidin possesses four active sites, not all of which react with the biotin residues associated with the antigen–biotinylated antibody complex. Thus, the remaining free active sites can operate as acceptors for another biotin-labeled protein secondarily added to the system.

Since the first papers on the avidin–biotin system appeared, the use of streptavidin as an alternative to the highly positively charged avidin has been introduced. Streptavidin is a biotin-binding protein extracted from bacteria of the species *Streptomyces* which has an affinity comparable to that of avidin[2] and has been shown to be as useful as avidin.[8] Streptavidin is less basic (p*I* 6.5) than avidin and has no carbohydrate residues,[2] thus limiting nonspecific reactions with acidic groups or lectins, respectively.

Experimental Procedures

Materials

Biotinyl-*N*-hydroxysuccinimide ester (BNHS)
Avidin from egg white
Streptavidin from *Streptomyces avidinii*
Peroxidase, β-galactosidase, alkaline phosphatase, glucose oxidase
Double-distilled glycerol
Dimethylformamide
β-Galactosidase anti-mouse Ig conjugate, prepared by the one-step glutaraldehyde method[9,10]

Buffers and Solutions

Phosphate-buffered saline (PBS): 0.15 *M* NaCl, in 10 m*M* potassium phosphate buffer (pH 7.4)
Saturated ammonium sulfate solution: dissolve by boiling 750 g of solid ammonium sulfate in 1 liter of distilled water and allow the solution to cool
0.1 *M* BNHS solution: dissolve 1 mg of biotinyl-*N*-hydroxysuccinimide ester in 30 μl of dimethylformamide

Biotinylation of Polyclonal Purified Antibodies

Antibodies are biotinylated using BNHS as detailed in Table I.

[8] F. M. Finn, N. Iwata, G. Titus, and K. Hofmann, *Hoppe-Seyler's Z. Physiol. Chem.* **362,** S679 (1981).
[9] S. Avrameas, *Immunochemistry* **6,** 43 (1969).
[10] S. Avrameas, T. Ternynck, and J.-L. Guesdon, *Scand. J. Immunol.* **8,** suppl. 7, 7 (1981).

TABLE I
GENERAL PROCEDURE FOR BIOTINYLATION OF ANTIBODIES

Step	Procedure
1	Prepare a solution of 2 mg of antibodies in 1 ml of sodium bicarbonate solution and dialyze overnight at 4° against the same solution
2	Add 10 μl of 0.1 *M* BNHS (in dimethylformamide)
3	Incubate 1 hr at room temperature
4	Dialyze against PBS overnight at 4°
5	Add an equal volume of glycerol (for storage)
6	Store at 4 or −20°

Biotinylation of Monoclonal Antibodies in Hybridoma Culture Supernatant

Precipitate 10–20 ml of hydridoma culture supernatant with an equal volume of saturated ammonium sulfate solution. Centrifuge (3000 *g*, 20 min, 4°) then wash the pellet once with a 40% solution of saturated ammonium sulfate in distilled water. Centrifuge (3000 *g*, 20 min, 4°) and suspend the pellet in 1 ml of distilled water and dialyze overnight at 4° against PBS and then against 0.1 *M* bicarbonate solution. Proceed as in step 2 (Table I).

Comments. We have shown (Ref. 7 and Table II) that (1) the number of biotin molecules fixed per molecule of antibody depends on the molar ratio of BNHS per amino group during the coupling reaction; (2) the capacity of the biotin-labeled antibody preparation to bind to avidin depends on the number of biotin moieties introduced into the molecule; and (3) compared to native antibodies, the antigen-binding capacity of biotin-labeled antibodies is not significantly modified by the number of biotin molecules incorporated (see Table II, the enzyme immunoassay results with 11.4 and 114 μl of biotin). Figure 2 shows the reactivity of monoclonal antihapten antibody, prepared against the trinitrophenyl (TNP) group, in relation to the degree of biotinylation. It is evident from Fig. 2 that a maximum level of reactivity is already obtained with a concentration of 1 m*M* BNHS (curve 3 in Fig. 2) and that the use of higher concentrations does not further modify the reactivity of the biotinylated antibody.

In addition, these preparations give the same signal as the unlabeled preparation when revealed by an anti-mouse immunoglobulin conjugate (data not shown). For these reasons, there is no risk in adding an excess of biotin and thus in working without knowing the exact IgG content of

TABLE II

ANTIGEN-BINDING CAPACITY OF SHEEP ANTI-RABBIT IMMUNOGLOBULIN ANTIBODIES BIOTINYLATED WITH VARIOUS AMOUNTS OF BNHS[a]

BNHS[b] (μl)	BNHS/amino group (molar ratio)	Biotin labeling[c] (%)	Activity[d] (OD)
0	0	0	0
1.14	0.1	26	0.44
11.4	1	42	1.04
45.6	4	95	n.d.[e]
114	10	100	1.12

[a] Data from Ref. 7.

[b] Biotin-*N*-hydroxysuccinimide ester at 34 mg/ml (0.1 *M* solution) in dimethylformamide. The quantities are given for 2 mg of purified polyclonal anti-mouse Ig antibodies in 1 ml of 0.1 *M* bicarbonate solution.

[c] Percentage of blocked amino groups determined according to Habeeb's method [A. F. S. A. Habeeb, *Anal. Biochem.* **14,** 328 (1966)].

[d] Optical densities obtained by incubating solutions of biotinylated antibodies at 1 μg/ml with Ig-coated plates followed by avidin labeled with β-galactosidase, read at 414 nm.

[e] Not determined.

the preparation. However, as far as monoclonal antibodies are concerned, it is well known that their physicochemical properties vary from one antibody to another and that one cannot predict in advance whether the binding capacity of one monoclonal antibody will be modified or totally abolished by chemical modifications, such as those produced by the labeling. In this case, preliminary assays must be performed to assess the remaining antibody activity after the biotin labeling of a monoclonal antibody by testing its antigen-binding capacity using a second labeled antibody (usually enzyme-labeled polyclonal rabbit or sheep anti-mouse immunoglobulin antibodies).

Biotinylation of Antigens

Dialyze 1 ml of protein at 2 mg/ml in 0.1 *M* bicarbonate solution overnight at 4°. To each 0.25-ml fraction of protein, add 20, 10, 5, or 2 μl of 10 m*M* BNHS (1 mg of BNHS in 300 μl of dimethylformamide). Proceed as in step 3 (Table I). Test by enzyme-immunoassay using antibody-coated plates (see Fig. 3).

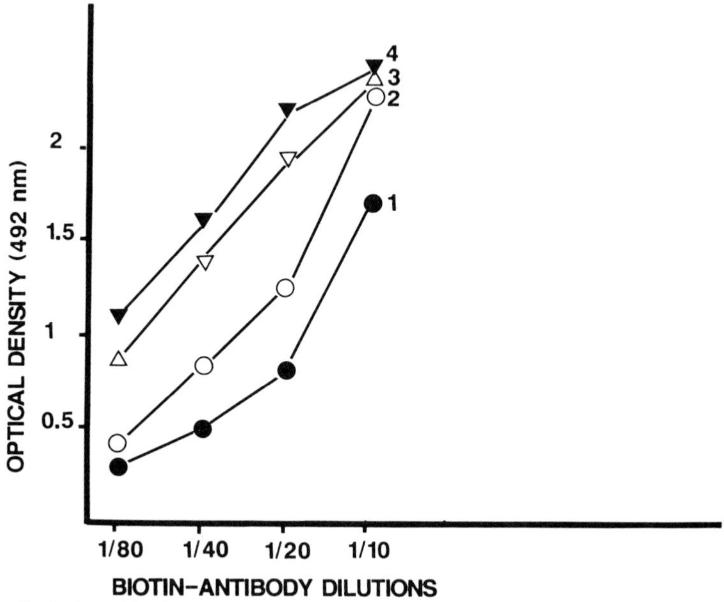

FIG. 2. Antigen-binding capacity of various preparations of biotinylated anti-TNP monoclonal antibodies. The culture supernatant of a monoclonal anti-TNP antibody was precipitated as described in the text. Biotinylation was performed by adding 5 (curve 1), 10 (2), 20 (3), or 40 μl (4) of 10 mM BNHS to 200-μl samples (20–100 μg/ml antibody). Samples were tested by enzyme immunoassay on TNP–ovalbumin-coated plates and revealed with peroxidase-labeled avidin (Table IV).

Comments. As we have shown for bovine serum albumin (BSA),[7] the antibody-binding capacity of an antigen may decrease with increasing biotin substitutions. In addition, since biotin substitution depends on the BNHS/amino group molar ratio and since the number of amino groups in a given protein is not always cited in the literature (this could be difficult to estimate in nonspecialized laboratories), it is therefore suggested that each antigen be tested to determine the optimal amount of BNHS needed for the coupling reaction following the general scheme given above. Figure 3 gives an example corresponding to the biotinylation of actin. It can be seen that the greater the biotin substitution, the higher the signal, and in this particular case the extent of biotinylation of actin does not modify its antibody-binding capacity.

Biotinylation of Enzymes

Dissolve 2 mg of peroxidase or glucose oxidase in 0.5 ml of 0.1 M carbonate–bicarbonate buffer (pH 9.5) or dialyze 2 mg of β-galactosidase

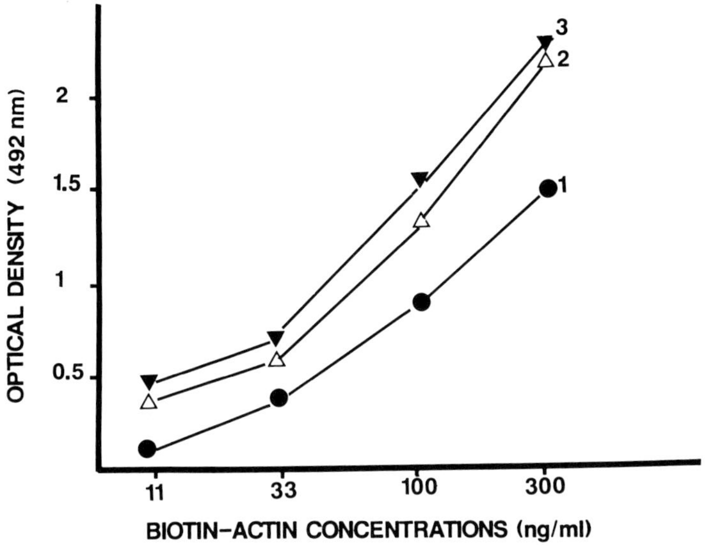

FIG. 3. Antibody-binding capacity of various biotinylated actin preparations. Biotinylation was performed by adding 5 (curve 1), 10 (2), or 20 μl (3) of 10 mM BNHS to 250-μl samples of actin at 2 mg/ml. Preparations were tested by enzyme immunoassays on antiactin monoclonal antibody-coated plates with peroxidase-labeled avidin (Table IV).

(usually kept in saturated ammonium sulfate) against this buffer. Add 20 μl of 0.1 M BNHS and proceed as in step 4 (Table I).

Comments. We have shown (Ref. 7 and data summarized in Table III) that the enzymes commonly used in immunoenzymatic technique (i.e., peroxidase, glucose oxidase, alkaline phosphatase, and β-galactosidase) can be labeled with biotin. Our results indicate, however, that some enzymes lose some of their catalytic activity after biotinylation. Indeed, the activity of alkaline phosphatase (from *Escherichia coli*) is highly affected even at low substitution: 35% of the activity is lost after addition of 3.4 biotin molecules per molecule, whereas that of alkaline phosphatase from calf intestine remains unchanged. The same loss of activity is observed when β-galactosidase is 80% substituted. At this level of substitution, neither glucose oxidase nor peroxidase activity appears to be modified (Table III). Thus, for enzyme assays, glucose oxidase, peroxidase, β-galactosidase, or alkaline phosphatase from calf intestine are recommended.

Coupling of Avidin or Streptavidin to Enzymes

Avidin or streptavidin is coupled to enzymes by the one-step glutaraldehyde procedure.[9] Mix 4 mg of either peroxidase or β-galactosidase

TABLE III
CATALYTIC ACTIVITY OF ENZYMES BIOTINYLATED WITH VARIOUS AMOUNTS OF BNHS[a]

	BNHS/amino group[b] (molar ratio)							
Enzyme and parameters	0.05	0.1	0.5	1	2	5	10	100
Glucose oxidase								
Labeling (%)[c]		4	35	50	71	84		
Activity (%)[d]			n.d.[e]	100	100	100	100	
Peroxidase								
Labeling (%)						68	72	80
Activity (%)						100	100	100
Alkaline phosphatase (*E. coli*)								
Labeling (%)	1.4	2.8	3.4		90	100	100	
Activity (%)	n.d.	95	65		35	35	35	
β-Galactosidase (*E. coli*)								
Labeling (%)				55	74	84		
Activity (%)				95	78	60		

[a] Data from Ref. 7.
[b] Biotin-*N*-hydroxysuccinimide ester at 34 mg/ml (0.1 *M* solution) in dimethylformamide. The quantities are given for 2 mg of purified polyclonal anti-mouse Ig antibodies in 1 ml of 0.1 *M* bicarbonate solution.
[c] Percentage of blocked amino groups determined according to Habeeb's method [A. F. S. A. Habeeb, *Anal. Biochem.* **14,** 328 (1966)].
[d] Enzyme activity versus native enzyme.
[e] Not determined.

with 2 mg of avidin (or streptavidin) in a final volume of 1 ml of 0.1 *M* phosphate buffer (pH 6.8). Add 50 μl of a 1% glutaraldehyde solution in the same buffer and incubate for 3 hr at room temperature (below 24°). Add 50 μl of 2 *M* lysine in phosphate buffer for 30 min and dialyze overnight against PBS at 4°. Centrifuge for 10 min in an Eppendorf centrifuge (~10,000 *g*). Add an equal volume of glycerol and store at 4 or −20°.

Comments. The coupling of avidin to enzymes using glutaraldehyde, which interacts with protein amino groups, decreases its positively charged groups and thus its nonspecific binding (see below). However, as in all protein coupling procedures, the conditions described here should be followed carefully because extensive cross-linking can occur, which often decreases the biological activity of the macromolecules.

Chemical Modification of Avidin

Mix 1 ml of avidin (2 mg/ml) in 0.1 *M* phosphate buffer (pH 7.4) with 1 ml of 4% (w/v) paraformaldehyde in the same buffer. Incubate for 2 hr at room temperature and dialyze against PBS overnight at 4°. Add 2 ml of glycerol.

Comments. Avidin is a basic protein (p*I* 10) and thus could bind to other macromolecules by electrostatic interactions, leading to nonspecific reactions. Treatment with formaldehyde (which blocks amino groups) results in a preparation of avidin possessing low nonspecific binding characteristics. Furthermore, the formaldehyde-treated avidin preparation is stable in solution, whereas native avidin is not and must be kept frozen in small aliquots. Glutaraldehyde treatment of avidin also decreases its nonspecific binding but at the same time can decrease its specific biotin binding. We have shown that incubation with biotin-labeled compounds under alkaline (pH 9) or high ionic strength (0.5 *M* NaCl) conditions decreases nonspecific reactions.[7] However, such nonphysiological conditions could effect the specific binding in some antigen–antibody systems.

To overcome the disadvantage of the nonspecific binding of avidin, the use of streptavidin was introduced. Streptavidin has the same biotin-binding properties as avidin[8] and can be labeled with biotin or enzymes in exactly the same manner as avidin.

Quantitative Enzyme Immunoassay

Media and Substrates

Washing medium: PBS–Tween 20 (0.1%)

Saturating medium: PBS–gelatin (0.5%) or PBS–BSA (1%)

Diluting medium: PBS–Tween–gelatin (0.5%)

Alkaline phosphatase substrate: 1 mg/ml solution of *p*-nitrophenylphosphate (Sigma Chemical Co., St. Louis, MO) in 0.1 *M* Tris-HCl buffer (pH 8.2 for *E. coli* phosphatase or pH 9.8 for calf intestinal phosphatase), containing 1.5 *M* NaCl; stop the reaction with 1 *N* NaOH and read at 414 nm.

β-Galactosidase substrate: 0.8 mg/ml of *o*-nitrophenyl-β-D-galactopyranoside (Sigma) in 0.1 *M* phosphate buffer (pH 7.4) containing 0.1 *M* mercaptoethanol; stop the reaction with 2 *M* sodium carbonate and read at 414 nm.

Peroxidase substrate: 0.5 mg/ml of *o*-phenylenediamine (Sigma) in sodium citrate–citric acid buffer (pH 5.5) containing 0.03% H_2O_2; stop the reaction with 2 *N* sulfuric acid and read at 492 nm.

Procedure. Enzyme immunoassay (EIA) using the avidin–biotin system is performed using the protocol described in Table IV.

Comments. The coating of plates is usually performed with proteins diluted in 0.1 *M* carbonate–bicarbonate buffer (pH 9.5) at a concentration

TABLE IV
GENERAL PROCEDURE FOR ENZYME IMMUNOASSAY

Step	Procedure
1	Coat the plates with 100 μl of antigen or antibody preparation (usually at 5 μg/ml) in 0.1 *M* carbonate–bicarbonate buffer (pH 9.5), 2 hr at 4°
2	Keep the plates at 4° until used
3	Discard the coating solution and incubate with the saturating medium, 30–60 min at 37°
4	Wash 3 times
5	Incubate with the first antibody (or antigen) diluted in diluting medium, 60–120 min at 37°
6	Wash 3 times
7	Incubate with the biotinylated second antibody, 60 min at 37°
8	Wash 3 times
9	Proceed with one of the following two avidin–biotin systems
9.1	LAB (labeled avidin–biotin) procedure: Incubate with enzyme-labeled avidin (1 μg/ml), 30 min at 37°
9.2	BRAB (bridged avidin–biotin) procedure: Incubate with the paraformaldehyde-treated avidin (2–5 μg/ml), 30 min at 37°, wash 3 times, incubate with biotinylated enzyme (1 μg/ml), 30 min at 37°
10	Wash
11	Incubate with the appropriate substrate solution until the desired color intensity is obtained.
12	Stop the enzymatic activity and read the optical density at 492 nm

of 5 μg/ml. However, the optimal coating concentration, especially for antigens, may vary considerably. In such cases, this concentration has to be determined for each system. Furthermore, the carbonate–bicarbonate buffer (pH 9.5) can be replaced by PBS if some proteins are found to be unstable at alkaline pH.

The BRAB procedure using native avidin does not present any particular advantage. It does not significantly increase the sensitivity, while it includes an additional step. The specific type of enzyme used for labeling avidin is not critical and is rather a matter of personal choice. As in all these immunoassays, the first antibody determines the sensitivity of the assay rather than the detection system used.

EIAs performed with the avidin–biotin system are at least as sensitive as those using conventional enzyme–antibody conjugates. The use of biotinylated antigen has been shown to be useful in competitive assays and has good sensitivity.[11]

[11] R. Rappuoli, P. Leoncini, P. Tarli, and P. Neri, *Anal. Biochem.* **118,** 168 (1981).

Immunocytochemical Assays

Media and Substrate

Washing medium: PBS

Diluting medium: PBS–gelatin (0.5%) or PBS–BSA (1%)

Substrate for enzyme detection

Alkaline phosphatase (calf intestine) substrate: mix 1 volume of 0.4 mg/ml of naphthol AS-MX phosphate (sodium salt, Sigma) in 0.2 *M* Tris-HCl buffer (pH 8.2), with 1 volume of 6 mg/ml of Fast Red TR salt (Sigma) in distilled water and incubate for 20 min; the reaction product is red

Peroxidase substrate: prepare a solution of diaminobenzidine (Sigma) at 0.5 mg/ml in 0.1 *M* Tris-HCl buffer (pH 7.6) containing 0.03% H_2O_2 and incubate for 7–10 min; the reaction product is brown

Procedure. Following the general procedure described in Table V the avidin–biotin system can be applied to the immunocytochemical detection of antigens. The substrate solutions are prepared, filtered, and used

TABLE V
GENERAL PROCEDURE FOR IMMUNOCYTOCHEMICAL ASSAYS

Step	Procedure
1	Prepare tissue sections or cells using standard methods
2	Fix the tissue (with paraformaldehyde, acetone, alcohol, etc.) or deparaffinize and hydrate tissue sections of wax-embedded material exactly as described in other procedures; treat the sections to reduce the endogenous enzyme activity
3	Add appropriate dilutions of the first antibody and incubate, 30–60 min
4	Wash
5	Incubate with biotinylated second antibody (10 μg/ml for purified antibody or previously determined dilutions for hybridoma supernatants), 30–60 min
6	Wash
7	Proceed with one of the following two avidin–biotin systems:
7.1	LAB (labeled avidin–biotin) procedure: Incubate with the avidin–enzyme preparation (10–20 μg/ml), 30 min
7.2	BRAB (bridged avidin–biotin) procedure: Incubate with paraformaldehyde-treated avidin (10 μg/ml), 15–30 min, wash, incubate with biotinylated enzyme (20 μg/ml), 15–30 min
8	Wash
9	Stain with appropriate substrate solution
10	Counterstain with suitable histological stain
11	Mount with a suitable medium

immediately.[9] After the incubation, slides are washed with distilled water and mounted with a microscope mounting medium.

Comments. As for EIA, streptavidin often gives better results than avidin, although formaldehyde treatment of avidin consistently decreases the nonspecific background. The LAB procedure is more rapid than that using BRAB, because an additional step is not required. However, the BRAB procedure often allows more sensitive immunocytochemical detection of antigen. To shorten the BRAB procedure, the use of preformed avidin–biotinylated enzyme complexes (ABC) has been proposed.[12] This procedure, in our hands, results in less sensitive detection of antigens than the sequential BRAB procedure. This is probably due to the size of the complex, which would penetrate with increased difficulty into the tissues. Furthermore, the preparation of the ABC is critical, and it is preferable to use a commercially available preparation.

As was pertinently commented,[13,14] some points need to be stressed for the use of the biotin–avidin system in immunocytochemistry. Biotin is an ubiquitous compound; it is a vitamin required by all living cells, and thus it is present in all tissues and body fluids either in a circulating form or bound to proteins. Consequently, high levels of "nonspecific" binding of avidin or streptavidin can be obtained. Furthermore, avidin is a glycoprotein and thus could bind to lectins sometimes present in tissues.

Immunoblotting

Media and Substrate

Saturating medium: PBS–BSA (1%)
Washing medium: PBS–Tween 20 (0.1%)
Staining solutions for transferred proteins: 0.2% Ponceau S (Serva) in 3% trichloroacetic acid
Substrates for enzyme detection: the same as for immunocytochemical assays; the incubation time indicated can, however, be increased until the desired intensity is obtained

Procedure. The avidin–biotin system has been successfully used for detecting antigens, after their electrophoretic separation and transfer onto nitrocellulose sheets either by electrophoresis[15] or by diffusion under pressure,[16] using the protocol outlined in Table VI.

[12] S. M. Hsu, L. Raine, and H. Fanger, *J. Histochem. Cytochem.* **29,** 577 (1981).
[13] E. A. Bayer and M. Wilchek, *Methods Biochem. Anal.* **26,** 1 (1980).
[14] R. E. Morris and C. B. Saelinger, *Immunol. Today* **5,** 127 (1984).
[15] H. Towbin, T. Staehelin, and F. Gordon, *Proc. Natl. Acad. Sci. U.S.A.* **72,** 313 (1974).
[16] G. Peltre, J. Lapeyre, and B. David, *Immunol. Lett.* **5,** 127 (1982).

TABLE VI
GENERAL PROCEDURE FOR IMMUNOBLOTTING

Step	Procedure
1	Separate proteins by electrophoresis either according to their molecular weights [sodium dodecyl sulfate (SDS)–polyacrylamide gel] or according to their p*I* (isoelectrofocusing)
2	Transfer the proteins from the gel onto nitrocellulose filters either electrophoretically[15] (polyacrylamide gel) or by diffusion under pressure[16] (agarose gel)
3	Assess the protein transfer by staining with Ponceau S
4	Saturate the nitrocellulose sites by incubation in the saturating solution, 30 min at 37°
5	Wash
6	Incubate the sheets with the biotinylated antibody (or antigen) at the optimal concentration previously determined (1–20 μg/ml), 60 min at 37°
7	Wash
8	Incubate with enzyme-labeled avidin (or streptavidin) (1–10 μg/ml), 15–30 min at 37°
9	Wash
10	Incubate with the appropriate substrate until the desired color intensity is obtained
11	Wash with distilled water
12	Keep dry

Comments. This procedure has been used with success with biotinylated monoclonal antibodies to characterize their reactivity on electrophoretically separated proteins and with biotinylated antigens to determine the p*I* of antibodies after separation by isoelectrofocusing.

[56] Two-Site and Competitive Chemiluminescent Immunoassays

By CHRISTIAN J. STRASBURGER and FORTUNE KOHEN

Introduction

Conventional immunoassay procedures for haptens and for peptide hormones require the preparation, purification, and characterization of the antigen or antibody labeled with a radioisotope or enzyme, thus rendering each particular assay tedious to develop and limited in scope. Furthermore, the short half-life and radiolysis-induced damage of the ^{125}I-labeled antibody or antigen limit assay sensitivity and impose a time limit

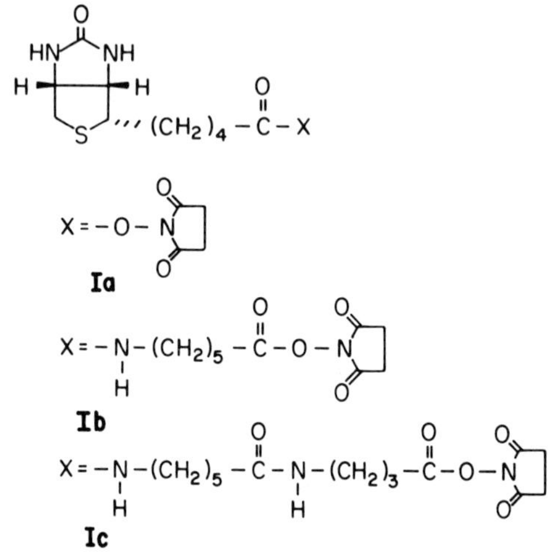

FIG. 1. Structure of different *N*-hydroxysuccinimide ester derivatives of biotin with varying alkyl chains: **Ia**, biotin-*N*-hydroxysuccinimide ester; **Ib**, biotin-ε-aminocaproyl-*N*-hydroxysuccinimide ester; **Ic**, biotin-ε-aminocaproyl-γ-butyryl-*N*-hydroxysuccinimide ester.

on the usefulness of a kit. In addition, the potential health hazards associated with the use and disposal of radioactive compounds are incompatible with the development of simple assay procedures in many clinical laboratories and in developing countries.

To avoid these drawbacks while retaining the specificity and sensitivity of an immunoassay, we explored the use of biotinylated probes (antibodies, antigens, or enzymes) in the development of immunoassays for polypeptide hormones and haptens. The biotinylated probes were prepared under mild conditions by conjugating *N*-hydroxysuccinimide ester derivatives of biotin (Fig. 1) to the various proteins (antibodies, antigens, or enzymes) via a peptide linkage. The biotinylated antibodies or enzymes were then evaluated in a solid-phase ELISA, using alkaline phosphatase as the enzyme and colorimetry at 405 nm as an end point. As the *N*-hydroxysuccinimide ester derivative of biotin possessing a long alkyl chain between the ureido ring of biotin and the carboxy terminus (compound **Ic**, see Fig. 1) led to the highest signal intensity, this derivative was chosen for all further biotinylations of proteins.[1]

The biotinylated probes serve as primary labeled components in immunoassay procedures. Two types of formats have been adopted in our

[1] F. Kohen, Y. Amir-Zaltsman, C. J. Strasburger, E. A. Bayer, and M. Wilchek, *in* "Complementary Immunoassays" (W. P. Collins, ed.), p. 57. Wiley, New York, 1987.

studies: two-site or sandwich-type immunoassays[1-4] and competitive-type immunoassays.[4] In both types of formats the high affinity for biotin and the four biotin-binding sites on avidin or streptavidin are exploited in order to amplify the sensitivity of the assay.

The properties of the avidin–biotin complex have enabled the development of two different approaches in two-site immunoassays for polypeptide hormones. In both cases an immobilized antibody and a biotinylated antibody preparation are employed, each of which recognizes a different epitope of the antigen. In one approach, following immunological reaction of the immobilized and the biotinylated antibody with the antigen, conjugates comprising avidin or streptavidin covalently labeled with a given probe are employed.[1,2] The latter probe can be an enzyme (e.g., alkaline phosphatase or horseradish peroxidase), a chemiluminescent agent, or a fluorescent marker. In the second approach, the use of conjugates is replaced by that of secondary probes. In this approach, unconjugated avidin or streptavidin is first applied followed by the addition of biotinylated enzyme (e.g., alkaline phosphatase,[1] penicillinase,[3] glucose-6-phosphate dehydrogenase[5,6]). In both procedures, depending on the label used, the end point is determined by colorimetry, luminometry, disappearance of color, or time-resolved fluorescence. Assays based on these principles have been developed for human growth hormone (hGH)[1,2,4] and human chorionic gonadotropin (hCG).[5,6]

In the competitive-type immunoassay for haptens,[4] a solid-phase antigen and the specific homologous biotin-labeled antibody are used. After the immunological reaction, labeled streptavidin is added, and the end point is determined depending on the type of marker attached to streptavidin. Figure 2 shows the approach used for haptens. Assays based on these principles have been developed for cortisol[4] and for estradiol.

The use of the avidin–biotin interaction as a mediator in two-site sandwich-type and competitive immunoassays has several advantages: (1) biotinylation of proteins can be achieved with great facility under mild

[2] C. J. Strasburger, Y. Amir-Zaltsman, and F. Kohen, 69th Endocrine Society Meeting, Indianapolis, Indiana, June, 1987, Abstr. No. 228.

[3] Y. Amir-Zaltsman, B. Gayer, E. A. Bayer, M. Wilchek, and F. Kohen, *in* "Non-Isotopic Immunoassay" (T. T. Ngo, ed.), p. 117. Plenum, New York, 1988.

[4] C. J. Strasburger, Y. Amir-Zaltsman, and F. Kohen, *in* "Non-Radiometric Assays" (B. Albertson and F. Hazeltine, eds.), p. 79. Alan R. Liss, New York, 1988.

[5] F. Kohen, E. A. Bayer, M. Wilchek, G. Barnard, J. B. Kim, W. P. Collins, I. Beheshti, A. Richardson, and F. McCapra, *in* "Analytical Applications of Bioluminescence and Chemiluminescence" (L. Kricka and T. P. Whitehead, eds.), p. 149. Academic Press, New York, 1984.

[6] G. Barnard, E. A. Bayer, M. Wilchek, Y. Amir-Zaltsman, and F. Kohen, this series, Vol. 133, p. 284.

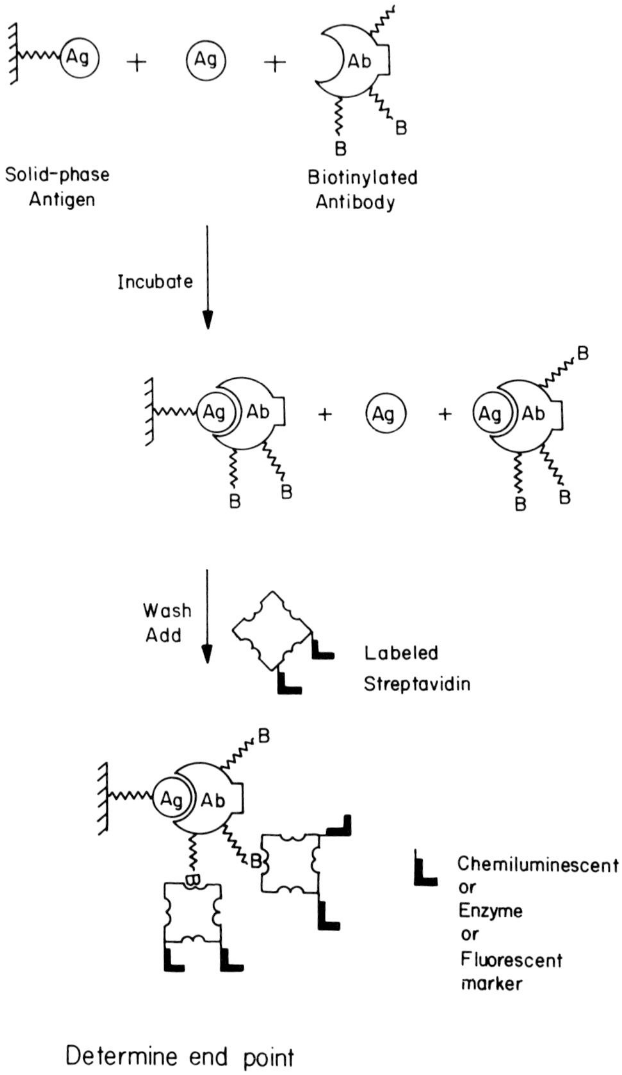

FIG. 2. Schematic of a competitive immunoassay procedure for haptens using biotinyl antibody and labeled streptavidin as probes.

conditions, and the resulting conjugates show very little or no loss of activity[1,4]; (2) the end point in the assay is versatile (luminometry, colorimetry, time-resolved fluorescence); (3) mediation via the avidin–biotin interaction serves to amplify the sensitivity of the assay; and (4) reactive biotinyl derivatives, biotinylated enzymes, and enzyme-labeled avidin and streptavidin conjugates are all commercially available.

Owing to the commercial availability of reactive biotinyl derivatives (e.g., various *N*-hydroxysuccinimide ester derivatives of biotin), the biotinylation reaction can be performed with ease in any laboratory. In addition, the use of labeled streptavidin in the assay allows the design of a uniform immunoassay methodology that includes both haptens and polypeptide hormones. The choice of the label on streptavidin depends on the sensitivity desired and on the instrumentation at the disposal of the particular investigator. In the development of immunoassays for haptens and peptide hormones, we have mainly used chemiluminescent-labeled streptavidin in conjunction with measurement of the light yield in a luminometer as the end point. In order to compare this label with others, we also used enzyme- or fluorescent-labeled streptavidin conjugates.

As representative examples, we report here a two-site immunochemiluminometric assay for human growth hormone (hGH) and a competitive-type immunoassay for cortisol based on the use of a chemiluminescent-labeled streptavidin as a probe. The use of streptavidin or acetylated avidin and biotinyl enzymes as probes in two-site assays has been described previously in this series[6] and in other publications.[1,3–5]

Methods

Materials. Avidin–alkaline phosphatase conjugate, polyphenylalanine–polylysine (1 : 1), biotin-*N*-hydroxysuccinimide, biotin-ε-aminocaproyl-*N*-hydroxysuccinimide, and biotin-ε-aminocaproyl-γ-butyryl-*N*-hydroxysuccinimide (Cat. No. 7304-1) were purchased from Bio-Makor (Rehovot, Israel); streptavidin from Cell-Tech (London); alkaline phosphatase (Type VII, from bovine intestine), Tween 20, *p*-nitrophenyl phosphate, and microperoxidase (MP-11) from Sigma Chemical Co. (St. Louis, MO); etched polystyrene balls from Northumbria Biologicals (England); Sepharose–protein A and disposable Sephadex G-25 PD-10 columns from Pharmacia (Uppsala, Sweden); 8% glutaraldehyde from Ted Bell Inc. (Tustin, CA); and 6-[*N*-(4-aminobutyl-*N*-ethyl)-2,3-dihydrophthalazine-1,4-dione] hemisuccinamide (ABEI-H) was a gift from LKB-Wallac (Turku, Finland). The following equipment was used to design and perform the assays described here under optimal conditions: UV monitor, e.g., LKB Uvicord 2138; luminometer, e.g., Berthold Clinilumat or Lumac M2500; reaction trays (20 or 60 wells) from Abbott; multiwash unit, e.g., Pentawash from Abbott; and horizontal shaker, e.g., Heidolph, 170–200 rpm.

Preparation of Antibodies

Specific polyclonal antibodies to steroids conjugated to bovine serum albumin are raised in rabbits and characterized in terms of titer, affinity, and specificity by radioimmunoassay (RIA) procedures. The hybridoma technique of Köhler and Milstein[7] is used to generate monoclonal antibodies to steroids[8] and to peptide hormones. Polyclonal and monoclonal antibodies belonging to the IgG class are purified by affinity chromatography on Sepharoase–protein A.

Purification of Antibodies on Sepharose–Protein A

Mouse IgG_1 binds to protein A at pH 8.0, whereas mouse IgG of other subclasses as well as IgG from polyclonal rabbit antiserum are bound at pH 7.2. A Sepharose–protein A column (5 ml) is equilibrated with 0.1 *M* sodium phosphate buffer (pH 7.2 or 8.0), and 1 ml of ascitic fluid or 0.5 ml of antiserum, diluted with 0.5 ml of the respective buffer, is applied and allowed to react for a period of 30–60 min. The column is rinsed with the same buffer until baseline absorbance (A_{280}) is regained in the effluent. For elution of the IgG fraction from the protein A column, the pH is lowered gradually by replacing the phosphate buffer with 0.1 *M* citrate buffers of pH 6, 4.5, and 3. The pooled IgG-containing peak is dialyzed against phosphate-buffered saline [(PBS) 10 m*M* phosphate, 150 m*M* NaCl (pH 7.2)] and concentrated to 1–2 mg protein/ml over a P10 membrane in an Amicon concentrator. The preparation is stored at $-20°$ until use.

Biotinylation of Antibodies

Biotin-ε-aminocaproyl-γ-butyryl-*N*-hydroxysuccinimide ester is dissolved in dry, amine-free dimethylformamide (DMF) at a concentration of 25 mg/ml (10 μl of this solution is sufficient to label 1 mg of IgG, with a 75-fold molar excess of biotin per mole of IgG in the reaction). The purified antibodies in PBS are made slightly basic (to pH 8.2–8.4) by adding 45 μl of 1 *M* Na_2HPO_4 (pH 9) for every milliliter of antibody solution. The above-described biotin-containing reagent (250 μg in 10 μl of DMF) is then added to the antibody solution (1 mg protein/ml). The reaction mixture is stirred at room temperature for 1–6 hr or overnight at 4°.

The reaction mixture is then applied to a Sephadex G-25 PD-10 column equilibrated with 25 m*M* Tris-HCl buffer (pH 7.1; Tris buffer), and the elution profile is monitored at 280 nm using an LKB Uvicord. The biotinylated antibody is eluted in the first peak (recovery ~95%), and the second peak is discarded. The Sephadex G-25 column can be reused up to at least 20 times by regenerating it after use with 50 ml of 0.5 *M* sodium

[7] G. Köhler and C. Milstein, *Eur. J. Immunol.* **6,** 511 (1976).

[8] F. Kohen and S. Lichter, *in* "Monoclonal Antibodies: Basic Principles, Experimental and Clinical Applications in Endocrinology" (G. Forti, M. B. Lipsett, and M. Serio, eds.), p. 87. Raven, New York, 1986.

chloride followed by 50 ml of Tris buffer. The column is stored after regeneration in PBS containing 0.05% sodium azide.

For storage, 5 μl of 20% bovine serum albumin and 2 μl of 10% sodium azide are added per milliliter of biotinylated antibody. The biotinylated antibodies are then stored in aliquots of 1 ml at −20°. As refreezing of the biotinylated antibody is not recommended, the unused portion of the biotinylated antibody is stored at 4°.

Preparation of Active Ester of ABEI-H

The active ester of ABEI-H is prepared by dissolving the hemisuccinamide derivative of ABEI (3.76 mg, 10 μmol) in 140 μl of dry DMF. *N*-Hydroxysuccinimide (1.15 mg, 10 μ*M*) in 20 μl of DMF is added, and after 5 min dicyclohexylcarbodiimide (3.09 mg, 15 μmol) in 40 μl of DMF is added to the solution containing ABEI-H. The vial is then covered with aluminum foil for light protection and kept overnight at 4°. The activated ester is used in the next step without any further purification. When stored at 4°, the active ester preparation can be used for at least 1 week.

Conjugation of Chemiluminescent Marker ABEI-H to Streptavidin

Streptavidin (2 mg, 0.027 μmol) is dissolved in 1 ml of 50 m*M* phosphate buffer (pH 8.3). To this solution the active ester derivative of ABEI-H in DMF (40 μl, 2 μmol) is added. The reaction mixture is left overnight at 4°. The precipitate formed is separated by centrifugation, and the supernatant is transferred to a Sephadex G-25 PD-10 column equilibrated with Tris buffer. On elution in the same buffer, three peaks absorbing at 280 nm are obtained. The first peak represents the conjugate.

The yield of chemiluminescent-labeled streptavidin is over 60%, and the incorporation ratio is 11–13 mol of the chemiluminescent marker per mole of streptavidin. The conjugate is stored at −20° in Tris buffer containing 0.1% bovine serum albumin and 0.05% sodium azide. After thawing, the remaining conjugate is stored at 4°. At an excess level of reagent, usually 10–20 ng of the streptavidin–ABEI-H conjugate is required per assay tube in our immunoassay procedures.

Preparation of Steroid–Protein Conjugates

Steroid–protein conjugates are prepared in two steps. In the first step of the reaction, activated *N*-hydroxysuccinimide esters of carboxy derivatives of steroids (e.g., cortisol 21-hemisuccinate or cortisol-3-carboxymethyloxime) are prepared as described previously.[9] The activated ste-

[9] F. Kohen, J. de Boever, and J. B. Kim, this series, Vol. 133, p. 387.

roid derivative is then coupled to a protein carrier (ovalbumin or bovine serum albumin) using the above-described conditions for the preparation of chemiluminescent-labeled streptavidin. The conjugates are purified by gel filtration on Sephadex G-25 columns.

Immobilization of Proteins on Polystyrene Balls

The following procedure is a modification of the methodology of Wood *et al.*[10,11]

Activation of Polystyrene Balls. The copolymer poly(phenylalanine–lysine) is dissolved in water at 60° and stored at 1 mg/ml at 4°. One thousand balls are transferred to a 500-ml Erlenmeyer flask and covered with 125 ml of water containing 3.5 mg of the copolymer (3.5 μg polymer/ball). The coating process is allowed to proceed for 4 days at room temperature with occasional shaking (2 to 3 times/day). After coating, the balls are washed once with 0.15 *M* NaCl, washed 3 times with water, and dried under a stream of compressed air. The coated dry balls can be stored at room temperature for at least 6 months.

Covalent Coupling of Proteins to Activated Balls

Glutaraldehyde activation. One hundred poly(phenylalanine–lysine)-coated balls (equivalent to 14.3 g in weight) are covered in an Erlenmeyer flask with 20 ml of 0.4% glutaraldehyde in 50 m*M* phosphate buffer (pH 8.0). The reaction, during which aldehyde functions are formed on the ε-amino groups of the lysine residues, is allowed to proceed for 30 min at room temperature, with stirring of the flask every 5 min. After activation, the balls are washed twice with water, washed once with 0.15 *M* NaCl, and used immediately for the coupling of proteins.

Protein coupling. The washed balls are poured into a solution of the desired protein (0.1–0.3 mg/100 balls) in 20 ml of 50 m*M* phosphate buffer (pH 8.0). The reaction is allowed to proceed for 2–3 hr at room temperature with occasional shaking (every 20 min) and overnight at 4°, after which 10 μg bovine serum albumin per ball is added as the saturation reagent from a 200 mg/ml stock solution in 0.15 *M* NaCl (saline). After a further 90 min at room temperature, the coupling solution is aspirated.

Stabilization of glutaraldehyde linkages. The Schiff bases present in the balls prepared as above are reduced for 20 min at room temperature with a solution (20 ml) of freshly prepared 75 m*M* sodium borohydride in water. The formation of hydrogen bubbles indicates that the reaction is taking place, and the flask is shaken every 5 min. The sodium borohydride

[10] W. G. Wood and A. Gadow, *J. Clin. Chem. Clin. Biochem.* **21,** 789 (1983).

[11] W. G. Wood, J. Braun, and U. Hantke, this series, Vol. 133, p. 354.

solution is aspirated, and the balls are washed 4 times with water. For storage the balls are covered with Tris buffer containing 2.5% bovine serum albumin and 0.05% sodium azide. At 4° in this solution the balls are stable for at least 2 years.

Coating of Microtiter Plates. A solution of hGH (100 μl, 2.5 μg/ml) in 50 m*M* carbonate buffer (pH 9.6) is distributed into each well of a microtiter plate. The plate is incubated overnight at 4° and drained. A solution (100 μl) of 0.3% bovine serum albumin in saline is the added to each well to block additional binding sites, and the plate is left at room temperature for 2 hr. After draining and 1 wash step with 0.05% Tween 20 in saline, the plate is covered and can be stored dry at −70° for at least 2 months.

Reagent Solutions

Buffer A: 50 m*M* Tris-HCl (pH 7.4) containing, per liter, 5 g bovine serum albumin (BSA) and 0.5 g sodium azide

Buffer B: buffer A, containing in addition, 0.5 g bovine γ-globulin, 0.1 ml Tween 20, and 9 g NaCl per liter

Coating buffer: 50 m*M* sodium carbonate (pH 9.6) containing 0.1 g sodium azide per liter

Wash solution: 9 g/liter NaCl containing 0.05% (v/v) Tween 20

Heme catalyst: a stock solution of 1 mg/ml of microperoxidase (MP-11, Sigma) is prepared in water; the stock solution can be stored for up to 6 months at 4°, and a working solution is prepared freshly before each assay by diluting the stock solution 1 : 100 with distilled water

Oxidant solution: a working solution is prepared freshly by mixing 0.1 ml of 30% hydrogen peroxide solution (Merck, Darmstadt, FRG) with 15 ml of distilled water

Evaluation of Biotinylated Proteins: Effect of Spacer Length Exemplified by Detection of Enzyme Activity

The effect of the spacer length in various biotinyl derivatives (Fig. 1) on the signal in immunoassays involving avidin–biotin amplification was investigated by the following experiment. The biotin derivatives **Ia** (short chain) and **Ic** (long chain) are used to derivatize both alkaline phosphatase and a monoclonal antibody against human growth hormone (hGH) (clone 69, see below). The conjugates obtained are evaluated by incubating serial dilutions of the two biotinyl antibody derivatives in hGH-coated microtiter plates overnight 4°, followed by sequential incubations for 30 min each at room temperature first with 100 ng of underivatized streptavidin and then with one of the two different biotinyl derivatives of alkaline phospha-

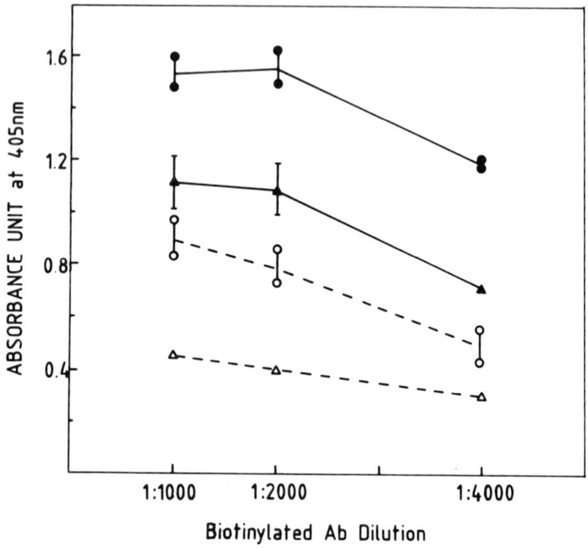

FIG. 3. Effect of spacer length of the biotinylated proteins on the signal intensity of the color development reaction. hGH-coated microtiter plates were incubated sequentially with different combinations of short- and long-chain derivatives of antibody 69 and of alkaline phosphatase. For experimental details, see text. Filled symbols represent the use of alkaline phosphatase derivatized with long-chain biotin (compound **Ic**, Fig. 1), whereas open symbols indicate the use of short-chain (**Ia**) biotinyl enzyme. Circles indicate long-chain (**Ic**) derivatized antibody, triangles short-chain (**Ia**) biotinylated antibody 69.

tase at 1 : 1000 dilution. The incubation steps are carried out in a 0.1-ml volume of buffer B and separated by washing 3 times with 0.05% Tween 20 in saline. Substrate solution [1 mg/ml *p*-nitrophenyl phosphate in 10% diethanolamine (pH 9.6) containing 5 m*M* $MgCl_2$] is added, and the yellow color that develops is measured at 405 nm. The results (Fig. 3) indicate that significantly higher specific signals are obtained with the long alkyl chain derivatives of biotin. We therefore use the biotin derivative **Ic** in all subsequent experiments.

Two-Site Immunochemiluminometric Assay for Human Growth Hormone Using Labeled Streptavidin as Probe[1,2,4]

Reagents

Monoclonal antibodies to hGH which bind to different epitopes on the HGH molecule: clone 518, affinity constant $K_A = 8 \times 10^{10}\ M^{-1}$; clone 69, affinity constant $K_A = 2 \times 10^9\ M^{-1}$

hGH reference preparation: WHO International Laboratory for Biological Standards, Hampstead, London NW3 6RB, England

Streptavidin-ABEI-H

Serum specimens: from normal individuals or from patients undergoing growth hormone stimulation tests or submitted to the laboratory to assess growth hormone pituitary reserve; sera are stored at −20° until assayed

Purified monoclonal antibody (clone 518) coupled to polystyrene balls (as described above) serves as the capture antibody

Monoclonal antibody to hGH (clone 69) biotinylated with compound **Ic** (Fig. 1)

Streptavidin labeled with ABEI-H (see text) is used as the chemiluminescent probe

Assay buffer (buffer B) and reagents for initiating the light reaction are described above

Procedure. A schematic diagram of the procedure is shown in Fig. 4, and details are presented in the legend. After the final incubation step of the reaction with ABEI-H-labeled streptavidin, the balls are washed 3 times. Each individual ball is then placed in a Lumacuvette, and sodium hydroxide (2 *N*, 250 μl) is added to each tube. The tubes are incubated at 60° for 30 min. After cooling to room temperature and placing the tubes in the luminometer, 100 μl of diluted microperoxidase and 100 μl of oxidant solution (0.2% H_2O_2) are injected sequentially into each individual tube. The chemiluminescence signal is integrated and recorded for 10 sec.

Evaluation. A typical dose–response curve (Fig. 5) reflects the analog recording of the signal emanating from the photomultiplier tube. The sensitivity of the method (defined as the zero-standard signal plus 2 standard deviations) is 1.5 pg of hGH/tube or 0.03 mIU hGH/liter. This technique provides a working range of 0.04–100 mIU hGH/liter. The intraassay precision is 5.5%, and the interassay precision is 7.4%. The dynamics of hGH concentrations in basal plasma levels or under stimulation tests in normal patients are determined by conventional RIA and the two-site immunochemiluminometric assay. A good correlation between the two methods is consistently obtained ($r = 0.97$, $n = 88$). As an example, Fig. 6 shows the circadian profile of hGH levels in a female patient as determined by RIA and immunochemiluminometric assay. In addition, this method has been used to detect basal secretion and minor bursts in the circadian profile of hypopituitary patients whose plasma hGH levels are undetectable by conventional RIA methods.[1,2,4] This method shows a gain in sensitivity by a factor of 10 in comparison with existing radio- or enzyme-immunoassay methods.

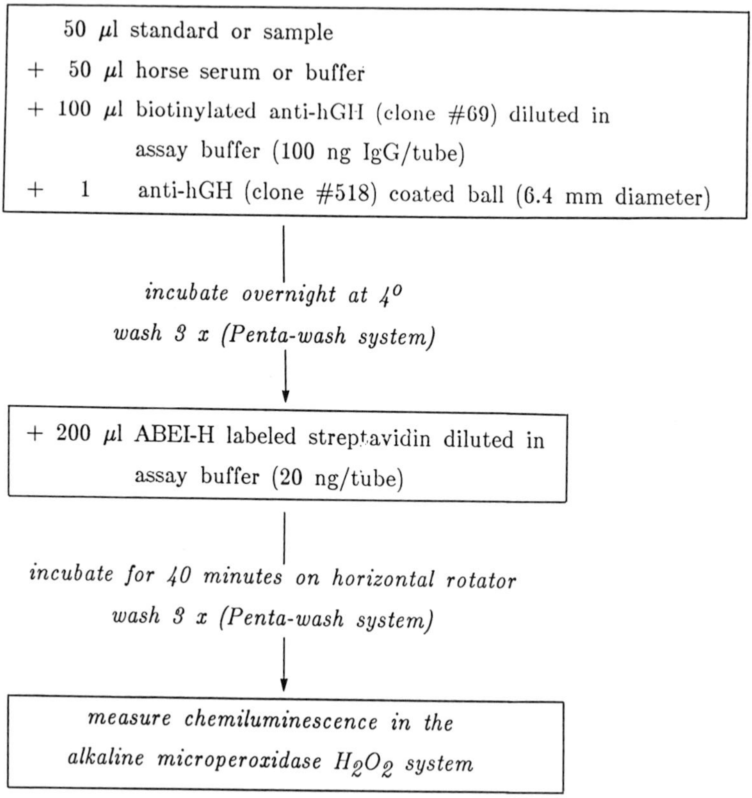

50 μl standard or sample
+ 50 μl horse serum or buffer
+ 100 μl biotinylated anti-hGH (clone #69) diluted in assay buffer (100 ng IgG/tube)
+ 1 anti-hGH (clone #518) coated ball (6.4 mm diameter)
incubate overnight at 4°
wash 3 x (Penta-wash system)
+ 200 μl ABEI-H labeled streptavidin diluted in assay buffer (20 ng/tube)
incubate for 40 minutes on horizontal rotator
wash 3 x (Penta-wash system)
measure chemiluminescence in the alkaline microperoxidase H2O2 system

FIG. 4. Flow diagram of a two-site immunochemiluminometric assay for hGH. The biotinylated antibodies and chemiluminescent-labeled streptavidin are diluted in buffer B. The hGH standards (50 μl/well) are prepared in buffer A, and 50 μl of horse serum is added to each well to compensate for matrix effects. The volume of the serum samples was 50 μl/well, and each well received 50 μl of buffer A.

Competitive-Type Immunoassays for Haptens Based on Biotinyl Antibody and Labeled Streptavidin as Probes Exemplified by Immunoassay of Salivary Cortisol

Reagents

Polyclonal rabbit antiserum to cortisol-3-carboxymethyloxime–bovine serum albumin conjugate
Ovalbumin–cortisol-3-carboxymethyloxime conjugate
Streptavidin–ABEI-H

The antibodies are purified on Sepharose–protein A and biotinylated using compound **Ic** (see Fig. 1) as described in the text. Cortisol-3-car-

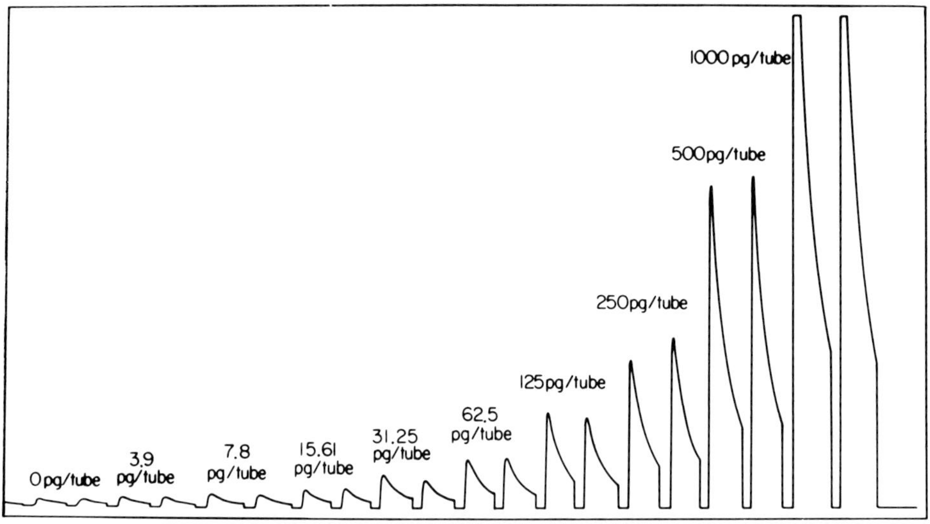

FIG. 5. Analog recording of a dose–response curve for hGH using biotinyl antibodies as primary probe, chemiluminescent-labeled streptavidin as secondary probe, and luminometry as an end point. The light generated on introduction of each sample is integrated for 10 sec and recorded in arbitrary light units.

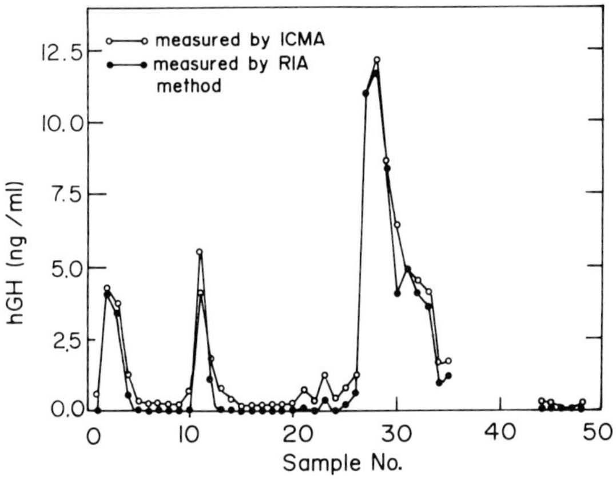

FIG. 6. Circadian plasma hGH pattern in a patient (5 years accelerated growth) as determined by RIA and immunochemiluminometric assay. The RIA data were measured at the laboratory of A. Kowarski (Baltimore, MD). The low values plotted as zero for the RIA method represent measurements of less than 0.25 ng/ml, the detection limit of the RIA method.

10 μl sample or standard
+ 200 μl biotinylated polyclonal anti-cortisol antibodies, diluted in assay buffer (30 ng IgG/tube)
+ 1 Cortisol-coated ball (6.4 mm diameter)

↓

incubate for 40 minutes on horizontal rotator
wash 3 x (Penta-wash system)

↓

+ 200 μl ABEI-H labeled streptavidin diluted in assay buffer (20 ng/tube)

↓

incubate for 40 minutes on horizontal rotator
wash 3 x (Penta-wash system)

↓

measure chemiluminescence in the alkaline microperoxidase H_2O_2 system

FIG. 7. Flow diagram for a competitive-type immunoassay for salivary cortisol using biotinyl antibody and chemiluminescent-labeled streptavidin as probes.

boxymethyloxime–ovalbumin conjugate is coupled to polystyrene balls and serves as the solid-phase antigen. Streptavidin labeled with ABEI-H is used as the chemiluminescent probe. For hapten assays buffer B (see above) is used throughout, and all the other reagents are the same as those used for the hGH assay.

Immunoassay Procedure. A schematic diagram of the method is shown in Fig. 7. After the final incubation step, the balls are treated as described for the hGH assay. The light generated is integrated for 10 sec and recorded.

Evaluation. A typical dose–response curve is shown in Fig. 8. The sensitivity of the method is less than 3 pg/tube and is comparable to that obtained by conventional RIA methods. For comparative purposes we have used two other end points in the cortisol assay (fluorometry or colorimetry, respectively, using streptavidin–β-galactosidase or avidin–alkaline phosphatase as the markers). Table I shows a comparison of the

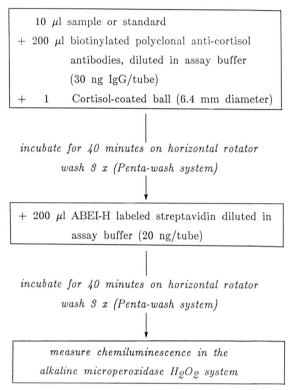

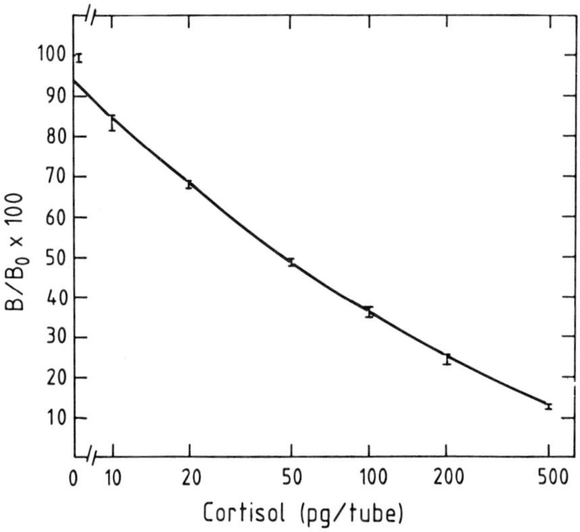

FIG. 8. Dose–response curve for cortisol using chemiluminescence as an end point. The value of B/B_0 represents the ratio of the "bound" signal to the zero-dose bound signal.

three methods. We would like to stress that all these techniques can easily be set up in most laboratories and that the end point in a given assay will depend on the available instrumentation.

Discussion

The use of biotinylated probes in immunoassays for haptens and peptide hormones is appealing for a variety of reasons: (1) the biotinylated enzymes and antibodies are stable reagents, and they can be prepared

TABLE I
COMPARISON OF THREE DIFFERENT END POINTS FOR MEASUREMENT OF CORTISOL BY SOLID-PHASE ANTIGEN LUMINESCENCE TECHNIQUE (SPALT) METHOD

Variable	Chemiluminescence[a]	End-point fluorometry[b]	Colorimetry[c]
Detector response (light units or OD)	23191–290	684–24.6	1.138–0.072
Sensitivity (μg/dl)	0.1	0.2	0.2
Working range (μg/dl)	0.1–100	0.2–100	0.2–100
Reagent stability		More than 2 years	

[a] System used: components of Fig. 2 in which L is a chemiluminescent marker.
[b] System used: components of Fig. 2 in which L is β-galactosidase.
[c] System used: components of Fig. 2 in which L is alkaline phosphatase.

easily since the reactive biotinyl derivatives are commercially available; (2) the end-point reaction in the assay can be varied according to the character of the avidin or streptavidin conjugate used in the method (colorimetry, fluorescence, or luminescence); and (3) the use of labeled avidin or streptavidin enables a uniform immunoassay methodology that comprises both small and large molecules.

We have described here a two-site immunochemiluminometric assay for hGH and a competitive-type assay for cortisol. In both methods, chemiluminescent-labeled streptavidin serves as the marker. The methods can be applied for the measurement of other polypeptide hormones (e.g., luteinizing hormone, hCG, or thyroid-stimulating hormone) or haptens (e.g., estradiol). The two-site technique requires only a capture antibody and a biotin-labeled antibody directed against a different epitope of the antigen. On the other hand, the competitive-type assay for haptens requires a solid-phase antigen and a specific biotinylated homologous antibody. The present avidin–biotin-mediated immunochemiluminometric assay for hGH has a lower limit of sensitivity below 75 amol (attomoles) of hGH per sample.

The dynamic range of the signal (zero-dose versus high-dose) obtained in the cortisol assay using a biotinylated polyclonal antibody shows a ratio of approximately 5 : 1. However, when using the same assay technique in an estradiol assay employing a biotinylated monoclonal antibody,[12] the dynamic range of the signal was increased to 20 : 1. This result suggests that the slope of a hapten assay can be improved by using biotinylated monoclonal antibodies instead of biotinylated polyclonal antibodies.

Acknowledgments

We are grateful to Drs. M. Wilchek, E. A. Bayer, G. Barnard, G. Messeri, J. Kostyo, Z. Zadik, Y. Zaltsman, and Mr. L. Toldo for permission to quote collaborative work; to Mr. B. Resch and Dr. F. Berthold for the loan of a Clinilumat luminometer; to Dr. M. Pazzagli for the gift of rabbit anticortisol serum; to Dr. N. Moav for a generous gift of the monoclonal antibodies to hGH; and to Mrs. M. Kopelowitz for excellent secretarial assistance. C. J. S. is a Minerva Fellow at the Weizmann Institute of Science.

[12] C. J. Strasburger and F. Kohen, unpublished results.

[57] Biotinylated Protein A in Immunoassay

By IKUO TAKASHIMA, NOBUO HASHIMOTO, and HSHI-CHI CHANG

Enzyme-linked immunoassay (ELISA) has been used extensively in serological diagnostic tests. In order to apply ELISA for serodiagnosis of zoonotic viral diseases such as Japanese encephalitis (JE), the conjugates must be prepared specifically to individual animal species involved in the transmission cycle of the virus. Since Guesdon *et al.*[1] introduced avidin–biotin into ELISA, this system has been used widely to detect or quantify immunoglobulins[2,3] or antigens.[4,5] Protein A is known to bind IgG in several species of animals via the Fc portion of the IgG molecule.[6] In this chapter, we describe the use of biotin-labeled protein A ELISA for simultaneous detection of JE antibody in the sera of humans, swine, and several other vertebrates using the same reagents.

Materials and Reagents

Antigen: purified JE virus JaGAr-01 strain vaccine (total N 96.6 μg/ml) from Biken (Osaka, Japan)[7]

Sera: swine sera were collected from endemic and nonendemic areas of Japan; human, cattle, horse, monkey, dog, and pigeon sera were collected from an endemic area; and rabbit, rat, and mice sera were obtained from individuals experimentally infected with the JE virus

Protein A: salt-free lyophilized powder (E-Y Laboratories, San Mateo, CA)

Horseradish peroxidase-labeled avidin (HRP–avidin): commercially available product of E-Y Laboratories

Substrate: 0.2 m*M* 2,2′-azinodi(3-ethylbenzthiazolinesulfonic acid) (ABTS; Sigma Chemical Co., St. Louis, MO) and 0.004% H_2O_2 in 50 m*M* citrate buffer (pH 4.0)

[1] J. L. Guesdon, T. Ternynck, and S. Avrameas, *J. Histochem. Cytochem.* **27,** 1131 (1979).
[2] S. Jackson, J. A. Sogn, and T. T. Kindt, *J. Immunol. Methods* **48,** 229 (1982).
[3] D. V. Subba Rao, N. L. McCartney-Francis, and D. D. Metcalfe, *J. Immunol. Methods* **57,** 71 (1983).
[4] C. Kendall, I. Ionescu-Mediu, and G. R. Dreesman, *J. Immunol. Methods* **56,** 329 (1983).
[5] R. H. Yolken, F. J. Leister, L. S. Whitcomb, and M. Santosham, *J. Immunol. Methods* **56,** 319 (1983).
[6] A. Forsgren and J. Sjoquist, *J. Immunol.* **97,** 882 (1966).
[7] K. Takaku, T. Yamashita, T. Osanai, I. Yoshida, M. Kato, H. Goda, M. Takagi, T. Hirota, T. Amano, K. Fukai, N. Kunita, K. Inoue, K. Shoji, A. Igarashi, and I. Ito, *Biken J.* **11,** 25 (1968).

Coating buffer: 50 m*M* sodium carbonate–bicarbonate buffer (pH 9.6)

Rinsing buffer: phosphate-buffered saline (PBS, pH 7.4) containing 0.05% Tween 20 (Tween 20–PBS)

Serum or biotin–protein A diluent buffer: PBS containing 1% Tween 80 (Tween 80–PBS)

HRP–avidin diluent buffer: 50 m*M* phosphate buffer (pH 8.0) containing 0.5 *M* NaCl and 0.1% Tween 20 (Tween 20–0.5 *M* NaCl PB)

Biotin Labeling of Protein A

The procedure is essentially the same as that described by Guesdon *et al.*[1] Protein A (1 mg/ml) is dialyzed against 0.1 *M* $NaHCO_3$ at 4°. After dialysis, 1 ml of the protein A solution is mixed with 60 μl of biotin-*N*-hydroxysuccinimide ester (E-Y Laboratories; 1 mg/ml in dimethyl sulfoxide). The solution is kept at room temperature for 4 hr and dialyzed overnight against PBS at 4° with several changes of PBS.

Procedure for ELISA

The procedure followed is that previously described.[8] Briefly, the optimal dilutions of the reagents are determined by checkerboard titration. JE antigen is diluted with coating buffer to a concentration of 8 units. One-hundred microliters of the antigen is delivered into U-shaped polystyrene microtiter plates (96 U, PS, SH, Nunc, Denmark). The plates are incubated at 37° for 3 hr to fix the antigen onto the wells. The solution is then discarded, and the wells are rinsed once with Tween 20–PBS. After tapping the plates, serum samples (diluted serially at 2-fold with Tween 80–PBS) are delivered to each well at a volume of 50 μl. The plates are kept at 37° for 40 min and then rinsed 3 times with Tween 20–PBS. Biotin–protein A in Tween 80–PBS (50 μl containing 4 units) is delivered into each well, and the plates are incubated at 37° for 40 min. The solution is again discarded, and the wells are washed 3 times with Tween 20–0.5 *M* NaCl PB. The HRP–avidin in Tween 20–0.5 *M* NaCl PB (50 μl at 8 units) is delivered into each well, and the plates are incubated at 37° for 40 min. Next, the contents of the plates are discarded, and the wells are washed 3 times with Tween 20–0.5 *M* NaCl PB. Finally, 100 μl of freshly prepared ABTS substrate solution is added to each well.

[8] H.-C. Chang, I. Takashima, J. Arikawa, and N. Hashimoto, *J. Virol. Methods* **9,** 143 (1984).

Following an incubation for 1 hr at 37°, the degree of color development is measured by a microplate photometer (Corona MTP-12, Corona Electric, Katsuda, Japan) at a wavelength of 405 nm. The cutoff point is set at an average of the OD values of control wells (without serum) plus an additional factor of 0.1. End-point titers of serum are expressed as a reciprocal of the highest serum dilution at which the OD value is above the cutoff point. The ELISA titers are compared with the titers of the hemagglutination inhibition (HI) test described by Clarke and Casals.[9]

Application

ELISA antibody titers to JE virus were determined in swine sera collected from nonendemic (Hokkaido) and endemic (Shizuoka) areas of Japan (Fig. 1A). ELISA titers of 106 swine sera from Hokkaido were lower than 1:5. HI titers of all these sera from Hokkaido were negative (<1:10). Of the 68 HI-negative sera from Shizuoka, 67 (98.5%) had ELISA titers lower than 1:5 and 1 serum had an ELISA titer of 1:5. Therefore, the cutoff point was set at a serum dilution of 1:5 for positive ELISA antibody. By these criteria, a positive antibody rate by ELISA was determined to be 43.8% (53/121), and the same rate was achieved by the HI test. There was no correlation between ELISA and HI titers ($r = 0.40$). Twelve sera had HI titers equal to or higher than 4 times ELISA titers. These 12 sera had IgM as the predominant types of antibody to the JE virus, determined by ELISA using anti-swine IgM–peroxidase conjugates.

ELISA titers to the JE virus were determined in 95 human sera from Miyazaki Prefecture, an endemic area of JE, and the results were compared with HI titers (Fig. 1B). All of the 45 HI-positive sera had ELISA titers higher than 1:5. Forty of 50 (80%) HI-negative sera had ELISA titers lower than 1:5. Thus, the cutoff point for this assay was also set at a serum dilution of 1:5 for positive antibody. By these criteria, 55 out of 95 sera (57.9%) were judged as positive in ELISA. Ten out of 50 HI-negative sera were positive in ELISA. These 10 HI-negative sera were examined using the 50% plaque-reduction neutralization test and showed neutralization titers ranging from 1:2.5 to 1:10 (data not shown). There was a close correlation both between the ELISA and HI titers ($r = 0.916$).

Using this system, ELISA antibody titers to JE were successfully detected in sera from rhesus monkey, dog, horse, and mouse but not in sera from cattle, pigeon, and rat (data not shown).

[9] D. H. Clarke and J. Casals, *Am. J. Trop. Med. Hyg.* **7**, 561 (1958).

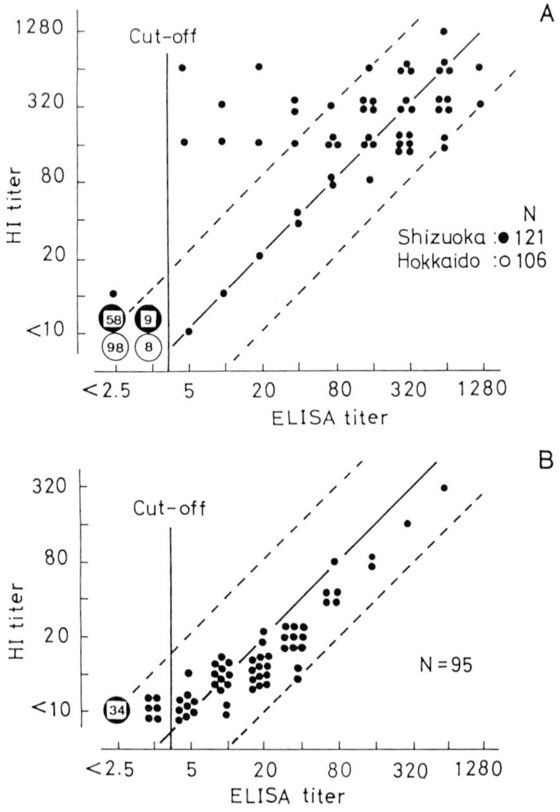

FIG. 1. (A) Comparison of JE antibody titers by ELISA and HI tests in swine sera from Shizuoka and northern Hokkaido. Numbers in black circles indicate numbers of swine sera from Shizuoka recording a particular value, and those in white circles represent those from Hokkaido. Each symbol shows the results obtained for an individual swine serum. (B) Comparison of JE antibody titers by ELISA and HI tests in human sera from Miyazaki Prefecture. Each symbol represents values for an individual serum sample. The number in the black circle indicates the number of human sera recording the given result. (Modified from Ref. 8, with permission.)

Comments

Since nonspecific reactions in the ELISA were very low in the sera of swine and humans, the cutoff titer for the assay could be set at 1 : 5 for positive antibody. Biotin-labeled protein A was found not to bind IgM antibody. The results suggest that the ELISA system as described here is not able to detect IgG antibody of certain animal species (e.g., cattle, pigeon, and rat).

[58] Slide Immunoenzymatic Assay for Human Immunoglobulin E

By EVERLY CONWAY DE MACARIO, ROBERT J. JOVELL, and ALBERTO J. L. MACARIO

The slide immunoenzymatic assay (SIA)[1] is a modular system (U.S. Patent No. 4,682,891) with multiple laboratory and field applications for measuring antigens, antibodies, and other analytes.[2,3] It is based on the principles of the enzyme-linked immunosorbent assay.[4] The advantages and limitations of SIA-related technology have been discussed.[2,3] Briefly, SIA affords a simple, rapid, and low-cost means for accurately measuring and storing[5] particulate and nonparticulate (molecular) analytes. Results can be read with the naked eye or instrumentally. The latter can be done manually or automatedly using commercially available spectrophotometers and SIA adapters. Unique to SIA is the geometry of the reaction area,[2] which endows the system with peculiar and convenient properties, not the least important being the virtual absence of background readings.

Descriptions of the components and variations of SIA have been published,[2,3,5–8] including specific applications in hybridoma technology[3,6,7] and in the quantification of immunoglobulins (SIA-Ig).[9] Pertinent to this chapter is the application of SIA to the measurement of human IgE in biological fluids. A basic SIA-IgE procedure has been described.[10] Here, improvements are reported which make use of biotin and streptavidin[11,12]

[1] E. Conway de Macario, A. J. L. Macario, and R. J. Jovell, *J. Immunol. Methods* **59,** 39 (1983).

[2] E. Conway de Macario, R. J. Jovell, and A. J. L. Macario, *BioTechniques* **3,** 138 (1985).

[3] E. Conway de Macario, A. J. L. Macario, and R. J. Jovell, this series, Vol. 121, p. 509.

[4] E. Engvall and P. Perlmann, *Immunochemistry* **8,** 871 (1971).

[5] E. Conway de Macario, R. J. Jovell, and A. J. L. Macario, *J. Immunol. Methods* **99,** 107 (1987).

[6] A. J. L. Macario and E. Conway de Macario, *in* "Monoclonal Antibodies against Bacteria" (A. J. L. Macario and E. Conway de Macario, eds.), Vol. 2, p. 213. Academic Press, Orlando, Florida, 1985.

[7] S. A. Kumar and A. J. L. Macario, *Hybridoma* **4,** 297 (1985).

[8] E. Conway de Macario, R. J. Jovell, and A. J. L. Macario, *J. Immunoassay* **8,** 283 (1987).

[9] E. Conway de Macario, A. J. L. Macario, and R. J. Jovell, *J. Immunol. Methods* **68,** 311 (1984).

[10] E. Conway de Macario, A. J. L. Macario, and R. J. Jovell, *J. Immunol. Methods* **90,** 137 (1986).

[11] B. Falini and C. R. Taylor, *Arch. Pathol. Lab. Med.* **107,** 105 (1983).

[12] D. A. Fuccillo, *BioTechniques* **3,** 494 (1985).

(see also Zymed's Streptavidin–Biotin system booklet). A considerable increase in sensitivity has been obtained, inasmuch as $A_{450\ nm}$ readings were enhanced and lower levels of IgE were detected as compared with the basic procedure described earlier. Four protocols were compared: protocols I and II represent the basic procedure whereas protocols III and IV are the new developments.

A series of general principles applicable to all protocols is as follows: (1) All incubations are done inside a humid chamber at 23°. (2) All washes are done with 5 drops (20 μl each) of distilled water or other washing solution per SIA slide circle. (3) Drying is done by placing the SIA slide face up onto a slide warmer (Fisher Scientific, Springfield, NJ). (4) Optimal dilutions of reagents (e.g., antisera, purified antibodies, biotin, streptavidin) must be determined when a new batch is used for the first time. Optimal dilutions (i.e., those producing negligible background and reproducible readings at levels within the most accurate range of the instrument being used) vary with the batch and source of each reagent. (5) SIA slides (Cel-Line Associates, Newfield, NJ) with three rows of 8 circles each are used. Circles are 3 mm in diameter and have a distribution pattern compatible with instrumental automation (Fig. 1). (6) All reagent solutions are applied as a drop of the specified volume onto each SIA slide circle.

Protocol I

1. Cover each SIA circle with 5 μl of goat IgG anti-human IgE (E-Y Laboratories, San Mateo, CA) at a dilution of 1 : 5 in phosphate-buffered saline (PBS, pH 7.2).
2. Dry.
3. Wash each circle first with PBS and then with H_2O.
4. Dry.
5. Cover each circle with 10 μl of the fluid in which IgE is to be measured. Sample dilutions are in PBS.
6. Incubate for 15 min.
7. Wash with H_2O.
8. Dry.

FIG. 1. SIA slides and SIA slide carrier (U.S. Patent No. 4,682,891) for automated reading with a vertical-beam spectrophotometer. (A) Four SIA slides, each with 24 reaction areas (circles, bottom), SIA slide carrier (middle), and spectrophotometer support (top). (B) The four SIA slides are inserted in the carrier. The 96 circles are distributed in a pattern matching that of the wells of a standard microtitration plate. (C) SIA slide carrier with slides mounted on support ready for spectrophotometric reading.

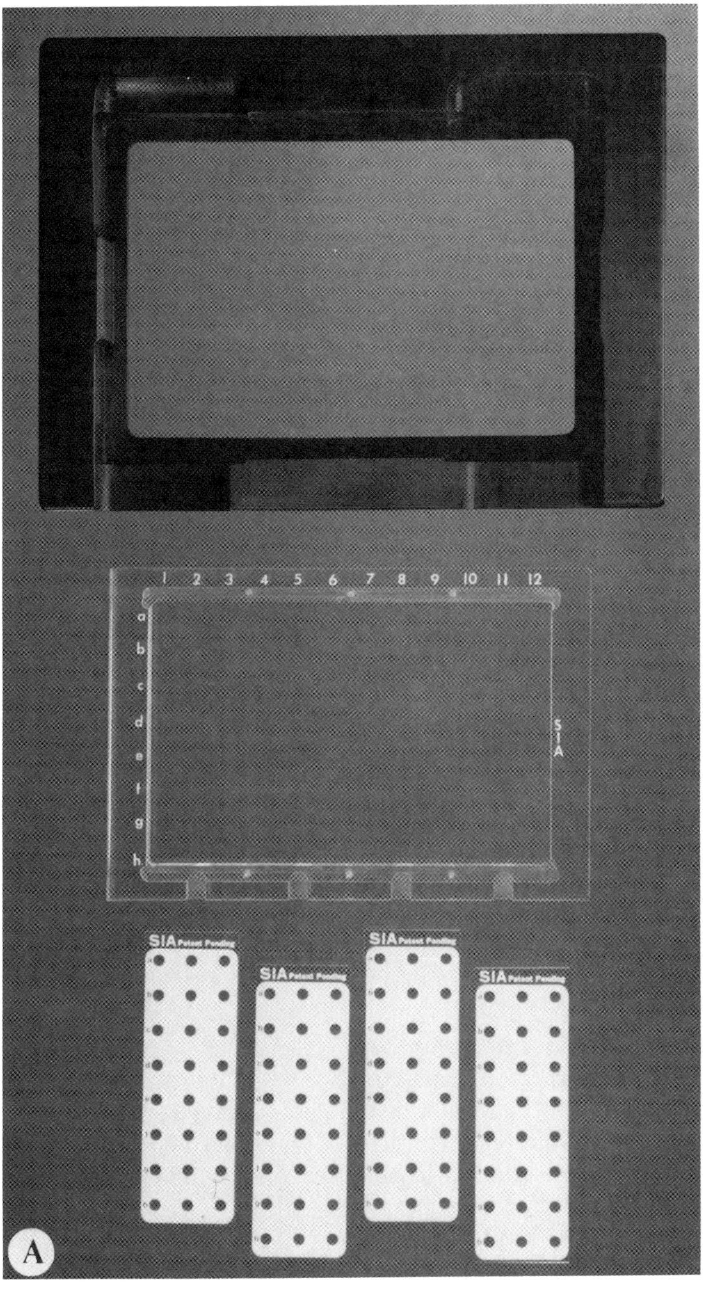
1 2 3 4 5 6 7 8 9 10 11 12
a
b
c
d
e
f
g
h
SIA
SIA Patent Pending
SIA Patent Pending
SIA Patent Pending
SIA Patent Pending
A

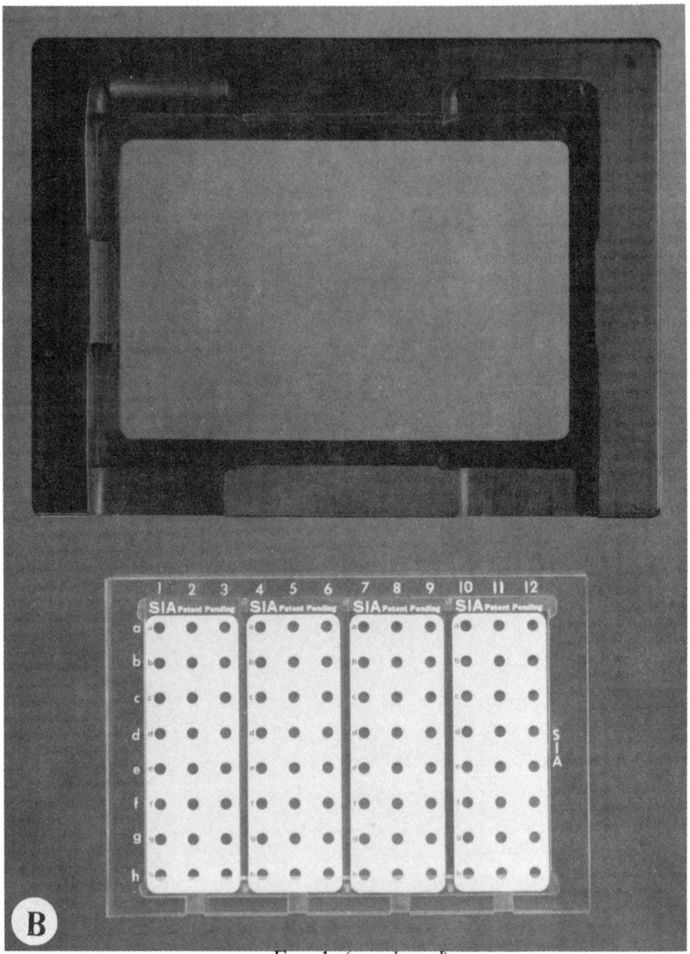

FIG. 1. (*continued*)

9. Cover each circle with 10 μl of goat IgG–peroxidase conjugate anti-human IgE (Cappel, Cochranville, PA) at a dilution of 1 : 50 in CM+ [Dulbecco's modified Eagle's medium, Gibco Laboratories, Grand Island, NY, with 15% (v/v) normal goat serum, Wadsworth Center for Laboratories and Research, Albany, NY].
10. Incubate for 15 min.
11. Wash with distilled water.
12. Dry.
13. Cover each circle with 10 μl of substrate solution prepared immediately before use by mixing 1 μl of 30% H_2O_2 with 10 ml of

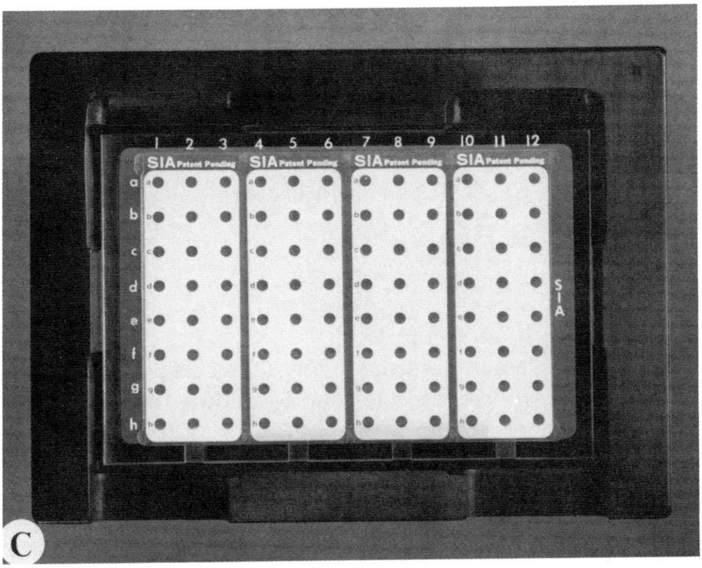

FIG. 1. *(continued)*

o-phenylenediamine (OPD; Sigma Chemical Co., St. Louis, MO) solution. OPD solution is 1 mg OPD in 1 ml of buffer (0.1 *M* citric acid in distilled water, adjusted to pH 4.5 with NaOH). This OPD solution can be prepared beforehand and stored frozen for 1 week at most.

14. Keep slide with drops of substrate solution on circles in a humid chamber at 23°. Read $A_{450\text{ nm}}$ values with a vertical beam spectrophotometer (e.g., Mini Reader II, Dynatech Instruments, Alexandria, VA) using an SIA slide carrier at 5 and 15 min. Additional readings, if necessary, are done every 15 min.

Protocol II

1. Cover each circle with 5 μl of biotinylated goat IgG anti-human IgE (Tago-Immunodiagnostic Reagents, Burlingame, CA) at a dilution of 1 : 5 in PBS.

2–14. Same as Steps 2–14 in Protocol I.

Protocol III

1 to 8. Same as Steps 1 to 8 in Protocol I.

9. Cover each circle with 10 μl of biotinylated goat IgG anti-human IgE (Tago) at a dilution of 1 : 25 in CM+.

10. Incubate for 15 min.
11. Wash with distilled water.
12. Dry.
13. Cover each circle with 10 μl of streptavidin–peroxidase (Amersham Corp., Arlington Heights, IL).

14–16. Same as Steps 10–12 above (this protocol).

17 and 18. Same as Steps 13 and 14 in Protocol I.

Protocol IV

1–13. Same as Steps 1–13 in Protocol III.

14. Incubate for 7.5 min.
15. Without removing drops or washing, add onto each drop already lying on the circles one 10-μl drop of biotin–peroxidase (E-Y Laboratories) at a dilution of 1 : 10,000 in CM+ with Tween 20 (1 : 300, v/v).
16. Incubate for 7.5 min.
17. Wash with distilled water.
18. Dry.

19 and 20. Same as Steps 13 and 14, respectively, in Protocol I.

Comments

An increase in sensitivity of SIA-IgE was observed while progressing from Protocol I through IV, the latter giving the highest $A_{450\ nm}$ readings

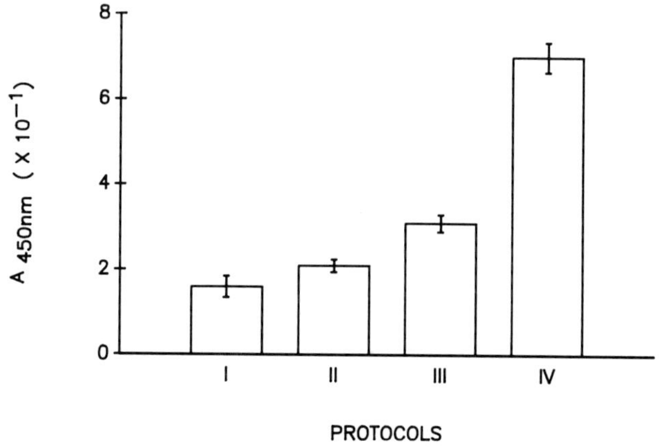

FIG. 2. Examples of data comparing Protocols I–IV in a single run. Bars represent arithmetic mean values at the peak of the reading curve, usually reached at 15 min. Brackets represent ranges (n = 3). The concentration of IgE used here was 200 IU/ml.[13]

TABLE I
REPRODUCIBILITY OF RESULTS WITH PROTOCOL IV TO MEASURE VARIOUS LEVELS OF IgE IN DIFFERENT-DAY RUNS

IgE (IU/ml)	Test; day run (peak $A_{450\ nm}$)[a]		
	1	2	3
200	0.75 [0.71–0.77][b]	0.77 [0.73–0.79]	0.68 [0.64–0.72]
20	0.38 [0.35–0.40]	0.32 [0.28–0.36]	0.39 [0.36–0.41]
2	0.27 [0.24–0.29]	0.19 [0.16–0.24]	0.26 [0.24–0.28]
0.2	0.10 [0.09–0.11]	0.07 [0.05–0.09]	0.08 [0.06–0.10]
0.02	n.d.[c]	0.03 [0.02–0.04]	0.03 [0.02–0.05]

[a] Tests were run in triplicate in different days. Values shown represent the peak of the reading curve after addition of substrate for peroxidase.
[b] Arithmetic mean [range; $n = 3$].
[c] Not done.

(Fig. 2).[13] This was a consistent trend confirmed in different runs performed on different days using either the same or different reagent batches. For example, the mean factor of increase obtained with Protocol IV compared with I was 3.20 ($n = 4$). Protocol IV also consistently detected low levels of IgE (Table I).

[13] IgE levels are expressed in International Units (IU), based on a World Health Organization reference serum. One International Unit is equivalent to 2.4 ng IgE.

[59] Quantification of Carcinoembryonic Antigen in Serum and Analysis of Epitope Specificities of Monoclonal Antibodies

By CHRISTOPH WAGENER, LUCIA WICKERT, and JOHN E. SHIVELY

The gene coding for the carcinoembryonic antigen (CEA) is a member of the immunoglobulin supergene family.[1] Similar to other members of this family, the CEA gene is a member of a subfamily of genes coding for glycoproteins of close structural relationship. Monoclonal antibodies

[1] R. J. Paxton, G. Mooser, H. Pande, T. D. Lee, and J. E. Shively, *Proc. Natl. Acad. Sci. U.S.A.* **84,** 920 (1987).

METHODS IN ENZYMOLOGY, VOL. 184

(MAbs) induced against CEA may be directed against CEA-specific epitopes or more or less common epitopes present on CEA as well as on structurally related antigens.[2] In this regard, a logical sequence of experimental steps can be designed in order to use monoclonal antibodies for the quantification of a single antigen or possibly several antigens in body fluids. The following strategy has thus been used for the characterization of monoclonal anti-CEA antibodies: (1) selection of MAbs with different epitope specificities[3,4]; (2) binding of the selected MAbs to different antigens of the CEA family, e.g., in Western blots or related techniques[5,6]; (3) determination of affinity constants of antibodies with interesting antigen-binding patterns[7]; (4) establishment of monoclonal antibody-based immunoassays.

For many techniques involved in the approach outlined above, either antigen or antibody have to be tagged by a label. Radiolabels such as ^{125}I have found wide application. However, because of the usually high number of antibodies to be handled and the time course of the experiments, the instability of radiolabels represents a significant drawback. In addition, radioactive hazards have to be taken into account. We have found the biotin label to be adequate or superior to radioactive labels in several aspects. Once antibodies and antigen are labeled by biotin, all of the experiments involved in the above approach can be performed, including initial antibody screening, determination of affinity constants, and double monoclonal, sandwich-type immunoassays. In this context, the use of avidin–enzyme conjugates for solid-phase immunoassays and the complementary application of avidin for the precipitation of either antigen or antibody in solution-phase assay[7] are particularly helpful.

Here we describe the use of the avidin–biotin system for the initial screening for epitope specificities of MAbs and for their ultimate use in double-monoclonal immunoassays for the sensitive detection of antigens in biological fluids. The precipitation of biotin-labeled antibodies or antigens from solution for the determination of affinity constants and epitope specificities is described elsewhere in this volume.[7]

[2] J. E. Shively and J. D. Beatty, *CRC Crit. Rev. Oncol./Hematol.* **2,** 355 (1985).

[3] C. Wagener, Y. H. J. Yang, F. G. Crawford, and J. E. Shively, *J. Immunol.* **130,** 2308 (1983).

[4] C. Wagener, U. Fenger, B. R. Clark, and J. E. Shively, *J. Immunol. Methods* **68,** 269 (1984).

[5] M. Neumaier, U. Fenger, and C. Wagener, *J. Immunol.* **135,** 3604 (1985).

[6] M. Neumaier, U. Fenger, and C. Wagener, *Mol. Immunol.* **22,** 1273 (1985).

[7] C. Wagener, B. R. Clark, K. J. Rickard, and J. E. Shively, *J. Immunol.* **130,** 2302 (1983); C. Wagener, U. Krüger, and J. E. Shively, this volume [60].

General Procedures

Monoclonal Antibodies. The production and characterization of MAbs against CEA are described in detail by Wagener *et al.*[3] In the original publications, the following abbreviations for the monoclonal antibodies were used: MAb 1, CEA.66-E3; MAb 2, T84.1-E3; MAb 3, CEA.41C-12.1-D8; MAb 4, T84.66-A3.1-H11; MAb 5, CEA.281-H5; MAb 6, CEA.11-H5. The IgG fractions of MAbs are purified from ascitic fluids over protein A–Sepharose.[8] The purified IgG can be dialyzed against water or a low ionic strength buffer (1 m*M* sodium phosphate, pH 7.0), lyophilized, and stored at −20°. The A_{280} of a 10 mg/ml solution of mouse IgG is 14.[9] CEA is purified according to Coligan *et al.*[10]

Determination of Epitope Specificities Using Biotinylated Monoclonal Antibodies in Competitive Solid-Phase Enzyme Immunoassay

Materials

PBS: 50 m*M* sodium phosphate (pH 7.2) containing 0.15 *M* NaCl and 0.1% NaN_3
Na_2CO_3, 0.2 *M* (pH 9.3–9.4)
BSA–PBS: PBS containing 1% (w/v) bovine serum albumin (Sigma Chemical Co., St. Louis, MO)
PBR (phosphate buffer–rabbit serum): 0.2 *M* potassium phosphate (pH 6.5) containing 20% (v/v) normal rabbit serum
APC: avidin–peroxidase conjugate (Sigma) diluted 1 : 100 in PBR
3 *M* HCl: dilute 260 ml concentrated HCl in 1 liter of distilled water
6 m*M* H_2O_2: dilute 68 μl of 30% H_2O_2 (w/v) in 100 ml of distilled water
Citrate buffer: 0.1 *M* sodium citrate (pH 5.0)
Substrate 1: 6 m*M* H_2O_2 (freshly prepared) and 40 m*M* *o*-phenylenediamine (Sigma) in citrate buffer
BNHS: *N*-hydroxysuccinimidobiotin (Sigma)
DMF: dimethylformamide (J. T. Baker, Phillipsburg, NJ), distilled from ninhydrin

Preparation of Biotin-Labeled MAb. A modification of the method of Clark and Todd[11] is employed. An equal volume of MAb (70 mg/ml in

[8] P. L. Ey, S. J. Prowse, and C. R. Jenkin, *Immunochemistry* **15,** 429 (1978).
[9] H. N. Eisen, E. S. Simms, and M. Potter, *Biochemistry* **7,** 4126 (1968).
[10] J. E. Coligan, J. T. Lautenschleger, M. L. Egan, and C. W. Todd, *Immunochemistry* **9,** 377 (1972).
[11] B. R. Clark and C. W. Todd, *Anal. Biochem.* **121,** 257 (1982).

PBS, 0.1 ml) is added to a freshly prepared solution of BNHS (1 mg/ml, 0.1 ml).[12] The BNHS solution is immediately added to the MAb dropwise with constant stirring, allowed to react 2 hr at room temperature, diluted to 1 ml with PBS, and dialyzed versus 3 to 4 changes of PBS. The degree of biotinylation can be changed by increasing or decreasing the amount of BNHS by a factor of 2–3 if desired. Aliquots of the biotinylated MAb are stored at −20° until used.

EIA Protocol. The protocol is similar to that described by Wagener *et al.*[4] Wells of a 96-well polyvinyl microtiter plate are coated with 100 μl of CEA (20 μg/ml) in sodium carbonate buffer for 12–18 hr at room temperature. Nonspecific binding is blocked by incubating antigen-coated wells with BSA–PBS for 2 hr at 37° followed by rinsing with PBS 5 times. A constant amount (50 μl; see below) of biotinylated MAb is added to 50 μl of serial 2-fold dilutions of unlabeled test MAb in PBS–BSA. Aliquots (50 μl) are added to the wells, after which the plate is incubated for 2 hr at 37° and washed 3 times with PBS. The wells are washed once with PBR, incubated with APC conjugate (100 μl of conjugate diluted 1 : 100 in PBR) for 2 hr at 37°, washed 5 times with 0.1 *M* citrate buffer (pH 5.0), and incubated with 100 μl of substrate for 30 min at room temperature in the dark. The reaction is stopped with 100 μl of HCl, and the absorbance is read at 492 nm. The dilution of biotinylated antibody was chosen to give a maximum A_{492} value of 0.8–1.5 in the absence of unlabeled MAb. Controls include wells not coated with antigen, incubations with no MAb, and incubations with normal mouse serum. Nonspecific binding should not exceed 0.05 absorbance units.

EIA Results. Sample results for epitope analysis of four MAbs are shown in Fig. 1. The results are equivalent to those published for a competitive solid-phase radioimmunoassay.[3] See next section for further discussion.

Determination of Epitope Specificities Using Biotinylated Antigen in Competitive Solid-Phase Enzyme Immunoassay

Materials

PBS: 10 m*M* sodium phosphate buffer (pH 7.4) containing 0.15 *M* NaCl

PBS–Tween: PBS plus 0.05% (v/v) Tween 20 (Sigma)

BSA–PBS: PBS containing 1% (w/v) bovine serum albumin (Sigma)

[12] The BNHS (2.0 mg) is added to 0.5 ml of DMF, stirred for 5 min, and diluted to 2.0 ml with 1.5 ml of water.

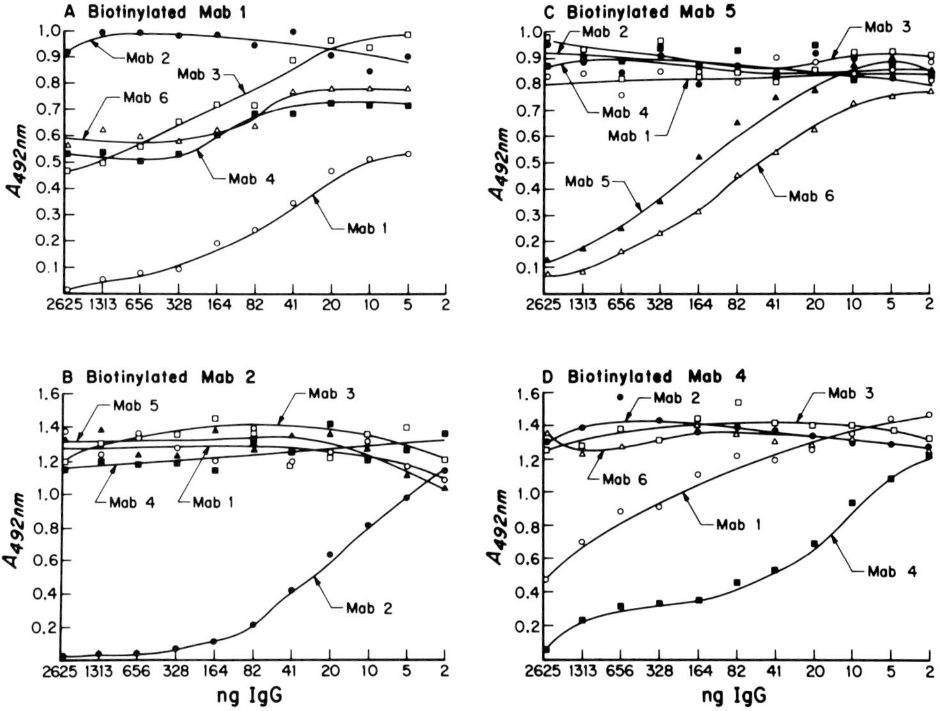

FIG. 1. Epitope analysis using a competitive avidin–biotin enzyme immunoassay. The wells of a microtiter plate were coated with CEA and incubated with a biotinylated MAb preparation and decreasing dilutions of the given unlabeled MAb. [From C. Wagener, U. Fenger, B. R. Clark, and J. E. Shively, *J. Immunol. Methods* **68,** 269 (1984), with permission.]

Reaction buffer: 20 m*M* sodium phosphate (pH 6.8)
Coating buffer: PBS containing 0.1% NaN_3
APC: avidin–peroxidase conjugate (Sigma) diluted 1 : 100 in PBS–Tween
1% H_2O_2: dilute 30% H_2O_2 1 : 30 in distilled water
Substrate 2: 1 mg/ml of 5-aminosalicylic acid (Sigma) in preheated (56°) reaction buffer; add 5 mg active charcoal per 100 ml and filter; add 100 μl of freshly prepared 1% H_2O_2 per 10 ml of above
3 *M* NaOH: dissolve 12 g of NaOH in 100 ml of distilled water
BNHS: *N*-hydroxysuccinimidobiotin (Sigma)
DMF: dimethylformamide, distilled from ninhydrin

EIA Protocol. The wells of a 96-well polystyrene microtiter plate (Costar, Cambridge, MA) are coated with MAb (5 μg MAb/ml in coating buffer) for 2 hr at 37° and overnight (8–12 hr) at 4°. Biotinylated CEA (50 μl of a 1.0 μg/ml solution) is added to separate test tubes containing 50 μl of the test MAb dilution (0.5–1000 ng of IgG is a good range), and the tubes containing the biotinylated CEA–MAb mixture are incubated overnight at 37°. The plates are washed 3 times with PBS, incubated for 1 hr at room temperature with 150 μl of BSA–PBS to block nonspecific binding, and washed 3 times with PBS. The biotinylated CEA–MAb mixture (100 μl) is added to the wells and incubated for 90 min at room temperature. The plates are washed 5 times with PBS–Tween, treated with 100 μl of APC, incubated for 2 hr at room temperature, and washed as follows: (1) rinse once with PBS; (2) fill wells to top with PBS, wait 5 min, discard, and repeat 2 times; (3) rinse wells with distilled water 3 times. The wells are then incubated with 200 μl of substrate 2 for 15 min, and the reaction is stopped with 100 μl of NaOH. Absorbance is read at 450 nm. The usual controls are run (see previous section). Nonspecific binding should be less than 0.01 absorbance units.

Comments. The inhibition analysis of five MAbs versus two MAbs coated on microtiter wells is shown in Fig. 2. The binding of biotinylated CEA to immobilized MAb 4 is inhibited by low amounts of the identical unlabeled antibody (Fig. 2A). At higher amounts of IgG, MAb 1 also quantitatively inhibits the binding of biotinylated CEA to MAb 4. MAb 2 shows partial binding inhibition at high IgG excess. Except for the latter inhibition by MAb 2, the results are comparable with those of the inhibition assay using biotinylated MAb 4 (Fig. 1D). The binding of biotinylated CEA to wells coated with MAb 2 is not inhibited by MAbs 1, 3, 4, or 5 (Fig. 2B). This result is equivalent with the result shown in Fig. 1B for the competition assay with biotinylated MAb 2.

Discussion of Inhibition EIA Results for Epitope Analysis of Monoclonal Antibodies. The results of the above inhibition assays are influenced by several factors, such as the affinity of different MAbs for the respective epitopes; the relative position of the respective epitopes; the number of epitopes recognized by the different MAbs; the effect of immobilization on the antigen–antibody interaction; and the effect of labeling on the antigen–antibody interaction.

Negative results in the inhibition assays do not necessarily mean that the antibodies in question are directed against unrelated epitopes. If the affinity of the labeled antibody (first assay version) or of the immobilized antibody (second assay version) is much higher than the affinity of the competing antibody, negative or inconclusive results may be obtained even at high excess of the low-affinity antibody. For this reason, labeled and competing antibodies should be exchanged against each other in the first assay version. The exchange of competing and immobilized antibod-

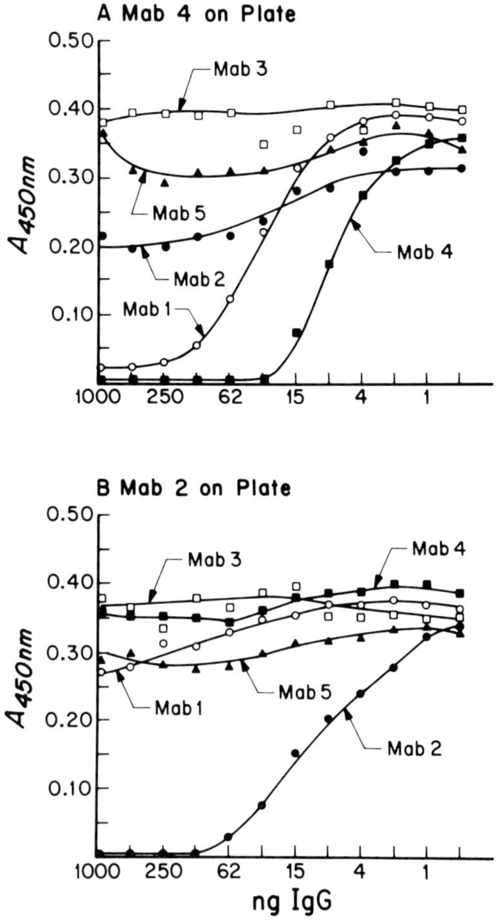

FIG. 2. Epitope analysis using biotinylated antigen. Microtiter wells coated with monoclonal antibody IgG were incubated with constant amounts of biotinylated CEA and decreasing dilutions of MAbs. The bound biotinylated CEA was detected with the avidin–peroxidase conjugate. (From J. E. Shively, C. Wagener, and B. R. Clark, this series, Vol. 121, p. 459.)

ies in assay version 2 does not demand additional labeling procedures. Occasionally, MAbs do not bind efficiently to the solid phase. In these cases, assay version 1 is more appropriate.

The relatively complex inhibition patterns obtained with labeled MAb 1 (Fig. 1A) and MAb 4 (Fig. 1D) or with immobilized MAb 4 (Fig. 2A), can be explained as follows. MAbs 1 and 2 are directed against repetitive epitopes on CEA, whereas MAbs 4, 5, and 6 are directed against singular epitopes.[13] (The epitope specificies of MAbs 5 and 6 are identical, see Fig. 1C.) The presence of more than one epitope on the protein moiety of CEA is probably due to the fact that CEA is built up of three highly homologous repeat units.[14] Though MAbs 1, 4, and 5 (6) exhibit totally different antigen specificities, one of the MAb 1 epitopes overlaps with the epitopes of MAbs 4 and 5 (6). For this reason, the binding of labeled MAb 1 to immobilized CEA is partially inhibited by MAbs 4 and 6 (Fig. 1A). In contrast, MAb 1 approaches 100% binding inhibition at high antibody excess when MAb 4 is used either as biotinylated MAb (Fig. 1D) or as immobilized MAb (Fig. 2A).

A possible objection to an immunoassay requiring the labeling of an antigen or antibody is that the labeling process may affect binding. In solid-phase assays, a further objection is that either an antigen or an antibody may undergo a conformational change affecting binding. By reversing the roles of antigen and antibody in the two tests described above, it is possible to probe the effect of both parameters on the formation of the antigen–antibody complex.

Determination of CEA and Related Antigens in Serum Using Biotinylated Monoclonal Antibodies in Additive Solid-Phase Enzyme Immunoassays

Materials

Coating buffer: 20 m*M* sodium acetate buffer (pH 5.5)
PBS: 50 m*M* sodium phosphate buffer (pH 7.2) containing 0.15 *M* NaCl and 0.01% (w/v) NaN_3
BSA–PBS: PBS containing 1% (w/v) BSA
Incubation buffer: 0.1 *M* potassium phosphate (pH 7.3) containing 20% (v/v) fetal calf serum (FCS), 0.02% (v/v) Tween, and 4% (v/v) normal mouse serum
APC: avidin–peroxidase conjugate (Sigma)
APC buffer: 0.2 *M* potassium phosphate buffer (pH 6.5) containing 20% biotin-free fetal calf serum
Washing buffers: 0.1 *M* potassium phosphate buffer (pH 6.5) and 0.1 *M* sodium citrate buffer (pH 5.0)
Substrate solution: 6 m*M* H_2O_2 (freshly prepared) and 40 m*M* *o*-phenylenediamine (Sigma) in 0.1 *M* citrate buffer (pH 5.0)
DMF and BNHS: see above

[13] B. M. Giannetti, M. Neumaier, and C. Wagener, *Fresenius Z. Anal. Chem.* **324,** 253 (1986).

[14] S. Oikawa, H. Nakazato, and G. Kosaki, *Biochem. Biophys. Res. Commun.* **142,** 511 (1987).

Preparation of Biotin-Labeled MAbs. Monoclonal antibody IgG is biotin-labeled as described above.[11] As antibodies, MAbs 4 and 6 are used. Molar ratios of BNHS to IgG in the range of 10 to 600 were tested in order to determine which ratio yields the highest sensitivity in the solid-phase EIA. Using MAb 4, a ratio of 60–70 was found to be optimal.

Preparation of $F(ab)_2$ Fragments. The IgG_1 fraction of MAb 2 is purified from ascitic fluid over protein A–Sepharose.[8] IgG_1 (6–12 mg/ml) is digested with pepsin (3%, w/v) in 0.1 *M* sodium acetate buffer (pH 4.1) for 18 hr at room temperature. The solution is then brought to pH 8.1 by the addition of 0.1 *M* Tris-HCl (pH 9.0). The reaction mixture is passed over a protein A–Sepharose column with a 0.1 *M* sodium phosphate solution (pH 8.1) as column and elution buffer. The nonbound fraction is characterized by gel-permeation HPLC (TSK 3000, LKB, Gräfelfing, FRG) and SDS–PAGE under reducing and nonreducing conditions.

Protocol of Additive Solid-Phase EIAs. The quantification of CEA in serum samples is performed by double monoclonal, sandwich-type immunoassays. MAb 2, which is a broadly cross-reactive, high-affinity antibody,[5–7] is immobilized on a solid support. As second antibodies, either biotinylated MAb 4 or MAb 6 is used. Each of the three antibodies binds to distinct epitopes on CEA. MAb 4 binds CEA with high affinity and specificity.[3–7] MAb 6, which has a moderate affinity for CEA, binds a 128-kDa tumor-associated CEA variant in addition to CEA.[5] Neither MAb 4 nor MAb 6 binds to nonspecific cross-reacting antigens present in normal tissues and body fluids. Solid-phase antibody and solution-phase antibodies are incubated in a single incubation step. The incubation time depends on the affinity of the respective solution-phase antibody. In a second incubation step, the biotin-labeled antibodies bound to the antigen(s) are reacted with avidin–peroxidase conjugate, followed by the addition of a suitable substrate.

Polystyrene beads with a diameter of 1/4 inch (Spherotech, Fulda, FRG) are washed several times in absolute ethanol and then sonicated for 60 min in PBS. The PBS is exchanged after 30 min. Subsequently, the beads are washed with PBS. The beads are coated with the IgG_1 fraction or $F(ab)_2$ fragments of MAb 2. Per 100 beads, 30 ml of coating buffer containing 10 μg/ml of either antibody (IgG) or $F(ab)_2$ fragments, respectively, are added. Coating is performed for 30 min at room temperature with shaking. Blocking of nonspecific binding sites and storage of the beads are performed in a solution of 2% BSA in PBS.

The CEA standard is dissolved in BSA–PBS. The biotinylated antibodies are diluted with incubation buffer to 2.66 μg/ml in the case of MAb 4 and 5.7 μg/ml in the case of MAb 6. The upper concentration limit of biotinylated antibody depends on the nonspecific binding to the beads which, in the present assays, does not exceed A_{492} values of 0.05.

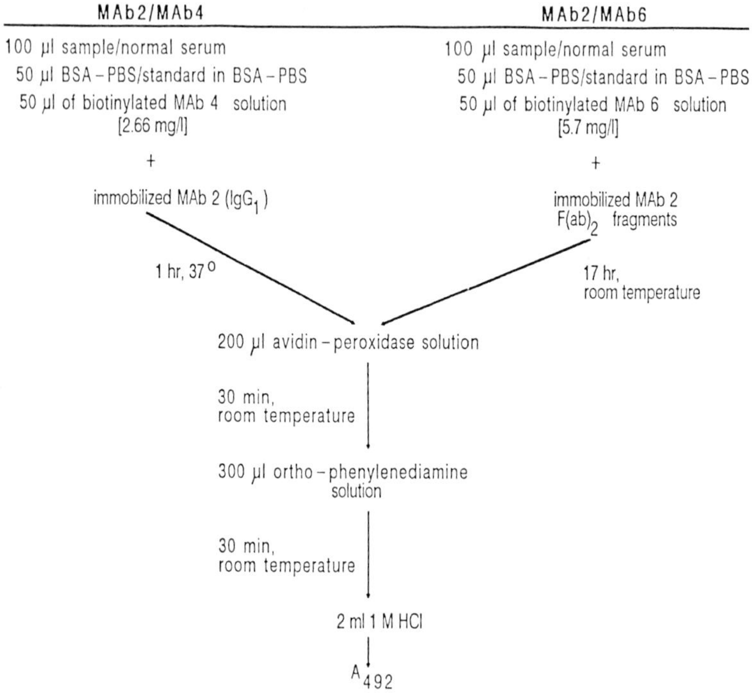

FIG. 3. Protocols of two additive, solid-phase enzyme immunoassays for CEA based on the avidin–biotin system.

The assay protocols are given in Fig. 3. The assays are performed in special reaction trays (Abbott, Wiesboden, Federal Republic of Germany). For the analysis of serum samples, 100 μl of serum is mixed with 50 μl of the solution of biotinylated MAb and 50 μl of BSA–PBS. For the determination of CEA standards, 100 μl of normal serum (checked for low CEA content) is mixed with CEA standard solution in BSA–PBS and 50 μl of antibody solution. After addition of the antibody-coated beads, the MAb 2/MAb 4 immunoassay is incubated for 1 hr at 37°, and the MAb 2/MAb 6 immunoassay for 17 hr at room temperature, respectively. Subsequently, the beads are washed 4 times with 0.1 *M* potassium phosphate buffer (pH 6.5). The concentration of avidin–peroxidase conjugate is chosen as the maximum concentration which yields negligible nonspecific binding. Compared with 1–5% BSA or 1% gelatin, 20% FCS most efficiently reduces background staining. However, FCS may contain free biotin. Therefore, new batches of FCS are checked for biotin by adsorption to avidin–Sepharose. The serum is tested in the CEA immunoassay prior to and after absorption. If interfering amounts of biotin are present, the FCS has to be adsorbed with avidin–Sepharose prior to its use in the immunoassay. Maximum binding of the avidin–peroxidase conjugate is reached after an incubation time of 20 min at room temperature. An incubation time of up to 80 min does not increase the sensitivity of the assay. In the present assay, an incubation time of 30 min is chosen. The beads are then washed 3 times in 0.1 *M* citrate buffer (pH 5.0). Subsequently, the substrate solution is incubated with the beads in the dark for 30 min at room temperature. The reaction is stopped with 2 ml of 1 *M* HCl, and the OD is read at 492 nm within 30 min.

Discussion of Assay Results. The assay protocols of the two immunoassays differ in two major aspects (Fig. 3). Owing to the lower affinity of MAb 6, the incubation time of the MAb 2/MAb 6 assay had to be longer than that of the MAb 2/MAb 4 assay in order to reach comparable sensitivities. The increase of incubation time, however, has an undesirable side effect: nonspecific anti-mouse IgG-binding substances that are occasion-

TABLE I
ASSAY CHARACTERISTICS OF TWO DOUBLE-MONOCLONAL, AVIDIN–BIOTIN-BASED ENZYME IMMUNOASSAYS FOR CEA

	Immunoassays	
Assay parameter	MAb 2/MAb 4	MAb 2/MAb 6
Sensitivity[a] (μg CEA/liter)	0.55	0.63
Measuring range (μg CEA/liter)	0.55–64.0	0.63–64.0
Recovery		
CEA added (μg/liter)	4.8	4.8
CEA found (μg/liter)	5.7	4.5
Recovery (%)	119	94
CEA added (μg/liter)	19.6	19.6
CEA found (μg/liter)	19.3	16.6
Recovery (%)	99	84
Between-run precision		
Mean CEA (μg/liter)	4.8	5.2
SD	0.55	0.80
CV (%)	11.5	15.3
n	8	15
Mean CEA (μg/liter)	17.2	18.5
SD	1.50	2.50
CV (%)	8.7	13.4
n	8	15
Upper limit, normal range 95% percentile (μg CEA/liter)	4.2	6.5

[a] CEA concentration corresponding to the zero standard +3 SD.

ally present in human sera may lead to false-positive signals when intact IgG is used.[15] Generally, such interference can be blocked by the addition of 1% mouse serum when high-affinity antibodies and, as a consequence, short incubation times are used; with prolonged incubation times, however, the addition of mouse serum may be insufficient. For this reason, $F(ab)_2$ fragments of MAb 2 are used in the MAb 2/MAb 6 immunoassay.

The characteristics of the two immunoassays are given in Table I. Considering 4.2–6.5 μg/liter as the upper limit of normal range, the sensitivity of both assays is sufficiently high for the measurement of CEA in serum samples. The assay characteristics of the MAb 2/MAb 4 EIA are quite comparable to those of immunoassays not based on the avidin–biotin system. The assay correlates well with commercial CEA immunoassays. The lower precision of the MAb 2/MAb 6 immunoassay is probably due to the lower affinity of MAb 6.

Acknowledgment

This research was supported by grants from the Deutsche Forschungsgemeinschaft, Wa-473/4-1, and from the National Large Bowel Program, National Cancer Institute, Grant CA 37808. We wish to thank Birgit Esser, Frances Crawford, and Karen Rickard for expert technical assistance, and Dr. Y. H. Joy Yang for production of monoclonal antibodies.

[15] C. Wagener, *Ann. Clin. Biochem.* **24,** Suppl. 2, 208 (1987).

[60] Selective Precipitation of Biotin-Labeled Antigens or Monoclonal Antibodies by Avidin for Determining Epitope Specificities and Affinities in Solution-Phase Assays

By CHRISTOPH WAGENER, ULRICH KRÜGER, and JOHN E. SHIVELY

Solid-phase assays in which either antigen or antibody are adsorbed to a solid support and the second reactant is present in solution may impose several problems in the analysis of the antigen–antibody interaction. The degree of adsorption to the solid matrix depends on the properties of the individual proteins.[1] In addition to the potential denaturation of proteins,[2]

[1] J. D. Andrade, *in* "Surface and Interfacial Aspects of Biomedical Polymers, Protein Adsorption" (J. D. Andrade, ed.), Vol. 2, p. 1. Plenum, New York, 1985.
[2] M. E. Soderquist and A. G. Walton, *J. Colloid Interface Sci.* **75,** 386 (1980).

adsorption to solid supports may modify or obscure epitopes of native antigens[3] or may create neoepitopes not present on the native antigen.[4,5] As the solid-phase support is effectively a ternary component of the reaction, the nonspecific interaction of antigens or antibodies with the solid phase may interfere both in binding and in inhibition experiments, especially when the affinity of the antigen–antibody interaction is moderate or low.[1]

The above problems do not apply to homogeneous assays in which both reactants are present in solution phase. However, a prerequisite for binding and inhibition studies in solution phase is the effective separation of bound from free antigen or antibody. Methods used for the precipitation of polyclonal antibodies may not be suitable for monoclonal antibodies. Precipitation by second antibodies is limited by the formation of soluble complexes between the monoclonal IgG and anti-mouse IgG antiserum at high monoclonal antibody excess. The precipitation of antigen–antibody complexes by ammonium sulfate or polyethylene glycol is complicated by high levels of nonspecific binding and incomplete precipitation of the complexes.[6] In order to perform binding and inhibition experiments in solution phase using a constant amount of antigen and variable amounts of antibody, a general method for the selective precipitation of antigen would be desirable.

Here we describe the use of the avidin–biotin system to selectively and quantitatively precipitate either biotin-labeled antigen or antibody from solution by the addition of avidin at optimal ratios of avidin to biotinylated reactant. Using this technique, inhibition studies and saturation experiments can be performed in solution phase at constant concentrations of biotinylated antibody or antigen.

General Procedures

Monoclonal antibodies (MAbs) against the carcinoembryonic antigen (CEA) are produced as described.[7] MAbs are elicited by immunizations either with purified CEA (MAbs 1, 3, and 5) or with the transplanted colon carcinoma cell line T84 (MAbs 2 and 4). CEA is purified according to

[3] F. J. Stevens, J. Jwo, W. Carperos, H. Köhler, and M. Schiffer, *J. Immunol.* **137,** 1937 (1986).

[4] A. D. Smith and J. E. Wilson, *J. Immunol. Methods* **94,** 31 (1986).

[5] H. C. Vaidya, D. N. Dietzler, and J. H. Ladenson, *Hybridoma* **4,** 271 (1985).

[6] C. Wagener, B. R. Clark, K. J. Rickard, and J. E. Shively, *J. Immunol.* **130,** 2302 (1983).

[7] C. Wagener, L. Wickert, and J. E. Shively, this volume [59]; C. Wagener, Y. H. J. Yang, F. G. Crawford, and J. E. Shively, *J. Immunol.* **130,** 2308 (1983).

Coligan *et al.*,[8] and the T84 cell line is that of Murakami and Masui.[9] IgG fractions of MAbs are purified from ascites fluid by protein A affinity chromatography according to Ey *et al.*[10] The IgG concentration is calculated from the absorbance of the solutions at 280 nm, assuming an extinction coefficient [1% (w/v); 1 cm] of 14.2.[11] CEA concentrations and concentrations of Fab fragments are determined by amino acid analysis.

Binding and Inhibition Studies with Biotin-Labeled Antibodies and Avidin as Precipitating Agent

Materials

PBS: 50 mM sodium phosphate buffer (pH 7.0) containing 0.15 M sodium chloride and 0.1% NaN_3

Avidin–PEG: avidin (Sigma Chemical Co., St. Louis, MO) 0.1 mg/ml in PBS containing 5% (w/v) polyethylene glycol (PEG 6000, Sigma)

BNHS: *N*-hydroxysuccinimidobiotin (Sigma)

DMF: dimethylformamide (J. T. Baker, Phillipsburg, NJ), distilled from ninhydrin

Preparation of ^{125}I-Labeled Antigen. Purified CEA is ^{125}I-labeled by the chloramine-T procedure of Hunter and Greenwood.[12]

Biotinylation Procedures. MAb is biotinylated as described elsewhere in this volume.[7] As a biotinylated carrier protein, either normal rabbit or goat serum can be used. After dialysis against PBS, carrier serum (2 ml) is treated with 2 ml of freshly made aqueous BNHS, prepared by dissolving 20 mg of BNHS in 0.5 ml of DMF and diluting to 2 ml with 1.5 ml of distilled water. The remainder of the procedure is as described for preparing the biotinylated MAbs.

Assay Overview. The protocol is similar to that described by Clark and Todd[13] and Wagener *et al.*[6] Determination of K_{aff} is performed in two steps. First, binding curves for each MAb are determined. A constant amount of radiolabeled antigen (0.5 ng of [^{125}I]CEA) is titered against serial 2-fold dilutions of MAb. The concentration of MAb IgG yielding half-maximal binding is determined. This concentration is used in the second step. The second step comprises a competitive RIA involving the

[8] J. E. Coligan, J. T. Lautenschleger, M. L. Egan, and C. W. Todd, *Immunochemistry* **9,** 377 (1972).

[9] H. Murakami and H. Masui, *Proc. Natl. Acad. Sci. U.S.A.* **77,** 3464 (1980).

[10] P. L. Ey, S. J. Prowse, and C. R. Jenkin, *Immunochemistry* **15,** 429 (1978).

[11] H. N. Eisen, E. S. Simms, and M. Potter, *Biochemistry* **7,** 4126 (1968).

[12] W. M. Hunter and F. C. Greenwood, *Nature (London)* **194,** 495 (1962).

[13] B. R. Clark and C. W. Todd, *Anal. Biochem.* **121,** 257 (1982).

addition of increasing amounts of unlabeled antigen (0.5–100 ng of CEA) which competes with a fixed amount of radiolabeled antigen (0.5 ng of [^{125}I]CEA).

The present RIA version required the separation of free from bound radiolabeled antigen. This is accomplished by precipitation of the biotinylated MAb with avidin. Since only a minute amount of MAb is present, the precipitation is driven to completion by adding a biotinylated carrier protein mixture.

Binding RIA Protocol with Biotinylated Antibodies. The RIA is performed in 400–μl polypropylene tubes (Bio–Rad, Richmond, CA) which can be tightly capped and centrifuged in a Beckman microcentrifuge. Serial 2-fold dilutions of biotinylated MAb (50 μl; 0.01–1,000 ng) in biotinylated carrier serum (1 : 800 dilution in either underivatized normal rabbit or goat serum diluted 1 : 40 in PBS) are incubated with 200 μl of PBS containing 6.5% PEG and 72 mM EDTA and with 10 μl of [^{125}I]CEA (0.5 ng in PBS; ~10^4 cpm) for 18 hr at 37°. The [^{125}I]CEA cocktail may include ^{57}Co as a volume marker if desired. Biotinylated MAb and biotinylated MAb–antigen complexes are coprecipitated with biotinylated carrier proteins by addition of 10 μl of avidin (1.0 mg/ml in PBS containing 5% PEG 6000), and the mixture is incubated for 1 hr at room temperature. The tubes are centrifuged, the supernatant fluids are removed, and the precipitates are counted.

The binding of constant amounts of [^{125}I]CEA by decreasing amounts of monoclonal antibody IgG is shown in Fig. 1. At high IgG excess, MAb 3 binds over 90% of labeled antigen, thus indicating quantitative precipitation of the antibody. The broad plateau regions of the binding curves for MAbs 2 and 4 show that nonprecipitable soluble complexes between biotin-labeled monoclonal antibody and avidin were not observed when excess IgG concentrations were used. Nonspecific precipitation of labeled CEA as determined by the use of biotin-labeled carrier without specific antibody was in the range of 1–3%.

CEA Inhibition Assay Protocol. The assay tubes, centrifugation protocol, and reagents are the same as in the previous section. To each assay tube (in duplicate) are added 10 μl of ^{125}I-labeled CEA (0.5 ng in PBS) and 50 μl of biotinylated MAb at concentrations indicated in the legend to Fig. 2. The dilution of MAbs is performed in a 1 : 800 dilution of biotinylated goat serum in either underivatized normal goat or rabbit serum (1 : 40 in PBS) and 200 μl of a CEA inhibitor solution (0.5–100 ng in PBS with 6.5% PEG). The mixture is incubated, precipitated with 10 μl of avidin (1.0 mg/ml in PBS with 5% PEG), and counted as before.

For inhibition experiments, IgG concentrations corresponding to 40–60% of the maximum binding of [^{125}I]CEA (see plateau regions in Fig. 1)

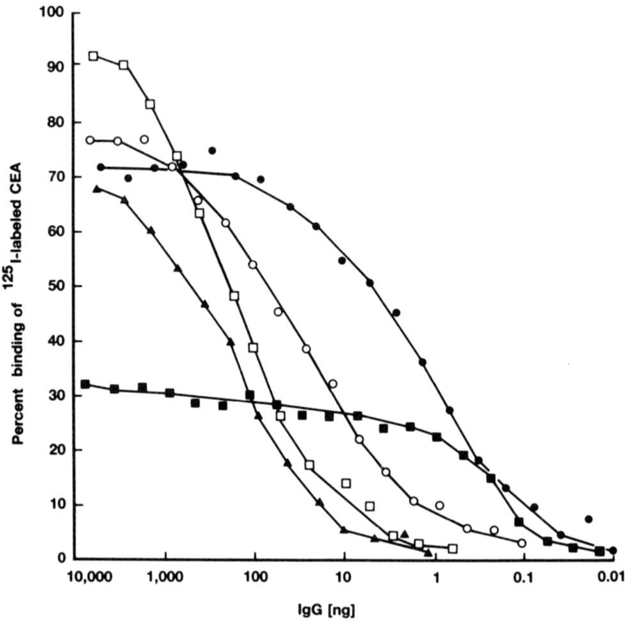

FIG. 1. Binding curves for radiolabeled antigen to MAb in a solution-phase, avidin–biotin-based RIA. Radiolabeled CEA was titered with decreasing dilutions of biotinylated MAbs. The biotinylated MAbs and MAb–CEA complexes were precipitated with avidin in the presence of 5% PEG. ○, MAb 1; ●, MAb 2; □, MAb 3; ■, MAb 4; ▲, MAb 5. [From C. Wagener, B. R. Clark, K. J. Rickard, and J. E. Shively, *J. Immunol.* **130,** 2308 (1983), with permission.]

are chosen for MAbs 1, 2, 3, and 5. Because of the low binding plateau of MAb 4, inhibition experiments are performed (in this case) at a higher relative binding of [^{125}I]CEA. The inhibition curves for MAbs 1–5 are shown in Fig. 2.

Calculation of Affinity Constants. Affinity constants are calculated from inhibition experiments by using a general form of the equation derived by Müller.[14] The original equation refers to 50% inhibition of tracer binding, whereas the general form of the equation can be applied to a broader range of inhibitor concentrations. Thus,

$$K_{\text{aff}} = \frac{b(1 - r) - r(1 - b)}{(1 - r)(1 - b)[r(T_t + I_t) - bT_t]}$$

where T_t is the total concentration of tracer, I_t the total concentration of

[14] R. Müller, *J. Immunol. Methods* **34,** 345 (1980).

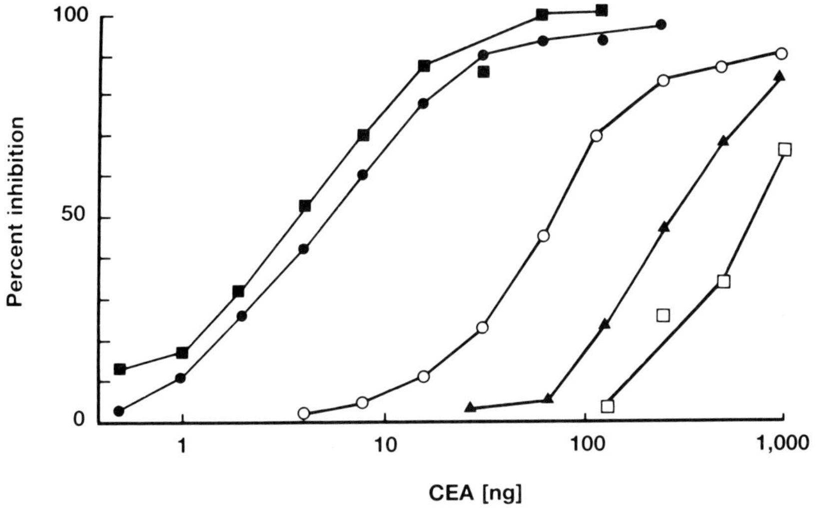

FIG. 2. Inhibition curves for calculating affinity constants of five MAbs. Constant amounts of radiolabeled CEA and biotinylated MAb were incubated with increasing amounts of unlabeled CEA. The biotinylated MAb and MAb–CEA complexes were precipitated with avidin in the presence of 5% PEG. The IgG concentrations of MAbs used were as follows: ○, MAb 1, 26.5 ng IgG per tube; ●, MAb 2, 1.4 ng IgG per tube; □, MAb 3, 96.9 ng IgG per tube; ■, MAb 4, 0.9 ng IgG per tube; ▲, MAb 5, 171.6 ng IgG per tube. [From C. Wagener, B. R. Clark, K. J. Rickard, and J. E. Shively, *J. Immunol.* **130,** 2308 (1983), with permission.]

inhibitor, b the fraction of tracer bound in the absence of inhibitor, and r the fraction of tracer bound in the presence of inhibitor.

A model calculation is given for MAb 2. Inhibition experiments were performed with 0.5 ng of [^{125}I]CEA per tube (260 μl). In the absence of inhibitor, 31.1% of labeled CEA was bound ($b = 0.311$). In the presence of 1.94 ng of unlabeled CEA, 22.9% of labeled CEA was precipitated ($r = 0.229$). This corresponds to a percent binding inhibition of $(1 - r/b) \times 100 = 26.4\%$ (see Fig. 2). Assuming an M_r of 180,000 for CEA, the concentration of tracer is 10.7×10^{-12} M, and the concentration of inhibitor is 41.5×10^{-12} M. Using the above formula, $K_{aff} = 1.8 \times 10^{10}$ M^{-1}. In the presence of 7.7 ng of unlabeled CEA (164.5×10^{-12} M), 12.1% of the tracer was precipitated ($r = 0.121$), corresponding to a percent inhibition of 61.1% (see Fig. 2). From these values, $K_{aff} = 1.8 \times 10^{10}$ M^{-1} is calculated.

As shown in Fig. 1, the fraction of radiolabeled CEA bound at high IgG excess is different for each of the MAbs. Since over 90% of unlabeled CEA is bound by the different MAbs,[6] the decrease of the immunoreactiv-

TABLE I
AFFINITY CONSTANTS OF MONOCLONAL ANTI-CEA ANTIBODIES[a,b]

MAb	Affinity constants (M^{-1})
1	3.7×10^{9}
2	1.8×10^{10}
3	1.0×10^{8}
4	2.6×10^{10}
5	3.8×10^{8}

[a] For calculation, see text.
[b] Data from C. Wagener, B. R. Clark, K. J. Rickard, and J. E. Shively, *J. Immunol.* **130,** 2302 (1983).

ity of radiolabeled CEA is most probably due to partial radioiodination damage of individual epitopes. If only 70% immunoreactivity of labeled CEA is assumed for MAb 2 (see plateau region in Fig. 1), the calculated affinity constants increase to 2.2×10^{10} to 2.5×10^{10} M^{-1}, depending on the inhibitor concentration used for calculation. For MAb 4, the affinity constants obtained for 30% immunoreactivity of tracer are in the range of 4.9×10^{10} to 5.7×10^{10} M^{-1}.

The affinity constants calculated for 5 MAbs from inhibition experiments without corrections for immunoreactive tracer are shown in Table I. The K_{aff} values for MAbs 1, 2, and 4 are sufficiently high to warrant their use in an immunoassay for CEA. The final choice of the optimum MAb is also dependent, however, on the required epitope specificity. A discussion of this topic and the problem of cross-reacting antigens is given by Wagener *et al.*[6]

Determination of Epitope Specificities and K_{aff} Using Biotinylated CEA and Avidin as Precipitating Agent

Reagents

PBS: see above

Avidin–PEG: avidin (Sigma) 5 mg/ml in PBS containing 6.5% (w/v) PEG 6,000 (Sigma)

BNHS and DMF: see above

Preparation of ^{125}*I-Labeled Fab fragments*. Fab fragments are prepared from MAbs 2 and 4 according to the procedure described by

Parham.[15] Molecular weight determinations and purity control are performed by SDS–PAGE using the Phast System (Pharmacia) with preformed gradient gels of 8–25% acrylamide. Under nonreducing conditions, molecular weights of 48K are obtained for the Fab fragments of both antibodies. The procedure is not generally applicable to all monoclonal antibodies as MAb 1 is completely digested by pepsin treatment.

For iodination, the concentration of Fab fragments is determined by amino acid analysis. The Fab fragments are labeled with ^{125}I by the chloramine-T method according to Bolton and Hunter.[16] The protein iodination is stopped by the addition of excess free tyrosine. The specific activity is approximately 2 MBq/μg for both Fab fragments.

Preparation of Biotinylated CEA and Carrier Protein. An equal volume of CEA (1 mg/ml in PBS, 0.1 ml) is added to a freshly prepared solution of BNHS (3.5 mg/ml, 0.1 ml). The preparation of the BNHS solution and the biotinylation procedure are performed as described for the monoclonal antibodies.[7] Biotin labeling of carrier protein (normal rabbit serum) is performed as described above.

RIA Overview. The RIA procedure employed here represents an inverse version of the RIA described above. Prior to epitope analysis and determination of K_{aff}, titration curves are established. Constant amounts of radiolabeled Fab fragments are titered against serial 2-fold dilutions of biotinylated CEA. From the plateau region, the fraction of immunologically active radiolabeled Fab fragment is determined. For inhibition experiments, the CEA concentration yielding 15–27% binding of labeled Fab is determined. This concentration is used for competitive inhibition studies involving the addition of increasing amounts of unlabeled monoclonal antibody IgG or Fab fragments respectively, which compete for the binding of a fixed amount of labeled Fab fragment at constant concentrations of biotinylated CEA. In case of the present RIA version, separation of bound and free antibody is required. This is accomplished by precipitation of the biotinylated antigen with avidin in the presence of a biotinylated carrier protein. The method allows the determination of epitope specificities and affinity constants in a single experiment.

Binding Assay Protocol. To 200 μl of 6.5% PEG (w/v) in PBS containing biotinylated CEA at 2-fold serial dilutions, an aliquot (50 μl) of PBS containing labeled Fab fragments ($\sim 10^4$ cpm, corresponding to $\sim$0.1 ng),

[15] P. Parham, *in* "Handbook of Experimental Immunology, Immunochemistry" (D. M. Weir, L. A. Herzenberg, C. Blackwell, and L. A. Herzenberg, eds.), Vol. 1, p. 14.1. Blackwell, Oxford, 1986.

[16] A. E. Bolton and W. M. Hunter, *in* "Handbook of Experimental Immunology, Immunochemistry" (D. M. Weir, L. A. Herzenberg, C. Blackwell, and L. A. Herzenberg, eds.), Vol. 1, p. 26.1. Blackwell, Oxford, 1986.

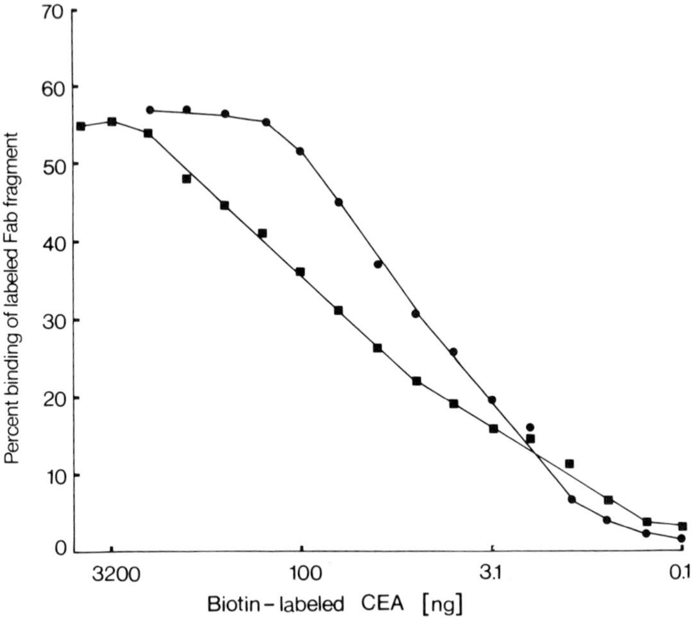

FIG. 3. Binding curve of radiolabeled Fab fragments of MAbs to antigen in a solution-phase, avidin–biotin-based RIA. Radiolabeled monoclonal Fab fragments were titered with increasing dilutions of biotinylated CEA. The biotinylated CEA and CEA complexes were precipitated with avidin in the presence of 5% PEG. ●, MAb 2; ■, MAb 4. [From U. Krüger, L. Wickert, and C. Wagener, *J. Immunol. Methods* **117,** 25 (1989), with permission.]

normal rabbit serum (1 : 40 dilution in PBS), and biotin-labeled normal goat serum (1 : 100 dilution in PBS) is added. The carrier serum is prepared as described in the previous section. The assay mixture is incubated for 18 hr at 37°. Subsequently, 10 μl of 6.5% PEG (w/v) in PBS containing 50 μg egg-white avidin (Sigma) is added. After 1 hr at room temperature, the mixture is centrifuged (10 min, 15,000 *g*). A sample (200 μl) of the supernatant is removed. The remaining volume is counted in a γ scintillation counter. The percentage of radioactivity precipitated is calculated from the total counts and the volume removed, thus obviating the addition of a volume marker.

Two-fold serial dilutions of biotinylated CEA are performed in the range of 6400–0.012 ng per tube (Fig. 3). For the MAb 2 Fab fragment a plateau was reached with 400–1600 ng of biotinylated CEA and with 3200–6400 ng CEA for the MAb 4 Fab fragment. Depending on individual Fab preparations and radioiodination, 40–60% of the radiolabeled Fab fragment is bound in the plateau region. The finding that 100% binding is not reached at high antigen excess may be due to the following: (1) the presence of nonimmune IgG in the ascitic fluid, (2) partial degradation during the preparation of Fab fragments, and (3) damage by the radioiodination procedure. The possibility that soluble complexes between biotinylated CEA and avidin may be responsible for the binding of less than 100% of the labeled Fab fragments has been excluded for the concentration range of biotinylated CEA used for the binding studies. Less than 10% of total biotinylated CEA is present in the supernatant when levels of 5,000 ng per tube are used. At higher amounts of antigen, soluble complexes between avidin and CEA–biotin may develop. The nonspecific binding is less than 3%.

Inhibition Assay Protocol. For inhibition experiments, concentrations of CEA–biotin of 5–25 ng per tube are chosen. Under these conditions, 15–27% of the total radioactivity is bound. To 200 μl of 6.5% PEG (w/v) in PBS containing the appropriate amount of biotinylated CEA and 2-fold serial dilutions of unlabeled monoclonal antibody IgG or Fab fragments, 50 μl of PBS containing labeled Fab fragments ($\sim 10^4$ cpm), normal rabbit serum (1 : 40 dilution), and biotin-labeled normal goat serum (1 : 100 dilution) is added. The rest of the procedure is identical to that described above for the binding assay protocol.

A representative inhibition curve is shown in Fig. 4. Using labeled Fab fragments of MAb 2, increasing concentrations of unlabeled MAb 2 Fab fragments result in complete binding inhibition. The remaining MAbs do not inhibit binding of the labeled Fab fragments. These results are comparable with those obtained by competitive solid-phase assays.[7,17]

Calculation of Affinity Constants. In addition to epitope specificities, K_{aff} can be calculated by using the general form of the equation derived by Müller[14] given above. In the case of inhibition experiments performed with constant amounts of biotin-labeled antigen and ^{125}I-labeled Fab fragments in the presence of increasing amounts of unlabeled Fab fragments, I_t is the concentration of unlabeled Fab fragments, T_t the concentration of labeled Fab fragments, b the fraction of tracer bound in the absence of inhibitor, and r the fraction of tracer bound in the presence of inhibitor. In the inhibition experiment presented in Fig. 4, $T_t = 0.17$ ng per tube. In the absence of inhibitor, 26.1% of the radioactivity is precipitated. In the presence of 8.85 ng per tube of unlabeled Fab fragments, 14.8% of the tracer is precipitated, corresponding to a percent binding of 56.7% (see Fig. 4). Given a reaction volume of 260 μl and an M_r for Fab fragments of 48,000, $T_t = 13.8 \times 10^{-12}\ M$, $I_t = 709 \times 10^{-12}\ M$, $b = 0.261$, and $r = 0.148$. From these values, $K_{aff} = 1.7 \times 10^9\ M^{-1}$.

[17] C. Wagener, U. Fenger, B. R. Clark, and J. E. Shively, *J. Immunol. Methods* **68,** 269 (1984).

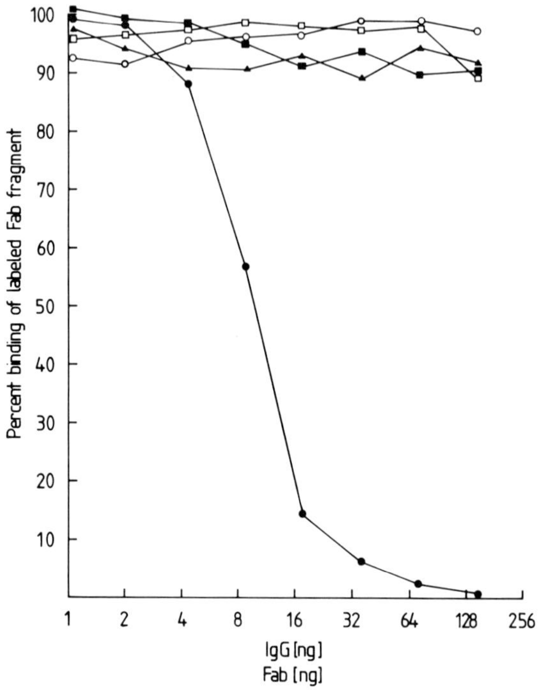

FIG. 4. Inhibition curves for determination of epitope specificities and affinity constants of monoclonal antibodies. Constant amounts of radiolabeled Fab fragments of MAb 2 and biotinylated CEA were incubated with increasing amounts of unlabeled intact MAbs or Fab fragments, respectively. Biotinylated CEA and CEA–antibody complexes were precipitated with avidin in the presence of 5% PEG. The total amount of radiolabeled Fab fragments was 0.1 ng per tube, and the concentration of biotinylated CEA was 6.25 ng per tube. ●, Fab fragment of MAb 2; the rest of the symbols are as in Fig. 1. [From U. Krüger, L. Wickert, and C. Wagener, *J. Immunol. Methods* **117,** 25 (1989), with permission.]

As shown in Fig. 3, only 58% of labeled Fab fragments from MAb 2 is precipitated at high excess of biotin-labeled CEA. It therefore follows that only part of the labeled Fab is immunoreactive. As discussed above, the reason for partial immunoreactivity may be the presence of nonimmune IgG in the IgG preparation from ascitic fluid, partial damage during the preparation of Fab fragments, or damage during radioiodination. Since the immunoreactive fraction of the unlabeled Fab fragments is unknown, K_{aff} can be estimated assuming partial immunoreactivity either of labeled Fab fragments alone or of both labeled and unlabeled Fab fragments. Assuming 58% immunoreactivity of the labeled fragment and 100% immunoreactivity of the unlabeled fragment, $T_t = 8.02 \times 10^{-12}\,M$, $I_t = 709 \times 10^{-12}\,M$, $b = 0.261/0.58 = 0.45$, and $r = 0.148/0.58 = 0.255$. From these values, $K_{aff} = 2.7 \times 10^9\ M^{-1}$. If 58% immunoreactivity of both tracer and inhibitor is assumed, then $I_t = 411 \times 10^{-12}\ M$, and $K_{aff} = 4.6 \times 10^9\ M^{-1}$.

The calculated affinity constants are lower than that shown for the intact antibody in Table I. This finding may be due to a variety of reasons, including the use of Fab fragments instead of intact antibodies, the presence of nonimmune IgG in the ascitic fluid, and/or the corrections made for the immunoreactive fraction of the radiolabeled Fab fragments or antigen.

Acknowledgments

This research was supported by grants from the Deutsche Forschungsgemeinschaft, Wa-473/4-1, and from the National Large Bowel Program, National Cancer Institute, Grant CA 37808. We wish to thank Karen Rickard for expert technical assistance, and Dr. Y. H. Joy Yang for production of monoclonal antibodies.

[61] Immunoassays for Diagnosis of Infectious Diseases

By ROBERT H. YOLKEN

Introduction

Traditionally, the diagnoses of infectious diseases have been based on the isolation of the infecting microorganisms in pure culture. Recently, however, infectious disease diagnoses have relied more heavily on the direct identification of infecting organisms in blood and other body fluids of the ill individual. The diagnosis of infectious diseases by the direct detection of microorganisms has been particularly important for the identification of viral agents since these agents are particularly difficult to detect in short periods of time by means of standard cultivation methods.[1]

Direct detection methods are also highly useful in the detection of bacterial, parasitic, and fungal pathogens that are fastidious or difficult to cultivate under standard laboratory conditions.[2] Most of these assays have relied on the measurement of the binding of antigens to defined antimicrobial antibodies. Immunoassays have a number of advantages for infectious disease diagnosis. These advantages are based on the sensitiv-

[1] D. A. Fuccillo, I. C. Shekarchi, and J. L. Sever, *in* "Manual of Clinical Laboratory Immunology" (N. R. Rose, H. Friedman, and J. L. Fahey, eds.), 3rd ed., p. 489. American Society for Microbiology, Washington, D. C., 1986.

[2] N. J. Schmidt, *Med. Clin. North Am.* **67,** 953 (1983).

ity and specificity inherent in antigen–antibody interactions. While there are a number of immunoassay systems that can be used for the direct detection of infectious antigens in human body fluids, solid-phase immunoassay techniques have attained widespread usage for this purpose.[3] The use of enzymatic methods offers the possibility of high degrees of sensitivity owing to the amplificatory nature of the enzyme–substrate reaction. Furthermore, the stability and safety of many enzyme-substrate systems allow for the application of enzyme immunoassays to a wide range of clinical and laboratory environments.[4-7]

Microorganisms are capable of infecting humans and causing disease in low concentrations. It is thus necessary that efficient assay systems for the diagnosis of infectious diseases attain a high rate of assay sensitivity. While there are a number of factors that determine the sensitivity of solid-phase immunoassay systems, the most critical are related to the efficiency of the binding of labeled immunoreagents to microbial antigens.

Initially, enzyme immunoassays for the detection of infecting antigens have been performed in "direct" formats in which the antimicrobial immunoglobulin is coupled to an enzyme.[8] One problem in this type of enzyme immunoassay system is that chemical linkage can result in a variable loss of the antigen-binding capacity of the antibody. In addition, the large mass of the enzyme–immunoglobulin complexes can result in slower rates of diffusion and thus in less favorable reaction kinetics. Furthermore, the variability inherent in the interactions of macromolecular species in solutions makes it difficult to control the rate and extent of immunoglobulin–enzyme coupling reactions. These problems can limit the development of immunoassay systems with consistent performance characteristics.

There are a number of possible solutions to the problems inherent in the direct linkage of enzymes to immunoglobulins. For example, immunoassays can be performed in indirect formats in which an unlabeled primary antimicrobial immunoglobulin is reacted with immobilized antigens and the resultant complex is quantified by an enzyme–conjugated second antibody.[9] A single enzyme conjugate can thus be utilized to de-

[3] R. H. Yolken, *Rev. Infect. Dis.* **4,** 36 (1982).

[4] S. Avrameas, T. Ternynck, and J. L. Guesdon, *Scand. J. Immunol.* **8,** (Suppl. 7), 7

[5] G. B. Wisdom, *Clin. Chem.* **22,** 1243 (1976).

[6] A. Voller, A. Bartlett, and D. E. Bidwell, *Trans. R. Soc. Trop. Med. Hyg.* **70,** 98 (1976).

[7] R. H. Yolken, *Yale J. Biol.* **59** (1), 25 (1986).

[8] R. H. Yolken, H. W. Kim, T. Clem, R. G. Wyatt, A. R. Kalica, R. M. Chanock, and A. Z. Kapikian, *Lancet* **2,** 263 (1977).

[9] R. H. Yolken, R. Viscidi, F. Leister, C. Harris, and S-B. Wee, *in* "Manual of Clinical Laboratory Immunology" (N. R. Rose, H. Friedman, and J. L. Fahey, eds.), 3rd ed. p. 573. American Society for Microbiology, Washington, D. C., 1986.

tect a number of target antigens. The use of this format avoids the need to generate enzyme-labeled immunoglobulins for each target antigen. However, this format still relies on the direct conjugation of enzyme and immunoglobulin molecules and is thus subject to the above-mentioned disadvantages. More importantly, the species specificity of many antiimmunoglobulin conjugates is not absolute. Nonspecific binding can therefore generate high background levels or, in certain cases, false-positive reactions.[10] The same problem is inherent in the use of other immunoglobulin-binding macromolecules such as staphylococcal protein A.[11]

Another approach to this problem is to link the antimicrobial immunoglobulin to a low molecular weight molecule which can then be detected specifically with an enzyme-labeled macromolecular counterpart. Such a system makes it less likely that the antigen-binding capacity of the immunoglobulin will be altered by macromolecular cross-linking. In addition, low molecular weight agents can be added in great excess to immunoglobulins, thus ensuring that coupling reactions will proceed to completion in relatively short periods of time. While a number of low molecular weight agents can be bound to immunoglobulin, biotin has been most widely used for the detection of microbial antigens in solid-phase formats.[12,13] The high affinity of avidin for biotin and biotin-substituted macromolecules ensures high degrees of assay sensitivity.

There are a number of methods that can be utilized for the detection of biotin-labeled immunoreagents. For example, avidin can be covalently coupled with enzyme and reacted to the biotinylated antibody on the solid-phase surface. Chemically modified avidin or bacterial streptavidin can be used to minimize nonspecific binding which sometimes discourages the use of the system.[14] Another limitation, the need for precise chemical coupling of avidin to enzyme, can also be minimized by forming complexes between unlabeled avidin and biotin-substituted avidin. This approach can be optimized to generate more enzyme-catalyzed signal than methods using enzyme-linked avidin.[15]

In terms of the labeling of immunoglobulins, we have found that standard biotinylation reagents such as biotin-*N*-hydroxysuccinimide ester can be used to label virtually all polyclonal antibodies over a wide range of ester and antibody concentrations (Fig. 1). We have also found that

[10] R. H. Yolken and P. J. Stopa, *J. Clin. Microbiol.* **10,** 703 (1979).

[11] R. H. Yolken and P. J. Stopa, *J. Clin. Microbiol.* **11,** 546 (1980).

[12] E. A. Bayer, M. Wilchek, and E. Skutelsky, *FEBS Lett.* **68,** 240 (1976).

[13] J. L. Guesdon, T. Ternynck, and S. Avrameas, *J. Histochem. Cytochem.* **27,** 1131 (1979).

[14] L. S. Nerurkar, N. R. Miller, M. Namba, M. Monzon, G. Brashears, G. Scherba, and J. L. Sever, *J. Clin. Microbiol.* **25**(1), 128 (1987).

[15] R. H. Yolken, F. J. Leister, L. S. Whitcomb, and M. Santosham, *J. Immunol. Methods* **56,** 319 (1983).

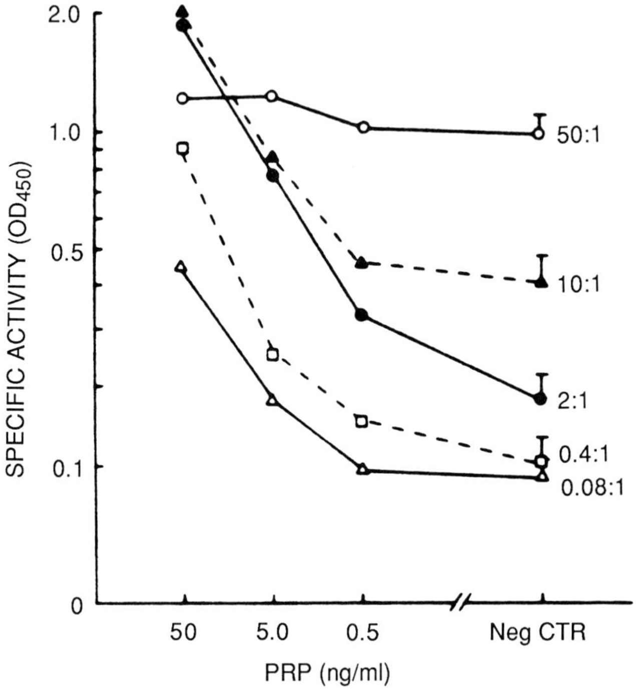

FIG. 1. Activity of biotin conjugates for the measurement of *Haemophilus influenzae* polyribitol phosphate (PRP). Numbers indicate the ratio (mg/mg) of biotin-*N*-hydroxysuccinimide ester to anti-PRP IgG utilized to formulate the conjugate. The specific activity was measured by means of a solid-phase assay performed in a manner similar to that described in the text. (From Ref. 15.)

monoclonal antibodies can be labeled with this method; however, some monoclonal antibodies appear to lose much of their antigen-binding capacity. This phenomenon is probably related to the linkage of biotin to an amino group located near the antigen-combining site of the immunoglobulin molecule. As an alternative to linkage of biotin to amino groups, biotin hydrazide can be coupled to carboxyl groups of immunoglobulin molecules by reaction with carbodiimide.[16] Polyclonal chicken antirotavirus antibodies labeled with biotin by this method performed as well in enzyme immunoassays for detection of rotavirus antigens in stool specimens as the same immunoglobulins labeled by the *N*-hydroxysuccinimide ester method. Other means of biotin linkage, such as reaction of biotin hydrazide with periodate-oxidized sugar residues, might also be utilized in

[16] D. J. O'Shannessy, M. J. Dobersen, and R. H. Quarles, *Immunol. Lett.* **8,** 273 (1984).

cases in which other linkages result in a diminution of antigen-binding activity.

We have devised a number of biotin-based assays for the detection of microbial antigens in the body fluids of infected humans.[15] We have found, for example, that assays utilizing biotinylated antibodies and a complex of avidin and biotinylated peroxidase can be used for the sensitive detection of cell wall polysaccharides of pathogenic bacteria such as *Haemophilus influenzae* and *Streptococcus pneumoniae* (Fig. 2). These assays were somewhat more sensitive than analogous assays that utilized immunoglobulins directly labeled with enzyme. We have also devised similar systems for the direct detection of viral antigens in body fluids. Furthermore, other enzymes can be used in addition to horseradish peroxidase to accomplish the enzyme–substrate indicator reaction. More

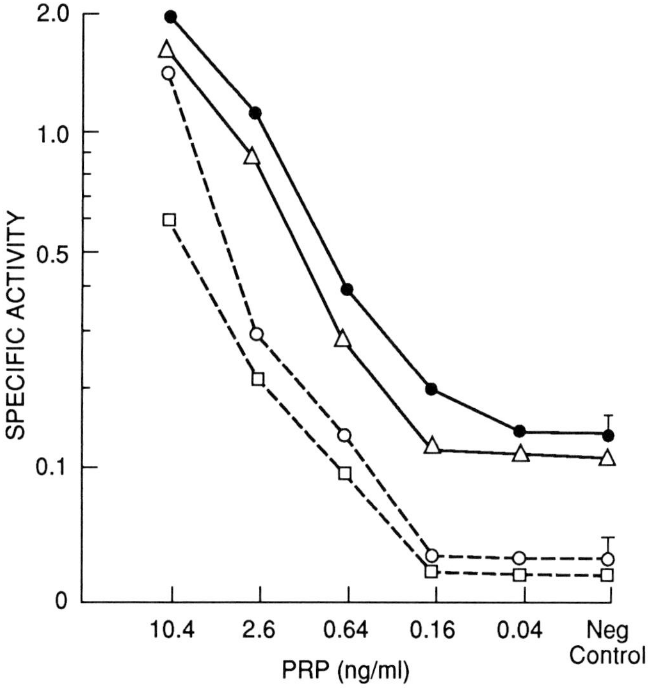

FIG. 2. Sensitivity of enzyme immunoassays for the measurement of *Haemophilus influenzae* PRP. The conjugate systems utilized were as follows: ●, biotin-labeled rabbit anti-PRP, enzyme-labeled biotin–avidin complex; △, biotin-labeled rabbit anti-PRP, enzyme-labeled avidin; □, unlabeled rabbit anti-PRP, fluorescein-labeled anti-IgG, enzyme-labeled antifluorescein; and ○, unlabeled rabbit PRP, enzyme-labeled anti-rabbit IgG. (From Ref. 15.)

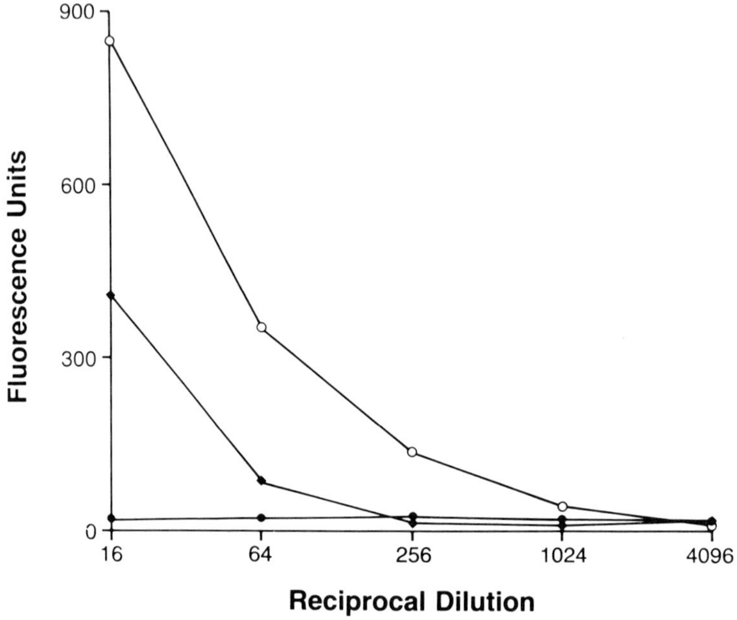

FIG. 3. Detection of HIV antigens in serial dilutions of cell culture fluids. Samples from a cell culture inoculated with an HIV isolate (○) and from a control uninoculated culture (●) were tested in an avidin–biotin-enhanced immunoassay for HIV antigens. Samples from the HIV cultures were also incubated with a monoclonal antibody to HIV-1 p24 protein and were then tested in the immunoassay (◆). The assays utilized streptavidin labeled with β-galactosidase and a fluorescent substrate for that enzyme. (From Ref. 18.)

recently, we have utilized biotinylated immunoglobulins for the detection of rotavirus antigenic variants[17] and human immunodeficiency viruses.[18] In both cases the biotin-based immunoassays proved to be at least as sensitive as analogous immunoassays utilizing immunoreagents labeled with other markers (Fig. 3).

One group of enzymes particularly useful for the diagnosis of infectious diseases are the bacterial β-lactamases.[19,20] In addition to possessing

[17] S. L. Vonderfecht, R. L. Miskuff, J. J. Eiden, and R. H. Yolken, *J. Clin. Microbiol.* **22,** 726 (1985).

[18] R. Viscidi, H. Farzadegan, F. Leister, M. L. Francisco, F. Polk, and R. Yolken, *J. Clin. Microbiol.* **26**(3), 453 (1987).

[19] U. Joshi, V. Rashavan, G. Zemse, A. Sheth, P. S. Borkar, and S. Ramachandran, *in* "Enzyme Labeled Immunoassay of Hormones and Drugs" (S. B. Pal, ed.), p. 233. de Gruyter, Berlin, 1978.

[20] P. B. Geetha, A. A. Koshy, C. N. Dandawate, B. K. Shaikh, S. N. Ghosh, and P. S. Borkar, *Hind. Antibiot. Bull.* **24,** 34 (1982).

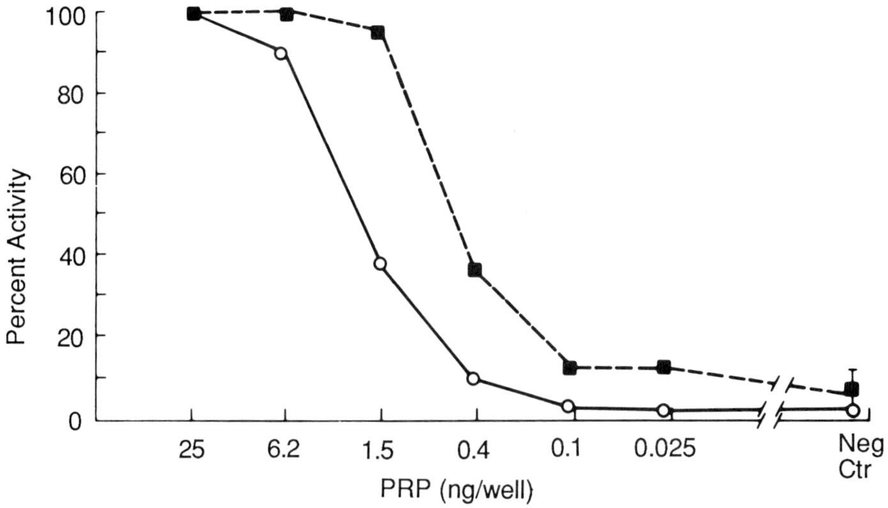

FIG. 4. Comparison of enzyme immunoassays for PRP. ■, Results with biotinylated β-lactamase; ○, results with biotinylated horseradish peroxidase. Percent activity was calculated from duplicate determinations as described in the text. The values for the negative control indicate the mean and standard deviation of reactivity in quadruplicate wells to which buffer was added in place of antigen dilution. (From Ref. 21.)

high turnover rates, these enzymes are stable, inexpensive, and can make use of a wide range of substrates. We have utilized biotinylated β-lactamase from *Bacillus cereus* and a starch–iodine penicillin substrate solution to devise immunoassays for the detection of rotavirus and adenovirus antigens in human body fluids.[21] The sensitivity of these assays was at least as favorable as analogous ones utilizing biotinylated horseradish peroxidase (Fig. 4).

In addition, biotin-labeled reagents have been utilized for the detection of a wide range of bacterial, viral, fungal, and parasitic antigens.[22–28] As-

[21] R. H. Yolken and S.-B. Wee, *J. Clin. Microbiol.* **19**(3), 356 (1984).

[22] J. C. Hierholzer, K. H. Johansson, L. J. Anderson, C. J. Tsou, and P. E. Halonen, *J. Clin. Microbiol.* **25**(9), 1662 (1987).

[23] S. Edwards, and G. C. Gitao, *Vet. Microbiol.* **13**(2), 135 (1987).

[24] L. C. Shekarchi, D. A. Fuccillo, R. Strouse, and J. L. Sever, *J. Clin. Microbiol.* **25**(2), 320 (1987).

[25] V. P. Kurup, *Zentralbl. Bakteriol. Mikrobiol. Hgy. Ser. A* **261**(4), 509 (1986).

[26] A. Belmaaza, J. Hamel, S. Mousseau, S. Montplaisir, and B. R. Brodeur, *J. Clin. Microbiol.* **24**(3), 440 (1986).

[27] P. Vilja, H. J. Turunen, and P. O. Leinikki, *J. Clin. Microbiol.* **22**(4), 637 (1985).

[28] A. Sutton, W. F. Vann, A. B. Karpas, K. E. Stein, and R. Schneerson, *J. Immunol. Methods* **82,** 215 (1985).

says making use of avidin–biotin interactions have also been utilized for the measurement of the immune response to infection with microbial antigens. While the degrees of sensitivity and specificity obtained in such assays varied depending on the binding characteristics of the immunoreagents, the utilization of avidin–biotin interactions has generally resulted in the attainment of the maximal degree of sensitivity allowed by immunoreagents. The use of avidin–biotin interactions has thus allowed for the development of a number of important assays for the diagnosis of infectious diseases. The efficient application of practical assays for microbial detection should allow for the improved clinical management of patients with suspected infections and for the containment of disease transmission in susceptible populations.

Preparation of Biotin-Labeled Immunoglobulins

Immunoglobulins or immunoglobulin fragments are purified from the serum to be labeled by standard methods and diluted to a concentration of 1 mg/ml in 60 m*M* carbonate buffer (pH 9.0). Biotin-*N*-hydroxysuccinimide ester is dissolved in dimethyl sulfoxide to make a stock solution of 1 mg/ml. Aliquots of biotin are added to the immunoglobulin (generally at a biotin to immunoglobulin molar ratio of 10 : 1), and the mixture is incubated for 3 hr at 25°. After incubation, unreacted biotin is removed by extensive dialysis against PBS or by fractionation with Sephadex G-25. The biotinylated antibody is aliquoted and stored at −20°. Preservatives such as sodium azide (20 mg/liter) can be added to prevent bacterial contamination.

Coat alternate rows of wells of the microtiter plate with a dilution of antiviral IgG and equal dilution of IgG from the same animal species that does not contain measurable antibody to rotavirus. Incubate the plate at least overnight at 4°. If the plate is not used the next day, it should be covered with Parafilm and stored at 4° until used.

Wash the plate 5 times with phosphate-buffered saline containing 0.05% Tween 20 (PBS–Tween). Add 50 μl of PBS–Tween to each of the wells. Buffer containing nonimmune animal sera can also be utilized to reduce the occurrence of nonspecific reactions. Add an equal amount of specimen to two wells coated with antiviral IgG and two wells coated with nonimmune IgG. Include a weakly positive control and four negative controls in each test.

Incubate the plate for 2 hr at 37° or overnight at 4°. Add antiviral antibody that has been labeled with biotin, and incubate the plate for 1 hr at 37°. Wash the plate 5 times with PBS–Tween.

Prepare the avidin–biotin complex by adding predetermined amounts of avidin and biotin-linked enzyme to a tube containing PBS–Tween. Incubate for 10 min at 37°, and add the complex to the wells. Incubate the plate for 30 min at 37°. Wash the plate 5 times with PBS–Tween.

Add appropriate substrate. Incubate the plate at 37° or room temperature until the weakly positive control has visible color equivalent to an optical density of approximately 0.1 measured at the appropriate wavelength.

Comments. For each sample calculate the virus-specific activity by subtracting the mean activity of the wells coated with the nonimmune IgG from that measured in wells coated with the antiviral IgG. To ensure accurate quantification, specimens giving readings of greater than 1.2 optical density units should be diluted 1 : 10 and retested. Calculate the mean and standard deviation of the virus activity of the negative controls. A specimen is considered positive if its mean activity is greater than 2 standard deviations above the mean activity of the negative controls. Alternatively, a specimen can be considered positive if its specific activity is greater than that of the weakly positive control.

If qualitative visual determinations are used, a specimen is considered positive if the degree of color in the antiviral IgG wells is different than the amount of color generated in the nonimmune wells and in the wells containing the weakly positive control samples. Note that if the β-lactamase starch–iodine system is utilized, a positive reaction will be manifested by a decrease in color, whereas in the case of most other enzymes, the binding of enzyme will result in an increase in color production.

Acknowledgment

This work was supported by contract N01-AI-52579 from the National Institute of Allergies and Infectious Diseases.

[62] Enzyme-Linked Immunosorbent Assay

By WILLIE F. VANN, ANN SUTTON, and RACHEL SCHNEERSON

Capsular polysaccharides are important in invasive bacterial disease as virulence factors and as immunogens. Since polysaccharides and oligosaccharides often do not adhere to a plastic solid phase as well as proteins, the measurement of antibodies to polysaccharides by enzyme-

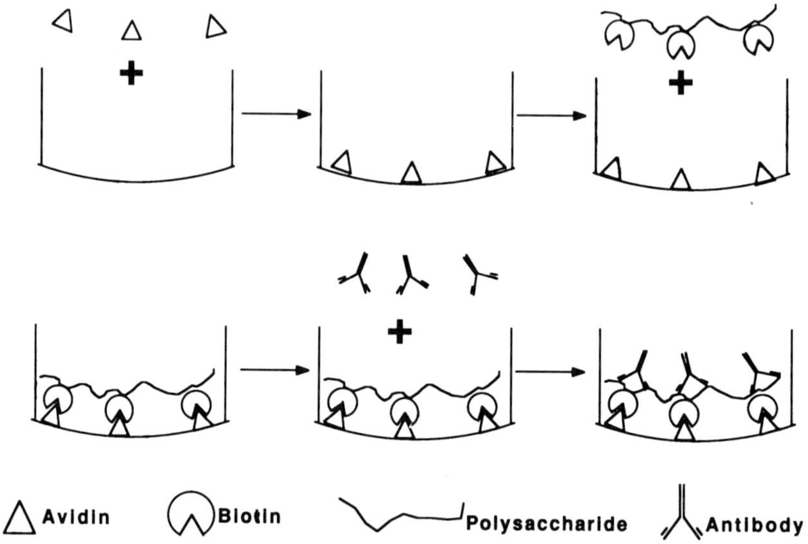

FIG. 1. Schematic illustration of avidin–biotin-based antipolysaccharide ELISA.

linked immunosorbent assay (ELISA) has been difficult.[1] Avidin–biotin complex formation has been used to enhance ELISA sensitivity by the preparation of biotinylated antibodies or indicator enzyme conjugates.[2,3] In the assay described here, the avidin–biotin complex is used to enhance the binding of the polysaccharide antigen to the solid phase.[4,5]

The polysaccharide is derivatized with biotin via an amino or hydrazide functional group.[4] The biotin moieties are then used to complex the polysaccharide antigen to an avidin-coated solid phase as illustrated in Fig. 1. The resulting solid phase can then be used in a routine ELISA to measure polysaccharide-specific antibody.[2,5] This assay has also been used to measure the ability of carbohydrate inhibitors to bind antibody specific for immobilized polysaccharide antigen.[5]

[1] L. T. Callahan, A. F. Woodhour, J. B. Meeker, and M. R. Hilleman, *J. Clin. Microbiol.* **10,** 459 (1979).

[2] P. Tijssen, *in* "Laboratory Techniques in Biochemistry and Molecular Biology" (R. H. Burdon and P. H. van Knippenberg, eds.), Vol. 15, p. 329. Elsevier, Amsterdam, 1985.

[3] J.-L. Guesdon, T. Ternynck, and S. Avrameas, *J. Histochem. Cytochem.* **27,** 1131 (1979).

[4] E. A. Bayer and M. Wilchek, *Methods Biochem. Anal.* **26,** 1 (1980).

[5] A. Sutton, W. F. Vann, A. B. Karpas, K. E. Stein, and R. Schneerson, *J. Immunol. Methods* **82,** 215 (1980).

TABLE I
METHODS OF PREPARING POLYSACCHARIDE HYDRAZIDES

Polysaccharide	Available function	Method of attaching hydrazide
Haemophilus influenzae b	—OH	CNBr
Streptococcus pneumoniae 12F	—COOH	RN=C=NR′
Staphylococcus aureus type 8	—COOH	RN=C=NR′
Streptococcus pneumoniae 6A	—OH	CNBr

Preparation of Biotinylated Polysaccharides

Polysaccharides are derivatized with biotin via adipic acid hydrazide groups. The method of attachment of the hydrazide or amino groups is determined by the structure of the polysaccharide (Table I).

Haemophilus influenzae b Polysaccharide Hydrazide. Adipic acid hydrazide groups are attached to *Haemophilus influenzae* b (Hib) polysaccharide[6] by cyanogen bromide activation of hydroxyl groups. A 10-ml solution of polysaccharide (5–10 mg/ml) is adjusted to pH 10.5 with 0.1 *N* NaOH. Cyanogen bromide (0.4 mg/ml polysaccharide) is added (1 g/ml prepared by dissolving 2 g CNBr in 1 ml acetonitrile to yield 2 ml), and the reaction mixture is maintained at 4° and pH 10.5. After 6 min, 10 ml of 0.5 *M* adipic dihydrazide in 0.5 *M* Na_2CO_3 (pH 8.5) is added, and the solution is stirred overnight at 4°. The reaction mixture is then desalted on Sephadex PD-10 and lyophilized.

Streptococcus pneumoniae Type 12F Polysaccharide Hydrazide. Polysaccharide carboxyl groups are derivatized with adipic acid dihydrazide by the carbodiimide method.[7] Polysaccharide (*Streptococcus pneumoniae* type 12F, 23 mg)[8] is dissolved in 10 ml of 0.25 *M* adipic acid dihydrazide and adjusted to pH 4.75 at 4°. Ethyldimethylaminopropylcarbodiimide (0.5 mmol) is added, and the pH is maintained with 0.1 *N* HCl for 3 hr. The reaction mixture is dialyzed against water and lyophilized. Yield, 20 mg.

The degree of substitution of polysaccharides by hydrazide using either method of activation is determined by the trinitrobenzenesulfonic

[6] R. M. Crisel, R. S. Baker, and D. E. Dorman, *J. Biol. Chem.* **250,** 4926 (1975).
[7] D. G. Hoare and D. E. Koshland, *J. Biol. Chem.* **242,** 2447 (1967).
[8] K. Leontein, B. Lindberg, and J. Lonngren, *Can. J. Chem.* **59,** 2081 (1981).

acid reaction.[9] The number of groups incorporated is dependent on the polysaccharide and usually varies between 100 and 400 nmol/mg.[5]

Biotinylation of Polysaccharide Hydrazides. All polysaccharide hydrazides are biotinylated by the following procedure. *Streptococcus pneumoniae* type 12F hydrazide (10.8 mg), dissolved in 1 ml of phosphate-buffered saline (PBS), is added to 0.5 ml of biotin-*N*-hydroxysuccinimide ester (8–10 mg/ml) dissolved in anhydrous amine-free dimethylformamide at 4°. The resulting solution is stirred overnight and desalted on a Sephadex PD-10 column. Polysaccharides are detected by capillary precipitation using the appropriate antisera. Yield, 5 mg.

ELISA

Reagents

Phosphate-buffered saline (PBS, pH 7.4)
Phosphate-buffered saline (pH 7.4) containing 1% bovine serum albumin, 0.1% Triton X-100, 0.1% sodium azide (PBS–BSA diluent)
Avidin, egg white (Sigma Chemical Co., St. Louis, MO), 4 μg/ml in PBS
Biotinylated polysaccharide antigen (2 μg/ml in PBS–BSA, prepared immediately before use, 10 ml/plate)
2% BSA, 0.2% sodium azide in PBS, sterile
Antiimmunoglobulin, species- and class-specific, conjugated to alkaline phosphatase (Kirkegaard & Perry Laboratories, Inc., Gaithersburg, MD)
p-Nitrophenyl phosphate (Sigma 104 substrate), 1 mg/ml in 1 *M* Tris-HCl, 0.3 m*M* $MgCl_2$ (pH 9.8)
Antisera, stock dilutions of test sera prepared in 2% BSA (usually 1 : 100 to 1 : 10,000 depending on antibody activity); antibody activity varies considerably and may require dilution outside this range

Procedure. Wells of U-well microtiter plates (Dynatech polyvinyl or Nunc Immunoplate II) are coated with 100 μl of avidin in PBS. Plates are sealed, incubated overnight at room temperature, and washed four times with PBS. Biotinylated polysaccharide is diluted to 2 μg/ml in PBS–BSA immediately before use, and 100 μl is added to avidin-coated plates. After incubation for 30–120 min at room temperature, the plates are washed with PBS. Test sera are serially diluted with PBS–BSA (2-fold) pipetting 100 μl/well. Plates are sealed, incubated overnight, and washed.

Diluted goat antiimmunoglobulin, conjugated to alkaline phosphatase (100 μl), is added, and the plates are incubated for 4 hr and washed. The

[9] A. F. S. A. Habeeb, *Anal. Biochem.* **14,** 328 (1966).

dilution of anti-Ig conjugate (usually 1 : 1000) must be determined for each lot such that the maximum absorbance of the lowest reference serum dilution is approximately 1.0 after color development.[2] The color is developed by addition of 100 μl *p*-nitrophenyl phosphate to the washed plate and incubation for 30–60 min. The optical density is read at 405 nm.

Comments

Several polysaccharides have been successfully biotinylated and used in this ELISA including *Escherichia coli* K1, *S. pneumoniae* 6A and 12F, dextran, and *Staphylococcus aureus* types 5 and 8. Some polysaccharides, however, either are insoluble and unusable after derivatization by this procedure (meningococcal Group C) or develop precipitates after storage at 4° (*S. aureus* type 5).

Inherent with any assay method involving chemical modification of polysaccharide antigens is the risk that antibody binding could be altered. We have not observed a noticeable change in the ability of *S. aureus* and *S. pneumoniae* polysaccharides to bind antibody; however, there is a noticeable difference in the ability of *H. influenzae* type b polysaccharide to inhibit antibody binding.[5]

No noticeable background was observed in experiments measuring antibody levels in mice or rabbits. Nonspecific binding of human sera to avidin-coated plates became evident in low-titered sera. This background reactivity can be controlled by addition of sufficient antigen to coat most of the avidin sites.

[63] Rapid Detection of Herpes Simplex Virus

By LATA S. NERURKAR

Herpes simplex virus (HSV) causes diverse clinical illnesses associated with its primary infection. Generalized infection during immunosuppression, neonatal infection, and HSV encephalitis (infection of the brain) are very devastating and in many instances fatal if not properly treated. Antiviral therapy that responds to both types (HSV type 1 and HSV type 2) is now available. This makes it essential and important to identify the virus so that chemotherapy can be initiated to save the life of the patient. In neonatal herpes, the baby often acquires the infection during birth from the asymptomatic mother who is shedding virus from the cervix or vagina. In this case, a laboratory viral culture is recommended as close to

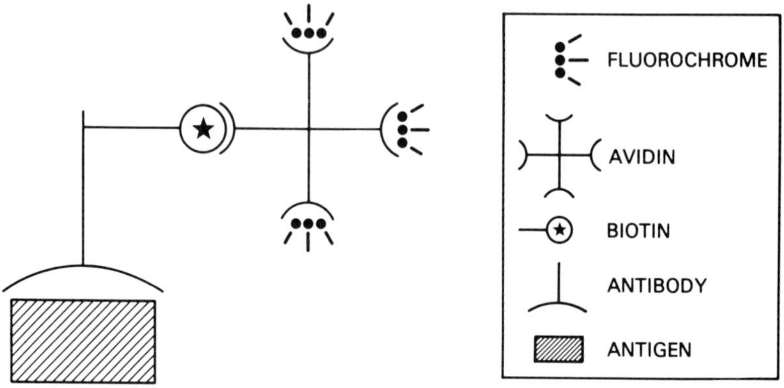

FIG. 1. Schematic of antigen detection using biotinylated primary antigen-specific antibody followed by avidin–fluorochrome.

the time of delivery as possible in order to detect the presence of virus. A cesarian section can be performed, if virus is detected, to avoid acquisition of infection by the neonate.

Virus is generally present in low titers in recurrent infections compared to the primary infections. This prolongs the detection by conventional tissue culture techniques, which can take up to 7 days. In order to rapidly detect the virus in clinical specimens, we have taken a 2-fold approach: (1) amplification of the virus by short-term (24 hr) tissue culture, followed by immunostaining with sensitive procedures employing biotin-labeled specific antibody and avidin fluorochrome reagents (Fig. 1) or (2) detection of viral antigens directly in the clinical specimens using a capture enzyme-linked immunosorbent assay employing biotin-labeled specific antibody and streptavidin–alkaline phosphatase detector systems (Fig. 2). Comparisons of time requirements, sensitivities, and specificities indicate that both methods are relatively more rapid with either no loss or only minor loss of sensitivity or specificity, respectively, compared to the conventional tissue culture. This suggests that a quick and reliable answer can be obtained using these techniques, which can be simultaneously confirmed by the conventional technique.

The importance of avidin and biotin in modern biotechnology was long recognized by Bayer and Wilchek,[1] but its application in immunoassays is evident only since the mid-1980s. This chapter details the use of avidin–biotin reagents for the detection and typing of herpes simplex virus. A flow chart (Fig. 3) is given to indicate the procedures used and time taken

[1] E. A. Bayer and M. Wilchek, *Methods Biochem. Anal.* **26,** 1 (1980).

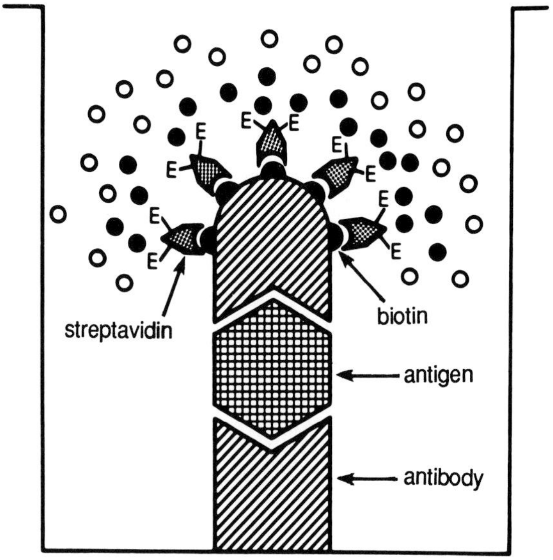

FIG. 2. Schematic of antigen capture B/SA ELISA for HSV. Symbols: ○, substrate; ●, product; E, enzyme.

for the detection of infectious virus, viral antigen, or virus type in the specimens received by the laboratory. Some of the basic methodology that forms the basis of comparison for new technology is included for those readers unfamiliar with the field of herpes virology.

Materials and Methods

Collection of Specimens. Specimens are collected by rubbing lesions with cotton swabs. The swabs are suspended in 3 ml of Hanks' balanced salt solution containing streptomycin, penicillin, mycostatin, and 0.5% gelatin (henceforth referred to as collection medium). Refrigeration before transport is essential, and, if samples are not studied immediately, quick freezing and storage at −70° are recommended.[2] For direct staining methods, scrapings of lesions are taken by nonabrasive metal spatulas on ethanol-moistened slides and dried at room temperature. The slides are stained immediately or saved frozen at −20 or −70° until studied.

Preparation of Stock Virus. HSV-1 and -2 are grown in monolayer cultures of owl monkey kidney cells maintained on complete Eagle's

[2] L. S. Nerurkar, A. J. Jacob, D. L. Madden, and J. L. Sever, *J. Clin. Microbiol.* **17,** 149 (1983).

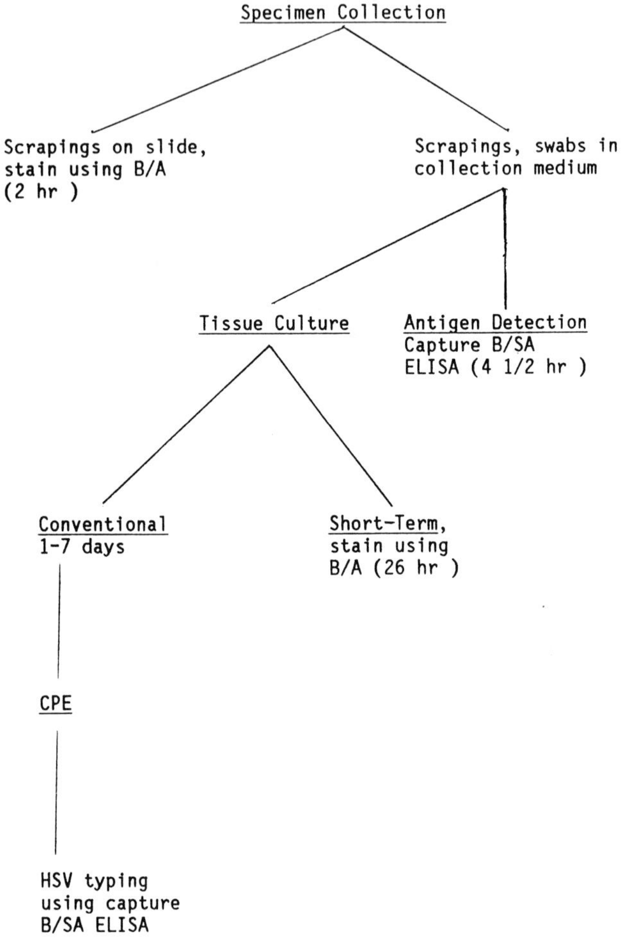

FIG. 3. Flow chart for HSV isolation, HSV antigen detection, and HSV typing.

minimum essential medium (EMEM; Microbiological Associates, Bethesda, MD) containing 2 m*M* glutamine (Gibco Laboratories, Madison, WI), penicillin (100 U/ml), and streptomycin (100 μg/ml) with 2% fetal bovine serum. The culture supernatants are harvested at 48 hr and concentrated 10-fold with an Amicon fiberglass concentrator (Amicon Corp., Lexington, MA). The virus is further purified by discontinuous density-gradient centrifugation with 20 and 60% sucrose in Tris buffer (20 m*M*; pH 7.8) with EDTA (1 m*M*) and NaCl (0.154 *M*) at 25,000 rpm using a

Beckman SW28 rotor. The virus is collected from the interphase of 20 and 60% sucrose and stored frozen in small portions at −70° until used.

Conventional Tissue Culture

Cell Lines and Virus Culture. Human foreskin fibroblasts (Flow 7000, obtained originally from Flow Laboratories, McLean, VA), rabbit kidney cell line MA111 (M. A. Bioproducts, Walkersville, MD), African green monkey kidney cells, WI-38, Vero cells (American Type Culture Collection, Rockville, MD), and owl monkey kidney cells (gift from Dr. M. Gravell, NINCDS, NIH), are all grown in EMEM containing 10% (inactivated and filtered) fetal bovine serum (FBS) (EMEM + 10% FBS), glutamine, and antibiotics. Monolayer glass tube cultures or plastic plate (Costar, 24-well plates) cultures of Flow 7000 are routinely used for conventional tissue culture and virus infection purposes. The specimens (0.2 ml/tube or well) are absorbed on monolayers for 1 hr, after which fresh EMEM with 2% FBS (EMEM + 2% FBS) is added (1.5 ml). The tube cultures are tightly capped and incubated at 37°. The plate cultures are incubated at 37° in 5% CO_2 atmosphere (humid).

All cultures are observed daily for cytopathic effect (CPE) until 7 days. The CPE includes characteristic rounding and ballooning of cells followed by degeneration. The foci of CPE are obvious throughout the monolayer, the number dependent on the infectivity of the inoculum. When stained, multinucleated giant cell formations and characteristic development of intranuclear inclusion bodies are evident. The cultures showing positive CPE are further passaged on Flow 7000 or rabbit kidney cell monolayers and confirmed for the presence of virus. When showing 80–90% CPE, cell monolayers are scraped and saved for determining the virus type (type 1 or 2, described later), or single cell suspensions are made and dried on slides to stain for HSV antigens.

Short-Term Tissue Culture for HSV on Lab-Tek Chamber Slides.[2] Human foreskin fibroblast cell cultures (Flow 7000) are grown on eight-well Lab-Tek chamber slides (Miles Scientific, Div. Miles Laboratories, Inc., Naperville, IL) as described previously.[2] In brief, the chamber slides are first rinsed and incubated overnight with EMEM +10% FBS to remove any toxic products (which are occasionally seen in certain batches of these chambers). Approximately 3×10^4 to 4×10^4 cells are seeded in each well and allowed to grow for 24–48 hr. It is recommended that each laboratory investigator optimize test conditions, e.g., dilutions of virus-containing specimens for inoculation, volume of inoculation, and time of tissue culture. In our laboratory, 100–200 μl of specimens is inoculated in duplicate. Two wells of each chamber slide are mock infected or infected

TABLE I
DETECTION OF HERPES SIMPLEX VIRUS BY TISSUE CULTURE AND SHORT-TERM TISSUE CULTURE WITH IMMUNOLOGICAL STAINING

Method	Time	Sensitivity	Specificity
TC-CPE (Conventional)	1–7 days	100	100
TC-B/A-FA	26 hr	100	100
TC-FA	26 hr	68.5	100
TC-IFA	26 hr	88.3	88.1

with laboratory-purified HSV-1 (McIntyre strain) or HSV-2 (MS strain) preparations, respectively, to provide negative and positive controls on the same slide. At the end of the incubation (24 hr), the medium is aspirated, the slides are washed 3 times with phosphate-buffered saline [(PBS) 6.7 m*M* phosphate (pH 7.2), 0.154 *M* NaCl], and the plastic chambers are removed and air dried. The slides are then fixed in chilled (4°) acetone for 10 min. The acetone is evaporated completely, and the slides are frozen at −70° until stained.

Determination of Infectivity of HSV. The infectivity of HSV in stock virus preparations and in clinical specimens is determined as described previously. In brief, Flow 7000 monolayer cultures are infected with 0.2-ml volumes of serial 10-fold dilutions of the virus preparation and observed for CPE. The end point is determined at 7 days and calculated as the 50% tissue culture infectious dose ($TCID_{50}$) by the method of Reed and Muench.[3]

Staining Procedures

Choice and Preparation of Antibody. After several trials and evaluations of different reagents on the market, our laboratory prefers to use a polyclonal rabbit anti-HSV immunoglobulin (IgG) antibody. A comparison of different staining procedures following short-term tissue culture (TC), e.g., (1) biotinylated rabbit anti-HSV IgG and fluoresceinated avidin (TC-B/A-FA), (2) rabbit anti-HSV IgG directly conjugated with fluorescein (TC-FA), and (3) rabbit anti-HSV IgG plus sheep anti-rabbit IgG conjugated with fluorescein (TC-IFA), has always resulted in choice of the biotin–avidin (B/A) reagents as in (1), and details are given below. Comparisons are described in Table I.[4]

The B/A staining procedure involves biotinylation of the rabbit anti-HSV IgG as follows. The antibody (obtained from Accurate Chemicals and Scientific Corp., Westbury, NY) is extensively dialyzed against PBS

[3] L. J. Reed and M. Muench., *Am. J. Hyg.* **27,** 493 (1938).
[4] L. S. Nerurkar, M. Namba, and J. L. Sever, *J. Clin. Microbiol.* **19,** 631 (1984).

to remove any traces of $(NH_4)_2SO_4$ present in the preparations. After sedimenting any insolubles, the antibody is dialyzed at 4° against 0.1 *M* $NaHCO_3$ (pH 8.2–8.6) overnight. It is then centrifuged (10 min at 2,000 rpm), and the protein concentration is estimated at 1 : 50 or 1 : 100 dilution of the antibody preparation according to the following formula[5]:

$$\text{Protein (mg/ml)} = 1.55(A_{280}) - 0.76(A_{260})$$

The antibody is adjusted to a concentration of 1 mg/ml with 0.1 *M* $NaHCO_3$ (pH 8.2–8.6) and mixed with freshly prepared biotin-*N*-hydroxysuccinimide ester (B-1000, Biosearch, San Rafael, CA), 1 mg/ml, in dimethyl sulfoxide in proportion (100 : 12) in a dropwise manner. The mixture is kept at 20° for 4 hr with occasional shaking. The biotinylated antibody is then extensively dialyzed against several changes of PBS to remove unconjugated biotin ester and dimethyl sulfoxide. The biotinylated antibody is then titrated and frozen in small volumes at −20°. The optimum working dilutions of this preparation are in the range of 1 : 20 to 1 : 40. The biotinylated anti-HSV-1 or -2 antibody preparations are found very stable even at 4° for 5–6 years, if stored properly.

Avidin. Fluoresceinated egg-white avidin (Cat. No. A-2001) is obtained from Vector Laboratories (Burlingame, CA).

Staining. The slides are rinsed in PBS for 2–5 min. This allows the dried monolayer to be rehydrated and the excess salt dried on the fixed cell monolayers to be washed off. One hundred microliters of properly diluted biotinylated antibody is then incubated with monolayers for 1 hr at 37° in humid chambers followed by 3 washings for 5 min each in PBS with gentle stirring. This is followed by incubation with avidin–fluorescein conjugate (1 : 50 dilution with PBS) in the dark for 30 min at 37° in humid chambers. The slides are then washed with PBS (3 washings for 5 min each) and counterstained with Evans blue (0.6 mg%). While using the 8-well Lab-Tek chambers, the soft plastic bonding material is peeled off at this stage (slides must be moist), and the slides are blotted, air dried, mounted with glycerol–PBS (90 : 10), and viewed under a fluorescent microscope. Typical pictures of stained infected and uninfected monolayers are given in Fig. 4.

HSV-Antigen Detection without Tissue Culture Amplification[6]*: Capture Biotin–Streptavidin ELISA*

The schematic diagram of the antigen capture biotin–streptavidin (B/SA) enzyme-linked immunosorbent assay is shown in Fig. 2.

[5] E. Layne, this series, Vol. 3, p. 447.

[6] L. S. Nerurkar, M. Namba, G. Brashears, A. J. Jacob, Y. J. Lee, and J. L. Sever, *J. Clin. Microbiol.* **20**, 109 (1984).

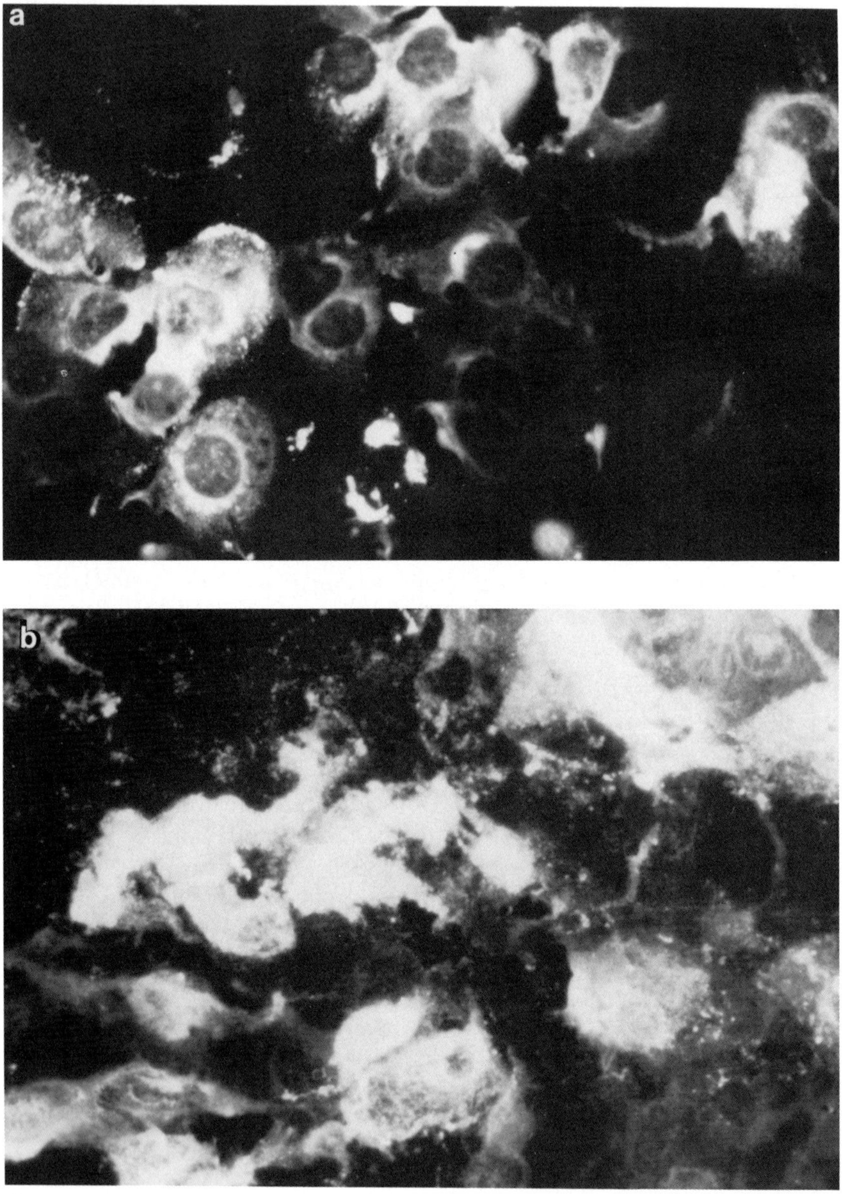

FIG. 4. Monolayers of owl monkey kidney cells infected with HSV-2 (a), WI-38 cells infected with HSV-2 (b), Flow 7000 cells infected with HSV-2 (c), and uninfected Flow 7000 cells (d) stained with biotin-labeled anti-HSV-2-specific rabbit IgG and egg avidin–fluorescein conjugate. Magnification: ×400.

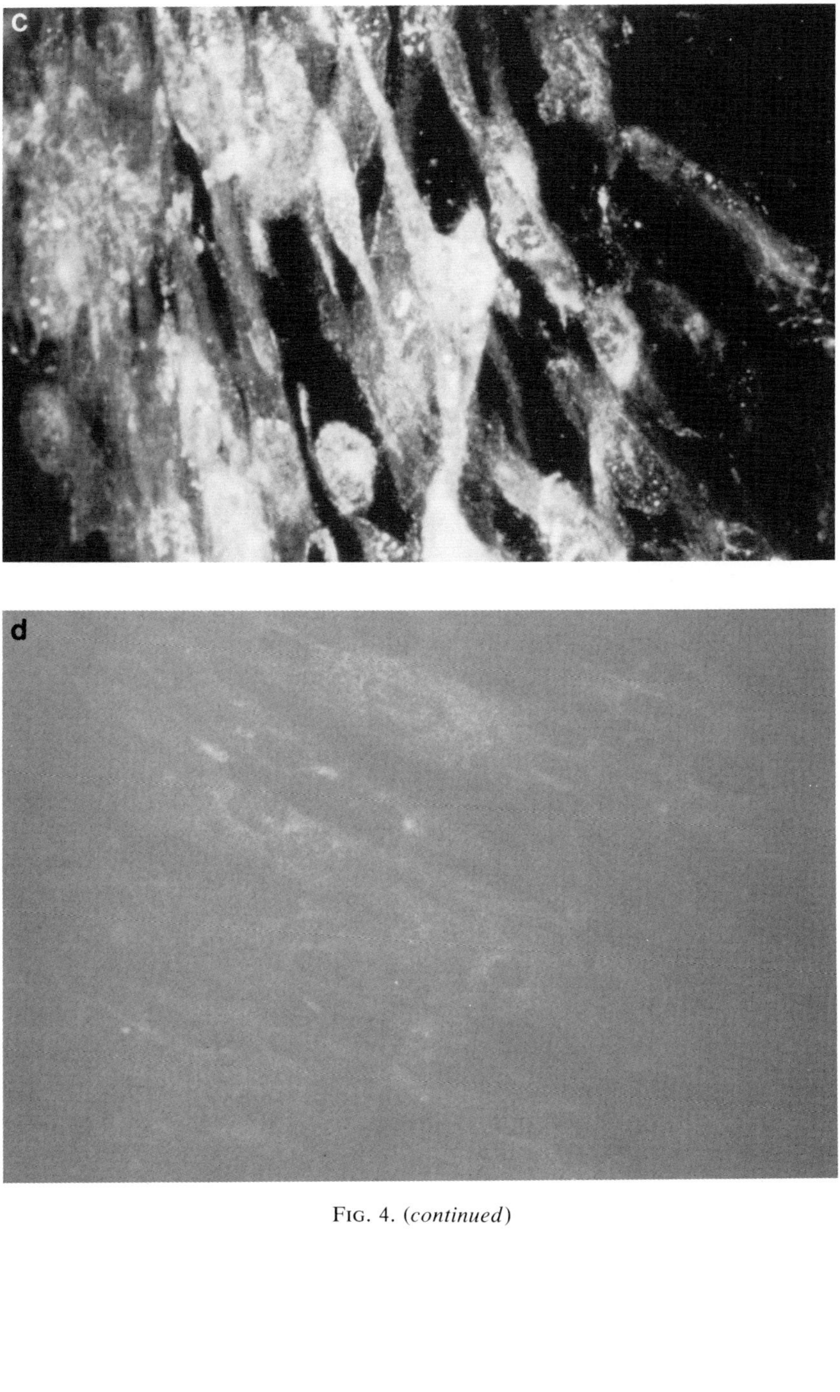

FIG. 4. (*continued*)

Choice of Plastic. Several 96-well microtiter plates are available, but only a few are recommended as solid-phase support. Initial testing included Immulon I and Immulon II plates (Dynatech Laboratories, Inc., Alexandria, VA), polyvinyl chloride plates (Dynatech), and Costar polystyrene plates (Costar, Cambridge, MA). Only the UV-irradiated plates, e.g., Immulon II and Costar polystyrene plates, are suitable for this assay. Many more brands are now available in the market but have not been systematically tested for the purpose.

Preparation of Biotinylated Anti-HSV Antibody and Streptavidin–Enzyme Conjugates. The biotinylation of anti-HSV immunoglobulin is described above. Streptavidin (Cat. No. 5532LA) is obtained from Bethesda Research Laboratories, Inc. (Gaithersburg, MD). Purified alkaline phosphatase prepared from bovine calf intestine is obtained from Sigma Chemical Co. (St. Louis, MO). The streptavidin–alkaline phosphatase conjugates are prepared by a one-step glutaraldehyde conjugation procedure as described previously.[7] Briefly, centrifuge the alkaline phosphatase suspension (Sigma, type VII from calf intestine, specific activity ~1000 U/mg protein) containing 5 mg enzyme at 2000 rpm for 10 min. Discard the supernatant. Dissolve 2 mg of streptavidin in 2 ml of PBS. Add the streptavidin solution to the 5 mg alkaline phosphatase precipitate. Mix well at room temperature, and dialyze extensively at 4° with several changes of PBS (minimum 3 changes). Add 10% glutaraldehyde (electron microscopy grade) in PBS to the enzyme–streptavidin mixture to give a final concentration of 0.2%. Incubate at room temperature with occasional gentle stirring for 2 hr. Dialyze against several changes of PBS at 4°. The streptavidin–alkaline phosphatase conjugate is titrated to determine the optimal concentration to be used in the ELISA experiments. During initial trials, commercially available egg avidin–enzyme conjugates were compared with streptavidin–enzyme conjugates prepared by us. The streptavidin conjugates are preferable, particularly in the ELISA experiments, in terms of lowering the absorbances of the negative controls.

Performance of Capture B/SA ELISA. All tests are run against anti-HSV-1 and -2 antibody reagents simultaneously. The capture antibodies are identical to the detecting antibodies except that the latter are linked with biotin. The microtiter plates are coated with an antibody dilution of 1 : 250, which is found to be sufficient to capture the levels of viral antigen present in clinical specimens. Microtiter plates are sensitized with 300 μl

[7] A. Voller, D. Bidwell, and A. Bartlett, *in* "Manual of Clinical Immunology" (N. R. Rose and H. Friedman, ed.), 2nd Ed., p. 359. American Society for Microbiology, Washington, D.C., 1980.

of antibody diluted in sodium carbonate–bicarbonate buffer [1.59 g Na_2CO_3, 2.93 g $NaHCO_3$, 0.2 g sodium azide per liter (pH 9.6)] in a humid chamber at 4° overnight.[6,7] The plates are washed 3 times with PBS containing 0.05% Tween 20 (Sigma), 0.1% bovine serum albumin (Miles Laboratories, Inc., Elkhart, IN), and 0.02% sodium azide (Sigma), henceforth referred to as ELISA wash. Stock virus or clinical specimens are then added, at least in duplicate (100 μl per well), and the plates are incubated at 37° in a 5% CO_2 incubator. The optimum time of incubation is determined to be 2 hr. Unbound material is removed by washing the plates 3 times with ELISA wash, and then pretitrated (1 : 40) biotin-linked antibody diluted in PBS containing 0.05% Tween 20, 1.0% bovine serum albumin, and 0.02% sodium azide (100 μl per well) is added to the plates. After incubation for 1 hr at 37° in a humid chamber, the plates are washed 3 times with ELISA wash. Pretitrated streptavidin–alkaline phosphatase conjugate (1 : 500, 100 μl per well) is then added, followed by incubation for 0.5 hr at 37° in a humid chamber. The unbound conjugate is removed by washing thoroughly, and 100 μl of freshly prepared *p*-nitrophenyl phosphate substrate (Sigma) is added to each plate [1 mg of substrate per ml of 0.5 m*M* diethanolamine buffer (pH 9.8) containing 100 mg of $MgCl_2 \cdot 6H_2O$ and 200 mg of sodium azide per liter, stored in the dark]. The plates are incubated for 30 min in a humid chamber at 37°, and the reaction is stopped by adding 50 μl of 3 *M* NaOH to each well. The absorbance measurements are made at 405 nm with a Dynatech ELISA reader model MR580. The addition of NaOH is optional, as in some samples very low levels of antigen are present and longer incubations may be necessary.

Several replicate readings of viral collection medium and control buffer must be incorporated in the assay to serve as negative controls. A known dilution of stock virus serves as a positive control. The use of the fluorogenic substrate methylumbelliferyl phosphate has not improved the sensitivity of detection of HSV antigen in clinical specimens or in purified virus preparations.[8]

Evaluation of Results. The absorbance readings on at least duplicate specimens and four to eight replicates of viral control medium are averaged, and the standard deviation (SD) of the control average in a given experiment is calculated. An average absorbance value for control buffer (Tris buffer used for virus purification) and the specimen collection medium ranges between 0.00 and 0.050 ± 0.020. The culture negative clinical specimens show absorbance values in the same range. The specimens which have an average absorbance value equal to or above the predicted

[8] I. C. Shekarchi, J. L. Sever, L. Nerurkar, and D. Fuccillo, *J. Clin. Microbiol.* **21,** 92 (1985).

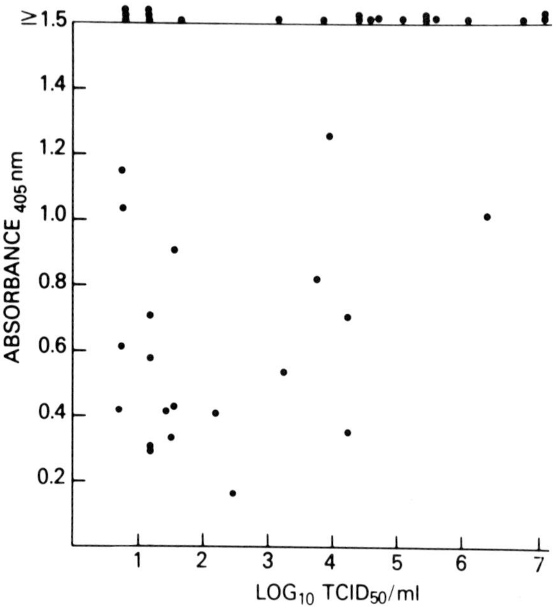

FIG. 5. Comparison of antigen capture B/SA ELISA and tissue culture detection of HSV in genital herpes specimens.

limit of absorbance values for a positive score are considered positive. The predicted limit of absorbance values for a positive score is determined by using the 95% prediction bound formula $\bar{X} + \{SD_{\bar{x}}\, C_{f.95\%}\, [(n + 1)/n]^{1/2}\}$, where $\bar{X}$ and $SD_{\bar{x}}$ are the average and the standard deviation, respectively, of n observations of control buffer absorbance values and $C_{f\cdot 95\%}$ is the t value at 95% probability for $f\,(=n - 1)$ degrees of freedom.[6]

The sensitivity of detection of gradient-purified HSV-1 and -2 is different using the biotin–streptavidin ELISA method.[6] In both the conventional tissue culture method and the short-term tissue culture followed by B/A staining method, there is an amplification due to virus replication ultimately capable of detecting very small numbers of infectious virus particles (1–10 particles). In the ELISA method, the sensitivity of HSV antigen detection appears to be better in clinical specimens, probably owing to the presence of free or cell-associated antigen available for reaction (Fig. 5) although no correlation is recognized between tissue culture infectivity titer and antigen level estimated by the ELISA absorbance values in clinical specimens. In contrast, loss of infectivity caused by improper storage does not drastically affect the viral antigen detection as it does the outcome of the tissue culture methods. Absorbance values

TABLE II
DETECTION OF HERPES SIMPLEX VIRUS BY TISSUE CULTURE AND NONTISSUE CULTURE METHODS

Method	Time	Sensitivity	Specificity
TC-CPE	1–7 days	100	100
Direct staining of cells[a] (B/A-FA)	2 hr	88.9	62.3
Ag detection by ELISA (B/SA)	4.5 hr	95.6	91.4[b]

[a] Obtained by scraping the lesions.
[b] The specificity increases to approximately 98–99% when the presence of antigen is confirmed in those specimens which are tissue culture HSV negative/ELISA HSV antigen positive using a blocking technique.[6]

appear to range between the minimum and the maximum for specimens of low infectivity titer as shown in Fig. 5 and indicate considerable variation in detectable levels of antigen at a given level of infectivity. This may allow an expedited study of follow-up specimens to look for the presence of infectious virus.

The method has good sensitivity and specificity as shown in Table II. The final specificity of antigen detection, as determined by positively blocking with unlabeled antibody and negatively blocking with preimmune rabbit sera, has been previously described.[6] The technique described here uses polyclonal anti-HSV antibody instead of monoclonal antibody, as used elsewhere,[9] to ensure better detection of low levels of viral antigen. The specific detection of HSV is confirmed by testing in the assay cytomegalovirus, *Candida albicans,* and *Staphylococcus aureus,* which are common pathogens in a variety of urinogenital infections or constituents of the normal flora in the genital region.

Typing of HSV Using Biotin–Streptavidin ELISA[10]

The test procedure for typing HSV employing the ELISA procedure using polyclonal and monoclonal antibodies is essentially the same as described earlier.[10] HSV isolates (not the clinical specimens) are studied by using (1) homologous polyclonal anti-HSV-1–anti-HSV-1 and homologous anti-HSV-2–anti-HSV-2 sandwiches and (2) a heterologous poly-

[9] K. Adler-Storthz, C. Kendall, R. C. Kennedy, R. D. Henkel, and G. R. Dressman, *J. Clin. Microbiol.* **18,** 1329 (1983).
[10] L. S. Nerurkar, N. R. Miller, M. Namba, M. Monzon, G. Brashears, G. Scherba, and J. L. Sever, *J. Clin. Microbiol.* **25,** 128 (1987).

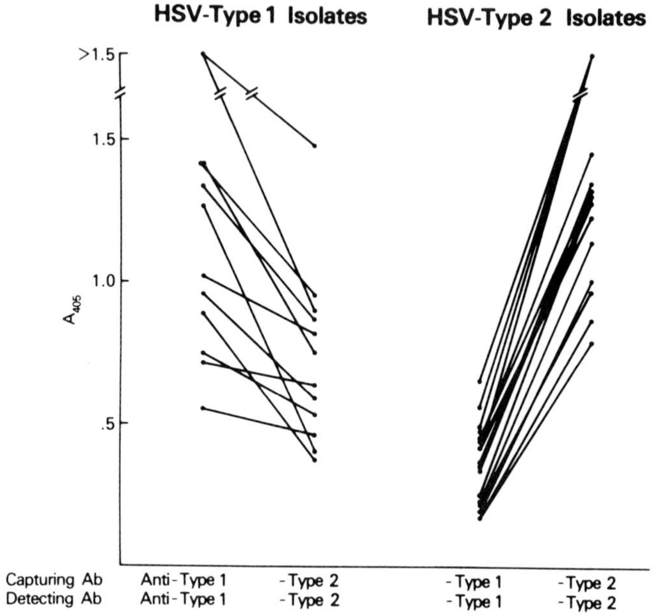

FIG. 6. Differences in A_{405} values using the homologous polyclonal antibody sandwich method for detection of HSV type in capture B/SA ELISA.

clonal anti-HSV-1–monoclonal anti-HSV-1 sandwich or heterologous polyclonal anti-HSV-2–monoclonal anti-HSV-2 sandwich. In the former, the ratio of the A_{405} of the type 1 sandwich to that of the type 2 sandwich is obtained and denoted as the typing index. It is noted that the clinical HSV-1 isolates or laboratory-purified HSV-1 always gives higher absorbances on the homologous anti-HSV-1–anti-HSV-1 sandwich. The reverse is true for clinical HSV-2 isolates or laboratory-purified HSV-2 using the homologous anti-HSV-2–anti-HSV-2 sandwich. The results of some of the experiments (Fig. 6) include comparisons of the absorbances for different isolates. The typing index for HSV-1 isolates is consistently above 1.0, and for HSV-2 isolates it is consistently below 1.0. In these experiments, the average and range values for the typing indices for HSV-1 are 1.74 and 1.12–3.27, respectively, and the typing indices for HSV-2 are 0.24 and 0.18–0.43, respectively. It is clear, based on such data, that polyclonal reagents can be successfully used in the ELISA for HSV typing.

A positive identification of HSV-1 or HSV-2 can be performed using the heterologous polyclonal anti-HSV-1–monoclonal anti-HSV-1 sand-

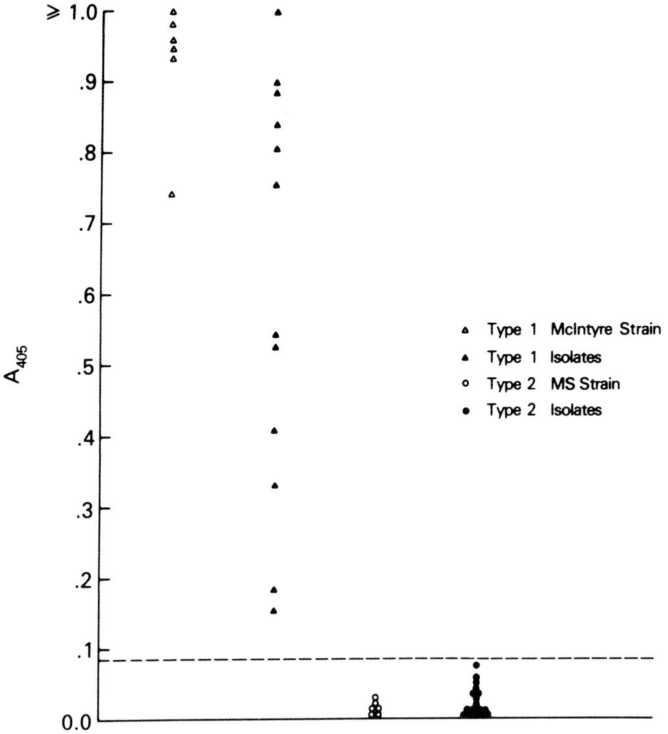

FIG. 7. Identification of HSV type using monoclonal HSV-1-specific antibody in capture B/SA ELISA. Capturing antibody, rabbit polyclonal anti-HSV-1; detecting antibody, biotin-labeled mouse monoclonal HSV-1 specific.

wich or polyclonal anti-HSV-2–monoclonal anti-HSV-2 sandwich, respectively. The clinical HSV-1 isolates and laboratory-purified HSV-1 show good, positive signals, with absorbance values above the cutoff limits (Fig. 7). The cutoff values for the identification of HSV-1 need to be established for each experiment based on the reactions of purified HSV-1 and HSV-2. This allows the distinction between HSV-1 and HSV-2 to be very clear and objective. Confirmation of HSV type can be done using restriction enzyme analysis or by staining with monoclonal antibodies.[10–12]

[11] D. M. Lonsdale, *Lancet* **1,** 849 (1979).

[12] E. Peterson, O. W. Schmidt, L. C. Goldstein, R. C. Nowinski, and L. Corey, *J. Clin. Microbiol.* **17,** 92 (1983).

Acknowledgments

This work was performed by the author while at Infectious Diseases Branch, National Institute of Neurological and Communicative Disorders and Stroke, National Institutes of Health. The continuous support, at the time, by Drs. John L. Sever and David L. Madden is greatly appreciated.

[64] Sensitive Immunoassay for Human Immunodeficiency Viral Core Proteins

By KENNETH R. HUSKINS, KEVIN J. REAGAN, JOHN A. WEHRLY, and JEANNE E. NEUMANN

The use of avidin–biotin technology for immunochemical research applications has been gaining steady acceptance. In particular, the methodology has been used for immunoaffinity chromatography,[1,2] for assessment of antigen epitope specificities,[3] and for infectious disease immunoassays.[4–7] Nearly all of the immunoassay applications have utilized enzyme-coupled signal generation to correlate with the specific analyte, although recently Hart and Taaffe[8] reported a chemiluminescent reporter system for immunoassay which utilizes acridinium ester-labeled streptavidin.

The development of sensitive assays is a critical need in improving the detection, diagnosis, and treatment of disease. It is particularly important in diseases with such devastating ramifications as acquired immunodeficiency syndrome (AIDS). The currently used blood-screening tests are capable of detecting antibodies to human immunodeficiency virus (HIV), the causative viral agent of AIDS. However, these serological tests cannot determine whether or not the specimen actually contains viral pro-

[1] T. V. Updyke and G. L. Nicolson, *J. Immunol. Methods*. **73,** 83 (1984).
[2] D. R. Gretch, M. Suter, and M. F. Stinski, *Anal. Biochem.* **163,** 270 (1987).
[3] C. Wagener, U. Fenger, B. R. Clark, and J. E. Shively, *J. Immunol. Methods* **68,**269 (1984).
[4] J.-L. Guesdon, T. Ternynck, and S. Avrameas, *J. Histochem. Cytochem.* **27,** 1131 (1979).
[5] K. Adler-Storthz, C. Kendall, R. C. Kennedy, R. D. Henkel, and G. R. Dreesman, *J. Clin. Microbiol.* **18,** 1329 (1983).
[6] L. S. Nerurkar, M. Namba, G. Brashears, A. J. Jacob, Y. J. Lee, and J. L. Sever, *J. Clin. Microbiol.* **20,** 109 (1984).
[7] C. M. Nielsen, K. Hansen, H. M. K. Andersen, J. Gerstoft, and B. F. Vestergaard, *J. Virol. Methods* **16,** 195 (1987).
[8] R. C. Hart and L. R. Taaffe, *J. Immunol. Methods* **101,** 91 (1987).

TABLE I
DU PONT HIV p24 CORE ANTIGEN ELISA

Step	Features
Antigen capture	200 μl Triton® X-100-treated sample
	Microwells coated with rabbit anti-p24 IgG, incubation at room temperature overnight (standard assay) or at 37° for 2 hr (short assay)
Antigen detection	Biotinylated second rabbit anti-p24 IgG, incubation at 37° for 2 hr
Signal amplification	Streptavidin–HRP conjugate, incubation at 37° for 30 min
	o-Phenylenediamine/H_2O_2 substrates, 30-min end-point absorbance (492 nm)

teins. For this reason, we have developed a sensitive, specific, and rapid immunoassay to measure HIV antigen. In this chapter we describe the assay format and detail its performance in various clinical settings.

HIV p24 protein is immunologically distinct from the core protein of most other retroviruses. The preparation of monospecific antisera with high specificity and affinity for this viral protein and the use of a streptavidin–biotin amplification system have allowed Du Pont to develop an extremely sensitive ELISA for HIV p24 antigen. The Du Pont HIV p24 Core ELISA test achieves immunological detection of p24 released after sample lysis with 0.5% Triton® X-100 under conditions gentle enough to retain reactivity. Highly specific rabbit polyclonal antisera provide the capture and detector immunoglobin reagents. Inactivated viral lysate (calibrated against immunoaffinity purified p24) serves as the standard. A biotin–streptavidin–horseradish peroxidase (HRP) couple provides the probe in a microtiter plate sandwich ELISA format. A description of the test format is outlined in Table I.

Using this protocol, the amount of assay color generated is directly proportional to the quantity of HIV p24 captured from the specimen. The sensitivity of the p24 ELISA is 30 pg/ml; using a 200-μl sample, the assay detection limit is 6 pg of p24. When the assay is used to detect viral antigen in supernatants of amplified cultures of peripheral blood lymphocytes, direct comparisons with the reverse transcriptase assay have shown an excellent correlation with additional advantages of increased sensitivity (up to 1000-fold) and reproducibility.

A major reason for the excellent sensitivity of the p24 ELISA is the use of the biotin–streptavidin–HRP couple for signal detection. As shown in Table II, a 14-fold increase in specific signal was observed for the

TABLE II
COMPARISON OF RABBIT ANTI-p24 CONJUGATES IN ANTIGEN ASSAY SENSITIVITY

p24 Antigen in sample (ng/ml)	Signal-to-noise ratio[a]	
	Biotinylation[b]	Direct HRP[c]
10	37.17	2.76
5	32.30	2.29
1.0	11.03	1.24
0.5	7.00	1.10
0.25	4.34	1.04

[a] Conjugates were diluted in PBS–50% normal rabbit serum containing 0.01% thimerasol.
[b] Biotinylated antibodies were subsequently reacted with streptavidin–HRP conjugates and *o*-phenylenediamine substrate.
[c] HRP–antibody conjugate.

biotinylated rabbit anti-p24 in comparison to the direct antibody–enzyme conjugate. Other attempts at chemical conjugations involving alkaline phosphatase also failed to achieve the sensitivity of the biotin–streptavidin system. It was concluded that the highest analytical sensitivity for HIV p24 core protein required signal amplification using the indirect biotin–avidin methodology. When biotinylated antibodies are employed as immunoassay reagents, assessment of the isoelectic point is critical to the optimization of these reagents as ELISA components.[9]

The specificity of a positive color response generated in the HIV p24 ELISA is confirmed by specific inhibition of signal by preincubating the sample with human antibody to HIV p24. This procedure forces putative antigen in the sample into immune complexes that are not captured by the anti-p24 coated microplate wells. Then, on retesting in the p24 ELISA, reduction of the signal by at least 50% constitutes evidence of a true positive result.

The p24 ELISA also detects free viral antigen without culture amplification from serum, plasma, cell lysates, and cerebrospinal fluid of individuals at high risk for developing AIDS. The significance of the presence of antigen in such human samples is the subject of ongoing research that is directed towards establishing diagnostic or prognostic links to clinical disease staging or patient management.

[9] J. J. Wadsley and R. M. Watt, *J. Immunol. Methods* **103,** 1 (1987).

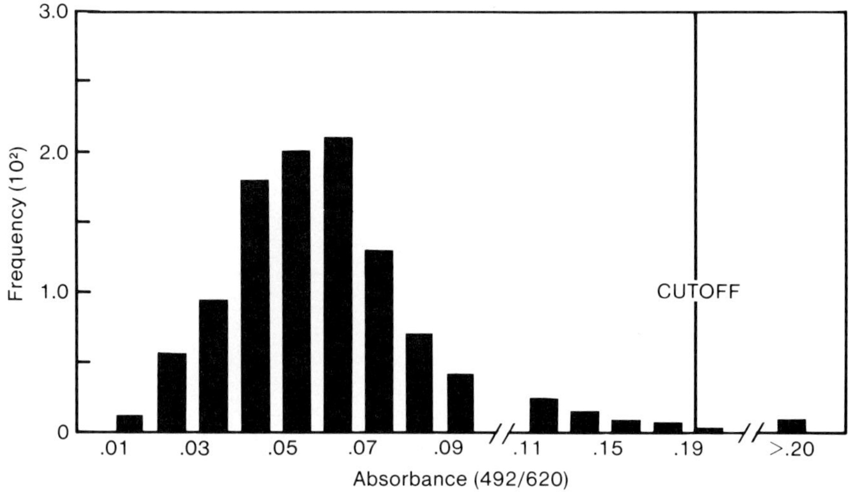

FIG. 1. HIV p24 core antigen ELISA reactivity in a normal donor population [n = 1049; 14 initial reactives (1.3%), 12 repeat reactives (1.1%)]. None of the reactive samples were true positives based on the confirmatory assay methodology (see text). The assay cutoff sensitivity is 0.3 ng/ml of HIV p24.

Our experience with large numbers of seronegative serum and plasma samples obtained from healthy normal donors is shown in Fig. 1. These data were collected using the standard (overnight) format and illustrate a tight absorbance distribution for the large majority of negative samples. An estimate of 1.1% (12 of 1049) for the assay false-positive rate was obtained from the repeat reactive samples that subsequently did not confirm. Extremely hemolysed, lipemic, and icteric samples may be assayed with no interference bias, although clear, nonhemolysed specimens are preferable whenever possible. A low percentage of rheumatoid factor and systemic lupus erythematosus plasma and serum specimens have given false-reactive results.

Table III summarizes the results from a preclinical evaluation of the HIV p24 ELISA as a method for detecting serum antigen in individuals either known to be infected or at high risk for infection with HIV. In this limited, coded sample study, the prevalence of antigenemia in patients with AIDS or AIDS-related complex (ARC) was seen to be greater than 40%. Note that nearly all of these individuals were HIV-antibody seropositive (as determined by the Du Pont HIV Antibody ELISA). In other studies of late-stage AIDS patients with markedly depressed HIV antibody titers, the presence of antigen was found in greater than 70% of such

TABLE III
HIV ANTIGENEMIA IN AIDS, ARC, AND HIGH-RISK INDIVIDUALS

Disease/risk	Number	HIV Western blot positive	Confirmed p24 positive
AIDS	29	29	12 (41%)
ARC	24	23	11 (46%)
High-risk symptomatics	131	101	15 (11%)
High-risk asymptomatics	117	42	2 (1.7%)

specimens. These results suggest that detection of circulating antigen in peripheral blood by immunoassay is more likely in seropositive individuals with advanced disease where reduced antibody titer likely limits the masking of antigen by endogenous immune complex formation.

In conclusion, the use of the biotin–streptavidin–horseradish peroxidase couple for signal detection has contributed to the excellent sensitivity of the Du Pont HIV p24 core antigen ELISA. This assay has demonstrated clinical utility in a variety of settings and provides physicians with another tool in the overall effort for diagnosis and treatment of AIDS.

[65] Gene Probes

By MEIR WILCHEK and EDWARD A. BAYER

Owing to its potential for replacement of radioactive DNA probes, the use of the avidin–biotin system as DNA probes in nucleic acid research is very broad, as can be seen from Chapter [3]. Nonetheless, the number of new methods for introducing biotin into DNA is quite limited, and many of these are included in this volume (Chapters [66]–[72]). Most of the previous studies have used the procedure introduced by David Ward and co-workers in the early 1980s.[1] Using this approach, a biotinyl derivative of dUTP is incorporated into a suitable DNA probe via nick translation. Following hybridization of the biotinylated probe with the target DNA, an avidin-associated probe is used for localization, isolation, and so on.

The major difference in most of these studies is the target DNA, i.e., the organism involved, and for this reason we have not included such

[1] P. R. Langer, A. A. Waldrop, and D. C. Ward, *Proc. Natl. Acad. Sci. U.S.A.* **78,** 6633 (1981).

examples, many of which will eventually find their way into other volumes in this series via other routes. Many studies have been published in which different enzymatic modes of incorporation of a biotinylated nucleotide have been described. In addition, various chemical methods are accumulating in the literature for incorporating the biotin moiety into DNA. We hope that, in the future, additional methods for labeling DNA will be developed and that the system will also become applicable for *in vivo* DNA diagnosis.

[66] Biotinylated Nucleotides for Labeling and Detecting DNA

By LEONARD KLEVAN and GULILAT GEBEYEHU

The emergence of nucleic acid hybridization as a primary tool for research in molecular biology and clinical diagnostics has intensified the search for safe and effective nonradioactive nucleic acid labeling and detection methods. The sensitivity, selectivity, and ease of use of nucleic acid probes and the availability of efficient protocols for the isolation and cloning of specific DNA sequences have led to the development of a wide selection of probes for biomedical and clinical applications.[1] However, the use of DNA hybridization in other than research applications has been limited by problems associated with the handling and disposal of radioisotope-labeled probes. For this reason, the ability to label DNA or RNA probes with stable, nonradioactive reporter groups and to detect the annealed probe–target hybrid with a safe and efficient detection system is a topic of intense current interest.

In the basic hybridization protocol, a labeled nucleic acid probe is annealed to a complementary DNA or RNA target sequence, which is either in solution or immobilized on an inert support. The labeled nucleic acid probe is used to determine the presence or absence of the target sequence in the reaction mixture. The detection system, which is usually composed of an amplification and a visualization component, must recognize and signal the presence or absence of the probe–target hybrid with a high degree of sensitivity and selectivity. Examples of nonradioactive amplification/visualization systems that have been used in DNA detection

[1] P. Zwadyk, Jr., and R. C. Cooksey, *in* "Critical Reviews in Clinical Laboratory Sciences" (J. Batsakis and J. Savory, eds.), p. 71. CRC Press, Boca Raton, Florida, 1987.

protocols include enzyme-catalyzed reduction of a dye to an insoluble precipitate,[2] enzyme-catalyzed enhanced chemiluminescence,[3] time-resolved fluorometry,[4] and bacterial transformation leading to the formation of colored bacterial colonies on an agar plate.[5]

An effective nonradioactive detection system should recognize the annealed nucleic acid probe with a degree of precision comparable to that obtained in the primary hybridization reaction. For this reason, the high binding affinity and specificity of the avidin (streptavidin)–biotin interaction have been used to direct the amplification component of nonradioactive detection systems to hybridized nucleic acid probes. In recent years, procedures have been developed to incorporate biotin directly into DNA and RNA probes in a manner that allows subsequent detection by a streptavidin-conjugated system. The biotinylated probes are stable and may be stored for several years, present no radiation hazard, are potentially easier to use than radioisotope-labeled probes, and may be used in situations where radioisotopes are difficult to obtain.

Synthesis and Characterization of Biotin–Nucleotide Analogs

General Considerations

In 1981, Langer *et al.*[6] reported the chemical synthesis and properties of dUTP and UTP nucleotide analogs containing biotin attached to the 5 position of the uridine base through an allylamine linker (Fig. 1). These biotinylated nucleotide analogs were shown to be effective substrates for a variety of DNA and RNA polymerases, including *Escherichia coli* DNA polymerase, T4 DNA polymerase, and T7 RNA polymerase. It was also demonstrated that nucleic acid probes with approximately 5% of the dTMP residues replaced with biotinylated dUMP hybridized in solution to complementary sequences with reassociation rates comparable to nonbiotinylated probes. The synthesis of additional biotinylated dUTP analogs was reported by Brigati *et al.*,[7] who prepared nucleotides with linkers of various lengths. These compounds were designated as biotin-*n*-dUTP,

[2] J. J. Leary, D. J. Brigati, and D. C. Ward, *Proc. Natl. Acad. Sci. U.S.A.* **80,** 4045 (1983).

[3] J. A. Matthews, A. Batki, C. Hynds, and L. J. Kricka, *Anal. Biochem.* **151,** 205 (1985).

[4] P. Dahlen, *Anal. Biochem.* **164,** 78 (1987).

[5] J. L. Hartley, M. Berninger, J. A. Jessee, F. R. Bloom, and G. F. Temple, *Gene* **49,** 295 (1986).

[6] P. R. Langer, A. A. Waldrop, and D. C. Ward, *Proc. Natl. Acad. Sci. U.S.A.* **78,** 6633 (1981).

[7] D. J. Brigati, D. Myerson, J. J. Leary, B. Spalholz, S. Z. Travis, C. K. Y. Fong, G. D. Hsiung, and D. C. Ward, *Virology* **126,** 32 (1983).

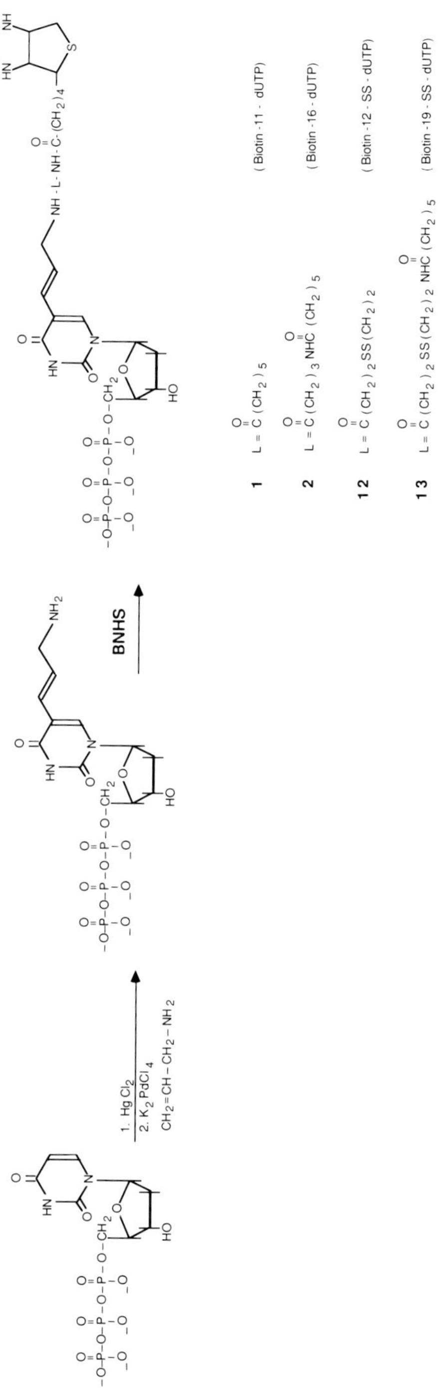

FIG. 1. Synthetic scheme for biotinylated dUTP analogs.

where n refers to the number of atoms between the biotin residue and the 5 position of uridine. Leary *et al.*[2] showed that DNA probes labeled with biotin-11-dUTP **(1)** or biotin-16-dUTP **(2)** could be detected via avidin-linked biotinylated alkaline phosphatase with a high degree of sensitivity in mixed-phase hybridizations.

We have recently reported the synthesis of biotinylated dATP and dCTP nucleotide analogs containing biotin linked to the N-6 and N-4 positions of the bases, respectively, through linkers which vary in length between 3 and 17 atoms.[8] We have also shown that these nucleotides can be incorporated into DNA by standard nick-translation protocols and that probes labeled with these biotinylated nucleotides may be effectively employed in sensitive DNA detection protocols. Identical Southern blots containing an *Eco*RI digest of human genomic DNA were hybridized to either biotin-7-dATP **(6)** or biotin-7-dCTP **(10)** labeled pBR322 plasmid DNA which contained a 1.1-kb human β-globin gene fragment (Fig. 2). In both cases, the expected 5.2-kb β-globin gene band was detected on the filters from 2.0 μg genomic DNA when detection was performed with a sensitive streptavidin–alkaline phosphatase conjugate [the BluGENE System, Bethesda Research Laboratories (BRL), Gaithersburg, MD].

An alternative method for modifying DNA with biotin or other reporter groups is to first incorporate the dATP analog N^6-(ω-aminoalkyl)dATP **(5)** into DNA and then react the amine-labeled DNA with an activated reporter group, e.g., biotin-*N*-hydroxysuccinimide ester (BNHS). Synthesis of the nucleotide analog N^6-(6-aminohexyl)dATP has been previously described,[8] and it has been shown that the nucleotide is an efficient substrate for a variety of DNA polymerases. A general method for the synthesis of biotinylated nucleotides and N^6-(ω-aminoalkyl)dATP analogs is presented in Fig. 3. 6-Chloropurine-2′-deoxyriboside 5′-monophosphate **(3)** is first converted to the corresponding nucleoside triphosphate by treatment with carbonyldiimidazole and tributylammonium pyrophosphate.[9] Treatment of the 6-chloropurine-2′-deoxyriboside 5′-triphosphate **(4)** with diaminoalkane then gives N^6-(ω-aminoalkyl)dATP. For example, when **4** was treated with 1,6-diaminohexane, N^6-(6-aminohexyl)dATP (**5**, $n = 6$) is obtained in good yield. The amine-labeled nucleotide triphosphate may be combined with amine-reactive reporter groups to construct a variety of modified nucleotide triphosphates.

[8] G. Gebeyehu, P. Y. Rao, P. SooChan, D. A. Simms, and L. Klevan, *Nucleic Acids Res.* **15,** 4513 (1987).

[9] D. E. Hoard and D. G. Otts, *J. Am. Chem. Soc.* **87,** 1785 (1965).

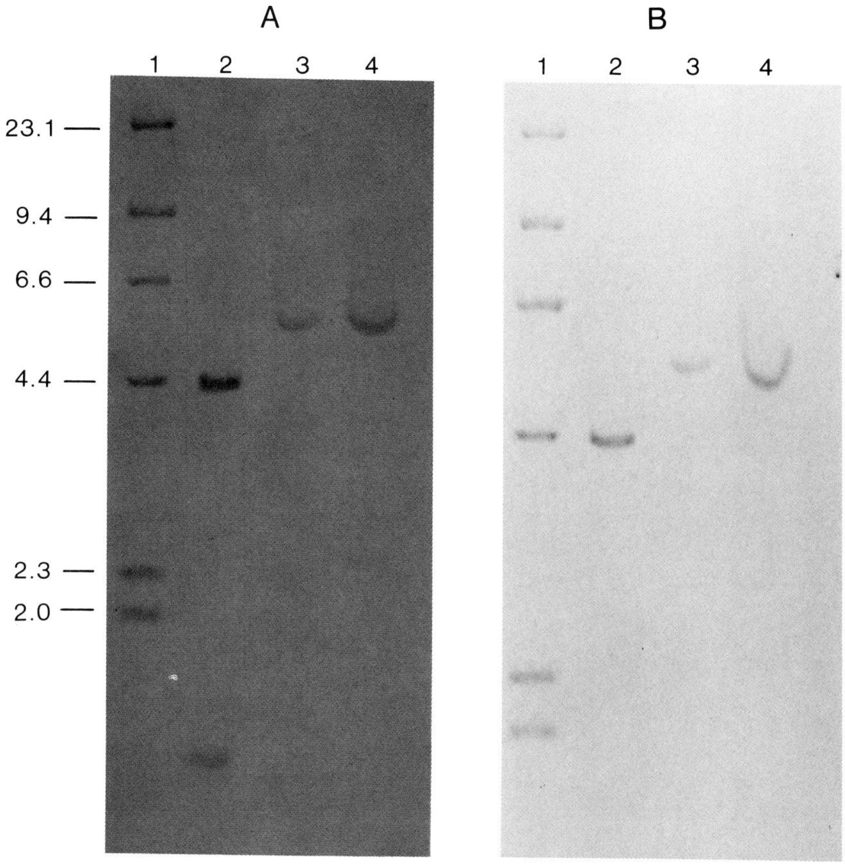

FIG. 2. Detection of single-copy genes. The plasmid pBR322, containing a 1.1-kb *Mst*II fragment of the human β-globin gene in the *Eco*RI site, was nick translated with biotin-7-dATP (A) or biotin-7-dCTP (B) and hybridized to a Southern blot containing human genomic DNA electrophoresed on a 1% agarose gel. The probe concentration was 100 ng/ml, and hybridization was performed in a buffer containing 45% formamide for 18 hr at 42°. The hybrids were detected by an alkaline phosphatase–streptavidin detection system. Lane 1 contains 0.1 μg biotinylated λ *Hin*dIII markers, lane 2 contains 5 pg of the *Eco*RI-digested plasmid DNA, and lanes 3 and 4 contain *Eco*RI-cut human DNA, 2 and 10 μg, respectively. (From Gebeyehu *et al.*[8])

The methods described above may be used to prepare a wide range of possible dATP nucleotide analogs (Table I). For example, the length of the linker arm between biotin and the nucleotide base may be varied [biotin-7-dATP **(6)** versus biotin-14-dATP **(7)**] or a biotinylated analog with a cleavable linker [biotin-15-SS-dATP **(8)**] may be prepared. Starting

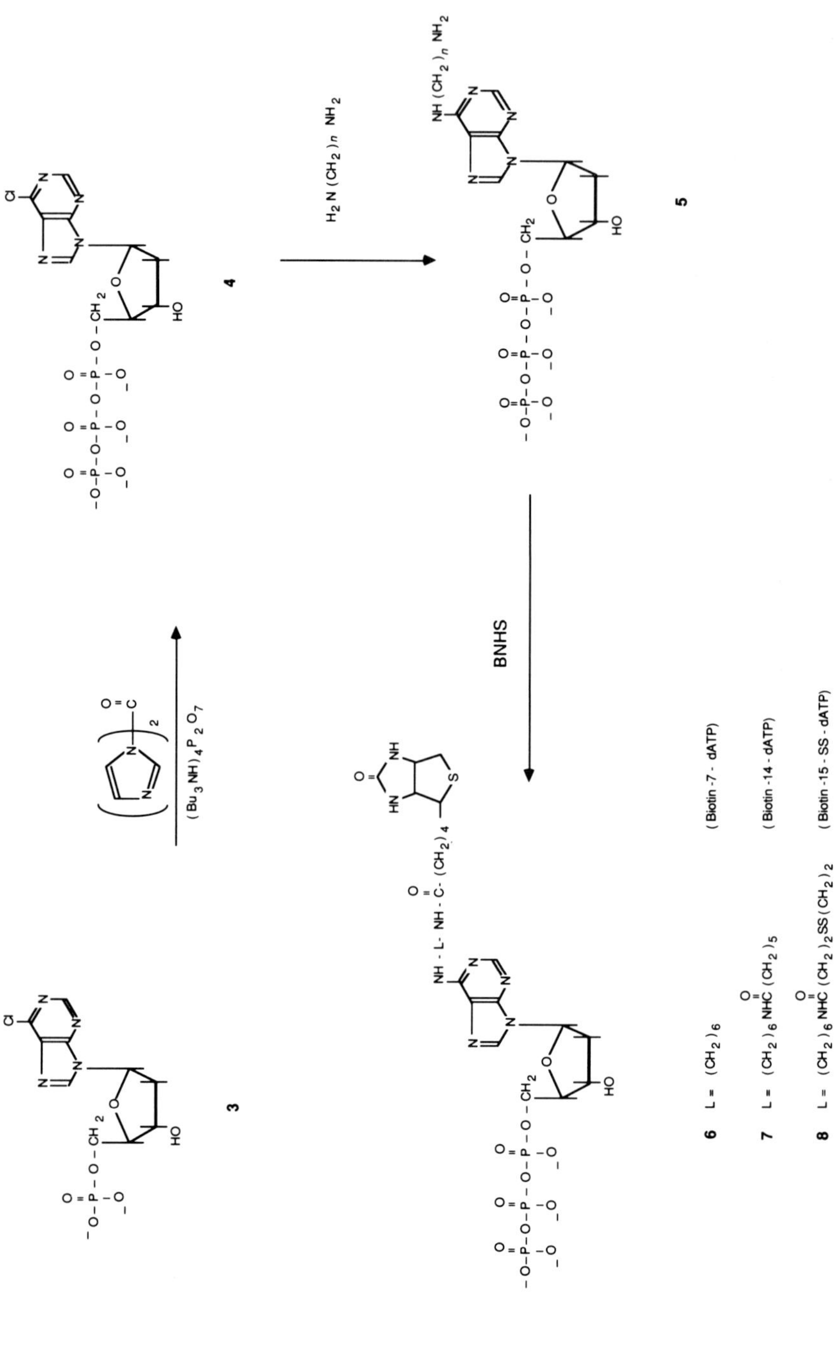

FIG. 3. Synthetic scheme for biotinylated dATP analogs.

TABLE I
CHARACTERIZATION OF BIOTIN NUCLEOTIDES AND PRECURSORS

Nucleotide analog	Molecular formula[a]	Molecular weight[a]	λ_{max}[b]	HPLC (retention time, min)[c]
N^6-(6-Aminohexyl)dATP	$C_{16}H_{29}N_6O_{12}P_3$	590.3	265	7.6
Biotin-7-dATP **(6)**	$C_{26}H_{43}N_8O_{14}P_3S$	816.6	265	9.0
Biotin-14-dATP **(7)**	$C_{32}H_{54}N_9O_{15}P_3S$	929.8	265	10.9
Iminobiotin-7-dATP	$C_{26}H_{44}N_9O_{13}P_3S$	815.6	266	10.6
Biotin-15-SS-dATP **(8)**	$C_{31}H_{52}N_9O_{15}P_3S_3$	979.9	265	11.4
5-(3-Amino)allyl-dUTP	$C_{12}H_{20}N_3O_{14}P_3$	523.2	286	2.6
Biotin-11-dUTP **(1)**	$C_{28}H_{45}N_6O_{17}P_3S$	862.7	288	8.7
Biotin-12-SS-dUTP **(12)**	$C_{27}H_{43}N_6O_{17}P_3S_3$	912.8	288	9.5
N^4-(6-Aminohexyl)dCTP	$C_{15}H_{29}N_4O_{13}P_3$	566.2	270	3.1
Biotin-7-dCTP **(10)**	$C_{25}H_{43}N_6O_{15}P_3S$	792.6	270	9.0
Biotin-14-dCTP **(11)**	$C_{31}H_{54}N_7O_{16}P_3S$	905.8	270	9.8

[a] Molecular formula and molecular weight are for the free acid.
[b] All UV spectra were recorded on a Beckman 25 dual-beam spectrophotometer.
[c] HPLC analysis was performed on a Waters Associates Chromatograph using a Radial-PAK C_{18} column with a Z-module radial compression system. A 10-min gradient from 90% 50 mM $NH_4H_2PO_4$ and 10% methanol–water (7:3) to 5% 50 mM $NH_4H_2PO_4$ and 95% methanol–water (7:3) was used for elution. The flow rate was maintained at 2 ml/min.

with the 6-chloropurine-riboside 5′-monophosphate, biotinylated ATP analogs may also be prepared by the synthetic methods outlined in Fig. 3. Biotin-7-ATP and biotin-14-ATP have been prepared in this manner.

A general procedure for the synthesis of biotinylated dCTP analogs is given in Fig. 4. dCTP is first subjected to a transamination reaction with sodium bisulfite and a diaminoalkane to give N^4-(ω-aminoalkyl)dCTP **(9)**.[10] N^4-(ω-Aminoalkyl)dCTP is then treated with the *N*-hydroxysuccinimide ester of either biotin or a biotin analog to obtain the desired compound. Biotin-7-dCTP **(10)** and biotin-14-dCTP **(11)** are prepared by this procedure. CTP may be treated in the same manner to obtain biotin-7-CTP or biotin-14-CTP.

The biotinylated nucleotides described above are characterized by UV absorption and HPLC (Table I). The biotinylated dATP and ATP analogs exhibit maximum absorbance (λ_{max}) at 265 nm whereas biotinylated dCTP and CTP analogs have a λ_{max} of 270 nm as expected. Biotin-7-dATP and biotin-7-ATP can also be characterized by negative ion fast atom bom-

[10] D. E. Draper, *Nucleic Acids Res.* **12,** 989 (1984).

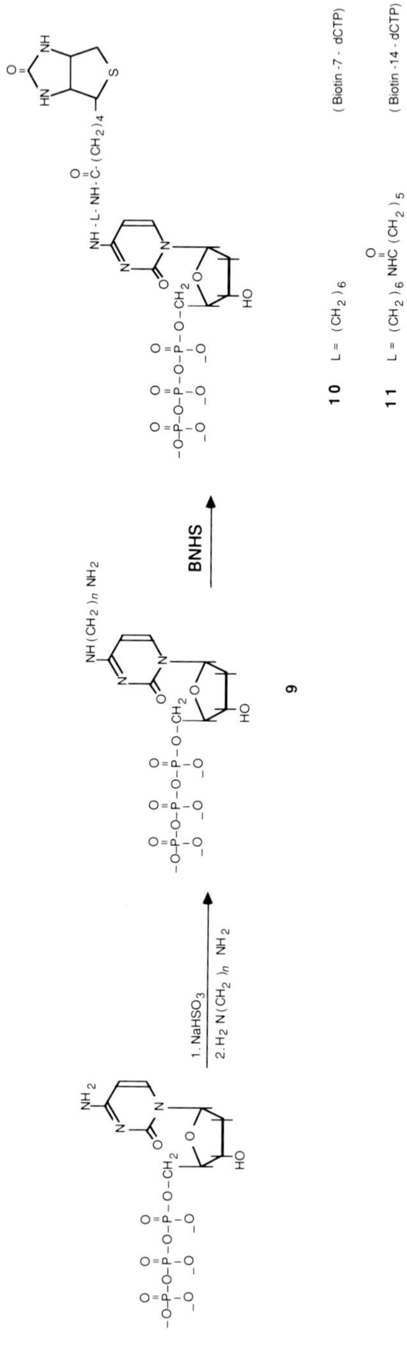

FIG. 4. Synthetic scheme for biotinylated dCTP analogs.

bardment mass spectra; M − H peaks of 815 and 831 m/z are observed, respectively, consistent with the expected molecular weights of 816 and 832 for the free acids. In addition, the characteristic fragmentation patterns corresponding to the di- and monophosphates are observed.

The desired products in each step of the synthesis of the biotinylated nucleotides described above are isolated by ion-exchange chromatography and their purity assayed by HPLC. We have found that reversed-phase HPLC is useful both for determining the purity and for following the progress of the synthesis. As demonstrated in Table I, compounds elute in the predicted manner when analyzed by this method. For example, compounds that become more lipophilic by addition of a biotin moiety will generally exhibit longer retention times in reversed-phase systems [compare, e.g., N^6-(6-aminohexyl)dATP and biotin-7-dATP]. The presence of either a free amine or biotin may also be ascertained by chemical analysis. Compounds containing primary amines give positive color development (pink) when spotted on TLC plates and sprayed with ninhydrin. The presence of biotin may be determined by a positive color reaction (orange-purple), when the compound is spotted on a TLC plate and sprayed with *p*-dimethylaminocinnamaldehyde in ethanolic H_2SO_4.[11]

Releasable Biotin–Nucleotide Analogs

The biotin–avidin/streptavidin interaction is reversible only under very harsh conditions (e.g., 6.0 *M* guanidine-HCl, pH 1.5).[12] However, two approaches have been used to synthesize derivatized nucleotides in which the high degree of specificity and selectivity of the biotin–avidin interaction may be reversed under milder reaction conditions. The first approach uses iminobiotin, an analog of biotin in which the ureido group has been replaced with a guanido group. Green has demonstrated that iminobiotin binds to avidin with a binding constant that increases with increasing pH and that nonprotonated iminobiotin binds efficiently to avidin.[12] Thus, macromolecules labeled with iminobiotin may be bound to streptavidin–agarose at basic pH (e.g., pH 11) while dissociation is achieved at acidic pH (e.g., pH 4). This principle has been successfully exploited for the purification of streptavidin from *Streptomyces* culture broth using an iminobiotin affinity column.[13] We have synthesized the releasable nucleotide analog iminobiotin-7-dATP by condensing N^6-(6-aminohexyl)dATP with iminobiotin-*N*-hydroxysuccinimide ester. This

[11] D. B. McCormick and J. A. Roth, *Anal. Biochem.* **34,** 226 (1970).

[12] N. M. Green, *Biochem. J.* **101,** 774 (1966).

[13] K. Hofmann, S. W. Wood, C. C. Brinton, J. A. Montibeller, and F. M. Finn, *Proc. Natl. Acad. Sci. U.S.A.* **77,** 4666 (1980).

nucleotide can be incorporated into DNA by standard labeling methods, such as nick translation, when reversible binding of the DNA to streptavidin is desired.

The second approach for construction of a releasable biotin–streptavidin system utilizes a biotin-labeling reagent containing a cleavable linker arm attached to biotin. For example, a macromolecule may be labeled with biotin through a linker arm containing a disulfide bond, and the labeled macromolecule may be bound to streptavidin–agarose under standard conditions. When the release of the macromolecule from the resin is desired, the complex may be treated with dithiothreitol (pH 8) to cleave the disulfide, effectively freeing the nucleotide.

Herman and co-workers[14] synthesized biotin-12-SS-dUTP (**12**, Fig. 1) and biotin-19-SS-dUTP (**13**, Fig. 1) and demonstrated that these nucleotides are efficiently incorporated into DNA by nick translation. The DNA thus obtained binds to streptavidin- and/or avidin–agarose and can be released from the resin with dithiothreitol. They also demonstrated that the efficiency of the cleavage reaction is dependent on the size of the linker between biotin and the nucleotide base. Thus, release of biotin-19-SS-labeled DNA from streptavidin–agarose was 10 times faster than for DNA labeled with the biotin-12-SS analog. This principle was employed by Rigas *et al.*[15] for the development a rapid plasmid isolation method using RecA protein-coated biotinylated probes that had been labeled with biotin-19-SS-dUTP. The dATP analog biotin-15-SS-dATP (**8**, Fig. 3), synthesized by the reaction of sulfosuccinimidyl 2-(biotinamido)ethyl-1,3-dithiopropionate with N^6-(6-aminohexyl)dATP, has also been successfully used for the reversible attachment of biotinylated probes to a streptavidin–agarose matrix.[16]

Synthesis of Biotin-7-dATP (6)

Materials. The nucleosides and nucleoside triphosphates used for organic synthesis are from Sigma Chemical Company (St. Louis, MO); Sephadex A-25 is purchased from Pharmacia (Piscataway, NJ). All other reagents are products of Aldrich (Milwaukee, WI). 6-Chloropurine-2′-deoxyriboside 5′-monophosphate is prepared as described previously.[8] Biotin-*N*-hydroxysuccinimide ester (BNHS) and caproylamidobiotin-*N*-hydroxysuccinimide ester (CAB-NHS) are from BRL.

[14] T. M. Herman, E. Lefever, and M. Shimkus, *Anal. Biochem.* **156,** 48 (1986).

[15] B. Rigas, A. A. Welcher, D. C. Ward, and S. M. Weissman, *Proc. Natl. Acad. Sci. U.S.A.* **83,** 9591 (1986).

[16] J. M. D'Alessio, A. W. Hammond, and D. K. Chatterjee, *Gene* **71,** 49 (1988).

6-Chloropurine-2′-deoxyriboside 5′-Triphosphate **(4).** 6-Chloropurine-2′-deoxyriboside 5′-monophosphate (triethylammonium salt, 0.3 mmol) is dissolved in 4 ml of dimethylformamide (DMF). Carbonyldiimidazole (240 mg, 1.5 mmol) is added, and the reaction mixture is stirred for 15 min. HPLC analysis shows the starting material is completely consumed. Methanol (99 μl) is added, and stirring is continued for 30 min to decompose the excess carbonyldiimidazole. Tri-*n*-butylammonium pyrophosphate (669 mg in 3 ml DMF) is then added, and the reaction mixture is placed on a magnetic stirrer overnight. The mixture is filtered and concentrated. The gummy residue which results is dissolved in 10 ml of water and loaded onto a Sephadex A-25 column (HCO_3^- form, 80 ml) which is equilibrated with 0.1 *M* triethylammonium bicarbonate (TEAB). The column is washed with 200 ml of 0.1 *M* TEAB and then eluted with a gradient from 0.1 to 1 *M* TEAB (500 ml each). The desired fractions are combined, concentrated, and coevaporated with ethanol to give the desired compound in 63% yield. The compound displays a single peak when analyzed by HPLC (retention time 6.1 min).

N^6-(6-Aminohexyl)dATP **(5).** 1,6-Diaminohexane (714 mg) is dissolved in water (48 ml), and the pH is adjusted to approximately 10 with carbon dioxide. The 6-chloropurine-2′-deoxyriboside 5′-triphosphate (0.614 mmol) in water (5 ml) is added, and the reaction mixture is heated at 55°. HPLC analysis after 1.5 hr reveals that the starting material is completely consumed. The reaction mixture is cooled to room temperature, diluted to 250 ml with water and loaded onto a Sephadex A-25 column (HCO_3^- form, 130 ml) equilibrated with 0.1 *M* TEAB. The column is washed with 250 ml 0.1 *M* TEAB followed by a gradient of 0.1 to 1.0 *M* TEAB (500 ml each). The fractions containing the desired material are combined, concentrated, and coevaporated with ethanol for a final yield of 71%. The compound displays a single peak when analyzed by HPLC (retention time 7.6 min) and gives a positive color development when spotted on a TLC plate and sprayed with ninhydrin.

Biotin-7-dATP **(6).** BNHS (52 mg, 0.15 mmol) is dissolved in DMF (1.5 ml) and added to a solution of N^6-(6-aminohexyl)dATP (0.077 mmol) in 0.1 *M* sodium borate (pH 8.5, 10 ml). The mixture is stirred for 1.5 hr, at which time HPLC analysis shows that the starting material is consumed. The reaction mixture is then concentrated to remove the DMF. The gummy material obtained is dissolved in water and loaded onto a Sephadex A-25 column (HCO_3^- form, 70 ml) which is equilibrated with 10 m*M* TEAB. The column is washed with 200 ml of 10 m*M* TEAB and then eluted with a gradient of 10 m*M* to 1.0 *M* TEAB (400 ml each). The desired fractions are combined, concentrated, and coevaporated with ethanol. A yield of 82% is obtained. The compound appears homogeneous when analyzed by HPLC (retention time 9.0 min). The compound gives a positive color development when spotted on a TLC plate and sprayed with *p*-dimethylaminocinnamaldehyde in ethanolic sulfuric acid,[11] which is indicative of the presence of biotin.

Incorporation of Biotinylated Nucleotides into DNA

General Considerations

The strategy for preparation of biotinylated DNA probes is often different from that employed for construction of radioisotope-labeled probes. In general, the sensitivity obtained in autoradiography or other radioisotope detection procedures is directly related to the efficiency of incorporation of radioactive nucleotides into DNA. However, the sensitivity of detection with biotin-labeled probes may be limited by the physical constraints of the nonisotopic detection system, as shown by the results in Table II. Biotinylated probes are prepared by nick translation (see below) in the presence of biotin-7-dATP and the three other unlabeled nucleoside triphosphates. Aliquots of the reaction mixture are removed at specific time intervals, purified, and applied to nitrocellulose membranes to immobilize the DNA. In these experiments, incorporation of biotin into DNA beyond the level of 7–35 biotins/kb does not increase the sensitivity of detection by streptavidin and biotinylated alkaline phosphatase. Likewise, labeling with more than one biotinylated nucleotide

TABLE II
SENSITIVITY OF DETECTION OF BIOTINYLATED DNA[a]

Time (min)	Number of biotin-7-dATP molecules/kb	Concentration of DNA (pg)						
		50	20	10	5	2	1	0
90	98	+	+	+	+	+	+	−
20	35	+	+	+	+	+	+	−
10	7	+	+	+	+	+	−	−

[a] The nick-translation reaction volume was 1 ml. At 10, 30, and 90 min; 200 μl of the reaction mix was removed and quenched with 30 m*M* EDTA and the DNA was purified on a Sephadex G-50 column. One to 50 pg of DNA was spotted onto a nitrocellulose filter and detected with an alkaline phosphatase–streptavidin detection system. + or − indicates presence or absence of color in dot spots after detection. The filter was developed in the dye mixture for 2 hr. From Gebeyehu *et al.*[8]

will not result in higher levels of sensitivity in subsequent detection protocols. This threshold for detection of biotinylated DNA observed at relatively low levels of biotin incorporation may be due to steric constraints on the binding of the streptavidin or biotinylated enyzme to the DNA.

Enzymatic Incorporation of Biotinylated Nucleotides

In general, biotinylated nucleotides are compatible with most enzymatic labeling protocols but exhibit lower levels of incorporation than corresponding canonical nucleotides. It has been shown, however, that high levels of biotin incorporation can adversely affect the reassociation kinetics of the labeled probe[17] and are usually not required to achieve high sensitivity in subsequent detection protocols. An efficient method for incorporating biotinylated nucleotides into double-stranded DNA is nick translation, which makes use of the combined enzymatic activities of DNase I and *E. coli* DNA polymerase I. In the standard nick translation reaction, the nicks generated by the action of DNase I on the duplex DNA are "translated" along the DNA template by the $5' \rightarrow 3'$-exonuclease activity and $5' \rightarrow 3'$-polymerase activity of DNA polymerase I. If a biotinylated deoxynucleoside triphosphate (e.g., biotin-7-dATP) is present in the reaction mixture, it will become incorporated into the newly synthesized DNA. The final size distribution of the DNA probe can be controlled by the ratio of DNA polymerase I to DNase I in the reaction mixture and the incubation time of the reaction. Generally, the enzyme ratio should be adjusted to yield probes in the size range of 500 to 1500 bases[18] when assayed on alkaline agarose gels.[19]

Alternatively, for small probes (<500 base pairs), DNA may be efficiently labeled by a modification of the random priming method of Feinberg and Vogelstein.[20] The random priming labeling method is an effective method for incorporation of biotinylated nucleotides into small DNA fragments and DNA isolated from agarose gels. The procedure relies on the hybridization of a mixture of random sequence oligodeoxyribonucleotides to heat-denatured DNA and efficient initiation of polymerization from the annealed template–primer complexes by the large fragment (Klenow fragment) of *E. coli* DNA polymerase I. Other enzymatic labeling methods that have been successfully used to incorporate biotinylated nucleotides

[17] R. P. Viscidi, C. J. Connelly, and R. H. Yolken, *J. Clin. Microbiol.* **23,** 311 (1986); R. P. Viscidi, this volume [70].

[18] J. Meinkoth and G. Wahl, *Anal. Biochem.* **138,** 267 (1984).

[19] M. W. McDonell, M. N. Simon, and F. W. Studier, *J. Mol. Biol.* **110,** 119 (1977).

[20] A. P. Feinberg and B. Vogelstein, *Anal. Biochem.* **132,** 6 (1983).

into DNA include 3′-end labeling with Klenow fragment[21] and replacement synthesis with T4 DNA polymerase.[22]

Nick Translation. A commercial preparation of DNA polymerase I–DNase I for nick translation and the nucleotide analog biotin-7-dATP (0.4 m*M* solution) are available from BRL, and nick-translation reagent systems are available from several commercial sources. To achieve efficient labeling by nick translation, all four deoxyribonucleoside triphosphates should be present at concentrations of approximately 20 μM.

Pipet the following into a 1.5-ml microcentrifuge tube (on ice): 5 μl nucleotide mixture [0.2 m*M* dCTP, 0.2 m*M* dGTP, 0.2 m*M* dTTP; in 500 m*M* Tris-HCl (pH 7.8), 50 m*M* $MgCl_2$, 100 m*M* 2-mercaptoethanol, and 100 μg/ml nuclease-free bovine serum albumin (BSA)], 1.0 μg DNA [in 10 m*M* Tris-HCl (pH 7.5), 120 m*M* NaCl, and 0.1 m*M* Na_2EDTA or similar buffer], 2.5 μl of 0.4 m*M* biotin-7-dATP, and distilled water to a final volume of 45 μl. Add 5 μl polymerase I–DNase I enzyme mixture [e.g., nick-translation grade reagent from BRL: 0.4 U/μl DNA polymerase I, 40 pg/μl DNase I in 50 m*M* Tris-HCl (pH 7.5), 5.0 m*M* magnesium acetate, 1.0 m*M* 2-mercaptoethanol, 0.1 m*M* phenylmethylsulfonyl fluoride, 50% glycerol, and 100 μg/ml nuclease-free BSA]. Mix thoroughly but gently, centrifuge briefly to bring liquid to the bottom of the tube, and incubate at 15° for 90 min.

After adding 5.0 μl of 300 m*M* Na_2EDTA (pH 8.0) and 1.25 μl of 5% (w/v) SDS, load the nick translation reaction onto a 5- to 10-ml Sephadex G-50 (Pharmacia) gel filtration column equilibrated with 1× SSC [0.15 *M* NaCl, 15 m*M* sodium citrate (pH 7.0)] and 0.1% (w/v) SDS to separate labeled DNA from unincorporated nucleotides. The presence of SDS in the elution buffer reduces nonspecific adsorption of the biotinylated DNA to the column and tubes. The probe may be tracked by addition of 0.5% Blue Dextran to the reaction mixture prior to fractionation, since both DNA and dextran will elute in the void volume of the column. Biotin-labeled probes may be stored in solution at −20° for at least 1 year.

Random Priming. Random sequence oligonucleotides and complete reagent systems for random priming are commercially available. However, standard labeling protocols should be modified[23] for incorporation of biotinylated dATP as described below.

Denature 25 ng of linear DNA dissolved in 5–20 μl dilute buffer in a microcentrifuge tube by heating in a boiling water bath for 5 min, then cool on ice. Add the following to the microcentrifuge tube sitting on ice:

[21] A. Murasugi and R. B. Wallace, *DNA* **3,** 269 (1984).

[22] J. D'Alessio, *Focus* **7,** 13 (1985).

[23] S. Bromley and R. Pless, unpublished results.

5 μl nucleotide mix [0.2 m*M* each dATP, dTTP, dCTP, dGTP; in 3 m*M* Tris-HCl (pH 7.0), 0.2 m*M* Na_2EDTA], 15 μl buffer [0.67 *M* HEPES, 0.17 *M* Tris-HCl, 17 m*M* $MgCl_2$, 33 m*M* 2-mercaptoethanol, 1.33 mg/ml nuclease-free BSA, 18 OD_{260} units/ml random sequence hexanucleotides (pH 6.8)], 1.2 μl of 0.2 m*M* biotin-7-dATP, and distilled water to a final volume of 49 μl. Mix briefly, add 1.0 μl Klenow fragment (3 U/μl), incubate the reaction mixture for 2 hr, add an additional 3 U of Klenow fragment, and continue the incubation for an additional 6 hr or longer. Add 5 μl of 0.2 *M* Na_2EDTA to stop the reaction and purify the DNA by column chromatography as described above. The random priming labeling method for incorporation of biotin-7-dATP may be scaled up in a linear fashion to produce larger quantities of the labeled probe.

Direct Incorporation of Biotin into DNA

Several chemical methods have been introduced for the direct attachment of biotin to the nucleotide bases of intact DNA probes. In 1985, Forster *et al.*[24] described the novel biotinylation reagent "photobiotin," which contains the biotin moiety attached to a photoreactive arylazide group through a positively charged linker. On photoactivation with visible light, the aryl azide is rapidly converted to the highly reactive aryl nitrene, which then incorporates the biotin group into the nucleic acid. However, DNA to be photobiotinylated must be highly purified since the aryl azide will also react with contaminants in the DNA probe preparation, including RNA, proteins, and organic buffers. Photobiotinylated DNA probes containing 5–10 biotins/kb DNA have been successfully used for the detection of single-copy genes in Southern blots of human genomic DNA and for detection of low-abundance mRNAs in Northern blots of total cytoplasmic RNA.[25]

An alternative method for direct chemical labeling of DNA probes is the transamination reaction of cytosine residues in denatured DNA with sodium bisulfite and a diaminoalkane.[17] This procedure introduces primary amino groups into the DNA which may then be labeled with an activated reporter group such as caproylamidobiotin-*N*-hydroxysuccinimide ester. It was demonstrated that DNA probes prepared in this manner with 2.9–10.3% of bases substituted with biotin can serve as hybridization probes and detect 1–2 pg DNA in dot-blot assays using an alkaline phosphatase-based detection system. Unpaired cytosine residues

[24] A. C. Forster, J. L. McInnes, D. C. Skingle, and R. H. Symons, *Nucleic Acids Res.* **13,** 745 (1985); see also J. L. McInnes, A. C. Forster, D. C. Skingle, and R. H. Symons, this volume [69].

[25] J. L. McInnes, S. Dalton, P. D. Vize, and A. J. Robins, *Bio/Technology* **5,** 269 (1987).

in DNA have also been biotinylated in a one-step transamination reaction using sodium bisulfite and biotin hydrazide.[26] DNA plasmids labeled by this procedure were shown to be effective as hybridization probes when detected with a sensitive streptavidin–alkaline phosphatase conjugate.

Applications of Biotinylated DNA Probes

Biotinylated DNA probes have been used in a wide variety of research applications and hybridization formats. For example, the specificity and selectivity of the biotin–streptavidin interaction have been successfully employed for the affinity selection of plasmids from a cDNA library using RecA protein,[15] for the development of a solution hybridization assay for rRNA from bacterial cell lysates,[27] and for purification of mRNA splicing complexes containing biotinylated pre-mRNA.[28] Biotinylated probes have also been successfully used in sandwich hybridization assays with europium-labeled streptavidin and a chemiluminescent enhancement system,[29] and for detection of single-copy restriction fragment length polymorphisms (RFLPs) in restriction fragment digests of human genomic DNA.[30–32] The strong binding interaction of biotin-labeled probes to streptavidin–agarose resins has also been applied to the isolation and separation of complementary strands from DNA restriction fragments that had been 3′-end-labeled with biotinylated nucleotides.[33]

Biotinylated probes have been incorporated into nonisotopic filter-based detection systems for a wide variety of infectious agents, including cytomegalovirus,[34,35] bovine herpesvirus,[36] enterotoxigenic *E. coli*,[37] peri-

[26] A. Reisfeld, J. M. Rothenberg, E. A. Bayer, and M. Wilchek, *Biochem. Biophys. Res. Commun.* **142,** 519 (1987); see also M. Wilchek, J. M. Rothenberg, A. Reisfeld, and E. A. Bayer, this volume [71].

[27] C. O. Yehle, W. L. Patterson, S. J. Boguslawski, J. P. Albarella, K. F. Yip, and R. J. Carrico, *Mol. Cell. Probes* **1,** 177 (1987).

[28] P. J. Grabowski and P. A. Sharp, *Science* **233,** 1294 (1986); see also P. J. Grabowski, this volume [34].

[29] P. Dahlen, A.-C. Syvanen, P. Hurskainen, M. Kwiatkowski, C. Sund, J. Ylikoski, H. Soderlund, and T. Lovgren, *Mol. Cell. Probes* **1,** 159 (1987).

[30] D. Dykes, J. Fondell, P. Watkins, and H. Polesky, *Electrophoresis* **7,** 278 (1986).

[31] J. Koch, N. Gregersen, S. Kolvraa, and L. Bolund, *Nucleic Acids Res.* **14,** 7133 (1986).

[32] G. J. Garbutt, J. T. Wilson, G. S. Schuster, J. J. Leary, and D. C. Ward, *Clin. Chem.* **31,** 1203 (1985).

[33] H. Delius, H. van Heerikhuizen, J. Clarke, and B. Koller, *Nucleic Acids Res.* **13,** 5457 (1985).

[34] N. S. Lurain, K. D. Thompson, and S. K. Farrand, *J. Clin. Microbiol.* **24,** 724 (1986).

[35] G. F. Buffone, C. M. Schimbor, G. J. Demmler, D. R. Wilson, and G. D. Darlington, *J. Infect. Dis.* **154,** 163 (1986).

[36] M. A. Dorman, C. D. Blair, J. K. Collins, and B. J. Beaty, *J. Clin. Microbiol.* **22,** 990 (1985).

odontal pathogens,[38] and *Leishmania*.[39] Biotinylated probes have also been used for *in situ* cytohybridization assays for detection of viral infections (e.g., human papillomavirus[40,41] and bovine herpesvirus[42]) and to determine levels of mRNA expression in cells or tissues.[43,44] New applications for biotinylated probes in molecular biology and clinical diagnostics will continue to emerge as nonisotopic detection techniques acquire the speed, sensitivity, and reproducibility to compete effectively with radioisotopes.

Acknowledgments

We thank Dr. Dietmar Rabussay for continued support and Dr. Reynaldo Pless and Dr. Susan Bromley for critical reading of the manuscript. We also thank Kathleen Blair for help in typing the manuscript.

[37] H. Bialkowska-Hobrzanska, *J. Clin. Microbiol.* **25,** 338 (1987).
[38] C. K. French, E. D. Savitt, S. L. Simon, S. M. Eklund, M. C. Chen, L. C. Klotz, and K. K. Vaccaro, *Oral Microbiol. Immunol.* **1,** 58 (1986).
[39] P. G. Trejo, *Focus (Bethesda Res. Lab./Life Technol.)* **9,** 11 (1987).
[40] A. M. Beckmann, D. Myerson, J. R. Daling, N. B. Kiviat, C. M. Fenoglio, and J. K. McDougall, *J. Med. Virol.* **16,** 265 (1985).
[41] K. Milde and T. Loning, *J. Oral Pathol.* **15,** 292 (1986).
[42] D. C. Dunn, C. D. Blair, D. C. Ward, and B. J. Beaty, *Am. J. Vet. Res.* **47,** 740 (1986).
[43] R. H. Singer, J. B. Lawrence, and C. Villnave, *BioTechniques* **4,** 230 (1986).
[44] P. Liesi, J.-P. Julien, P. Vilja, F. Grosveld, and L. Rechardt, *J. Histochem. Cytochem.* **34,** 923 (1986).

[67] Biotinylated Psoralen Derivative for Labeling Nucleic Acid Hybridization Probes

By COREY LEVENSON, ROBERT WATSON, and EDWARD L. SHELDON

Nucleic acid hybridization probes are widely used biochemical reagents with applications in research and diagnostics. Such probes require at least two structural features: a single-stranded region of nucleic acid capable of annealing to a complementary "target" sequence and one or more detectable moieties (labels). The label(s) may be incorporated in the single-stranded hybridizing region as long as its presence does not compromise the ability of the probe to anneal to the target sequence. Nonradioactive labels are preferable owing to their ease of handling and disposal and their long shelf life. One of the preferred labels for non-

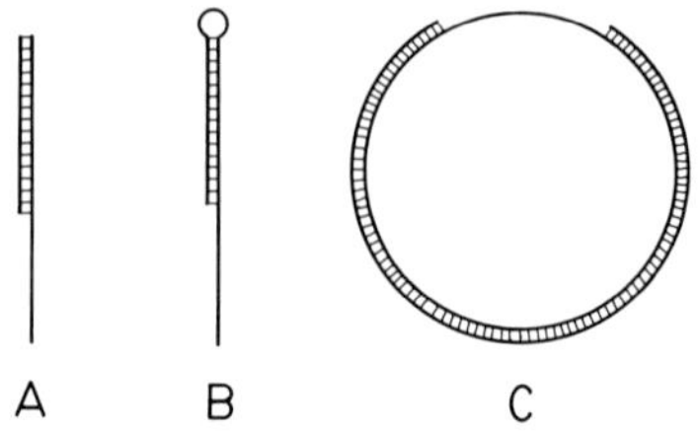

FIG. 1. Probe constructs that are suitable substrates for labeling using biotinylated psoralen: (A) two complementary single-stranded oligonucleotides, (B) a partially self-complementary oligonucleotide, and (C) an M13 phage-derived "gapped circle" [V. Courage-Tebbe and B. Kemper, *Biochim. Biophys. Acta* **697,** 1 (1982)]. Note: These representations are not drawn to scale.

isotopically labeled probes is biotin.[1,2] Conjugates of various enzymes with avidin or streptavidin are employed for colorimetric detection.[3,4]

Nonisotopic labels can be introduced into probes using appropriately modified derivatives of psoralens. Psoralens are planar compounds that are capable of intercalating into double-stranded nucleic acids. The intercalated complexes are photoactivated by long-wavelength (360 nm) ultraviolet light and form interstrand cross-links. Figure 1 depicts three types of hybridization probes that are suitable substrates for reaction with psoralen labeling reagents. Each probe consists of a single-stranded hybridizing region linked to a double-stranded region containing preferred psoralen-binding sites (5′-TpA-3′). Once these probes have been labeled with a psoralen derivative, the double-stranded region is cross-linked and therefore becomes incapable of being denatured; the duplex region serves only as a carrier for the label(s).

The psoralen-derived labeling reagents have three domains: the trimethylpsoralen ring system, a long, flexible, hydrophilic "spacer arm," and the nonisotopic label. For the reagent described herein, the psoralen ring system is derived from trimethylpsoralen (trioxsalen), the spacer arm is derived from tetraethylene glycol, and the label is biotin. The synthesis of the biotinylated psoralen labeling reagent is shown in Fig. 2. Reaction of tetraethylene glycol with *p*-toluenesulfonyl chloride in pyridine yields

[1] J. E. Manning, N. D. Hershey, T. R. Broker, M. Pellegrini, H. K. Mitchell, and N. Davidson, *Chromosoma* **53,** 107 (1975).

[2] P. R. Langer, A. A. Waldrop, and D. C. Ward, *Proc. Natl. Acad. Sci. U.S.A.* **78,** 6633 (1981).

[3] J. J. Leary, D. J. Brigati, and D. C. Ward, *Proc. Natl. Acad. Sci. U.S.A.* **80,** 4045 (1983).

[4] E. L. Sheldon, D. E. Kellogg, R. Watson, C. Levenson, and H. Erlich, *Proc. Natl. Acad. Sci. U.S.A.* **83,** 9085 (1985).

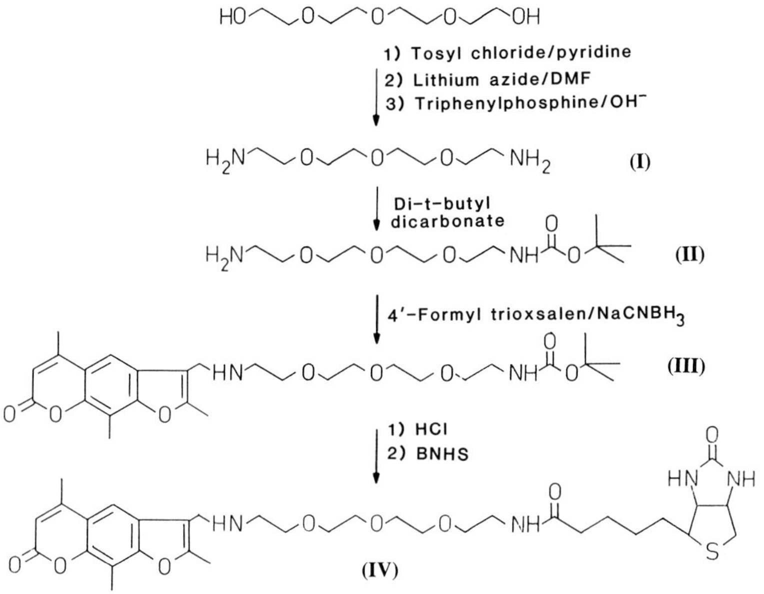

FIG. 2. Synthesis of the biotinylated psoralen labeling reagent.

bistosylate. Reaction of the tosylate with lithium azide in DMF yields bisazide. The azide is reduced to the bisamine **(I)** by triphenylphosphine–ammonium hydroxide. The bisamine is converted to the mono-*tert*-butyloxycarbonyl (BOC) derivative **(II),** which is purified by silica gel column chromatography. The mono-BOC-protected psoralen derivative **(III)** is produced via a reductive amination reaction employing compound **II** and 4′-formyltrioxsalen. Treatment of compound **III** with HCl in ethyl acetate removes the BOC group, and subsequent reaction with biotin-*N*-hydroxysuccinimide ester (BNHS) yields the biotinylated psoralenamine **(IV).** For the purpose of quantifying levels of incorporation, compound **IV,** may also be synthesized using tritiated BNHS.

Synthesis of Reagents

Materials. The following chemicals are obtained from Aldrich (Milwaukee, WI): tetraethylene glycol, *p*-toluenesulfonylchloride, triphenylphosphine, di-*tert*-butyl dicarbonate, and dimethylaminocinnamaldehyde. Lithium azide is obtained from Kodak (Rochester, NY). 4′-Formyltrioxsalen is obtained from HRI Associates (Berkeley, CA). *N*-Hydroxysuccinimide ester of biotin (BNHS) is obtained from Pierce

Chemical Co. (Rockford, IL). Tritiated BNHS is obtained from Amersham (Arlington Heights, IL). Analytical thin-layer chromatograms are run on Bakerflex IB2-F plates (J. T. Baker, Phillipsburg, NJ). Preparative TLC is performed using uniplate silica gel GF tapered layer plates (Analtech, Newark, DE). Biotinylated compounds are detected on TLC plates using a spray reagent consisting of equal volumes of 2% (v/v) sulfuric acid in ethanol and 0.2% (w/v) *p*-dimethylaminocinnamaldehyde in absolute ethanol.[5] NMR spectra are obtained using a Varian FT80, and shifts are reported as parts per million (ppm) downfield from tetramethylsilane as an internal standard. Mass spectral data are provided by the mass spectrometry laboratory at the University of California at Berkeley, Department of Chemistry.

Bistosyl-3,6,9-trioxaundecane-1,11-diol. To a chilled solution of tetraethylene glycol (42 g, 216 mmol) in 500 ml of dry pyridine is added *p*-toluenesulfonyl chloride (100 g, 525 mmol). The solution is stirred at 4° for 18 hr. To the solution is added 100 ml of methanol, and stirring is continued for an additional hour. The solvent is removed under reduced pressure and the residue partitioned between 300 ml of ethyl acetate and 300 ml of 0.5 *M* citric acid. The organic layer is washed with saturated NaCl (2 times, 300 ml each). The organic extract is dried over magnesium sulfate, filtered, and concentrated to a syrupy residue (92 g) which may be used directly for the next reaction. If desired, the material may be purified by silica gel column chromatography using dichloromethane–methanol (97:3) as eluant (TLC, R_f = 0.54, detect with iodine vapor). NMR (DMSO-d_6): 2.41(s), 3.46(s), 3.55(m), 4.14(m), 7.46(d, J = 8.3 Hz), 7.82 (d, J = 8.3 Hz).

1,11-Diazido-3,6,9-trioxaundecane. To a solution of bistosyl-3,6,9-trioxaundecane-1,11-diol (50.3 g, 100 mmol) in 250 ml of dry dimethylformamide (DMF) is added lithium azide (30 g, 613 mmol). The mixture is heated to 75° with stirring until TLC on silica gel (ethyl acetate) reveals no starting material remaining. The solvent is removed under vacuum and the residue partitioned between 500 ml ethyl acetate and 250 ml of saturated NaCl. The organic layer is dried over magnesium sulfate, filtered, and concentrated to an oil (21 g). This material can be either reduced directly in the following reaction or purified by column chromatography on silica gel using ethyl acetate as eluant (TLC, R_f = 0.71, iodine vapor detection). NMR (DMSO-d_6): 3.34(m), 3.57(s), 3.65(m). IR (neat): 2100 cm^{-1}.

1,11-Diamino-3,6,9-trioxaundecane **(I).** Triphenylphosphine (89 g, 339 mmol) is added to a chilled, stirred solution of 1,11-diazido-3,6,9-

[5] D. B. McCormick and J. A. Roth, this series, Vol. 18, p. 383.

trioxaundecane (24.4 g, 100 mmol) in 250 ml pyridine. Nitrogen bubbles evolve. The solution is stirred with cooling on ice for 45 min and then allowed to warm to room temperature. After 45 min at room temperature, 100 ml of concentrated ammonium hydroxide is added and the mixture stirred overnight. The solvent is removed under reduced pressure, and the mixture is partitioned between 500 ml of 0.5 *M* citric acid and 250 ml of ethyl acetate. The aqueous phase is washed with ethyl acetate (2 times 250 ml each) to remove triphenylphosphine oxide. The aqueous phase is saturated with sodium chloride and exhaustively extracted with *n*-butanol. The alcohol extract is concentrated under reduced pressure and the residue taken up in 100 ml of absolute ethanol. Concentrated hydrochloric acid (15 ml) is added, and the solvent is removed under reduced pressure. The residue is crystallized from ethanol–ether. (Note: the product will not crystallize unless it is anhydrous.) The hygroscopic crystals (21 g) are dried under vacuum. TLC, R_f = 0.28 [silica gel, dichloromethane–methanol–acetic acid (70 : 30 : 5), ninhydrin detection]. NMR (DMSO-d_6): 3.39(s), 3.57(s), 3.64(s), 8.10(br).

1 - Amino - 12 - tert - butyloxycarbonylamino - 3,6,9,- trioxaundecane **(II).** To a solution of **I** (13.26 g, 50 mmol) and triethylamine (8 ml, 5.8 g, 57.4 mmol) in 100 ml of methanol is added a solution of di-*tert*-butyl dicarbonate (12 g, 55 mmol) in 10 ml of methanol. Carbon dioxide bubbles evolve, and the reaction is slightly exothermic. Stirring is continued until gas evolution ceases. The solvent is removed under reduced pressure, and the residue is adsorbed onto 20 g of silica gel and fractionated on a column using dichloromethane–methanol–acetic acid (70 : 30 : 5) as eluant. The fractions containing product are pooled and concentrated to yield the mono-BOC (butyloxycarbonyl) derivative as a syrup. TLC, R_f = 0.67 [dichloromethane–methanol–acetic acid (70 : 30 : 5), ninhydrin positive].

12 - tert - Butyloxycarbonylamino - 1 - [(4,5′,8 - trimethylpsoralen - 4′-methylenyl)amino]-3,6,9-trioxaundecane **(III).** 4′-Formyltrioxsalen (1 g, 3.9 mmol) and 1 g of 3 Å molecular sieves are added to a solution of compound **II** (2.28 g, 7.80 mmol) and sodium cyanoborohydride (246 mg, 3.9 mmol) in 50 ml of anhydrous methanol. The suspension is adjusted to pH 7 by addition of triethylamine and is stirred in the dark at room temperature until TLC [silica gel, dichloromethane–methanol (8 : 1)] shows no formyltrioxsalen remaining. The mixture is filtered and evaporated to a residue which is purified by silica gel column chromatography using the TLC solvent system as eluant. Fractions containing the fluorescent product are pooled and evaporated to dryness to yield 1.8 g. TLC, R_f = 0.41 [dichloromethane–methanol (8 : 1); fluorescent under long-wavelength UV; ninhydrin positive after spraying the plate with 3 *M* HCl in ethyl acetate]. NMR (DMSO-d_6): 1.36(s), 2.39(s), 2.44(s), 2.65(t, J = 5.5 Hz), 3.07(t, J = 5.5 Hz), 3.48(s), 3.81(s), 6.25(s), 6.73 (br t), 7.73(s).

12-Biotinylamino-1-[(4,5',8-trimethylpsoralen-4'-methylenyl)amino]-3,6,9-trioxaundecane **(IV).** To a solution of **III** (71.2 mg, 133.7 μmol) in 1 ml of dry dioxane is added 0.5 ml of a saturated solution of HCl in dioxane (~12 M). The solution becomes cloudy, and a precipitate is formed which begins to crystallize. The mixture is checked by TLC [silica gel, 2-butanone–acetic acid–water (70 : 30 : 25)]. When the BOC-protected material is no longer evident (~30 min), the solvent is removed under a stream of air. The crystalline residue is taken up in 0.5 ml of methanol and taken to dryness. The residue is reconstituted in 0.4 ml of dry DMF, and 120 μl (689 μmol) of diisopropylethylamine is added. This solution is added to a suspension of BNHS (50 mg, 146.5 μmol) in 0.2 ml of dry DMF. All material dissolves. After 2 hr, the mixture is taken to dryness and reconstituted in 1 ml of methanol. The solution is applied to two 20 × 20 cm preparative TLC plates, and the plates are developed with 2-butanone–acetic acid–water (70 : 30 : 25). The product band (R_f = 0.77, fluorescent and biotin positive) is scraped and eluted from the silica gel with 20 ml of methanol. The extract is filtered, and the concentration of product is determined spectrophotometrically using an extinction coefficient of 25,000 at 250 nm. Yield, 87.6 μmol of **IV** (66%). Fast atom bombardment high-resolution mass spectroscopy: M + H = 659.3109 ($C_{33}H_{47}N_4O_8S$ = 659.829).

Probe Preparation and Biotinylation

Partially double-stranded probe constructs (as depicted in Fig. 1) are biotinylated by mixing the labeling reagent with the probe DNA at a molar ratio (reagent to base pairs) of 2 : 1 and a DNA concentration of 100 μg/ml in 100 mM NaCl, 10 mM Tris (pH 7.5), and 1 mM EDTA. The mixture is irradiated with 360-nm UV light (B-100A Black-Ray Lamp, UV Products, San Gabriel, CA) at a flux of 30 mW/cm^2 for 10 min, and the biotinylated gapped circles are purified from the reaction mixture by chromatography on a 0.7 × 30 cm Sephacryl S-200 column in 10 mM NaCl, 10 mM Tris, 1 mM EDTA. These conditions result in the incorporation of one biotinylated psoralen labeling reagent for every 10 base pairs (10%). Levels of label incorporation are assayed by scintillation counting of probes that had been photoreacted with tritiated biotinylated psoralen.

Comments

Biotinylated nucleic acid probe constructs have previously been made by enzymatic incorporation of biotinylated nucleotides into double-stranded DNA,[2] by chemical synthesis of oligonucleotides using biotin derivatives,[6,7] or by reaction of the probe with photoreactive biotin derivatives.[8] The disadvantages of the enzymatic method of biotinylation include the expense of reagents and difficulties involved with control and scale-up of the reaction. Biotinylated oligonucleotides are difficult to prepare and have the disadvantage of delivering very few biotins to the site of hybridization. Photoreactive biotin derivatives previously reported are based on arylazides that generate arylnitrenes upon photoactivation. These nitrenes do not discriminate between single- or double-stranded nucleic acids or between nucleic acids and other molecules.

The attribute of biotinylated psoralen that makes it attractive as a labeling reagent is its ability to discriminate between single- and double-stranded nucleic acids; psoralens are known to interact preferentially with the duplex form. It was demonstrated previously[9] that M13 probes could be treated (cross-linked) with trioxsalen without compromising the ability of the probe to hybridize to target nucleic acids. We have demonstrated that partial duplex probes are also suitable substrates for reaction with psoralen derivatives.[4] The presence of a large substituent at the 4′ position of the psoralen ring system does not prevent the molecule from intercalating and subsequently photoreacting with the duplex region. The level of incorporation (psoralens per 100 base pairs) is comparable for both 4′-aminomethyltrioxsalen and the biotinylated psoralen described herein. It is possible to bind one psoralen per 4–5 base pairs for either compound.[10] We have found that this level of incorporation is unnecessary for optimal signal detection. Because of the size of the enzyme conjugates used for detection, it is probably unnecessary to have more than approximately one biotinylated psoralen per 10 base pairs. Higher levels of incorporation lead to higher levels of background, presumably owing to aggregation of the heavily biotinylated probe or increased nonspecific adherence to the membrane. The extent of incorporation of biotinylated psoralen with nucleic acids is easily controlled by adjusting the base pair to psoralen stoichiometry and has been found to be insensitive to scale-up. Milligram quantities of probe DNA can be labeled in the single reaction.

Acknowledgments

The authors would like to acknowledge the creative input of Kary Mullis and Henry Rapoport, and the technical assistance of Diana Ho, David Kellogg, Todd Smith, Rick Snead, and Dragan Spasic.

[6] J. Kempe, W. I. Sundquist, F. Chow, and S. L. Ho, *Nucleic Acids Res.* **13,** 45 (1985).

[7] A. Chollet and E. H. Kawashima, *Nucleic Acids Res.* **13,** 1529 (1985).

[8] A. C. Forster, J. L. McInnes, D. C. Skingle, and R. H. Symons, *Nucleic Acids Res.* **13,** 745 (1985).

[9] D. M. Brown, J. Frampton, P. Goelet, and J. Karn, *Gene* **20,** 139 (1982).

[10] S. T. Isaacs, C. J. Shen, J. E. Hearst, and H. Rapoport, *Biochemistry* **16,** 1058 (1977).

[68] Chemically Cleavable Biotin-Labeled Nucleotide Analogs

By TIMOTHY M. HERMAN and BARBARA J. FENN

The ability to incorporate biotin-labeled nucleotide analogs into DNA and RNA has stimulated interest in the use of the avidin–biotin affinity system to isolate protein–DNA[1,2] and protein–RNA complexes.[3] However, the high affinity of avidin for biotin has resulted in one major limitation of this approach, namely, the inability to dissociate the biotinylated complex from the avidin affinity column under conditions that preserve the protein–nucleic acid as well as protein–protein interactions present in the complex. Although several approaches have been taken to solve this problem,[4,5] none have been entirely satisfactory.

The development of the chemically cleavable biotinylated nucleotide Bio-12-SS-dUTP[6] appears to be a general solution to this problem. Bio-12-SS-dUTP contains a disulfide bond in the 12-atom linker arm joining biotin to the 5-carbon of uridine. Following binding of biotinylated DNA to an avidin affinity column, a reducing agent such a dithiothreitol is used to cleave the linker arm and release the DNA. Because this cleavage occurs under gentle, nondenaturing conditions, the recovered complexes should be amenable to further investigation of the protein–protein and protein–nucleic acid interactions found within the native complex.

We describe here a procedure to synthesize the chemically cleavable biotinylated nucleotide Bio-19-SS-dUTP (Fig. 1). This analog is a modification of Bio-12-SS-dUTP in which an additional 7 atoms are present in the linker arm between the disulfide bond and biotin.[7] This modification makes it possible to use Bio-19-SS-dUTP with streptavidin, an acidic protein with no nonspecific DNA-binding properties. The procedure is based on that originally described by Langer et al.[8] The corresponding

[1] M. S. Kasher, D. Pintel, and D. C. Ward, *Mol Cell. Biol.* **6,** 3117 (1986).
[2] T. M. Herman, this series, Vol. 170, p. 41.
[3] P. J. Grabowski and P. A. Sharp. *Science* **233,** 1294 (1986).
[4] R. A. Gravel, K. F. Lam, D. Mahuran, and A. Kronis, *Arch. Biochem. Biophys.* **201,** 669 (1980).
[5] K. Hoffmann, S. W. Wood, C. C. Brinton, J. A. Montibeller, and F. M. Finn, *Proc. Natl. Acad. Sci. U.S.A.* **77,** 4666 (1980).
[6] M. L. Shimkus, J. Levy, and T. M. Herman, *Proc. Natl. Acad. Sci. U.S.A.* **82,** 2593 (1985).
[7] T. M. Herman, E. Lefever, and M. Shimkus, *Anal. Biochem.* **156,** 48 (1986).
[8] P. R. Langer, A. A. Waldrop, and D. C. Ward, *Proc. Natl. Acad. Sci. U.S.A.* **78,** 6633 (1981).

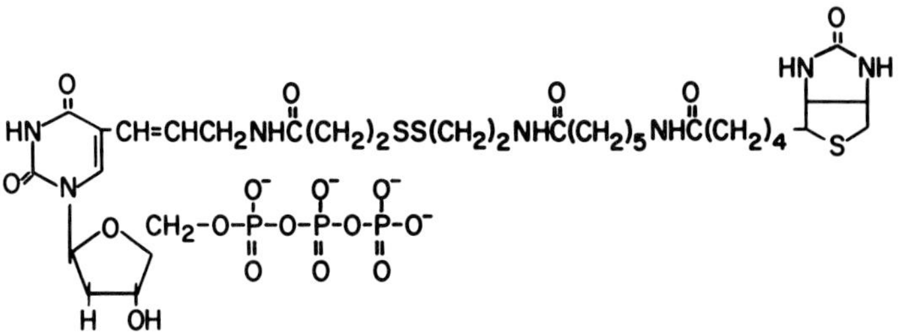

FIG. 1. Structure of Bio-19-SS-dUTP.

ribonucleotide analog can be prepared by the same procedure, substituting UTP for dUTP as the starting material.

Synthesis of Bio-19-SS-dUTP

The synthesis of Bio-19-SS-dUTP consists of three distinct steps. First, dUTP is reacted with mercury acetate to modify the 5-carbon of uracil. Second, allylamine is added to this position on the pyrimidine ring in the presence of a palladium catalyst. Third, allylamine-dUTP is reacted with an *N*-hydroxysuccinimide ester of biotin to produce the final product, Bio-19-SS-dUTP. The entire procedure requires 4 days and involves the use of two ion-exchange columns and a final purification of the product by reversed-phase HPLC. A detailed description of each step of the procedure follows.

Synthesis of 5-Mercurated dUTP (Hg-dUTP)

Dissolve 50 mg (91 μmol) of deoxyuridine 5′-triphosphate (Sigma Chemical Co., St. Louis, MO) in 10 ml of 0.1 *M* sodium acetate (pH 6.0). Add a 5-fold molar excess (142 mg) of mercury(II) acetate to the dUTP solution. Incubate at 50° for 4 hr. Cool the reaction on ice. Add 34 mg of LiCl (9-fold molar excess over dUTP). Remove $HgCl_2$ by extracting the reaction 6 times with an equal volume of ice-cold ethyl acetate. A brief centrifugation (1,000 *g* for 3 min at 4°) is used to separate the upper ethyl acetate phase from the lower aqueous phase containing the Hg-dUTP. The aqueous phase remains clear during these extractions only if it is kept at 4°. On warming, the aqueous phase becomes cloudy.

Precipitate the Hg-dUTP by adding 3 volumes of cold absolute ethanol. A white fluffy precipitate is immediately visible. Cool for 1 hr at −20°. Collect the precipitate by centrifugation (5,000 *g* for 10 min). Wash the precipitate twice with cold absolute ethanol and once with cold diethyl ether. Air dry. Dissolve the mercurated nucleotide in 0.1 *M* sodium acetate (pH 5.0) (~2 ml) and adjust to a concentration of 20 m*M* based on the millimolar extinction coefficient of 10.2 at 260 nm.

Comments. The above procedure routinely results in the mercuration of at least 90% of the dUTP. Although methods have been reported to assess the efficiency of the ethyl acetate extraction procedure[9] and to measure the extent of mercuration,[10] such procedures are not normally necessary. Instead, the Hg-dUTP obtained above is used directly in the next step in the synthesis procedure to produce allylamine-dUTP.

Synthesis of 5-(3-Amino)allyldeoxyuridine Triphosphate (AA-dUTP)

Prepare a fresh solution of 2.0 *M* allylamine by adding 1.5 ml of 13.3 *M* allylamine (Aldrich, Milwaukee, WI) slowly to 8.5 ml of ice-cold 4 *N* glacial acetic acid. Adjust the solution to pH 7.0 with 5 *N* NaOH. Add 480 μl (960 μmol) of the 2.0 *M* allylamine solution to 80 μmol Hg-dUTP (4 ml of the 20 m*M* solution in 0.1 *M* sodium acetate, pH 5.0) and 80 μmol of the catalyst K_2PdCl_4 (Aldrich; 0.64 ml of a 125 m*M* solution in water). Incubate the reaction for 18 hr (overnight) at room temperature. A black precipitate will form during this reaction. Filter the reaction several times through a 0.45-μm nylon filter to remove the precipitate. Dilute the clear filtrate with 5 volume of water.

Load the diluted filtrate onto a 35 ml DEAE-Sephadex A-25 column (1.5 × 20 cm) equilibrated with 0.1 *M* sodium acetate (pH 5.5). Wash the loaded column with 150 ml of equilibration buffer. Elute the aminoallyl dUTP with a 180-ml linear gradient of 0.1 to 0.6 *M* sodium acetate (pH 8.5). Collect 3-ml fractions. Monitor the absorbance of the column fractions at 260 nm. Aminoallyl-dUTP elutes from this column at approximately 0.4 *M* sodium acetate.[11,12]

Pool the fractions containing the aminoallyl-dUTP and concentrate by addition of 3 volumes of cold absolute ethanol. A heavy white precipitate forms immediately. Cool the solution for 1 hr at −20°. Collect the precipitate by centrifugation and resuspend in 3 ml of 0.1 *M* sodium borate (pH

[9] A. J. Christopher, *Analyst* **94,** 392 (1969).

[10] R. M. K. Dale, D. C. Ward, D. C. Livingston, and E. Martin, *Nucleic Acids Res.* **2,** 915 (1975).

[11] M. L. Shimkus, P. Guaglianone, and T. M. Herman, *DNA* **5,** 247 (1986).

[12] Fractions containing aminoallyl-dUTP are easily identified by their absorbance spectrum. The addition of the exocyclic double bond to the pyrimidine ring alters the absorbance properties of the nucleotide such that it now shows an absorbance minimum at 262 nm and absorbance maximums at 240 and 288 nm. The ratio of absorbances at 262 and 288 nm is approximately 1.4. In contrast, both dUTP and Hg-dUTP exhibit a single absorbance maximum at 260 nm.

8.5). Determine the concentration of aminoallyl-dUTP spectrophotometrically using a millimolar extinction coefficient of 7.1 at 288 nm.

Comments. The yield of aminoallyl-dUTP at this stage in the procedure is normally 15–20% of the starting dUTP.

Synthesis of Bio-19-SS-dUTP

Add 14.5 μmol of sulfo-NHS-LC-SS-biotin (Pierce Chemical Co., Rockford, IL) to 14.5 μmol of aminoallyl-dUTP in 2.7 ml of 0.1 *M* sodium borate (pH 8.5). Incubate the reaction for 2 hr at room temperature. Load the reaction onto a 35-ml DEAE-Sephadex A-25 column (1.5 × 20 cm) equilibrated in 0.1 *M* triethylamine–carbonate (pH 7.5).[13] Following application of the sample, wash the column with 2 volumes of equilibration buffer. Elute the Bio-19-SS-dUTP with a 180-ml linear gradient of 0.1 *M* triethylamine–carbonate (pH 7.5) to 0.9 *M* triethylamine–carbonate (pH 8.0). Wash with an additional 100 ml of the 0.9 *M* buffer to assure the complete elution of the product. Collect 3-ml fractions and monitor the absorbance at 288 nm. Pool fractions containing Bio-19-SS-dUTP (last major peak to elute from the column at approximately 0.7 *M* triethylamine–carbonate). Concentrate the Bio-19-SS-dUTP by rotary evaporation. Following several additions of methanol to the Bio-19-SS-dUTP, the solution is evaporated to dryness.

The identity of the pooled DEAE-Sephadex A-25 fractions containing Bio-19-SS-dUTP can be verified by reversed-phase HPLC analysis of an aliquot of each pool. Two-microliter aliquots of each pool, diluted to 200 μl with 50 m*M* triethylamine–carbonate (pH 7.5) are analyzed as described below. Dissolve Bio-19-SS-dUTP in 2.0 ml of 50 m*M* triethylamine–carbonate (pH 7.0)[13] in preparation for final purification by reversed-phase HPLC. Inject 200-μl aliquots of Bio-19-SS-dUTP onto a Bio-Sil ODS-5S column (250 × 4 mm; Bio-Rad, Richmond, CA) equilibrated with 50 m*M* triethylamine–carbonate (pH 7.0)–15% acetonitrile.[14]

[13] The triethylamine–carbonate buffer is a volatile buffer that facilitates the concentration of the final product by rotary evaporation. The pH of this buffer is adjusted by bubbling CO_2 through the solution. This buffer must be prepared just before use and care taken to avoid an increase in pH as the CO_2 is lost from the solution.

[14] Several precautions must be taken in the preparation of the buffer used in this step. Fifty millimolar triethylamine is first prepared and filtered through a 0.45-μm filter to remove any particulate impurities. The buffer is then adjusted to pH 7.0 with CO_2. Finally, acetonitrile is added to a final concentration of 15%. This final buffer should not be filtered or degassed at this stage. Attempts to do so will alter both the final pH and the acetonitrile concentration of the buffer. This in turn will alter the retention time of the biotinylated nucleotide. In addition, buffers with a pH greater than 7.0 should be avoided when using silica-based HPLC resins.

Bio-19-SS-dUTP is eluted isocratically with a retention time of approximately 10 min (flow rate 1.0 ml/min).[15] Bio-19-SS-dUTP can be identified in two ways. First, it is the most hydrophobic species present in the reaction and therefore is the last peak to elute. Second, simultaneous monitoring of the effluent at both 262 and 288 nm will reveal the characteristic ratio at these two wavelengths of 1.4.

Concentrate the Bio-19-SS-dUTP by rotary evaporation to dryness with several additions of methanol. Dissolve in 10 m*M* Tris-HCl (pH 7.5) and determine the concentration using the millimolar extinction coefficient of 7.1 at 288 nm. Store the nucleotide at −80°. Bio-19-SS-dUTP can be stored for at least 6 months in this way without any noticeable degradation.

Comments. The yield in the last step in the synthesis procedure has ranged from 25 to 75% of the aminoallyl-dUTP being converted to Bio-19-SS-dUTP. This variability is most likely due to differences in the reactivity of the *N*-hydroxysuccinimide-activated biotin ester.

Acknowledgment

This work was supported by National Science Foundation Grant DMB 8616956.

[15] The retention time of Bio-19-SS-dUTP in this system is critically dependent on the concentration of acetonitrile in the elution buffer. Therefore, the acetonitrile concentration can be easily altered to achieve the desired retention time on individual reversed-phase HPLC columns.

[69] Preparation and Uses of Photobiotin

By JAMES L. MCINNES, ANTHONY C. FORSTER, DEREK C. SKINGLE, and ROBERT H. SYMONS

The development of nonradioisotopic methods for the labeling of nucleic acid probes has created great interest worldwide. The basis for this approach is to introduce into the nucleic acid a modifying group which can subsequently be detected by means of the production of an insoluble or soluble colored product or by some other nonradioactive method. Such modifying groups can be introduced enzymatically[1–6] or chemically.[7–18] A

[1] P. R. Langer, A. A. Waldrop, and D. C. Ward, *Proc. Natl. Acad. Sci. U.S.A.* **78,** 6633 (1981).

[2] J. J. Leary, D. J. Brigati, and D. C. Ward, *Proc. Natl. Acad. Sci. U.S.A.* **80,** 4045 (1983).

more direct approach is to couple a detecting enzyme to the nucleic acid probe itself. Such an approach has been recently reported for oligonucleotide probes by Li *et al.*[19] and Jablonski *et al.*[20] Probes labeled by these nonradioisotopic procedures have advantages over radioactively labeled probes in terms of stability, safety, detection time, and cost.

In recent years the most popular nonradioactive label has been biotin, owing to its high affinity for avidin (or streptavidin).[21] Colorimetric detection can be carried out enzymatically via an avidin–enzyme (or streptavidin–enzyme) conjugate to visualize the probe–target hybrid. In 1981, Langer *et al.*[1] reported an approach whereby biotin-11-dUTP was incorporated into DNA by nick translation. This compound, an analog of dTTP, is dUTP with a biotin moiety covalently attached to the C-5 position of the pyrimidine by an allylamine arm. This arm, 11 atoms in length, spaces the biotin away from the DNA, thus allowing it to be more available for detection.[2] These biotinylated probes allow detection of target sequences in the range of 1–5 pg DNA and have been utilized in standard

---

[3] G. Gebeyehu, P. Y. Rao, P. SooChan, D. A. Simms, and L. Klevan, *Nucleic Acids Res.* **15,** 4513 (1987).

[4] S. McCracken, *Focus (Bethesda Res. Lab./Life Technol.)* **7**(2), 5 (1985).

[5] J. D'Alessio, *Focus (Bethesda Res. Lab./Life Technol.)* **7**(4), 13 (1985).

[6] R. W. Richardson and R. I. Gumport, *Nucleic Acids Res.* **11,** 6167 (1983).

[7] J. E. Landegent, N. Jansen in de Wal, R. A. Baan, J. H. J. Hoeijmakers, and M. van der Ploeg, *Exp. Cell Res.* **153,** 61 (1984).

[8] P. Tchen, R. P. P. Fuchs, E. Sage, and M. Leng, *Proc. Natl. Acad. Sci. U.S.A.* **81,** 3466 (1984).

[9] A. C. Forster, J. L. McInnes, D. C. Skingle, and R. H. Symons, *Nucleic Acids Res.* **13,** 745 (1985).

[10] E. L. Sheldon, D. E. Kellogg, R. Watson, C. H. Levenson, and H. A. Erlich, *Proc. Natl. Acad. Sci. U.S.A.* **83,** 9085 (1986).

[11] R. P. Viscidi, C. J. Connelly, and R. H. Yolken, *J. Clin. Microbiol.* **23,** 311 (1986).

[12] A. Reisfeld, J. M. Rothenberg, E. A. Bayer, and M. Wilchek, *Biochem. Biophys. Res. Commun.* **142,** 519 (1987).

[13] M. Renz, *EMBO J.* **2,** 817 (1983).

[14] A.-C. Syvänen, M. Alanen, and H. Söderlund, *Nucleic Acids Res.* **13,** 2789 (1985).

[15] A. H. Al-Hakim and R. Hull, *Nucleic Acids Res.* **14,** 9965 (1986).

[16] E. D. Sverdlov, G. S. Monastyrskaya, L. I. Guskova, T. L. Levitan, V. I. Scheichenko, and E. I. Budowsky, *Biochim. Biophys. Acta* **340,** 153 (1974).

[17] A. H. N. Hopman, J. Wiegant, G. I. Tesser, and P. Van Duijn, *Nucleic Acids Res.* **14,** 6471 (1986).

[18] M. Renz and C. Kurz, *Nucleic Acids Res.* **12,** 3435 (1984).

[19] P. Li, P. P. Medon, D. C. Skingle, J. A. Lanser, and R. H. Symons, *Nucleic Acids Res.* **15,** 5275 (1987).

[20] E. Jablonski, E. W. Moomaw, R. H. Tullis, and J. L. Ruth, *Nucleic Acids Res.* **14,** 6115 (1986).

[21] N. M. Green, *Adv. Protein Chem.* **29,** 85 (1975).

filter hybridization studies with a sensitivity comparable to ^{32}P-labeled DNA probes.[2]

In an effort to provide a simpler and more convenient method for introducing biotin into nucleic acids, we have developed a rapid, chemical labeling procedure involving a photoactivatable analog of biotin, *N*-(4-azido-2-nitrophenyl)-*N*′-(*N*-*d*-biotinyl-3-aminopropyl)-*N*′-methyl-1,3-propanediamine (photobiotin).[9] The procedure is versatile in that stable single- or double-stranded DNA or RNA probes can be produced on either a small or large scale. This chapter describes the synthesis of photobiotin, its use as a labeling reagent for the preparation of both DNA and RNA probes, and examples of their use.

Principle of Method

Photobiotin acetate (Fig. 1) consists of biotin attached by a charged linker arm to a photoreactive arylazide group. When a mixture of nucleic acid and photobiotin acetate is exposed to strong visible light for 15–20 min under defined conditions, the arylazide group is converted to an extremely reactive arylnitrene, which allows the formation of linkages to the nucleic acid. Although the precise nature and site of the linkages are unknown, the linkage is stable under standard hybridization conditions and is presumably covalent.

The presence of the linker arm containing a positively charged tertiary amino group (at neutral pH) enables photobiotin to be electrostatically attracted to the nucleic acid, thus allowing a high local concentration for reaction. A relatively long (9-atom) linker arm was chosen to allow the biotinyl group of the biotin-labeled nucleic acid to penetrate the biotin-binding sites within avidin (or streptavidin) efficiently.[21] Thus, steric hindrance by the nucleic acid in ligand detection is minimized. We have estimated that the resulting probes possess one biotin molecule per 100–

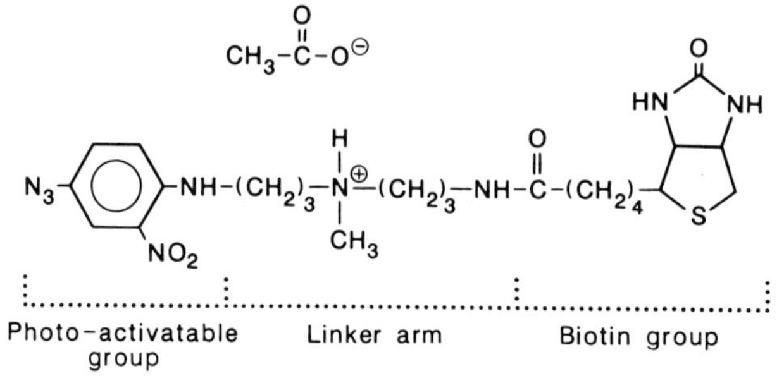

FIG. 1. Structure of photobiotin acetate.

150 bases of nucleic acid.[9] Such an extent of labeling is unlikely to interfere with the hybridization of the biotinylated probe to complementary target sequences.

Synthesis of Photobiotin Acetate

The procedure outlined below is a summary of the previously published method.[9] However, photobiotin acetate can be readily obtained from commercial sources (see below).

Materials. *N*-(3-Aminopropyl)-*N*-methyl-1,3-propanediamine is obtained from Tokyo Kasei. 4-Fluoro-3-nitrophenyl azide is synthesized by the method of Fleet *et al.*[22] *d*-Biotinyl-*N*-hydroxysuccinimide ester is synthesized from *d*-biotin by the DCC (*N,N'*-dicyclohexylcarbodiimide) method of Bayer and Wilchek[23] except that the ester is isolated by precipitation from ether and used without further purification. Pyridine is distilled twice from ninhydrin and once from CaH_2. Kieselgel 60 F_{254} and aluminum oxide 60 F_{254} TLC plates are obtained from Merck.

Procedure

The three-step synthesis of photobiotin acetate (Fig. 2) allows the rapid synthesis of gram amounts of the reagent.

Step 1: Synthesis of N-(3-Aminopropyl)-N'-(4-azido-2-nitrophenyl)-N-methyl-1,3-propanediamine (**III**). All reactions involving the formation and use of arylazides are carried out in the dark. A solution of 4-fluoro-3-nitrophenylazide (**I**; 0.91 g, 5.0 mmol) in dry ether (10 ml) is added with stirring to a solution of *N*-(3-aminopropyl)-*N*-methyl-1,3-propanediamine (**II**; 3.2 ml, 20 mmol) in dry ether (20 ml), and the mixture is stirred for 30 min. TLC on alumina with methanol is carried out to show the reaction to be complete. The solvent is removed, and the red oil is dissolved in water (25 ml). Following the addition of 1 *M* NaOH (25 ml), the product is extracted into ethyl acetate (2 times, 50 ml each). Although the aqueous phase is still dark red, less than 5% of the product remains. Analysis of the extract by TLC on silica with acetic acid–water (1 : 9, v/v), using ninhydrin to detect amines, should show the product to be free of compound **II**. The pooled ethyl acetate extracts are dried over Na_2SO_4, and the solvent is removed to give the crude product **III**, which is used without further purification.

Step 2: Synthesis of N-(4-Azido-2-nitrophenyl)-N'-(N-d-biotinyl-3-aminopropyl)-N'-methyl-1,3-propanediamine (Photobiotin, **V***).* *N*-(3-

[22] G. W. J. Fleet, J. R. Knowles, and R. R. Porter, *Biochem. J.* **128,** 499 (1972).

[23] E. A. Bayer and M. Wilchek, *Methods Biochem. Anal.* **26,** 1 (1980).

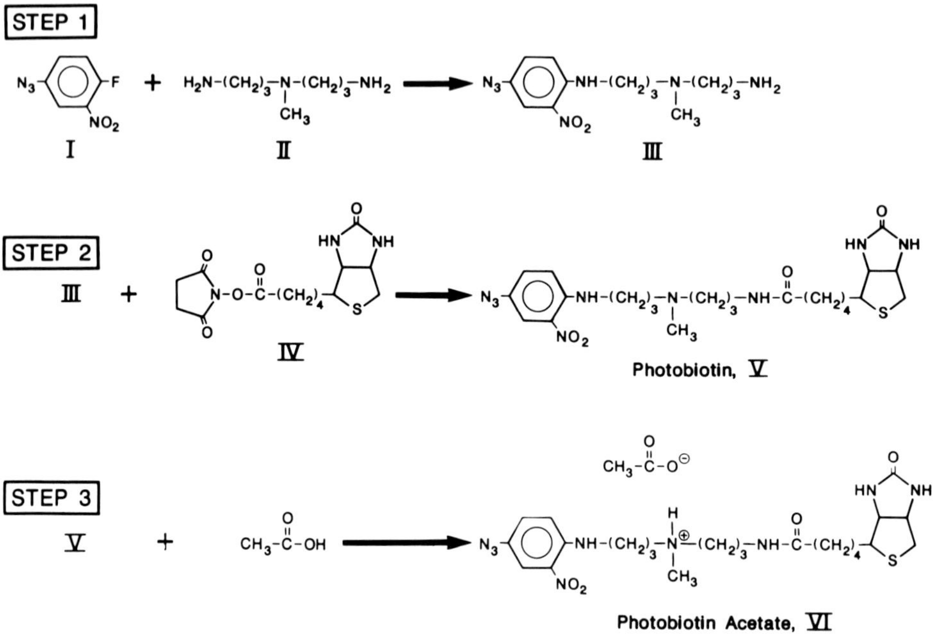

FIG. 2. Synthetic route to photobiotin acetate.

Aminopropyl)-*N'*-(4-azido-2-nitrophenyl)-*N*-methyl-1,3-propanediamine (**III**; ~5 mmol) is dissolved in a solution containing *d*-biotinyl-*N*-hydroxysuccinimide ester (**IV**; 1.7 g, 5.0 mmol) in pyridine–water (7 : 3, v/v, 50 ml), and the solution is incubated at 37° for 2 hr. TLC on alumina with tetrahydrofuran–water (19 : 1, v/v) is carried out to show the reaction to be complete. The solvent is removed, and the residue is dissolved in a mixture of 0.5 *M* $NaHCO_3$ (100 ml) and 2-butanol (100 ml). The large volume is necessary to visualize the interface between the phases. The 2-butanol layer is removed, washed successively with water (100 ml) and saturated NaCl (100 ml), and then slowly added to ether (900 ml). The suspension is stirred, the red precipitate is allowed to settle, and the supernatant is decanted. The precipitate is washed with ether (100 ml) and dried *in vacuo* to give the product (50% yield from 4-fluoro-3-nitrophenylazide, mp 114.5–115°). TLC on alumina with methanol should show the product to be a single spot of R_f 0.75 (the symmetrical disubstitution product formed from compounds **I** and **II** has R_f 0.6). TLC on silica with acetonitrile–water (19 : 1, v/v), using *p*-dimethylaminocinnamalde-

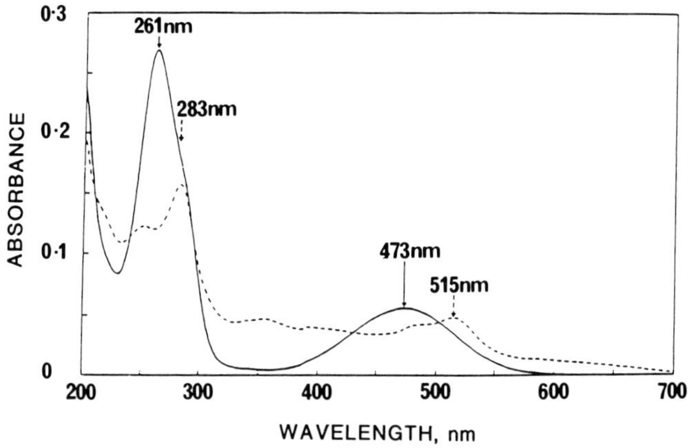

FIG. 3. Absorbance spectra of photobiotin acetate before and after photolysis. An aqueous solution of photobiotin acetate (0.5 μg/μl, 20 μl) was diluted with water to a volume of 1.2 ml for spectral analysis before (solid line) and after (dashed line) photolysis. [From A. C. Forster, J. L. McInnes, D. C. Skingle, and R. H. Symons, *Nucleic Acids Res.* **13,** 745 (1985).]

hyde spray to detect biotin derivatives,[24] should show the product to be free of compound **IV**. Photobiotin is poorly soluble in water.

Step 3: Preparation of Acetate Salt of Photobiotin (Photobiotin Acetate, **VI***).* A solution of photobiotin (**V**; 0.11 g, 0.21 mmol) in 1 *M* acetic acid (0.3 ml) is lyophilized to remove excess acetic acid. The resulting hygroscopic, red solid (**VI**) is very water soluble, with λ_{max} (water) values of 261 and 473 nm (ε_M = 19,200 and 3,900 M^{-1} cm^{-1}, respectively; Fig. 3).

Labeling of Nucleic Acids with Photobiotin Acetate

Photobiotin acetate, in addition to labeling RNA, labels single-stranded DNA and linear or supercoiled double-stranded DNA.[9,25] Since photobiotin acetate will react with any organic material, it is essential that nucleic acids to be used as probes be *highly purified* and free from any contaminating DNA, RNA, proteins, agarose, or buffers such as Tris. Inorganic salts may also inhibit the biotinylation of nucleic acids. Ideally, the nucleic acid to be labeled with photobiotin acetate should be dissolved in water or 0.1 m*M* EDTA.

[24] D. B. McCormick and J. A. Roth, *Anal. Biochem.* **34,** 226 (1970).

[25] J. L. McInnes, S. Dalton, P. D. Vize, and A. J. Robins, *Bio/Technology* **5,** 269 (1987).

Materials. Photobiotin acetate is obtainable from Bethesda Research Laboratories (Life Technologies, Inc., Gaithersburg, MD), Vector Laboratories, Inc. (Burlingame, CA), and Bresatec Ltd. (G.P.O. Box 498, Adelaide, South Australia 5000, Australia). The reagent is light sensitive and should be protected from direct light and should be handled only in subdued light for as short a period as practicable. The solid may be stored dry at 4° for at least 12 months without any detectable deterioration. Photobiotin acetate can be dissolved in distilled water and is stable for at least 12 months when stored at −20°.

A strong source of white light is required for the photoactivation reaction. A 250- to 500-W mercury vapor discharge lamp containing a tungsten filament with a built-in reflector to direct the output of light is the most suitable. Such lamps are normally available from major suppliers of industrial lighting. Tungsten lamps are not recommended because of their high heat output.

Procedure

The following protocol is suggested for the labeling of 1–25 μg of nucleic acid with photobiotin acetate. Prior to the photoactivation step (see below), the initial step of mixing the nucleic acid with photobiotin should be carried out in *subdued light* (e.g., by working in a darkened room with the door ajar to allow minimum light necessary to carry out the initial manipulations). All subsequent steps can be carried out under normal light.

In a sterile Eppendorf tube, a solution of photobiotin acetate (1 μg/μl) in water is added to an equal volume of nucleic acid (0.5–1.0 μg/μl) in water or 0.1 m*M* EDTA. The Eppendorf tube, with its lid open, is placed in crushed ice approximately 10 cm beneath a suitable light source (see above) and irradiated for 15–20 min. A solution (50 μl) of 0.1 *M* Tris-HCl (pH 9.0), 1 m*M* EDTA is added, and the volume is increased to 100 μl, if necessary, with water. 2-Butanol (100 μl) is added to the above solution and mixed well. After centrifuging for 1 min in a microcentrifuge, the upper, red 2-butanol phase is carefully removed and discarded. The 2-butanol extraction is repeated; the aqueous phase should now be colorless and concentrated to 30–35 μl. The biotin-labeled nucleic acid is precipitated by the addition of 5 μl of 3 *M* sodium acetate (pH 5.2) and 2.5 volumes (100 μl) ice-cold ethanol followed by chilling at −70° for 15 min or at −20° overnight. Centrifugation at 16,000 *g* for 15 min at 4° in a microcentrifuge yields a pellet which, when visible, is orange-brown in color and is indicative of effective photobiotin labeling. (A white nucleic acid pellet at this stage is indicative of the reaction having proceeded poorly or not at all.) The pellet is washed with cold 70% (v/v) ethanol, dried *in vacuo*, and dissolved in 0.1 m*M* EDTA. The probe concentration may be determined by measuring the absorbance at 260 nm. The recovery of biotin-labeled nucleic acid achieved using this protocol is approximately 70%. Storage of the biotin-labeled probe is at −20°.

Hybridization Conditions

Biotin-labeled nucleic acid probes can be used in standard dot-blot and Southern blot hybridizations in the same way as radioisotope-labeled probes. Hybridization parameters and the stringency of posthybridization washings should be optimized for the particular system and probe being used.[26–28] The majority of our hybridization studies have used nitrocellulose membranes. We routinely achieve lower backgrounds with nitrocellulose in comparison to nylon-based membranes. Acceptable backgrounds can be achieved with nylon membranes provided a stronger blocking procedure is followed after hybridization.[29]

The following hybridization procedure, adapted from Thomas,[30] is used routinely in our laboratory for DNA–DNA hybridizations. The nitrocellulose filter is baked *in vacuo* for 2 hr at 80° following Southern or dot-blotting. After transferring to a heat-sealable Polythene bag, the filter is prehybridized overnight at 42° with prehybridization buffer (80 μl of solution/cm^2 of nitrocellulose). The prehybridization buffer contains 50% (v/v) deionized formamide, 5× SSC (SSC: 0.15 *M* NaCl, 15 m*M* sodium citrate), 5× Denhardt's solution [1.0 mg/ml each of bovine serum albumin, Ficoll 400, and poly(vinylpyrrolidone) (M_r 40,000)], 50 m*M* sodium phosphate (pH 6.5), 0.25 mg/ml sonicated, denatured salmon sperm DNA, and 5 m*M* EDTA. The filter is then hybridized for 20–24 hr at 42° with gentle agitation in hybridization buffer. The hybridization buffer contains 50% (v/v) deionized formamide, 5× SSC, 1× Denhardt's solution, 20 m*M* sodium phosphate (pH 6.5), 50 μg/ml sonicated, denatured salmon sperm DNA, 5 m*M* EDTA, 10% (w/v) dextran sulfate (Pharmacia), and 20–100 ng/ml biotin-labeled DNA probe. Biotin-labeled double-stranded DNA probes, previously sonicated (see Comments), are dena-

[26] B. D. Hames and S. J. Higgins, "Nucleic Acid Hybridization: A Practical Approach." IRL Press, Oxford, 1985.

[27] J. Meinkoth and G. Wahl, *Anal. Biochem.* **138,** 267 (1984).

[28] T. Maniatis, E. F. Fritsch, and J. Sambrook, "Molecular Cloning: A Laboratory Manual." Cold Spring Harbor Laboratory, Cold Spring Harbor, New York, 1982.

[29] *Focus (Bethesda Res. Lab./Life Technol.)* **7**(2), 11 (1985).

[30] P. S. Thomas, this series, Vol. 100, p. 255.

tured by heating at 95° for 5 min and quick cooling on ice or by brief treatment (10 min at room temperature) with an equal volume of 100 mM NaOH, prior to addition to the hybridization buffer. Single-stranded probes (M13 DNA or RNA) require no pretreatment before addition to the hybridization buffer. After hybridization the filter is washed with frequent agitation in 300 ml of 2× SSC, 0.1% sodium dodecyl sulfate (SDS) (3 times, 15 min each, room temperature) and in 300 ml of 0.1× SSC, 0.1% SDS (3 times, 20 min each, 65°). The filter is gently blotted and placed in a heat-sealable bag ready for colorimetric detection.

Colorimetric Detection

The following method, adapted from Leary *et al.*,[2] is used routinely in our laboratory for the colorimetric detection of biotin-labeled probes following hybridization. The filter, inside a heat-sealable Polythene bag, is incubated for 20–30 min at 42° in TSMT buffer [0.1 M Tris-HCl (pH 7.5), 1 M NaCl, 2 mM $MgCl_2$, 0.05% (v/v) Triton X-100] containing 30 mg/ml bovine serum albumin (fraction V, Sigma). We recommend 80 μl of solution/cm^2 of filter. This blocking buffer is removed and replaced with 1 μg/ml of avidin–alkaline phosphatase (Sigma, Cat. No. A 2527) in TSMT buffer, and the filter is incubated for 10–15 min at room temperature with occasional gentle agitation. The filter is removed from the Polythene bag and washed 3 times at room temperature for 20 min each in TSMT buffer followed by 2 washes at room temperature for 10 min each in 0.1 M Tris-HCl (pH 9.5), 1 M NaCl, 5 mM $MgCl_2$.

For color development, the filter is transferred to a new Polythene bag and incubated at room temperature with substrate solution consisting of 0.1 M Tris-HCl (pH 9.5), 100 mM NaCl, 5 mM $MgCl_2$, 0.30 mg/ml nitroblue tetrazolium (NBT), and 0.20 mg/ml 5-bromo-4-chloro-3-indolyl phosphate (BCIP). To prepare 5 ml of substrate solution, add 20 μl NBT solution (75 mg/ml in 70% *N,N*-dimethylformamide) to 5 ml of the above buffer and gently mix. To this mixture add 20 μl BCIP solution (50 mg/ml in *N,N*-dimethylformamide) and again gently mix. The substrate solution may be freshly prepared just prior to use, or it can be stored as aliquots for 1–2 weeks at −20°. Color development should be carried out in the dark in order to decrease nonspecific background color. Maximum color development is normally obtained within 3–4 hr, although the optimal period of color development will vary, depending on the amount of biotin-labeled probe annealed to the target nucleic acid, and may range from a few minutes to 16 hr (overnight). Overnight incubations may result in background problems. Color development is terminated by removing the filter from the Polythene bag and placing it in 10 m*M* Tris-HCl (pH 7.5), 1 m*M* EDTA. The filter is best photographed while moist.

Comments

The photobiotin labeling procedure outlined here is simple, rapid, and reliable. Large quantities (e.g., 100 μg to milligram amounts) of biotinylated probe can be economically prepared. Once biotinylated, single-stranded RNA probes, M13 DNA probes, and double-stranded DNA probes are stable for at least 12 months when stored at $-20°$. Photoactivation can alternatively be performed in sealed siliconized glass microcapillary tubes (e.g., 10–100 μl), which have been placed beneath the surface of distilled water in a shallow tray, cooled in an ice bath. If necessary, an ultraviolet–visible spectral analysis (Fig. 3) may be used to ensure that the photolysis reaction has gone to completion. The use of molar ratios of photobiotin acetate to nucleic acid higher than those recommended may result in precipitation of the nucleic acid.

As stated previously, our laboratory routinely uses avidin–alkaline phosphatase for colorimetric detection. The biotin-labeled probe bound to target nucleic acid on nitrocellulose may alternatively be detected by a streptavidin–alkaline phosphatase conjugate (e.g., Bethesda Research Laboratories, BluGENE Nonradioactive Nucleic Acid Detection System, Cat. No. 8279SA). This, and other detection systems (e.g., streptavidin–biotinylated acid phosphatase complex, ENZO Biochemicals), may allow greater sensitivity. Nitrocellulose filters, unfortunately, cannot be reused after color detection since the precipitated dyes from the NBT–BCIP color reaction appear to bind irreversibly. However, these dyes may be removed from *nylon* membranes by heating in formamide, thus allowing reprobing.[3]

Photobiotin-labeled DNA (and, to a lesser extent, RNA) probes have been rigorously tested in our laboratory in different hybridization analyses. These studies have shown that photobiotin can replace the use of radioactivity in preparing gene probes for Southern, Northern, and dot-blot analyses.[25,31–33] In each case, the detection of target nucleic acids is at least as sensitive as with probes labeled with ^{32}P. An interesting observation is that brief sonication of photobiotin-labeled double-stranded DNA

[31] N. Habili, J. L. McInnes, and R. H. Symons, *J. Virol. Methods* **16,** 225 (1987).

[32] J. L. McInnes, A. C. Forster, and R. H. Symons, *in* "Methods in Molecular Biology" (J. M. Walker, ed.), Vol. 4, p. 401. Humana Press, Clifton, New Jersey, 1988.

[33] J. L. McInnes and R. H. Symons, *in* "Nucleic Acid Probes" (R. H. Symons, ed.), p. 113. CRC Press, Boca Raton, Florida, 1989.

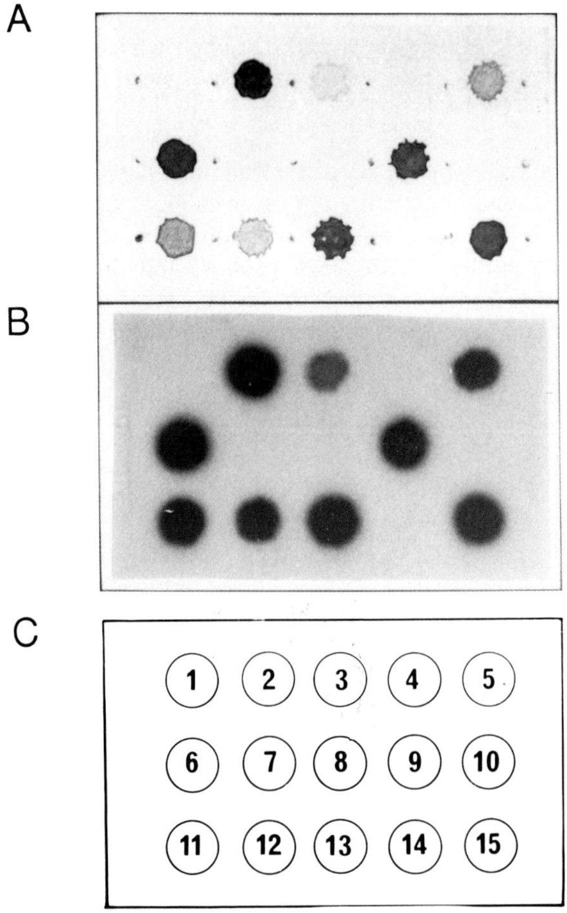

FIG. 4. Detection of chrysanthemum stunt viroid (CSV) in extracts of chrysanthemum leaves by dot-blot hybridization analysis.[34] (A) Sonicated, photobiotin-labeled recombinant DNA probe containing a full-length cloned monomer insert (353 nucleotides) of the Beltsville strain of CSV in plasmid pUC9.[35] The probe concentration was 100 ng/ml, and the detection time was 1 hr. (B) Nick-translated ^{32}P probe (specific activity 7×10^7 cpm/μg DNA). Autoradiography was for 7 hr at $-70°$ using an intensifying screen. (C) Location of samples on nitrocellulose filters. Samples 1–13, field samples; sample 14, healthy chrysanthemum tissue; sample 15, CSV-infected chrysanthemum tissue as control.

probes can result in a 5-fold increase in sensitivity over nonsonicated probes.[31] We recommend that sonication be carried out at a concentration of 20 μg/ml (250-μl aliquots) in the presence of carrier nucleic acid (100 μg/ml sonicated, denatured salmon sperm DNA) for 1 min (4 times, 15 sec each) using a Sonifier Cell Disrupter (Branson Sonic Power, Danbury, CT, U.S.A.) with an output setting control of 4 and a duty cycle of 50%.

TABLE I
APPLICATIONS OF PHOTOBIOTIN ACETATE

Application	Example	Hybridization	Ref.
ingle-copy gene detection	Chicken histone H5 gene	Southern blot	*a*
	Human metallothionein IIA gene	Southern blot	*b*
	Human β-globin gene	Southern blot	*c*
RNA detection	Chicken histone H5	Northern blot	*b*
	Oncogenes (c-*myc*, c-*sis*, and c-*abl*)	*In situ* hybridization	*d*
enomic DNA analysis	Transgenic mice/human metallothionein	Dot-blot	*b*
	IIA promoter–pig growth hormone fusion gene	Southern blot	*e*
iroid RNA detection	Avocado sunblotch viroid (ASBV)	Northern blot, dot-blot	*f*
	Potato spindle tuber viroid (PSTV)	Dot-blot	*g*
	Chrysanthemum stunt viroid (CSV)	Dot-blot	*h*
iral RNA detection	Barley yellow dwarf virus (BYDV)	Dot-blot	*e, i*
	Flavivirus	Dot-blot	*j*
	Soybean dwarf virus (SDV)	Dot-blot	*g*
iral DNA detection	Human papilloma virus (HPV)	*In situ* hybridization	*k*
larker DNA labeling	*Hin*dIII digest of λ DNA; 2–23 kb fragments	Southern blot	*a, b*
	*Hae*III digest of φX174 RF DNA: 271–1353 bp fragments	Southern blot	*c*
	*Eco*RI digest of SPP-1 bacteriophage DNA, 0.9–7.8 kb fragments	Southern blot	*e*
ffinity-based hybrid collection	Plasmid DNA/streptavidin–agarose matrix	Sandwich	*l*
elective enrichment	cDNA sequences/copper-chelate agarose	Subtractive	*m*
rotein labeling	Bovine intestinal alkaline phosphatase	—	*f*
	Sheep brain tubulin	—	*n*

[a] J. L. McInnes, A. C. Forster, and R. H. Symons, *in* "Methods in Molecular Biology" (J. M. Walker, ed.), Vol. 4, p. 401. Humana Press, Clifton, New Jersey, 1988.

[b] J. L. McInnes, S. Dalton, P. D. Vize, and A. J. Robins, *Bio/Technology* **5,** 269 (1987).

[c] Photobiotin Labeling System; Instruction manual, Bethesda Research Laboratories/Life Technologies, Inc., Gaithersburg, Maryland.

[d] J. Bresser and M. J. Evinger-Hodges, *Gene Anal. Tech.* **4,** 89 (1987).

[e] J. L. McInnes, P. D. Vize, N. Habili, and R. H. Symons, *Focus (Bethesda Res. Lab./Life Technol.)* **9**(4), 1 (1987).

[f] A. C. Forster, J. L. McInnes, D. C. Skingle, and R. H. Symons, *Nucleic Acids Res.* **13,** 745 (1985).

[g] J. L. McInnes and R. H. Symons, *in* "Nucleic Acid Probes" (R. H. Symons, ed.), p. 113. CRC Press, Boca Raton, Florida, 1989.

[h] This chapter.

[i] N. Habili, J. L. McInnes, and R. H. Symons, *J. Virol. Methods* **16,** 225 (1987).

[j] A. M. Khan and P. J. Wright, *J. Virol. Methods* **15,** 121 (1987).

[k] K. Milde and T. Löning, *J. Oral Pathol.* **15,** 292 (1986).

[l] A.-C. Syvänen, M. Laaksonen, and H. Söderlund, *Nucleic Acids Res.* **14,** 5037 (1986).

[m] A. A. Welcher, A. R. Torres, and D. C. Ward, *Nucleic Acids Res.* **14,** 10027 (1986).

[n] E. Lacey and W. N. Grant, *Anal. Biochem.* **163,** 151 (1987).

Of recent note, we have made use of photobiotin to prepare a biotinylated nonradioactive recombinant DNA probe for the detection of an important plant pathogen, chrysanthemum stunt viroid (CSV). This probe contains a full-length cloned monomer insert (353 nucleotides) of the Beltsville strain of CSV in plasmid pUC9. Figure 4A shows a typical dot-blot hybridization study for CSV RNA detection. Extracts prepared from infected leaf tissues of field samples give positive hybridization signals, which can be seen to range from weak to strong, indicating different levels of CSV infection. Extracts from healthy leaf material show no hybridization. An identical result (Fig. 4B) is obtained using the same recombinant DNA probe, labeled with ^{32}P by nick translation.

In addition to the above-mentioned results from our laboratory, photobiotin is proving to be an attractive reagent for other research groups. Table I lists applications of photobiotin. As can be seen, the usage of photobiotin has been extended to the detection of viral RNA and DNA, marker DNA labeling, *in situ* hybridization, affinity-based hybrid collection, a selective enrichment protocol, and protein labeling.

Acknowledgments

This work was supported by the Australian Research Grants Scheme and by a Commonwealth Government Grant to the Adelaide University Centre for Gene Technology in the Department of Biochemistry. We thank Dr. R. A. Owens for the CSV cDNA clone and Jane Moran for chrysanthemum leaf extracts.

[34] Partially purified nucleic acid extracts were prepared from healthy and infected leaves [R. K. Horst and S. O. Kawamoto, *Plant Dis.* **64,** 186 (1980)] by J. Moran, Department of Agriculture and Rural Affairs, Burnley, Victoria, Australia. Samples (3 μl) were spotted directly onto duplicate nitrocellulose membranes.[9] Each spot contains extract from approximately 120 mg of leaf tissue. The filters were prehybridized overnight at 42° and hybridized for 23 hr at 55°.

[35] R. A. Owens, U.S. Department of Agriculture, Beltsville, MD; personal gift.

[70] Modification of Cytosine Residues on DNA

By RAPHAEL P. VISCIDI

Nucleic acid hybridization techniques have found wide application in the biological sciences for the identification and quantification of specific nucleotide sequences in complex mixtures of nucleic acids. Most current hybridization assay systems use polynucleotide probes labeled with radioisotopes. However, the inherent instability of radiolabels and the potential health hazard posed by the preparation, handling, and disposal of radioactive material have stimulated interest in the development of nonradioactively labeled nucleic acid probes. In recent years a large number of methods have been developed for the linkage to nucleic acids of biotin and other markers that can be measured by means of enzymatic and fluorescent detection systems. A method that involves the selective chemical modification of cytosine residues on DNA is described below.[1] The preparation of biotinylated DNA probes and their use in mixed-phase hybridization assays demonstrate one application of this method.

Principle

Unpaired cytosine residues on deoxyribonucleic acids can be modified by the addition of sodium bisulfite to the 5,6-double bond of the pyrimidine base.[2,3] The exocyclic amino group of the cytosine–bisulfite adducts can undergo substitution with various amines.[4] In the presence of ethylenediamine, transamination forms an N^4-substituted cytosine derivative that has a side chain terminating in a primary amino group.[5] *N*-Hydroxysuccinimide ester derivatives of biotin will react under mild conditions with bisulfite–ethylenediamine-modified cytosine residues to yield biotin-labeled DNA (Fig. 1).

Procedures

Materials

λ phage DNA from Bethesda Research Laboratories (Gaithersburg, MD)
Sodium metabisulfite from Sigma Chemical Co. (St. Louis, MO)
Ethylenediamine, Gold Label from Aldrich (Milwaukee, WI)
N-Biotinyl-ε-aminocaproic acid *N*-hydroxysuccinimide ester from ENZO Biochemicals (New York, NY)
Phosphodiesterase I (*Crotalus adamanteus* venom) from Pharmacia (Piscataway, NJ)
Deoxyribonuclease I (bovine pancreas) from Pharmacia
Streptavidin from Bethesda Research Laboratories

[1] R. P. Viscidi, C. J. Connelly, and R. H. Yolken, *J. Clin. Microbiol.* **23,** 311 (1986).
[2] R. Shapiro, R. E. Servis, and M. Welcher, *J. Am. Chem. Soc.* **92,** 422 (1970).
[3] H. Hayatsu, Y. Wataya, and K. Kai, *J. Am. Chem. Soc.* **92,** 724 (1970).
[4] R. Shapiro and J. M. Weisgras, *Biochem. Biophys. Res. Commun.* **40,** 839 (1970).
[5] D. E. Draper, *Nucleic Acids Res.* **12,** 989 (1984).

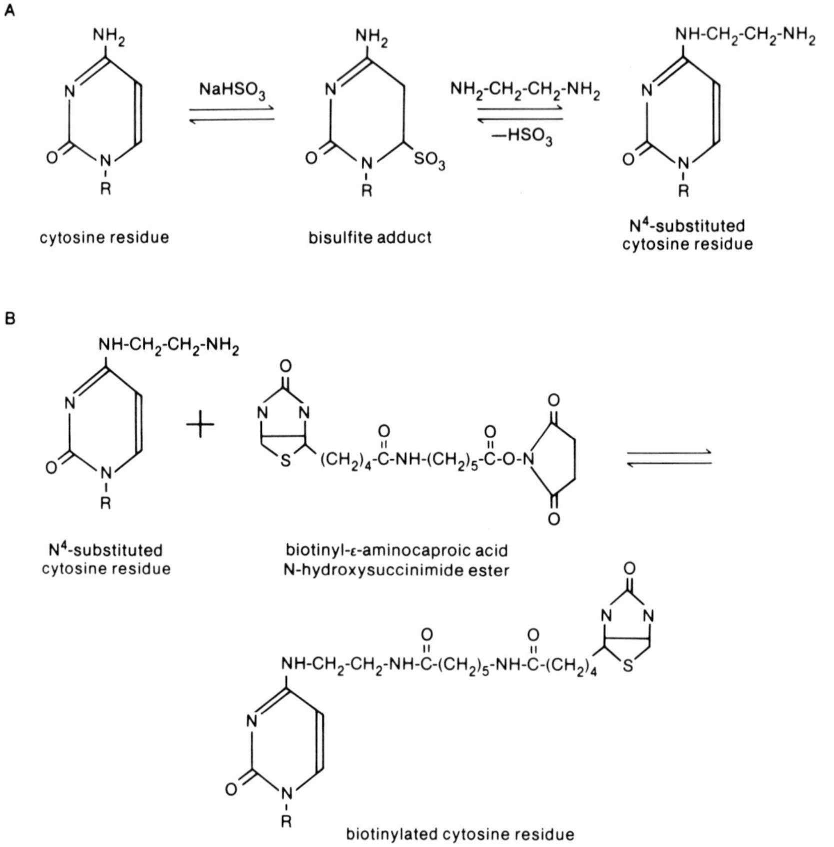

FIG. 1. Scheme for labeling DNA with biotin by the sodium bisulfite–ethylenediamine method. (A) Bisulfite-catalyzed transamination reaction with ethylenediamine. (B) Biotinylation reaction. R, Deoxyribonucleic acids. (From Viscidi *et al.*,[1] courtesy of ASM Publications.)

Calf intestinal alkaline phosphatase labeled with biotinyl-ε-aminocaproic acid from Calbiochem-Behring (San Diego, CA)
Nitroblue tetrazolium, grade III crystalline (NBT), from Sigma
5-Bromo-4-chloro-3-indolyl phosphate, *p*-toluidine salt (BCIP), from Sigma

Modification of DNA with Sodium Bisulfite and Ethylenediamine

A series of solutions of 1 *M* sodium bisulfite and 3 *M* ethylenediamine are prepared by the sequential addition to a polypropylene tube, on ice, of

1 ml ethylenediamine, 1 ml distilled water, 1 ml concentrated HCl, and 0.475 g $Na_2S_2O_5$. The pH of the solutions is adjusted to 6.0, 6.5, or 7.0, respectively, by titration with concentrated HCl, and the final volume of each is brought to 5 ml with distilled water. A 2 *M* sodium bisulfite–ethylenediamine solution is similarly prepared by the addition of 0.95 g of $Na_2S_2O_5$, and the pH of this preparation is adjusted to 5.5. A 1 mg/ml portion of hydroquinone (dissolved first in a small volume of 95% ethanol) is added to the bisulfite–ethylenediamine solutions to scavenge free radicals. Bisulfite solutions are always prepared fresh on the day of use. Since the bisulfite-catalyzed transamination reaction of cytosine residues is single-strand specific, nucleic acid samples (5–50 μg) in 10 m*M* Tris-HCl (pH 8.0), 1 m*M* EDTA are heat-denatured at 100° for 3 min and quickly cooled on ice. The reaction is initiated by adding 9 volumes of a bisulfite–ethylenediamine solution to 1 volume of single-stranded nucleic acid sample. The reaction mixture is incubated for 3 hr at 42°. After dialysis against 3 exchanges of 5 m*M* sodium phosphate buffer (pH 8.5), the samples are concentrated by placing the dialysis bag (Spectrapor #2, MW cutoff 12,000–14,000) in a plastic dish at room temperature and covering the bag with polyethylene glycol (MW 15,000–20,000) until the sample volume is reduced by diffusion to 200 μl.

Biotinylation of DNA

Modified nucleic acids (1–10 μg) in 0.1 *M* sodium phosphate buffer (pH 8.5) are biotinylated by incubation for 1 hr at 25° with 10 m*M* *N*-biotinyl-ε-aminocaproic acid *N*-hydroxysuccinimide ester. The ester is prepared freshly as a 0.2 *M* stock solution in *N′*,*N′*-dimethylformamide and is added dropwise to the modified nucleic acids to start the reaction. The unreacted biotin derivative is removed by extensive dialysis against 150 m*M* NaCl, 10 m*M* sodium phosphate buffer (pH 7.0), 1 m*M* EDTA. Labeled probes are stored at −20° until used.

Analysis of dCMP Residues of Modified DNA

λ phage DNA labeled with deoxycytidine 5′-[α-^{32}P]triphosphate by nick translation is modified with each of the bisulfite–ethylenediamine solutions described above, and the modified nucleic acids are biotinylated. A portion of each preparation is digested with DNase I (20 μg/ml) for 30 min at 25° in 50 m*M* Tris-HCl (pH 7.5), 5 m*M* $MgCl_2$. The reaction mixture is adjusted to 100 m*M* Tris-HCl (pH 9.3), 20 m*M* $MgCl_2$, 100 m*M* NaCl and incubated for an additional 4 hr at 37° with snake venom phosphodiesterase (0.25 mg/ml). An aliquot (1 μl) of the resulting mixture of nucleotide 5′-monophosphates is spotted on a polyethyleneimine-cellu-

TABLE I
DEOXYCYTIDINE COMPOSITION OF λ PHAGE DNA MODIFIED BY SODIUM BISULFITE AND ETHYLENEDIAMINE AND EFFICIENCY OF BIOTINYLATION REACTION WITH MODIFIED CYTOSINE DERIVATIVES

Bisulfite concentration (mol/liter)	pH	dCMP (%)	N^4-substituted dCMP (%)	Efficiency of biotinylation reaction (%)	Total nucleotides biotinylated (%)[a]
2	5.5	44	56	77	10.3
1	6.0	67	33	79	6.2
1	6.5	80	20	90	4.3
1	7.0	88	12	100	2.9

[a] Calculation based on a 24% cytosine content for λ phage DNA.

lose TLC plate. The plate is developed with 65% 2-propanol–2 *N* HCl solvent. The R_f values of the nucleotides are determined from autoradiograms of the plate, and the percentages of modified and unmodified cytosine residues are calculated by counting in a Beckman scintillation counter the appropriate spots dissolved in Econofluor. The approximate R_f values of unmodified dCMP, the N^4-substituted deoxycytidine derivative, and the biotinylated nucleotide are 0.71, 0.33, and 0.94, respectively.

Table I shows the percentages of modified and unmodified cytosine residues. The extent of formation of the N^4-substituted cytosine residue varies from 12 to 56%, depending on the pH and bisulfite concentration of the reaction mixture. The efficiency of the reaction between bisulfite–ethylenediamine-modified cytosine residues and *N*-biotinyl-ε-aminocaproic acid *N*-hydroxysuccinimide ester ranges from 77% for the most extensively modified nucleic acids to 100% for the lightly modified nucleic acids. Since the cytosine content of λ phage DNA is 24%, the percentage of total bases that can be labeled with biotin ranges from 2.9 to 10.3%. Unmodified λ phage DNA and undenatured DNA cannot be labeled with biotin.

Bisulfite-Modified, Biotinylated DNA Probes for Detection of Target Sequences Immobilized on Nitrocellulose

The modified nucleic acids, prepared by reaction with the different bisulfite–ethylenediamine solutions described above, are evaluated in a routine dot-blot hybridization assay to determine their performance characteristics as probes to detect complementary sequences immobilized on nitrocellulose. Serial 2-fold dilutions of λ phage DNA, denatured by boiling for 2 min in 0.2 *N* NaOH, 6× SSC (1× SSC is 150 m*M* NaCl, 15 m*M* sodium citrate), are spotted onto nitrocellulose paper (BA-85, Schleicher & Schuell, Keene, NH). The paper is neutralized in 2× SSC and baked overnight at 65° in a vacuum oven. Paper strips with a set of dilutions are individually prehybridized in sealed plastic bags for 4 hr at 37° in 45% formamide, 3× SSC, 10× Denhardt's solution, 0.1% sodium dodecyl sulfate (SDS), 0.1 m*M* EDTA, 1 m*M* sodium pyrophosphate, and 10 m*M* HEPES buffer (pH 7) containing 100 μg/ml of salmon sperm DNA. The hybridization reaction is performed in the same solution containing 10% dextran sulfate and 0.1 or 1 μg/ml of a heat-denatured biotinylated λ phage DNA probe. After an overnight incubation at 37°, the strips are washed 3 times each with 2× SSC, 0.1% SDS (10 min at 25°, 0.1× SSC, 0.1% SDS (30 min at 45°), and 0.1× SSC (10 min at 25°).

For detection of biotinylated probes, the strips of nitrocellulose paper are incubated for 1 hr at 37° with a mixture of 0.25 μg/ml of streptavidin and 0.2 U/ml of calf intestine alkaline phosphatase labeled with biotinyl-ε-aminocaproic acid diluted in phosphate-buffered saline (pH 7.2, PBS) containing 0.5% fetal calf serum. After washing the strips 3 times with PBS containing 0.05% Tween 20, they are developed with a precipitable, colorimetric substrate composed of 0.33 mg/ml of NBT and 0.17 mg/ml of BCIP diluted in 100 m*M* Tris-HCl (pH 9.3), 0.1 *M* NaCl, 5 m*M* $MgCl_2$.[6] Strips are incubated with the substrate solution for 1 hr in the dark at 25°. The reaction is terminated by washing the strips with tap water and blotting them dry.

With each of the four biotinylated probes, 1–2 pg of filter-bound DNA can be detected. In addition, the detection limit is the same with probe concentrations of 0.1 and 1 μg/ml (Fig. 2). Since all the probes provide an equivalent end-point level of sensitivity, biotinylation of more than 3–4% of the bases is not necessary. On the other hand, although accurate quantification is not possible, the intensity of the colored spots appears greater for probes containing a higher biotin content, and this increase in signal is not accompanied by a significant increase in background noise.

Tests have also been performed to determine the ability of bisulfite-modified, biotinylated probes to visualize nucleotide sequences in a Southern blot format. Various amounts of *Hin*dIII-digested λ phage DNA are electrophoresed on 1% agarose and then transferred bidirectionally to nitrocellulose filters by the method of Smith and Summers.[7] The filters are hybridized with 1 μg/ml of a λ phage probe with a 4% biotin content.

[6] J. J. Leary, D. J. Brigati, and D. C. Ward, *Proc. Natl. Acad. Sci. U.S.A.* **80,** 4045 (1983).
[7] G. E. Smith and M. D. Summers, *Anal. Biochem.* **109,** 123 (1980).

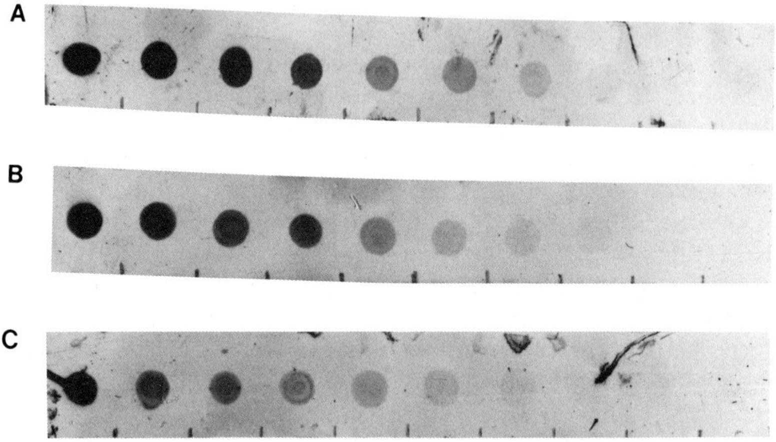

FIG. 2. Detection of λ phage DNA in dot-blot hybridization assays, using biotinylated DNA probes and a colorimetric detection method. Serial 2-fold dilutions of λ DNA (from 256 to 1 pg) were spotted on nitrocellulose paper. The last dot on the right represents 10 ng of salmon sperm DNA. The strips were hybridized with a λ DNA probe containing 10.3% of the total nucleotides biotinylated at a concentration of 1 μg/ml (a) or 0.1 μg/ml (b) or with 1 μg/ml of a λ DNA probe containing 2.9% of the total nucleotides biotinylated (c). (From Viscidi *et al.*,[1] courtesy of ASM Publications.)

Bands containing as little as 2.5 pg of DNA can be detected by the enzymatic method (Fig. 3).

Comments

The bisulfite-catalyzed transamination reaction of deoxyribonucleic acids with ethylenediamine and biotinylation of the N^4-substituted cytosine derivatives is a practical method to prepare nonisotopic probes for nucleic acid hybridization assays. The principal advantages of this labeling technique are the low cost of the reagents, the simplicity of the laboratory procedures, and the excellent performance characteristics of the labeled probes in standard mixed-phased hybridization assays. In addition, the extent of modification of nucleic acids can be controlled easily by adjusting the pH and bisulfite concentration of the reaction. This may be advantageous if the optimal biotin content varies for different probes. Another advantage of the method is the potential versatility offered by the selective introduction of a reactive amino group on DNA and the specificity of the bisulfite reaction for single-stranded cytosine residues. The reactive amine provides a site for attachment not only of biotin but also of

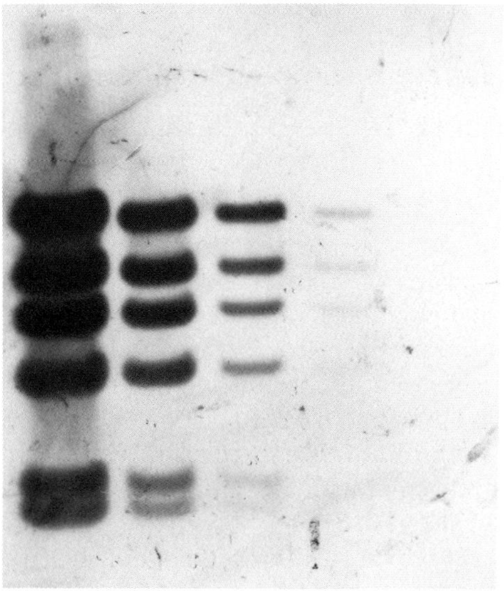

FIG. 3. Detection of λ phage DNA in a Southern blot, using a biotinylated DNA probe and a colorimetric detection method. A bidirectional Southern transfer was made from serial 10-fold dilutions of a *Hin*dIII digest of λ DNA. The agarose gel electrophoresis lanes contained from 100 to 0.01 ng of nucleic acid, and thus the lanes of the blot show from 50 to 0.005 ng. The blots were hybridized with 1 μg/ml of a λ DNA probe containing 4.3% of the total nucleotides biotinylated. (From Viscidi *et al.*,[1] courtesy of ASM Publications.)

other haptens such as dinitrophenol, of "reporter" molecules such as fluorescein, and of various *N*-hydroxysuccinimide esters of protein affinity-labeling reagents.[8,9] The specificity for unpaired cytosine residues can be used to selectively label cytosine-containing tails on DNA probes or to label specific regions on a double-stranded nucleic acid by creating deletion loops at a site of interest.[10]

Compared to some methods for the preparation of biotin-labeled DNA probes, the bisulfite–ethylenediamine method has one unique feature. Unlike methods that use nick-translation protocols, a nonenzymatic labeling procedure is less influenced by impurities such as phenol, salts, or agarose present in some nucleic acid preparations. For example, without extraction from the agarose matrix, DNA fragments that are in low-melting agarose can be labeled directly with biotin by the bisulfite–ethylenediamine method.

[8] D. E. Draper and L. Gold, *Biochemistry* **19,** 1774 (1980).

[9] A. K. Sarkar and L. H. Schulman, this series, Vol. 59, p. 156.

[10] K. W. C. Peden and D. Nathans, *Proc. Natl. Acad. Sci. U.S.A.* **79,** 7214 (1982).

[71] Direct Incorporation of Biotin into DNA

By MEIR WILCHEK, JEFFREY M. ROTHENBERG, AVI REISFELD, and EDWARD A. BAYER

As shown in other chapters of this volume, biotin is currently the nonradioactive probe of choice. For gene probes, biotin-containing mono-, oligo-, and polynucleotides have been prepared by various enzymatic and chemical means. Nevertheless, the utility of biotin as a probe can be further increased by introducing other biotin-containing reagents that react on the DNA at different positions which are not important for hybridization. In the following discussion, we describe the use of two classes of such reagents, namely, hydrazide derivatives of biotin and *p*-diazobenzoylbiocytin (DBB).

Biotin hydrazide (BHZ) and biocytin hydrazide (BCHZ) label cytidine residues in a one-step transamination reaction (Fig. 1),[1] and DBB labels guanidine residues at position 8 (Fig. 2).[2] Neither of the labeling procedures interferes with subsequent hybridization, since an NH group remains on the cytidine, while position 8 of guanidine is not involved in hybridization. Biotin-dCTP and biotin-dGTP have also been prepared and were shown to be suitable substrates for DNA polymerases. Both derivatives can be introduced into DNA by nick translation.

Interaction of BHZ with DNA

Reagents

Denatured[3] or single-stranded DNA preparation

Biotin hydrazide[4] (BHZ), 10 mg/ml in 0.1 *M* sodium acetate buffer (pH 4.5)

[1] A. Reisfeld, J. M. Rothenberg, E. A. Bayer, and M. Wilchek, *Biochem. Biophys. Res. Commun.* **142,** 519 (1987). For synthesis of BHZ and BCHZ, see M. Wilchek and E. A. Bayer, this volume, [13].

[2] J. M. Rothenberg and M. Wilchek, *Nucleic Acids Res.* **16,** 7197 (1988). For synthesis of diazobenzoyl biocytin, see M. Wilchek and E. A. Bayer, this volume [13].

[3] The DNA preparation is denatured immediately prior to biotinylation by heating to 100° for 10 min followed by rapid cooling in an ice–NaCl bath.

[4] To solubilize BHZ, the aqueous solution may be boiled. Owing to the relatively low solubility of BHZ, the reproducibility of the interaction can be improved by employing a more water-soluble hydrazide derivative, e.g., BCHZ. The extended lysyl chain in BCHZ would also serve to improve subsequent interaction with the avidin-based detection system.

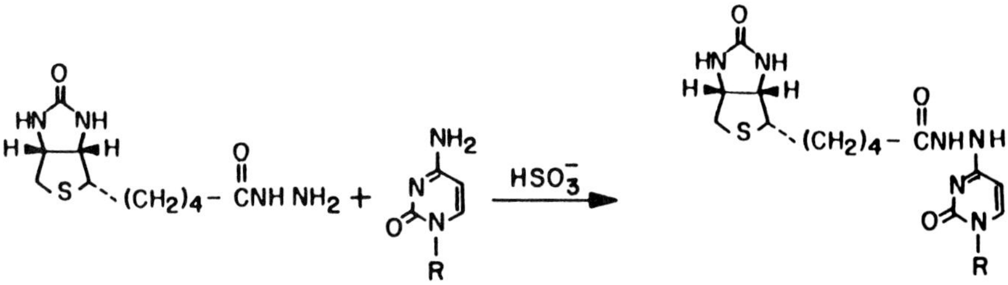

FIG. 1. Reaction of biotin hydrazide with cytidine derivatives.

Sodium bisulfite
Distilled water
Tris–EDTA buffer [10 mM Tris-HCl buffer (pH 7.6) containing 0.1 mM EDTA]

Procedure. The DNA sample (50 μg in 50 μl solution) is brought to 0.5 ml with the biotin hydrazide solution, and sodium bisulfite is added to a final concentration of 1 M. The reaction is allowed to proceed for 24 hr, after which the solution is dialyzed against distilled water for 24 hr at 4°. The solution is changed 3 times during the dialysis. The contents of the dialysis bag are collected, placed into an Eppendorf tube, and concentrated to dryness in a SpeedVac Concentrator (Savant Instruments Inc., Hicksville, NY). The DNA is resuspended in the original volume of Tris–EDTA buffer and stored at 4°.

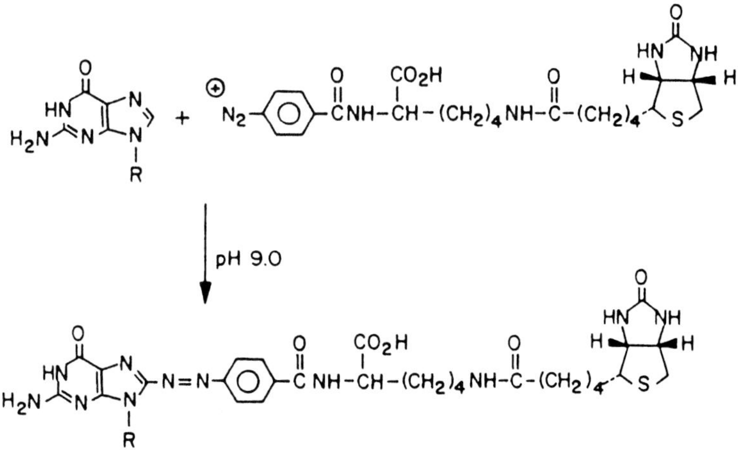

FIG. 2. Reaction of p-diazobenzoylbiocytin with guanidine derivatives.

Interaction of DBB with DNA

Reagents

Denatured[3] or single-stranded DNA preparation, 10 μg in 10 μl
Diazobenzoylbiocytin (DBB),[5] 8.8 mg/ml (14.6 m*M*) in 260 μl of 0.1 *M* borate buffer (pH 9)
3 *M* Sodium acetate buffer (pH 5.5)
Ethanol
Tris–EDTA buffer

Procedure. The DNA sample is added to the DBB solution, and the reaction is allowed to proceed for 30 min at room temperature. The DNA is precipitated by adding successively 0.1 volume of acetate buffer and 2 volumes of ice-cold ethanol. The sample is kept at −20° (1 hr to overnight), and the DNA is collected by a 15-min centrifugation in an Eppendorf microfuge. The pellet is then treated with 70% cold ethanol, and the washing procedure is repeated 4 times to remove excess reagent. The DNA is then resuspended in Tris–EDTA buffer (10 μl) and stored at −20°.

Dot-Blot Hybridization

Reagents

Target DNA
Biotinylated DNA probe
Prehybridization solution: 50% formaldehyde solution containing 5× SSC [1× SSC is 0.15 *M* NaCl and 15 m*M* sodium citrate buffer (pH 7)], 2× Denhardt's solution [50× Denhardt's solution is 10 g poly(vinylpyrrolidone), 10 g Ficoll 70, and 10 g bovine serum albumin dissolved in 1 liter distilled water], 50 m*M* sodium phosphate, and 300 μg/ml shredded sonicated salmon sperm DNA
Hybridization solution: same as prehybridization solution except that the solution is brought to 0.3% sodium dodecyl sulfate (SDS) and contains 500 μg/ml salmon sperm DNA
Tris–EDTA buffer

Direct Analysis of Biotinylated DNA. Nitrocellulose membrane filters are spotted with heat-denatured, biotin hydrazide-labeled DNA in sequential half-logarithmic dilutions. The filters are prehybridized for 4 hr at 42° in prehybridization solution. The filters are then subjected to enzyme-based detection as described below.

[5] DBB is prepared *in situ* from the aminobenzoyl-biocytin precursor immediately prior to use.

Analysis of Target DNA. Nitrocellulose membrane filters are spotted with sequential half-logarithmic dilutions of target DNA samples. The filters are prehybridized for 4 hr at 42° in prehybridization solution and then hybridized with 0.5 μg/ml of the biotinylated probe dissolved in hybridization buffer. The reaction is carried out for a minimum of 12 hr at 42°, after which the filters are washed as described earlier. The filters are then subjected to enzyme-based detection as described below.

Enzyme-Mediated Colorimetric Detection

Reagents

Quenching buffer: 0.1 *M* Tris-HCl buffer (pH 7.5) containing 0.15 *M* NaCl
Quenching solution: 3% bovine serum albumin in quenching buffer
Streptavidin-conjugated alkaline phosphatase[6] (BRL BlueGENE detection kit; Bethesda Research Laboratories, Gaithersburg, MD) diluted to 2 μg/ml in quenching buffer
Substrate buffer: 0.1 *M* Tris-HCl buffer (pH 9.5) containing 0.1 *M* NaCl and 50 m*M* $MgCl_2$
Substrate solution[7]: 5-bromo-4-chloro-3-indolyl phosphate (50 mg/ml in 33 μl of dimethylformamide) and nitroblue tetrazolium (75 mg/ml in 40 μl of 70% dimethylformamide) in 10 ml substrate buffer

Procedure. The dried filters containing the biotinylated DNA samples are incubated with quenching solution for 1 hr at 65°. The filters are then incubated at room temperature with the conjugate for 10 min. The filters are washed twice in the quenching buffer and then incubated for 10 min in substrate buffer. The reaction with substrate solution is carried out for 1–3 hr in sealed plastic bags (Dazey Seal-a-Meal).

Comments

Using the above-described procedures, we have succeeded in labeling picogram quantities of DNA on dot blots. The major problem with the two reagents described in this chapter, however, is their low solubility under conditions required to modify the DNA. Other biotin-containing reagents

[6] Other avidin- or streptavidin-containing conjugates or complexes (see E. A. Bayer and M. Wilchek, this volume [18]) can be used in place of the streptavidin-conjugated alkaline phosphatase reagent used in this instance.

[7] Other suitable substrate solutions may be used instead. For most of our blotting studies with alkaline phosphatase, we regularly use 10 mg naphthol AS-MX phosphate dissolved in 200 μl of dimethylformamide and mixed with a solution containing 30 mg Fast Red dissolved in 100 ml of 0.1 *M* Tris-HCl (pH 8.4).

that are based on the same principles but designed to comprise elements of increased solubility in aqueous solution should be developed. The use of such reagents at higher concentrations would increase the incorporation of biotin, which will eventually lead to higher sensitivity than that achieved with the reagents described in this chapter.

[72] Colorimetric-Detected DNA Sequencing

By STEPHAN BECK

Since their introduction in 1981,[1] biotin-labeled polynucleotides have found widespread application as sensitive affinity probes for the detection of nucleic acids and have proved to be a real alternative to the convenient but hazardous detection using radioisotopes.[2] Their use for the detection of DNA sequencing bands, however, has been described only recently.[3] Two problems, specific to the DNA-sequencing technique, have complicated its application. First, for DNA sequencing (for review, see Refs. 4 and 5), relatively large gels are used to separate the sequencing reactions (0.1–0.3 mm in thickness, 40–100 cm in length, and 20–40 cm in width). Since a biotin-based detection requires accessibility to the separated sequencing bands for the actual visualization reaction, the band pattern has to be transferred from the very fragile gels onto an immobilizing matrix. For this purpose, special techniques and equipment had to be developed beforehand.[6,7]

The second problem is a consequence of the nature of the biotinylated nucleotides. Traditionally DNA-sequencing bands are labeled by enzymatic incorporation of radioactive nucleotides, such as [^{32}P]dATP or [^{35}S]dATP, and are visualized by autoradiography.[8,9] Although a variety of biotinylated nucleotides have been synthesized[1,10] and are commer-

[1] P. R. Langer, A. A. Waldrop, and D. C. Ward, *Proc. Natl. Acad. Sci. U.S.A.* **78,** 6633 (1981).

[2] M. Wilchek and E. A. Bayer, this volume [2].

[3] S. Beck, *Anal. Biochem.* **164,** 514 (1987).

[4] J. C. Moores, *Anal. Biochem.* **163,** 1 (1987).

[5] R. Wu, ed., this series Vol. 155.

[6] G. M. Church and W. Gilbert, *Proc. Natl. Acad. Sci. U.S.A.* **81,** 1991 (1984).

[7] S. Beck and F. M. Pohl, *EMBO J.* **3,** 2905 (1984).

[8] A. M. Maxam and W. Gilbert, *Proc. Natl. Acad. Sci. U.S.A.* **74,** 560 (1977).

[9] F. Sanger, S. Nicklen, and A. R. Coulson, *Proc. Natl. Acad. Sci. U.S.A.* **74,** 5463 (1977).

[10] G. Gebeyehu, P. Y. Rao, P. SooChan, D. A. Simms, and L. Klevan, *Nucleic Acids Res.* **15,** 4513 (1987).

METHODS IN ENZYMOLOGY, VOL. 184

cially available, they are not yet suitable for use in DNA sequencing for two reasons. First, they appear to be poorer substrates, especially for the large fragment of DNA polymerase I (Klenow fragment) widely used in the dideoxy chain termination sequencing method.[1,10] This does not seem to affect their suitability for DNA labeling by nick translation, but during the elongation of a dideoxy sequencing reaction it enhances the likelihood of nonspecific chain terminations which result in artificial extra bands. Second, when using biotinylated nucleotide analogs for DNA sequencing, unpredictable band shifts in mobility are likely to occur during electrophoresis, which consequently results in an unreadable sequence ladder. This shift is due to the random incorporation of the biotinylated analog, which has a different molecular weight than its parent compound. The same band shift problem also applies to labeling of DNA-sequencing bands using a photoactivatable analog of biotin.[11]

Recent advances in transfer or blotting technology and in DNA synthesis have made it possible to overcome these problems. Church and Gilbert[6] have developed a technique for electroblotting the band pattern of a sequencing gel onto nylon membranes, and Beck and Pohl[7] have developed a direct blotting electrophoresis (DBE) method. The DBE technique combines electrophoretic separation and electroblotting in a single run by electroeluting the DNA bands, during their separation, onto an immobilizing matrix that is driven across the bottom of a very short sequencing gel by a conveyor belt.[12] In the field of DNA synthesis, Agrawal *et al.*[13] and others have developed a method to covalently attach nonradioactive labels, such as fluorophores and biotin, to the 5′ terminus of a synthetic oligonucleotide (primer) that can be used in a primer-extension sequencing reaction and overcomes the bandshift problem. In the following section a protocol is described for the colorimetric detection of DNA-sequencing bands using such a biotinylated primer, direct blotting electrophoresis, and a phosphatase-catalyzed color reaction.

Materials and Methods

The mixes for the DNA sequencing reactions are prepared according to Refs. 14 and 15. The 17-nucleotide-long primer, biotinylated at the

[11] A. C. Forster, J. L. McInnes, D. C. Skingle, and R. H. Symons, *Nucleic Acids Res.* **13,** 745 (1985).

[12] F. M. Pohl, German Patent 3022 527 (1980).

[13] S. Agrawal, C. Christodoulou, and M. J. Gait, *Nucleic Acids Res.* **14,** 6227 (1986).

[14] A. T. Bankier, K. M. Weston, and B. G. Barrell, this series, Vol. 155, p. 51.

[15] A. T. Bankier and B. G. Barrell, *in* "Techniques in the Life Sciences" (R. A. Flavell, ed.), p. 1. Elsevier, Amsterdam.

5′ terminus {5′-biotin-NH$(CH_2)_2O$-d[pGTAAAACGACGGCCAGT]-3′}, was synthesized and kindly provided by Agrawal *et al.*[13] Streptavidin, 5-bromo-4-chloro-3-indolyl phosphate (BCIP), nitroblue tetrazolium (NBT), and biotinylated alkaline phosphatase were from Bethesda Research Laboratories (BRL, Gaithersburg, MD; Kit No. 8239SA).

Reagents

T mix: 25 μl 0.5 m*M* dTTP, 500 μl 0.5 m*M* dCTP, 500 μl 0.5 m*M* dGTP, 50 μl 10 m*M* ddTTP, and 1000 μl 10 m*M* Tris-HCl (pH 8), 0.1 m*M* Na_2EDTA

C mix: 500 μl 0.5 m*M* dTTP, 25 μl 0.5 m*M* dCTP, 500 μl 0.5 m*M* dGTP, 8 μl 10 m*M* ddCTP, and 1000 μl 10 m*M* Tris-HCl (pH 8), 0.1 m*M* Na_2EDTA

G mix: 500 μl 0.5 m*M* dTTP, 500 μl 0.5 m*M* dCTP, 25 μl 0.5 m*M* dGTP, 16 μl 10 m*M* ddGTP, and 1000 μl 10 m*M* Tris-HCl (pH 8), 0.1 m*M* Na_2EDTA

A mix: 500 μl 0.5 m*M* dTTP, 500 μl 0.5 m*M* dCTP, 500 μl 0.5 m*M* dGTP, 1 μl 10 m*M* ddATP, and 500 μl 10 m*M* Tris-HCl (pH 8), 0.1 m*M* Na_2EDTA

Biotin primer mix: 1.5 μl T, C, G, or A mix, 0.25 μl 0.2 μ*M* biotin primer, and 0.25 μl 100 m*M* Tris-HCl (pH 8), 10 m*M* $MgCl_2$ in a total volume of 2 μl[16]

Reaction mix: 0.25 μl 0.1 *M* dithiothreitol, 0.5 μl 5 μM dATP, 0.5 U Klenow fragment (Boehringer Mannheim, FRG), and 0.75 μl H_2O in a total volume of 2 μl[16]

Chase mix: 0.25 m*M* dTTP, dCTP, dGTP, and dATP in 10 m*M* Tris-HCl (pH 8), 0.1 m*M* Na_2EDTA

Stop mix: 10 ml deionized formamide, 10 mg xylene cyanol, 10 mg bromphenol blue, 200 μl 0.5 *M* Na_2EDTA

PBS–HT buffer (pH 7): 7.3 g/liter NaCl, 2.36 g/liter Na_2HPO_4, 1.31 g/liter $NaH_2PO_4 \cdot 2H_2O$, 15000 U/liter heparin, and 0.5% Tween 20

TMS buffer: 100 m*M* Tris-HCl (pH 9.5), 100 m*M* NaCl, and 50 m*M* $MgCl_2$

Developer: 330 μg/ml NBT and 166 μg/ml BCIP; always make up fresh with TMS buffer and filter through a 0.45-μm disposable filter before use.

Protocol for Biotin–Streptavidin-Detected DNA Sequencing

Sequencing Reactions. Sequencing reactions are performed using the reagents as tabulated below. In the procedure, template and biotin primer mix are incubated in sealed tubes for 30 min at 55°. Then reaction mix is

[16] Volumes are calculated per sequencing reaction.

added, and the tubes are resealed and incubated for 15 min at 37°. Chase mix is added, and the sealed tubes are incubated for 15 min at 37°. The reaction is stopped by the addition of stop mix and incubation of the tubes, unsealed, for 20 min at 80° (end volume ~2 μl).

	T	C	G	A
Template (~0.5 μg/2 μl)	2 μl	2 μl	2 μl	2 μl
Biotin primer mix	2 μl	2 μl	2 μl	2 μl
Reaction mix	2 μl	2 μl	2 μl	2 μl
Chase mix	2 μl	2 μl	2 μl	2 μl
Stop mix	2 μl	2 μl	2 μl	2 μl
Total volume	10 μl	10 μl	10 μl	10 μl

Electrophoresis/Blotting. Run approximately 2 μl per lane on a standard sequencing gel and transfer bands onto an immobilizing matrix by electroblotting,[6] or run/blot approximately 2 μl per lane with DBE onto an immobilizing matrix using a short sequencing gel.[3,7,17,18]

Membrane Pretreatment. Dry the membrane. UV-cross-link DNA to the membrane for 3 min, and bake the membrane at 80° for 30 min.

Colorimetric Detection. Block the membrane in PBS–HT buffer for 30 min at room temperature in a plastic tray. Incubate the wet, rolled-up membrane in PBS–HT buffer (1–2 ml per 100 cm^2) containing 1 μg/ml streptavidin in a glass culture tube for 10 min on a horizontal roller. Wash in 30 volumes of PBS–HT buffer for 5 min at room temperature in a plastic tray with moderate shaking. Repeat twice. Incubate the wet, rolled-up membrane in PBS–HT buffer (1–2 ml per 100 cm^2) containing 1 μg/ml biotinylated alkaline phosphatase in a glass culture tube for 10 min on a horizontal roller. Wash twice in 30 volumes of PBS–HT buffer and once in TMS buffer in a plastic tray with moderate shaking for 5 min each. Add developer (1–2 ml per 100 cm^2) to the wet, rolled-up membrane in a glass culture tube and incubate at room temperature until bands become visible (~1 hr). Dry the membrane at 80° for 5 min. The glass culture tube can be replaced by a heat-sealable plastic bag.

Concluding Remarks and Future Prospects

Although the development of the biotin–streptavidin-mediated, colorimetric detection of DNA-sequencing bands is still at a very early stage, it

[17] S. Beck, *in* "Electrophoresis '86" (M. J. Dunn, ed.), p. 173. VCH Verlagsgemeinschaft Weinhem, 1986.

[18] F. M. Pohl and S. Beck, this series, Vol. 155, p. 250.

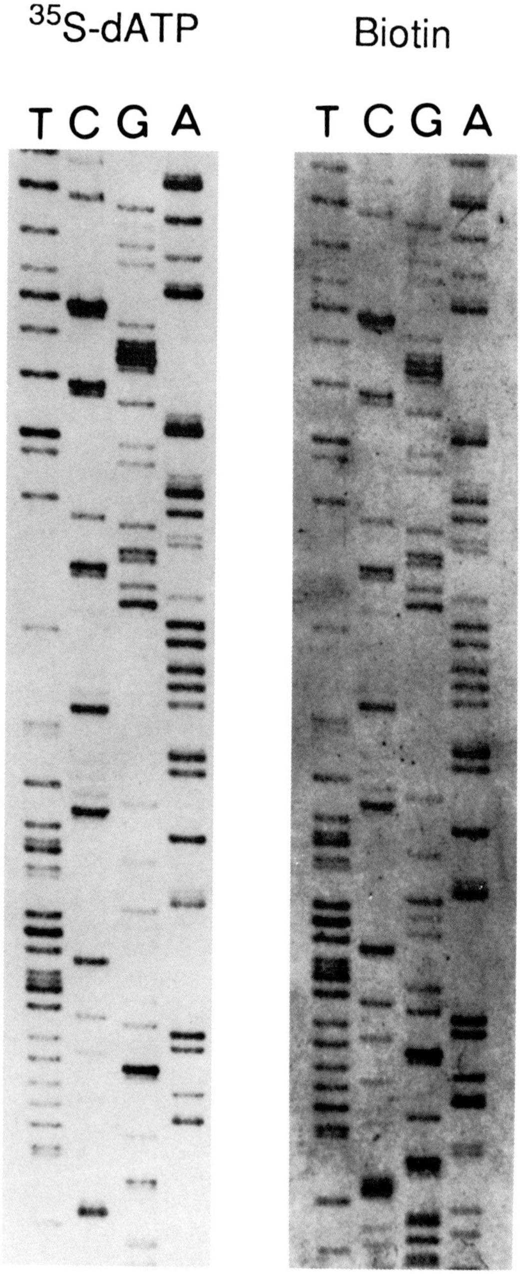

FIG. 1. Comparison of an autoradiographic ([35S]dATP) and colorimetric (biotin) detected band pattern of colabeled, standard sequencing reactions that were separated and blotted onto Nytran membranes (Schleicher & Schuell, Keene, NH) via direct blotting

clearly shows potential to become a real alternative to radioactive detection (Fig. 1). The sensitivity of the biotin–streptavidin-mediated detection compares to that of [^{35}S]dATP in a standard sequencing reaction, and the biotinylated probes exhibit very high reagent stability. Multiple freezing and thawing, formamide (up to 100%), 8 *M* urea, and heat up to 80° do not affect their reactivity. Besides some obvious advantages (such as lower costs in the long run, the possibility to keep the probes frozen without deterioration arising from radiolysis/decay or bleaching, and no requirement of special safety regulations), the colorimetric detection bears great potential for new developments. In combination with the direct blotting electrophoresis, the whole detection procedure can possibly be largely automated, since all reactions can be carried out at room temperature and the membrane can easily be moved from one incubation chamber into another by a conveyor belt. The two-step reaction, including the streptavidin and alkaline phosphatase incubation, can possibly be reduced into a single-step reaction by using, for instance, a bispecific, monoclonal antibody. Such an antibody, directed against biotin and peroxidase, has already been described and has been successfully used for the detection of immunodots, Western, and Southern blots.[19]

Another very promising approach is to synthesize biotinylated dideoxynucleotides similar to those reported for fluorescent-tagged chain-terminating dideoxynucleotides.[20] As described for these fluorescent terminators, this approach is likely to solve two problems simultaneously. First, similar to using a biotinylated primer, the band-shift problem is solved by the incorporation of exactly one label molecule per DNA chain. Second, since each DNA molecule becomes specifically labeled only at the 3′ terminus, artificially occurring extra bands (arising from nonspecific chain termination) are no longer a problem because they remain undetectable. Like the fluorescent dideoxy analogs, the biotinylated counterparts are likely to be a very bad substrate or would not react at all with conventional DNA polymerases used for DNA sequencing. A chemically modified, bacteriophage T7 DNA polymerase,[21] however, shows lower discrimination against nucleotide analogs and incorporates the fluorescent-tagged dideoxynucleotides sufficiently for DNA sequencing.[20,22]

[19] M. M. Leong, C. Milstein, and R. Pannell (1986). *J. Histochem. Cytochem.* **34,** 1645 (1986).

[20] J. M. Prober, G. L. Trainor, R. J. Dam, F. W. Hobbs, C. W. Robertson, R. J. Zagursky, A. J. Cocuzza, M. A. Jensen, and K. Baumeister, *Science* **238,** 336 (1987).

[21] S. Tabor and C. C. Richardson, *Proc. Natl. Acad. Sci. U.S.A.* **84,** 4767 (1987).

[22] For additional information see Beck *et al., Nucleic Acids Res.* **17,** 5115 (1989).

electrophoresis. The [^{35}S]dATP-labeled band pattern corresponds to a 15-hr exposure to an X-ray film (Fuji safety), and the biotin-labeled band pattern corresponds to a colorimetric detection of approximately 3 hr. (Reprinted with permission from Ref. 3.)

[73] Composite Applications

By Meir Wilchek and Edward A. Bayer

The avidin–biotin system was initially introduced with the intention of applying the technique in a unified fashion for analysis by various disciplines. For example, if a specific target molecule is under investigation in a given experimental system, then a single biotinylated binder can be used with subsequent analysis by a variety of different avidin-associated probes. In such an application, the target molecule can be quantified by immunoassay using an avidin-conjugated enzyme. At the cellular level, the distribution of the target can be analyzed using a fluorescent form of avidin combined with flow cytometry. At the ultrastructural level of resolution, the localization of the molecule of interest can be accomplished under the electron microscope using avidin–gold conjugates. Avidin–Sepharose can be employed for the isolation of the target molecule, and avidin–enzyme or radioactive avidin can be used with blotting techniques for information at the molecular level of resolution.

Another form of composite application would be if a variety of different biotinylated binders were combined with a single avidin-associated probe for comparative studies. An example of this would be the employment of a series of lectins with different sugar specificities or a collection of antibodies specific for different antigens for assay in a given experimental system using a single avidin–enzyme conjugate. Of course, the epitome of such application would be to have a group of well-characterized biotinylated binders and a series of well-characterized avidin-conjugated (or biotinylated) probes for routine work in different areas (for isolation, localization, quantification, etc.).

This final section of the volume includes a collection of chapters exemplifying such studies ([74]–[79]). Unfortunately, only few such detailed studies have been performed. It seems that the interests and technological expertise of a given laboratory are often limited to a particular area (i.e., the major concern of some laboratories is in localization, in others the interest lies in immunoassay, whereas in other laboratories the research is concentrated on isolation of a given target molecule). It appears that very few laboratories have devoted their research to composite studies involving avidin–biotin technology. We have been able, therefore, to include only a representative sample of such a combined approach. We hope that more studies of this nature will be published in the near future.

Because of the limited number of these studies, we have also included in this part of the book some miscellaneous or special applications that do not strictly fall into the other categories. These include the preparation of complex neoglycoproteins, the probing of cellular receptors, and the targeted production of hybridomas.

[74] Third Component of Human Complement: C3

By MELVIN BERGER

The third component of human complement (C3) is of great importance not only because it provides a subunit for three different enzymes in the complement-activation cascade, but also because it is the major opsonic protein of the serum complement system.[1] Native C3 is a two-chain protein with a molecular weight of 195,000. Its concentration in normal serum, approximately 1.2 mg/ml, is the highest of any complement component. The opsonic function of C3 is due to its ability to attach covalently to the target of complement activation, providing bound fragments which can interact noncovalently with specific receptors on phagocytic cells.[2]

Activation of C3 is caused by proteolytic cleavage at arginine-77 of the α chain, as shown schematically in Fig. 1. This is accompanied by release of the 8000-Da C3a fragment and by conformational changes in the C3b fragment that allow acyl transfer from an internal thiol ester[3] (composed of the thiol of cysteine-262 and a γ-carboxyl group derived from glutamine-265 of the remaining α' chain)[4] to a suitable acceptor such as a free amino group or hydroxyl on the target.[5] The resulting covalently bound C3b fragment may interact with specific C3b receptors (CR1) on phagocytic cells or with other complement components to continue the activation cascade. Further proteolytic cleavage of C3b, carried out by Factor I (C3 inactivator), forms $C3b_i$, which is inactive in the complement cascades. This partially cleaved molecule is still opsonic, however, and it interacts with a different cellular receptor, CR3, rather than CR1. Addi-

[1] E. J. Brown, K. A. Joiner, and M. M. Frank, *in* "Fundamental Immunology" (W. Paul, ed.), p. 645. Raven, New York, 1984.

[2] G. D. Ross and M. E. Medof, *Adv. Immunol.* **37,** 217 (1985).

[3] B. F. Tack, R. A. Harrison, J. Janatova, M. L. Thomas, and J. W. Prahl, *Proc. Natl. Acad. Sci. U.S.A.* **77,** 5764 (1980).

[4] M. H. L. De Bruijn and G. H. Fey, *Proc. Natl. Acad. Sci. U.S.A.* **82,** 708 (1985).

[5] S. K. Law and R. P. Levine, *Proc. Natl. Acad. Sci. U.S.A.* **74,** 2701 (1977).

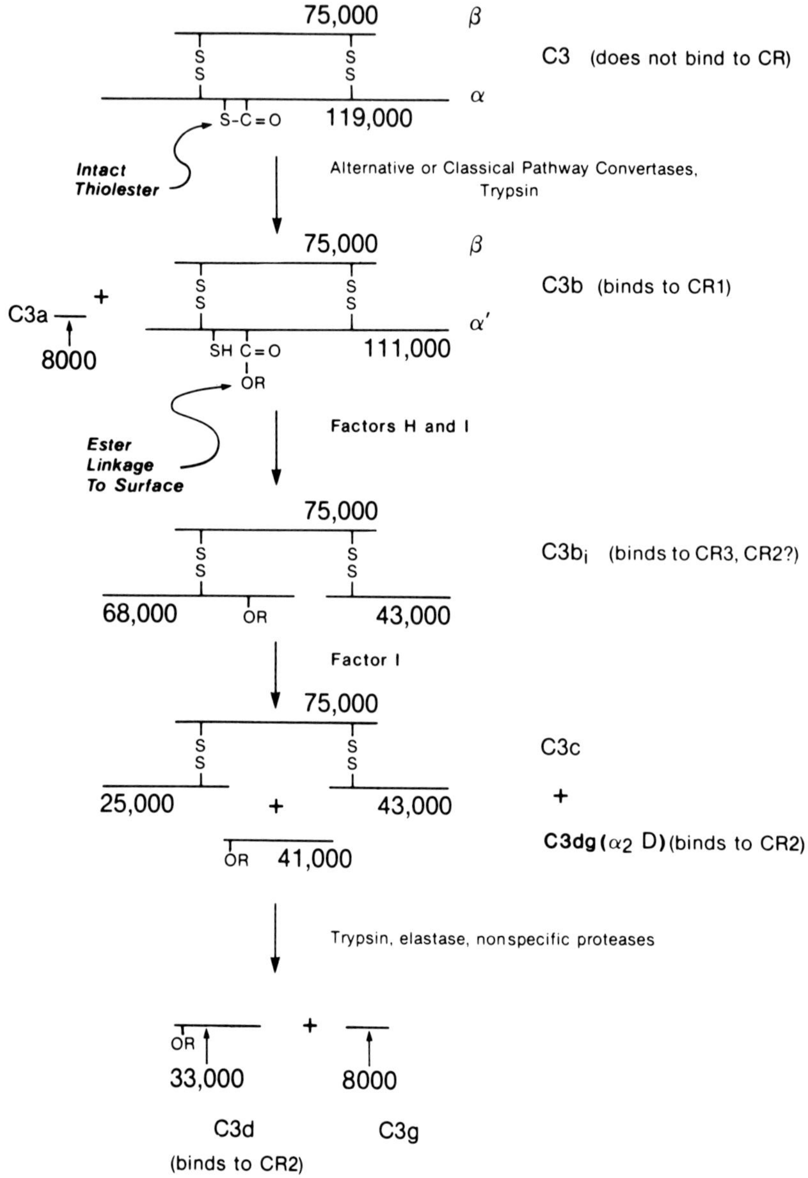

FIG. 1. Schematic diagram of C3 protein chain cleavage reactions and fragments that bind to cellular complement receptors (CR). Note that positions and number of disulfide bonds are not precise.

tional cleavage steps result in release of C3c and further alterations in the receptor-binding specificity of the remaining C3dg fragment.

Several factors, including the chemical nature of the C3b acyl acceptor and the localization and distribution of C3 fragments on the target, determine the rate of degradation of C3b and the opsonic activity of the fragments. The demonstration that native C3 could be biotinylated without loss of activity[6] and the commercial availability of several avidin derivatives suggested that these reagents could be employed to answer many questions about C3 functions. In particular, the resistance of the avidin–biotin bond to conventional denaturants including sodium dodecyl sulfate (SDS) suggested that this reagent system could be used with great advantage in the isolation and identification of the acceptors for acylation during C3 activation and in defining the role of C3 fragments in opsonization.

Biotinylation of Hemolytically Active C3

Reagents

Buffer: 10 m*M* phosphate-buffered 0.14 *M* NaCl (pH 7.4, PBS)
Na_2CO_3, 5% (w/v) in distilled water
(+)-Biotin-*N*-hydroxysuccinimide ester (Calbiochem, La Jolla, CA)
Dimethylformamide (DMF)
Avidin, 11 U/mg (Worthington-Millipore, Freehold, NJ)
Pronase (Boehringer-Mannheim, Indianapolis, IN)
o-[(*p*-Hydroxyphenyl)azo]benzoic acid (HABA) (Eastman Kodak, Rochester, NY)
Purified human C3,[7] 4–5 mg/ml in PBS
Complement reagents for hemolytic assay of C3 (see Ref. 7)

Procedure. Although the thiol ester of intact C3 is susceptible to nucleophilic attack, which causes loss of hemolytic activity, this group is apparently protected in a hydrophobic pocket, and the native molecule remains stable at alkaline pH for brief periods of time. Reaction of C3 with biotin-*N*-hydroxysuccimide (BNHS) is carried out as described by Bayer *et al.*,[8] using a stock solution of 1.7 mg/ml (5 m*M*) in DMF. C3, stored at 4–5 mg/ml in the presence of protease inhibitors such as phenylmethylsulfonyl fluoride, is titrated to pH 8.5 by adding small aliquots of 5% Na_2CO_3. Depending on the desired number of biotins per C3 molecule (see below), the necessary amount of BNHS solution, usually 20–50 μl/ml

[6] M. Berger, T. A. Gaither, P. M. Cole, T. M. Chused, C. H. Hammer, and M. M. Frank, *Mol. Immunol.* **19,** 857 (1982).

[7] C. H. Hammer, G. H. Wirtz, L. Renfer, H. D. Gresham, and B. F. Tack, *J. Biol. Chem.* **256,** 3996 (1981).

[8] E. A. Bayer, E. Skutelsky, and M. Wilchek, this series, Vol. 62, p. 308.

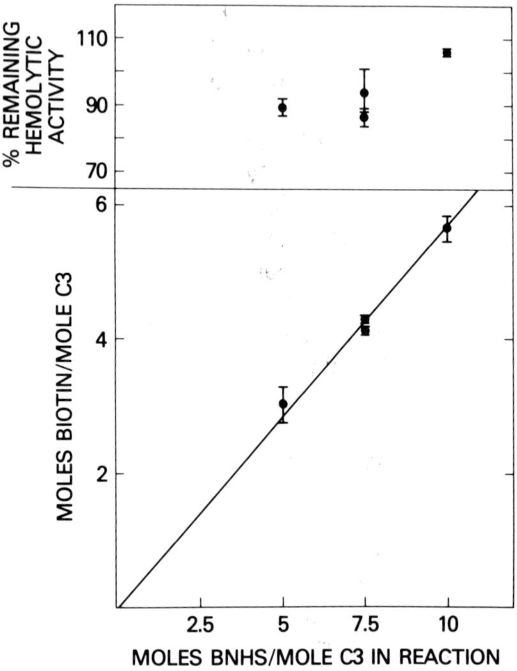

FIG. 2. Incorporation of biotin and retention of hemolytic activity. All determinations were performed on two to four replicates. Means ± SEMs are indicated. (From Ref. 6, with permission of Pergamon Press Ltd.)

of C3, is added to the alkalinized C3 and allowed to react at room temperature for 4 hr, then dialyzed at 4° overnight against PBS, with 1 or 2 buffer changes to remove excess biotin and released succinimide.

To determine the stoichiometry of biotinylation, 1-mg aliquots of derivatized C3 are denatured by heating at 56° for 10 min, allowed to cool to room temperature, and treated with pronase in a final weight ratio of 1% relative to the exact amount of C3 present. (Protein concentrations of purified C3 solutions are determined from absorbance at 280 nm using $E_{1\,cm}^{1\%} = 9.7$).[9] The digestion mixtures are stirred at room temperature overnight, then spun for 2 min in a microcentrifuge. Aliquots of the supernatant are used for biotin determination by the avidin–HABA dye exclusion assay as described by Green,[10] adapted to a total volume of 0.55 ml per determination. A standard curve with free biotin is run with each set of assays. As shown in Fig. 2, as the input of BNHS is varied from a ratio

[9] B. F. Tack and J. W. Prahl, *Biochemistry* **15,** 4513 (1976).
[10] N. M. Green, this series, Vol. 18, p. 418.

of 5 to 10 mol per mole of C3, the biotin incorporation increases linearly from 3 to 6 mol per mole of conjugate.[6] Control experiments demonstrated that digests of nonderivatized C3 will not interfere with the HABA-binding assay. Determination of the hemolytic activity of the biotinyl C3, using erythrocytes presensitized with antibody and complement components C1 and C4 (in the presence of excess amounts of the remaining components C2 and C5 to C9),[7] have indicated excellent preservation of hemolytic activity, with a mean recovery greater than 90%. For routine preparations of biotinyl C3 an input ratio of 7.5 mol of BNHS per mole of C3 is used, which generally results in the incorporation of 4 biotins per C3 molecule.

Localization of Biotin in C3 Molecule. In order to determine which regions of the protein carry biotin, an aliquot of biotinyl C3 was incubated under sterile conditions for 8 days at 37° in the serum of a patient genetically deficient in C3 so that fragmentation by Factor I would be complete. Initial experiments had revealed that fresh biotinyl C3 had normal mobility on agarose gel immunoelectrophoresis and that aged biotinyl C3 showed typical C3c and C3d immunoprecipitation patterns. Adsorption with avidin–Sepharose removes both the C3c and C3d fragments but has no effect on unmodified C3.[6] This result indicates that biotin is present in both fragments of the derivatized molecule and provides assurance that even if fragmentation by Factor I in whole serum continues to completion, a biotinylated fragment (C3d) will remain bound to the target of complement activation.

Effects of Avidin on Biotinyl C3.[6] Although biotinylation of C3 does not interfere with the activities of the native molecule or its fragments, several functions of biotinyl C3 are inhibited by avidin. Incubation of intact biotinyl C3 with avidin leads to loss of hemolytic activity and may also cause precipitation, depending on the ratios used. In contrast, avidin has no effect on unmodified C3. Also, avidin does not form precipitates if the C3 has first been fragmented by factor I. Depending on the concentration used, avidin treatment of sheep erythrocytes sensitized with biotinyl C3 may result in blocking of the sites necessary for interaction of the bound C3b fragments with their receptors, thereby inhibiting immune adherence or opsonic functions. In addition, avidin treatment of such cells can cause agglutination by cross-linking the biotinyl C3b moieties on different erythrocytes. Fluorescent avidin derivatives can be used to quantify cell-bound biotinyl C3 fragments by flow cytometry, showing an excellent correlation with uptake of ^{125}I-labeled C3.[6] In addition, ferritin- or gold-conjugated avidin can be used to localize bound C3 fragments on targets of complement activation. This approach has been used to determine the topography (i.e., clustered versus randomly distributed) and site (i.e., cell wall versus capsule) of C3b deposition on various targets and thus to explain differences in opsonic activity of C3b fragments at different sites.[6,11]

Use of Avidin–Sepharose and Biotinyl C3 to Identify C3 Acceptors

Reagents

Hemolytically active biotinyl C3
Complement-activating system: purified components or active serum
Avidin–Sepharose
1% sodium dodecyl sulfate (SDS) in 37.5 m*M* Tris-HCl (pH 8.0)
1 *M* NH_2OH in 0.05% SDS–20 m*M* Tris (pH 10.5)

As noted above, the chemical nature of the acceptor for acylation by C3 is a primary determinant of whether the C3b fragments will serve to continue complement activation by the alternative and/or classical pathways or become inactivated by Factor I and its cofactor, H. The well-known detergent and chemical resistance of the avidin–biotin bond suggested that this reagent system could be used to isolate membrane constituents and immunological reactants to which C3 fragments had become bound, even under conditions where strong hydrophobic and antibody–antigen bonds would be broken. This reagent system is particularly useful when putative acceptors for C3 can be selectively labeled and their retention on avidin–Sepharose quantified. In the examples described here, ^{125}I-labeled antipneumococcal antibodies are used to demonstrate that the antibody-mediated classical pathway activation is accompanied by binding of C3b to the antibodies themselves. *Escherichia coli,* intrinsically labeled in the capsular polysaccharides or outer membrane proteins, is used to demonstrate alternative pathway-mediated binding of C3b to these structures.

Binding of C3b to Antipneumococcal Antibodies.[12] Strain R36a pneumococci, grown to log phase in trypticase soy broth, are washed in PBS and heated to 56° for 30 min to inactivate autolysin. The cells are then incubated with ^{125}I-labeled IgG or IgM antibodies and washed with PBS. The classical pathway C3 convertase is assembled on these organisms by incubation with purified human C1, C4, and C2. Biotinylated human C3 is then reacted with the organisms for 60 min at 30°, and the cells are washed extensively with PBS. The sensitized organisms (2×10^9) are then incubated for 30 min in 1% SDS–37.5 m*M* Tris (pH 8), which removes most of the antibody but fails to cause the release of intrinsically labeled cell wall or internal protein constituents. The pH 8 SDS extract is

[11] E. J. Brown, K. A. Joiner, R. M. Cole, and M. Berger, *Infect. Immun.* **39,** 403 (1983).
[12] E. J. Brown, M. Berger, K. A. Joiner, and M. M. Frank, *Infect. Immun.* **42,** 594 (1983).

then mixed with 0.5 ml of avidin–Sepharose by slowly rotating at room temperature for 12 hr to allow maximal binding of the biotinyl C3b-bearing constituents. This mixture is then washed 5 times with 2-ml aliquots of the same buffer to remove nonspecifically bound material. The amount of ^{125}I retained by the adsorbent is then determined.

Using the above procedure, it was found that 14.6% of the IgM and 9.2% of the IgG were bound through the interaction of covalently attached biotinyl C3b with the avidin–Sepharose. The specificity of this binding was confirmed in parallel control experiments using either nonbiotinylated C3 or avidin–Sepharose presaturated with free biotin, which showed less than 1.5% binding of ^{125}I-labeled IgG or IgM. In reciprocal experiments in which the antipneumococcal antibodies were biotinylated and the C3 radiolabeled with ^{125}I, retention of ^{125}I-labeled C3b on the avidin–Sepharose was 14% with biotinyl-IgM, 29% with biotinyl-IgG, but less than 0.4% with unmodified antibodies. To determine the nature of the bond between the ^{125}I-labeled antibodies and the biotinyl C3b adsorbed to the avidin–Sepharose, the adsorbent bearing these complexes was treated with 1 *M* NH_2OH in 0.5% SDS–20 m*M* Tris (pH 10.5), since such treatment had previously been shown to cleave the ester bonds formed between C3b and its targets. The NH_2OH pH 10.5 treatment released 39% of the IgM and 53% of the IgG, indicating that this fraction of the C3b–antibody linkages represented ester bonds and that the remainder reflected amide bonds, which would not be hydrolyzed under these conditions. Again, the reciprocal experiments using ^{125}I-labeled C3 and biotinylated antibodies gave similar results. In addition, inclusion of reducing agents such as 50 m*M* dithiothreitol in the wash buffers allowed us to demonstrate that the C3b fragment attaches covalently to both heavy and light chains.

Binding of C3b to Polysaccharide Constituents of E. coli.[13] *Escherichia coli* 0111B4, strain CL 99, is grown with [^{14}C]galactose so that this sugar will be incorporated into its capsular O antigen. The bacteria [10^9 colony-forming units (cfu)/ml] are then sensitized with an equal volume of 10% antibody-depleted human serum to which radiolabeled biotinyl C3 (by reductive methylation with NaB^3H_4) is added to comprise about 50% of the total C3 present. Complement activation is allowed to proceed at 37° for 45 min with intermittent shaking, and the bacteria are then washed 3 times and resuspended in buffer containing 1 m*M* phenylmethylsulfonyl fluoride. The washed bacterial suspension is lysed by 3 passages through a French press at 16,000 psi, and a supernatant fraction

[13] K. A. Joiner, R. Goldman, M. Schmetz, M. Berger, C. H. Hammer, M. M. Frank, and L. Leive, *J. Immunol.* **132,** 369 (1984).

containing the capsular antigens is prepared by centrifugation at 113,000 *g* at 4° for 2 hr.

SDS is added from a stock solution to bring the final concentration to 2%, and the solution is heated to 100° for 5 min to further inactivate proteases. After cooling to room temperature, the solution is diluted to a final SDS concentration of 1% and mixed with avidin–Sepharose at an 8 : 1 ratio by volume of solution to adsorbent. The suspension is mixed by rotating at room temperature for 4 hr and loaded into a 10-ml disposable column (Econo column, Bio-Rad, Richmond, CA) containing an additional 0.5 ml of fresh avidin–Sepharose. The column is then washed with buffer containing 1% SDS until the effluent contains less than 0.01% of the input radioactivity, and the column outlet is clamped.

Under these conditions, 40% of the biotinyl ^{3}H-labeled C3 and 0.41% of the applied [^{14}C]galactose was found to be specifically bound to the avidin–Sepharose. To determine the nature of the C3–capsular polysaccharide bond, the adsorbent is resuspended in 4 volumes of 1% SDS–1 *M* NH_2OH (pH 10.5) and placed in a 37° incubator for 30 min; elution is then continued with this buffer. This treatment does not affect the avidin–biotinyl C3 binding, as only 3.6% of the bound ^{3}H could be eluted. In contrast, however, the NH_2OH at pH 10.5 cleaves the C3–polysaccharide linkage, and essentially 100% of the bound [^{14}C]galactose can be eluted. This demonstrates quite clearly that the C3 is bound to the capsular polysaccharide by ester linkages which are cleaved by the NH_2OH pH 10.5 treatment, and that this can be accomplished under conditions where the avidin–biotin bond remains stable.

Avidin-Linked C3b Oligomers with Enhanced Affinity for C3b Receptors

Reagents

Purified C3b[14]
(+)-Biotin-*N*-hydroxysuccinimide ester
Avidin
BioGel A 1.5m or A 5m

Although the receptors on phagocytic cells for C3b and $C3b_i$ play critical roles in the host defense and in clearance of immune complexes, studies of their binding interactions and functions have been hampered by low affinities for soluble monomeric ligands. For example, binding of monomeric C3b to CR1 is easily demonstrable only at reduced ionic

[14] M. Berger, T. A. Gaither, C. H. Hammer, and M. M. Frank, *J. Immunol.* **127,** 1329 (1981).

strength, and the K_a has been estimated at only 10^6–10^7 L/M.[15,16] For this reason, and to eliminate variables inherent in the use of large opsonized particles, stable oligomers are preferable reagents for studies of receptor–ligand interactions. Such oligomers may be generated by avidin cross-linking of biotinyl-C3b.

Procedure. C3b is prepared by treating purified human C3 with 0.1% (w/w) trypsin for 5 min at 37°, then stopping the reaction with excess soybean trypsin inhibitor.[14] C3b is isolated from the reaction mixture by chromatography on BioGel A 0.5m in 1 *M* NaCl, 0.1 *M* sodium acetate (pH 5.6). To avoid excessive conformational changes that might be caused by multiple biotinylated C3b molecules wrapping themselves around individual tetravalent avidin molecules, it is desirable to incorporate only one molecule of biotin per molecule of C3b. This may be accomplished quite reproducibly with the procedure outlined above for biotinylation of active C3 but using only a 2- to 2.5-fold molar excess of BNHS. The actual biotin content of the conjugate may be determined as described above.

An amount of avidin (2 mg/ml solution) exactly sufficient to bind the total amount of biotin present in the conjugate is then added to the biotinyl C3b (5 mg/ml), and the solution is rotated in the cold overnight to assure complete equilibration. In general, the amount of avidin on a molar basis will be about one-fourth of the amount of biotinyl C3b. The mixture is then chromatographed over a 1.5 × 90 cm BioGel A 1.5m or A 5m column equilibrated with PBS, and the protein eluting at the void volume is used as a mixture of C3b trimers and tetramers. Additional peaks, corresponding to monomers and dimers are also generally found but may be incompletely resolved. Brief centrifugation in an airfuge (100,000 *g*) should always be performed before these oligomers are utilized in cell-binding assays.

When the binding affinities of these products for human neutrophils were compared to that of unmodified monomeric C3b using a rosette inhibition assay, it was found that monomeric biotinyl C3b was 84 ± 5% as effective as unmodified C3b on a weight basis, while the oligomers were 10–20 times more effective than monomers, depending on the density of C3b on the indicator particles. Avidin-linked biotinyl C3b dimers were 4–5 times more effective than unmodified monomers and were closely comparable to dimers formed spontaneously during the trypsinization of C3. An interesting variation on this theme was employed by

[15] M. A. Arnaout, N. Dana, J. Melamed, R. Medicus, and H. R. Colten, *Immunology* **48,** 229 (1983).

[16] M. Berger and A. S. Cross, *Immunology* **51,** 431 (1984).

Okada and Brown,[17] who used a biotin derivative containing a disulfide spacer to conjugate C3b and prepare avidin-linked oligomers. This allowed the complexes to be easily cleaved by the addition of dithiothreitol, in which case the monomeric C3b was released from CR1. With this reagent, then, susceptibility to release by the reducing agent could be used to distinguish between surface-bound and internalized oligomers.

Comments

The ease with which C3 can be biotinylated without loss of hemolytic activity allows a number of avidin derivatives to be employed for studies that help define the function of this complement component in the host defense. In particular, since C3 and antibody may often be found together in immune complexes or on opsonized particles, it is desirable to have another reagent system available for localization and quantification of C3 fragments. C3 shares with only C4 and α_2-macroglobulin the ability to bind covalently to its targets by transacylation from an internal thiol ester. The acceptor for C3 binding is a critical determinant of the subsequent fate of the molecule and its biological function.

The extraordinary affinity of the avidin–biotin interaction and its stability to detergents suggested that this reagent system would be uniquely suited for using solid-phase avidin adsorbents to retain biotinyl C3–acceptor complexes under conditions where antibody–antigen bonds and even lipid–lipid interactions could be dissolved, i.e., by 2% SDS. This provided the original impetus for studying the biotinylation of C3. As shown by the examples cited here, acceptors such as polysaccharides can be selectively released by cleaving the ester linkage between the C3 fragment and the sugar hydroxyl. In addition, this system allowed us to demonstrate that approximately one-half of the C3b–immunoglobulin binding is characterized by amide linkages. In any situation where the potential acceptors for acylation by C3 can be selectively labeled or otherwise identified after release, this reagent system should provide a powerful tool. Other complement components may also be biotinylated without loss of hemolytic activity so that avidin–Sepharose may be used to study their interactions, i.e., in detergent-solubilized membrane attack complexes[18] or other multimolecular structures.

[17] K. Okada and E. J. Brown, *J. Immunol.* **140,** 878 (1988).
[18] E. R. Podack and H. J. Muller-Eberhard, *J. Biol. Chem.* **256,** 3145 (1981).

[75] Purification and Characterization of Membrane Proteins

By JEAN-PIERRE KRAEHENBUHL, and CLAUDE BONNARD

Introduction

The interaction of avidin or streptavidin with the coenzyme biotin has been used to isolate polypeptide hormone receptors[1–6] and has simplified the problem of linking tracers to biological probes[7–10] such as hormones, antibodies, or nucleic acids. Streptavidin, like avidin, binds biotin with one of the strongest noncovalent bonds[5] yet discovered ($K_A \sim 10^{15}\ M^{-1}$).[11] This 60,000-Da nonglycosylated protein with a neutral isoelectric point, which is produced by *Streptomyces avidinii,*[12] is stable to denaturation by urea, guanidine-HCl, and heat.[12,13] The absence of a carbohydrate moiety and its neutral isoelectric point confer on streptavidin advantages over avidin, inasmuch as nonspecific interactions between the protein and cellular components are minimized. These properties make streptavidin an ideal reagent for the binding of biotinylated ligands. Conjugation of biotin to biological macromolecules is usually a simple procedure, and several biotinylating reagents are now commercially available (Table I).[14–21]

[1] K. Hofmann and Y. Kiso, *Proc. Natl. Acad. Sci. U.S.A.* **73,** 3516 (1976).
[2] K. Hofmann, F. M. Finn, H. J. Friesen, C. Diaconescu, and H. Zahn, *Proc. Natl. Acad. Sci. U.S.A.* **74,** 2697 (1977).
[3] G. A. Orr, *J. Biol. Chem.* **256,** 761 (1981).
[4] M. T. Haeuptle, M. L. Aubert, J. Djiane, and J. P. Kraehenbuhl, *J. Biol. Chem.* **258,** 305 (1983).
[5] G. Redeuilh, C. Secco, and E. E. Baulieu, *J. Biol. Chem.* **260,** 3996 (1985).
[6] M. A. Dunand, J. P. Kraehenbuhl, B. C. Rossier, and M. L. Aubert, *Am. J. Physiol.* (in press) (1988).
[7] M. Wilchek and E. A. Bayer, this volume [13]; E. A. Bayer and M. Wilchek, this volume [14].
[8] H. Heitzmann and F. M. Richards, *Proc. Natl. Acad. Sci. U.S.A.* **71,** 3537 (1974).
[9] J. E. Manning, N. D. Hershey, T. R. Broker, M. Pellegrini, H. K. Mitchell, and N. Davidson, *Chromosoma* **53,** 107 (1975).
[10] D. S. Papermaster, B. G. Schneider, M. A. Zorn, and J. P. Kraehenbuhl, *J. Cell Biol.* **77,** 196 (1978).
[11] N. M. Green, *Adv. Protein Res.* **29,** 85 (1975).
[12] L. Chaiet and F. J. Wolf, *Arch. Biochem. Biophys.* **106,** 1 (1964).
[13] L. Chaiet, T. W. Miller, F. Tansig, and F. J. Wolf, *Antimicrob. Agents Chemother.* **3,** 28 (1963).
[14] K. Hofmann, F. M. Finn, and Y. Kiso, *J. Am. Chem. Soc.* **100,** 3585 (1978).
[15] C. Bonnard, D. S. Papermaster, and J. P. Kraehenbuhl, *in* "Immunolabelling for Electron Microscopy" (J. M. Polak and I. M. Varndell, eds.), p. 95. Elsevier, New York, 1984.

METHODS IN ENZYMOLOGY, VOL. 184

TABLE I
BIOTINYLATING REAGENTS

Reagent	Specificity	Comments	Source	Refs.
N-Hydroxysuccinimidobiotin	R—NH_2	Spontaneous reaction at pH 7.0–9.0	Pierce	8, 14
N-Hydroxysuccinimidyl 6-(biotinamido)hexanoate	R—NH_2	Long-chain analog of *N*-hydroxysuccinimido-biotin	ENZO Biochem	15, 16
Sulfo-*N*-hydroxysuccinimido-biotin	R—NH_2	Water soluble	Pierce	17
Sulfo-*N*-hydroxysuccinimidyl 6-(biotinamido)hexanoate	R—NH_2	Water soluble, long-chain analog of sulfo-*N*-hydroxysuccinimido-biotin	Pierce	18, 19
Sulfosuccinimidyl-2-(biotinamido)ethyl 1,3′-dithiopropionate	R—NH_2	Water soluble, cleavable by thiols or reducing agents	Pierce	20
N-Iodoacetyl-*N*′-biotinyl-hexylenediamine	R—SH	Active halogen reaction at pH 7.5–8.5	Pierce	21

The biotin–streptavidin bridge technique applied to the purification of membrane proteins offers several advantages over conventional affinity chromatography procedures that require immobilization of the ligand. When a ligand is linked directly or via a linker to a solid matrix, it becomes difficult to evaluate its biological activity. Coupling may cause impairment or loss of biological activity or binding capacity. These parameters can be determined independently for the biotinylated ligand in a fluid phase and compared to those of the native molecule. Furthermore, the number of biotin molecules cross-linked to the ligand can be controlled by varying the stoichiometry of biotinylated reagent added to the ligand. The degree of biotinylation can be monitored by using radiolabeled biotinylating reagents. A further advantage is that the binding sites (receptors, antigens) can be visualized at the cell or tissue level with the biotinylated ligand (hormone, factor, antibody) and streptavidin coupled to

[16] S. M. Costello, R. T. Felix, and R. W. Giese, *Clin. Chem.* **25,** 1572 (1979).
[17] W. T. Loc and D. H. Conrad, *J. Exp. Med.* **159,** 1790 (1984).
[18] J. J. Leary, D. J. Brigati, and D. C. Ward, *Proc. Natl. Acad. Sci. U.S.A.* **80,** 4045 (1983).
[19] K. Hofmann, G. Titus, J. A. Montibeller, and F. M. Finn, *Biochemistry* **21,** 978 (1982).
[20] M. Shimkus, T. Levy, and T. Herman, *Proc. Natl. Acad. Sci. U.S.A.* **82,** 2593 (1985).
[21] K. Sutoh, K. Yamamoto, and T. Wakabayashi, *J. Mol. Biol.* **178,** 323 (1984).

fluorochromes, enzymes, or colloidal gold. Thus, the same biotinylated ligand can serve, first, to identify and purify a membrane constituent that specifically interacts with the ligand using immobilized streptavidin and, second, to localize the membrane protein in cells or tissues using labeled streptavidin.

In this chapter, we first describe the various biotinylation reaction schemes that have been used in our laboratory and then present protocols for the immobilization of streptavidin followed by the various steps involved in the affinity purification of hormone receptors from rabbit liver or mammary gland[4] and from toad kidney[6] using the streptavidin–biotin bridge system. Finally, the use of the biotin–streptavidin system for the immunocytochemical localization of two plasma membrane proteins[22] involved in Na^+ transport across tight epithelia is documented in amphibian tissues.

Biotinylating Reagents

In order to utilize the biotin–streptavidin affinity to the best advantage, it is essential that the streptavidin-binding sites for biotin bound to macromolecular carriers be fully accessible with minimal steric hindrance. Since the biotin-binding site in avidin, and presumably in streptavidin, resides within an approximately 1-nm depression,[23] a spacer between the biotin and the macromolecule can be expected to increase the affinity between the biotinylated ligand and streptavidin.[23] This has been tested experimentally in our laboratory, and the results are given below.

If the very strong noncovalent association between streptavidin and biotin can be maintained after derivatization, it should be possible to retrieve from crude mixtures in one step a high yield of biotinylated ligand–receptor complexes. Once isolated, however, the biotinylated ligand cannot be gently released from streptavidin, and dissociation requires harsh conditions deleterious to most receptor proteins. Guanido analogs of biotin, such as 2-iminobiotin, have thus been synthesized and shown to bind avidin at alkaline pH and dissociate at pH values of 4 or lower.[19] However, the weak affinity of iminobiotin for avidin, its sensitivity to steric effects, and the rather unphysiological proton concentration required for maximal binding argue against the use of these analogs.[19,24]

[22] F. Verrey, P. Kairouz, E. Schaerer, P. Fuentes, K. Geering, B. C. Rossier, and J. P. Kraehenbuhl, *Amer. J. Physiol.* **256,** 1034 (1989).

[23] N. M. Green, L. Konieczny, E. J. Toms, and R. C. Valentine, *Biochem. J.* **125,** 781 (1971).

[24] K. Hofmann, S. W. Wood, C. C. Brinton, J. A. Montibeller, and F. M. Finn, *Proc. Natl. Acad. Sci. U.S.A.* **77,** 4666 (1980).

Biotinylation usually occurs in an aqueous phase. Thus, the biotinylating agent should be soluble in a solvent miscible with the aqueous buffer. However, the most commonly used solvent [*N,N*-dimethylformamide (DMF)], used to dissolve *N*-hydroxysuccinimidobiotin, is rather toxic to cells even at concentrations as low as 0.1% DMF (C. Bonnard and J.-P. Kraehenbuhl, unpublished observation). Lysis, which has been observed with erythrocytes, will allow the biotinylation of non-plasma-membrane proteins of the cell. The addition of a sulfonate group on the *N*-hydroxysuccinimido moiety of the biotinylation reagent represents a major improvement for the selective labeling of plasma membrane proteins. Cleavable analogs containing a disulfide bond have been synthesized and are now commercially available (Table I).[20] Such reagents should allow the recovery of membrane proteins or receptors under mild conditions, but so far their use has not been reported.

Biotinylation Reaction Schemes

Four major reaction schemes have been developed, which allow control of where substitution should occur on the ligand. These involve esters, maleimides, active halogens, and hydrazides. The properties of the reagents that have been used in our laboratory are summarized in Table I and described in detail elsewhere in this volume.[7]

When biotinylating a ligand, it is often useful to monitor the degree of biotinylation. This is possible using radiolabeled biotinylating reagents, which can easily be synthesized or are commercially available (Amersham International, plc, Amersham, England). *N*-Hydroxysuccinimido[^{14}C]-biotin is synthesized using [^{14}C]biotin (Amersham), according to the method of Heitzmann and Richards.[8] *N*-Hydroxysuccinimido[^{3}H]biotin is commercially available (Amersham). *N*-Hydroxysuccinimidyl 6-([^{3}H]biotinamido)hexanoate, synthesized according to Costello *et al.*,[16] is also available from Amersham.

The coupling of radiolabeled biotinylating reagents to ligands is performed according to the supplier's instructions, and the various steps are summarized as follows.

Protocol for Coupling of Radiolabeled Biotinylating Reagents to Ligands

Solutions

Ligand: dissolve 30–300 nmol of ligand in 1 ml of 10 m*M* sodium phosphate buffer (pH 7.5) containing 150 m*M* NaCl (PBS)

Biotinylating reagent: dissolve 12 mg *N*-hydroxysuccinimido-[^{14}C]biotin (specific activity 60 μCi/mmol) or 6 mg *N*-hydroxysuccimido-6-([^{3}H]biotinamido)hexanoate (specific activity 0.34 mCi/mg) in 1 ml dry dimethylformamide (DMF)

Procedure. The coupling is carried out with a 100- to 1000-fold molar excess of the biotinylating reagent over the ligand. Mix the appropriate volume of biotinylating reagent (100 μl) in DMF to 1 ml of ligand solution and stir for 1 hr at room temperature. Remove uncoupled biotin by gel filtration of the reaction mix on Sephadex G-25 equilibrated with 25 m*M* Tris-HCl (pH 7.5). Determine the optical density and radioactivity in 0.5-ml fractions. Pool the fractions containing the ligand and determine the ligand concentration and the radioactivity. From these values calculate the number of biotin residues on the ligand. Usually 5–20 residues are coupled. Higher values have been obtained with immunoglobulins, which did, however, alter the antibody activity. In such cases, the molar ratio of biotinylating reagent per ligand should be reduced empirically until an optimum ratio is obtained where both the number of biotin residues coupled and the biological activity of the ligand remain high.

Binding Constant of Interaction between Biotinylated Ligand and Immobilized Streptavidin

Using two biotinylating reagents and immobilized streptavidin, we have performed binding studies and Scatchard plot analysis to determine the effect of a spacer arm inserted between the biotin and the ligand on the interaction with streptavidin. Bovine serum albumin (BSA) is derivatized with *N*-hydroxysuccinimido[^{3}H]biotin or with *N*-hydroxysuccinimidyl 6-([^{3}H]biotinamido)hexanoate to monitor the degree of biotin cross-linking. The biotinylated BSA (on average 4–10 biotin molecules are cross-linked) is radioiodinated and incubated with nonradioiodinated BSA and streptavidin as indicated in Table II.[25] The results, summarized in Table II, indicate that coupling biotin to a ligand drastically reduces its affinity for streptavidin. The K_A is decreased 5–6 orders of magnitude from 10^{15} to approximately 10^{8} M^{-1}. With a spacer arm between the ligand and biotin, the association constant is increased one order of magnitude to around 10^{9} M^{-1}. Interestingly, a second high-capacity, low-affinity site can also be detected. When biotin coupled to red blood cells is incubated with soluble streptavidin, nonlinear Scatchard plots are generated, suggesting strong positive cooperativity. The affinity between biotinylated ligand with or

[25] P. J. Munson and D. Rodbard, *Anal. Biochem.* **107,** 220 (1980).

TABLE II
BINDING CONSTANTS OF DERIVATIZED BSA AND IMMOBILIZED STREPTAVIDIN[a]

Derivatized ligand	Temperature (°)	K_A (M^{-1}) First site	K_A (M^{-1}) Second site
Biotin–BSA	4	4.3×10^8	—
	23	4.6×10^8	—
	37	5.2×10^8	—
Biotin–spacer–BSA	4	3.0×10^9	1.2×10^8
	23	3.4×10^9	1.6×10^8
	37	4.1×10^9	2.3×10^8

[a] In the assay 5 pmol of ^{125}I-labeled biotinylated BSA was added to 50 μg insolubilized streptavidin in a volume of 100 μl. Cold biotinylated BSA was added in increasing concentrations ranging between 5 pmol and 1 nmol. The mixture was incubated at the indicated temperature for 12 hr until equilibrium was reached. Free ligand was separated from streptavidin by centrifuging the mixture through a cushion of 0.5 *M* sucrose. The pellets were counted and the data analyzed using the Scatfit program developed by Munson and Rodbard.[25]

without a spacer arm and streptavidin increases with acidic pH (data not shown).

Immobilization of Streptavidin

Affi-Gel 10 (Bio-Rad, Richmond, CA) (2 ml) is washed once with 2-propanol and 3 times with cold distilled water. The washed beads are incubated for 4 hr at room temperature with 0.6 mg of streptavidin [Amersham; Bethesda Research Laboratories (BRL), Gaithersburg, MD] in 1 ml of 0.1 *M* $NaHCO_3$ and 10^6 cpm of ^{125}I-labeled streptavidin to monitor the coupling efficiency. The beads are then incubated for 2 hr at room temperature with 2.0 *M* glycine buffer (pH 9.0) to quench free reactive groups. Between 0.5 and 2.0 mg of streptavidin are coupled per milliliter of packed Affi-Gel. Immobilized streptavidin is available from Pierce Chemical Co. (Rockford, IL) at 1–2 mg/ml of gel.

Purification of Plasma Membrane Receptors

Lactogenic and somatogenic hormone-binding sites from rabbit liver or mammary membranes are solubilized with the nonionic detergent Triton X-100 and purified by affinity chromatography using biotinylated human growth hormone (hGH) and immobilized streptavidin.[4,6]

Detergent Solubilization of Membrane Proteins. Liver or mammary glands are removed from 5-month-old outbred New Zealand White rabbits at day 15 of gestation. The tissue is homogenized in the presence of protease inhibitors [1 m*M* of phenylmethylsulfonyl fluoride (Serva, Heidelberg, FRG) and 0.2 μg/ml of pepstatin, leupeptin, and antipain (Sigma Chemical Co., St. Louis, MO)]. A microsomal fraction enriched in plasma membrane enzyme is prepared as described elsewhere.[26]

The membrane suspension is centrifuged for 2 hr at 200,000 g_{av} and suspended at a concentration of 1–5 mg/ml in lysis buffer [50 m*M* Tris-HCl (pH 7.5) containing 100 m*M* NaCl, 10 m*M* $MgCl_2$, 0.02% sodium azide, and 0.5% (w/v) Triton X-100 (Sigma) or 10 m*M* 3-[(3-cholamidopropyl)dimethylammonio]-1-propane sulfonate (CHAPS, Calbiochem, San Diego, CA)]. The samples are agitated for 30 min at room temperature. Insoluble material is removed by centrifugation for 2 hr at 200,000 g_{av} at 4°. On solubilization about 50% of the original binding sites are generally recovered in the supernatant as determined by a radioreceptor assay.[27,28] Usually, 0.5 and 1.0 ng of hGH is bound per milligram of solubilized mammary or liver membrane proteins, respectively. The optimal solubilization conditions and the nature of the detergent must be determined for each plasma membrane protein. For instance, solubilization of lactogenic hormone receptor cannot be achieved in the presence of Triton X-100, because the hormone undergoes aggregation. In this case, one can use CHAPS.[29]

Affinity Purification of Membrane Proteins. Four steps are required for the purification of membrane receptors using biotinylated ligand and immobilized streptavidin. The detailed protocol is given below. Briefly, the solubilized membrane proteins are incubated with the biotinylated ligand until equilibrium is reached (12 hr using hGH and lactogenic or somatogenic receptors). Then, the unbound biotinylated ligand has to be separated from the hormone complex. This can be achieved by polyethylene glycol precipitation when the mass of the hormone–receptor complex is significantly larger than that of the free ligand. A high M_r protein such as immunoglobulin G serves as a carrier. Third, the hormone–receptor complex is adsorbed on immobilized streptavidin. Finally, the receptor is

[26] L. Kuehn and J. P. Kraehenbuhl, *J. Biol. Chem.* **254,** 11072 (1979).

[27] R. P. Shiu and H. G. Friesen, *Biochem. J.* **140,** 301 (1974).

[28] Y. M. L. Suard, J. P. Kraehenbuhl, and M. L. Aubert, *J. Biol. Chem.* **254,** 10466 (1979).

[29] D. S. Liscia and B. K. Vonderhaar, *Proc. Natl. Acad. Sci. U.S.A.* **79,** 5930 (1982).

recovered from the matrix by elution conditions that vary depending on the type of ligand.

Affinity Purification of Hormone Receptors Using Biotinylated Ligand and Immobilized Streptavidin

Step 1: Binding of Biotinylated Ligand to Solubilized Receptor. Solubilized membrane proteins (1–4 ml, 1–5 mg/ml) in lysis buffer [50 m*M* Tris-HCl (pH 7.5), 100 m*M* NaCl, 10 m*M* $MgCl_2$, 0.02% sodium azide, and 0.5% Triton X-100 or 10 m*M* CHAPS] are incubated overnight at room temperature with 25–100 μg of biotinylated ligand. Radioiodinated biotinylated ligand (10^7 cpm) is added to the mixture to monitor the recovery and yield of receptors during subsequent purification.

Step 2: Separation of Unbound Biotinylated Ligand. Ice-cold lysis buffer containing 0.1% rabbit γ-globulin (0.35–.5 ml) is added as a carrier to the solubilized membranes. Add an equal volume of a polyethylene glycol 6000 (Siefried, Zofingen, Switzerland) solution (25% if the membranes are solubilized with Triton X-100, 32% if solubilized with CHAPS) and vortex vigorously. Let sit on ice 10 min and then spin for 30 min at 3000 g_{av} at 4°. Remove the supernatant by vacuum aspiration and drain the tubes. Count the tubes to determine the amount of hormone bound. Dissolve the resulting pellet in lysis buffer containing 1% Triton X-100 or 10 m*M* CHAPS and allow to stand for 16 hr at 4° in order to solubilize the ligand–receptor complex.

Step 3: Binding of Hormone–Receptor Complex to Immobilized Streptavidin. Wash 50–200 μl of streptavidin–Affi-Gel 3 times with 1 ml lysis buffer containing 1% Triton X-100 or 10 m*M* CHAPS in an Eppendorf tube. Add the washed beads to the solubilized hormone–receptor complex mixture and incubate for 15–30 min at 4° with constant shaking. Centrifuge the beads, wash the pellet 3 times with lysis buffer containing 0.1% Triton X-100, and count the pellet.

Step 4: Elution of Receptor. The elution conditions depend on the type of membrane receptor that is purified. The lactogenic or somatogenic hormone–receptor complex is dissociated with 5.0 *M* $MgCl_2$.[4] Incubate the beads with elution buffer [25 m*M* Tris-HCl (pH 7.5), 5.0 m*M* $MgCl_2$, 30 m*M* β-octylglycoside (Sigma)] and dialyze the eluate in lysis buffer. The binding properties of the purified receptor can be determined in lysis buffer.[4] For analytical purposes, the receptor can be eluted with 2% sodium dodecyl sulfate (SDS).

Comments. Using this affinity purification, a 35-kDa polypeptide was isolated from the mammary gland, which based on its binding properties corresponds to the lactogenic hormone receptor. In liver, a second 67-

kDa protein was isolated which corresponds to the somatogenic hormone receptor. The primary sequence of the two receptors, deduced from cDNA clones, has recently been established. The rat prolactin receptor consists of a 291-amino acid polypeptide[30] and the hGH receptor of a 620 amino acid protein.[31] Thus, the protocol described in this chapter allows the purification of the monomeric form of each receptor, the mass of which corresponds to the predicted sequence. The larger M_r forms reported for the receptor from rat ovary or liver using different purification protocols[32,33] have not been observed with this procedure.

Plasma Membrane Protein Localization

Plasma membrane proteins, including receptors,[34] ion pumps, or channels,[22] have been localized immunocytochemically in cells or tissues using biotinylated antibodies and labeled streptavidin. The use of streptavidin–colloidal gold in electron microscope immunocytochemistry is described elsewhere in this volume.[35] In this chapter, we restrict ourselves to light microscope immunocytochemistry.

Preparation of Tissue Sections. It is important to select a fixative that does not significantly alter antigenicity. The effect of fixation on antigenicity can easily be tested on dot blots. The antigen in increasing concentration is applied to nitrocellulose using a dot-blot or a slot-blot device. The nitrocellulose membranes are then treated with common fixatives (paraformaldehyde–glutaraldehyde,[34] paraformaldehyde–lysine–periodate[36]) known to preserve antigenicity. Following incubation with primary and biotinylated antibodies, the sheets are treated with enzyme-coupled streptavidin. The amount of reaction product generated by the samples treated with fixatives can be compared to that of control samples.

Biotinylation of Antibodies. The biotinylation protocol is the same as described above. It is convenient to biotinylate an antibody directed against the primary antibody, rather than the primary antibody itself. Such biotinylated antibodies directed against mouse, rabbit, rat, goat, or sheep immunoglobulins are now commercially available (Amersham,

[30] J. M. Boutin, C. Jolicoeur, H. Okamura, J. Gagnon, M. Edery, M. Shirota, D. Banville, I. Dusauter-Fourt, J. Djiane, and P. Kelly, *Cell* **53,** 69 (1988).

[31] D. W. Leung, S. A. Spencer, G. Cachianes, R. G. Hammonds, C. Collins, W. Henzel, R. Barnard, M. J. Waters, and W. I. Woods, *Nature (London)* **330,** 537 (1987).

[32] J. S. Bonifacio and M. L. Dufau, *J. Biol. Chem.* **259,** 4542 (1984).

[33] L. A. Haldosen and J. A. Gustafson, *J. Biol. Chem.* **262,** 7404 (1987).

[34] R. Solari and J. P. Kraehenbuhl, *Cell* **36,** 61 (1984).

[35] D. S. Papermaster, B. G. Schneider, and I. Nir, this volume [41].

[36] J. W. McLean and P. K. Nakane, *J. Histochem. Cytochem.* **22,** 1077 (1974).

BRL). For double-labeling experiments[37] it is not possible to use biotinylated secondary antibodies if the primary antibodies are raised in the same species. In this case, the primary antibody should be biotinylated directly.

Labeled Streptavidin Systems. Fluorescent (fluorescein isothiocyanate Lissamine, Texas Red, phycoerythrin) or enzyme (alkaline phosphatase, β-galactosidase, horseradish peroxidase) conjugates of streptavidin are now commercially available (Amersham). Either the fluorochromes or enzymes are directly coupled to streptavidin, or streptavidin serves as a bridge between the biotinylated antibodies and the biotinylated enzymes.

Protocol for Light Microscope Immunocytochemistry

Step 1: Fixation and Tissue Sectioning. The two fixatives that preserve antigenicity and yield good tissue preservation are fixative a [freshly depolymerized 3% paraformaldehyde, 0.5% glutaraldehyde in 0.1 *M* sodium cacodylate buffer (pH 7.5)] and fixative b [freshly depolymerized 2% paraformaldehyde, 150 m*M* lysine, 10 m*M* sodium periodate in 75 m*M* sodium phosphate buffer (pH 7.5) containing 15 m*M* glucose[30]]. The tissue is fixed for 30 min–1 hr in fixative a at room temperature and for 4–6 hr in fixative b.[30] The fixed tissue can be embedded in hydrophilic resins[35] or frozen in liquid isopropane. Standard 5- to 10-μm or 0.5-μm thick[32] frozen sections are harvested on polylysine-coated microscope slides. When an aldehyde is used as fixative, the free aldehyde groups are blocked by incubating the sections with 100 m*M* glycine or lysine. The Schiff bases, which tend to autofluoresce, can be reduced to amide bonds by a 15-min incubation in 10 m*M* sodium cyanoborohydride dissolved in PBS.

Step 2: Incubation with Primary Antibody. The sections are washed 3 times with PBS and incubated for 15 min with PBS containing 0.1% BSA. The sections are incubated with primary antibody at the appropriate dilution (in 0.1% BSA–PBS). The incubation time with primary antibody of 1–2 hr is suitable in most instances. The characteristics of some antibodies, however, are such that longer incubation periods may be required. Evaporation losses can be minimized in this case by keeping the slides in a moist chamber. Wash the slides 3 times with PBS.

Step 3: Detection using Labeled Streptavidin. Proceed with one of the streptavidin detection systems. Incubate for 15–30 min with FITC-, lissamine-, Texas Red-, or phycoerythrin-labeled streptavidin diluted in 0.5% BSA–PBS, as suggested by the supplier, or incubate for 15–30 min with streptavidin–enzyme complexes, enzyme-labeled streptavidin, or

[37] K. T. Tokuyasu and S. J. Singer, *J. Cell Biol.* **71,** 894 (1976); A. H. Dutton, M. Adams, and S. J. Singer, this volume [37].

streptavidin bridge reagents (Amersham). With streptavidin bridge reagents, an additional 15-min incubation period with the biotinylated enzyme is required. Wash the slides with PBS 3 times. Apply the enzyme–substrate solution to sections and incubate for 10 min at room temperature when enzyme-labeled streptavidin is used. In this case, the sections can be counterstained with a suitable histological stain.

Step 4: Mounting Sections. For fluorescent reagents, selected mounting solutions that prevent fluorescence fading are available. For FITC, Citifluor (Department of Chemistry, The City University of London, England) has been found to give good results by drastically reducing the fluorescence fading. A special mountant is supplied with the phycoerythrin–streptavidin kit; for Texas Red, buffered glycerol is a suitable mountant.

Comments. The best results are obtained using 0.5-μm frozen sections as illustrated in Figs. 1 and 2. A polyclonal antibody directed against the catalytic α subunit of the amphibian Na^+,K^+-ATPase cross-reacts with one of the Na^+ channel subunits.[22] The antibody labels both the apical cell surface and the basolateral plasma membrane in epithelial cells of the urinary bladder (Fig. 1), the distal colon, and a cell line derived from the bladder of the toad *Bufo marinus*. In contrast, the apical cell surface of most cells in the small intestine (Fig. 2) is not recognized by the antibody. These data suggest that the Na^+ channel associated with the apical cell

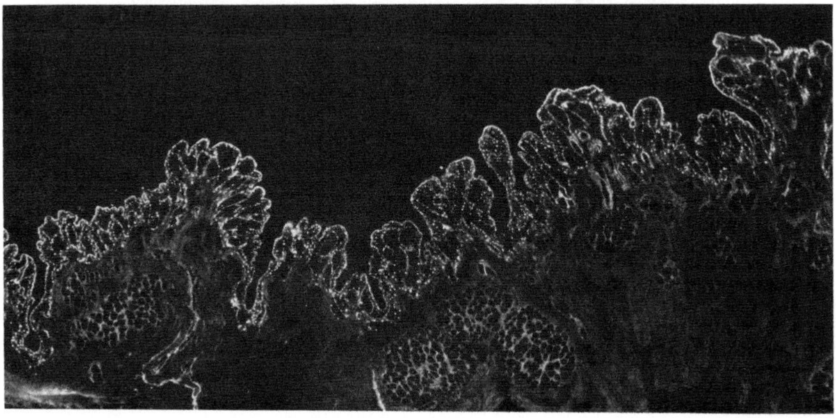

FIG. 1. Toad (*Bufo marinus*) bladder mucosa. Frozen sections (0.5 μm) were incubated with biotinyl α subunit-specific antibodies and FITC–streptavidin. An antibody directed against the α subunit of amphibian Na^+,K^+-ATPase labels both the apical and basolateral membranes of the superficial cells as well as the entire surface of the basal cells. The plasma membrane of smooth muscle cell also reacts with these antibodies. Magnification: ×456.

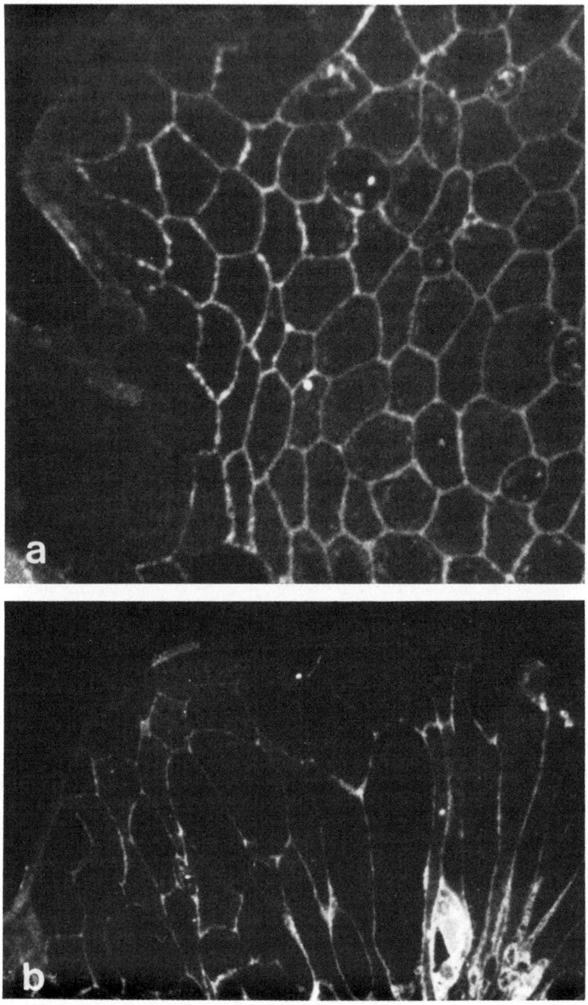

FIG. 2. Toad (*Buto marinus*) jejunal mucosa. Frozen sections (0.5 μm) were incubated first with biotinyl α subunit-specific antibodies and then with FITC–streptavidin. Magnification: $\times 600$. (a) Section tangential to the axis of the jejunal epithelial cells. The labeling of the lateral cell surface generates a honeycomb pattern. The brush border of most cells remains unlabeled. (b) Section perpendicular to the axis of the villus. The labeling is restricted to the lateral and basal cell surface of the absorptive cells, and the brush border remains unlabeled with the exception of one cell.

surface of tight epithelia is recognized by the anti-α subunit in the distal colon and the urinary bladder, but not in jejunal cells. In contrast, the sodium pump is expressed on the basolateral cell surface of all three types of epithelia. The sodium channel and the α subunit of the sodium pump share an epitope that is recognized by the polyclonal antibody.[22]

Conclusion

Using the streptavidin–biotin bridge system, the number of steps required for the purification of rare cellular proteins is reduced and their tissue localization is improved. Most reagents are now commercially available, which greatly facilitates the use of the system.

Acknowledgments

We wish to thank Dr. Michel Aubert from the Department of Pediatrics of the University of Geneva who performed the mathematical analysis of the binding studies. We are grateful to Dr. Jean Wilson who kindly revised the manuscript and to Mme. Cesco who typed it. This work was supported by grants from the Swiss National Science Foundation (3.398-0.86) and from Amersham International, plc, Amersham, England.

[76] Facilitated Cell Fusion for Hybridoma Production

By MARY K. CONRAD and MATHEW M. S. LO

For many years hybridomas have been produced for a variety of purposes by fusion of two cell types in the presence of viruses[1] or polyethylene glycol (PEG).[2-3] There are, however, disadvantages to the method of chemical fusion. One is the fact that the fusion between cells is a random event, causing problems when cell populations cannot be specifically selected in advance. As a result, large numbers of growing hybridoma colonies need to be screened in order to find the usually rare colonies of interest arising from the fusion.

More recently, attention has been given to the use of intense, short-duration electric pulses to achieve cell–cell fusion.[4-6] Cell fusion using

[1] G. Kohler and C. Milstein, *Nature* (*London*) **256,** 495 (1975).
[2] G. Galfre, S. C. Howe, and C. Milstein, *Nature* (*London*) **266,** 550 (1977).
[3] G. Galfrè and C. Milstein, this series, Vol. 73, p. 3.
[4] E. Neumann, G. Gerisch, and K. Opatz, *Naturwissenschaften* **67,** 414 (1980).
[5] T. Y. Tsong, *Biosci, Rep.* **3,** 487 (1983).
[6] U. Zimmermann, J. Vienken, J. Halfmann, and C. C. Emeis, *Adv. Biotechnol. Processes* **4,** 79 (1985).

this method is usually 10–100 times more efficient than PEG-mediated fusion.[7] Also of advantage is the fact that exogenous chemicals need not be introduced into the cell suspensions to be fused. The key to successful fusion is cell–cell contact[8] upon application of the electric field. However, the same disadvantages that are found in PEG fusions also apply to this method and are, in fact, multiplied many times over because of the enhanced fusion efficiency. Therefore, unless specific cells of interest can be stringently isolated in advance, potentially millions of randomly fused hybridomas would need to be screened for the rare one of interest.

When cell populations can be isolated in advance, there are a number of techniques that can be used to bring cells into contact prior to electrofusion.[6,9–12] In order to benefit from the advantages of electrofusion, a method is required for selectively achieving contact between cells of interest within a bulk suspension of mixed cell types. Previously it has been reported that fusion between B cells and antigen-coated myeloma cells can be promoted by such targeting techniques alone, without the use of fusogens.[13] This method, however, has never been in widespread use, suggesting difficulties in its practical application. The technique illustrates the fact, though, that B cells possess convenient surface antibodies directed against one specific antigen, the use of which can facilitate targeted preselection in cell suspensions.

Targeting

We have used the high-affinity interaction between antigen and surface immunoglobulin receptors on B cells in conjunction with the tight binding between avidin and biotin to achieve specific preselected cell–cell contact in suspensions. The procedures we developed result in high-affinity monoclonal antibody production from hybrids formed during electrofusion.[14] To accomplish this, myeloma cells are coated with either biotin or avidin, and the antigen of interest is conjugated to the affinity counterpart (Fig. 1). A mixture of these with spleen cells, followed by electrofu-

[7] U. Karsten, G. Papsdorf, G. Roloff, P. Stolley, H. Abel, I. Walther, and H. Weiss, *Eur. J. Cancer Clin. Oncol.* **21,** 733 (1985).

[8] A. Sowers, *J. Cell Biol.* **102,** 1358 (1986).

[9] H. A. Pohl, "Dielectrophoresis." Cambridge Univ. Press, Cambridge, 1978.

[10] J. Vienken and U. Zimmermann, *FEBS Lett.* **137,** 11 (1982).

[11] R. Bischoff, R. M. Eisert, I. Schedel, J. Vienken, and U. Zimmermann, *FEBS Lett.* **147,** 64 (1982).

[12] M. K. Conrad, M. M. S. Lo, T. Y. Tsong, and S. H. Snyder, *in* "Cell Fusion" (A. Sowers, ed.), p. 427. Plenum, New York, 1987.

[13] R. B. Bankert, C. DesSoye, and L. Power, *Transplant. Proc.* **12,** 443 (1980).

[14] M. M. S. Lo, T. Y. Tsong, M. K. Conrad, S. M. Strittmatter, L. H. Hester, and S. H. Snyder, *Nature (London)* **310,** 792 (1984).

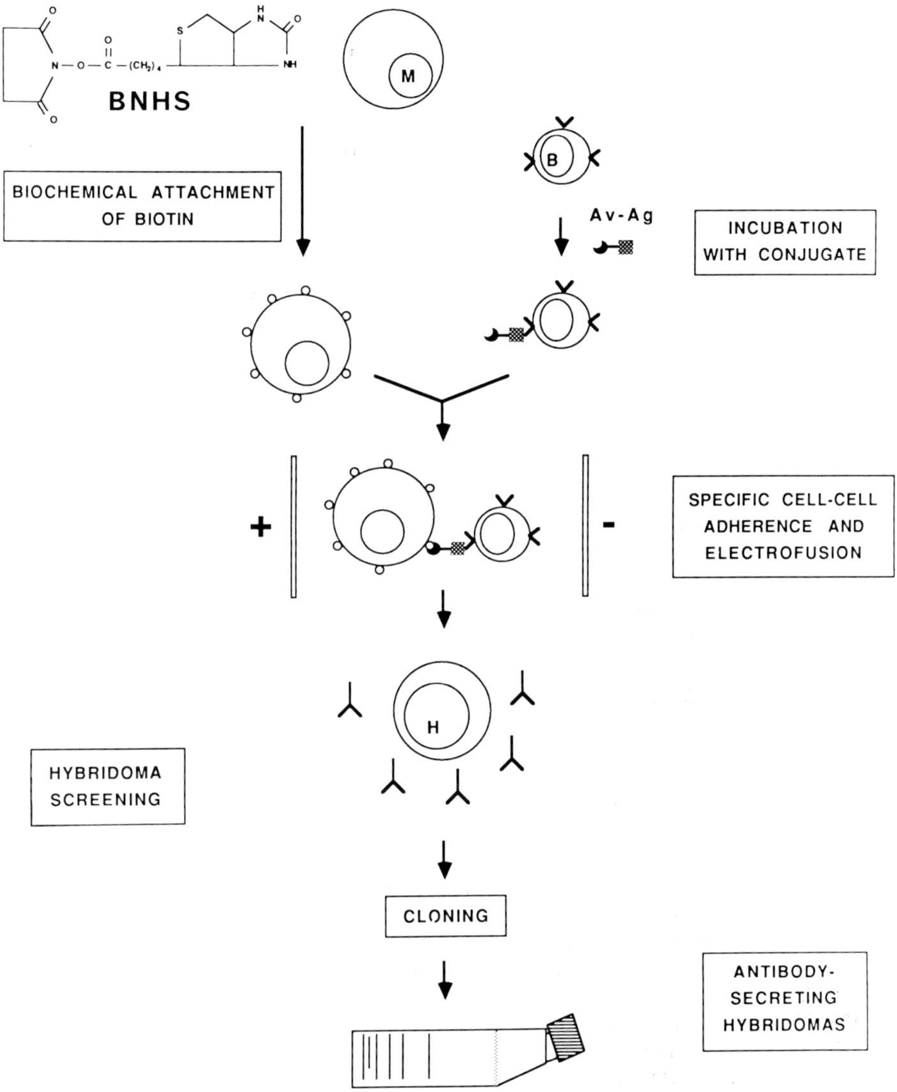

FIG. 1. Basic protocol for bioselective cell–cell contact and electrofusion to form monoclonal antibody-secreting hybridomas. Myeloma and spleen cells are prepared as described in the text. Following coincubation, electrofusion is performed on the bulk cell suspension containing adherent pairs of interest. Resulting hybridomas are screened, and those producing the desired antibodies are cloned. See text for details. M, Myeloma cell; B, B-type spleen cell; H, hybridoma cell; Av-Ag, avidin–antigen conjugate. (From Ref. 12.)

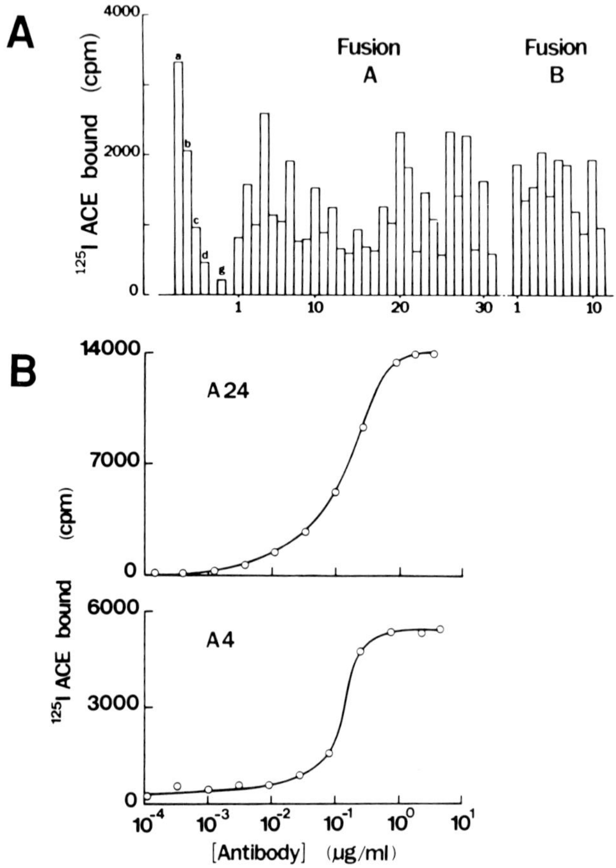

FIG. 2. Radioimmunoassays for (A) initial screening for antiangiotensin-converting enzyme (ACE) antibody and (B) affinity determinations for two monoclonals. (A) Two fusions using spleen cells of mice immunized with ACE resulted in 31 and 11 growing colonies, respectively. Supernatants from the colonies were screened with a radioimmunoassay.[14] All colonies produced anti-ACE antibody titers. Bars a–d are 3,000-, 10,000-, 30,000-, and 100,000-fold dilutions of a goat polyclonal antiserum raised against ACE. Bar g is a mouse IgG_{2a} monoclonal antibody raised against Russell viper venom. (B) Known quantities of monoclonal antibodies were serially diluted for incubation with ^{125}I-labeled ACE in the radioimmunoassay. Results for two monoclonals are shown. The affinity is equal to the concentration of antibody that binds 50% of added antigen. Affinities (K_d) ranged between 1 and 0.1 nM. (Figure 2A was originally published in Ref. 14.)

sion, results in relatively few growing colonies, a very high proportion of which secrete high-affinity antibodies directed against the antigen used (Fig. 2). Uninteresting hybridomas are rarely produced, thus saving resources in the screening steps.

Construction of Avidin–Antigen Conjugates

The chemical conjugation of antigen to avidin is complex, and a variety of chemical cross-linking procedures (similar to those used to prepare enzyme–antibody or enzyme–macromolecule conjugates) may be employed.[15] Conjugation to avidin is preferred over conjugation to biotin because of the relative ease of characterizing the avidinylated conjugate and the greater efficiency of biotinylating myeloma cells.

The essential steps in preparing avidin–antigen conjugates consist of chemical activation of avidin (or antigen) and subsequent reaction with antigen (or avidin), followed by purification of the conjugate from other reaction products by chromatographic separation techniques based on either size or charge differences. Cross-linking reactions may be performed in solution or in the solid phase by immobilization of avidin onto iminobiotin–Sepharose.

We have previously described the use of a small homobifunctional cross-linker 1,5-difluoro-2,4-dinitrobenzene (DFDNB)[16,17] in a solid-phase procedure.[14] In this procedure, avidin, bound to iminobiotin–Sepharose, is reacted with a large excess of DFDNB, and the resin is washed to remove unreacted reagent. As a consequence, the immobilized avidin is chemically activated, and subsequent addition of proteins or peptides results in their covalent conjugation to avidin through free amino groups.

Preparation of Iminobiotin–Sepharose. Ethylenediamine is combined with CNBr-activated Sepharose CL-4B,[18] and the product is allowed to react subsequently with the *N*-hydroxysuccinimide derivative of iminobiotin.[19,20] The free (unreacted) amines of the iminobiotin–Sepharose preparation are blocked by reaction with Sanger's reagent (2,4-dinitrofluorobenzene). Typically, the iminobiotin concentration is about 10–30 μmol/ml of packed Sepharose resin.

Preparation of Avidin-Conjugated Antigen. Avidin (70 μg) is incubated with iminobiotin–Sepharose (0.1 ml packed resin) in a solution (0.5 ml) consisting of 0.1 *M* sodium borate buffer (pH 10.5). The incubation and subsequent reaction are carried out in a small 1-ml column. After a 1-hr incubation period at 23°, the resin is washed with 2 ml of sodium borate buffer and reacted for 30 min at 23° with a solution (0.5 ml) containing 2

[15] P. Tijssen, "Practice and Theory of Enzyme Immunoassays," Vol. 26. Elsevier, New York, 1985.

[16] H. S. Tager, *Anal. Biochem.* **71,** 367 (1976).

[17] E. E. Golds and P. E. Braun, *J. Biol. Chem.* **253,** 8162 (1978).

[18] P. Cuatrecasas and I. Parikh, *Biochemistry* **11,** 2291 (1972).

[19] K. Hoffmann, S. W. Wood, C. C. Brinton, J. M. Montibeller, and F. M. Finn, *Proc. Natl. Acad. Sci. U.S.A.* **77,** 4666 (1980).

[20] G. A. Orr, *J. Biol. Chem.* **256,** 761 (1981); See also B. F. Goldin and G. A. Orr, this volume [17].

mM DFDNB in 0.1 M sodium borate buffer (pH 8.5). The resin is washed with 5 ml of sodium borate buffer (pH 10.5) to remove unreacted reagent. Antigen (0.1 nmol) is dissolved in sodium borate buffer (pH 10.5) and added to the resin now containing the chemically activated avidin. These are reacted for 18 hr at 23°. The resin is washed with 2 ml of sodium borate buffer (pH 10.5) and 1 ml sodium borate buffer containing 1 mM lysine or glycine to block unreacted DFDNB groups. The avidin–antigen conjugate is eluted with 0.1 M citrate buffer (pH 3.5). A carrier protein such as bovine serum albumin (BSA) is usually added to the conjugate, which is then desalted by gel filtration on a Sephadex G-25 column and eluted with phosphate-buffered saline (PBS).

Comments. This method of conjugation is particularly effective with large proteins where modification of basic residues does not drastically affect the antigenicity of the protein of interest. However, smaller peptides or acidic proteins could become insoluble when conjugated in this manner.

Smaller proteins or peptides can be derivatized with DFDNB and conjugated to avidin in solution.[16] The molar ratio of the peptide to avidin should be reduced to less than 5 mol of peptide per mole of avidin to prevent formation of insoluble conjugate. The coupling efficiency of this method is very high. However, any amino groups present on the peptide will react with DFDNB and may form very large conjugated complexes, severely altering the antigenicity of the peptide.

We have also used a heterobifunctional cross-linker, *N*-maleimidobenzoyl-*N*-hydroxysuccinimide (MBS).[15,21,22] The advantage of using MBS is that conjugation occurs through sulfhydryl groups on the antigen. We have found this to be less detrimental to the antigenicity of proteins and peptides. Also, MBS reacts with lysine residues of avidin, which causes a beneficial reduction in the isoelectric point of avidin–MBS to a more neutral pH value. In contrast, native avidin is very basic and usually binds nonspecifically even mildly acidic proteins, such as those on cell surface membranes. Avidin is first reacted with MBS at a 5-fold molar excess of reagent for 1 hr at 23°. Free reagent is removed by gel filtration on a Sephadex G-25 column. The concentration of avidin–MBS is calculated from the absorbance of the avidin–MBS at 282 nm. The ratio of avidin to MBS is usually between 3 : 1 and 5 : 1. Antigen is reduced with 10 mM dithiothreitol (DTT) prior to reaction with avidin–MBS for 20 hr. Maleimide groups react with free sulfhydryl groups on cysteines or with those produced from reduction of cystine residues in proteins. In most cases, conjugates may be used without further purification. However,

[21] F. T. Lui, M. Finnecker, T. Hamaoka, and D. Katz, *Biochemistry* **18,** 690 (1979).

[22] R. J. Youle and D. M. Neville, Jr., *Proc. Natl. Acad. Sci. U.S.A.* **77,** 5483 (1980).

purification is easily accomplished by ion-exchange chromatography or by gel filtration.

Cross-linkers may not be suitable for conjugating haptens or carbohydrates to avidin. Many haptens or carbohydrates can be chemically activated either by synthesizing the appropriate *N*-hydroxysuccinimide derivative[15] or by periodate oxidation, and the activated substances are in turn conjugated to avidin. These reactions are usually very efficient, and the conjugate can be easily purified by ion-exchange chromatography, since reaction with amino groups on avidin reduces its isoelectric point. For such conjugations, avidin from egg white (which contains many lysines) is more useful than, and therefore preferred over, streptavidin from *Streptomyces avidinii*.[23]

Construction of Biotin–Antigen Conjugates

In some cases where shotgun experiments are conducted to raise antibodies to unknown antigens, biotinylation may offer advantages over avidinylation. For this purpose, biotinylation reactions are carried out with derivatives of biotin or iminobiotin. *N*-Hydroxysuccinimidyl (NHS) esters, reactive with amines on proteins, have been most commonly employed. The reactions are simple, quantitative, and are less likely to modify biochemical and antigenic properties. Other biotin-containing reagents that react with proteins and glycoconjugates may also be used.[24]

Testing of Conjugate

Before relying on the conjugate to bring about selective cell–cell cross-linking, the functionality of both the avidin (or biotin) and the antigen moieties must be ascertained. *In vitro* assays, including enzyme linked immunoassay, radioimmunoassay, and fluorescent immunoassay, are used for this purpose. The same assays can also provide some idea as to the amount of conjugate to use in the actual fusion procedure. Any of the assays requires a control antiserum to the antigen being used. Often one can use serum from the animals immunized for fusion purposes, in which detectable titers can often be found with enzyme or radioimmunoassays using native antigen.[15,25] Commercially available polyclonal or other monoclonal antibody preparations can also be used.

Once an appropriate standard antiserum is available, the binding activity of the conjugate is measured by incubating together biotinylated

[23] N. M. Green, *Adv. Protein Chem.* **29,** 85 (1975).

[24] M. Wilchek and E. A. Bayer, this volume [13]; E. A. Bayer and M. Wilchek, this volume [14].

[25] B. B. Mishell and S. M. Shiigi, "Selected Methods in Cellular Immunology." Freeman, San Francisco, California 1980.

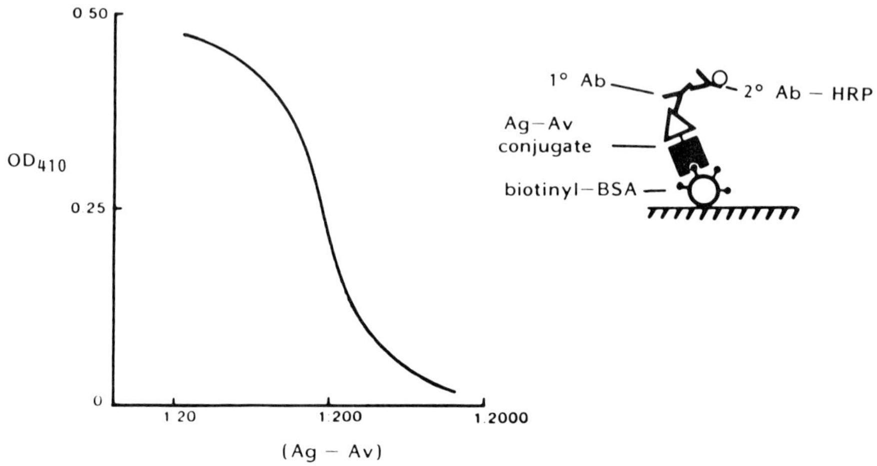

FIG. 3. Characterization of activity of antigen–avidin conjugates. An antigen–avidin conjugate was prepared by cross-linking with DFDNB to avidin.[16] The conjugate was analyzed by adsorption to biotinylated microtiter plates. The plates (Dynatek) were coated with biotinylated bovine serum albumin, conjugate was added and incubated for 40 min at 37°, and the wells were washed 5 times with PBS containing 0.05% (v/v) Tween 20. A polyclonal antiserum, raised against the antibody, was added, and the plates were incubated for 1 hr at 37°. Following another wash cycle, the plates were treated with a second antibody covalently attached to horseradish peroxidase; after incubation and washing, the plates were developed with a suitable substrate solution. The curve shows the amount of conjugate bound as measured by enzymatic conversion of substrate to a colored product at 410 nm. Conjugate activity can be detected to at least a 1:200 dilution of the original conjugate preparation. (From Ref. 12.)

myeloma cells, the avidin–antigen conjugate, a predetermined dilution of the antiserum, and a goat anti-mouse antibody that has been either radiolabeled or conjugated to peroxidase or fluorescent microspheres. Cells are then washed, and the degree of second antibody binding is quantified. Negative controls should include normal (unbiotinylated) myeloma cells, normal mouse serum (unimmunized), and conjugate replacement with native avidin or streptavidin. Titrating the conjugate or inhibiting with known concentrations of free antigen provides quantification of the degree of binding. An example of the type of result that may be obtained by enzyme immunoassay is shown in Fig. 3.

Preparation of Cells for Fusion

Myeloma Cells

Biotinylation. Biotinyl-*N*-hydroxysuccinimide ester (BNHS) is used to biotinylate myeloma cells. Log-phase cells are harvested and washed with Dulbecco's PBS, which contains no divalent cations. BNHS (0.1 *M*) is freshly dissolved in dimethylformamide (a minimal volume is used as this solvent lyses cells at high concentrations) and diluted with PBS. Cells are resuspended in 100 μM BNHS, incubated for 15 min at 23° without further agitation, and centrifuged through 10 ml of serum at 200 *g* for 8 min. Cells are washed twice in Dulbecco's modified Eagle's medium (DME), which is used because it is biotin free. Deoxyribonuclease I (DNase) is added to the DME (final concentration 10–50 μg/ml) to reduce cell aggregation caused by the release of intracellular components. In later stages the enzyme is removed by washing, as DNase would prove toxic if introduced into cells during fusion.

To test the degree of biotinylation, cells are stained with fluorescein-conjugated avidin. Two million P3×63/Ag8.653 myeloma labeled cells are analyzed with a fluorescence-activated cell sorter. Greater than 98% of the cells are biotinylated. Two fractions containing either heavily or lightly biotinylated cells are cloned in soft agar. There are no differences in viability between the two fractions. In four different cell lines tested (P3×63/Ag8.653, SP2/0, S194, and FOX-NY), the fluorescence intensity remains maximal up to 3 hr after biotinylation. P3×63/Ag8.653 cells retain approximately 10–30% of the original intensity after overnight incubation following biotinylation, while the other three lines lose virtually all of the label.

Model fusion experiments (using an appropriate cell surface marker) and growth experiments (assessed by trypan blue exclusion) have demonstrated that both the capacity to undergo membrane fusion and the viability of myeloma cells are not significantly altered by biotinylation using BNHS.

Avidinylation. Alternatively, myeloma cells can be avidinylated. The heterobifunctional cross-linker *N*-succinimidyl 3-(2-pyridylthio)propionate (SPDP) is used to covalently attach avidin to myeloma cell surfaces. Avidin is reacted with a 5-fold molar excess of SPDP for 1 hr at 23°, and free reagent is removed by gel filtration on a Sephadex G-25 column with PBS. The degree of conjugation is determined from the absorbance of the avidin–SPDP solution at 253 and 282 nm. A 4 : 1 ratio of avidin to SPDP is typical.[12]

About 10^7 log-phase myeloma cells grown in medium containing 0.1 m*M* 2-mercaptoethanol are reacted with 1 μM avidin–SPDP for 1 hr at 23° without agitation. Cells are collected by centrifugation through a cushion of solution made from equal volumes of isosmotic sucrose and DME. The degree of labeling by this method is measured by the binding of biotinylated fluorescein-conjugated lysozyme. The fraction of cells labeled is usually greater than 80%, but the intensity of staining is about 10-

fold lower compared with biotinylated myeloma cells stained with fluorescein-conjugated avidin.

The density of avidin on myeloma cell surfaces can be greatly increased by prior reaction of myeloma cells with reduced SPDP. This is prepared by dissolving 1 mg of SPDP and 0.5 mg of DTT in 30 μl of dimethylformamide and then diluting into 30 ml of PBS. Myeloma cells are then incubated in this solution for 15 min at 23°, centrifuged through sucrose–DME, and resuspended in PBS. The cells are then incubated with 1 μM of avidin–SPDP as described above. This method results in staining intensities equivalent to biotinylated cells stained with fluorescein-conjugated avidin.

Comments. Although more complex, avidinylating of myeloma cells permits the use of biotinylated antigen to cross-link spleen and myeloma cells for fusion. Often large or complex proteins are more easily biotinylated than avidinylated. We try to avoid bridging biotinylated antigen and biotinylated myeloma cells with avidin. This process is usually uncontrolled and is not reproducible. Large aggregates of cells are often formed. Therefore, when biotinylated antigen is used, it is preferable to use avidinylated myeloma cells.

Spleen Cells

For purposes of immunization, antigenic proteins are adsorbed to either bentonite or alumina. They may also be emulsified in Freund's adjuvant; however, we have found that this gives rise to additional problems with intraperitoneal adhesions. Small molecules, such as haptens or peptides, are first conjugated to keyhole limpet hemocyanin (KLH) before adsorption to the carrier substance. In some cases, animals, primed with injections of KLH alone before immunization with a KLH–antigen conjugate, develop serum titers to the antigen more rapidly.[25]

The animals commonly used for monoclonal antibody production are either C57BL/6 or BALB/c mice. The route of administration is usually by intraperitoneal injection. However, when the antigen used is very precious, the final immunization can be given by intravenous or intrasplenic injection of a very small amount of antigen 3 days prior to fusion.

The immunization schedule can be varied a great deal, depending on the availability of the antigen and considerations of time and experimental goals. It has been shown that a greater variety of higher-affinity antibodies are produced from hyperimmunized animals, as compared to those produced following acute exposure to the antigen.[26] A typical schedule may include 3 or 4 injections given 1–2 weeks apart. However, in our initial

[26] C. Berek, G. M. Griffiths, and C. Milstein, *Nature (London)* **316,** 412 (1985).

experiments, we were interested in raising antibodies to very rare entities that are extremely hard to isolate. We therefore immunize animals only twice, 1 week apart, with microgram quantities of antigen each time. Both injections are intraperitoneal, and the final one is given 3 days prior to fusion.

Three days after the final immunization, spleens are removed. Cells are dissociated on a fine wire mesh using a rubber policeman. Red blood cells may be lysed with freshly prepared ammonium chloride solution (0.84%); however, we find that the presence of red blood cells does not noticeably affect fusion in a high-voltage field. The lymphoid cells are collected in DME containing 50 μg/ml DNase. Incubation with the antigen–avidin (or antigen–biotin) conjugate is carried out for a minimum of 30 min up to 2 hr, at 4° (to prevent capping and internalization of the conjugate).

Combining Cells

Treated spleen cells are mixed with biotinylated (or avidinylated) myeloma cells at a ratio of 4 : 1[27] and centrifuged. The mixed cell pellet is loosened and spun for 10 sec at 50 *g*. The pellet is incubated for 30 min at either 23 or 37° and then diluted in DME. An aliquot containing up to 2×10^7 cells is underlaid with sucrose and centrifuged at 200 *g* for 6 min at 23°. Cells are then resuspended in 0.5 ml isosmotic sucrose (300–320 mOsm) and placed into the cell fusion chamber (Fig. 4).

It should be noted that harsh resuspension techniques (such as pipetting) should be avoided, since many steps in the preselection process involve mechanical agitation and disruption. Instead, gentle disruption by tapping or flicking the tube should be used. Since divalent cations promote nonspecific cell aggregation, they should be excluded.

Cell Fusion

After completion of the preparation steps described above, one is left with a suspension of treated myeloma and spleen cells in isosmotic sucrose. In order to avoid heating of the cells during high-voltage fusion, sucrose is used because of its low electrical conductance. The suspension contains the relatively few B cell–antigen–(avidin–biotin)–myeloma cell complexes. Typically, a 0.5-ml suspension, containing about 50 million cells, is introduced into the fusion chamber (Fig. 4). The cells are exposed to 2–4 pulses of 5 μsec duration at a field intensity of 3 kV/cm. The

[27] We find, however, that the ratio of cell types is not as critical in high-voltage fusion as it is in fusions using PEG (see Ref. 3).

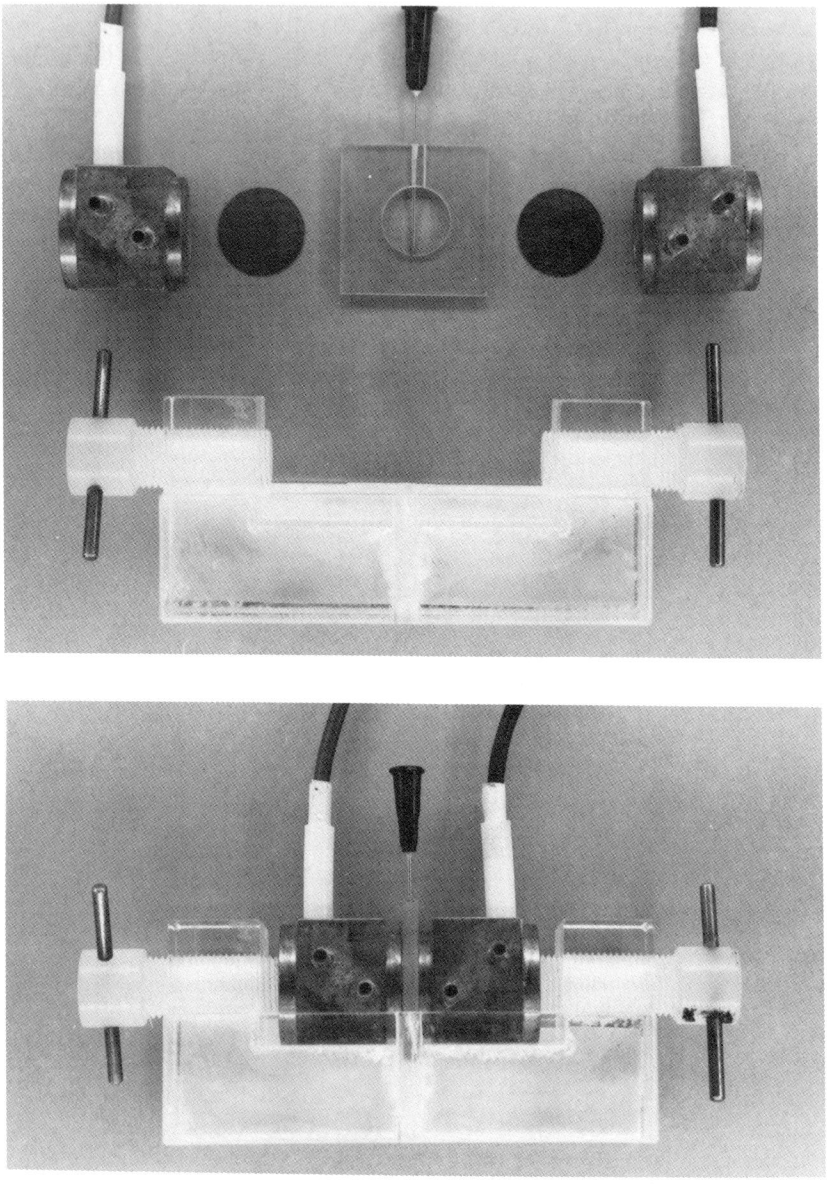

FIG. 4. Electrofusion chamber. Cells are fused in 0.5 ml of a suspension held in a chamber created by sandwiching two stainless steel disks around a well drilled into a 3 mm thickness of polycarbonate, which are then clamped between two brass electrodes. The suspension is introduced into the chamber through a 22-gauge port. The chamber is also vented with another small port to prevent the formation of air bubbles. (Top) Unassembled apparatus; (bottom) apparatus assembled and ready for use.

electrical pulse consists of a square wave generated by a commercial high-voltage pulse generator. The field is monitored with a storage oscilliscope connected to the electrodes through a 100× reduction probe. The field polarity is alternated between the pulses to reduce bulk cell transport toward one electrode or the other.

Following fusion, cells are maintained at 37° for 30 min in DME. Membrane pores generated during electrofusion undergo spontaneous repair, and this process is temperature dependent. Cells are then plated into 96-well tissue culture plates with DME containing aminopterin, which kills all unfused myeloma cells over the course of the next few days. Unfused spleen cells die off in a matter of days in culture. Growth of the hybridomas is usually detected from 1 to 6 weeks following electrofusion.

Comment. We were surprised to obtain predominantly IgG antibodies following the above acute immunization regimen, rather than those of the IgM class, which is what we had predicted. We hypothesize that the bioselective method, utilizing binding of antigen to surface antibody, facilitates fusion of those B cells stimulated to produce high-affinity antibodies (i.e., those of the IgG class), while the more weakly binding antibody-producing spleen cells are left behind.

[77] Complex Neoglycoproteins

By MING-CHUAN SHAO, LING-MEI CHEN, and FINN WOLD

Introduction

Biotinylated glycans bound to avidin or streptavidin represent useful glycoprotein models for the study of both glycan processing by Golgi enzymes and glycoprotein interactions with lectins/receptors. Although we have used primarily asparagine-linked glycans derived from pronase digestion of ovalbumin, in principle, any glycan isolated from a protease digest of a glycoprotein (with the amino acid or a short peptide still attached) should be amenable to the reaction of the free α-amino group with activated biotinyl derivatives and the subsequent tight binding to avidin or streptavidin. Once a particular avidin preparation has been characterized, the stoichiometry of binding and the stability of the neoglycoprotein complex are reproducible, constant features of that preparation, and after any experiment in which the glycans may have been altered, the noncovalent mode of binding permits quantitative recovery of the biotinylated glycan for chemical analysis. The major drawback of this neoglycoprotein model is that it does not directly provide the type of protein matrix environment that exists in natural glycoproteins. One way to overcome this problem is to insert chemical "extension arms" between the glycan and the biotin anchor, and thus alter the display of the glycan relative to the protein surface. These aspects of the avidin–biotinylglycan neoglycoprotein preparation and their use are considered in the following.

Preparation and Assay of Biotinylated Glycan Derivatives[1,2]

The *N*-hydroxysuccinimide esters of biotin and 6-biotinamidohexanoic acid, which are now available commercially, have been used for most preparations of the glycan derivatives used in our studies. If radiolabeled biotin is needed, its activated *N*-hydroxysuccinimide ester can be prepared as outlined elsewhere in this volume.[3] The radiolabeled derivative with the 6-carbon extension arm can be prepared by reaction of 6-aminohexanoate with the activated radiolabeled biotin derivative. The product can be further activated with *N*-hydroxysuccinimide. An alternative for the latter preparation is to modify the α-amino group of the glycopeptide with N-blocked aminohexanoic acid and then, after unblocking, to react with activated biotin.[4]

The glycopeptides and the glycosylasparagine derivatives are prepared by exhaustive proteolysis with pronase. For most of our work we have used the ovalbumin-derived glycans $Man_6GlcNAc_2$-Asn and $Man_5GlcNAc_2$-Asn as the main starting materials. The glycans are prepared according to the method of Huang *et al.*[5] with modifications suggested by other workers.[6,7] The desalted and lyophilized glycan derivatives are in general of sufficient purity (90%) to proceed directly with the biotinylation reaction. An additional gel filtration step on BioGel P-4 with 0.1 *M* ammonium acetate as eluent can be used to further improve the purity.

Biotinylation is carried out according to previously described procedures.[1,2] The glycosyl-Asn derivative (2 μmol) is dissolved in 0.25 ml of

[1] V. J. Chen, this series, Vol. 138, p. 418.
[2] V. J. Chen and F. Wold, *Biochemistry* **25,** 939 (1986).
[3] E. A. Bayer and M. Wilchek, this volume [14].
[4] S.-C. B. Yan, this series, Vol. 138, p. 413.
[5] C. C. Huang, H. E. Mayer, and R. Montgomery, *Carbohydr. Res.* **13,** 127 (1970).
[6] K. Yamashita, Y. Tachibana, and A. Kobata, *J. Biol. Chem.* **253,** 3862 (1978).
[7] J. Conchie and I. Strachan, *Carbohydr. Res.* **63,** 193 (1978).

dimethyl sulfoxide and mixed with 0.25 ml of a separate solution containing 5 μmol of the *N*-hydroxysuccinimide ester of biotin (or 6-biotinamidohexanoic acid); a drop (40 μl) of triethylamine is added, and the thoroughly mixed reaction mixture is stoppered and left at room temperature overnight. The solvent is then removed by lyophilization, and the residue is dissolved in 1 ml of 1 *M* triethylamine-HCl (pH 8) to hydrolyze the excess biotinylation reagent. After 4 hr at room temperature, the reaction mixture is subjected to gel filtration on a 1.5 × 100 cm column of Sephadex G-25 eluted with water. The eluate is monitored for carbohydrate content using the phenol–sulfuric acid test[8] and/or for biotin content by the method of Green[9] or by the method of McCormick and Roth.[10] The biotinylated derivative appears immediately after the void volume; after pooling the appropriate fractions and lyophilization, the product is generally obtained in a yield of about 80% and is free of unreacted starting material and salt.

A variety of other biotinylated glycans have been prepared. Owing to the difficulty in obtaining the appropriate pure glycosyl-Asn from glycoproteins, it has been necessary to prepare some of these derivatives from the major ovalbumin glycans by enzymatic modification using solubilized Golgi enzymes[11] from rat liver (detergent-treated Golgi membrane preparations which contain all the glycan-processing enzymes). For these preparations the starting biotinylated glycan (1 μmol) is incubated for 12 hr at room temperature with 0.8–1 mg of Golgi membrane proteins in 10 m*M* MES buffer (pH 6.7; Sigma Chemical Co., St. Louis, MO) containing 10 m*M* $MnCl_2$ and 5 μl/ml of Nonidet P-40. The sample is brought to a total volume of 800 μl with the same buffer, containing either 24 mg of bovine serum albumin (BSA) and 8 μmol of donor substrate (UDP-GlcNAc or UDP-Gal) in reactions involving transferases or the mannosidase II inhibitor swainsonine (40 μM) for reactions producing hybrid structures. The high concentration of BSA has been found to be important to stabilize the enzymes in the reaction mixture. Under these conditions the reactions that have been explored quantitatively are 75–100% complete in 8 hr.[12] If larger (or smaller) quantities are needed, the preparation can readily be scaled up (or down). After the incubation, the reaction mixture is heated for 10 min at 95° and then subjected to centrifugation. The clear supernatant solution is collected, the pellet is washed with 400

[8] M. Dubois, K. A. Gilles, J. K. Hamilton, P. H. Rebers, and F. Smith, *Anal. Chem.* **28,** 350 (1956).

[9] N. M. Green, this series, Vol. 18, p. 418.

[10] D. B. McCormick and J. A. Roth, this series, Vol. 18, p. 383.

[11] I. Tabas and S. Kornfeld, this series, Vol. 83, p. 416.

[12] M.-C. Shao and F. Wold, *J. Biol. Chem.* **262,** 2968 (1987).

μl of distilled water, and the combined solution and wash are lyophilized. The residue is finally dissolved in 400 μl of water and subjected to gel filtration on BioGel P-6 eluted with 0.1 *M* ammonium acetate. The pooled glycan peak is lyophilized and constitutes the final product.

To conserve material we have found the following method useful for monitoring the column: 2 μl of each column fraction in the product region is pipetted into the wells of a microtiter plate and treated with 10 μl of a reagent containing 0.25 m*M* 4-hydroxyazobenzene-2′-carboxylic acid and 0.3 mg/ml of avidin in 0.2 *M* phosphate buffer (pH 7.0). Fractions containing biotinyl derivatives are identified by their ability to bleach the brown reagent to yellow.[9] If incorporation of ^{14}C-labeled sugars is involved in the reaction, the samples are subsequently transferred directly from the microtiter plates to scintillation vials for counting. It is important to note that it is essential to use biotinylated substrates in these reactions. If glycans containing an unsubstituted Asn are used, the aspartylglycosylamine amidohydrolase which is found in most tissues,[13] and which is present in substantial amounts in Golgi enzyme preparations, will effectively hydrolyze the substrate to reducing glycan (glycosylamino) and free aspartate. Since the amidohydrolase is inactive on substrates with the α-amino group of Asn blocked, the biotinylated derivatives are not affected.

We have determined the purity and the nature of the various biotinylated glycans almost entirely by fast atom bombardment mass spectrometry (FAB-MS),[14] supported by incorporation of radiolabeled sugars. Although this methodology does not permit separation of positional isomers or distinction between different epimeric monosaccharides, the processing steps in and the glycan products from rat liver and oviduct are considered to be so well established,[15] that the structure of the products can be assigned with a good deal of confidence. Recent new developments in detection systems indicate that HPLC may provide a readily available method for the analysis of glycan isomers.[16]

The biotinylated glycans produced in this laboratory are listed in Table I along with their method of production and their purity as estimated by FAB-MS. The corresponding structures are illustrated in Fig. 1.

Preparation and Properties of Avidin–Biotinylglycan Neoglycoproteins

The avidin–biotinylglycan complexes are prepared by mixing the appropriate molar ratio of avidin (streptavidin) and biotinylglycan in the

[13] M. Kohno and I. Yamashina, this series, Vol. 28, p. 786.

[14] M.-C. Shao, C. C. Q. Chin, R. M. Caprioli, and F. Wold, *J. Biol. Chem.* **262,** 2973 (1987).

[15] A. Kobata, K. Yamashita, and S. Takasaki, this series, Vol. 138, p. 84.

[16] L.-M. Chen, M.-G. Yet, and M. C. Shao, *FASEB J.* **2,** 2819 (1988).

TABLE I
BIOTINYLATED GLYCANS

Glycan structure[a]		Method of preparation[b]: Isolated from ovalbumin	Substrate	Swainsonine	Avidin	Purity (%)
A	M_6-R	+				90–95
B	M_5-R	+				90–95
C	GnM_5-R		B + UDP-Gn	+	−	90–95
D	Gn_2M_5-R	+				90–95
E	Gn_3M_5-R	+				90–95
F	GGn_3M_5-R	+				90–95
G	$GGnM_5$-R		C + UDP-G	+	−	90–95
H	GnM_4-R		C + UDP-Gn	−	+	45–70[c]
I	Gn_3M_4-R	+				70
J	GnM_3-R		C	−	−	90–95
K	Gn_2M_3-R		J + UDP-Gn	−	−	90–95
L	$G_2Gn_2M_3$-R		K + UDP-G	−	−	80
M	Gn_4M_3-R	+				70

[a] The actual structures are given in Fig. 1. R is -Gn_2-(biotinyl)Asn and -Gn_2-(biotinamido-hexanoyl)Asn in all cases except for compound H, for which only the former has been prepared. M, Mannose (Man); Gn, *N*-acetylglucosamine (GlcNAc); G, galactose (Gal).

[b] The biotinylated glycans were either obtained by purification from pronase digests of ovalbumin followed by the biotinylation, or by enzymatic conversion of biotinylated products as indicated.

[c] The major contaminant in these preparations is the substrate, C.

buffer and at the pH at which they are to be used at a concentration of about 10^{-5} *M* (subunits). Under these conditions the complex forms rapidly, and, with an excess of binding sites, essentially all the glycan is bound. Dialysis for as long as 2 weeks does not give any significant loss of bound glycan.[1,2] The proper molar ratio must be established for each batch of avidin. Although avidin and streptavidin have four binding sites per tetramer, the commercial preparations have, in our hands, been found to express an average of only 2.8–3.2 sites for the biotinylglycans. We assume that the missing sites are already occupied by contaminating biotin, and for practical purposes we routinely use a molar ratio of 2.5 biotinylglycan per tetramer.

Our empirical evaluation of the stability of the complexes of biotinylated $Man_6GlcNAc_2$-Asn with avidin and streptavidin confirmed their expected stability.[2] At a concentration of 2×10^{-6} *M*, the complexes show less than 5 and 10% loss of bound ligand in 48 hr for avidin and streptavidin, respectively. In the presence of 1 m*M* biotin, the half-lives of the two

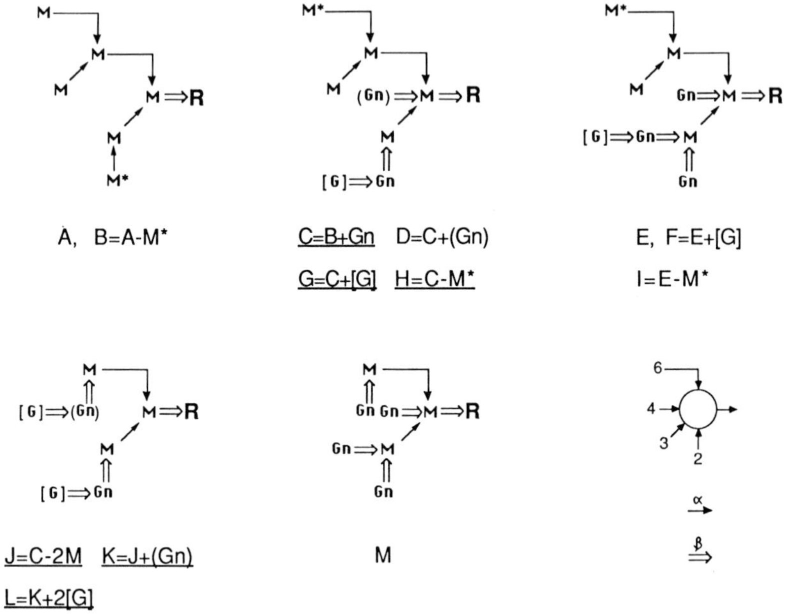

FIG. 1. Structural representations of the glycan derivatives listed in Table I. The anomeric configuration and the linkage positions are encoded as indicated in lower right-hand corner of the figure. *R* is either Gn ⇒ Gn ⇒ (N^{α}-biotinyl)Asn or Gn ⇒ Gn ⇒ [N^{α}-(6-biotinamido)hexanoyl]Asn. The assignment of the specific structures to these glycan derivatives, which were only characterized by mass spectrometry, is based on previous extensive structural analyses of these compounds[15] and should be correct.

complexes are significantly reduced (27 hr for avidin and 2–3 hr for streptavidin). Direct titration of the binding sites shows that the binding of biotinylglycan is not as tight as that of free biotin[2]; the fact that it is possible to essentially completely replace the biotinylglycan with a 10-fold excess of biotin[12] (heating at 95°) suggests that this difference in binding may approach a factor of 10. The latter method has been used routinely to liberate the bound glycan from avidin at the end of an experiment. The reaction mixture is treated with a 10-fold excess of biotin and heated at 95° for 10 min, and the resulting free biotinylglycans are separated from the avidin–biotin complex and from free biotin by gel filtration, as described above for biotinylglycan preparations.

Application of Avidin–Biotinylglycan Neoglycoproteins

Because of the high stability of the complex over a broad range of pH and temperature in the absence of free biotin, the neoglycoproteins can be used in a variety of studies as if they were covalent derivatives. They behave like multiglycosylated glycoproteins in their interaction with receptors[17] and show characteristic properties of glycoproteins as substrates for glycosidases and glycosyltransferases. Thus, it is well established that endoglycosidase H (endo H), an enzyme which catalyzes the cleavage between the two GlcNAc residues in protein-associated oligomannose glycans (such as glycans A and B in Table I), will act on some but far from all such structures in intact glycoproteins. The resistant glycans will become available to endo H action after denaturation or proteolytic cleavage of the glycoprotein, thus showing that the availability of the glycan on a protein surface is an important specificity determinant for this enzyme. When biotinylglycan (A or B in Table I) was used as substrate for endo H, the avidin-bound glycan was completely resistant to the enzyme under conditions in which the free substrate was hydrolyzed. The avidin-bound derivative with the extension arm (biotinamidohexanoyl) showed a slow cleavage under the same conditions.[18] Similarly, exposure of the free and bound biotinylated glycans to α-mannosidase showed significant effects of the protein matrix on both the rate and the end products of the mannosidase action.[2] A major area of use for these derivatives has been in the study of the effect of the protein matrix on the individual processing enzymes from Golgi membranes. Many of the various substrates in Table I have been prepared with these enzymes and have subsequently been used as substrates, free or bound to avidin (streptavidin), to permit a direct evaluation of the effect of the protein matrix on individual processing steps.[12,14] The facile, quantitative removal of the glycans from the complexes after the reactions have been carried out makes these noncovalent yet stable neoglycoproteins unique tools in the study of glycoprotein processing.

Acknowledgments

The studies leading to the methods discussed here were supported by a U.S. Public Health Service grant (GM 31305) and a Robert A. Welch Foundation grant (AU-916). We gratefully acknowledge the contributions of Theresa Domany in preparing the manuscript.

[17] Y. Ohsumi, V. J. Chen, S.-C. B. Yan, F. Wold, and Y. C. Lee, *Glycoconjugate J.* **5**, 99 (1987).

[18] V. J. Chen, S.-C. B. Yan, and F. Wold, *in* "Microbiology 1986," p. 297. Amer. Soc. Microbiol. Washington, 1986.

[78] Probing β-Adrenergic Receptors

By KATHRYN E. MEIER and ARNOLD E. RUOHO

Adrenergic receptors mediate responses to endogenous catecholamines. These receptors are subdivided into two major types according to their relative affinities for catecholamine agonists.[1] For α-adrenergic receptors, the order of agonist potency is epinephrine > norepinephrine > isoproterenol. For β-adrenergic receptors, the order is isoproterenol > epinephrine > norepinephrine. β-Adrenergic receptors have been further subdivided into β_1 and β_2 subtypes on the basis of their relative affinities for agonists and antagonists.[2] Agonist occupation of β-adrenergic receptors results in activation of adenylate cyclase, leading to increased intracellular levels of cyclic AMP. β-Adrenergic receptors are intrinsic membrane glycoproteins. The complete amino acid sequences of the mammalian β_2-adrenergic receptor[3] and the avian β-adrenergic receptor[4] have been deduced from their cDNA nucleotide sequences.

Cellular events involved in the regulation of receptor expression and function can be studied by means of microscopy techniques that allow visualization of the receptors on the cell surface and in intracellular compartments. β-Adrenergic receptors can be sequestered into an intracellular compartment following agonist binding and are down-regulated in response to prolonged exposure to agonist.[5,6] However, these conclusions have been reached through studies of radioligand binding to the receptor, rather than by microscopy. Adrenergic ligands are small molecules that cannot be directly visualized, and antibodies to the receptor have been difficult to obtain. Furthermore, the receptors are generally present in low densities on the surface of responsive cells (<10,000 receptors/cell); this fact has complicated attempts at their localization. In order to visualize

[1] R. P. Ahlquist, *Am. J. Physiol.* **153,** 586 (1948).

[2] A. M. Lands, A. Arnold, J. P. McAuliff, F. P. Ludvena, and T. G. Brown, *Nature (London)* **214,** 597 (1967).

[3] R. A. F. Dixon, B. K. Kobilka, D. J. Strader, J. L. Benovic, H. G. Dohlman, T. Freille, M. A. Bolanowski, C. D. Bennett, E. Rands, R. E. Diehll, R. A. Mumford, E. E. Slater, I. S. Sigal, M. G. Caron, R. J. Lefkowitz, and C. D. Strader, *Nature (London)* **321,** 75 (1986).

[4] Y. Yarden, H. Rodriguez, S. K.-F. Wong, D. R. Brandt, D. C. May, J. Burnier, R. N. Harkins, E. Y. Chen, J. Ramachandran, A. Ullrich, and E. M. Ross, *Proc. Natl. Acad. Sci. U.S.A.* **83,** 6795 (1986).

[5] T. K. Harden, *Pharmacol. Rev.* **35,** 5 (1983).

[6] L. S. Mahan, R. M. McKernan, and P. A. Insel, *Annu. Rev. Pharmacol.* **27,** 215 (1987).

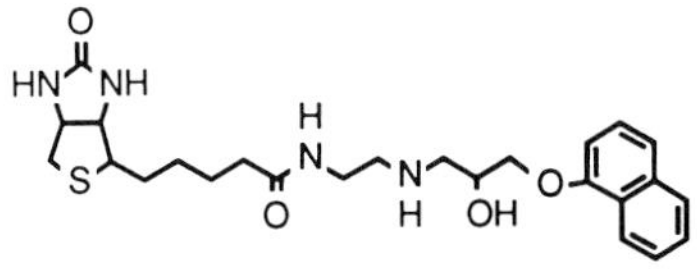

BN

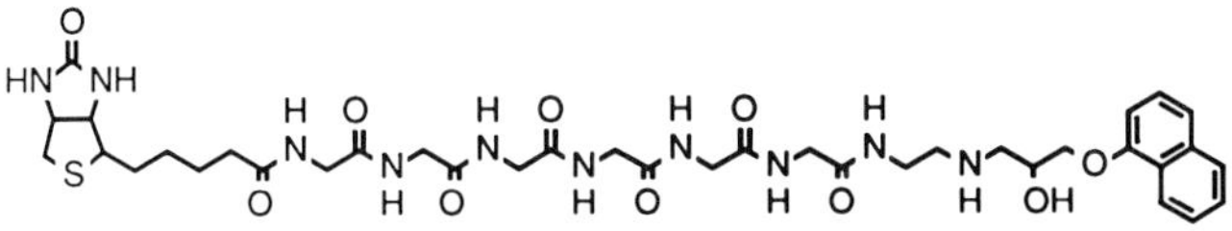

BGN

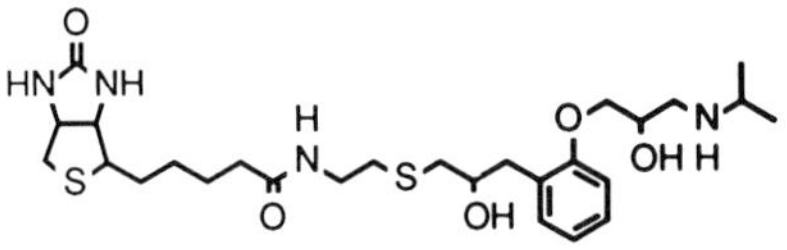

BCA

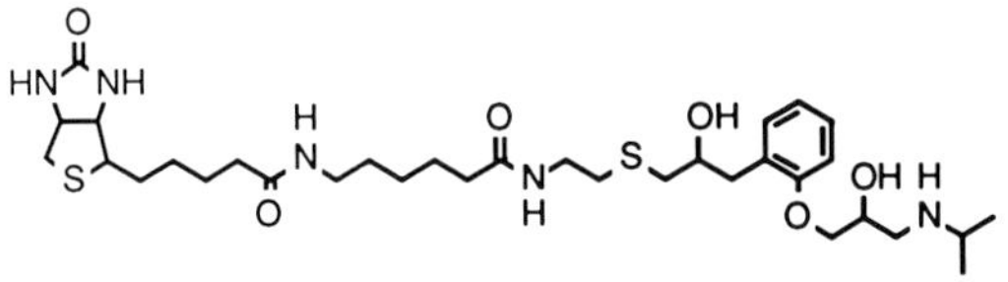

BCCA

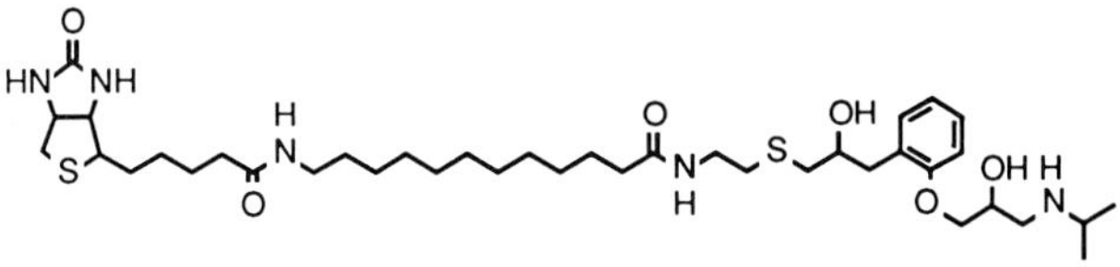

BDCA

FIG. 1. Structures of biotinylpropranolol and biotinylalprenolol derivatives.

the receptor by the techniques of light and electron microscopy, it is therefore necessary to create receptor-specific ligands that can be linked to a fluorescent or electron-dense marker.

We synthesized a series of biotinylated derivatives of propranolol and alprenolol, two β-adrenergic antagonists that bind with high affinity to the receptor (Fig. 1).[7–9] The goal of this work was to obtain derivatives that

[7] K. E. Meier and A. E. Ruoho, *J. Supramol. Struct.* **9,** 243 (1978).

[8] K. E. Meier and A. E. Ruoho, *Biochim. Biophys. Acta* **761,** 257 (1983).

[9] K. E. Meier, *Diss. Abst. Int. B* **42,** 4036 (1982).

would bind simultaneously to both the β-adrenergic receptor and to avidin. The interaction between avidin and biotin is essentially irreversible. The interaction between β-antagonists and the receptor is of sufficiently high affinity (equilibrium K_D 1–10 n*M*) to allow excess ligand to be washed from the cell surface prior to fixation. Avidin can be linked to fluorescent, electron-dense, or macromolecular marker molecules. Using biotinylated antagonists, it should therefore be possible to visualize β-adrenergic receptors on the cell surface. Success in achieving this goal will be dependent on the receptor density on the cell surface, as well as on the sensitivity of the method used to detect the label. The remainder of this chapter is concerned with the preparation and characterization of biotinylpropranolol and biotinylalprenolol derivatives. Discussion focuses on biotinyldodecanoylcysteaminylalprenolol (BDCA) because this compound possesses the optimal characteristics desired for a bifunctional reagent.

Methodology

General Methods for Detection and Quantification of Biotin. The presence of biotin in reaction products is detected using *p*-dimethylaminocinnamaldehyde as previously described by McCormick and Roth.[10] A solution containing 0.2% (w/v) of the reagent in 1% sulfuric acid–ethanol is applied as a spray to thin-layer chromatography plates containing separated reaction products. Biotin-containing compounds appear as pink spots following this treatment.

The amount of biotin present in solutions of biotinylated products is determined by a colorimetric avidin titration technique described by Green.[11] This procedure involves spectrophotometric titration (500 nm) of a known quantity of biotin-binding sites in an avidin solution by competition with 10 m*M* hydroxyazobenzene-2′-carboxylic acid. The observation of a sharp end point to the titration curve is indicative of high-affinity binding of the biotinylated compound to avidin.

The stabilities of the complexes between biotin derivatives and avidin are assessed as follows. A 50-μl aliquot of a 1 m*M* solution of the biotinyl derivative is added to 500 μl of 0.1 m*M* avidin in 50 m*M* Tris-HCl (pH 7.4). This solution is incubated for 10 min at 25°. A 50-μl aliquot of a solution of 10 m*M* [^{14}C]biotin (50 μCi/ml) is then added. The incubation is continued at 4°, 25°, or 37°. At various times, 100-μl aliquots of the mixture are removed and diluted with 100 μl of 50 m*M* Tris-HCl (pH 7.5). The

[10] D. B. McCormick and J. A. Roth, this series, Vol. 18, p. 383.

[11] N. M. Green, this series Vol. 18, p. 418.

diluted samples are applied to a 7 × 80 mm column of Sephadex G-50 and then eluted with 50 m*M* Tris-HCl (pH 7.5). Two 0.9-ml fractions are collected; the ^{14}C content of each fraction is determined by liquid scintillation spectrometry. The avidin–biotin complex elutes in the first fraction (void volume); free biotin is contained in the second fraction. Data are normalized for column recovery by expressing the bound [^{14}C]biotin as a percentage of the total [^{14}C]biotin eluted from each column.

Methods for Assessment of Binding Affinity of Derivatives for β-Adrenergic Receptor. Several assay systems can be used to determine the affinity of biotinylated antagonists for the β-adrenergic receptor. The ability of the compounds to inhibit isoproterenol-induced increases in cyclic AMP levels in intact duck erythrocytes or HeLa cells is assessed using a commercially available radioimmunoassay method[7] or a binding protein assay[8] to measure intracellular cyclic AMP. Similar studies have been carried out with duck erythrocyte membranes using [α-^{32}P]ATP as substrate for adenylate cyclase activity.[7] The K_D values of the compounds for β-adrenergic receptor binding are determined using radioligand binding assays with membrane fractions (or solubilized membrane fractions) prepared from avian erythrocytes[8] or cultured mammalian cells (see footnote to Table I).[9,12] These determinations are based on the abilities of the derivatives to compete for binding of a radiolabeled antagonist ([^3H]dihydroalprenolol or [^{125}I]iodohydroxybenzylpindolol) to the receptor preparation.

Synthesis of Biotinyl-N-hydroxysuccinimide. Biotinyl-*N*-hydroxysuccinimide (BNHS) is prepared by the method of Jasiewicz *et al.*[13]; this reagent is now commercially available under the designation *N*-hydroxysuccinimidobiotin (Sigma Chemical Co., St. Louis, MO).

Synthesis of Biotinylpropranolol Derivatives. Biotinylpropranolol compounds are prepared from an aminopropranolol derivative, 1-(1-naphthoxy)-3-*N*-(ethylamino)propan-2-ol (NEDA), as previously described.[7] NEDA is reacted directly with BNHS to give biotinyl-NEDA (BN). This compound, whose synthesis and characterization has independently been reported by Atlas and co-workers,[14] is now commercially available under the designation "biotinylpropranolol analog" (Sigma). A second derivative is obtained by first reacting BNHS with hexaglycine to give biotinylhexaglycine.[7] This intermediate is then reacted with NEDA to give biotinylhexaglycyl-NEDA (BGN).

[12] K. E. Meier and A. E. Ruoho, *Biochem. Biophys. Acta* **804,** 331 (1984).

[13] M. L. Jasiewicz, D. R. Schoenberg, and G. C. Mueller, *Exp. Cell Res.* **100,** 213 (1976).

[14] D. Atlas, D. Yaffe, and E. Skutelsky, *FEBS Lett.* **95,** 173 (1978).

Synthesis of 1-[2-Hydroxy-3-(2-propyl)amino]propyloxy-2-[2-hydroxy-3-[(2-amino)ethyl]thio]propyl Benzene (Cysteaminylalprenolol). Bromoalprenolol is synthesized from *N*-bromosuccinimide (Sigma) and (−)-alprenolol (+)-tartrate (Sigma) as described by Vauquelin *et al.*,[15] using a reaction time of 4 hr. The reaction mixture, containing 2 mmol of bromoalprenolol, is adjusted to pH 9.0 with Na_2CO_3. This mixture is extracted 5 times with 200 ml of ether. The ether extract, containing bromoalprenolol, is reduced to 10 ml using a rotary evaporator. This solution is added to 320 mg NaOH (8 mmol) and 454.4 mg cysteamine (4 mmol) in 50 ml of ethanol. The mixture is stirred for 30 min under N_2 in the dark. The N_2 source is then removed, and the reaction is allowed to proceed for 18 hr at 25° with stirring. The mixture is then dried under vacuum. The residue is dissolved in 25 ml of water and is extracted 6 times with 25 ml of ethyl acetate. The ethyl acetate extract is reduced to 50 ml using a rotary evaporator. This solution is extracted 3 times with 50 ml of water that had been adjusted to pH 6.0 with HCl. The aqueous extract is then adjusted to pH 10 with Na_2CO_3. This solution is extracted 6 times with 50 ml of ethyl acetate. These extractions are monitored by thin-layer chromatography on silica gel using chloroform–methanol–acetic acid (1 : 5 : 1, v/v); a ninhydrin spray [0.29% (w/v) in acetone] is used to visualize the amine-containing starting material and product. The final ethyl acetate extract is dried under vacuum. This residue is dissolved in 15 ml of methanol, filtered, and dried under vacuum to a yellow oil (411 mg, 1.2 mmol, 60% yield). The product gives a single component (R_f 0.6) that can be visualized using UV light or ninhydrin spray following thin-layer chromatography on silica gel with the solvent system described above.

Synthesis of Biotinylaminododecanoic Acid. Biotinylaminododecanoic acid is prepared by a modification of a method described by Bayer *et al.*[16] A solution of 12-aminododecanoic acid (258.4 mg, 1.2 mmol, Aldrich, Milwaukee, WI) in 12 ml of 0.2 *M* $NaHCO_3$ is added to a solution of BNHS (341 mg, 1 mmol) in 10 ml of dimethylformamide (DMF). The reaction is allowed to proceed for 9 hr at 25° with stirring. The reaction mixture is then filtered to remove the resulting white precipitate. The precipitate is washed with water followed by water acidified to pH 2.0 with HCl. The washed precipitate is dried under vacuum to a white powder (380 mg, 86% yield, mp 216°).

Synthesis of Biotinyldodecanoylcysteaminylalprenolol. A solution of biotinyldodecanoic acid (88.2 mg, 0.2 mmol) in 5 ml of DMF is heated to 95°; 1,1′-carbonyldiimidazole (32.4 mg, 0.2 mmol, Sigma) is then added.

[15] G. Vauquelin, P. Geynet, J. Hanoune, and A. D. Strosberg, *Proc. Natl. Acad. Sci. U.S.A.* **74,** 3710 (1977).

[16] E. A. Bayer, T. Viswanatha, and M. Wilchek, *FEBS Lett.* **60,** 309 (1975).

The reaction is allowed to proceed for 5 min at 95°, followed by 30 min at 25°. A solution of cysteaminylalprenolol (137 mg, 0.4 mmol) in 10 ml of DMF is then added. The reaction mixture is stirred overnight at 25°. The solvent is removed under vacuum. The residue is dissolved in 5 ml of chloroform–methanol–acetonitrile–acetic acid (9 : 9 : 9 : 2, v/v). This solution is applied to a silica gel column (60–200 mesh, 3 × 14 cm, Sigma). The column is eluted with the solvent mixture described above with collection of 5-ml fractions. Recovery of the product (BDCA) is monitored by thin-layer chromatography of the column fractions on silica gel, using the same solvent system; dimethylaminocinnamaldehyde spray is used to detect the biotinylated product. Fractions 20–36 are pooled. The solvent is removed under vacuum, leaving a yellow oil (0.052 nmol by avidin titration, 26% yield). The product appears as a single component (visualized with either UV light or dimethylaminocinnamaldehyde spray) in two different thin-layer chromatography systems (silica gel): chloroform–methanol–acetic acid, 1 : 5 : 1, R_f 0.81; chloroform–methanol–acetonitrile–acetic acid 9 : 9 : 9 : 2, R_f 0.81. BDCA can also be detected as a single component (visualized by iodine vapor) using reversed-phase thin-layer chromatography plates with methanol as solvent (R_f 0.23). BDCA is stored as a solution in methanol and is stable in this form for at least 1 year. The concentration of the stock solution is determined by avidin titration, as described above.

Synthesis of Biotinylcysteaminylalprenolol and Biotinylcaproylcysteaminylalprenolol. Biotinylcysteaminylalprenolol (BCA) is synthesized by reaction of BNHS with cysteaminylalprenolol.[8,9] Biotinylcaproylcysteaminylalprenolol (BCCA) is synthesized by reaction of biotinyl ε-aminocaproic acid (prepared by analogy to biotinylaminododecanoic acid, as described above) with cysteaminylalprenolol.[8,9]

Comments

Our initial studies of two biotinylpropranolol derivatives (BN and BGN, Fig. 1) led to the following conclusions. BN, which did not contain a spacer group between the biotin and propranolol residues, was a potent β-adrenergic antagonist in the absence of avidin, but it bound to the receptor with very low affinity in the presence of avidin (Table I). A hexaglycyl spacer group was introduced into BGN in order to allow simultaneous binding of the ligand to avidin and to the receptor. This goal was achieved at the expense of substantial loss of affinity of the ligand for the receptor (Table I). This reduced affinity is presumed to be due to the addition of a bulky side chain to the amino portion of the parent compound.

A second series of derivatives were synthesized from alprenolol, an-

TABLE I
BINDING CONSTANTS FOR BIOTINYLPROPRANOLOL AND BIOTINYLALPRENOLOL DERIVATIVES

Ligand	K_D (nM)[a]
Propranolol	11
BN	3.9
Avidin–BN	>650
BGN	47
Avidin–BGN	320
Alprenolol	3.5
BCA	12
Avidin–BCA	59
BCCA	13
Avidin–BCCA	13
BDCA	2.0
Avidin–BDCA	2.0

[a] The apparent equilibrium dissociation constants for each ligand are determined at 4° using a digitonin-solubilized duck erythrocyte membrane preparation. The radioligand used is [^{3}H]dihydroalprenolol, and the binding constants are determined by competition for radioligand binding by the indicated antagonists. The avidin complexes of the ligands are prepared with a 10% molar excess of biotin-binding sites over biotinylated ligand. Similar trends in affinity values are obtained using a duck erythrocyte membrane preparation (binding and adenylate cyclase assays), except that the differences between the values for BN versus avidin–BN and BCA versus avidin–BCA are more pronounced.

other β-adrenergic antagonist that possesses high affinity for the receptor. In the case of alprenolol, it was possible to derivatize a different portion of the parent molecule than had been derivatized in the biotinylpropranolol series. The biotinylalprenolol derivatives contained spacer groups of varying chain length. All of the compounds bound with high affinity to the β-adrenergic receptor in the absence of avidin, with BDCA possessing the highest affinity (Table I). In the presence of avidin, BCCA and BDCA retained their affinities for the receptor, while BCA bound with much lower affinity. These data, consistent with the data obtained for the biotinylpropranolol compounds, indicate that it is necessary to introduce a spacer sequence between the two ends of these bifunctional reagents in order to allow simultaneous binding of the compounds to both the β-adrenergic receptor and to avidin. Similar requirements for a spacer sequence were observed in the studies of Pitha *et al.*,[17] who found that such a sequence is required for binding of β-adrenergic receptors to alprenolol which had been linked to an affinity column support resin.

All of the derivatives described here bind with high affinity to avidin. Their dissociation rates from avidin are slightly higher than that of biotin, e.g., 20% dissociation of BDCA in 20 hr at 25° as compared to 3% dissociation of biotin.[9] This rate is not rapid enough to interfere with use of the complex in biological assay systems. The avidin–BDCA complex is always prepared shortly before each use.

We have attempted to use BDCA as a receptor visualization reagent with cultured mammalian cell model systems (HeLa epithelial cells, L6 myoblast cells).[9] Using binding and activity assays, we found that BDCA is capable of binding to the β-adrenergic receptor of intact HeLa cells in the presence of ferritin–avidin conjugates. HeLa and L6 cells were incubated with ferritin–avidin–BDCA, washed, fixed, embedded, and thin sectioned. Examination of the sections revealed that the level of labeling was too low to allow positive identification of β-adrenergic receptor sites, despite low levels of nonspecific binding of ferritin–avidin to the cell surface in the absence of BDCA. Similar results were obtained with various incubation temperatures, with various incubation times, with avidin linked to horseradish peroxidase, with various types of fixatives, and by prelabeling the cells with BDCA prior to addition of ferritin–avidin. Subsequent analysis of receptor content, showed that, in contrast to earlier reports of high levels of receptor expression, these cells expressed approximately 5000 receptors/cell.[9] With this level of receptor expression, one would expect to visualize between 10 and 30 receptors in a 75-nm thin section of a cell. At this level of labeling, it was not possible to distinguish nonspecific binding of ferritin–avidin from specific binding of the ferritin–avidin–BDCA complex.

Several new developments in this field may lead to the successful visualization of β-adrenergic receptor sites on intact cells using the biotin–avidin approach developed here. First, the techniques of molecular biology have made it possible to transfect cells with the nucleotide sequence encoding the receptor. Such cells can theoretically be made to express greatly elevated levels of receptor protein. Second, improved techniques for detection of fluorescent marker molecules have enabled

[17] J. Pitha, J. Zjawiony, R. J. Lefkowitz, and M. G. Caron, *Proc. Natl. Acad. Sci. U.S.A.* **77,** 2219 (1980).

the detection of very low levels of ligand binding to cell surfaces. The application of such approaches to visualization studies utilizing the high-affinity biotinylalprenolol probe described here should result in interesting new information regarding the location, expression, and regulation of β-adrenergic receptors.

[79] Probing Adenosine Receptors Using Biotinylated Purine Derivatives

By KENNETH A. JACOBSON

Extracellular adenosine acts as a neuromodulator through two subtypes of membrane-bound receptors. Agonist activation of A_1- or A_2-adenosine receptors is generally linked to an inhibitory or stimulatory effect, respectively, on the enzyme adenylate cyclase, which converts intracellular 5′-ATP to cyclic 3′,5′-AMP. Many N^6-alkyl- and N^6-aryl-substituted adenosine analogs have been synthesized as potent agonists having nanomolar K_i values at the high-affinity A_1 site.[1] At each receptor subtype, xanthines act as competitive antagonists. Nonselective antagonists caffeine and theophylline have K_i values in the 10 μM range, and at A_1 receptors some synthetic 1,3-dialkyl-8-substituted analogs have K_i values close to 1 nM.[2] Adenosine antagonism is the principal mechanism by which caffeine acts as a stimulant *in vivo*.

The avidin–biotin complex has been utilized in studies directed toward the isolation, histochemical localization, and microscopic structural probing of the receptor protein.[3] Both adenosine and xanthine analogs have been coupled to biotin,[4,5] forming bifunctional probes to serve as noncovalent cross-linkers between adenosine receptors and avidin. Structure–activity relationships for adenosine agonists and antagonists were studied initially to identify potential sites for derivatization. Since neither adenosine nor theophylline contains a readily derivatized functional group that is nonessential for biological activity, a "functionalized conge-

[1] K. A. Jacobson, K. L. Kirk, W. L. Padgett, and J. W. Daly, *J. Med. Chem.* **28,** 1341 (1985).

[2] K. A. Jacobson, K. L. Kirk, W. L. Padgett, and J. W. Daly, *J. Med. Chem.* **28,** 1334 (1985).

[3] G. L. Stiles, *Trends Pharmacol. Sci.* **7,** 486 (1986).

[4] K. A. Jacobson, K. L. Kirk, W. Padgett, and J. W. Daly, *FEBS Lett.* **184,** 30 (1985).

[5] K. A. Jacobson, D. Ukena, W. Padgett, K. L. Kirk, and J. W. Daly, *Biochem. Pharmacol.* **36,** 1697 (1987).

ner'' methodology[1,2] was used. In this approach, a chemically functionalized chain (terminating in an amine or carboxylic acid) is incorporated in the molecular structure. Insensitive structural sites for attachment of chains were located, and the effect on activity of substituents present in a sequentially extended chain were investigated. In brief, an N^6-*p*-(carboxymethyl)phenyl substituent on adenosine or an 8-*p*-(carboxymethyloxy)phenyl substituent on 1,3-dipropylxanthine provided sites for functionalization tolerated at the receptor. Furthermore, the enhanced affinity of certain related amide derivatives for adenosine receptors indicated accessory binding sites at or near the receptor.

A range of chain lengths separating biotin and adenosine pharmacophore moieties were synthesized (Table I). Spacer groups consisting of ε-aminocaproic acid and oligoglycine were utilized. The affinity-enhancing effect noted previously[1,2] of an amino group present on the chains of congeners in the series of ADAC and XAC suggested substitution of an additional amino group on the ε-aminocaproyl spacer, as in the D-lysyl residue[5] of compound **3**. The ε-biotinyl-D-amino acid isomer (equivalent to D-biocytin) was included to minimize proteolytic cleavage, a potential problem in biological assay systems. N^{α}-(9-Fluorenemethyloxycarbonyl)-D-biocytin was prepared as a synthetic intermediate. The base-sensitive FMOC protecting group, removable in 10% diethylamine in dimethylformamide, was used because of the acid lability of purine nucleosides.

Synthetic Methods

ADAC and XAC (amine congeners, see Table I) are available from Research Biochemicals, Inc. (Natick, MA).

N^{α}-(9-Fluorenemethyloxycarbonyl)-D-biocytin.[5] N^{α}-(9-fluorenemethyloxycarbonyl)-*N*-ε-benzyloxycarbonyl-D-lysine[5] (0.70 g, 1.4 mmol) is treated with 30% HBr–acetic acid for 1 hr. Most of the solvent is removed under a nitrogen stream, and the residue is purified by preparative thin-layer chromatography (silica, chloroform–methanol–acetic acid, 70 : 25 : 5, v/v) to give N^{α}-(9-fluorenemethyloxycarbonyl)-D-lysine as an amorphous glass (0.14 g, 27% yield).

FMOC-D-lysine (83 mg, 0.23 mmol) is combined with succinimidylbiotin (88 mg, 0.23 mmol) and diisopropylethylamine (39 μl, 0.23 mmol) in 5 ml of dimethylformamide–tetrahydrofuran (2 : 1) and stirred overnight. Ethyl acetate and 0.1 *M* HCl are added, and the phases are mixed. The upper phase is washed with water and evaporated, and the residue is triturated with ethyl acetate and petroleum ether. The solid product, N^{α}-(9-fluorenemethyloxycarbonyl)-D-biocytin, is collected. Yield, 63 mg (43%). mp 139–143°. $[\alpha]_D$ at 23° in dimethylformamide, 36.5°. The product is then coupled to ADAC[5] and deprotected to give compound **3**.

TABLE I
BIOTINYL CONJUGATES OF ADENOSINE RECEPTOR LIGANDS[a]

B =
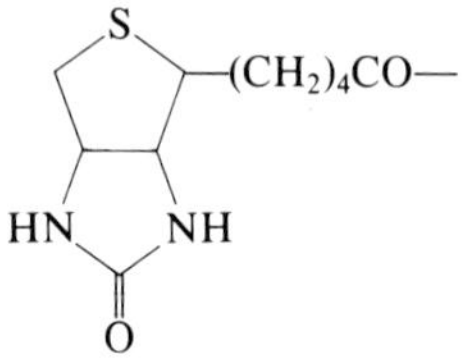

ADAC =
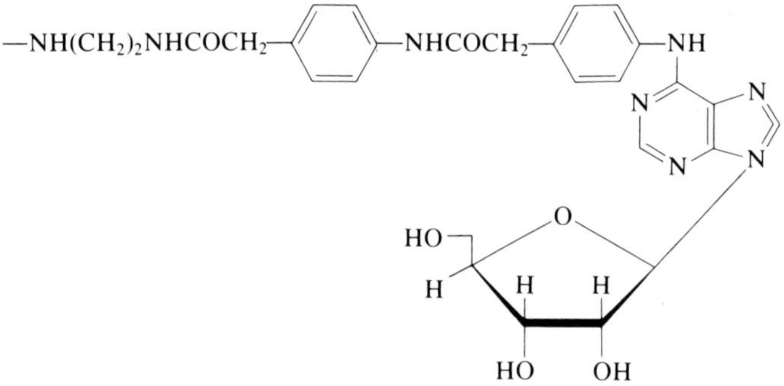

XAC =
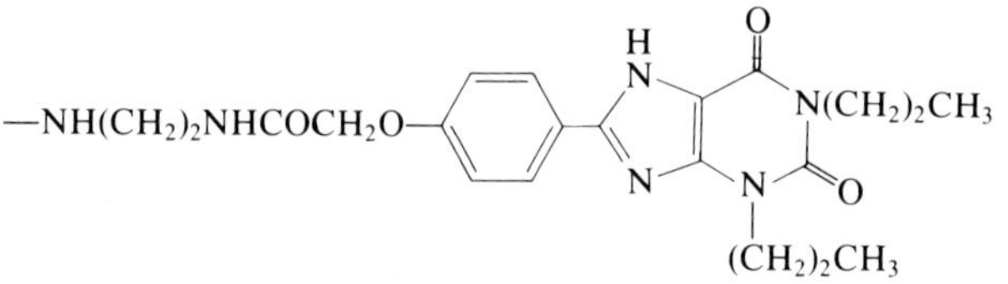

Biotinyl (B) derivative	K_i (no avidin)	K_i (+ avidin)
(1) B–ADAC	11.4	36
(2) B–$NH(CH_2)_5CO$–ADAC	18	35
(3) B–$NH(CH_2)_4CHNH_2CO$–ADAC (D-configuration)	8.9	n.d.[b]
(4) B–XAC	54	>500
(5) B–$NH(CH_2)_5CO$–XAC	50	>500
(6) B–$NH(CH_2)_5CO$–Gly_3–XAC	50	260

[a] K_i values (nanomolar) are for inhibition of the binding of N^6-[^{3}H]phenylisopropyladenosine or N^6-[^{3}H]cyclohexyladenosine to rat brain membrane A_1-adenosine receptors in the presence or absence of precomplexed avidin (1 μM). Data from Refs. 4 and 5.

[b] n.d., not determined.

Biotinyl-ε-aminocaproyl-ADAC (**2**). ADAC[5] (an adenosine amine congener, 17 mg, 29 μmol) is suspended in 0.5 ml of dimethylformamide and treated with succinimidyl 6-(biotinamido)hexanoate (Pierce Chemical Co., Rockford, IL; 17 mg, 36 μmol). Solubilization of the reactants is followed by gradual precipitation of the product. Methanol (1 ml) and dry ether (2 ml) are added. The product is obtained by filtration. Yield, 25 mg (93%). mp 195–198°.

Other biotin conjugates, **1** and **4–6**, are prepared similarly by acylation of a xanthine or adenosine amine congener[1,2,5] using biotinyl-*N*-hydroxysuccinimide ester. The pure biotin conjugates are recrystallized from dimethylformamide–ether and characterized by thin-layer chromatography [chloroform–methanol–acetic acid, 85 : 10 : 5, v/v; biotin derivatives turn pink after spraying with a fresh 0.1% solution of 4-(dimethylamino)cinnamaldehyde in ethanolic sulfuric acid, 1%, then heating], elemental analysis, NMR, and mass spectroscopy. The low volatility precludes using standard chemical ionization mass spectrometry. Instead, californium plasma desorption mass spectrometry[6] can be used and generally gives intense positive ion peaks corresponding to $(M + H)^+$ and $(M + Na)^+$.

Biochemical and Physiological Characterization

Affinity at A_1-adenosine receptors is determined in a competitive binding assay using a tritiated agonist radioligand (see Table I).[4,5] The potencies of both free ligand and that which is precomplexed to avidin[4,7] are measured. All of the adenosine conjugates (**1–3**) bind simultaneously to A_1-adenosine receptors and to avidin. Thus, based on estimates of the extended chain lengths of the conjugates and the depth of the biotin binding site, the adenosine agonist binding site must be located within 12 Å from the receptor protein surface. In the form of avidin complexes, the xanthine conjugates lose their affinity for the receptor, which suggests conformational or topographical differences between agonist- and antagonist-occupied receptors.

Adenosine acts as a potent vasodilator through vascular A_2 receptors. When infused directly into canine coronary arteries,[8] compounds **1** and **2** are 3.0 ± 0.2-fold and 10.2 ± 7.3-fold, respectively, more potent than adenosine. In the form of avidin complexes (purified by gel filtration), only the longer chain derivative **2** is active (relative potency 2.1 ± 0.4), with a slow onset of action and fast washout. Compound **1** is active in stimulating adenylate cyclase[9] through A_2-adenosine receptors in human platelets (EC_{50} 2.43 μM).

[6] K. A. Jacobson, L. K. Pannell, K. L. Kirk, H. M. Fales, and E. A. Sokoloski, *J. Chem. Soc., Perkin Trans. 1*, 2143 (1986).

[7] R. A. Kohanski and M. D. Lane, *J. Biol. Chem.* **260,** 5014 (1985).

[8] K. A. Jacobson, N. Yamada, K. L. Kirk, W. L. Padgett, J. W. Daly, and R. A. Olsson, *Biochem. Biophys. Res. Commun.* **136,** 1097 (1986); K. A. Jacobson, N. Yamada, K. L. Kirk, W. L. Padgett, J. W. Daly, and R. A. Olsson, *Biochem. Biophys. Res. Commun.* **139,** 375 (1986).

[9] D. Ukena, R. A. Olsson, and J. W. Daly, *Can. J. Physiol. Pharmacol.* **65,** 365 (1987).

Author Index

Numbers in parentheses are footnote reference numbers and indicate that an author's work is referred to although the name is not cited in the text.

A

B

C

D

E

H

I

J

K

L

M

O

P

S

T

U

V

W

Y

Z

Subject Index

A

B

C

D

H

I

J

M

O

P

T

U

V

W

X

Y

Z